T0304633

IDEAL MHD

Comprehensive, self-contained, and clearly written, this successor to *Ideal Magnetohydrodynamics* (Plenum Press, 1987) describes the macroscopic equilibrium and stability of high-temperature plasmas – the basic fuel for the development of fusion power.

Now fully updated, this book discusses the underlying physical assumptions for three basic MHD models: ideal, kinetic, and double-adiabatic MHD. Included are detailed analyses of MHD equilibrium and stability, with a particular focus on three key configurations at the cutting-edge of fusion research: the tokamak, stellarator, and reversed field pinch. Other new topics include continuum damping, MHD stability comparison theorems, neoclassical transport in stellarators, and how quasi-omnigeneity, quasi-symmetry, and quasi-isodynamic constraints impact the design of optimized stellarators.

Including full derivations of almost every important result, in-depth physical explanations throughout, and a large number of problem sets to help master the material, this is an exceptional resource for graduate students and researchers in plasma and fusion physics.

JEFFREY FREIDBERG is Korea Electric Power Professor of Nuclear Science and Engineering, and former Head of Nuclear Science and Engineering, at MIT, and a former Associate Director of MIT's Plasma Science and Fusion Center. He is a Fellow of the APS and the AAAS, and the author of *Plasma Physics and Fusion Energy* (2007).

IDEAL MHD

JEFFREY FREIDBERG

Massachusetts Institute of Technology

CAMBRIDGE
UNIVERSITY PRESS

Shaftesbury Road, Cambridge CB2 8EA, United Kingdom

One Liberty Plaza, 20th Floor, New York, NY 10006, USA

477 Williamstown Road, Port Melbourne, VIC 3207, Australia

314–321, 3rd Floor, Plot 3, Splendor Forum, Jasola District Centre, New Delhi – 110025, India

103 Penang Road, #05–06/07, Visioncrest Commercial, Singapore 238467

Cambridge University Press is part of Cambridge University Press & Assessment,
a department of the University of Cambridge.

We share the University's mission to contribute to society through the pursuit of
education, learning and research at the highest international levels of excellence.

www.cambridge.org
Information on this title: www.cambridge.org/9781107006256

First published 2014

A catalogue record for this publication is available from the British Library

Library of Congress Cataloging-in-Publication data
Freidberg, Jeffrey P., author.
[Ideal magnetohydrodynamics]
Ideal MHD / Jeffrey Freidberg.
pages cm
Updated version of: Ideal magnetohydrodynamics. 1987.
ISBN 978-1-107-00625-6 (Hardback)
1. Magnetohydrodynamics–Mathematical models.   2. Fluid dynamics–Mathematical models.
3. Turbulence–Mathematical models.   4. High temperature plasmas.   5. Plasma (Ionized gases)
6. Fusion reactors.   I. Title.
QC718.5.M36F74 2014
538´.6–dc23   2014002053

ISBN   978-1-107-00625-6   Hardback

To Karen

Contents

Preface

It has been over 25 years since my original textbook on *Ideal Magnetohydrodynamics* was published. The book, I believe, was well received by the fusion community but, unfortunately, the publisher has long since gone out of business, making it quite difficult to obtain copies. As a result I have often been asked by students and colleagues to write an updated version of the original book and this volume is the result of that effort. The second volume describing extended MHD will be published in the future.

In writing the book I have found some similarities with my original MHD book but many differences as well. One might hope and expect that a considerable amount of new ideas and discoveries will have been developed over the past 25 years. The overall result is that about a third of the book is closely related to my original textbook, about a third has similar subject titles but has been entirely rewritten, and the final third is completely new. The material is largely based on an evolving course on MHD that I have been teaching at MIT for nearly 25 years.

The basic structure of the book is similar to the original volume. It still makes good sense to talk first about the MHD model, then MHD equilibrium, and finally MHD stability. The completely new topics include an extensive discussion of alternate MHD models, specifically the Chew–Goldberger–Low double adiabatic model and the kinetic MHD model. There is also a detailed discussion of continuum damping. A major topic discussed in far greater detail than in the original volume is the equilibrium and stability of stellarators. Included in this discussion is a reasonably complete but still physically intuitive description of neoclassical transport and how it affects the design of stellarators.

To keep the volume to a reasonable length attention is focused on the three concepts leading the charge towards fusion: the tokamak, stellarator, and reversed field pinch. Thus certain concepts, such as Scyllac and the Elmo Bumpy Torus, covered in the original volume are omitted from this volume. Also, other concepts, such as the field reversed configuration, the spheromak, and the gas dynamic trap, are not included since they are not well described by ideal MHD. Extended MHD physics is required.

Is this a good time to write such a book? I believe so. MHD is one of the basic building blocks of fusion science. It plays a primary role in the design and physics goals of major fusion devices such as ITER. Equally important there are still basic questions to be answered, for example the simultaneous optimization of a stellarator with respect to both MHD and neoclassical transport. In short, the current and future generation of fusion scientists will need to understand MHD to help propel the program forward.

Who is the intended audience? The textbook is aimed at first-year graduate students and new scientists and engineers entering the field. It is also appropriate for undergraduate seniors assuming they have a strong enough background in physics. The text is almost completely self-contained. It includes in-depth physical explanations and full derivations of almost every important result. Hopefully the discussions will help clarify some of the mathematical mysteries of plasma physics. The overall goal is to present the material in a sufficiently complete manner so that readers will not require a hoard of accompanying references in order to master the physics.

What background is required? Readers should have a good background in undergraduate mathematics and physics. Specifically, in mathematics a knowledge up to and including partial differential equations is important. In terms of physics the requirements include mechanics, basic fluid dynamics, and electromagnetic theory (i.e., electrostatics, magnetostatics, and wave propagation). Experience has shown that an undergraduate degree in physics or most engineering disciplines provides satisfactory preparation. Note that since the material is largely self-contained, a basic course in plasma physics is not actually required. Still such a course would be helpful to put MHD in proper perspective with respect to the overall field of fusion plasma physics.

A note to instructors using the book for one of their courses: the book contains a large number of problem sets to help master the material, although some require considerable effort. Also, there is more than enough material to cover a one-semester subject. Thus, instructors have the option of picking and choosing among various topics to assemble a coherent one-semester subject.

In the end it is my hope that the textbook will help educate the current and next generation of fusion scientists and engineers in the important subject of MHD, a crucial area of research for the ultimate achievement of fusion energy.

Acknowledgements

The material for this book has evolved over many years of research and teaching. A large number of friends, colleagues, and students, too numerous to mention, have contributed in a significant way to my knowledge of the field, making this book possible. I acknowledge my deep appreciation for their collaboration, cooperation, and camaraderie.

A number of people at MIT deserve special thanks. Peter Catto, Matt Landremann, and Felix Parra for helping me understand neoclassical transport in stellarators. Martin Greenwald provided great insight on tokamak density limits. Similarly Bob Granetz clarified many of the experimental issues associated with disruptions and ELMs. Jesus Ramos was an invaluable resource on the topic of alternate MHD models.

Friends and colleagues from the general fusion community have also generously provided advice and suggestions for parts of the book. Mike Zarnstorff, Per Helander, and Harold Weitzner contributed substantially to my knowledge of stellarators. Riccardo Betti and John Menard provided many insights into standard tokamaks and spherical tokamaks. Hans Goedbloed's beautiful books were a great help on basic MHD theory. I have also received considerable help and insight on the reversed field pinch concept through discussions with Piero Martin and Emilio Martines.

Two other friends and colleagues stand out for their contributions to the content of the book, not only through many discussions but by actually collaborating and carrying out specific calculations, many of which appear in the book: Antoine Cerfon (tokamaks and stellarators) and Luca Guazzotto (the reversed field pinch).

In terms of preparation of the manuscript I have had spectacular help from Liz Parmelee for invaluable administrative and organizational support during the entire project. Also Mary Pat MacNally carried out the daunting task of creating over 100 figures appearing in the book. For this I am most grateful. Several students, Aaron Bader, Arturo Dominguez, and Nathan Howard, produced many of the figures for the early chapters. The dramatic cover, which illustrates the NCSX experiment, was created by Princeton Plasma Physics Laboratory Stellarator Team.

As always it has been a pleasure working with the team at Cambridge University Press. Thanks to Simon Capelin (publishing director), Elizabeth Horne (assistant editor), Emma Walker (senior production editor), and David Hemsley (copy editor).

Last, but absolutely not least, I would like to thank my wife Karen for her unending patience, support, and encouragement while I prepared the manuscript. Now that it is complete I hope to renew our acquaintance.

Jeff Freidberg,
August 2013

1

Introduction

1.1 The role of MHD in fusion energy

Magnetohydrodynamics (MHD) is a fluid model that describes the macroscopic equilibrium and stability properties of a plasma. Actually, there are several versions of the MHD model. The most basic version is called "ideal MHD" and assumes that the plasma can be represented by a single fluid with infinite electrical conductivity and zero ion gyro radius. Other, more sophisticated versions are often referred to as "extended MHD" or "generalized MHD" and include finite resistivity, two-fluid effects, and kinetic effects (e.g. finite ion gyro radius, trapped particles, energetic particles, etc.). The present volume is focused on the ideal MHD model.

Most researchers agree that MHD equilibrium and stability are necessary requirements for a fusion reactor. If an equilibrium exists but is MHD unstable the result is almost always very undesirable. There can be a violent termination of the plasma known as a major disruption. If no disruption occurs, the result is likely to be a greatly enhanced thermal transport which is highly detrimental to fusion power balance. In order to avoid MHD instabilities it is necessary to limit the regimes of operation so that the plasma pressure and current are below critical values. However, these limiting values must still be high enough to meet the needs of producing fusion power. In fact it is fair to say that the main goal of ideal MHD is the discovery of stable, magnetically confined plasma configurations that have sufficiently high plasma pressure and current to satisfy the requirements of favorable power balance in a fusion reactor.

1.1.1 The plasma pressure in a fusion reactor

To put the role of MHD in context with respect to fusion it is useful to quantify the value of plasma pressure required in a reactor. This is easily done by considering

1

simple power balance in a deuterium–tritium (D–T) fusion plasma where the heating power produced by fusion alpha particles balances the thermal conduction losses due to classical collisions and plasma turbulence.[1] This balance must be achieved at an optimum temperature that maximizes fusion energy production. The resulting "ignited" plasma is self-sustaining, requiring no external heating sources. The power balance condition is given by

$$\text{alpha heating} = \text{thermal loss}$$

$$\frac{E_\alpha}{4} n^2 \langle \sigma v \rangle = \frac{3}{2} \frac{p}{\tau_E} \tag{1.1}$$

where $E_\alpha = 3.5$ MeV, n is the electron number density, $\langle \sigma v \rangle$ is the velocity averaged D–T fusion cross section, p is the plasma pressure, and τ_E is the thermal conduction energy confinement time. For a plasma with equal temperatures $T_D = T_T = T_e \equiv T$, the plasma pressure is equal to $p = 2nT$, where T is measured in units of energy. Some simple manipulations allow Eq. (1.1) to be rewritten in terms of one version of the Lawson (1957) parameter as follows:

$$p\tau_E = \frac{24}{E_\alpha} \frac{T^2}{\langle \sigma v \rangle} \tag{1.2}$$

For many years this fundamental requirement has divided fusion research into three main areas of study: heating, transport, and MHD. The reasoning for this division starts with the recognition that the function $T^2/\langle \sigma v \rangle$ has a minimum at approximately $T = 15$ keV. It is important to operate at this temperature or else p and/or τ_E would have to raised, both of which lead to increased costs. It is the job of the heating community to provide ways to heat the plasma to about 15 keV.

At this temperature ignition requires

$$(p\tau_E)_{\min} \approx 8 \text{ atm-sec} \tag{1.3}$$

Learning how to produce a plasma with a sufficiently long τ_E is the job of the transport community. Learning how to produce plasmas with a sufficiently large p is the job of the MHD community. For many years these three areas of research were reasonably separated. As fusion research has progressed, longer duration, high-performance plasmas have been produced and these three areas have started to overlap. The reason is that plasma–wall interactions have become increasingly important and have a large, simultaneous impact on heating, transport, and MHD. For the moment it is, nonetheless, still useful to think of the three separate plasma requirements for an ignited plasma.

---

[1] Readers unfamiliar with fusion reactor power balance should refer to the Further reading at the end of the chapter.

One might think on the basis of Eq. (1.3) that it might be possible to make tradeoffs between p and τ_E in order to reach ignition in as easy a way as possible. In practice there is not much room for tradeoffs. The reason is that if one wants to construct a standard base-load reactor with a power output of 1 GWe as economically possible, this actually requires a specific value of p. The reasoning behind this conclusion is based on (1) the intuition that "most economical" translates into smallest size and (2) the smallest size is set by the maximum neutron flux passing through the first wall. The maximum allowable neutron wall loading as set by material limitations is typically assumed to be $P_W \approx 4\,\text{MW/m}^2$. The condition that the neutron flux not exceed the wall loading limit in a toroidal reactor is given by

$$\text{fusion neutron flux} = \text{wall loading}$$

$$\frac{E_n}{16}p^2 \frac{\langle \sigma v\rangle}{T^2}\left(2\pi^2 R_0 a^2\right) = P_W\left(4\pi^2 R_0 a\right) \tag{1.4}$$

Here, $E_n = 14.1$ MeV, R_0 is the major radius of the torus, and a is the minor radius. Solving for p yields

$$p = \left(32\frac{T^2}{\langle \sigma v\rangle}\frac{P_W}{E_n a}\right)^{1/2} \tag{1.5}$$

The minor radius of the plasma appearing in Eq. (1.5) can be accurately approximated by assuming that most of the electric power is produced by the fusion neutrons with a conversion efficiency $\eta \approx 0.4$. Thus, Eq. (1.4) can be rewritten as

$$\text{electric power} = \eta(\text{neutron power})$$

$$P_E = \eta P_n = \eta P_W\left(4\pi^2 R_0 a\right) \tag{1.6}$$

Now, the minor radius a can be rewritten in terms of the dimensionless inverse aspect ratio a/R_0

$$a = \left(\frac{1}{4\pi^2}\frac{a}{R_0}\frac{P_E}{\eta P_W}\right)^{1/2} \tag{1.7}$$

Typically $R_0/a \sim 3$. The exact value is not too critical since it enters the value of the pressure as a fourth root. For the parameters under consideration one finds $a \approx 2.3$ m, which when substituted into Eq. (1.5) leads to

$$p \approx 7 \text{ atm} \tag{1.8}$$

The conclusion is that a fusion plasma must have a pressure of about 7 atm and a corresponding energy confinement time equal to 1.1 sec. In general there is some, but not a lot, of flexibility in these values.

1.1.2 The dimensionless pressure, β

The analysis of MHD is almost always carried out in terms of a dimensionless pressure denoted by β. There are various detailed definitions in the literature, the most important of which are discussed in the text. All definitions involve the ratio of plasma pressure to applied magnetic pressure:

$$\beta \equiv \frac{p}{B^2/2\mu_0} \tag{1.9}$$

In configurations with a large toroidal magnetic field and an aspect ratio $R_0/a \sim 3$, the corresponding reactors typically require $\beta \sim 5-10\%$, values that have been already achieved experimentally. In tighter aspect ratio devices, higher stable β values are attainable, but often the pressure is not higher because, for engineering and geometric reasons, the magnetic field is smaller. Other concepts do not rely on a large toroidal magnetic field, which is an important engineering advantage. As a result their required and achieved MHD β values are higher. However, such configurations typically have poorer MHD stability behavior leading to enhanced thermal transport. Almost all discussions of MHD in the literature involve β, but readers should stay alert to the fact that it is pressure that is the critical parameter for a fusion reactor.

1.1.3 A variety of fusion concepts

What is the best magnetic geometry for a fusion reactor from the point of view of MHD? Over the years many ideas have been tried. A list is given below:

Belt pinch	Reversed field pinch
Cusp	Screw pinch
Elmo bumpy torus	Spherical tokamak
Field reversed configuration	Spheromak
Force-free pinch	Stellarator
Heliac	Stuffed caulked cusp
High β stellarator	Tandem mirror
Levitated dipole	Theta pinch
Mirror	Tokamak
Octopole	Tormac
Perhapsatron	Z-pinch
Plasma focus	Z-pinch – hard-core

Clearly there has not been a shortage of imagination in inventing new concepts. Of this long list two concepts have risen to the top, largely because of superior overall plasma physics performance. These are the tokamak and the stellarator. It should be noted that while these configurations have the best plasma physics performance, they may not be the optimized choice from an engineering point of view. Both of these concepts have a large toroidal magnetic field which adds to the cost and complexity of a fusion reactor. Still, unless other concepts can overcome the plasma physics challenges their more desirable engineering features cannot be utilized. So far, while progress has been made, they have not yet been able to overcome these challenges, thereby explaining why tokamaks and stellarators remain at the top of the list.

1.1.4 Structure of the textbook

The basic structure of the textbook is straightforward. The discussion begins with a description of the ideal MHD model and some of its general properties. This is followed by a discussion of MHD equilibrium in simple and general geometries. The last main topic discussed involves MHD stability.

There are many examples presented, although the bulk of the actual applications involve tokamaks and stellarators. There is also a substantial discussion of the reversed field pinch, a concept that is not as yet quite as advanced as tokamaks and stellarators in terms of performance. Still, it does hold some promise and its relatively simple geometric properties make it an ideal example to help understand MHD equilibrium and stability.

The overall purposes of the textbook are to provide both a qualitative and quantitative understanding of ideal MHD theory as applied to magnetic fusion. The specific goals are to discover concepts capable of achieving MHD stable, high-pressure, fusion-grade plasmas.

1.2 Units

The basic units used throughout the textbook are the usual SI units. The one exception is temperature, which always appears in conjunction with Boltzmann's constant, k. This constant is always absorbed into the temperature which then has the units of energy: $kT \rightarrow T$.

In the course of the text a number of relations are derived in terms of practical units as defined below:

Number density	n	10^{20} m^{-3}
Temperature	T	keV
Magnetic field	B	T (tesla)
Current	I	MA (megamperes)
Minor radius	a	m
Major radius	R_0	m

References

Lawson, J. D. (1957), *Proceedings of the Physical Society* **B70**, 6.

Further reading

The history of fusion

Dean, Stephen O. (2013). *Search for the Ultimate Energy Source*. New York: Springer Press.
Fowler, T. Kenneth (1997). *The Fusion Quest*. Baltimore, MD: The John Hopkins University Press.

Power balance in a fusion reactor

Dolan, T. J. (1982). *Fusion Research*. New York: Pergamon Press.
Freidberg, Jeffrey (2007). *Plasma Physics and Fusion Energy*. Cambridge: Cambridge University Press.
Kikuchi, Mitsuru, Karl Lackner, and Minh Quang Tran, editors (2012). *Fusion Physics*. Vienna: International Atomic Energy Agency.
Stacey, Weston M. (2005). *Fusion Plasma Physics*. Weinhein, Germany: Wiley-VCH.
Wesson, John (2011). *Tokamaks*, 4th edn. Oxford: Oxford University Press.

2

The ideal MHD model

2.1 Introduction

The goal of Chapter 2 is to provide a physical understanding of the ideal MHD model. Included in the discussion are (1) a basic description of the model, (2) a derivation starting from a more fundamental kinetic model, and, most importantly, (3) an examination of its range of validity.

In particular, it is shown that ideal MHD is the simplest fluid model that describes the macroscopic equilibrium and stability properties of a plasma. The claim of "simplest" is justified by a discussion of the large number of important plasma phenomena *not* covered by the model. However, in spite of its simplicity it is still a difficult model to solve analytically or even computationally because of the geometrical complexities associated with the two and three dimensionality of the configurations of fusion interest.

The derivation of the MHD model follows from the standard procedure of starting with a more fundamental and inclusive kinetic description of the plasma which describes the behavior of the electron and ion distribution functions. The mass, momentum, and energy moments of the kinetic equations are then evaluated. By introducing the characteristic length and time scales of ideal MHD, and making several corresponding ordering approximations, one is then able to close the system. The end result is the set of ideal MHD fluid equations.

The validity of the model is then assessed by examining the ordering assumptions used for closure to see whether or not they are consistent with the actual properties of fusion plasmas. This is a crucial step since ideal MHD is widely used in the design and interpretation of fusion experiments and one must be sure to understand the limits on the validity of the model. The assessment shows that while the basic derivation of MHD is straightforward there are several hidden surprises and subtleties.

Questions arise for two reasons. First, one of the basic assumptions used in the derivation, i.e., that the plasma is collision dominated, is *never* satisfied in plasmas of fusion interest. Even so, there is overwhelming empirical evidence that MHD provides an accurate description of macroscopic plasma behavior. This apparent good fortune is not a lucky coincidence but the consequence of some subtle physics; namely, those parts of the MHD model that are not valid because of violation of the collision dominated assumption are not directly involved in many if not most phenomena of interest. In other words, the model is only incorrect when it is unimportant. An attempt is made to clarify these issues in Chapter 9 by the introduction of several more sophisticated, low-collisionality plasma models whose regimes of validity are more closely aligned with actual experimental operating conditions. These models are more difficult to solve mathematically. However, several general equilibrium and stability comparison theorems are derived in Chapter 10 that help explain why ideal MHD works as well as it does.

The second subtle MHD issue concerns the following. Ideal MHD is an asymptotic model in the sense that specific length and time scales must be assumed for the derivation to be valid. In addition certain naturally appearing dimensionless parameters involving the MHD length and time scales must be ordered as small, medium, or large in order to close the system. For instance, high collisionality is represented by one such parameter. The issue here is that the multiple criteria defining the regime of validity arise from the need to simultaneously satisfy each assumption used in the derivation. However, a certain subset of phenomena described by the model requires only a corresponding subset of criteria to be satisfied, and consequently can have a much wider range of validity. One important example is MHD equilibrium. This important and useful information is discussed as the analysis progresses.

With these subtleties in mind attention is now focused on providing an in-depth description of the ideal MHD model.

2.2 Description of the model

The ideal MHD model provides a single-fluid description of long-wavelength, low-frequency, macroscopic plasma behavior. To put the model in perspective, it is perhaps useful to first discuss those plasma phenomena *not* described by ideal MHD.

Regarding physics in general, it has been pointed out that the three major discoveries of modern physics during the last two centuries, namely:

- Maxwell's equations with wave propagation
- relativity
- quantum mechanics

are each eliminated in the derivation of MHD.

Within the narrower confines of plasma physics itself, there are a variety of phenomena important in fusion plasmas. Among them are:

- radiation
- RF heating and current drive
- resonant particle effects
- micro instabilities
- classical and anomalous transport
- plasma–wall interactions
- resistive instabilities
- α-particle behavior.

Similarly, none of these phenomena is adequately described by ideal MHD.

Although the apparent lack of physical content is humbling, the one crucial phenomenon simply but accurately described by the model is the effect of magnetic geometry on the macroscopic equilibrium and stability of fusion plasmas. Specifically, ideal MHD answers such basic questions as: How does a given magnetic geometry provide forces to hold a plasma in equilibrium? Why are certain magnetic geometries more stable against macroscopic disturbances than others? Why do fusion configurations have such technologically undesirable shapes as a torus or a toroidal-helix?

One should be aware that in spite of the simplicity implied by its limited physical content, the ideal MHD model is still too difficult to solve in most geometries of interest. This will become evident as the text progresses by noting the many sophisticated expansions required to obtain analytic insight into the MHD behavior of various magnetic configurations. Attempts to solve similar problems using more comprehensive kinetic models are extremely difficult, even numerically, in realistic two- and three-dimensional geometries.

With this perspective the ideal MHD model is given by

$$\text{Mass:} \quad \frac{\partial \rho}{\partial t} + \nabla \cdot (\rho \mathbf{v}) = 0$$

$$\text{Momentum:} \quad \rho \frac{d\mathbf{v}}{dt} = \mathbf{J} \times \mathbf{B} - \nabla p$$

$$\text{Energy:} \quad \frac{d}{dt}\left(\frac{p}{\rho^\gamma}\right) = 0 \qquad (2.1)$$

$$\text{Ohm's law:} \quad \mathbf{E} + \mathbf{v} \times \mathbf{B} = 0$$

$$\text{Maxwell:} \quad \nabla \times \mathbf{E} = -\frac{\partial \mathbf{B}}{\partial t}$$

$$\nabla \times \mathbf{B} = \mu_0 \mathbf{J}$$

$$\nabla \cdot \mathbf{B} = 0$$

In these equations, the electromagnetic variables are the electric field $\mathbf{E}$, the magnetic field $\mathbf{B}$, and the current density $\mathbf{J}$. The fluid variables are the mass density ρ, the fluid velocity $\mathbf{v}$, and the pressure p. Also, $\gamma = 5/3$ is the ratio of specific heats and $d/dt = \partial/\partial t + \mathbf{v} \cdot \nabla$ is the convective derivative.

Observe that in ideal MHD the electromagnetic behavior is governed by the low-frequency, pre-Maxwell equations. The MHD fluid equations describe the time evolution of mass, momentum, and energy.

The mass equation implies that the total number of plasma particles is conserved; phenomena such as ionization, recombination, charge exchange, and unfortunately fuel depletion by fusion reactions, are negligible to a high order of accuracy on the MHD time scale.

The basic physics of the momentum equation corresponds to that of a fluid with three interacting forces: the pressure gradient force ∇p, the magnetic force $\mathbf{J} \times \mathbf{B}$, and the inertial force $\rho d\mathbf{v}/dt$. In static equilibrium it is the $\mathbf{J} \times \mathbf{B}$ force that balances the ∇p force, thereby confining the plasma. Dynamically, one must examine the stability of any such equilibrium to determine whether or not the plasma remains in place.

The energy equation expresses an adiabatic evolution characterized by a ratio of specific heats, $\gamma = 5/3$. The remaining relation is Ohm's law, which implies that in a reference frame moving with plasma the electric field is zero; that is, the plasma is a perfect conductor. It is the perfect conductivity assumption of Ohm's law that gives rise to the name "ideal" MHD.

As stated previously, the conditions for validity of the ideal MHD model imply that the phenomena of interest correspond to certain length and time scales. For macroscopic behavior the characteristic length scale is that of the overall plasma dimension. Denoting this dimension by a, then typically, for present day high-performance experiments, $a \sim 1$ m. The characteristic speed with which MHD phenomena occur is the thermal velocity of the plasma ions: $V_{Ti} = (2T_i/m_i)^{1/2}$, where T_i is the ion temperature and m_i is the ion mass. This gives rise to a characteristic MHD time $\tau_M \equiv a/V_{Ti}$. For m_i equivalent to deuterium and $T_i = 3$ keV then $\tau_M \sim 2$ μsec. The MHD length and time scales are compared with those of other basic plasma physics phenomena in Tables 2.1 and 2.2. In computing these values it has been assumed that $a = 1$ m, $T_e = T_i = 3$ keV, $B = 5$ T, $n = 10^{20}$ m^{-3} (particle number density) and m_i equivalent to deuterium. Also the Coulomb logarithm has been set to $\ln \Lambda = 19$.

The richness of plasma physics is clearly evidenced by the large number and wide range of length and time scales. Among these, ideal MHD lies midway between a variety of high-frequency microscopic phenomena and low-frequency collisional transport phenomena. This is the regime of macroscopic equilibrium and stability.

Table 2.1 *Comparison of the characteristic MHD time with those of other basic plasma physics phenomena. In practical formulas $T_e = T_i \equiv T$ is expressed in keV and n in $10^{20}\, m^{-3}$.*

Plasma physics time scales	Formulas	Values (sec)
Electron gyro period	$\tau_{ce} = 2\pi/\omega_{ce} = 2\pi m_e/eB$	7.1×10^{-12}
Electron plasma period	$\tau_{pe} = 2\pi/\omega_{pe} = 2\pi(m_e\varepsilon_0/ne^2)^{1/2}$	1.1×10^{-11}
Ion plasma period	$\tau_{pi} = (m_i/m_e)^{1/2}\tau_{pe}$	6.7×10^{-10}
Ion gyro period	$\tau_{ci} = (m_i/m_e)\tau_{ce}$	2.8×10^{-8}
MHD time	$\tau_M = a/V_{Ti}$	1.9×10^{-6}
Electron–electron collision time	$\tau_{ee} = 6.7 \times 10^{-6}\, T^{3/2}/n$	3.0×10^{-5}
Ion–ion collision time	$\tau_{ii} = (2m_i/m_e)^{1/2}\tau_{ee}$	2.6×10^{-3}
Energy equilibration time	$\tau_{eq} \approx (m_i/2m_e)\tau_{ee}$	5.5×10^{-2}
Energy confinement time for ignition	$\tau_E = 1.7/n$	1.7
Resistive diffusion time	$\tau_D = \mu_0 a^2/\eta = 40\, a^2 T^{3/2}$	2.1×10^2

Table 2.2 *Comparison of the characteristic MHD length with those of other basic plasma physics phenomena.*

Plasma physics length scales	Formulas	Values (m)
Electron gyro radius	$r_{Le} = V_{Te}/\omega_{ce}$	3.7×10^{-5}
Debye length	$\lambda_D = V_{Te}/\omega_{pe}$	5.8×10^{-5}
Electron skin depth	$\delta_e = c/\omega_{pe}$	5.3×10^{-4}
Ion gyro radius	$r_{Li} = (m_i/m_e)^{1/2} r_{Le}$	2.2×10^{-3}
Ion skin depth	$\delta_i = (m_i/m_e)^{1/2}\delta_e$	3.2×10^{-2}
MHD length	a	1
Ion–ion mfp	$\lambda_{ii} = V_{Ti}\tau_{ii}$	1.4×10^3
Electron–electron mfp	$\lambda_{ee} = \lambda_{ii}$	1.4×10^3

Observe that while typical MHD times (on the order of microseconds) are much shorter than typical experimental times (on the order of seconds), this does not imply that MHD can be applied only during a small fraction of the time of interest. The widest use of MHD involves the repeated calculation of equilibrium and stability over many small time increments during the slow evolution of a discharge.

For example, illustrated in Fig. 2.1 is a curve of the time evolution of the current in a tokamak. Also shown is a narrow time increment Δt during which MHD is valid. If at the beginning of Δt the external fields and the plasma current and pressure are given, then the MHD model can be used to calculate the equilibrium and stability of the system. This procedure can then be repeated many times over a continuing sequence of small time increments as the pressure and current slowly evolve on the transport time scale, an evolution whose physics is not described by

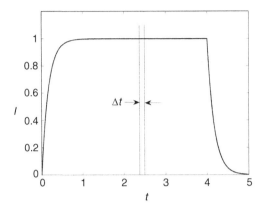

Figure 2.1   Diagram of a 5 sec current pulse in a typical tokamak. The interval $\Delta t \sim 50\,\mu\mathrm{sec}$ corresponds to one of a continual sequence of time slices during which ideal MHD is valid.

ideal MHD. In other words, the plasma conditions at the start of each MHD increment must be provided by experimental measurements or by theoretical transport modeling.

There is a critical implication, however, that during each successive time increment Δt, the plasma is MHD stable. If not, the plasma properties would likely change in a rapid and unfavorable way (e.g. a large increase in anomalous transport or complete destruction of the plasma) within several MHD times because of the strength of the instabilities. Stated differently, it is often more important to learn how to avoid MHD instabilities during the long time transport evolution of a plasma rather than to learn the precise details of what happens after an instability is excited – invariably such an instability leads to a major degradation in performance.

In summary, although many important plasma physics processes are neglected in the derivation of ideal MHD, the one critical phenomenon that remains is the self-consistent treatment of macroscopic equilibrium and stability in multidimensional magnetic geometries.

2.3 Derivation of the ideal MHD model

2.3.1 Starting equations

In order to more fully appreciate the physics content of ideal MHD as well as the subtleties involved, a derivation of the model is presented starting from basic principles. A number of such derivations exist in the literature and each, including the one presented here, follows the same general procedure: fluid moments are calculated from a general kinetic model, and then a number of assumptions are made to obtain closure of the system.

Many would agree that Braginskii's (1965) classic calculation represents one of the earliest and most rigorous derivations of plasma fluid equations in the collision dominated regime. The analysis presented here is noticeably simpler because the end goal is much narrower in scope; that is, the derivation is focused solely on the ideal MHD model with little attention given to the general question of determining transport coefficients in different collisionality regimes. Even so, some of the results of Braginskii are quoted to demonstrate the smallness of certain terms neglected in the ideal MHD model.

The starting point for the present derivation is the full set of Maxwell's equations coupled to the Boltzmann kinetic model for the plasma. Specifically, the kinetic model for each species and its coupling to Maxwell's equations is given by

$$\frac{df_\alpha}{dt} \equiv \frac{\partial f_\alpha}{\partial t} + \mathbf{v} \cdot \nabla f_\alpha + \frac{Z_\alpha e}{m_\alpha}(\mathbf{E} + \mathbf{v} \times \mathbf{B}) \cdot \nabla_v f_\alpha = \left(\frac{\partial f_\alpha}{\partial t}\right)_c$$

$$\nabla \times \mathbf{E} = -\frac{\partial \mathbf{B}}{\partial t}$$

$$\nabla \times \mathbf{B} = \mu_0 \mathbf{J} + \frac{1}{c^2}\frac{\partial \mathbf{E}}{\partial t} \tag{2.2}$$

$$\nabla \cdot \mathbf{E} = \frac{\sigma}{\varepsilon_0}$$

$$\nabla \cdot \mathbf{B} = 0$$

with the current and charge densities defined as

$$\mathbf{J} = \sum_\alpha Z_\alpha e \int \mathbf{v} f_\alpha d\mathbf{v}$$

$$\sigma = \sum_\alpha Z_\alpha e \int f_\alpha d\mathbf{v} \tag{2.3}$$

Here, $f_\alpha(\mathbf{r}, \mathbf{v}, t)$ is the distribution function for each species α, and is in general a function of spatial coordinate, velocity, and time. It is assumed that the plasma is fully ionized and consists of two species, electrons and singly charged ions. Hence, $\alpha = e, i$ and $Z_i = 1, Z_e = -1$. (A derivation of the starting kinetic equation can be found in many textbooks. See the Further reading at the end of the chapter. A heuristic derivation is given in Appendix A.)

In the kinetic description there are two types of forces that act on the particles. First there are the long-range Lorentz forces, $Z_\alpha e\, \mathbf{E}$ and $Z_\alpha e\, \mathbf{v} \times \mathbf{B}$, in which $\mathbf{E}$ and $\mathbf{B}$ are smoothly behaved fields calculated from the averaged current and charge densities as indicated in Eq. (2.3). Second, the right-hand side of the kinetic equation represents the forces due to short-range interactions, or collisions. In the derivation presented here, the details of the collision operator are not of major importance. Only certain global conservation relations are needed. For plasmas of fusion interest, the dominant collisions are elastic Coulomb collisions between both like and unlike particles. The conservation laws for elastic collisions can be summarized as follows. If the collision operator is defined in the usual way:

$$\left(\frac{\partial f_\alpha}{\partial t}\right)_c = \sum_\beta C_{\alpha\beta} \tag{2.4}$$

where $C_{\alpha\beta}$ represents collisions of particles of species α with particles of species β, then:

- Conservation of particles between like and unlike particle collisions implies

$$\int C_{ee}d\mathbf{v} = \int C_{ii}d\mathbf{v} = \int C_{ei}d\mathbf{v} = \int C_{ie}d\mathbf{v} = 0 \tag{2.5}$$

- Conservation of momentum and energy between like particle collisions implies

$$\int m_e \mathbf{v} C_{ee}d\mathbf{v} = \int m_i \mathbf{v} C_{ii}d\mathbf{v} = 0$$
$$\int \frac{1}{2} m_e v^2 C_{ee}d\mathbf{v} = \int \frac{1}{2} m_i v^2 C_{ii}d\mathbf{v} = 0 \tag{2.6}$$

- Conservation of total momentum and energy between unlike particle collisions implies

$$\int (m_e \mathbf{v} C_{ei} + m_i \mathbf{v} C_{ie})d\mathbf{v} = 0$$
$$\int \frac{1}{2} (m_e v^2 C_{ei} + m_i v^2 C_{ie})d\mathbf{v} = 0 \tag{2.7}$$

More information on collisions in a plasma can be obtained from many excellent textbooks listed at the end of the chapter.

The full set of kinetic-Maxwell equations provides a detailed and complete description of plasma behavior. At one end of the spectrum, it contains microscopic information about the orbits of individual charged particles on the very short gyro time scale and gyro radius length scale. At the other end, it accurately describes the macroscopic behavior of large plasma experiments including MHD equilibrium and stability as well as very slow transport phenomena. Not surprisingly, the complexity arising from this breadth of information makes it virtually impossible to solve, even numerically, the kinetic-Maxwell system of equations in any non-trivial geometry. This realization has led to the development of several simpler models with narrower physics content. Ideal MHD is one such model.

2.3.2 Two-fluid equations

The derivation of the ideal MHD equations begins by taking moments of the kinetic equation. This calculation transforms a single equation for f_α in seven variables $(\mathbf{r}, \mathbf{v}, t)$ into an infinite set of fluid equations in four variables $(\mathbf{r}, t)$. The procedure is straightforward although somewhat tedious. Once the fluid equations are obtained, physical variables such as density, velocity, and pressure are introduced. Then, the physical regime of interest is defined, allowing various approximations and expansions to be made, ultimately leading to truncation and closure of the infinite set of equations. Since the truncated system is by design focused on a specific regime of physics its total information content is much less than that of the original kinetic equation. Even so, the virtue of the fluid equations is that they are enormously simpler to solve.

To derive the ideal MHD model, the moments that are required correspond to mass, momentum, and energy; that is, starting with the kinetic equation one evaluates

$$\int g_i \left[\frac{df_\alpha}{dt} - \left(\frac{\partial f_\alpha}{\partial t} \right)_c \right] d\mathbf{v} = 0 \tag{2.8}$$

for $i = 1{-}3$ with $g_i(\mathbf{v})$ given by

$$\begin{aligned} g_1 &= 1 && \text{(mass)} \\ g_2 &= m_\alpha \mathbf{v} && \text{(momentum)} \\ g_3 &= m_\alpha v^2/2 && \text{(energy)} \end{aligned} \tag{2.9}$$

After a straightforward calculation the fluid equations for each species can be written as

$$\frac{\partial n_\alpha}{\partial t} + \nabla \cdot (n_\alpha \mathbf{u}_\alpha) = 0$$

$$\frac{\partial}{\partial t} (m_\alpha n_\alpha \mathbf{u}_\alpha) + \nabla \cdot (m_\alpha n_\alpha \langle \mathbf{v} \mathbf{v} \rangle) - Z_\alpha e n_\alpha (\mathbf{E} + \mathbf{u}_\alpha \times \mathbf{B}) = \int m_\alpha \mathbf{v} \, C_{\alpha\beta} d\mathbf{v} \tag{2.10}$$

$$\frac{\partial}{\partial t} \left(\frac{1}{2} m_\alpha n_\alpha \langle v^2 \rangle \right) + \nabla \cdot \left(\frac{1}{2} m_\alpha n_\alpha \langle v^2 \mathbf{v} \rangle \right) - Z_\alpha e n_\alpha \mathbf{u}_\alpha \cdot \mathbf{E} = \int \frac{1}{2} m_\alpha v^2 \, C_{\alpha\beta} d\mathbf{v}$$

Here, n_α and $\mathbf{u}_\alpha$ are the macroscopic number density and fluid velocity defined by

$$\begin{aligned} n_\alpha &= \int f_\alpha d\mathbf{v} \\ \mathbf{u}_\alpha &= \frac{1}{n_\alpha} \int \mathbf{v} f_\alpha d\mathbf{v} \end{aligned} \tag{2.11}$$

and the only non-zero contributions to the collision terms are from unlike particle interactions, $\alpha \neq \beta$. The quantities $\langle \mathbf{vv} \rangle$, $\langle v^2 \rangle$, and $\langle v^2 \mathbf{v} \rangle$ are higher moments of the distribution function defined by the general relation

$$\langle h \rangle = \frac{1}{n_\alpha} \int h f_\alpha d\mathbf{v} \tag{2.12}$$

The next step in the derivation is to introduce a new independent velocity variable, $\mathbf{w} = \mathbf{v} - \mathbf{u}_\alpha(\mathbf{r}, t)$, representing the random thermal motion of the particles, which by definition satisfies $\langle \mathbf{w} \rangle = 0$. By introducing this variable into Eq. (2.10) one can write the fluid equations in terms of more physical macroscopic quantities. In particular, the quantities of interest are the scalar pressure,

$$p_\alpha = \frac{1}{3} m_\alpha n_\alpha \langle w^2 \rangle \tag{2.13}$$

the total pressure tensor,

$$\mathbf{P} = m_\alpha n_\alpha \langle \mathbf{ww} \rangle \tag{2.14}$$

the anisotropic part of the pressure tensor,

$$\mathbf{\Pi}_\alpha = \mathbf{P}_\alpha - p_\alpha \mathbf{I} \tag{2.15}$$

the temperature,

$$T_\alpha = p_\alpha / n_\alpha \tag{2.16}$$

the heat flux due to random motion,

$$\mathbf{q}_\alpha = \frac{1}{2} m_\alpha n_\alpha \langle w^2 \mathbf{w} \rangle \tag{2.17}$$

the mean momentum transferred between unlike particles due to the friction of collisions,

$$\mathbf{R}_\alpha = \int m_\alpha \mathbf{w} C_{\alpha\beta} d\mathbf{w} \tag{2.18}$$

and the heat generated and transferred between unlike particles due to collisional dissipation,

$$Q_\alpha = \int \frac{1}{2} m_\alpha w^2 C_{\alpha\beta} d\mathbf{w} \tag{2.19}$$

Substituting these definitions into Eq. (2.10) and making use of the mass continuity relation leads to the following form of the moment equations:

$$\left(\frac{dn_\alpha}{dt}\right)_\alpha + n_\alpha \nabla \cdot \mathbf{u}_\alpha = 0$$

$$m_\alpha n_\alpha \left(\frac{d\mathbf{u}_\alpha}{dt}\right)_\alpha - Z_\alpha e n_\alpha (\mathbf{E} + \mathbf{u}_\alpha \times \mathbf{B}) + \nabla \cdot \mathbf{P}_\alpha = \mathbf{R}_\alpha$$

$$n_\alpha \left[\frac{d}{dt}\left(\frac{1}{2}m_\alpha u_\alpha^2 + \frac{3}{2}T_\alpha\right)\right]_\alpha - Z_\alpha e n_\alpha \mathbf{u}_\alpha \cdot \mathbf{E} + \nabla \cdot (\mathbf{u}_\alpha \cdot \mathbf{P}_\alpha + \mathbf{q}_\alpha) = Q_\alpha + \mathbf{u}_\alpha \cdot \mathbf{R}_\alpha$$

$$(2.20)$$

where

$$\left(\frac{d}{dt}\right)_\alpha \equiv \frac{\partial}{\partial t} + \mathbf{u}_\alpha \cdot \nabla \qquad (2.21)$$

is the convective derivative for the species α.

A further reduction can be made leading to a simpler form of the energy equation. This form is obtained by evaluating the dot product of the momentum equation with $\mathbf{u}_\alpha$ and subtracting the result from the energy equation. Upon replacing the energy equation with the simplified form one obtains the final set of two-fluid equations given by

$$\left(\frac{dn_\alpha}{dt}\right)_\alpha + n_\alpha \nabla \cdot \mathbf{u}_\alpha = 0$$

$$m_\alpha n_\alpha \left(\frac{d\mathbf{u}_\alpha}{dt}\right)_\alpha - Z_\alpha e n_\alpha (\mathbf{E} + \mathbf{u}_\alpha \times \mathbf{B}) + \nabla \cdot \mathbf{P}_\alpha = \mathbf{R}_\alpha$$

$$\frac{3}{2}n_\alpha \left(\frac{dT_\alpha}{dt}\right)_\alpha + \mathbf{P}_\alpha : \nabla \mathbf{u}_\alpha + \nabla \cdot \mathbf{q}_\alpha = Q_\alpha$$

$$(2.22)$$

$$\nabla \times \mathbf{E} = -\frac{\partial \mathbf{B}}{\partial t}$$

$$\nabla \times \mathbf{B} = \mu_0 e(n_i \mathbf{u}_i - n_e \mathbf{u}_e) + \frac{1}{c^2}\frac{\partial \mathbf{E}}{\partial t}$$

$$\nabla \cdot \mathbf{E} = \frac{e}{\varepsilon_0}(n_i - n_e)$$

$$\nabla \cdot \mathbf{B} = 0$$

where, as stated, the ions are singly charged so that $Z_i = -Z_e = 1$.

This set of equations is exact if not very useful since there is as yet no prescription for closing the system: there are more unknowns than equations. The particular prescription that leads to the ideal MHD equations consists of the following steps. First, certain asymptotic orderings are introduced that eliminate the very high-frequency, short-wavelength information in the model. These orderings are well satisfied for the phenomena of interest, i.e., those involving the macroscopic behavior of fusion plasmas. Next, the equations are rewritten as a set of single-fluid equations by the introduction of appropriate single-fluid variables. The system still has more unknowns than equations. A crucial step follows. The plasma is assumed to be collision dominated, thereby enabling the higher-order moments to be approximated in terms of the basic variables by means of Braginskii's well-known transport theory (1965). The result is a complicated but closed set of equations. It is at this point that the characteristic MHD length and time scales are introduced. The various terms appearing in the equations can now be ordered as small, medium, or large in terms of several dimensionless parameters. One can then define a set of validity conditions in terms of these parameters such that the remaining, greatly simplified set of equations corresponds to ideal MHD.

A remarkable feature of the analysis is that while the model is derived on the assumption that the plasma is collision dominated, the final ideal MHD equations do not explicitly depend on the details of the collisions. The main purpose of the collisional transport theory is to provide the appropriate scaling of the transport coefficients. The corresponding terms are then systematically ordered out of the model by comparing them with other terms dominated by the length and time scales associated with macroscopic MHD behavior.

2.3.3 *Low-frequency, long-wavelength, asymptotic expansions*

There are two important approximations that can be made, leading to substantial simplifications in the two-fluid equations. These approximations serve to eliminate the high-frequency, short-wavelength information from the model. Each approximation has the form of an asymptotic expansion which eliminates a leading-order time derivative and thus alters the basic mathematical structure of the model.

The first approximation represents the transformation of the full Maxwell's equations to the low-frequency pre-Maxwell's equations. This limit is formally accomplished by letting $\varepsilon_0 \to 0$. As a result, the displacement current $\varepsilon_0 \partial \mathbf{E}/\partial t$ and the net charge $\varepsilon_0 \nabla \cdot \mathbf{E}$ can both be neglected. The resulting set of equations is easily shown to be Galilean invariant.

The neglect of the displacement current requires that the electromagnetic waves of interest have phase velocities much slower than the speed of light (i.e., $\omega/k \ll c$), and that the characteristic thermal velocities be non-relativistic (i.e., $V_{Ti} \ll V_{Te} \ll c$). The neglect of the net charge restricts attention to plasma behavior whose characteristic frequency is much less than the electron plasma frequency (i.e., $\omega \ll \omega_{pe}$)

and whose characteristic length is much longer than the Debye length (i.e., $a \gg \lambda_D$). An examination of Tables 2.1 and 2.2 shows that these assumptions are satisfied by a large margin when applied to the MHD behavior of fusion plasmas.

The neglect of $\varepsilon_0 \nabla \cdot \mathbf{E}$ implies that

$$n_i = n_e \equiv n \tag{2.23}$$

This relationship is called the quasineutral approximation. It should be emphasized that Eq. (2.23) does not imply that $\mathbf{E} = 0$ or $\nabla \cdot \mathbf{E} = 0$, only that $\varepsilon_0 \nabla \cdot \mathbf{E}/en \ll 1$. For any low-frequency macroscopic charge separation that tends to develop, the electrons have more than an adequate time to respond, creating an electric field whose direction is such as to cancel the charge imbalance. This maintains the plasma in local quasineutrality.

As an example, consider an electrostatic problem with $\mathbf{E} = -\nabla\phi$ and $n_e = n_e(\phi, \mathbf{r})$, $n_i = n_i(\phi, \mathbf{r})$. Equating n_e to n_i and inverting the relationship then gives $\phi = \phi(\mathbf{r})$. Quasineutrality implies that the electric field calculated from $\mathbf{E} = -\nabla\phi$ satisfies $\varepsilon_0 \nabla \cdot \mathbf{E}/en \ll 1$.

The second asymptotic assumption neglects electron inertia in the electron momentum equation and is accomplished formally by letting $m_e \to 0$. This implies that on time scales of MHD interest the electrons have an infinitely fast response time because of their small mass. Specifically, time scales long compared to those of the electron plasma frequency, ω_{pe}, and the electron cyclotron frequency, ω_{ce}, are required. Similarly, the length scales must be long compared to the Debye length, λ_D, and the electron gyro radius, r_{Le}. As before, an examination of Tables 2.1 and 2.2 shows that these conditions are easily satisfied for macroscopic phenomena in fusion plasmas.

2.3.4 The single-fluid equations

Using the asymptotic approximations just described, one can derive a set of single-fluid equations by introducing a new set of fluid variables. To begin, it is customary to introduce a mass density rather than a number density. Since $m_e \to 0$ and $n_i = n_e \equiv n$, the mass density corresponds to that of the ions and is defined as

$$\rho = m_i n \tag{2.24}$$

Likewise, the momentum of the fluid is also carried by the ions, so that the appropriate definition of the fluid velocity is given by

$$\mathbf{v} = \mathbf{u}_i \tag{2.25}$$

The current density is proportional to the difference in flow velocity between electrons and ions,

$$\mathbf{J} = en(\mathbf{u}_i - \mathbf{u}_e)$$
$$\mathbf{u}_e = \mathbf{v} - \mathbf{J}/en \tag{2.26}$$

The final definitions required are for the total pressure and temperature,

$$p = p_i + p_e = 2nT$$
$$T = (T_i + T_e)/2 \tag{2.27}$$

Equations (2.24)–(2.27) relate the single-fluid variables ρ, $\mathbf{v}$, $\mathbf{J}$, p, T to the two-fluid variables n, $\mathbf{u}_i$, $\mathbf{u}_e$, p_i, p_e, T_i, T_e.

Two points require discussion before proceeding with the single-fluid equations. First, care must be exercised to conserve the total information content of the starting equations. For instance, several of the single-fluid equations are obtained by combining various equations of the two-fluid model. If these manipulations involve two equations from the starting model, then two equivalent equations must appear in the final model.

This is not as trivial as it might seem, as evidenced by the second point of discussion: namely, that if one counts fluid variables then the two-fluid model has 11 unknowns while the single-fluid model has 9 unknowns. A careful accounting of the information content in the two-fluid equations is shown to explain and eliminate this imbalance.

Consider now the derivation of the single-fluid model. The first equation is obtained from the mass conservation equations. Multiplying the ion mass equation by m_i one finds

$$\frac{\partial \rho}{\partial t} + \nabla \cdot (\rho \mathbf{v}) = 0 \tag{2.28}$$

which is identical to the ideal MHD mass conservation relation. The other information contained in these equations is obtained by multiplying the electron and ion equations by e and then subtracting. The result is

$$\nabla \cdot \mathbf{J} = 0 \tag{2.29}$$

Equation (2.29) is redundant with charge conservation in the low-frequency form of Maxwell's equations: $\nabla \cdot (\nabla \times \mathbf{B} - \mu_0 \mathbf{J}) = \mu_0 \nabla \cdot \mathbf{J} = 0$.

The next set of single-fluid equations follows from the momentum equations. The electron and ion equations are first added together. Making use of the fact that $\mathbf{R}_e = -\mathbf{R}_i$ leads to the relationship

$$\rho \frac{d\mathbf{v}}{dt} - \mathbf{J} \times \mathbf{B} + \nabla p = -\nabla \cdot (\mathbf{\Pi}_i + \mathbf{\Pi}_e) \tag{2.30}$$

Here, $d/dt = \partial/\partial t + \mathbf{v} \cdot \nabla$ now represents the convective derivative moving with the (ion) fluid. The left-hand side of Eq. (2.30) corresponds to the ideal MHD momentum equation. The condition under which the right-hand side is negligible defines one constraint on the parameter regime for which the MHD equations are valid and will be discussed shortly. Note that there is no electric force, $\sigma\mathbf{E}$, acting on the fluid, since $\sigma = 0$ as a result of the quasineutral approximation.

The second piece of information contained in the two-fluid momentum equations is obtained by simply rewriting the electron equation in terms of the single-fluid variables. This leads to the following relation:

$$\mathbf{E} + \mathbf{v} \times \mathbf{B} = \frac{1}{en}(\mathbf{J} \times \mathbf{B} - \nabla p_e - \nabla \cdot \mathbf{\Pi}_e + \mathbf{R}_e) \qquad (2.31)$$

The left-hand side of Eq. (2.31) corresponds to the ideal MHD Ohm's law. The conditions under which the right-hand side is negligible determine additional constraints on the parameter regime for the validity of MHD and are also discussed shortly.

The third set of single-fluid equations follows from the two-fluid energy equations. After some straightforward algebra one obtains

$$\frac{d}{dt}\left(\frac{p_i}{\rho^\gamma}\right) = \frac{2}{3\rho^\gamma}(Q_i - \nabla \cdot \mathbf{q}_i - \mathbf{\Pi}_i : \nabla\mathbf{v})$$

$$\frac{d}{dt}\left(\frac{p_e}{\rho^\gamma}\right) = \frac{2}{3\rho^\gamma}\left[Q_e - \nabla \cdot \mathbf{q}_e - \mathbf{\Pi}_e : \nabla\left(\mathbf{v} - \frac{\mathbf{J}}{en}\right)\right] + \frac{1}{en}\mathbf{J} \cdot \nabla\left(\frac{p_e}{\rho^\gamma}\right)$$

$$(2.32)$$

where $\gamma = 5/3$. The last term in the electron component of Eq. (2.32) results from the convective derivative relation $(d/dt)_e = (d/dt) - (\mathbf{J}/en) \cdot \nabla$. The left-hand sides of Eq. (2.32) are closely related to the ideal MHD equation of state. There are a number of terms on the right-hand side, which in order to be neglected define additional conditions on the parameter regime of validity. Furthermore, since MHD is a single-fluid model, other assumptions must be made to couple the individual electron and ion temperatures and pressures into single-fluid variables.

The final equations of the single-fluid model are the remaining low-frequency Maxwell equations given by

$$\nabla \times \mathbf{E} = -\frac{\partial \mathbf{B}}{\partial t}$$

$$\nabla \times \mathbf{B} = \mu_0\mathbf{J} \qquad (2.33)$$

$$\nabla \cdot \mathbf{B} = 0$$

As stated previously, Eq. (2.29) is consistent with the low-frequency form of Ampere's law.

The single-fluid model is described by Eqs. (2.28)–(2.33). No assumptions other than the asymptotic approximations have been made at this point. Although the left-hand sides of the fluid equations are identical to the MHD model, the full set of equations is still not closed because of the presence of the as yet undefined higher moments and the unresolved coupling of the electron and ion energies.

2.3.5 The ideal MHD limit

This section contains a description of the critical assumptions required for the single-fluid model to reduce to ideal MHD; in particular, quantitative conditions are derived that determine when the right-hand sides of Eqs. (2.30)–(2.32) are negligible.

The basic requirement for the validity of ideal MHD is that both the electrons and ions be collision dominated. This is the usual requirement for a fluid model to be useful. If there are sufficient collisions, a given particle remains reasonably close to its neighboring particles during the time scales of interest. In this case the division of the plasma into small identifiable fluid elements provides a good description of the physics.

The question as to whether a given model describes collision dominated or collisionless behavior rests largely in the evolution of the pressure tensor; that is, if one considers the exact mass and momentum equations it is only the pressure tensor that remains as an undefined higher moment. A comparison between Eq. (2.30) and the ideal MHD momentum equation indicates that for the models to be equivalent, $\mathbf{\Pi}_i$ and $\mathbf{\Pi}_e$ must be negligible. The implication is that the full pressure tensor $\mathbf{P}$ reduces to a simple scalar isotropic pressure p.

One intuitively expects an isotropic pressure to arise in systems where many collisions take place on a time scale which is short compared to those of interest. The collisions rapidly randomize the distribution function into a Maxwellian form giving rise to an isotropic pressure. This is the basic assumption of the scalar pressure ideal MHD model. In collisionless models closure of the pressure tensor is more complicated, requiring the solution of reduced kinetic equations. A perhaps surprising feature of MHD is that for most problems involving equilibrium and stability the results are not overly sensitive to the specific model used to describe the evolution of $\mathbf{P}$. It is this fact that is ultimately responsible for the unexpectedly reliable predictions of ideal MHD well outside its regime of validity.

The collision dominated assumption enters into the single-fluid equations as follows. In this limit the distribution functions for electrons and ions are nearly locally Maxwellians. As a consequence, one can refer to well-established theories, such as given by Braginskii (1965), in order to obtain expressions for the higher-order fluid moments in terms of appropriate transport coefficients. By defining the characteristic length and time scales of ideal MHD it is then possible to compare the MHD terms with the transport terms. This then determines a set of conditions for the right-hand sides of the single-fluid equations to be negligible. These are the validity conditions for ideal MHD.

A convenient place to begin the analysis is to recall the characteristic MHD length and time scales. Since the main goal of ideal MHD is the investigation of macroscopic phenomena, the length scales of interest correspond to the macroscopic dimensions of the plasma denoted by a. The typical time scale of MHD interest corresponds to a/V_{Ti}, the ion thermal transit time across a macroscopic plasma dimension. This time scale is characteristic of many MHD plasma instabilities and represents the fastest time scale in which macroscopic plasma motion can occur. In determining the scaling relations it is helpful to introduce the characteristic MHD frequency ω and wave number k as follows:

$$
\omega \sim \frac{\partial}{\partial t} \sim \frac{V_{Ti}}{a}
$$
$$
k \sim |\nabla| \sim \frac{1}{a}
\tag{2.34}
$$

and, similarly, the resulting velocity

$$
\frac{\omega}{k} \sim |\mathbf{v}| \sim V_{Ti}
\tag{2.35}
$$

The next step in the analysis is to consider the conditions for the collision-dominated transport theory to be valid. There are two such conditions. The first requires that during the MHD time scale of interest, each species has sufficient collisions to make the distribution function nearly Maxwellian. For the ions the dominant collision mechanism is due to ion–ion interactions, characterized by a collision time τ_{ii}. The electrons become Maxwellian by colliding with either other electrons or ions. Since $\tau_{ee} \sim \tau_{ei}$ it is not important to make this distinction. Hence, the condition that each species be collision dominated is given by

$$
\begin{array}{lll}
\text{Ions} & \omega\tau_{ii} \sim V_{Ti}\tau_{ii}/a \ll 1 & \\
\text{Electrons} & \omega\tau_{ee} \sim (m_e/m_i)^{1/2}V_{Ti}\tau_{ii}/a \ll 1 &
\end{array}
\tag{2.36}
$$

Here, use has been made of the fact that $\tau_{ee} \sim (m_e/m_i)^{1/2}\tau_{ii}$ when $T_i \sim T_e$. As might be expected, this condition is more restrictive for ions than electrons.

The second condition for the collision dominated theory to be valid requires that the macroscopic scale length be much longer than the mean free path. Noting that the mean free path for each species is given by $\lambda_\alpha \sim V_{T\alpha}\tau_{\alpha\alpha}$, the condition $\lambda_\alpha \ll a$ reduces to $V_{Ti}\tau_{ii}/a \ll 1$ for both electrons and ions; that is, for the ions both the time scale and length scale requirements yield the same condition for high collisionality. For the electrons, the length scale requirement is more restrictive and yields the same condition as for the ions. Thus, the conditions for a collision dominated plasma can be summarized as follows:

$$V_{Ti}\tau_{ii}/a \sim V_{Te}\tau_{ee}/a \ll 1 \tag{2.37}$$

The collision dominated assumption is now used to estimate the higher-order moments, which enter as transport terms, in the single-fluid equations. (Readers unfamiliar with the evaluation of transport coefficients should attempt Problems 2.2–2.4, which demonstrate a simple procedure for deriving the basic scaling of such coefficients.) For present purposes the transport coefficients used in the analysis are those derived by Braginskii (1965) and are summarized for convenience in Appendix B.

The analysis begins with Eq. (2.31), the momentum equation. The matrix elements for $\mathbf{\Pi}_i$ and $\mathbf{\Pi}_e$ are a rather complicated series of terms. The leading-order effect, however, is viscosity. Moreover, the ion viscosity coefficient is larger than that of the electrons by a factor $(m_i/m_e)^{1/2}$. From Braginskii (1965) it follows that the largest elements of the $\mathbf{\Pi}_i$ tensor have the form (in rectangular coordinates)

$$\Pi_{jj} \sim \mu\left(2\nabla_\parallel \cdot \mathbf{v}_\parallel - \frac{2}{3}\nabla \cdot \mathbf{v}\right) \sim \mu\frac{V_{Ti}}{a} \tag{2.38}$$

where the viscosity coefficient μ is given by

$$\mu \sim nT_i\tau_{ii} \tag{2.39}$$

If the right-hand side of Eq. (2.30) is now compared with the ∇p term one finds

$$|\nabla \cdot \mathbf{\Pi}_i/\nabla p| \sim V_{Ti}\tau_{ii}/a \ll 1 \tag{2.40}$$

Therefore, if the collision dominated assumption is satisfied, the viscosity is negligible and the momentum equation reduces to that of ideal MHD.

The next equation to consider is Ohm's law, given by Eq. (2.31). From the argument just given it is clear that the $\mathbf{\Pi}_e$ term can be neglected compared to the ∇p_e term. From the momentum equation it also follows that the $\mathbf{J} \times \mathbf{B}$ and ∇p_e terms are comparable. The $\mathbf{J} \times \mathbf{B}$ term in Ohm's law represents the Hall effect, while the ∇p_e term represents the effect of the electron diamagnetic drift. Comparing either of these terms with the $\mathbf{v} \times \mathbf{B}$ term yields

$$\frac{|\nabla p_e/en|}{|\mathbf{v} \times \mathbf{B}|} \sim \frac{r_{Li}}{a} \tag{2.41}$$

where $r_{Li} = V_{Ti}/\omega_{ci}$ is the ion gyro radius. Thus, if one makes the additional assumption that

$$r_{Li}/a \ll 1 \tag{2.42}$$

which is well satisfied in fusion experiments, then the $\mathbf{J} \times \mathbf{B}$ and ∇p_e terms can be neglected in the Ohm's laws. Note that Eq. (2.42) implies that

$$\omega/\omega_{ci} \sim r_{Li}/a \ll 1 \qquad (2.43)$$

MHD frequencies are much lower than the ion gyro frequency.

The remaining term on the right-hand side of Eq. (2.31), $\mathbf{R}_e/en$, represents the momentum transfer due to the friction of collisions between electrons and ions. The dominant contribution to $\mathbf{R}_e$ is electrical resistivity. Using the results from Braginskii, one can express $\mathbf{R}_e$ as

$$\mathbf{R}_e/en \sim \eta \mathbf{J} \qquad (2.44)$$

where the electrical resistivity η is given by

$$\eta \sim \frac{m_e}{ne^2 \tau_{ei}} \qquad (2.45)$$

Substituting the scaling relation (from the momentum equation) $|\mathbf{J}| \sim |\nabla p_e|/|\mathbf{B}|$ leads to the following requirement for the $\eta \mathbf{J}$ term to be negligible compared to the $\mathbf{v} \times \mathbf{B}$ term:

$$\frac{|\eta \mathbf{J}|}{|\mathbf{v} \times \mathbf{B}|} \sim \left(\frac{m_e}{m_i}\right)^{1/2} \left(\frac{r_{Li}}{a}\right)^2 \left(\frac{a}{V_{Ti}\tau_{ii}}\right) \ll 1 \qquad (2.46)$$

Equation (2.46) implies that while τ_{ii} must be sufficiently small for the collision dominated approximation to hold it cannot be so small that the plasma will be dominated by resistive diffusion. Alternatively, one can view Eq. (2.46) as a requirement to make the macroscopic dimension a large enough so that the resistive diffusion time is long compared to the characteristic MHD time. Thus, in order for the ideal MHD Ohm's law to apply, the small gyro radius and small resistivity conditions must both be satisfied. Practically, the condition given by Eq. (2.46) is well satisfied in all fusion experiments.

Before proceeding to the energy equations there are several subtleties related to the electric field and Ohm's law that should be discussed. (1) The ideal MHD Ohm's law $\mathbf{E} + \mathbf{v} \times \mathbf{B} = 0$ implies that $|\mathbf{E}_\perp| \sim V_{Ti}B$. However, in many MHD stability analyses the equilibrium ion flow is assumed to be zero (i.e., $\mathbf{v} = 0$). In this case the equilibrium ion pressure is confined by an electrostatic electric field $\mathbf{E} = -\nabla \phi$ that scales as $|\mathbf{E}_\perp| \sim |\nabla p_i|/en \sim (r_{Li}/a)V_{Ti}B$. The equilibrium $\mathbf{E}_\perp$ is incorrectly omitted from ideal MHD equilibrium analyses (since it is being compared to zero) because of the $r_{Li}/a \ll 1$ assumption. (2) The electric field associated with fast time-varying MHD motions is treated correctly. Specifically, in terms of the vector potential it follows that $\mathbf{E}_\perp \sim i\omega \mathbf{A}_\perp$ and $\mathbf{B} \sim i\mathbf{k} \times \mathbf{A}$ which, using the MHD ordering, implies that $|\mathbf{E}_\perp| \sim V_{Ti}|\mathbf{B}|$. This is consistent with the ideal MHD

Ohm's law. The situation can be summarized as follows: ideal MHD treats the inductive part of the electric field correctly and the electrostatic part incorrectly. The error, nevertheless, is not important since the only way the electric field is used in ideal MHD is in Faraday's law, which requires the evaluation of $\nabla \times \mathbf{E}$. Therefore, even if the electrostatic part of $\mathbf{E}$ (i.e., the $\nabla \phi$ part) is calculated incorrectly the error is annihilated by the curl operation.

Lastly, consider the energy conservation relations given by Eq. (2.32). With the assumptions already made, most of the terms on the right-hand sides are already negligible. In particular,

$$\frac{\mathbf{\Pi}_e : \nabla(\mathbf{J}/en)}{\partial p_e/\partial t} \sim \left(\frac{m_e}{m_i}\right)^{1/2} \left(\frac{r_{Li}}{a}\right) \left(\frac{V_{Ti}\tau_{ii}}{a}\right) \ll 1$$

$$\frac{(\mathbf{J}\cdot\nabla p_e)/en}{\partial p_e/\partial t} \sim \left(\frac{r_{Li}}{a}\right) \ll 1$$

$$\frac{\mathbf{\Pi}_e : \nabla\mathbf{v}}{\partial p_e/\partial t} \sim \left(\frac{m_e}{m_i}\right)^{1/2} \left(\frac{V_{Ti}\tau_{ii}}{a}\right) \ll 1$$

$$\frac{\mathbf{\Pi}_i : \nabla\mathbf{v}}{\partial p_i/\partial t} \sim \left(\frac{V_{Ti}\tau_{ii}}{a}\right) \ll 1$$

(2.47)

The remaining terms contain the heat flux $\mathbf{q}_\alpha$ and the collisional heating Q_α. The largest contribution to $\mathbf{q}_\alpha$ is due to thermal conductivity. There are separate conductivity coefficients parallel and perpendicular to the field and for electrons and ions. By far, the largest coefficients are those parallel to the field so that $\mathbf{q}_\alpha \approx -\kappa_{\|\alpha}\nabla_\| T_\alpha$. The main contributions to the collisional heating Q_α are joule heating and electron and ion energy equilibration. If the condition to neglect resistive diffusion given by Eq. (2.46) is satisfied, then joule heating is also negligible. Therefore, what remains of the energy equations is as follows:

$$\frac{d}{dt}\left(\frac{p_i}{\rho^\gamma}\right) = \frac{2}{3\rho^\gamma}\left[\nabla_\| \cdot \left(\kappa_{\|i}\nabla_\| T_i\right) + \frac{n(T_e - T_i)}{\tau_{eq}}\right]$$

$$\frac{d}{dt}\left(\frac{p_e}{\rho^\gamma}\right) = \frac{2}{3\rho^\gamma}\left[\nabla_\| \cdot \left(\kappa_{\|e}\nabla_\| T_e\right) - \frac{n(T_e - T_i)}{\tau_{eq}}\right]$$

(2.48)

where τ_{eq} is the energy equilibration time. As they now stand, these equations describe the separate time evolution of the electrons and ions. In order to obtain the MHD energy equation a further assumption is required which couples the electron and ion energies together. This assumption corresponds to the condition that the energy equilibration time be short compared to the characteristic time of interest, so that $T_e \approx T_i$. This can be expressed as $\omega \tau_{eq} \ll 1$, or, using the relation $\tau_{eq} \sim (m_i/m_e)^{1/2} \tau_{ii}$,

$$\omega \tau_{eq} \sim \left(\frac{m_i}{m_e}\right)^{1/2} \left(\frac{V_{Ti}\tau_{ii}}{a}\right) \ll 1 \tag{2.49}$$

Because the energy equilibration time is long compared to the momentum exchange time, Eq. (2.49) is even more restrictive in terms of collisionality than Eq. (2.37), the basic condition for the collision dominated expansion to be valid.

If Eq. (2.49) is satisfied, then

$$T_i \approx T_e \equiv T$$
$$p_i \approx p_e = nT \equiv p/2 \tag{2.50}$$

Note that the strong collisionality assumption resolves the problem of the unequal number of fluid variables in the two-fluid and ideal MHD models as only a single temperature and pressure rather than two separate ones for each quantity are now required.

Both the electron and ion energy equations yield the identical piece of information – namely that $T_i \approx T_e$ in the limit of small τ_{eq}. The second piece of independent information follows from annihilating the $T_i \approx T_e$ redundancy by adding the equations and setting $T_i = T_e = T$. The result is

$$\frac{d}{dt}\left(\frac{p}{\rho^\gamma}\right) = \frac{2}{3\rho^\gamma} \nabla_\parallel \cdot \left[(\kappa_{\parallel i} + \kappa_{\parallel e})\nabla_\parallel T\right] \tag{2.51}$$

The ideal MHD equation of state follows when the right-hand side of Eq. (2.51) can be neglected. Braginskii's transport theory shows that the parallel electron thermal conductivity $\kappa_{\parallel e} \sim nT_e\tau_{ee}/m_e$ is larger by $(m_i/m_e)^{1/2}$ than that of the ions. Consequently, the right-hand side is negligible when

$$\frac{\nabla_\parallel \cdot (\kappa_{\parallel e}\nabla_\parallel T)}{\partial p/\partial t} \sim \left(\frac{m_i}{m_e}\right)^{1/2} \left(\frac{V_{Ti}\tau_{ii}}{a}\right) \ll 1 \tag{2.52}$$

which is identical to Eq. (2.49).

This completes the derivation of the ideal MHD equations. In summary, the validity of ideal MHD imposes several conditions on the plasma: (1) high collisionality; (2) characteristic dimensions much larger than an ion gyro radius; and (3) sufficient size that resistive diffusion is negligible, despite the high collisionality.

2.4 Region of validity

2.4.1 Overall criteria

With the somewhat tedious analysis completed, the next important issue is to focus on the most restrictive assumptions required for the derivation and to discuss their regions of validity. Furthermore, there is additional useful information to be extracted by examining the conditions for validity, one equation at a time. This provides insight into which specific phenomena are not accurately described by ideal MHD; perhaps more importantly, it also indicates those phenomena that will still be reliably treated even if a particular validity condition is violated.

These goals are accomplished in several steps. The starting point is the introduction of dimensionless variables

$$
\begin{aligned}
y &= \left(\frac{r_{Li}}{a}\right) \\
x &= \left(\frac{m_i}{m_e}\right)^{1/2}\left(\frac{V_{Ti}\tau_{ii}}{a}\right)
\end{aligned}
\tag{2.53}
$$

As just discussed, there are three independent conditions that must be satisfied for ideal MHD to be valid. They are

$$
\begin{aligned}
&(1)\ \text{High collisionality} \quad x \ll 1 \\
&(2)\ \text{Small gyro radius} \quad\ y \ll 1 \\
&(3)\ \text{Small resistivity} \qquad y^2/x \ll 1
\end{aligned}
\tag{2.54}
$$

These conditions are illustrated in Fig. 2.2. In the region labeled "ideal MHD" all three conditions are simultaneously satisfied. Although a significant number of approximations have been made in the derivation, there is a substantial region of parameter space where all the assumptions are satisfied and the ideal MHD equations are valid.

The next question to be asked is whether plasmas of fusion interest lie in the region of MHD validity. This can be answered by transforming the (x, y) diagram into a (T, n) diagram and observing whether values of n and T of fusion interest lie in the region of validity.

The first step is to define the parameter range of fusion interest. Past experiments and extrapolations to future fusion reactors indicate that the densities and temperatures of fusion plasmas lie in the range

$$
\begin{aligned}
10^{18}\ \text{m}^{-3} &< n < 10^{22}\ \text{m}^{-3} \\
0.5\ \text{keV} &< T < 50\ \text{keV}
\end{aligned}
\tag{2.55}
$$

These conditions describe a rectangle in the (n, T) diagram.

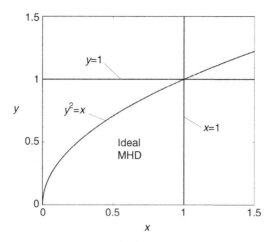

Figure 2.2   Region of validity for the ideal MHD model in terms of the normalized variables $y = r_{Li}/a$ and $x = (m_i/m_e)^{1/2}(V_{Ti}\tau_{ii}/a)$. In the region labeled "Ideal MHD" the validity conditions are satisfied.

The second step is to rewrite the conditions in Eq. (2.54) in terms of n and T. Since the B field explicitly appears, a prescription is needed to specify how B varies with n and T. A reasonable choice is to assume that $\beta = 4\mu_0 nT/B^2$ is held fixed. The parameter β measures the ratio of plasma energy to magnetic energy. It is a dimensionless quantity whose value is important in fusion reactor designs and is often limited by MHD instabilities. Consequently, treating β and the scale length a as parameters leads to the following expressions for the three MHD criteria of validity:

(1) High collisionality $x = 9.2 \times 10^3 (T^2/an) \ll 1$
(2) Small gyro radius $y = 2.3 \times 10^{-2}(\beta/na^2)^{1/2} \ll 1$ (2.56)
(3) Small resistivity $y^2/x = 5.6 \times 10^{-8}(\beta/aT^2) \ll 1$

In these expressions the units are a (m), T (keV), and n (10^{20} m^{-3}). The characteristic collision times are chosen as $\tau_{ee} \approx \tau_e$ and $\tau_{ii} \approx \tau_i$ where τ_e and τ_i are given by Braginskii (see Appendix B). Also, the Coulomb logarithm has been set to $\ln \Lambda = 19$, and the ion mass corresponds to deuterium. Equation (2.56) is illustrated in Fig. 2.3 for the case $a = 1$ m and $\beta = 0.05$. Observe that the conditions of small gyro radius and small resistivity are well satisfied for plasmas of fusion interest. Note, however, that the high collisionality assumption is never satisfied. *Thus, the region in which ideal MHD is valid completely excludes plasmas of fusion interest!* This disconcerting conclusion contradicts the overwhelming empirical experimental evidence which demonstrates that ideal MHD provides a very accurate description of most macroscopic plasma behavior.

The question may now be asked whether or not this is coincidence or, alternatively, the result of some subtle and perhaps unexpected physics? The resolution

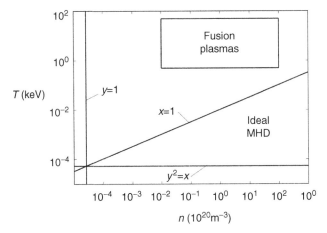

Figure 2.3   Region of validity for the ideal MHD model in (n, T) parameter space for the case $\beta = 0.05$ and $a = 1$ m. In the region labeled "Ideal MHD" the validity conditions are satisfied.

follows from an examination of the validity conditions, one equation at a time. The results show that there are indeed subtle physics issues at play which are only rigorously resolved by examining certain general stability predictions of more realistic but complicated models. Several such models are described in Chapter 9 and the stability predictions are covered in Chapter 10.

2.4.2 Conservation of mass

To begin, consider the conservation of mass relation given in Eq. (2.1). Since none of the MHD validity conditions is required to obtain this equation, its range of applicability extends far beyond that of ideal MHD. It is valid for collisionless or collision dominated systems regardless of the size of the ion gyro radius or the influence of plasma resistivity. Modifications are only required when phenomena such as recombination, ionization, charge exchange, fusion fuel depletion, or source terms due to gas puffing or pellet injection are present.

2.4.3 Momentum equation

The validity of the momentum equation is also much wider than the ideal MHD conditions would imply, although the reasons involve considerably more subtle physics. Recall that the collision dominated assumption is required to neglect Π_i and Π_e. In a collisionless plasma the magnetic field in a certain sense plays the role of collisions for the perpendicular motion; that is, perpendicular to the field, particles are confined to the vicinity of a given field line executing nearly (two-dimensional) isotropic motion if their gyro radius is much smaller than the characteristic plasma dimension. In fact, a calculation of $\Pi_{\perp i}$ in the collisionless regime (see, for instance, Bowers and Haines (1971)) shows that for the MHD ordering $\Pi_{\perp i}/p_i \sim r_{Li}/a \ll 1$. Therefore, the perpendicular motion is fluid-like, implying that the perpendicular components of the momentum equation provide an excellent description of plasma behavior in either the collision dominated or collisionless regimes.

The general situation parallel to the field is much more complicated and it is here that ideal MHD treats the physics very inaccurately. In a collisionless plasma there is no reason a priori to assume that any simple relationship exists between the perpendicular and parallel pressures, $p_\perp$ and $p_\parallel$; perpendicular to the **B** field, particles execute well-confined gyro orbits, while along the field they execute free-streaming kinetic motion.

Even so, in practice the situation parallel to the field simplifies considerably for the following subtle reason. Consider the application of the MHD model over the narrow increment Δt as discussed in connection with Fig. 2.1. By assumption, the plasma has evolved through a series of macroscopically stable states over a long period of time up until the beginning of Δt. During this period there is more than adequate time for collisions to isotropize the pressure. Consequently, at the beginning of Δt it is a good approximation to set $p_\perp = p_\parallel$ as an *initial condition*. However, as time progresses during the short MHD increment Δt, the two pressures will in general evolve quite differently because of the different physics involved perpendicular and parallel to the field. During this short increment there is not enough time for collisions to re-isotropize the plasma.

Now, it turns out that a not immediately obvious property of most MHD instabilities is that the plasma pressure neither compresses nor expands as the modes grow; instead, the perpendicular and parallel pressures are simply convected with the fluid motion. The reason is that compression and expansion correspond to very stable motions and would suppress most instabilities if excited to any appreciable level. This crucial fact of plasma behavior, demonstrated explicitly in Chapter 8, implies that neither $p_\perp$ nor $p_\parallel$ changes significantly from its initial value during the increment Δt. For fast MHD instabilities in a collisionless plasma, $p_\perp$ and $p_\parallel$ thus remain isotropic if they are so initially.

Under the situation just described, it is accurate to use the ideal MHD momentum equation in both the collision dominated and collisionless regimes. In the collisionless regime that part of the physics treated inaccurately by ideal MHD does not matter: the nature of MHD instabilities is such that the dynamical motions where the inaccuracies would be important are not excited.

Finally, as a caution, it should be noted that this conclusion applies to most but not all fusion configurations of interest. Specifically, for magnetic configurations with closed field lines certain modes are strongly affected by plasma compression and for these cases the ideal MHD and collisionless models can give substantially different results.

2.4.4 Energy equation

The ideal MHD adiabatic energy equation is never valid for plasmas of fusion interest. Like the parallel motion in the momentum equation, it too, for similar reasons, has little effect on the applicability of the model.

Even assuming that evolution over the long period prior to Δt leads to temperature equilibration (i.e., $T_e = T_i = T$), one still requires negligible parallel thermal conductivity for the adiabatic energy relation to be valid. Substituting numerical values from fusion experiments one can show that it would be far more accurate to treat the parallel thermal conductivity as infinite rather than zero. Therefore, a more appropriate energy equation follows from setting $\nabla_\parallel \cdot (\kappa_\parallel \nabla_\parallel T) = 0$ for both electrons and ions, yielding

$$\mathbf{B} \cdot \nabla T = 0 \tag{2.57}$$

If the assumption is again made that MHD instabilities are incompressible for most fusion configurations of interest, it can then be proven that the "incorrect" adiabatic energy relation yields the same results as the "infinite parallel thermal conductivity" constraint given by Eq. (2.57). The proof of this statement is as follows.

By definition, if the plasma is incompressible the density is just convected with the fluid as it moves under the action of an MHD instability. Density convection implies that

$$\frac{d\rho}{dt} = \frac{\partial \rho}{\partial t} + \mathbf{v} \cdot \nabla \rho = 0 \tag{2.58}$$

The conservation of mass equation shows that this relation is equivalent to the well-known result that incompressible motions are characterized by the constraint

$$\nabla \cdot \mathbf{v} = 0 \tag{2.59}$$

The adiabatic energy equation, before incompressibility is assumed, can with the aid of the mass conservation relation be written as

$$\frac{dp}{dt} + \gamma p \nabla \cdot \mathbf{v} = 0 \tag{2.60}$$

Invoking incompressibility gives

$$\frac{dp}{dt} = 0 \tag{2.61}$$

implying that the energy as well as the mass is convected with the fluid.

The same result follows from the collisionless equation of state. Taking the total time derivative of Eq. (2.57) and using Faraday's law yields

$$\frac{d}{dt}\left[\mathbf{B}\cdot\nabla\left(\frac{p}{\rho}\right)\right] = \mathbf{B}\cdot\nabla\left[\frac{d}{dt}\left(\frac{p}{\rho}\right)\right] - (\nabla\cdot\mathbf{v})\left[\mathbf{B}\cdot\nabla\left(\frac{p}{\rho}\right)\right] \qquad (2.62)$$

Now assume the motions are incompressible. Then $\nabla\cdot\mathbf{v} = 0$ and from mass conservation $d\rho/dt = 0$, implying that Eq. (2.62) reduces to

$$\frac{dp}{dt} = 0 \qquad (2.63)$$

The collision dominated and collisionless equations of state are identical when the plasma motions of interest are incompressible. The only hidden assumption in the analysis is the requirement that the operator $\mathbf{B}\cdot\nabla \neq 0$ over the entire plasma, a requirement that is satisfied by almost all fusion configurations of interest. The one exception involves certain modes in closed line systems in which case, as with the parallel momentum equation, the results from a more sophisticated kinetic model can differ substantially from ideal MHD. This too is discussed in Chapters 9 and 10.

A corollary of the analysis presented above is the following. Any calculation using the ideal MHD model in which the results depend explicitly on the ratio of specific heats γ is unreliable. Equation (2.60) implies that such results must involve compressibility. Therefore, they make explicit use of the collision dominated assumption, which is not valid for fusion plasmas.

2.4.5 Ohm's law

The final equation to consider is Ohm's law. Here, the main terms neglected (i.e., the Hall effect, the electron diamagnetic drift, and electrical resistivity) do not depend on the collision dominated assumption. Thus, Ohm's law is approximately valid in the collisionless regime for phenomena on the MHD time scale. Even so, there is one important subtlety that should be noted.

The subtlety involves the resistivity. Although its magnitude is very small it nonetheless can play an important role. This follows because the ideal MHD Ohm's law implies that $E_{\parallel} = 0$ for any situation. With resistivity included, $E_{\parallel} = \eta J_{\parallel}$. While $E_{\parallel}$ can be shown to be small (i.e., $E_{\parallel}/E_{\perp} \ll 1$) it must really be compared to zero since it occurs in the vector component of Ohm's law that does not contain $\mathbf{E}_{\perp}$ or $\mathbf{v}\times\mathbf{B}$.

The critical feature is that unlike the Hall and diamagnetic effects discussed earlier, the resistivity contribution does not arise from an electrostatic potential. As such, it has the important property that, while small, it allows a new class of motions to occur that are entirely excluded from ideal MHD: the tearing and reconnecting of magnetic field lines. These phenomena occur on a hybrid time scale related to a geometric-like mean of the short ideal MHD and long resistive diffusion time scales. Since the hybrid scale is much longer than the characteristic MHD time, resistivity has little effect on ideal MHD behavior. However, the hybrid time scale sets a well-defined upper limit to the incremental width of time Δt over which ideal MHD is valid. Typical numerical values for the hybrid scale are on the order of 1 ms.

2.4.6 Summary of validity conditions

The ideal MHD model follows from a straightforward, self-consistent closure of the two-fluid moment equations. The conditions for its validity are (1) high collisionality, (2) small ion gyro radius, and (3) small resistivity. Conditions (2) and (3) are well satisfied in fusion plasmas. In contrast, condition (1) is never satisfied. A more subtle examination of ideal MHD, one equation at a time, shows that in most situations the model is still reliable even though the collision dominated assumption is not satisfied. Crucial to this argument is the claim that for problems involving MHD equilibrium and stability, the plasma motions of interest are incompressible. Specifically, the conservation of mass and the perpendicular momentum equation are shown to be valid in the collisionless limit even when $\nabla \cdot \mathbf{v} \neq 0$. However, both the parallel momentum equation and the energy equation are incorrect in this regime. Nevertheless, when $\nabla \cdot \mathbf{v} = 0$ neither of these equations, which though incorrect, plays an important role.

An analysis of Ohm's law demonstrates that ideal MHD accurately calculates the inductive part of the electric field, but incorrectly predicts the electrostatic part. Even so, for the low-frequency Maxwell's equations, this has no effect on the evaluation of the remaining fluid and field variables since only $\nabla \times \mathbf{E}$ enters the analysis. Finally, resistivity, while small, can play an important role in that it allows new motions to occur that are prohibited in ideal MHD. These phenomena occur on the slower MHD-resistive diffusion hybrid time scale and thus set a practical upper limit on the maximum time period Δt over which ideal MHD can be considered reliable.

The overall conclusion is that ideal MHD predictions are actually more accurate than one might expect. The reasons are subtle and associated with stability results that are stated but not (at this point) proved. The proofs will be given in Chapters 9 and 10, after development of the general equilibrium and stability properties of ideal MHD which then serve as a point of reference.

2.5 Overall summary

The ideal MHD model is a single-fluid model that describes the effects of magnetic geometry on the macroscopic equilibrium and stability properties of fusion plasmas. The MHD length and time scales of interest are a, the macroscopic plasma dimension, and a/V_{Ti}, the ion thermal transit time across the plasma.

The model is derived in a straightforward manner by forming the mass, momentum, and energy moments of the kinetic-Maxwell equations. The moment equations reduce to ideal MHD with the introduction of three critical assumptions: high collisionality, small ion gyro radius, and small resistivity. An analysis of the validity conditions shows that the collision dominated assumption is never satisfied in plasmas of fusion interest. The remaining two conditions are satisfied by a wide margin.

A careful examination of the collision dominated assumption shows that those particular parts of ideal MHD treated inaccurately (i.e., the parallel momentum and energy equations) play little if any practical role in most MHD equilibrium and stability phenomena; that is, these equations primarily describe the compression and expansion of a plasma whereas most MHD instabilities involve incompressible motions. The question on plasma compressibility is examined in Chapters 9 and 10 by means of several alternate, collisionless models.

References

Bowers, E. C. and Haines, M. G. (1971). *Phys. Fluids* **14**, 165.
Braginskii, S. I. (1965). In *Reviews of Plasma Physics*, Vol. 1, ed. M. A. Leontovich. New York: Consultants Bureau.

Further reading

Most of the material covered in this chapter has been known for many years. Listed below are some early and more recent references covering the material of interest.

The Boltzmann kinetic equation

Krall, N. A. and Trivelpiece, A. W. (1973). *Principles of Plasma Physics*. New York: McGraw-Hill.
Montgomery, D. C. and Tidman, D. A. (1964). *Plasma Kinetic Theory*. McGraw-Hill, New York.

Derivation of fluid equations

Goedbloed, H. and Poedts, S. (2004). *Principles of Magnetohydrodynamics*. Cambridge: Cambridge University Press.
Grad, H. (1956). *Notes on Magnetohydrodynamics*. New York University Report NYO-6486-I, New York.
Hazeltine, R. D. and Meiss, J. D. (2003). *Plasma Confinement*. Redwood City, CA: Addison-Wesley.

Rose, D. J. and Clark, M. (1961). *Plasmas and Controlled Fusion*. Cambridge, MA: MIT Press.
Spitzer, L. (1962). *Physics of Fully Ionized Gases*. New York: Interscience.

Collisions and transport

Goldston, R. J. and Rutherford, P. H. (1995). *Introduction to Plasma Physics*. Bristol: Institute of Physics.
Helander, P. and Sigmar, D. J. (2002). *Collisional Transport in Magnetized Plasmas*. Cambridge: Cambridge University Press.

Problems

2.1 Starting from the kinetic-Maxwell equations, and following the steps outlined in the text, derive the general set of two-fluid equations given by Eq. (2.22).

2.2 This problem, and the ones below, demonstrate a simple procedure for estimating the transport coefficients in a collision-dominated plasma. The required calculations are based on a simplified form of the Krook collision operator discussed in Appendix A. Also, to minimize the algebra, only the dominant terms in the kinetic equation are maintained which focus on the transport coefficient of interest. The first problem involves the parallel plasma resistivity. The relevant kinetic equation is given by

$$-\frac{e}{m_e}E_z\frac{\partial f}{\partial v_z} = -v_{ei}(f - f_M)$$

$$f_M = n\left(\frac{m_e}{2\pi T}\right)^{3/2}\exp\left(-\frac{m_e v^2}{2T}\right)$$

Assume that n, T, E_z are constants. Here and below the collision frequencies are also treated as constants. Consider the collision dominated limit corresponding to large v_{ei} and solve for the distribution function by expanding as follows:

$$f = f_M + f_1 + f_2 + \cdots$$

where $f_n/f_M \sim O(1/v_{ei}^n)$.

(a) Using this expansion calculate f_1.
(b) Substitute the solution for f_1 into the definition of $J_z = -e\int v_z f\,d\mathbf{v}$. The result should be of the form $J_z = (1/\eta_{\parallel})E_z$. Find the parallel resistivity $\eta_{\parallel}$.

2.3 Consider next the parallel electron thermal conductivity which arises from both electron–electron and electron–ion collisions. In this case the relevant limit of the kinetic equation can be written as

$$v_z \frac{\partial f}{\partial z} = -(\nu_{ee} + \nu_{ei})(f - f_M)$$

$$f_M = n \left(\frac{m_e}{2\pi T} \right)^{3/2} \exp\left(-\frac{m_e v^2}{2T} \right)$$

Here it has been assumed that n is a constant and $T_e = T_i \equiv T(z)$.

(a) Calculate f_1 using the expansion described in Problem 2.2.
(b) Substitute the solution for f_1 into the definition for the heat flux. Show that the result is of the form $\mathbf{q}_e = -\kappa_\parallel (\partial T/\partial z)\mathbf{e}_z$. Find the parallel electron thermal conductivity $\kappa_\parallel$.

2.4 The last transport coefficient of interest is the parallel viscosity which is largest for the ions. The appropriate limit of the kinetic equation is given by

$$v_z \frac{\partial f}{\partial z} = -\nu_{ii}(f - f_M)$$

$$f_M = n \left(\frac{m_i}{2\pi T} \right)^{3/2} \exp\left[-\frac{m_i(\mathbf{v} - u_z \mathbf{e}_z)^2}{2T} \right]$$

Assume that n and T are constants and note that the distribution function corresponds to a shifted Maxwellian whose fluid velocity is given by $\mathbf{u}_i = u_z(z)\,\mathbf{e}_z$.

(a) Solve for f_1 using the expansion described in Problem 2.2.
(b) Substitute the solution into the definition of Π_{zz} and show that the result is of the form $\Pi_{zz} = -\mu_\parallel(\partial u_z/\partial z)$. Find the parallel viscosity $\mu_\parallel$.

2.5 Consider a system in which each species of particles is acted upon by an additional force arising from an external potential; that is, Newton's law for each particle is given by

$$m_\alpha \frac{d\mathbf{u}_\alpha}{dt} = Z_\alpha e(\mathbf{E} + \mathbf{v} \times \mathbf{B}) - \nabla \phi_\alpha$$

where $\phi_\alpha = \phi_\alpha(\mathbf{r})$ is a known fixed potential function.

(a) Using the same assumptions made in the derivation of ideal MHD, derive an equivalent set of closed, single-fluid equations which include the effects of ϕ_e and ϕ_i. In order to complete this derivation some assumptions must be made regarding the size ϕ_e and ϕ_i. Order ϕ_e and ϕ_i so that their contributions to the momentum equation are comparable to the $\mathbf{J} \times \mathbf{B}$ force. What ordering gives this result?
(b) Assume the external potential represents a fixed gravitational force in the $\mathbf{e}_x$ direction of magnitude g. What are the corresponding functional dependences

of ϕ_e and ϕ_i? Write down the modified MHD equations including gravity for $m_e \ll m_i$.

2.6 Does the local Maxwellian distribution function

$$f = n\left(\frac{m}{2\pi T}\right)^{3/2}\exp\left(-\frac{mv^2}{2T}\right)$$

$$n = n(\mathbf{r}, t)$$
$$T = T(\mathbf{r}, t)$$

satisfy the Vlasov equation (i.e. the kinetic equation neglecting collisions)?

2.7 Consider a system in steady state $\partial/\partial t = 0$, having y symmetry $\partial/\partial y = 0$. The ions are described by a distribution function of the form

$$f = n_0\left(\frac{m}{2\pi T}\right)^{3/2}\exp\left(-\frac{\varepsilon + \alpha p^2/m}{T_0}\right)$$

$$n_0, T_0, \alpha = \text{constants}$$
$$\varepsilon = \frac{1}{2}mv^2 + e\phi(x, z)$$

$$p = mv_y + eA_y(x, z)$$

Does this distribution function satisfy the Vlasov equation? Here, ϕ and A_y are the scalar and vector potential, respectively. For this distribution function calculate n, $\mathbf{u}$, p, $\mathbf{\Pi}$, and $\mathbf{q}$ as functions of ϕ and A_y.

2.8 This problem shows that a similar set of equilibrium equations can exist for an ideal MHD plasma and a Vlasov plasma despite the opposing limits on collisionality. To illustrate this point derive the following exact result. Consider a quasi-neutral plasma described by a hybrid model. The ions satisfy the Vlasov equation. The electrons satisfy the massless fluid equation $0 = -e(\mathbf{E} + \mathbf{u}_e \times \mathbf{B}) - \nabla p_e/n$.

(a) For static equilibria ($\partial/\partial t = 0$) show that any distribution function $f_i = f_i(\varepsilon)$ with $\varepsilon = m_i v^2/2 + e\phi(\mathbf{r})$ satisfies the Vlasov equation for an arbitrary three-dimensional geometry.
(b) Calculate $\mathbf{J} = e\int \mathbf{v}f_i d\mathbf{v} - en\mathbf{u}_e$ and show that

$$\mathbf{J} \times \mathbf{B} = \nabla p$$
$$p = p_e + \int \frac{mv^2}{3}f_i d\mathbf{v}$$

Note that this equation is identical to the ideal MHD static equilibrium relation.
(c) Compare the equilibrium electric field in ideal MHD and the hybrid model.

3

General properties of ideal MHD

3.1 Introduction

This chapter presents a discussion of the basic properties of the ideal MHD model. These properties include the general conservation laws satisfied by the model as well several types of boundary conditions that are of interest to fusion plasmas. The discussion demonstrates the physical foundations of ideal MHD while providing insight into its reliability in predicting experimental behavior.

The material is organized as follows. First, a short description is given of the three most common types of boundary conditions that couple the plasma behavior to the externally applied magnetic fields: (1) plasma surrounded by a perfectly conducting wall; (2) plasma isolated from a perfectly conducting wall by an insulating vacuum region; and (3) plasma surrounded by a vacuum with embedded external coils. The most complex of these provides a quite accurate description of realistic experimental conditions.

Second, it is shown that despite the significant number of approximations made in the derivation of the model, ideal MHD still conserves mass, momentum, and energy, both locally and globally. This is one basic reason for the reliability of the model.

Finally, a short calculation shows that as a consequence of the perfect conductivity assumption, the plasma and magnetic field lines are constrained to move together; that is, the field lines are "frozen" into the plasma. This leads to important topological constraints on the allowable dynamical motions of the plasma. In fact the property of "frozen-in" field line topology can be taken as the basic definition of "ideal" MHD.

3.2 Boundary conditions

In order to properly formulate an MHD problem a set of appropriate boundary conditions must be specified that couples the plasma behavior to the externally applied magnetic fields. For problems involving MHD equilibrium and stability,

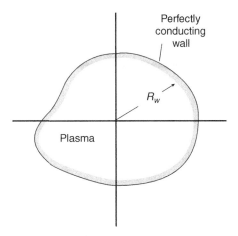

Figure 3.1   A plasma surrounded by a perfectly conducting wall whose surface is defined by $S_w(r,\,\theta,\,z) \equiv r - R_w(\theta,\,z) = 0$.

there are three types of boundary conditions, of varying complexity, that are often used. A discussion of each now follows.

3.2.1 Perfectly conducting wall

The first and simplest boundary condition assumes that the plasma extends out to a stationary, perfectly conducting wall whose shape is defined by $S_w(r,\,\theta,\,z) = 0$, as shown in Fig. 3.1. In this case the electromagnetic boundary conditions require that the tangential electric field and normal magnetic field vanish on the conducting wall:

$$\begin{aligned} \mathbf{n} \times \mathbf{E}|_{S_w} &= 0 \\ \mathbf{n} \cdot \mathbf{B}|_{S_w} &= 0 \end{aligned} \qquad (3.1)$$

Here, $\mathbf{n}$ is the outward-pointing normal vector. In time-varying systems the magnetic condition is redundant, a consequence of Faraday's law.

Next, from the ideal Ohm's law one sees that $\mathbf{n} \times \mathbf{E} + (\mathbf{n} \cdot \mathbf{B})\mathbf{v} - (\mathbf{n} \cdot \mathbf{v})\mathbf{B} = 0$; that is, the normal component of velocity also automatically vanishes on the wall:

$$\mathbf{n} \cdot \mathbf{v}|_{S_w} = 0 \qquad (3.2)$$

Thus, once appropriate initial data and the shape of the wall are specified, the conditions in Eq. (3.1) plus the condition given by Eq. (3.2) completely specify the problem. Note that there are no constraints imposed on the tangential velocity at the wall.

3.2.2 Insulating vacuum region

A more realistic set of boundary conditions assumes that the plasma is isolated from the conducting wall by a vacuum region as shown in Fig. 3.2. In most cases this model is more appropriate than the previous one for describing "confined" plasmas.

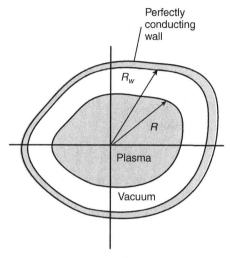

Figure 3.2 A plasma isolated from a perfectly conducting wall by a vacuum region. The plasma surface is defined by $S_p(r, \theta, z) \equiv r - R(\theta, z) = 0$ while the wall surface is defined by $S_w(r, \theta, z) \equiv r - R_w(\theta, z) = 0$.

In principle, one solves the combined plasma–vacuum system as follows. In the plasma region the ideal MHD equations apply, while in the vacuum region, where the fluid variables are not defined, the relevant equations are those determining the magnetic field:

$$\nabla \times \hat{\mathbf{B}} = 0$$
$$\nabla \cdot \hat{\mathbf{B}} = 0 \qquad (3.3)$$

Here, quantities with a ^ denote vacuum variables. Equation (3.3) implies that $\hat{\mathbf{B}} = \nabla \hat{V}$ with $\nabla^2 \hat{V} = 0$. Assume now that the relevant equations can be solved in each region and consider the boundary conditions. On the perfectly conducting wall $S_w(r, \theta, z) = 0$, the normal component of magnetic field must vanish:

$$\mathbf{n} \cdot \hat{\mathbf{B}}\big|_{S_w} = 0 \qquad (3.4)$$

Unlike the previous case, however, the plasma surface defined by $S_p(r, \theta, z, t) = 0$ is now free to move since the plasma is surrounded by vacuum. Hence, $\mathbf{n} \cdot \mathbf{v}\big|_{S_p}$ is arbitrary. There are, however, three non-trivial jump conditions that must be satisfied to connect the fields across the surface. These arise from the divergence $\mathbf{B}$ equation, Faraday's law, and the momentum equation. There are other jump conditions that can be derived but these in general are subsidiary relations not required to solve MHD problems.

Consider now the jump conditions. A convenient way to obtain the desired relations is to assume that the plasma surface is moving with a normal velocity

$v_n\mathbf{n} = (\mathbf{n} \cdot \mathbf{v})\mathbf{n}$. The jump conditions are straightforward to derive in a reference frame moving with the plasma surface. Once these conditions are obtained, all that is then required is to convert back to the laboratory frame using the corresponding Galilean transformation.

In the moving frame where the surface appears stationary, integrating the divergence **B** equation over a small closed volume through which the surface passes leads to the jump condition

$$\llbracket \mathbf{n} \cdot \mathbf{B}' \rrbracket_{S_p} = 0 \tag{3.5}$$

Here and below, primed quantities correspond to the moving frame and $\llbracket Q \rrbracket \equiv \hat{Q} - Q$. Similarly, by integrating Faraday's law over a small open surface area whose normal vector is tangent to the surface one obtains the jump condition

$$\llbracket \mathbf{n} \times \mathbf{E}' \rrbracket_{S_p} = 0 \tag{3.6}$$

The last jump condition is obtained by rewriting the momentum equation as follows:

$$\rho'\left(\frac{\partial \mathbf{v}'}{\partial t'} + \mathbf{v}' \cdot \nabla'\mathbf{v}'\right) = \frac{1}{\mu_0}\mathbf{B}' \cdot \nabla'\mathbf{B}' - \nabla'\left(p' + \frac{B'^2}{2\mu_0}\right) \tag{3.7}$$

Only the last term can exhibit a delta function behavior leading to a non-trivial jump condition. For finite accelerations the inertial terms at most have a step discontinuity across the surface. Similarly, since the operator $\mathbf{B}' \cdot \nabla'$ contains only surface derivatives the term $\mathbf{B}' \cdot \nabla'\mathbf{B}'$ also has at most a step discontinuity across the surface. Thus, integrating across a small distance perpendicular to the surface leads to the jump condition

$$\llbracket p' + B'^2/2\mu_0 \rrbracket_{S_p} = 0 \tag{3.8}$$

The desired form of the jump conditions is obtained by moving back to the laboratory reference frame by means of the Galilean transformation defined by $p' = p$, $\mathbf{B}' = \mathbf{B}$, $\mathbf{n} \times \mathbf{E}' = \mathbf{n} \times (\mathbf{E} + \mathbf{v} \times \mathbf{B}) = \mathbf{n} \times \mathbf{E} - (\mathbf{n} \cdot \mathbf{v})\mathbf{B}$. Here, the unprimed coordinates correspond to the laboratory reference frame. With these substitutions, the relevant jump conditions for ideal MHD can be written as

$$\llbracket \mathbf{n} \cdot \mathbf{B} \rrbracket_{S_p} = 0$$
$$\llbracket \mathbf{n} \times \mathbf{E} - (\mathbf{n} \cdot \mathbf{v})\mathbf{B} \rrbracket_{S_p} = 0$$
$$\llbracket p + B^2/2\mu_0 \rrbracket_{S_p} = 0 \tag{3.9}$$

A final simplification occurs by making use of the fact that the plasma is a perfect conductor: $\mathbf{E} + \mathbf{v} \times \mathbf{B} = 0$. This implies that in the plasma $[\mathbf{n} \cdot \mathbf{B}]_{S_p}$ and $[\mathbf{n} \times \mathbf{E} - (\mathbf{n} \cdot \mathbf{v})B]_{S_p}$ are both automatically zero. Therefore, Eq. (3.9) reduces to

$$\mathbf{n} \cdot \hat{\mathbf{B}}|_{S_p} = 0$$

$$\mathbf{n} \times \hat{\mathbf{E}} - (\mathbf{n} \cdot \mathbf{v})\hat{\mathbf{B}}|_{S_p} = 0 \tag{3.10}$$

$$[\![p + B^2/2\mu_0]\!]_{S_p} - 0$$

In the interesting situation where there are no surface currents (i.e., $[\![\mathbf{n} \times \mathbf{B}]\!]_{S_p} = 0$) and the pressure falls smoothly to zero at the plasma edge (i.e., $p|_{S_p} = 0$), the jump conditions reduce to

$$\mathbf{n} \cdot \hat{\mathbf{B}}|_{S_p} = 0$$

$$\mathbf{n} \times \hat{\mathbf{E}}|_{S_p} = 0 \tag{3.11}$$

$$[\![B^2/2\mu_0]\!]_{S_p} = 0$$

Although Eqs. (3.4) and (3.9) completely specify the boundary conditions, the plasma–vacuum problem is in practice difficult to solve. The reason is that a straightforward counting of boundary conditions suggests that the problem is over determined. To understand this point recall that for a vacuum field defined by $\hat{\mathbf{B}} = \nabla \hat{V}$ with $\nabla^2 \hat{V} = 0$, the two boundary conditions on $\mathbf{n} \cdot \hat{\mathbf{B}}$ on the wall and plasma surfaces, uniquely determine $\hat{V}$, implying that $\mathbf{n} \times \hat{\mathbf{B}}$ and hence $\hat{B}^2$ are also known quantities. However, there remains the pressure balance jump condition, which places an additional constraint on $\hat{B}^2$ apparently leading to an over-determined problem.

The problem is resolved as follows. One must treat the shape of the plasma $S_p(r, \theta, z, t)$ as an additional unknown to be self-consistently determined by the analysis. Herein lies the extra degree of freedom to make the problem well posed although difficult to solve. A problem for which $S_p(r, \theta, z, t)$ must be determined is known as a "free boundary" problem. In contrast, it is often far simpler, though less relevant, to specify the shape of the plasma surface $S_p(r, \theta, z, t)$ and then determine a self-consistent shape for the outer perfect conductor. When $S_p(r, \theta, z, t)$ is specified the problem is known as a "fixed boundary" problem.

This completes the discussion of the boundary and jump conditions for a system consisting of a plasma isolated from a conducting wall by a vacuum region.

3.2.3 Plasma surrounded by external coils

The most difficult but realistic set of boundary conditions corresponds to the situation where the plasma is confined by the magnetic fields created by a fixed set of external current-carrying conductors embedded in the vacuum region as shown in Fig. 3.3. This problem is more difficult than the previous one because the current-carrying conductors must by definition have spaces between them. Consequently, the problem must be solved in an infinite rather than finite domain.

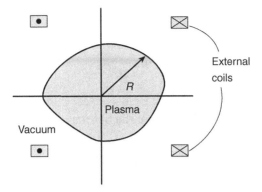

Figure 3.3 A plasma surrounded by a vacuum region into which are embedded external current-carrying coils. The plasma surface is defined by $S_p(r, \theta, z) \equiv r - R(\theta, z) = 0$.

In this regard note that the jump conditions at the plasma–vacuum interface remain unchanged: i.e., Eq. (3.9) still applies. However, with external coils the wall condition, Eq. (3.4), is no longer required. Instead, the vacuum field is written as $\hat{\mathbf{B}}(\mathbf{r}, t) = \mathbf{B}_a(\mathbf{r}) + \tilde{\mathbf{B}}(\mathbf{r}, t)$, where $\mathbf{B}_a(\mathbf{r})$ is the known steady state (or perhaps slowly varying) applied field due to the external coils and $\tilde{\mathbf{B}}(\mathbf{r}, t)$ is the induced field due to the plasma. The applied field is obtained by writing $\mathbf{B}_a = \nabla \times \mathbf{A}$ and then calculating $\mathbf{A}$ from the Biot–Savart law

$$\mathbf{A} = \frac{\mu_0}{4\pi} \sum_i \int \frac{\mathbf{J}_i}{|\mathbf{r} - \mathbf{r}'|} \, d\mathbf{r}' \qquad (3.12)$$

where the sum is over the external conductors. The induced field satisfies $\nabla \times \tilde{\mathbf{B}} = \nabla \cdot \tilde{\mathbf{B}} = 0$ again implying that $\tilde{\mathbf{B}} = \nabla \tilde{V}$ with $\nabla^2 \tilde{V} = 0$. The potential $\tilde{V}$ must be regular throughout the entire vacuum region thus leading to the following boundary conditions:

$$\tilde{\mathbf{B}}\big|_{\infty} = 0$$
$$\tilde{\mathbf{B}}\big|_{S_p} = -\mathbf{B}_a\big|_{S_p} \qquad (3.13)$$

The shape of the plasma surface is also an unknown for this class of "free boundary" problems and must be determined self-consistently from the analysis.

Solutions to the MHD equations satisfying the external coil boundary conditions provide an accurate description of plasma behavior in realistic experimental situations. Because of the complexity involved it is perhaps not surprising that most such applications require substantial numerical computations. The MHD equations plus any set of the boundary conditions just discussed constitute a well-posed formulation to investigate the macroscopic equilibrium and stability of fusion plasmas.

3.3 Local conservation relations

The original kinetic-Maxwell equations from which the MHD model has been derived conserve mass, momentum, and energy, not only macroscopically, but microscopically as well. Since a considerable number of assumptions have been made in the derivation of the MHD equations it is important to investigate whether the resulting model still satisfies these basic conservation laws. In this section the question of local conservation is treated by showing that the ideal MHD equations can be written in canonical conservation form. The implications regarding global conservation are discussed in the next section.

The canonical conservation form is given by

$$\frac{\partial}{\partial t}(\ ) + \nabla \cdot (\ ) = 0 \tag{3.14}$$

Once the mass, momentum, and energy equations can be written in this form, it is then straightforward to derive the global conservation relations.

3.3.1 Conservation of mass

To begin observe that the MHD mass equation is already in the desired conservation form:

$$\frac{\partial \rho}{\partial t} + \nabla \cdot \rho \mathbf{v} = 0 \tag{3.15}$$

The first term represents the gain in mass within a given volume element. This gain is due to a net inward flux of particles through the boundaries of the volume element as described by the second term.

3.3.2 Conservation of momentum

Consider now the momentum equation. If one makes use of the tensor identity $\nabla \cdot (\mathbf{AC}) = (\nabla \cdot \mathbf{A})\mathbf{C} + (\mathbf{A} \cdot \nabla)\mathbf{C}$ then a short calculation allows the momentum equation to be written in conservation form as follows:

$$\frac{\partial \rho \mathbf{v}}{\partial t} + \nabla \cdot \mathbf{T} = 0$$

$$\mathbf{T} = \rho \mathbf{vv} + \left(p + \frac{B^2}{2\mu_0}\right)\mathbf{I} - \frac{1}{\mu_0}\mathbf{BB} \tag{3.16}$$

The $\partial(\rho \mathbf{v})/\partial t$ term represents the increase in momentum within a volume element. This increase is generated by a net inward flux of momentum $\nabla \cdot \mathbf{T}$ through the boundaries of the volume element.

The contributions to **T** have the following physical interpretation. The $\rho\mathbf{v}\mathbf{v}$ term represents the Reynolds stress and is important in systems with large fluid flows. Often it is not too important in studies of plasma stability where the equilibrium flows are assumed to be small or zero. The remaining contributions to **T** include the effects of the pressure and magnetic field. In a locally orthogonal coordinate system in which one coordinate is aligned along **B** these contributions can be conveniently rewritten as

$$
\mathbf{T}_B = \begin{vmatrix} p_\perp & & \\ & p_\perp & \\ & & p_\parallel \end{vmatrix} \tag{3.17}
$$

where

$$
\begin{aligned}
p_\perp &= p + \frac{B^2}{2\mu_0} \\
p_\parallel &= p - \frac{B^2}{2\mu_0}
\end{aligned} \tag{3.18}
$$

The quantities $p_\perp$ and $p_\parallel$ represent the total pressures perpendicular and parallel to the magnetic field respectively. Equation (3.18) implies, as expected, that the particle pressure p acts isotropically perpendicular and parallel to the field. In contrast, the magnetic pressure $B^2/2\mu_0$ adds to the total pressure perpendicular to the field, while subtracting when parallel to the field. The "negative parallel magnetic pressure" actually corresponds to a tension along the magnetic field lines. This anisotropic behavior of the magnetic field (i.e., producing pressure perpendicular to **B** and tension parallel to **B**) is fundamental to the understanding of the equilibrium and stability properties of the magnetic geometries of fusion interest.

3.3.3 Conservation of energy

The last conservation equation of interest corresponds to energy. Several steps are required to obtain the desired form. To begin, form the dot product of the momentum equation with **v**:

$$
\rho\mathbf{v} \cdot \frac{d\mathbf{v}}{dt} = \mathbf{v} \cdot (\mathbf{J} \times \mathbf{B}) - \mathbf{v} \cdot \nabla p \tag{3.19}
$$

Next, by making use of the conservation of mass, one can rewrite the inertial terms as

$$
\rho\mathbf{v} \cdot \frac{d\mathbf{v}}{dt} = \frac{\partial}{\partial t}\left(\frac{1}{2}\rho v^2\right) + \nabla \cdot \left(\frac{1}{2}\rho v^2 \mathbf{v}\right) \tag{3.20}
$$

Now, by using the ideal MHD Ohm's law and Faraday's law the electromagnetic terms can be expressed as

$$\mathbf{v} \cdot \mathbf{J} \times \mathbf{B} = \frac{1}{\mu_0} \mathbf{E} \cdot \nabla \times \mathbf{B}$$

$$= -\frac{1}{\mu_0} \nabla \cdot (\mathbf{E} \times \mathbf{B}) + \frac{1}{\mu_0} \mathbf{B} \cdot \nabla \times \mathbf{E} \qquad (3.21)$$

$$= -\nabla \cdot \left(\frac{1}{\mu_0} \mathbf{E} \times \mathbf{B} \right) - \frac{\partial}{\partial t} \left(\frac{B^2}{2\mu_0} \right)$$

Lastly, the term $\mathbf{v} \cdot \nabla p$ can be rewritten by (1) combining the adiabatic energy equation with the conservation of mass yielding $\partial p/\partial t + \mathbf{v} \cdot \nabla p + \gamma p \nabla \cdot \mathbf{v} = 0$ and (2) using the definition $\nabla \cdot (p\mathbf{v}) = \mathbf{v} \cdot \nabla p + p \nabla \cdot \mathbf{v}$. Eliminating $\nabla \cdot \mathbf{v}$ from these two equations yields

$$\mathbf{v} \cdot \nabla p = \frac{1}{\gamma - 1} \frac{\partial p}{\partial t} + \frac{\gamma}{\gamma - 1} \nabla \cdot p\mathbf{v} \qquad (3.22)$$

Substituting Eqs. (3.20)–(3.22) leads to the energy equation in conservation form:

$$\frac{\partial w}{\partial t} + \nabla \cdot \mathbf{s} = 0$$

$$w = \frac{1}{2}\rho v^2 + \frac{p}{\gamma - 1} + \frac{B^2}{2\mu_0} \qquad (3.23)$$

$$\mathbf{s} = \left(\frac{1}{2}\rho v^2 + \frac{p}{\gamma - 1} \right) \mathbf{v} + p\mathbf{v} + \frac{1}{\mu_0} \mathbf{E} \times \mathbf{B}$$

In this equation w represents the total energy of the system, which consists of the kinetic, internal (i.e., $3p/2$ for $\gamma = 5/3$), and magnetic energies. The quantity $\mathbf{s}$ is comprised of the net flux of kinetic plus internal energy, the mechanical work done on the plasma through compression, and the flux of electromagnetic energy as given by the Poynting vector.

Summarizing, it has just been demonstrated that the ideal MHD equations can be written in local conservation form in which each of the terms has a simple physical interpretation.

3.4 Global conservation laws

By integrating the local conservation relations over an appropriate volume it is possible to obtain a set of global conservation laws for ideal MHD. These laws are exact and are valid for general, non-linear, multidimensional, time-dependent situations. The specific forms of the global laws, as well as the choice of the appropriate integration volume, depend on the boundary conditions to be applied. Consequently, a separate derivation is required for each type of boundary condition as presented below.

3.4.1 Perfectly conducting wall

For this case the global conservation laws are obtained by integrating the local conservation relations out to the perfectly conducting wall. Making use of the boundary conditions given by Eqs. (3.1) and (3.2) then leads to the following global conservation laws:

$$\frac{dM}{dt} = 0$$

$$\frac{d\mathbf{P}}{dt} = -\int_{S_w} \left(p + \frac{B^2}{2\mu_0} \right) \mathbf{n} dS = 0 \tag{3.24}$$

$$\frac{dW}{dt} = 0$$

where M is the total mass of the plasma

$$M = \int_{V_w} \rho \, d\mathbf{r} \tag{3.25}$$

$\mathbf{P}$ is the mechanical momentum of the plasma

$$\mathbf{P} = \int_{V_w} \rho \mathbf{v} \, d\mathbf{r} \tag{3.26}$$

and W is the sum of the kinetic, internal, and magnetic energies of the plasma

$$W = \int_{V_w} \left(\frac{1}{2} \rho v^2 + \frac{p}{\gamma - 1} + \frac{B^2}{2\mu_0} \right) d\mathbf{r} \tag{3.27}$$

The first contribution to Eq. (3.24) demonstrates that the total mass of plasma is conserved. The second equation represents the conservation of momentum. The boundary term represents the total force exerted by the walls on the plasma. If, as one might reasonably expect, the wall remains in place during the duration of experimental operation, then this force must exactly vanish: $d\mathbf{P}/dt = 0$; that is, mechanical momentum is conserved. The final contribution to Eq. (3.24) shows that the total energy of the system is conserved. From the definition of W given by

Eq. (3.27) one observes that during a given dynamical motion of the plasma (an instability, for instance) energy can in general be transferred between the magnetic field, the plasma internal energy, and the plasma kinetic energy, although the sum must be conserved.

3.4.2 Insulating vacuum region

When the plasma is surrounded by a vacuum region it is of particular interest to focus attention on the conservation of energy. This relation is slightly more complicated than for the case of the perfectly conducting wall since the plasma boundary is now allowed to move. The result is that it is not the individual, but the combined plasma–vacuum energy that is conserved.

To show this, first consider a global quantity G, defined by

$$G(t) = \int_V g(\mathbf{r}, t) \, d\mathbf{r} \tag{3.28}$$

The total time derivative of G in a closed volume $V(t)$ whose surface $S(t)$ is moving with a normal velocity $v_n(S)\mathbf{n} = (\mathbf{n} \cdot \mathbf{v})\mathbf{n}$ is given by

$$\frac{dG(t)}{dt} = \int_V \frac{\partial g}{\partial t} \, d\mathbf{r} + \int_S g\, \mathbf{n} \cdot \mathbf{v} \, dS \tag{3.29}$$

where $\mathbf{n}(S)$ is the outward unit normal to the surface. For readers unfamiliar with this result a derivation is given in Appendix C. Equation (3.29) is applied to the total plasma energy by setting $G(t) = W(t)$, $g(\mathbf{r}, t) = w(\mathbf{r}, t)$, plus choosing $V = V_p$ to be the plasma volume and $S = S_p$ the plasma surface. The normal boundary velocity is $\mathbf{n} \cdot \mathbf{v}(S_p) = \mathbf{n} \cdot \mathbf{v}(\mathbf{r},t)|_{S_p}$, corresponding to the motion of the plasma surface. Making use of the ideal Ohm's law, and the fact that on the plasma boundary $\mathbf{n} \cdot \mathbf{B}|_{S_p} = 0$, it follows that

$$
\begin{aligned}
\frac{dW}{dt} &= \int_{V_p} \frac{\partial w}{\partial t} \, d\mathbf{r} + \int_{S_p} w\, \mathbf{n} \cdot \mathbf{v} dS \\
&= -\int_{V_p} \nabla \cdot \mathbf{s} \, d\mathbf{r} + \int_{S_p} w\, \mathbf{n} \cdot \mathbf{v} dS \\
&= -\int_{S_p} (\mathbf{s} - w\, \mathbf{v}) \cdot \mathbf{n} \, dS \\
&= -\int_{S_p} \left(p + \frac{B^2}{2\mu_0} \right) \mathbf{n} \cdot \mathbf{v} dS
\end{aligned}
\tag{3.30}
$$

Note that if the plasma surface is moving (i.e., $\mathbf{n} \cdot \mathbf{v} \neq 0$), then the boundary term is in general non-zero.

Consider now the vacuum region. Here the total energy is given by

$$\hat{W} = \int_{\hat{V}} \frac{\hat{B}^2}{2\mu_0} \, d\mathbf{r} \tag{3.31}$$

where $\hat{V}$ is the volume of the vacuum region between the plasma and the wall. From Eq. (3.29) it follows that

$$\frac{d\hat{W}}{dt} = \int_{\hat{V}} \frac{1}{\mu_0} \left(\hat{\mathbf{B}} \cdot \frac{\partial \hat{\mathbf{B}}}{\partial t} \right) d\mathbf{r} - \int_{S_p} \frac{\hat{B}^2}{2\mu_0} \mathbf{n} \cdot \mathbf{v} dS + \int_{S_w} \frac{\hat{B}^2}{2\mu_0} \mathbf{n} \cdot \mathbf{v} dS \tag{3.32}$$

The surface integral consists of two contributions, one from the wall surface and one from the plasma surface. The wall surface contribution vanishes because the wall is not moving: $\mathbf{n} \cdot \mathbf{v}|_{S_w} = 0$. The plasma surface contribution has a minus sign (here and below) because $\mathbf{n}$ represents an inward normal vector to the vacuum region. The first term can be simplified by substituting Faraday's law and converting the volume integral to a surface integral by the divergence theorem. One also makes use of the facts that (1) $\nabla \times \hat{\mathbf{B}} = 0$ in vacuum and (2) $\mathbf{n} \cdot \hat{\mathbf{B}}|_{S_p} = 0$ and $\mathbf{n} \times \hat{\mathbf{E}} - (\mathbf{n} \cdot \mathbf{v})\hat{\mathbf{B}}|_{S_p} = 0$ from Eq. (3.10). This yields

$$\frac{d\hat{W}}{dt} = -\int_{\hat{V}} \frac{1}{\mu_0} \left(\hat{\mathbf{B}} \cdot \nabla \times \hat{\mathbf{E}} \right) d\mathbf{r} - \int_{S_p} \frac{\hat{B}^2}{2\mu_0} \mathbf{n} \cdot \mathbf{v} dS$$

$$= -\int_{S_p} \left(\frac{1}{\mu_0} \hat{\mathbf{E}} \times \hat{\mathbf{B}} + \frac{\hat{B}^2}{2\mu_0} \mathbf{v} \right) \cdot \mathbf{n} \, dS$$

$$= \int_{S_p} \frac{\hat{B}^2}{2\mu_0} \mathbf{n} \cdot \mathbf{v} \, dS \tag{3.33}$$

The final energy relation is obtained by adding Eqs. (3.30) and (3.33):

$$\frac{d}{dt} (W + \hat{W}) = \int_{S_p} \left(\frac{\hat{B}^2}{2\mu_0} - p - \frac{B^2}{2\mu_0} \right) \mathbf{v} \cdot \mathbf{n} \, dS = 0 \tag{3.34}$$

Here, the boundary term vanishes by virtue of the pressure balance jump condition given by Eq. (3.10).

Equation (3.34) implies that when an ideal MHD plasma is isolated from a perfectly conducting wall by a vacuum region, the combined energy of the plasma–vacuum system is conserved. The fact that only the sum is conserved indicates that, in general, energy can and will flow from one region to the other as the plasma moves.

3.4.3 Plasma surrounded by external coils

If the conducting wall in the vacuum is replaced by a series of current-carrying conductors, the energy of the system is no longer conserved. The reason is that with external sources present, energy can be supplied to, or extracted from, the system. Even so, it is still possible to derive a relatively simple energy balance relation for the system.

The procedure is almost identical to that given for the vacuum region bounded by a perfectly conducting wall. The main differences are that (1) the conducting wall is now moved to infinity and (2) $\nabla \times \hat{\mathbf{B}} \neq 0$ due to the presence of external conductors. For this situation there is an additional volume contribution to the energy in the region outside the plasma due to these conductors. The energy balance relation becomes

$$\frac{d}{dt}\left(W + \hat{W}\right) = P_{\text{ext}} \tag{3.35}$$

where

$$
\begin{aligned}
P_{\text{ext}} &= -\int_{\hat{V}} \frac{1}{\mu_0}\hat{\mathbf{E}} \cdot \nabla \times \hat{\mathbf{B}}\ d\mathbf{r} \\
&= -\sum_i \int_{V_i} \hat{\mathbf{J}}_i \cdot \hat{\mathbf{E}}_i\ d\mathbf{r}
\end{aligned}
\tag{3.36}
$$

represents the power delivered to the system by the external circuits. The integrals in P_{ext} are carried out over the volume of each external conductor. Equation (3.35) has the simple physical interpretation that the rate of increase of the combined plasma–vacuum energy is equal to the power delivered by the external circuits.

3.5 Conservation of flux: the "frozen-in field line" concept

The final conservation law to be considered concerns the magnetic flux. The basic result, which is a consequence of the perfect conductivity Ohm's law, is that the magnetic flux contained within an arbitrary open surface area moving with the plasma does not change; that is, the flux is "frozen" into the plasma.

To show this, one starts with the definition of the magnetic flux ψ passing through an open area S_p in the plasma as shown in Fig. 3.4:

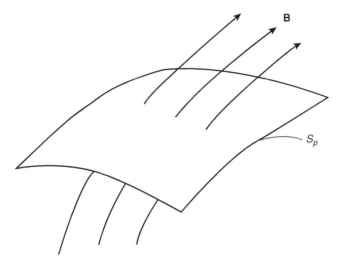

Figure 3.4   Magnetic flux passing through an open surface S_p whose surface normal vector is **n**.

$$\psi = \int_{S_p} \mathbf{B} \cdot \mathbf{n}\, dS \tag{3.37}$$

Assume now that the plasma contained within S_p is moving with a velocity **v**. As the surface moves the change in the flux passing through the area is given by

$$\frac{d\psi}{dt} = \int_{S_p} \frac{\partial \mathbf{B}}{\partial t} \cdot \mathbf{n}\, dS - \oint \mathbf{v} \times \mathbf{B} \cdot d\mathbf{l} \tag{3.38}$$

where $d\mathbf{l}$ is the arc length along the perimeter of the surface. For readers unfamiliar with this relation, a derivation is presented in Appendix C.

Substituting for $\partial \mathbf{B}/\partial t$ from Faraday's law and then converting the surface integral into a line integral by applying Stokes' theorem yields

$$\frac{d\psi}{dt} = -\oint (\mathbf{E} + \mathbf{v} \times \mathbf{B}) \cdot d\mathbf{l} \tag{3.39}$$

Clearly, if the plasma obeys the ideal MHD Ohm's law then

$$\frac{d\psi}{dt} = 0 \tag{3.40}$$

Since the derivation of Eq. (3.40) applies to any arbitrary surface area, it immediately follows that by setting S_p equal to the entire cross section of the plasma, the total flux contained within an ideal MHD plasma is conserved. An equally interesting case is to allow $S_p(l, t) = S_p(l_0, 0)$ to initially coincide with the

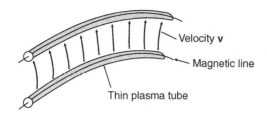

Figure 3.5   Thin tube of plasma containing a magnetic field line. As the plasma moves with a velocity **v**, the field line is "frozen" into the plasma motion.

cross section of a long thin flux tube (see Fig. 3.5). Assume now that the plasma within $S_p(l_0, 0)$ moves with a velocity $\mathbf{v}(\mathbf{r}, t)$ for a time Δt. After this period the area $S_p(l_0, \Delta t)$, which still contains the original plasma, will in general have a different shape and size (e.g., due to a non-uniform expansion of the plasma). However, Eq. (3.40) implies that the flux contained within the new cross section is the same as in the original cross section. Now, apply this result to a continuous sequence of cross sections along the original flux tube, $S_p(l, 0)$ with $l_0 \leq l \leq l_1$. Then, in the limit where $S_p \to 0$, one arrives at the well-known intuitive picture that in ideal MHD, magnetic lines move with the plasma; they are "frozen" into the fluid.

The conservation of flux relation has very important implications about the structure of the magnetic field. This follows because any allowable physical velocity **v** of a dissipationless ideal MHD plasma requires that neighboring fluid elements remain adjacent to one another; fluid elements are not allowed to tear or break into separate pieces. Since the magnetic lines move with the plasma, the field line topology must thus be preserved during any physically allowable MHD motion. This is a very strict requirement on the structure of the magnetic fields.

There are many configurations in plasma physics in which intuition suggests that it would be energetically favorable for field lines to break and reconnect, forming new configurations with lower potential energy. Such transitions are not allowed in ideal MHD because of the constraint on the topology. It is for this reason that the introduction of even a small resistivity can have a dramatic effect on plasma stability, much larger than indicated by simple dimensional arguments. A small dissipation breaks the topological constraint by allowing field lines to diffuse through the plasma. This permits a much wider class of motions to take place, although admittedly on a slower and presumably less dangerous time scale. These new motions allow the plasma to access the lower energy states that are prohibited in the ideal model. As stated earlier, the effects of resistivity lie beyond the scope of the present textbook.

To conclude, the essential distinguishing feature that serves as the definition of an "ideal" MHD plasma is that the magnetic field lines are frozen into the plasma during all allowable dynamical motions.

3.6 Summary

The results of this chapter show that the ideal MHD model conserves mass, momentum, energy, and magnetic flux. These conservation laws apply to general non-linear, time-dependent, multidimensional systems. The existence of such laws is a non-trivial consequence in view of the many assumptions made in the derivation of the model. However, having shown the existence of the conservation laws, one can proceed with some confidence to the problems of equilibrium and stability of magnetic fusion configurations knowing that the model should prove reliable and provide valuable insight because of its inherently sound foundation.

Further reading

Much of the work presented in this chapter has been derived from "classic" papers written during the early days of the fusion program. Some of these early papers are cited below as well as more recent contributions.

Boundary conditions and conservation laws

Bernstein, I.B., Frieman, E.A., Kruskal, M.D. and Kulsrud, R.M. (1958). *Proc. R. Soc. London*. **A244**, 17.

Goedbloed, J.P. (1983). *Lecture Notes on Ideal Magnetohydrodynamics*. Fom-lnstituut voor Plasmafysica, Nieuwegein, the Netherlands, Rijnhuizen Report.

Goedbloed, J.P. and Poedts, S. (2004). *Principles of Magnetohydrodynamics*. Cambridge: Cambridge University Press.

Hazeltine, R.D. and Waelbroeck, F.L. (1998). *The Framework of Plasma Physics*. Reading, MA: Perseus Books.

Basic properties of MHD

Alfven, H. (1950). *Cosmical Electrodynamics*. Oxford: Clarendon Press.

Bateman, G. (1978). *MHD Instabilities*. Cambridge, MA: MIT Press.

Goedbloed, J.P. and Poedts, S. (2004). *Principles of Magnetohydrodynamics*. Cambridge: Cambridge University Press.

Goldston, R.J. and Rutherford, P.H. (1995). *Introduction to Plasma Physics*. Bristol: IOP Publishing Ltd.

Grad, H. and Rubin, H. (1958). *Proceedings of the Second United Nations International Conference on the Peaceful Uses of Atomic Energy*. Geneva: United Nations, Vol. 31, 190.

Morozov, A.I. and Solov'ev, L.S. (1966). *Reviews of Plasma Physics* Vol. 2, ed. M. A. Leontovich. New York: Consultants Bureau.

Newcomb, W.A., (1958). *Ann. Phys. (NY)* **3**, 347.

Problems

3.1 An MHD plasma is surrounded by a rigid wall. For this problem the MHD model is generalized to include the dissipative effects of resistivity and thermal conduction. Specifically the Ohm's law and energy equation are replaced by

$$\mathbf{E} + \mathbf{v} \times \mathbf{B} = \eta \mathbf{J}$$

$$\frac{1}{\gamma - 1}\left(\frac{\partial p}{\partial t} + \mathbf{v} \cdot \nabla p\right) + \frac{\gamma}{\gamma - 1} p \nabla \cdot \mathbf{v} = \eta J^2 - \nabla \cdot \mathbf{q}$$

Here η is the resistivity and $\mathbf{q} = \mathbf{q}_i + \mathbf{q}_e$ is the total heat flux vector. Show that the global conservation of energy relation can be written as

$$\frac{dW}{dt} = -Q = -\int \mathbf{q} \cdot \mathbf{n}\, dS$$

where the energy W is given by Eq. (3.27) and Q represents the heat loss through the surface of the wall due to thermal conduction. What happened to the Ohmic heating term?

3.2 Consider an ideal MHD plasma surrounded by a perfectly conducting wall. The global angular momentum of the plasma is defined as

$$\mathbf{L} = \int \rho\, \mathbf{r} \times \mathbf{v}\, d\mathbf{r}$$

Prove that

$$\frac{d\mathbf{L}}{dt} = 0$$

This result demonstrates that the global angular momentum is a conserved quantity.

3.3 In this problem the global conservation of energy relation is generalized to include multiple ion species (e.g., D, T, plus impurities). The basic multiple species ideal MHD model is now written as

$$\frac{\partial n_\alpha}{\partial t} + \nabla \cdot n_\alpha \mathbf{u}_\alpha = 0$$

$$m_\alpha n_\alpha \left(\frac{\partial}{\partial t} + \mathbf{u}_\alpha \cdot \nabla\right)\mathbf{u}_\alpha = Z_\alpha e(\mathbf{E} + \mathbf{u}_\alpha \times \mathbf{B}) - \nabla p_\alpha$$

$$\left(\frac{\partial}{\partial t} + \mathbf{u}_\alpha \cdot \nabla\right)\frac{p_\alpha}{n_\alpha^\gamma} = 0$$

Here α denotes species (electrons and all ions), Z_α is the charge number, and $m_e = 0$.

For a plasma surrounded by a perfectly conducting wall show that the global conservation of energy relation has the form

$$\frac{dW}{dt} = 0$$

where

$$W = \int \left[\frac{B^2}{2\mu_0} + \sum_\alpha \left(\frac{1}{2} m_\alpha n_\alpha u_\alpha^2 + \frac{p_\alpha}{\gamma - 1} \right) \right] d\mathbf{r}$$

3.4 A plasma surrounded by a perfectly conducting wall satisfies the MHD equations with the exception of Ohm's law, which now includes resistivity: $\mathbf{E} + \mathbf{v} \times \mathbf{B} = \eta \mathbf{J}$. Assume the current density vanishes at the wall. Show that the total magnetic flux contained within the wall is a constant. However, show that for any open surface area interior to the wall the flux is no longer a conserved quantity, implying that the magnetic field is no longer frozen into the plasma.

3.5 This problem provides an alternate demonstration that in an ideal MHD plasma the field lines are frozen into the plasma. It can be easily shown that any divergence-free magnetic field can be written as

$$\mathbf{B} = \nabla \alpha \times \nabla \beta$$

where $\alpha = \alpha(\mathbf{r}, t)$, $\beta = \beta(\mathbf{r}, t)$ represent field line coordinates; that is, the intersection of the surfaces $\alpha = \alpha_0$ and $\beta = \beta_0$ defines a line that is everywhere tangent to $\mathbf{B}$, thus defining the field line. Compute $\partial \mathbf{B}/\partial t$ and substitute the result into Faraday's law using the ideal MHD Ohm's law $\mathbf{E} + \mathbf{v} \times \mathbf{B} = 0$. Show that the resulting relation is satisfied when the field lines move with the plasma:

$$\frac{d\alpha}{dt} = \frac{d\beta}{dt} = 0$$

with $d/dt = \partial/\partial t + \mathbf{v} \cdot \nabla$.

3.6 The requirement $d\alpha/dt = d\beta/dt = 0$ in Problem 3.5 does not lead to a unique solution, as it is possible to add various homogeneous solutions to α and β. To illustrate this point consider a magnetic field $\mathbf{B} = B_\theta(r) \, \mathbf{e}_\theta + B_z(r) \, \mathbf{e}_z$. Write $\mathbf{B} = \nabla \alpha \times \nabla \beta$ with $\alpha = r^2/2$ and $\beta = \theta f_1(r) + (z/r) f_2(r)$.

(a) Express $f_1(r)$ and $f_2(r)$ in terms of $B_\theta(r)$ and $B_z(r)$.
(b) On any surface $\alpha = \alpha_0$ show that replacing $\beta(r, \theta, z)$ with $\beta(r, \theta, z) + g(r, t)$ leaves the magnetic field unchanged but corresponds to either a rotation in θ or a translation in z of the field line. Note that such rotations and translations have no physical significance because of the cylindrical symmetry.
(c) Assume the plasma moves with a velocity $\mathbf{v} = v_\theta(r) \, \mathbf{e}_\theta + v_z(r) \, \mathbf{e}_z$ and $v_r = 0$.

Is this velocity field consistent with Faraday's law? Could the given magnetic field be consistent with any $v_r \neq 0$? Explain.

The conclusion is that the concept of "frozen-in" field lines, while intuitively appealing, is not unique because of the ambiguity in identifying a field line from one instant of time to the next. Even so, the "frozen-in" concept represents an extremely useful interpretation of the motion of plasma and field lines.

4

MHD equilibrium: general considerations

4.1 Introduction

The goal of MHD equilibrium theory is the discovery of magnetic geometries that confine and isolate hot plasmas from material walls and are macroscopically stable at sufficiently high values of β (where β is plasma pressure/magnetic pressure) to be attractive as potential fusion reactors.

Research starting in the 1950s has led to the discovery of several magnetic geometries possessing such attractive MHD properties. Chapter 4 focuses on the general features common to these configurations. Included in the discussion are (1) a description of the basic equilibrium equations, (2) the need for toroidicity, (3) the concept of magnetic flux surfaces, (4) the definition of the basic plasma parameters and figures of merit describing an MHD equilibrium, and (5) the fundamental conflict between the requirements for good equilibrium and good stability in toroidal geometry. Thus, a framework and perspective are developed that provide an overview of the nature of MHD equilibria. This background, once established, is the basis for the discussions of the specific applications presented in the chapters that follow.

4.2 Basic equilibrium equations

To begin, consider the MHD equilibrium equations given by

$$\mathbf{J} \times \mathbf{B} = \nabla p$$
$$\nabla \times \mathbf{B} = \mu_0 \mathbf{J} \qquad (4.1)$$
$$\nabla \cdot \mathbf{B} = 0$$

These are just the time-independent form of the full MHD equations with $\mathbf{v} = 0$; the equilibria of interest are static. Equilibrium is achieved by balancing the magnetic force $\mathbf{J} \times \mathbf{B}$ with the pressure gradient force ∇p.

It is worthwhile noting that while the full MHD equations including dynamics are valid for times comparable to the MHD time scale, the equilibrium equations themselves are actually valid over the much longer transport time scale. This follows because for static or very slowly moving systems, the issues of $E_\parallel$, collisions, and viscosity are either negligible or exactly cancel when adding the individual electron and ion momentum equations to obtain the single-fluid MHD momentum equation.

The question of static ($\mathbf{v} = 0$) vs. stationary ($\mathbf{v} \neq 0$) equilibria is also worth discussing. In many current fusion experiments substantial equilibrium flows are observed suggesting that stationary flows should be included in the analysis. Even so, such equilibrium flows are not included in the present textbook, primarily because of their mathematical complexity. Still, there is a wealth of information and insight that can be obtained by studying purely static MHD equilibria. This strategy thereby avoids the high level of complexity arising from the introduction of flow.

A final point of interest concerns the conservation of mass equation and the adiabatic equation of state. For static equilibria these relations are automatically satisfied. This implies that $\rho(\mathbf{r})$ is arbitrary and that $p(\mathbf{r})$ is decoupled from $\rho(\mathbf{r})$.

4.3 The virial theorem

The ideal MHD equilibrium equations satisfy a particular integral relation known as the virial theorem (Shafranov, 1966). A consequence of this theorem is the basic requirement that for an MHD equilibrium to exist, the plasma must be held in force balance by externally supplied currents; it is not possible to create a configuration confined solely by the currents flowing within the plasma itself. If this were possible, one could design a very attractive fusion reactor requiring only minimal "start-up" coils, which could be shut off once the plasma currents were established (see Fig. 4.1).

To demonstrate the virial theorem, recall that the equilibrium force balance equation can be written in conservation form,

$$\nabla \cdot \mathbf{T} = 0 \tag{4.2}$$

where the stress tensor $\mathbf{T}$ is given by

$$\begin{aligned}
\mathbf{T} &= p_\perp (\mathbf{I} - \mathbf{bb}) + p_\parallel \mathbf{bb} \\
p_\perp &= p + B^2/2\mu_0 \\
p_\parallel &= p - B^2/2\mu_0 \\
\mathbf{b} &= \mathbf{B}/B
\end{aligned} \tag{4.3}$$

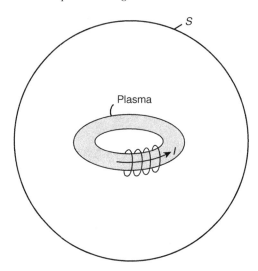

Figure 4.1   Geometry illustrating a plasma attempting to confine itself by its own currents. The surface S is the one referred to in the virial theorem.

Note that in a local orthogonal coordinate system with unit vectors $\mathbf{e}_1(\mathbf{r})$, $\mathbf{e}_2(\mathbf{r})$, $\mathbf{b}(\mathbf{r})$, $\mathbf{T}$ has the form

$$\mathbf{T} = \begin{vmatrix} p_\perp & & \\ & p_\perp & \\ & & p_\| \end{vmatrix} \tag{4.4}$$

indicating that $p_\perp$ and $p_\|$ represent the stresses perpendicular and parallel to $\mathbf{B}$, respectively.

The next step in the derivation is to integrate the identity

$$\nabla \cdot (\mathbf{r} \cdot \mathbf{T}) = \mathbf{r} \cdot (\nabla \cdot \mathbf{T}) + \mathrm{Trace}(\mathbf{T}) \tag{4.5}$$

over an arbitrary volume V, bounded by the surface S. Setting $\nabla \cdot \mathbf{T} = 0$ for equilibrium leads to

$$\int_V \left(3p + \frac{B^2}{2\mu_0} \right) d\mathbf{r} = \int_S \left[\left(p + \frac{B^2}{2\mu_0} \right) (\mathbf{n} \cdot \mathbf{r}) - \frac{B^2}{\mu_0} (\mathbf{r} \cdot \mathbf{b})(\mathbf{n} \cdot \mathbf{b}) \right] dS \tag{4.6}$$

where $d\mathbf{S} = \mathbf{n} dS$ and $\mathbf{n}$ is the outward-pointing normal vector to S.

Assume now that the virial theorem is false; confined equilibria do exist without external currents. Let S lie outside the confined plasma so that $p(S) = 0$ (see Fig. 4.1). If no external currents are present, then S can extend to infinity. Furthermore, if as assumed, the equilibrium currents are indeed confined to the plasma, then B must decrease with radius for large r at least as rapidly as a dipole field: $B(S) \leq K/r^3$ as $r \to \infty$. Consequently, for large r the integrand on the right-hand side of Eq. (4.6) scales as

$$\text{Integrand} \leq B^2 r dS \leq \frac{K^2 \sin\theta \, d\theta d\phi}{r^3} \tag{4.7}$$

Under these circumstances, as S extends to infinity, the right-hand side of Eq. (4.6) vanishes while the left-hand side approaches a non-zero positive constant. The resulting contradiction thus proves the virial theorem. When confined equilibria are surrounded by external conductors, Eq. (4.6) is not violated, since the right-hand side must now be evaluated over the surface of the conductors.

4.4 The need for toroidicity

The most obvious common feature of current magnetic fusion concepts is that, with one exception, each is constructed in the shape of a torus. The idea is to create configurations in which magnetic field lines remain contained within a closed toroidal volume; lines should not intersect the vacuum chamber.

The reason for building such technologically complicated toroidal systems rather than simpler, linear, open-ended systems is associated with the enormous difference in the energy loss rates perpendicular and parallel to the magnetic field. In an open-ended device, magnetic field lines leave the system and ultimately make contact with material walls (see Fig. 4.2a). Both particles and energy can be lost very quickly because charged particles move freely along magnetic field lines. Assuming that the potentially faster free streaming particle loss can be eliminated by magnetic trapping, the dominant parallel energy loss mechanism is due to collisional transport, in particular thermal conduction by electrons.

In contrast, for a toroidal device with contained magnetic lines (see Fig. 4.2b), cross field ion thermal conduction is the dominant loss mechanism. There are no

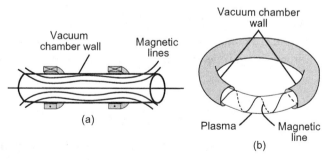

Figure 4.2   Magnetic field line trajectories in (a) an open-ended system and (b) a toroidal system.

parallel losses to the wall. The ratio of the parallel to perpendicular classical thermal conductivities is given by (Braginskii, 1965)

$$\frac{\kappa_{\|e}}{\kappa_{\perp i}} \approx \left(\frac{m_i}{m_e}\right)^{1/2} (\omega_{ci}\tau_i)^2 = 3.4 \times 10^{10} \left(\frac{B^2 T_k^3}{n_{20}^2}\right) \tag{4.8}$$

where ω_{ci} is the ion cyclotron frequency and τ_i is the characteristic ion–ion collision time for momentum exchange. It is also assumed that (1) the ions are singly charged deuterons, (2) the electrons and ions are at the same temperature, and (3) the Coulomb logarithm appearing in τ_i has been set to $\ln \Lambda = 19$. For plasma parameters corresponding either to current experimental operation or to reactor conditions this ratio is enormous. As an example, consider a tokamak with $T = 3\,\text{keV}$, $B = 5\,\text{T}$, and $n = 10^{20}\,\text{m}^{-3}$. In this case $\kappa_{\|e}/\kappa_{\perp i} = 2.3 \times 10^{13}$!

In practice the actual value of the perpendicular thermal conductivity in a toroidal plasma is substantially larger than the classical value. The reasons for this are twofold. First, the orbits of a small class of trapped particles are strongly modified by toroidal effects, leading to a disproportionately large contribution to $\kappa_{\perp i}$ (neoclassical transport). Second, and more importantly, anomalous effects due to plasma micro-instabilities increase the effective collision frequency, leading to greatly enhanced values of $\kappa_{\perp i}$ and $\kappa_{\perp e}$. However, experimental measurements of $\kappa_{\|e}$ indicate that its value is approximately classical. Even with these larger values of $\kappa_{\perp i}$ the ratio of $\kappa_{\|e}/\kappa_{\perp i}$ remains so large that there exists a common consensus that, to avoid such losses, the magnetic configuration must be toroidal.

The one exception to this consensus is the open-ended mirror confinement concept. Here, by a combination of innovative ideas and clever plasma operation, large gains have been made in reducing end loss. However, even with these improvements in confinement, mirrors, as of this writing, have yet to achieve comparable performance to the tokamak. As a result the worldwide mirror program has been dramatically reduced in size with respect to its peak years in the 1970s and 1980s. A main focus of today's mirror program is based on the idea of the gas dynamic trap, whose goal is the development of a 14 MeV neutron source, a very important facility for testing materials in a fusion environment. The mirror machine may be attractive for this purpose because of reduced plasma confinement requirements and the possibility of a simple, very compact, low cost design.

4.5 Flux surfaces

In general, for a fusion plasma the magnetic field lines lie on a set of closed nested toroidal surfaces. This follows from the equilibrium relation obtained by forming the dot product of MHD momentum equation with **B**:

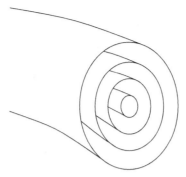

Figure 4.3 Contours of constant pressure in a well-confined toroidal equilibrium.

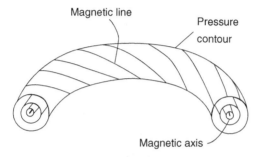

Figure 4.4 Toroidal flux surfaces showing the magnetic axis and the magnetic lines lying on a surface.

$$\mathbf{B} \cdot \nabla p = 0 \tag{4.9}$$

That is, for a well-confined equilibrium the pressure is maximum near the center of the poloidal cross section and is weakly varying around the toroidal direction. For such profiles the contours of constant pressure are nested toroidal surfaces (see Fig. 4.3). From Eq. (4.9) it follows that the magnetic lines lie on the $p = $ constant contours.

Consequently these contours are usually referred to as magnetic flux surfaces or simply just flux surfaces. The limiting flux surface, which approaches a single magnetic line where the pressure is a maximum, is called the magnetic axis (see Fig. 4.4).

A similar equilibrium relation is obtained by forming the dot product of the MHD momentum equation with $\mathbf{J}$,

$$\mathbf{J} \cdot \nabla p = 0 \tag{4.10}$$

which implies that the current lines also lie on the surfaces of constant pressure; the current flows between and not across flux surfaces. Note that while both $\mathbf{J}$ and $\mathbf{B}$ lie

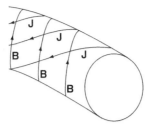

Figure 4.5   Toroidal flux surface showing magnetic lines and current lines lying on the surface. Note that the **B** and **J** lines are not in general purely parallel or perpendicular to each other.

on constant p contours, this does not imply that they are parallel. In general, the angle between **J** and **B** is arbitrary and the limits of being either purely parallel or perpendicular correspond to special cases. The general situation is illustrated in Fig. 4.5.

On any set of closed flux surfaces there are three important classes of magnetic field line trajectories that must be distinguished: rational, ergodic, and stochastic. A constant p contour on which all the lines exactly close on themselves after a finite number of toroidal circuits is known as a rational surface. If the field lines do not close, but instead wrap around indefinitely covering the entire constant p contour, this corresponds to an ergodic surface. Lastly, in those situations where the field line wanders around and actually fills a volume, the result is a region of stochasticity.

In general, toroidal magnetic geometries with ergodic or rational surfaces are the ones of interest and importance to fusion. Most configurations with desirable MHD confinement properties have both ergodic and rational surfaces. Usually the ratio of rational to ergodic surfaces is of measure zero. The tokamak, stellarator, reversed field pinch, and spheromak are configurations of this type. Concepts in which all field lines are rational on every flux surface are known as closed line systems. Two examples are the field reversed configuration and the levitated dipole.

Systems with regions of stochasticity are undesirable for fusion as they are characterized by poor confinement properties. Specifically, the equilibrium relation $\mathbf{B} \cdot \nabla p = 0$ implies that $p =$ constant over the stochastic volume, which is equivalent to a region with infinite transport. That is, the enormous parallel thermal conductivity of a plasma forces the temperature in the stochastic volume to rapidly equilibrate: $T =$ constant over the volume. Thus, to support a finite perpendicular heat flux $\mathbf{q}_\perp = -\kappa_\perp \nabla_\perp T$, assuming a flat temperature profile, one requires $\kappa_\perp \to \infty$. Stochastic regions can occur only in systems without geometric symmetry, for instance, in the vicinity of a separatrix in a 3-D stellarator or, more importantly, over substantial regions of plasma where there exists multidimensional

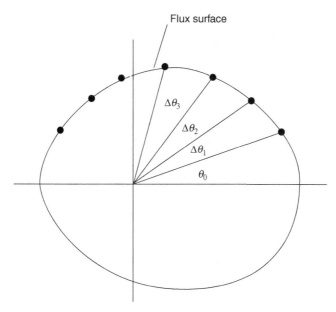

Figure 4.6   Magnetic field line projections used in the definition of rotational transform.

micro-turbulence involving a large number of different field harmonics. Because of their unattractive confinement properties, plasmas whose central core has regions of stochasticity should, to the maximal extent possible, be avoided experimentally. Consequently such systems are not treated here.

The usual way to characterize the two important classes of flux surfaces is in terms of the rotational transform, which is defined as follows. Imagine the projection of a magnetic line on a given poloidal cross section as shown in Fig. 4.6. The magnetic line starts off at a poloidal angle θ_0. After one toroidal transit around the torus the magnetic line returns to a slightly different angle $\theta_0 + \Delta\theta$ on the flux surface. In general, $\Delta\theta$ depends upon the poloidal angle θ_0, where the line started. The rotational transform ι is the average value of the angle $\Delta\theta$ after an infinite number of transits:

$$\iota = \lim_{N \to \infty} \frac{1}{N} \sum_{1}^{N} \Delta\theta_n \tag{4.11}$$

If ι is a rational fraction of 2π the line is closed. If it is not, the line is ergodic. The rotational transform plays an important role in both equilibrium and stability, and a practical procedure for its evaluation is discussed in Section 4.6. Note that ergodic surfaces can be mapped out either by plotting $p = $ constant contours or by tracing magnetic field line trajectories over many circuits of the torus. For closed line systems, however, the flux surfaces are defined as $p =$ constant surfaces since closed lines do not trace out complete surfaces.

4.6 Surface quantities: basic plasma parameters and figures of merit

By carrying out appropriate integrals over the flux surfaces one can define a number of quantities that are of importance to the study of MHD equilibria. These quantities, known as "surface quantities," describe the global properties of the equilibria. As such they are essential for distinguishing different configurations and for interpreting experimental data. Several quantities also serve as important figures of merit that provide insight into MHD stability. Furthermore, as discussed in the next section (Section 4.7), a precise number of freely chosen surface quantities must be specified a priori in order to formulate a well-posed equilibrium problem.

4.6.1 Fluxes and currents

Surface quantities are, in general, one-dimensional functions depending only upon the flux surface label. For example, each contour of the set of nested toroidal surfaces illustrated in Fig. 4.3 is labeled by a different value of the pressure p. If, as shown in Fig. 4.7, one now calculates the poloidal flux ψ_p contained within any given pressure contour

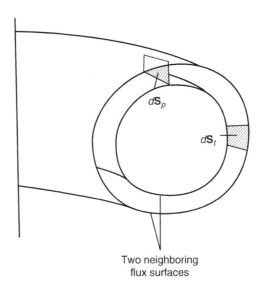

Two neighboring
flux surfaces

Figure 4.7 The poloidal and toroidal surface elements $d\mathbf{S}_p$ and $d\mathbf{S}_t$ used to calculate the corresponding magnetic fluxes.

$$\psi_p = \int \mathbf{B} \cdot d\mathbf{S}_p \quad \text{poloidal flux} \tag{4.12}$$

then

$$\psi_p = \psi_p(p) \tag{4.13}$$

The amount of poloidal flux contained within any given pressure contour depends only on the value of p labeling that contour.

Note that it makes physical sense for ψ_p to be a surface quantity since the magnetic lines lie in the constant p contours. Were the magnetic lines to cross the pressure surfaces, the definition of "contained flux" would become ambiguous. In fact, since $\mathbf{J}$ and $\mathbf{B}$ both lie on constant p contours, it follows that

$$\psi_t = \int \mathbf{B} \cdot d\mathbf{S}_t \qquad \text{toroidal flux}$$

$$I_t = \int \mathbf{J} \cdot d\mathbf{S}_t \qquad \text{toroidal current} \tag{4.14}$$

$$I_p = \int \mathbf{J} \cdot d\mathbf{S}_p \qquad \text{poloidal current}$$

are also surface quantities. In contrast, the components of magnetic field and current density vary with the poloidal and toroidal angles, θ and ϕ, around a flux surface and thus do not qualify as surface quantities, i.e., on any surface $p = $ constant, and B_p is a function $B_p(p, \theta, \phi)$ and not $B_p(p)$.

A further point concerns the labeling of the flux surfaces. Up until now it has been assumed that each flux surface has been labeled by a value of p. Clearly this is not unique. One could just as well label the surfaces with the values of ψ_p, or with any other surface quantity for that matter. In many MHD calculations appearing in the literature, either ψ_p, ψ_t, or a combination thereof is used as the label. The key point is that a properly defined surface quantity remains a surface quantity regardless of how the flux surfaces are labeled.

With this as background, one can now define several basic plasma parameters and figures of merit that measure the quality of an MHD equilibrium with respect to confinement efficiency and stability. All of the quantities introduced are discussed in detail in later sections of the textbook. For present purposes, they are simply introduced for convenience but with little accompanying discussion. Also, keep in mind that the figures of merit related to stability are in general neither necessary nor sufficient; instead, they serve as approximate guidelines for achieving overall favorable stability properties.

For a plasma parameter or figure of merit to be useful several requirements must be met: (1) the quantities must be relatively easy to evaluate but still accurately reflect the physics issue under consideration; (2) the quantities must be expressible in terms of simple physical parameters, easily related to experiment; and (3) the quantities must be defined in a sufficiently general manner so as to apply to all configurations of interest.

The plasma parameters and figures of merit are now described below.

4.6.2 Normalized plasma pressure, β

The quantity β is a global plasma parameter whose value is critical for a fusion reactor. It measures the efficiency of plasma confinement by the magnetic field. Interestingly, there is actually no unique definition of plasma β that is agreed upon by the entire fusion community. Various definitions are distinguished by different geometric factors whose choice is motivated by a given configuration's aspect ratio and cross sectional shape. Still, there is usually not a large difference in numerical values between the various definitions. The definition used here and throughout the textbook follows the strategy of mathematical simplicity and ease of comparison with experiment.

Qualitatively, β measures the ratio of plasma pressure to magnetic pressure:

$$\beta = \frac{\text{plasma pressure}}{\text{magnetic pressure}} \tag{4.15}$$

The definition of plasma pressure is straightforward. It is assumed to be the volume averaged value defined as

$$\langle p \rangle = \frac{1}{V_p} \int p d\mathbf{r} \tag{4.16}$$

where V_p is the volume of the plasma.

It is the magnetic pressure, $B^2/2\mu_0$, that is more subtle. In general, both the toroidal and poloidal magnetic pressures must be included in the definition; that is, $B^2 = B_t^2 + B_p^2$. A convenient choice for the toroidal magnetic pressure for any cross section is $B_t^2 = B_0^2$, where B_0 is the vacuum toroidal field at the geometric center of the chamber confining the plasma: $R = R_0$ as shown in Fig. 4.8.

For the poloidal magnetic pressure a good choice for a circular cross section plasma is $B_p^2 = (\mu_0 I / 2\pi a)^2$ where I is the total toroidal plasma current and a is the minor radius of the plasma also shown in Fig. 4.8. This definition can be generalized to include non-circular plasmas as follows: $B_p^2 = (\mu_0 I / C_p)^2$. Here, C_p is the poloidal circumference of the plasma surface, which for simplicity is approximated by

$$C_p \approx 2\pi a \left(\frac{1 + \kappa^2}{2}\right)^{1/2} \tag{4.17}$$

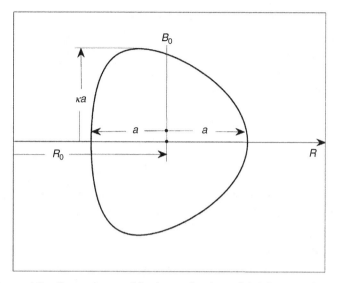

Figure 4.8   Geometry used in the evaluation of the figures of merit.

where κ is the plasma elongation as shown in Fig. 4.8. More accurate geometric approximations for the circumference can be made but they are more complicated mathematically and lead to only small quantitative changes in the value of C_p.

To summarize, the definition of β used throughout the textbook is given by

$$\beta = \frac{2\mu_0 \langle p \rangle}{B^2}$$

$$B^2 = B_0^2 + \left(\frac{\mu_0 I}{2\pi a}\right)^2 \frac{2}{1 + \kappa^2}$$

(4.18)

It is often useful to define separate toroidal and poloidal βs measuring plasma confinement efficiency with respect to each component of the magnetic field. These definitions have the form

$$\beta_t = \frac{2\mu_0 \langle p \rangle}{B_0^2}$$

$$\beta_p = \frac{4\pi^2 a^2 (1 + \kappa^2) \langle p \rangle}{\mu_0 I^2}$$

(4.19)

Note that

$$\frac{1}{\beta} = \frac{1}{\beta_t} + \frac{1}{\beta_p}$$

(4.20)

indicating that the smaller of the two quantities β_t or β_p dominates the overall magnetic confinement efficiency.

In general, high values of β are desirable for fusion reactor economics and technology. However, there is a maximum allowable value of β set by MHD equilibrium requirements and by MHD instabilities driven by the pressure gradient.

4.6.3 Kink safety factor, q*

The kink safety factor is a global plasma parameter that measures stability against dangerous long-wavelength modes driven by the toroidal plasma current. These are known as "kink modes." The kink safety factor is proportional to the ratio of toroidal field to toroidal current. High values provide "safety" against current-driven modes. The kink safety factor for a circular cross section plasma is defined by

$$q_* = \frac{aB_t}{R_0 B_p} = \frac{2\pi a^2 B_0}{\mu_0 R_0 I} \tag{4.21}$$

The definition is extended to include non-circular cross sections by replacing $a \to C_p/2\pi \approx a[(1 + \kappa^2)/2]^{1/2}$. Thus, the definition used throughout the text is given by

$$q_* = \frac{2\pi a^2 B_0}{\mu_0 R_0 I} \left(\frac{1 + \kappa^2}{2}\right) \tag{4.22}$$

Observe that with this definition of q_* there is a simple relationship between β_t and β_p which can be written as

$$\frac{\beta_t q_*^2}{\varepsilon} = \varepsilon \beta_p \left(\frac{1 + \kappa^2}{2}\right) \tag{4.23}$$

In many fusion applications it is desirable to operate at high current (i.e., low q_*), for improved confinement and ohmic heating. However, kink instabilities prevent this type of operation and set a lower limit on the minimum achievable q_* (i.e., the maximum toroidal current).

4.6.4 Rotational transform, ι, and the MHD safety factor, q

The concept of the rotational transform ι has already been discussed in Section 4.5. Recall that the rotational transform is a surface quantity whose profile plays an important role in the MHD equilibrium and stability of magnetic fusion concepts. Its magnitude and profile have important consequences for the maximum achievable β in a reactor. Usually the quantity $\iota/2\pi$ is directly used to describe 3-D configurations such as the stellarator. However, the inverse of $\iota/2\pi$, known as the safety factor q (not to be confused with the kink safety factor q_*), is the quantity that is typically used to describe 2-D axisymmetric configurations such as the tokamak and reversed field pinch. The relation between q and ι is thus given by

$$q = \frac{2\pi}{\iota} \qquad (4.24)$$

As its name implies the "safety factor" is a qualitative indicator of stability. High q is "good" for stability while low q is "bad."

Below a practical method is described for calculating the rotational transform in a multidimensional geometry. The starting assumptions are that a set of nested toroidal ergodic flux surfaces exist and that an MHD equilibrium has been calculated. Specifically, it is assumed that the magnetic fields $B_r(r,\theta,\phi)$, $B_\theta(r,\theta,\phi)$, and $B_\phi(r,\theta,\phi)$ are known quantities. Using this information one can calculate the field line trajectories $r(l)$, $\theta(l)$, and $\phi(l)$, where l is the arc length coordinate along the magnetic field. The trajectories are obtained by solving the following set of non-linear, first-order differential equations describing the tangent lines to the magnetic field:

$$\frac{dr}{dl} = \frac{B_r}{B}$$

$$\frac{d\theta}{dl} = \frac{B_\theta}{rB} \qquad (4.25)$$

$$\frac{d\phi}{dl} = \frac{B_\phi}{RB}$$

where $R = R_0 + r\cos\theta$. For initial conditions at $l = 0$ assume that $\theta(0) = 0$, $\phi(0) = 0$, and $r(0) = r_0$. In this case the quantity r_0 serves as the flux surface label. Solving these equations is a straightforward numerical task.

Assume now that the field line trajectories are known. Then, over a segment of field line of length L, the changes in poloidal angle $\Delta\theta$ and toroidal angle $\Delta\phi$ are given by

$$\Delta\theta = \int_0^{\Delta\theta} d\theta = \int_0^L \frac{d\theta}{dl}\, dl = \int_0^L \frac{B_\theta}{rB}\, dl$$

$$\Delta\phi = \int_0^{\Delta\phi} d\phi = \int_0^L \frac{d\phi}{dl}\, dl = \int_0^L \frac{B_\phi}{RB}\, dl$$

$$(4.26)$$

One now takes the limit $L \to \infty$. The rotational transform is just the average value of the poloidal excursion per single toroidal transit. Simple proportions then yield

$$\frac{\lim\limits_{L\to\infty} \Delta\theta}{\lim\limits_{L\to\infty} \Delta\phi} = \frac{\iota}{2\pi} \tag{4.27}$$

Equation (4.27) can conveniently be rewritten in terms of the safety factor

$$q = \frac{\lim\limits_{L\to\infty} \displaystyle\int_0^L \frac{B_\phi}{RB}\, dl}{\lim\limits_{L\to\infty} \displaystyle\int_0^L \frac{B_\theta}{rB}\, dl} \tag{4.28}$$

where the integrands $I_\phi = B_\phi/RB$ and $I_\theta = B_\theta/rB$ are evaluated along the field line trajectory. For example, $I_\phi(r,\theta,\phi) = I_\phi[r(l,r_0), \theta(l,r_0), \phi(l,r_0)] = I_\phi(l,r_0)$.

Equation (4.28) is the desired result. It defines the surface quantity $q(r_0)$ whose profile is important to know for all fusion concepts of interest. Simplified forms of $q(r_0)$ are derived in Chapters 5, 6, and 7 for 1-D, 2-D, and 3-D configurations. Qualitatively, a $q(r_0)$ profile that has a strong radial variation is said to possess magnetic shear, defined as $s(r_0) = (r_0/q)(dq/dr_0)$. In general, large shear is favorable for MHD stability.

Finally, it is worth noting that in certain special limits it just so happens that $q(\text{plasma edge}) = q_*$, a result that sometimes leads to confusion as to the role each plays. These difficulties are resolved in the discussion on tokamak stability.

4.6.5 Summary

A number of basic plasma parameters and figures of merit have been introduced whose purpose is to provide intuition as to the quality and desirability of a given MHD equilibrium for fusion applications. Of particular interest to the equilibrium confinement properties are the values of β and q_* as well as the profile of $\iota/2\pi$ or, equivalently q. In terms of stability attractive fusion concepts have a high maximum β, low minimum q_*, and a q profile with substantial shear.

4.7 Equilibrium degrees of freedom

In addition to their importance in describing the global properties of fusion plasmas, the surface quantities play a crucial role in the proper formulation of an MHD equilibrium problem. The reason is that while many different surface quantities can be defined, only two are independent. This can be seen intuitively by recognizing that there are two independent external circuits that control the properties of any given MHD equilibrium: one is the toroidal field circuit (e.g., regulate the toroidal flux evolution) and the other is the poloidal field circuit (e.g., regulate the toroidal current evolution). The existence of two arbitrary free surface functions has been demonstrated mathematically for one-, two- and general three-dimensional configurations.

The implication is that in order to formulate a well-posed MHD equilibrium problem it must be both possible and necessary to specify a priori two and only two independent surface quantities.

The existence of two free functions is on the one hand a great advantage of MHD equilibrium theory. This freedom allows the theory to describe a wide range of configurations. On the other hand, one must appreciate that the physics content of the ideal MHD model is insufficient to predict the specific functional dependence of the two free functions. Hence, for a given configuration the accuracy of the results depends upon the accuracy with which the free functions can be specified. For example, if $p(S)$ and $q(S)$ are chosen as the free functions, their dependence must be specified at the outset of the problem, based on physical intuition, experimental data, or the results of a transport code. Only after these functions are specified is the MHD equilibrium problem properly formulated, permitting a solution for the fields, flux surfaces, etc.

4.8 The basic problem of toroidal equilibrium

Qualitatively, the problem of producing a magnetically confined toroidal equilibrium separates into two parts. First, the magnetic configuration must provide radial pressure balance in the poloidal plane so that the pressure contours form closed nested surfaces (see Fig. 4.9). Both toroidal and poloidal fields can readily

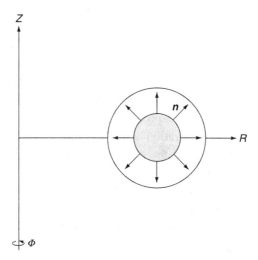

Figure 4.9   Schematic diagram of the forces required for radial pressure balance.

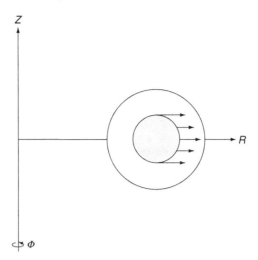

Figure 4.10   Schematic diagram of the outward toroidal expansion force inherent in all toroidal configurations.

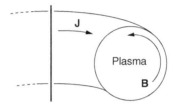

Figure 4.11   Toroidal configuration with a purely poloidal magnetic field.

accomplish this task although the manners in which they do so are quite different. Second, the configuration must balance the radially outward expansion force inherent in all toroidal geometries without sacrificing stability. This is known as toroidal force balance and is illustrated in Fig. 4.10. Although the forces associated with toroidal force balance are usually smaller than those corresponding to radial pressure balance, they are nevertheless more difficult to compensate. With this in mind it is helpful to examine two opposing limits that serve to illustrate the basic nature of the toroidal force balance problem.

To begin, consider a configuration with a purely poloidal magnetic field as shown in Fig. 4.11. There are two forces that cause the plasma to expand radially outward in the positive R direction: (1) the hoop force and (2) the tire tube force.

The hoop force arises because of toroidal current flowing in the plasma and is similar to the force produced by a current-carrying loop of wire. One can understand this force intuitively by examining the field pattern illustrated in Fig. 4.12a. Conservation of flux implies that $\psi_1 = \psi_2$. Since ψ_1 passes through a smaller area

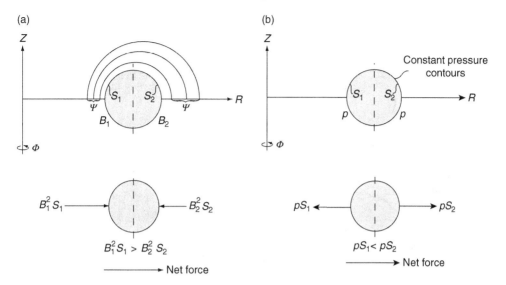

Figure 4.12 The toroidal expansion forces in a system with purely poloidal fields: (a) the hoop force and (b) the tire tube force.

than ψ_2, one expects the average magnetic field on the inner plasma surface S_1 to be greater than that on the outer surface S_2: $B_1 > B_2$. The net outward force on the plasma along R is proportional to $\mathbf{F}_R \propto \mathbf{e}_R(B_1^2 S_1 - B_2^2 S_2)/2\mu_0$. Even though S_1 is smaller than S_2 the quadratic dependence on B dominates, indicating that $\mathbf{F}_R > 0$: $\mathbf{F}_R$ represents the outward hoop force.

The tire tube force arises as follows. On a constant pressure contour the area S_1 on the inside of the surface is smaller than the area S_2 on the outside: $S_1 < S_2$ (see Fig. 4.12b). If the pressure on this contour is denoted by p then there is a net outward force along R proportional to $\mathbf{F}_R \propto -\mathbf{e}_R(pS_1 - pS_2)$. Since $S_1 < S_2$ then $\mathbf{F}_R > 0$: $\mathbf{F}_R$ represents an outward force very similar to that found in a rubber tire tube.

The combined outward force can be balanced in two different ways. First, if a perfectly conducting shell surrounds the plasma, then as the plasma expands along R in response to the outward force the poloidal magnetic flux outside the plasma is compressed, thereby increasing the magnetic pressure (Fig. 4.13a). Equilibrium is achieved when the plasma shifts sufficiently far out so that the increased magnetic pressure is large enough to balance the hoop plus tire tube forces.

A second way to balance the outward force is to replace the conducting wall with a set of fixed current-carrying vertical field coils (Fig. 4.13b). This is an important practical option. In actual experiments with conducting walls the materials used are usually copper, aluminum, or stainless steel, which have a high but not infinite conductivity. Hence, flux can only remain compressed for about a skin

(a)   (b)

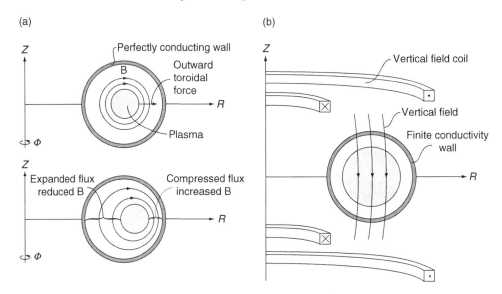

Figure 4.13   A configuration with purely poloidal fields held in toroidal equilibrium by means of (a) a perfectly conducting wall and (b) an externally applied
vertical field.

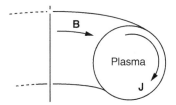

Figure 4.14   Toroidal configuration with a purely toroidal field.

time, which is with rare exceptions shorter than the experimental times of interest.
By a proper choice of magnitude and sign, the vertical field generated by the
external coils produces an inward compensating $\mathbf{J} \times \mathbf{B}_{\text{vert}}$ force for equilibrium:
$\mathbf{F}_R = \mathbf{e}_R 2\pi R_0 I B_{\text{vert}}$. Thus, configurations with purely poloidal magnetic fields can
be readily designed with good toroidal equilibrium properties.

This favorable equilibrium conclusion is negated by the fact that such configurations often develop catastrophic MHD instabilities leading to the destruction of
the plasma. (The stability of configurations with purely poloidal fields is discussed
in detail in Chapter 11.)

As a consequence, one is motivated to examine the opposite limit, a configuration with a purely toroidal field as shown in Fig. 4.14. In Chapter 11, it is
demonstrated that this configuration has inherently better stability properties than

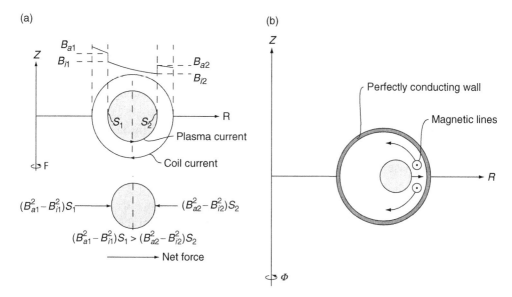

Figure 4.15 Toroidal force balance in a configuration with a purely toroidal field: (a) sharp boundary model showing the $1/R$ outward force and (b) the lack of toroidal equilibrium.

that corresponding to a purely poloidal field. Nevertheless, from the simple calculation outlined below, it follows that a purely toroidal configuration cannot be held in MHD equilibrium.

To show this, assume for simplicity that the plasma carries only a diamagnetic surface current: $B_{a1} > B_{i1}$ and $B_{a2} > B_{i2}$ (see Fig. 4.15b). The implication is that some toroidal field is excluded from the plasma interior. From Maxwell's equations it follows that the toroidal field in both the plasma and vacuum decrease inversely with R; that is $\mathbf{B} = B_\phi \mathbf{e}_\phi$ where

$$\hat{B}_\phi = \hat{K}/R$$
$$B_\phi = K/R$$
(4.29)

and $\hat{K} > K$.

Because of the $1/R$ dependence it is clear that the magnetic field on the inside of the torus is always greater than that on the outside: $B_1 > B_2$. This effect is partially compensated by the slightly smaller area on the inside $S_1 < S_2$, but the quadratic dependence of the magnetic pressure on B dominates and there remains a net outward force: $\mathbf{F}_R \propto \mathbf{e}_R \left[(B_{a1}{}^2 - B_{i1}{}^2)S_1 - (B_{a2}{}^2 - B_{i2}{}^2)S_2 \right] / 2\mu_0$. There is an additional outward force arising from the tire tube effect, identical to that in the purely poloidal case.

The end result is that a diamagnetic plasma confined solely by a toroidal field also experiences a radially outward toroidal expansion force. One can now ask whether or not this force can be balanced in a similar manner to the purely poloidal case. The answer is no! The explanation is as follows.

Since the magnetic field is in the toroidal direction, a conducting wall is not able to compensate the outward force; that is, as the plasma moves outward the magnetic lines simply slip around and let the plasma drift through (Fig. 4.16b); because of the topology, toroidal flux does not get trapped between the plasma and the wall. Likewise, a vertical field does not help because by symmetry the $\mathbf{J} \times \mathbf{B}_{vert}$ force does not point along R.

The conclusion from this discussion is that a configuration with a purely toroidal magnetic field cannot be held in MHD equilibrium because of the $1/R$ dependence of B_ϕ resulting from the toroidal geometry.

The basic problem of finding attractive magnetic geometries for confining fusion plasmas can thus be summarized as follows. On the one hand, toroidal systems with a purely poloidal magnetic field have good toroidal equilibrium properties but poor stability properties. On the other hand, for a pure toroidal field, stability is inherently much better, but serious equilibrium problems exist. Attempts to resolve this dilemma have led to the discovery of a number of different configurations whose strategy is to combine the desirable features of both toroidal and poloidal systems while to a reasonable extent suppressing the undesirable ones. Chapters 5–7 discuss the equilibrium properties of the currently most promising magnetic configurations.

4.9 A single particle picture of toroidal equilibrium

The final topic in this chapter is aimed at developing further intuition into the basic problem of toroidal equilibrium by re-examining the issues from the point of view of single particle guiding center motion.

Consider first the equilibrium problem in the case of a purely toroidal field. For a low β plasma with small diamagnetism the dominant drifts experienced by the particles are due to ∇B and curvature. These drifts arise because of the $1/R$ dependence of B and the centrifugal force, both toroidal effects. As illustrated in Fig. 4.16 the guiding center drift velocity is in the Z direction and is given by

$$\mathbf{V}_{g\perp} = \frac{1}{\omega_{c0} R_0} \left(\frac{w_\perp^2}{2} + w_\parallel^2 \right) \mathbf{e}_Z \tag{4.30}$$

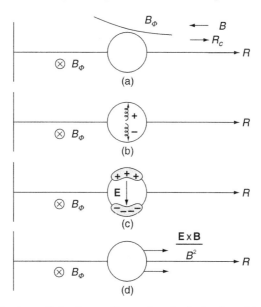

Figure 4.16   Single particle picture illustrating the lack of toroidal equilibrium in a system with purely toroidal field: (a) the geometry, (b) the guiding center drifts, (c) the induced electric field, and (d) the induced outward $\mathbf{E} \times \mathbf{B}$ drift.

where $\omega_{c0} = qB_0/m$. Note that the drift is in opposite directions for electrons and ions. As a result, positive and negative charges collect at the top and bottom of the plasma, respectively, as shown in Fig. 4.16. This charge sets up a Z directed electric field, which then propels the entire plasma outward along R with the $(\mathbf{E} \times \mathbf{B})/B^2$ drift velocity. Hence, there is no toroidal equilibrium for a configuration with purely toroidal field.

What happens when a poloidal field is superimposed on the toroidal field? In this case the magnetic field lines wrap around the plasma on helical trajectories similar to the stripes on a barber pole. Even so, the particles still experience the same up–down drift given by Eq. (4.30). One is now faced with the apparently paradoxical question of how the addition of a poloidal field can lead to single particle confinement for particles that are always drifting upward (or downward for the opposite charge).

The resolution is illustrated in Fig. 4.17. Here, the dashed lines represent flux surfaces. To leading order, the parallel motion of a particle simply follows a field line. Thus, the projection of the particle's guiding center orbit onto a poloidal cross section would exactly coincide with the flux surface if there was no vertical drift. However, when the drift is included, then as the particle spirals along the field line from point 1, it actually arrives at point 2, off the original surface, because of the

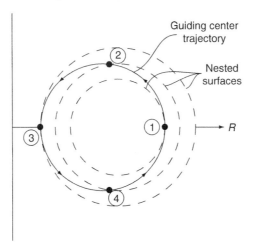

Figure 4.17   Single particle picture of toroidal confinement due to the presence of a poloidal magnetic field. Shown is a particle orbit including a uniform upward drift.

upward drift. From point 2 to 3 the drift off the surface continues to increase. In contrast, from point 3 to 4 and then back to 1, the upward drift actually causes the particle to drift back toward the surface, eventually returning to its starting position.

The point is that a constant upward drift corresponds to motion away from the surface half the time and toward the surface the other half of the time; on average the net motion due to the drift cancels and the particles remain confined. This simple picture demonstrates how a poloidal field leads to single particle confinement and macroscopic toroidal force balance.

4.10 Summary

Ideal MHD equilibrium theory has the goal of discovering magnetic geometries that stably confine hot plasmas at sufficiently high values of β to be of interest as fusion reactors. Over years of research many different concepts have been developed to achieve this goal. These different concepts, nevertheless, share a number of common features, which have been the subject of Chapter 4. A summary is given below.

- **Basic equations:** The equilibrium equations of interest correspond to the time-independent $\partial/\partial t = 0$, static $\mathbf{v} = 0$, limit of the MHD model, which describes a balance between the pressure gradient force ∇p and the magnetic force $\mathbf{J} \times \mathbf{B}$. The equilibrium model by itself is actually valid from the MHD time scale up until the very long transport time scale.

- **Virial theorem:** The virial theorem demonstrates that as desirable as it may be, it is not possible to construct a magnetic geometry in which the plasma is confined solely by its own currents; an external coil system must be supplied.

- **Toroidicity:** Because the thermal conduction loss rate parallel to the magnetic field is enormous compared to that perpendicular to the field, there is near consensus in the fusion community that magnetic geometries of fusion interest must be toroidal. The one important exception to this philosophy is the mirror concept.

- **Magnetic flux surfaces:** In a well-confined toroidal MHD equilibrium, the contours of constant pressure form a set of nested tori. Both the magnetic and current lines lie on these contours, which are known as flux surfaces.

- **Surface quantities and figures of merit:** For any given MHD equilibrium, it is possible to define a number of global quantities that depending only upon the flux surface label. These surface quantities represent important physical parameters which can be related to experiment and be used as figures of merit for measuring stability. Included are average beta β, toroidal beta β_t, poloidal beta β_p, kink safety factor q_*, rotational transform $\iota/2\pi$, and MHD safety factor q.

- **Equilibrium degrees of freedom:** For a well-posed MHD equilibrium problem it must be both possible and necessary to specify a priori two and only two independent surface quantities.

- **The basic problem of toroidal equilibrium:** The problem of MHD equilibrium separates into two parts: radial pressure balance and toroidal force balance. The latter is the more difficult problem and herein lies the basic dilemma of magnetic fusion geometries. Systems with a purely poloidal magnetic field are easy to maintain in toroidal force balance but have poor MHD stability properties. In contrast, systems with a purely toroidal field have poor toroidal force balance but much more favorable stability properties (in equivalent straight systems). The challenge then is to discover configurations which combine the favorable features and minimize the unfavorable features of these two limiting configurations.

References

Braginskii, S. I. (1965). In *Reviews of Plasma Physics*, Vol. 1, ed. M. A. Leontovich. New York: Consultants Bureau.

Shafranov, V. D. (1966). In *Reviews of Plasma Physics*, Vol. 2, ed. M. A. Leontovich. New York: Consultants Bureau.

Further reading

The references below all have detailed discussions concerning the basic properties of MHD
    equilibrium.
Bateman, G. (1978). *MHD Instabilities*. Cambridge, MA: MIT Press.
Bellan, P. M. (2006). *Fundamentals of Plasma Physics*. Cambridge: Cambridge University
    Press.
Goedbloed, J. P. and Poedts, S. (2004). *Principles of Magnetohydrodynamics*. Cambridge:
    Cambridge University Press.
Hazeltine, R. D. and Meiss, J. D. (2003). *Plasma Confinement*. Redwood City, CA:
    Addison-Wesley.
Morozov, A. I. and Solov'ev, L. S. (1966). In *Reviews of Plasma Physics*, Vol. 2, ed. M. A.
    Leontovich. New York: Consultants Bureau.
Shafranov, V. D. (1966). In *Reviews of Plasma Physics*, Vol. 2, ed. M. A. Leontovich. New
    York: Consultants Bureau.
Solov'ev, L. S. and Shafranov, V. D. (1970). In *Review of Plasma Physics*, Vol. 5, ed.
    M. A. Leontovich. New York: Consultants Bureau.
Wesson, J. (2011). *Tokamaks*, 4th edn. Oxford: Oxford University Press.

Problems

4.1 For each of the magnetic configurations given below calculate B_r and then
determine the magnetic field line trajectories and the shape of the flux surfaces.

(a) $B_\theta = B_{\theta 0}[ar/(r^2 + a^2)]$
   $B_z = B_{z0}$
   with $B_{\theta 0}$, B_{z0}, and a constants

(b) $B_\theta = 0$
   $B_z = B_{z0}(1 + \delta \cos kz)$
   with B_{z0} and δ constants

(c) $B_\theta = B_{\theta 0}(r + \Delta \cos \theta)/a$
   $B_z = B_{z0}$
   with $B_{\theta 0}$, B_{z0}, a, and Δ constants

4.2 Following the discussion associated with Fig. 4.17 sketch the orbit of a
particle starting off at point 1 with a parallel velocity in the opposite direction of
that illustrated.

4.3 Show that the MHD equilibrium equations for an arbitrary three-dimensional
geometry can be written as

$$\nabla_\perp \left(p + \frac{B^2}{2\mu_0} \right) - \frac{B^2}{\mu_0} \boldsymbol{\kappa} = 0$$

where $\boldsymbol{\kappa} = \mathbf{b} \cdot \nabla \mathbf{b}$ is the field line curvature and $\mathbf{b} = \mathbf{B}/B$.

4.4 The goal of this problem is to calculate the volume contained within a given flux surface. The starting point is the usual definition of volume given by

$$V = \int d\mathbf{r} = \int_0^{2\pi} d\phi \int_0^{2\pi} d\theta \int_0^{r_s} Rr dr$$

Here $R = R_0 + r_s \cos\theta$, $r_s = r_s(\theta,\phi,r_0)$ specifies the shape of the flux surface and r_0 is the label for the flux surface under consideration. See Fig. 4.8 for the geometry. Note that an alternate representation of the flux surface can be written as $S(r_s,\theta,\phi) = S_0$ where S_0 now represents the flux surface label.

(a) Make a change of variables from r,θ,ϕ to S,θ,ϕ and show that the volume of an arbitrary 3-D flux surface can be written as

$$V = \int_0^{S_0} dS \int_0^{2\pi} \int_0^{2\pi} \frac{Rr_s}{\partial S/\partial r_s} d\theta d\phi$$

(b) For a 1-D system with cylindrical symmetry set $S = r$ and $S_0 = r_0$. Show that the volume, as expected, is given by $V(r_0) = 2\pi^2 R_0 r_0^2$.

(c) Consider now a 2-D system with toroidal symmetry: $\partial/\partial\phi = 0$. In this case any divergence free poloidal magnetic field can be written as $\mathbf{B}_p = \nabla\psi \times \nabla\phi$ where $\psi = \psi(S)$ is an appropriately scaled flux surface label. Show that the expression for the volume reduces to

$$V(\psi_0) = 2\pi \int_0^{\psi_0} d\psi \oint \frac{dl_p}{B_p}$$

4.5 This problem investigates the approximation $C_p \approx 2\pi a[(1 + \kappa^2)/2]^{1/2}$ used in the definition of q_*. For the following model cross sections calculate and plot $C_p/2\pi a$ as a function of elongation κ over the range $1 \le \kappa \le 3$. The geometry for each cross section is shown in Fig. 4.18.

(a) An ellipse (this involves elliptic functions).
(b) A racetrack.
(c) A triangle plasma.
(d) A "D" shaped plasma.
(e) The approximation $C_p/2\pi a \approx [(1 + \kappa^2)/2]^{1/2}$.

How big are the errors between the approximation and the exact values for the various cross sections?

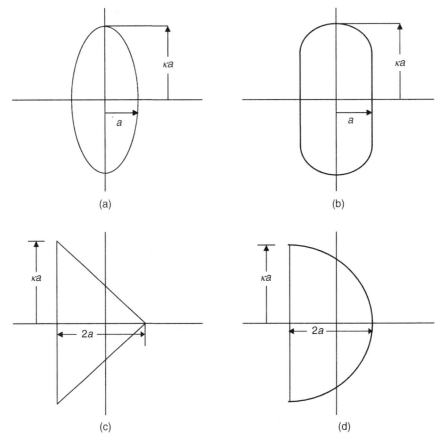

Figure 4.18   Geometry for various cross sections: (a) ellipse, (b) racetrack, (c) triangle plasma, and (d) "D" shaped plasma

5

Equilibrium: one-dimensional configurations

5.1 Introduction

Although the magnetic configurations of fusion interest are toroidal, one can begin to develop physical intuition by first investigating their one-dimensional cylindrically symmetric analogs: the θ-pinch, the Z-pinch, and the general screw pinch. These can be considered to be the basic building blocks of MHD equilibrium. Focusing on cylindrical systems allows the two basic problems of MHD equilibrium – radial pressure balance and toroidal force balance – to be separated, so that each can be studied individually.

The one-dimensional model focuses entirely on radial pressure balance. The question of toroidal force balance does not enter since by definition the geometry is a linear cylinder. For many configurations, once radial pressure balance is established, toroidicity can be introduced by means of an inverse aspect ratio expansion, from which one can then investigate toroidal force balance.

Chapter 5 provides a description of the basic one-dimensional configurations and how they provide radial pressure balance in a plasma. In particular, it is shown that both toroidal and poloidal fields as well as combinations thereof can easily accomplish this goal.

Included in the analysis are descriptions of two present day fusion concepts: the reversed field pinch, and the ohmic tokamak. These configurations are singled out since both their radial pressure balance and MHD stability are reasonably well described by the one-dimensional cylindrical model. Toroidal effects can be treated perturbatively and make small quantitative, but not qualitative, corrections to the cylindrical equilibrium and stability results.

5.2 The θ-pinch

The θ-pinch represents the one-dimensional analog of the toroidal configuration with purely toroidal field. The "equivalent" torus consists of a section $2\pi R_0$

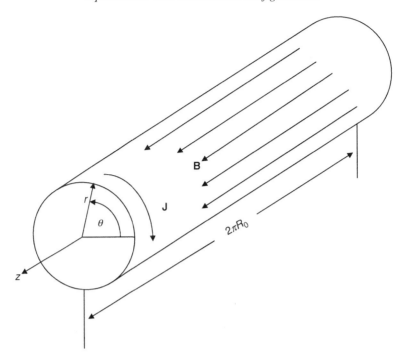

Figure 5.1   Linear θ-pinch geometry.

in length of the infinitely long cylindrically symmetric linear plasma column.
In a θ-pinch the only non-zero component of **B** is in the z direction: $\mathbf{B} = B_z(r)\mathbf{e}_z$.
It is applied externally and induces a large diamagnetic current in the θ
direction: $\mathbf{J} = J_\theta(r)\mathbf{e}_\theta$. The resulting $\mathbf{J} \times \mathbf{B}$ force confines the pressure $p(r)$.
See Fig. 5.1 for the geometry. The θ directed current is the origin of the name
θ-pinch.

The basic equilibrium relation for a θ-pinch is easily obtainable from the
radial component of the momentum equation. However, a slightly more formal
derivation is presented which demonstrates the general procedure used for all
one- and two-dimensional MHD configurations discussed in the textbook. Specifically, there is a natural sequence in which to solve the MHD equations which
provides a simple and direct path to the final equations of interest. This sequence
is given by (1) the $\nabla \cdot \mathbf{B} = 0$ equation, (2) Ampere's law, and (3) the momentum
equation:

- $\nabla \cdot \mathbf{B} = 0$: The equation $\nabla \cdot \mathbf{B} = 0$ is trivially satisfied for the θ-pinch because
  of symmetry; all quantities are only a function of the radial coordinate r.

$$\nabla \cdot \mathbf{B} = 0 \quad \rightarrow \quad \frac{\partial B_z}{\partial z} = 0 \qquad (5.1)$$

- **Ampere's law:** Ampere's law shows that only $J_\theta(r)$ is non-zero.

$$\mathbf{J} = \frac{1}{\mu_0} \nabla \times \mathbf{B} \quad \rightarrow \quad J_\theta = -\frac{1}{\mu_0} \frac{dB_z}{dr} \qquad (5.2)$$

- **Momentum equation:** The only non-trivial component of the momentum equation is in the radial direction and is given by

$$\mathbf{J} \times \mathbf{B} = \nabla p \quad \rightarrow \quad J_\theta B_z = \frac{dp}{dr} \qquad (5.3)$$

Here, the pressure $p = p(r)$. Eliminating J_θ by means of Eq. (5.2) leads to

$$\frac{d}{dr}\left(p + \frac{B_z^2}{2\mu_0}\right) = 0 \qquad (5.4)$$

This equation can be easily integrated, yielding

$$p + \frac{B_z^2}{2\mu_0} = \frac{B_0^2}{2\mu_0} \qquad (5.5)$$

where B_0 is the applied magnetic field.

Equation (5.4) is the basic radial pressure balance relation for a θ-pinch. Its integrated form given by Eq. (5.5) indicates that at any local value of r the sum of the local particle pressure plus local magnetic pressure is a constant, equal to the externally applied magnetic pressure. The conclusion is that in a θ-pinch radial pressure balance is achieved by the pressure exerted by the externally applied magnetic field.

Illustrated in Fig. 5.2 is a set of typical profiles given, for example, by

$$\frac{2\mu_0 p(r)}{B_0^2} = 1 - \left[1 - \hat{\beta}\left(1 - \rho^2\right)^2\right]^2$$

$$\frac{B_z(r)}{B_0} = 1 - \hat{\beta}\left(1 - \rho^2\right)^2$$

$$\frac{a\mu_0 J_\theta(r)}{B_0} = -4\hat{\beta}\rho\left(1 - \rho^2\right) \qquad (5.6)$$

$$\hat{\beta} = \frac{\beta_0}{1 + (1 - \beta_0)^{1/2}}$$

where $\rho = r/a$ and $\beta_0 = 2\mu_0 p_0 / B_0^2$ is defined as the beta on axis. For physical solutions $\beta_0 < 1$. The pressure peaks at $r = 0$ and vanishes at the plasma edge $r = a$ (i.e., $\rho = 1$). The profiles are chosen so that the current density and pressure

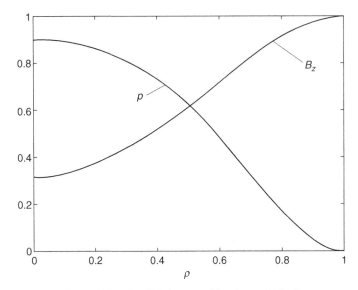

Figure 5.2   Equilibrium profiles for a θ-pinch.

gradient also vanish at $r = a$. The plasma is thus isolated from the containing wall indicating good radial confinement and closed, nested pressure contours.

Note that Eq. (5.4) implies that the θ-pinch has one free surface function, for example $B_z(r)$, from which it is then possible to calculate $p(r)$. The second free function (e.g., $B_\theta(r)$) has been set to zero because of the special symmetry associated with the θ-pinch.

The basic plasma parameters and figures of merit can now be easily evaluated. Consider first the value of beta. Since $B_\theta = 0$, then $\beta = \beta_t$ with the toroidal β given by Eq. (4.19)

$$
\begin{aligned}
\beta_t &= \frac{2\mu_0 \langle p \rangle}{B_0^2} \\
&= \frac{4\mu_0}{a^2 B_0^2} \int_0^a pr\,dr \\
&= 2 \int_0^1 \left(1 - \frac{B_z^2}{B_0^2}\right) \rho\,d\rho \\
&= \hat{\beta}\left(\frac{2}{3} - \frac{\hat{\beta}}{5}\right)
\end{aligned}
\tag{5.7}
$$

where the last expression corresponds to the profiles given by Eq. (5.6). Note that as $\beta_0 \to 0$ then $\hat{\beta} \approx \beta_0/2$ and $\beta_t \approx \beta_0/3$. Similarly, as $\beta_0 \to 1$ then $\hat{\beta} \to 1$ and $\beta_t \approx 7/15$.

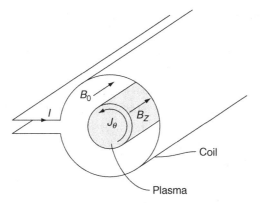

Figure 5.3   Schematic diagram of a θ-pinch experiment.

Equation (5.7) implies that, in general, physical profiles can exist for

$$0 < \beta_t < 1 \tag{5.8}$$

This wide range of variation including access to high values of β_t indicates that the θ-pinch is an excellent option for producing radial pressure balance in a fusion plasma.

Furthermore, since there is no z current in a θ-pinch (i.e., $I = 0$) then $\beta_p = \infty$ and as such is not a relevant quantity. Similarly, $q_* = \infty$ indicating favorable stability against current-driven kink modes, which is a trivial result since there is no toroidal current. The local safety factor $q(r) = \infty$, again a consequence of no z current.

The θ-pinch is more than a theoretical concept and in the early years of the fusion program a number of such devices were built. The typical experimental situation is illustrated in Fig. 5.3. The plasma fills an insulating discharge tube (usually made of quartz or Pyrex), which is surrounded by a single turn coil connected to a large capacitor bank. The gas is initially pre-ionized after which the switch on the capacitor bank is closed. As the capacitor discharges, current flows in the coil, producing a B_z field in the discharge chamber. The parameters of the high-voltage capacitor bank are chosen so that the rise time of the field is quite short, on the order of 2 μsec, while the overall pulse length is typically 10 μsec. The rapidly rising magnetic field acts like a piston, imparting a large impulse of momentum and energy to the particles as they are reflected off the piston face. This energy is ultimately converted to heat after repeated reflections off the converging piston.

The performance of θ-pinches in terms of fusion parameters represents one of the early successes of the fusion program. Using the implosion heating method, ion temperatures of 1–4 keV were routinely obtained at very high density, $n \sim 1$–$2 \times 10^{22}\,\mathrm{m}^{-3}$. The peak value of beta on axis was typically $\beta_0 \sim 0.7$–0.9, quite a high value, but corresponding to a volume averaged beta of only $\beta_t \sim 0.05$. The low average beta was a consequence of the large compression ratio of the plasma due to the implosion heating. This type of heating left a large volume of very low pressure plasma between the plasma core and the wall. As further evidence of good performance, radial profile measurements made by end-on holographic interferometry demonstrated excellent radial confinement, nested circular flux surfaces, and no indications of macroscopic instability. Because of the high temperatures and densities, the θ-pinch was the first laboratory device to successfully produce a substantial number of thermonuclear neutrons.

As might be expected, the biggest problem with the θ-pinch was end loss. Typically, little if anything was done to provide axial confinement of particles or energy; the plasma simply flowed out the end of the device along field lines in a characteristic time $\tau = L/V_{Ti}$, where L is the length of the magnet. For a 5-m device, $\tau \sim 10 \, \mu\text{sec}$, a very short time indeed. Because of the end-loss problem the θ-pinch has been replaced by more advanced toroidal concepts capable of much longer confinement times.

Before closing the discussion of θ-pinches, several comments should be made regarding toroidal equilibrium and stability. The cylindrical θ-pinch represents a highly degenerate configuration for the following reasons. A complete and detailed stability analysis indicates that the straight θ-pinch is marginally stable against the most dangerous MHD modes. It is never unstable. Also, as has been previously shown, the θ-pinch cannot be bent into a torus. Therefore, small additional fields, of order a/R_0, must be added to provide toroidal force balance. Since the basic straight configuration is marginally stable, the overall stability depends sensitively on these small, as yet unidentified, additional fields. Thus, the important problems have yet to be faced.

5.3 The Z-pinch

The Z-pinch is a one-dimensional model of the toroidal configuration with purely poloidal field and is in many ways orthogonal to the θ-pinch. In a Z-pinch, the only non-zero field component is $B_\theta(r)$ and consists entirely of the self-field induced by the longitudinal current $J_z(r)$ flowing in the plasma; hence the name Z-pinch.[1] Again the $\mathbf{J} \times \mathbf{B}$ force confines the pressure $p(r)$ (see Fig. 5.4). Unlike the θ-pinch, when the total plasma current vanishes there is no remaining background $\mathbf{B}$ field. The basic equation describing radial pressure balance in a Z-pinch is determined as follows:

---

[1] Note this does not violate the virial theorem since the current is not confined in that it extends from $-\infty < z < \infty$.

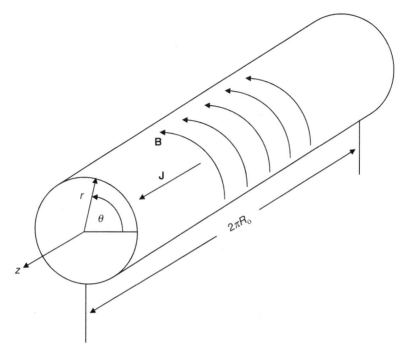

Figure 5.4   Linear Z-pinch geometry.

- $\nabla \cdot \mathbf{B} = 0$: As for the θ-pinch the $\nabla \cdot \mathbf{B} = 0$ equation is trivially satisfied because of symmetry; all quantities are only functions of r.

$$\nabla \cdot \mathbf{B} = 0 \quad \rightarrow \quad \frac{1}{r}\frac{\partial B_\theta}{\partial \theta} = 0 \tag{5.9}$$

- **Ampere's law:** Application of Ampere's law shows that only $J_z(r)$ is non-zero.

$$\mathbf{J} = \frac{1}{\mu_0}\nabla \times \mathbf{B} \quad \rightarrow \quad J_z = \frac{1}{\mu_0 r}\frac{d}{dr}(rB_\theta) \tag{5.10}$$

- **Momentum equation:** Only the radial component of the momentum equation yields non-trivial information

$$\mathbf{J} \times \mathbf{B} = \nabla p \quad \rightarrow \quad J_z B_\theta = -\frac{dp}{dr} \tag{5.11}$$

The radial pressure balance relation is obtained by substituting J_z from Eq. (5.10). The result is

$$\frac{dp}{dr} + \frac{B_\theta}{\mu_0 r}\frac{d}{dr}(rB_\theta) = 0 \tag{5.12}$$

which can be rewritten as

$$\frac{d}{dr}\left(p + \frac{B_\theta^2}{2\mu_0}\right) + \frac{B_\theta^2}{\mu_0 r} = 0 \tag{5.13}$$

Equation (5.13) is the basic radial pressure balance relation for a Z-pinch. The two terms in the derivative represent the particle pressure and the magnetic pressure. The last term represents the tension force generated by the curvature of the magnetic field lines. The connection between the tension force and the curvature is a general result obtained by substituting $\mathbf{J}$ from Ampere's law directly into the momentum equation for arbitrary 3-D magnetic configurations. Noting that $(\nabla \times \mathbf{B}) \times \mathbf{B} = -\nabla(B^2/2) + \mathbf{B} \cdot \nabla\mathbf{B}$ and defining $\boldsymbol{\kappa} = \mathbf{b} \cdot \nabla\mathbf{b}$ with $\mathbf{B} = B\mathbf{b}$ one finds

$$\nabla_\perp\left(p + \frac{B^2}{2\mu_0}\right) - \frac{B^2}{\mu_0}\boldsymbol{\kappa} = 0 \tag{5.14}$$

The quantity $\boldsymbol{\kappa}$ is known as the curvature vector and is related to the radius of curvature vector $\mathbf{R}_c$ by $\boldsymbol{\kappa} = -\mathbf{R}_c/R_c^2$. For readers unfamiliar with this relation, a derivation is presented in Appendix D. Now, since $\boldsymbol{\kappa} = \mathbf{e}_\theta \cdot \nabla\mathbf{e}_\theta = -\mathbf{e}_r/r$ in a Z-pinch, the connection between the tension force and curvature is apparent.

The tension force is very important. In contrast to a θ-pinch, for a Z-pinch it is the tension force and not the magnetic pressure gradient that provides radial pressure balance. This can be seen explicitly from an example. It can be easily verified that the profiles below, for which p, ∇p, and J_z all vanish at the plasma edge $r = a$, satisfy the Z-pinch pressure balance relation:

$$\frac{2\mu_0 p(r)}{B_{\theta a}^2} = \frac{2}{3}\left(5 - 2\rho^2\right)\left(1 - \rho^2\right)^2$$

$$\frac{B_\theta(r)}{B_{\theta a}} = 2\rho\left(1 - \rho^2/2\right) \tag{5.15}$$

$$\frac{a\mu_0 J_z(r)}{B_{\theta a}} = 4\left(1 - \rho^2\right)$$

Here, $\rho = r/a$, $B_{\theta a} \equiv B_\theta(a) = \mu_0 I/2\pi a$, and I is the total current flowing in the plasma. These profiles are illustrated in Fig. 5.5. Also illustrated are curves of $-p'$, $-(B_\theta^2/2\mu_0)'$, and $-B_\theta^2/\mu_0 r$ vs. r, representing the outward force density profiles of the three contributions to radial pressure balance. Note that in the outer region of the plasma, $r/a > (2/3)^{1/2}$, the plasma and magnetic pressure gradients are both pointing outward; the plasma is not confined by magnetic pressure. Instead it is the tension force that acts inwards, thereby providing radial pressure balance. If one thinks of the magnetic lines as rubber bands wrapped around the plasma column,

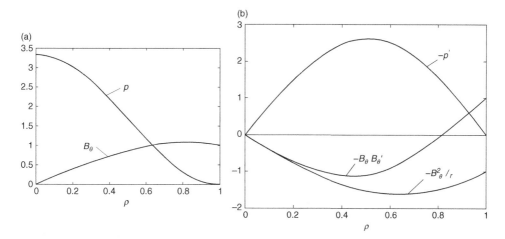

Figure 5.5 Equilibrium profiles for a Z-pinch: (a) $2\mu_0 p/B_{\theta a}^2$ and $B_\theta/B_{\theta a}$, (b) outward forces $-p'$, $-(B_\theta^2/2\mu_0)'$, and $-B_\theta^2/\mu_0 r$.

then the tension force is obvious. In analogy with the θ-pinch, the Z-pinch equilibrium has one free function, for instance $B_\theta(r)$, which then determines $p(r)$. The other free function, $B_z(r)$, has been set to zero because of the special Z-pinch symmetry.

From the basic Z-pinch relationships given by Eqs. (5.10)–(5.13) it is straightforward to calculate the plasma parameters and figures of merit. First, since $B_z = 0$ the plasma beta reduces to $\beta = \beta_p$. Using the form of β_p given by Eq. (4.19), it follows that β_p reduces to

$$\beta_p = \frac{2\mu_0 \langle p \rangle}{B_{\theta a}^2}$$
$$= \frac{4\mu_0}{a^2 B_{\theta a}^2} \int_0^a pr\,dr \qquad (5.16)$$
$$= 1$$

The result $\beta_p = 1$ is known as the Bennett pinch relation (1934) and is actually valid for any confined Z-pinch profile. While high β_p is desirable for confinement efficiency, the lack of flexibility in achieving small to moderate β_p is a disadvantage; that is, some classes of potentially dangerous MHD modes might be stabilized if β_p could be lowered. The fact that this cannot occur in a pure Z-pinch is one reason why its stability properties are so poor.

Since $B_z = 0$ for a Z-pinch, the quantities β_t, q_* and $q(r)$ approach limiting values. The toroidal beta has the value $\beta_t = \infty$ and as such is not a relevant quantity. The kink safety factor $q_* = 0$, indicating poor stability against current-driven kinks. Also, since the magnetic field line trajectories never progress along z it follows that $q(r) = 0$.

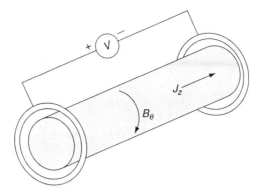

Figure 5.6   Schematic diagram of a linear Z-pinch experiment.

A number of linear Z-pinch experiments were constructed during the early years of the fusion program. A typical device is illustrated schematically in Fig. 5.6. Here, a capacitor bank is discharged across two electrodes located at each end of a cylindrical quartz or Pyrex tube. The high voltage ionizes the gas and produces a z current flowing along the plasma. These early experiments exhibited disastrous instabilities, often leading to a complete quenching of the plasma after 1–2 μsec. This behavior is discussed in detail in Chapter 11 and is consistent with the unfavorable figures of merit derived above.

Finally, with regard to toroidal equilibrium, recall that the Z-pinch can easily be bent into a torus, although its stability properties remain poor. An interesting innovation to the simple Z-pinch is the addition of a current-carrying conductor along the axis which greatly improves stability. This configuration is known as the hard-core Z-pinch.

5.4 The general screw pinch

5.4.1 General properties

It should come as no great surprise that the search for MHD stable toroidal equilibria has led to configurations that combine toroidal and poloidal fields. The hope, of course, is to combine the favorable features of each field while suppressing the unfavorable ones. The one-dimensional analogs of such systems have both $B_\theta(r)$ and $B_z(r)$ being non-zero (see Fig. 5.7). The basic pressure balance relation for a general screw pinch follows in a straightforward manner.

- $\nabla \cdot \mathbf{B} = 0$: Even with two field components present, cylindrical symmetry guarantees that $\nabla \cdot \mathbf{B} = 0$ is trivially satisfied.

$$\nabla \cdot \mathbf{B} = 0 \quad \rightarrow \quad \frac{1}{r}\frac{\partial B_\theta}{\partial \theta} + \frac{\partial B_z}{\partial z} = 0 \qquad (5.17)$$

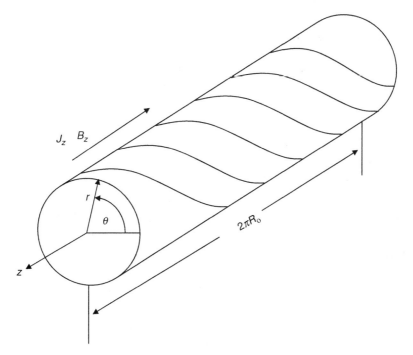

Figure 5.7   General screw pinch geometry.

- **Ampere's law:** Application of Ampere's law shows that two components of current, $J_\theta(r)$ and $J_z(r)$, are non-zero.

$$\mathbf{J} = \frac{1}{\mu_0} \nabla \times \mathbf{B} \quad \rightarrow \quad \mathbf{J} = \frac{1}{\mu_0 r} \frac{d}{dr}(rB_\theta)\mathbf{e}_z - \frac{1}{\mu_0} \frac{dB_z}{dr}\mathbf{e}_\theta \qquad (5.18)$$

- **Momentum equation:** As before, only the radial component of the momentum equation is non-trivial.

$$\mathbf{J} \times \mathbf{B} = \nabla p \quad \rightarrow \quad J_\theta B_z - J_z B_\theta = \frac{dp}{dr} \qquad (5.19)$$

Substituting Eq. (5.18) yields

$$\frac{d}{dr}\left(p + \frac{B_\theta^2 + B_z^2}{2\mu_0}\right) + \frac{B_\theta^2}{\mu_0 r} = 0 \qquad (5.20)$$

Equation (5.20) is the basic radial pressure balance relation for a general screw pinch. Note that even though the momentum equation is non-linear, the θ-pinch and Z-pinch forces superpose linearly, a consequence of the cylindrical symmetry. Although Eq. (5.20) is a relatively simple relation, it nevertheless exhibits many of the features and flexibility expected in more realistic, multidimensional, toroidal models. This is demonstrated by examining Eq. (5.20) and evaluating the corresponding plasma parameters and figures of merit.

- **Free functions:** There are two free functions available to specify the equilibrium; for example, $B_\theta(r)$ and $B_z(r)$. The θ-pinch and Z-pinch are special choices where B_θ or B_z is set to zero, respectively. In the general case, once both profiles are specified, the pressure is calculated from Eq. (5.20).
- **Flux surfaces:** The contours of constant pressure are given by $r = $ constant. Thus, the flux surfaces consist of closed concentric circles.
- **Beta:** The value of β can vary over a wide range if B_z is not zero. This can be seen from the general one-dimensional macroscopic force balance in the plasma. The relation is obtained by multiplying Eq. (5.20) by r^2 and integrating over the plasma from $0 < r < a$. A short calculation yields a volume averaged radial pressure balance relation which can be written in two equivalent forms

$$2\pi \int_0^a pr\,dr = \frac{\mu_0 I^2}{8\pi} + 2\pi \int_0^a \frac{B_0^2 - B_z^2}{2\mu_0} r\,dr$$

$$\langle p \rangle = \frac{B_{\theta a}^2}{2\mu_0} + \frac{1}{2\mu_0}\left(B_0^2 - \langle B_z^2 \rangle\right)$$

(5.21)

The second form is obtained by multiplying the first by $1/\pi a^2$ and using the definition of volume average.

The left-hand term in Eq. (5.21) represents the outward force due to the plasma pressure. The first right-hand term represents the inward force due to the magnetic tension in the poloidal field. The last term represents the net force due to the magnetic pressure of the toroidal field. It can be inward or outward depending on whether the plasma is diamagnetic or paramagnetic with respect to the toroidal field. If one now recalls the simplified definitions of β_t, β_p, and β from Eqs. (4.18)–(4.20) which are given by

$$\beta_t = \frac{2\mu_0}{B_0^2}\langle p \rangle$$

$$\beta_p = \frac{2\mu_0}{B_{\theta a}^2}\langle p \rangle$$

(5.22)

$$\beta = \frac{\beta_t \beta_p}{\beta_t + \beta_p}$$

it then follows that

$$\beta = \frac{2\mu_0 \langle p \rangle}{B_0^2 + B_{\theta a}^2}$$

(5.23)

Clearly $\beta < 1$ and can vary over a large range depending on the size of B_0.

- **Kink safety factor:** The kink safety factor is readily evaluated from Eq. (4.21) and is given by

$$q_* = \frac{2\pi a^2 B_0}{\mu_0 R_0 I} = \frac{aB_0}{R_0 B_{\theta a}} \qquad (5.24)$$

Its value can vary over a wide range depending on the ratio of toroidal field to toroidal current and the geometry a/R_0.

- **Rotational transform and safety factor:** The magnetic field lines wrap around the column along helical paths similar to the stripes on a barber pole (see Fig. 5.7). This gives rise to a non-zero rotational transform. The value of ι is calculated by noting that the angle $\Delta\theta$ defined in Eq. (4.26) is independent of the starting angle θ_0 because of the cylindrical symmetry. Consequently, to calculate ι it is only necessary to integrate the field line trajectory a distance $\Delta z = 2\pi R_0$ corresponding to one transit around the equivalent torus.

$$\iota = \Delta\theta = \int_0^{\Delta\theta} d\theta = \int_0^{2\pi R_0} \frac{d\theta}{dz}\, dz \qquad (5.25)$$

From the equations for the field line trajectories

$$\frac{dr}{dz} = \frac{B_r(r)}{B_z(r)} = 0$$

$$\frac{d\theta}{dz} = \frac{B_\theta(r)}{rB_z(r)} \qquad (5.26)$$

one sees from the first of these that $r =$ constant, indicating circular flux surfaces. After substituting the second relation into Eq. (5.25) it follows that the expression for ι given by Eq. (5.25) can be trivially integrated yielding

$$\iota(r) = \frac{2\pi R_0 B_\theta(r)}{rB_z(r)} \qquad (5.27)$$

The safety factor $q(r) = 2\pi/\iota(r)$ can thus be written as

$$q(r) = \frac{rB_z(r)}{R_0 B_\theta(r)} \qquad (5.28)$$

Again, with two free functions available, there is a wide range of flexibility in the $q(r)$ profile. In general, the magnetic shear will be non-zero. Also, note that for a screw pinch $q(a) = q_*$.

This completes the discussion of the plasma parameters and figures of merit for the screw pinch. By now it should be clear that a wide variety of fusion configurations are described by the general pressure balance relation given by Eq. (5.20). For purposes of illustration it is useful to describe two limiting conceptual examples of screw pinch equilibria which bracket the configurations of interest. These are the parallel pinch and the perpendicular pinch, and are discussed below. More interestingly, two actual present day fusion configurations, the reversed field pinch and the low β ohmic tokamak, are accurately described by the 1-D screw pinch equilibrium relation and these are discussed in Sections 5.5.2 and 5.5.3.

5.4.2 The parallel pinch

The parallel pinch is a configuration in which all the current flows parallel to the magnetic field. In other words, $\mu_0 \mathbf{J} = k(r)\mathbf{B}$, where $k(r)$ is an arbitrary scalar function. Since $\mathbf{J}$ is parallel to $\mathbf{B}$ there is no magnetic force $\mathbf{J} \times \mathbf{B}$ acting on the plasma and for obvious reasons such configurations are often called "force free." Furthermore, since $\mathbf{J} \times \mathbf{B} = 0$, the pressure $p(r) = 0$ everywhere. Force-free configurations are good approximations to very low β systems. For the parallel pinch it is instructive to calculate the magnetic field profiles as well as $q(r)$. The goal is to learn whether or not confined current density profiles exist with a high value of kink safety factor $q_* \sim 1$. The analysis proceeds as follows and is more complicated than one might expect.

The basic equilibrium equation for the parallel pinch is obtained by substituting $d(rB_\theta)/dr = krB_z$ into Eq. (5.20) yielding

$$\frac{r}{k}\frac{dB_z}{dr} = -rB_\theta \tag{5.29}$$

This equation is now differentiated with respect to r after which $d(rB_\theta)/dr$ is again eliminated. The result is a single differential equation for B_z given by

$$\frac{1}{rk}\frac{d}{dr}\left[\frac{r}{k}\frac{dB_z}{dr}\right] + B_z = 0 \tag{5.30}$$

One simple solution to this equation corresponds to the choice $k(r) = k_0 =$ constant. In this case the solutions are given by

$$B_z = \frac{\mu_0 J_z}{k_0} = B_0 J_0(k_0 r)$$
$$B_\theta = \frac{\mu_0 J_\theta}{k_0} = B_0 J_1(k_0 r) \tag{5.31}$$

where $B_0 =$ constant and J_0, J_1 (in Roman, not italic fonts) are Bessel functions. This solution is not entirely satisfactory since the current density is finite at the edge of the plasma $r = a$. It is not confined.

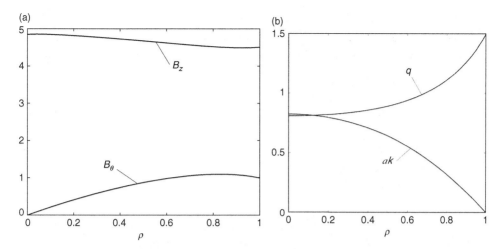

Figure 5.8 Equilibrium profiles for the parallel pinch for the case $\alpha = 2/9$ and $q_* = 3/2$. Illustrated are (a) $B_\theta/B_{\theta a}$ and $B_z/B_{\theta a}$, (b) ak and q.

However, Eq. (5.30) is difficult to solve analytically, even for simple choices of $k(r)$ describing confined equilibria. Such equilibria require $\mu_0 J_z(0) = k(0)B_z(0) =$ constant and $\mu_0 J_z(0) = k(a)B_z(a) = 0$. The corresponding $k(r)$ must satisfy $k(0) =$ constant and $k(a) = 0$. An alternate way to approach the problem is to specify a plausible profile for $\mu_0 J_z(r) = k(r)B_z(r)$ and then use Eq. (5.30) to calculate $k(r)$. A plausible choice for $J_z(r)$ is as follows:

$$J_z(r) = \frac{4B_{\theta a}}{\mu_0 a}\left(1 - \rho^2\right) \tag{5.32}$$

where again $\rho = r/a$. After a slightly tedious calculation the unintuitive $k(r)$ can be evaluated and then back substituted to obtain the desired profiles:

$$
\begin{aligned}
ak(\rho) &= \frac{4\alpha(1 - \rho^2)}{\left[1 + (2\alpha^2/3)(1 - \rho^2)^2(5 - 2\rho^2)\right]^{1/2}} \\
\frac{B_\theta(\rho)}{B_{\theta a}} &= \frac{\mu_0 J_\theta(\rho)}{B_{\theta a}k(\rho)} = \rho(2 - \rho^2) \\
\frac{B_z(\rho)}{B_{\theta a}} &= \frac{\mu_0 J_z(\rho)}{B_{\theta a}k(\rho)} = \frac{1}{\alpha}\left[1 + (2\alpha^2/3)(1 - \rho^2)^2(5 - 2\rho^2)\right]^{1/2} \\
q(\rho) &= \frac{q_*}{2 - \rho^2}\left[1 + (2\alpha^2/3)(1 - \rho^2)^2(5 - 2\rho^2)\right]^{1/2}
\end{aligned}
\tag{5.33}
$$

Here, $B_{\theta a} = B_\theta(a)$, $B_z(a) = B_0$, and $\alpha = B_{\theta a}/B_0 = \varepsilon/q_*$.

The profiles are illustrated in Fig. 5.8 for the case $q_* = 3/2$ and $\varepsilon = 1/3$ corresponding to $\alpha = 2/9$. Observe that $B_z(0)/B_z(a) = 1 + (10/3)\,\alpha^2 \approx 1.16 > 1$, showing that the toroidal field is paramagnetic. In fact it can be easily shown that any force-free equilibrium must be paramagnetic with respect to the toroidal magnetic field. This follows by setting $p = 0$ in the general force balance relation given by Eq. (5.21):

$$\text{Paramagnetism} \equiv 2\pi \int_0^a \frac{B_z^2 - B_0^2}{2\mu_0} r\,dr = \frac{\mu_0 I^2}{8\pi} > 0 \qquad (5.34)$$

The safety factor profile can either increase or decrease with radius. For the profiles in the example the transition point is $\alpha^2 = 9/10$ with smaller values corresponding to an increasing q profile. Specifically, one finds $q(0)/q(a) = [1 + (5/6)(\alpha^2 - 9/10)]^{1/2}$.

The conclusion is that force-free solutions can be found that describe equilibria with confined current densities and high values of the kink safety factor.

5.4.3 The perpendicular pinch

The perpendicular pinch is a configuration in which all the current flows perpendicular to the magnetic field. There is no parallel current: $\mathbf{J} \cdot \mathbf{B} = 0$. One might anticipate that this configuration confines the maximum amount of pressure since no current is "wasted" flowing parallel to the field, which would produce no force on the plasma. The goal of the example below is to calculate typical profiles for the perpendicular pinch and to see how large β can become assuming that the kink safety factor is large, of order unity.

The analysis begins by noting that $\mathbf{J} \cdot \mathbf{B} = 0$ implies that

$$\frac{1}{B_z}\frac{dB_z}{dr} = \frac{1}{rB_\theta}\frac{d}{dr}(rB_\theta) \qquad (5.35)$$

This equation can be easily integrated yielding

$$B_z(\rho) = B_0 \frac{\rho B_\theta(\rho)}{B_{\theta a}} \qquad (5.36)$$

Assume now that a confined $J_z(\rho)$ profile is specified. Plausible choices for $J_z(\rho)$ and the corresponding $B_\theta(\rho)$ are as follows:

$$\frac{a\mu_0 J_z(\rho)}{B_0} = \frac{4\varepsilon}{q_*}(1 - \rho^2)$$

$$\frac{B_\theta(\rho)}{B_0} = \frac{\varepsilon}{q_*}\rho(2 - \rho^2) \qquad (5.37)$$

where $B_{\theta a}/B_0 = \varepsilon/q_*$. The $B_z(\rho)$ and $J_\theta(\rho)$ profiles are now easily calculated from Eq. (5.36)

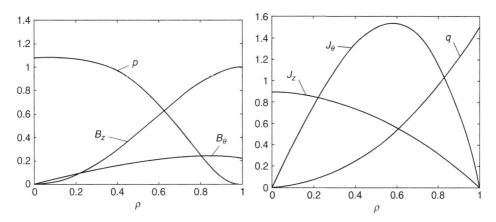

Figure 5.9   Equilibrium profiles for the perpendicular pinch for the case $q_* = 3/2$ and $\varepsilon = 1/3$: (a) B_z/B_0, B_θ/B_0, $2\mu_0 p/B_0^2$ and (b) $a\mu_0 J_z/B_0$, $a\mu_0 J_\theta/B_0$, q.

$$\frac{B_z(\rho)}{B_0} = \rho^2(2 - \rho^2)$$

$$\frac{a\mu_0 J_\theta(\rho)}{B_0} = 4\rho(1 - \rho^2)$$

(5.38)

Knowing B_z and B_θ, one can evaluate the safety factor

$$q(\rho) = q_* \rho^2$$

(5.39)

The last step is to determine $p(\rho)$ from Eq. (5.20), the radial pressure balance relation. A short calculation leads to

$$\frac{2\mu_0 p(\rho)}{B_0^2} = (1 - \rho^2)^2\left[1 + 2\rho^2 - \rho^4 + \frac{\varepsilon^2}{3q_*^2}(5 - 2\rho^2)\right]$$

(5.40)

The profiles are plotted in Fig. 5.9 for the case $q_* = 3/2$ and $\varepsilon = 1/3$. Note that $B_z(0) = 0$, indicating that the toroidal field is highly diamagnetic. Similarly $q(0) = 0$. Thus, while high values of q_* are allowed, one should be concerned that the safety factor on axis is always zero, a warning sign of potential internal instabilities.

The original question can now be addressed – how large can β become assuming $q_* \sim 1$? From the definition of β given by Eq. (5.23), one can show after a short calculation that

$$\beta = \frac{2\mu_0\langle p\rangle}{B_0^2 + B_{\theta a}^2} = \frac{1}{30}\left(\frac{14 + 15\varepsilon^2/q_*^2}{1 + \varepsilon^2/q_*^2}\right)$$

(5.41)

The value of β is almost independent of q_* and is approximately given by $\beta \approx 1/2$. This is indeed a high value, although it will be shown in Chapter 6 that the inclusion of toroidal force balance effects reduces the achievable β by a substantial amount when $q_* \sim 1$.

5.5 Inherently 1-D fusion configurations

There are two present day fusion configurations whose MHD equilibrium and stability are accurately described by a 1-D model: the reversed field pinch, and the low β ohmic tokamak with circular cross section. For these configurations toroidal effects enter only perturbatively when calculating toroidal force balance and do not have a qualitative impact on MHD stability. Radial pressure balance in other configurations such as the high β auxiliary heated tokamak and the stellarator can also be described by the 1-D screw pinch model but in these cases toroidicity qualitatively alters both toroidal force balance and MHD stability. For this reason discussions of the equilibrium properties of these configurations are postponed until multidimensional systems are analyzed in Chapters 6 and 7.

Consider now the MHD equilibrium properties of the two inherently 1-D configurations.

5.5.1 *The reversed field pinch*

Overview

The reversed field pinch (RFP) is an axisymmetric toroidal configuration with comparably sized toroidal and poloidal magnetic fields. The plasma pressure is high, similar in magnitude to the magnetic pressures. The aspect ratio $R_0/a \sim 5$ is relatively large compared to many other fusion configurations. An important feature of the RFP is that the toroidal field coils are very modest in size, a distinct technological advantage with respect to complexity and cost as one looks ahead to reactors.

Carefully tailored, and sometimes naturally occurring, pressure and current profiles are stable to macroscopic MHD modes localized internally within the plasma. For these "internal" modes the plasma surface does not move. However, "external" modes where the plasma surface is allowed to move are always MHD unstable and require a close fitting conducting shell around the plasma and/or feedback for stability. Research has also shown that under certain circumstances the plasma deforms from its "unstable" cylindrical shape into a "stable" helically shaped plasma. The present discussion is focused on cylindrical RFP equilibria.

There are several main RFP experiments currently in operation around the world: (1) the Madison Symmetric Torus (MST) at the University of Wisconsin

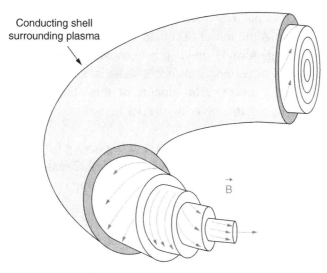

Figure 5.10   Schematic diagram of an RFP (courtesy of J. S. Sarff).

(USA); (2) the Reversed Field Experiment (RFX) operated by an Italian consortium consisting of the University of Padua, the government, and industry; and (3) the RFP at the AIST Laboratory in Tsukuba, Japan. There is also a smaller experiment, EXTRAP T2R, operating in Sweden.

The combination of comparably sized poloidal and toroidal magnetic fields, plus a large aspect ratio, imply that radial pressure balance and MHD stability can be accurately modeled by a straight cylindrical system. Toroidal effects make a small quantitative modification to the results but do not qualitatively affect the overall MHD behavior.

A schematic diagram of an RFP is illustrated in Fig. 5.10. Experimentally a typical RFP operates as follows. The discharge chamber is pre-filled with neutral deuterium. Next a small, uniform toroidal bias field, which homogeneously fills the chamber, is applied by the toroidal field coils. A large toroidal current is then induced in the plasma. This is accomplished by making the torus the secondary of a transformer. As the toroidal current rises it ionizes the deuterium and compresses both the resulting plasma and the toroidal bias field. In addition the toroidal current raises the plasma temperature by ohmic heating. At the end of the current rise the peak poloidal magnetic field due to the current and the on-axis compressed toroidal bias magnetic field are of comparable size. The follow-on "flat top" portion of the current cycle is where most of the interesting MHD and confinement physics takes place.

An interesting feature of the RFP is associated with the compression of the toroidal bias field. There is substantial compression so that at the end of the current ramp the residual edge toroidal field is much less than the central toroidal field as well as the initial bias field. Remarkably, under certain operating conditions the edge toroidal field spontaneously reverses direction, thereby motivating the name "reversed field pinch." The best performance of an RFP occurs when the field reverses. In support of this statement it is shown in Chapter 11 that without the reversal the RFP is unstable to internal MHD instabilities.

Current experimental programs are aimed at improving MHD and transport in RFPs by (1) feedback stabilizing external MHD modes, (2) external profile control, (3) operating in the single helicity state, and (4) learning how to drive long-lived toroidal currents without DC transformer action.

The analysis below considers radial pressure balance in a cylindrical RFP. The goal is to identify the critical equilibrium parameters of interest to experimental operation and to show that there is a wide range of parameter space where equilibria are possible. This range is narrowed considerably in Chapter 11 when MHD stability is analyzed.

The equilibrium p *and* B_z *profiles*

Radial pressure balance in an RFP satisfies the general screw pinch relation given by Eq. (5.20). For an RFP it is convenient to specify the pressure and toroidal field. One then uses the screw pinch pressure balance relation to determine the poloidal field.

The stability analysis in Chapter 11 shows that for internal MHD stability the pressure profile must be very flat on axis. Also, for good confinement it is desirable for the pressure and pressure gradient to vanish at the edge of the plasma $r = a$. A simple choice that captures these features is given by

$$p(\rho) = p_{\max}(1 - \rho^6)^3 \qquad (5.42)$$

where $\rho = r/a$ is the normalized radius. The quantity $p_{\max}$ is the maximum pressure that occurs on axis. It is related to the average pressure by the relation

$$\langle p \rangle = 2 \int_0^1 p\,\rho\,d\rho = \frac{81}{140} p_{\max} \qquad (5.43)$$

The pressure profile $p/p_{\max}$ is illustrated in Fig. 5.11.

Consider next the $B_z(\rho)$ profile. During the flat top period of operation the toroidal field is peaked on axis and reverses direction near the plasma edge. Also, for good confinement the corresponding current density must vanish at the plasma edge. A simple form of B_z, and the corresponding J_θ, having the desired properties can be written as

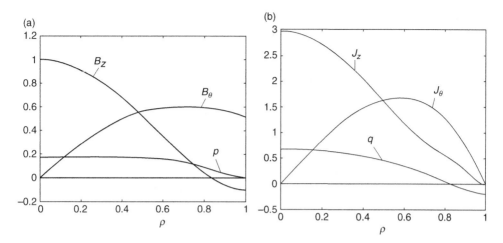

Figure 5.11   Equilibrium profiles for an RFP for the case $\alpha_z = 1.1$, $2\mu_0 \langle p \rangle / B_{z0}^2 = 0.1$, and $\varepsilon = 1/5$: (a) B_z/B_{z0}, B_θ/B_{z0}, $2\mu_0 p/B_{z0}^2$ and (b), $a\mu_0 J_\theta/B_{z0}$, $a\mu_0 J_z/B_{z0}$, q/ε.

$$\frac{B_z(\rho)}{B_{z0}} = 1 - 2\alpha_z \rho^2 + \alpha_z \rho^4$$

$$\frac{a\mu_0 J_\theta(\rho)}{B_{z0}} = -\frac{1}{B_{z0}}\frac{dB_z}{d\rho} = 4\alpha_z \rho \left(1 - \rho^2\right)$$

(5.44)

Here, B_{z0} is the field on axis and α_z is a parameter that measures the size of the edge toroidal field. Specifically,

$$\alpha_z = 1 - \frac{B_z(a)}{B_{z0}} = 1 - \frac{B_{za}}{B_{z0}}$$

(5.45)

Field reversal requires $\alpha_z > 1$. The B_z/B_{z0} and $a\mu_0 J_\theta/B_{z0}$ profiles are illustrated in Fig. 5.11 for the case $\alpha_z = 1.1$.

The B_θ profile

The B_θ profile is easily found from the following alternate form of the screw pinch pressure balance relation:

$$\frac{d}{d\rho}\left(\rho^2 B_\theta^2\right) = -\rho^2 \frac{d}{d\rho}\left(2\mu_0 p + B_z^2\right)$$

(5.46)

A straightforward integration yields the somewhat complicated expression

$$\frac{B_\theta^2(\rho)}{B_{z0}^2} = \frac{\alpha_p}{9}\left(35\rho^6 - 40\rho^{12} + 14\rho^{18}\right) + \frac{\alpha_z}{15}\left[30\rho^2 - 20(2\alpha_z + 1)\rho^4 + 45\alpha_z\rho^6 - 12\alpha_z\rho^8\right]$$

(5.47)

where $\alpha_p = 2\mu_0 \langle p \rangle / B_{z0}^2$ is a parameter related to the plasma beta. The corresponding current density is found from Ampere's law

$$\frac{a\mu_0 J_z(\rho)}{B_{z0}} = \frac{1}{B_{z0}}\left[\frac{1}{\rho}\frac{d}{d\rho}(\rho B_\theta)\right] = \frac{1}{2B_{z0}}\left[\frac{1}{\rho^2 B_\theta}\frac{d}{d\rho}(\rho^2 B_\theta^2)\right]$$

$$= \frac{140\alpha_p}{9}\left(\frac{\rho B_{z0}}{B_\theta}\right)\rho^4(1-\rho^6)^2 + 4\alpha_z\left(\frac{\rho B_{z0}}{B_\theta}\right)(1-\rho^2)(1-2\alpha_z\rho^2+\alpha_z\rho^4)$$

$$(5.48)$$

The B_θ/B_{z0} and $a\mu_0 J_z/B_{z0}$ profiles are also illustrated in Fig. 5.14 assuming that $\alpha_p = 0.1$.

Most of the desirable RFP profile properties just described were pointed out in the early analyses of Butt *et al.* (1958), Robinson (1971), and Bodin and Newton (1980).

Figures of merit

Consider now the safety factor profile. From the definition of q it follows that

$$q(\rho) = \varepsilon\frac{\rho B_z}{B_\theta} \sim \varepsilon \ll 1 \qquad (5.49)$$

where $\varepsilon = a/R_0 \ll 1$. This quantity is also plotted in Fig. 5.11 assuming that $\varepsilon = 0.2$. Observe that $q(\rho)$ decreases monotonically with radius and passes through zero at the reversal point. Particularly important is the fact that $q \sim \varepsilon$. The safety factor is small, suggesting that kink instabilities may be a problem for the RFP. For the case under consideration one finds that at the plasma edge $q_a = q_* \approx -0.039$. Stabilization requires a close fitting conducting shell and/or feedback. This conclusion is quantified in Chapter 11.

Global pressure balance

Global pressure balance in an RFP is described by a simplified form of the general screw pinch relation, obtained by making use of the fact that for typical operation, $|B_{za}| \ll B_{\theta a}$. It then follows that a useful definition of β in an RFP is given by

$$\beta = \frac{2\mu_0\langle p \rangle}{B_{za}^2 + B_{\theta a}^2} \approx \frac{2\mu_0\langle p \rangle}{B_{\theta a}^2} = \beta_p \qquad (5.50)$$

The plasma is held in equilibrium primarily by the poloidal magnetic field implying that $\beta \approx \beta_p$ in an RFP.

For the present profiles β_p can be evaluated by noting that from Eq. (5.47)

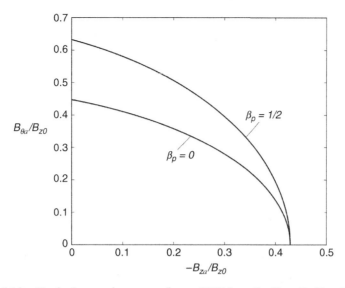

Figure 5.12   Typical operating space for an RFP in a $B_{\theta a}/B_{z0}$, B_{za}/B_{z0} diagram. The operating space lies below the curves.

$$B_{\theta a}^2 = 2\mu_0 \langle p \rangle + \frac{B_{z0}^2}{15} \alpha_z (10 - 7\alpha_z) \qquad (5.51)$$

leading to

$$\beta_p = \frac{15\alpha_p}{15\alpha_p + \alpha_z(10 - 7\alpha_z)} \qquad (5.52)$$

The operating regime of an RFP is thus defined by the two-dimensional (α_z, α_p) parameter space with the resulting β_p given by Eq. (5.52). For the example under consideration with $\alpha_z = 1.1$ and $\alpha_p = 0.1$, one finds that $\beta_p \approx 0.37$.

The (α_z, α_p) space is equivalent to a $B_{\theta a}/B_{z0}$, B_{za}/B_{z0} space. One boundary of the latter space has already been defined: $-B_{za}/B_{z0} > 0$ for reversal. Next, the requirement that $(B_{\theta a}/B_{z0})^2 > 0$ restricts B_{za}/B_{z0} to the range $-B_{za}/B_{z0} < 3/7$. Lastly, $0 < \beta_p < 1/2$ for normal operation of an RFP. The typical operating parameter space defined by these conditions is illustrated in Fig. 5.12. Operation outside these boundaries is possible but in general corresponds to degraded performance. There are further restrictions on the operation space that arise due to MHD stability as discussed in Chapter 11.

Summary of the RFP

The RFP is a relatively large aspect ratio axisymmetric toroidal configuration with a flat pressure profile on axis and a reversed B_z near the plasma edge. The particle pressure, and toroidal and poloidal magnetic pressures are all of comparable magnitude with the poloidal field providing the dominant force for radial pressure balance. The modest toroidal field system is a technological advantage from the reactor point of view but a sophisticated feedback system is needed to stabilize external MHD modes because of the small safety factor.

5.5.2 *The low* β *ohmic tokamak*

Overview

The tokamak is also an axisymmetric toroidal configuration. It is currently the leading contender to become the first power producing fusion reactor. Tokamaks have a large toroidal field and a small poloidal field with an aspect ratio typically on the order of $R_0/a \sim 3$. This combination of field ratio and aspect ratio leads to a safety factor satisfying $q \gtrsim 1$. The large safety factor leads to good MHD stability, even without a conducting wall or feedback, and the large toroidal field helps reduce energy and particle transport. The result is that the tokamak has exhibited the best plasma physics performance at the present time in terms of the Lawson parameter $p\tau_E$ and the ion temperature T_i as compared to all other fusion concepts. This is why it is the leading contender for a fusion reactor. However, the requirement of a large toroidal field introduces technological complexity into the configuration as well as increased capital cost.

　In any event from the perspective of MHD equilibrium and stability it is worth noting that there are two qualitatively different regimes of operation of a tokamak, depending on the method of heating. The first regime corresponds to early tokamaks heated entirely by the ohmic current induced in the plasma. Here, the plasma acts as the secondary of a transformer. In this regime the plasma β is very low and the toroidal field is slightly paramagnetic. Temperatures on the order of 1–3 keV have been achieved at densities of about $10^{20}\,\mathrm{m}^{-3}$. Since the plasma resistivity decreases with increasing electron temperature (i.e., $\eta \sim T^{-3/2}$) there is a practical upper limit to how high the temperature can be raised solely by ohmic heating: $T_{\mathrm{max}} \sim 3\text{–}5\,\mathrm{keV}$.

　With respect to stability, a low β ohmic tokamak is susceptible to MHD modes which limit the maximum toroidal current that can flow in the plasma. This corresponds to a minimum stable value of q. Violation of the current limit leads to violent MHD behavior that rapidly terminates the plasma and can in fact cause physical damage to the surrounding vacuum chamber. This catastrophic behavior is known as a "major disruption" and clearly must be avoided in a tokamak reactor. In fact many would agree that, alpha physics aside, the major plasma physics problems facing the tokamak reactor concept are the achievement of steady state and the avoidance of disruptions.

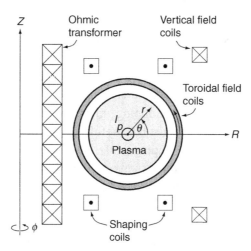

Figure 5.13   Schematic diagram of an ohmically heated tokamak with a circular cross section plasma.

Both radial pressure balance and the MHD current limit can be reasonably accurately calculated, at least for circular cross section plasmas, using the straight cylindrical model. Toroidal effects enter only as small corrections that do not affect the qualitative MHD behavior. It is the radial pressure balance of this configuration that is described in this subsection.

Before proceeding though it is worth describing the second regime of tokamak operation, which corresponds to the situation where a substantial amount of auxiliary power, provided either by neutral beams or RF waves, is injected into the plasma. This additional power raises the plasma temperature, and hence β as well. In the auxiliary heated regime toroidal effects are important, affecting both MHD equilibrium and stability. There is an equilibrium limit on βq^2 and separate stability limits on both β and q. The implication is that a high β auxiliary heated tokamak is inherently a 2-D configuration. Its equilibrium and stability properties are discussed in Chapters 6 and 12 respectively.

Return now to the low β circular tokamak. Typical operation is described in the schematic diagram illustrated in Fig. 5.13. Initially the toroidal field coils are energized, leading to a large, steady state B_z field. Neutral gas, usually deuterium, is injected into the vacuum chamber, and sometimes pre-ionized. The toroidal current is then ramped up by means of a changing flux in the primary of the ohmic transformer. After a short time, on the order of milliseconds, the plasma becomes fully ionized and achieves a steady state power balance between ohmic heating and thermal conduction losses. Radiation losses are usually small. The plasma is maintained in quasi-steady state as long the flux continues to swing at a constant rate in the primary of the transformer. This is the requirement to induce a constant (in time) toroidal electric field in the plasma (i.e., $2\pi R_0 E_z = -d\psi/dt \approx$ constant) which drives the quasi-steady state toroidal current. Once the transformer runs out of volt-seconds the toroidal current can no longer be maintained and the plasma decays away. The time period of quasi-steady state current is known as the "flat top period" and it is here that most of the interesting physics takes place.

Note that while the current is ramping up and during flat top operation the current in the vertical field coils must be carefully programmed to hold the plasma in toroidal force balance. The shaping coils are also carefully programmed to generate the desired cross sectional shape of the plasma which was chosen as circular in the early days of fusion research, but now usually corresponds to an elongated, outward pointing "D."

Qualitatively, operation of a low β tokamak and an RFP are similar. Quantitatively though the large toroidal field in the tokamak leads to a very different flat top plasma. The large B_z field in a tokamak greatly inhibits the toroidal flux compression so that the flat top plasma is only slightly paramagnetic: $B_z(0) \approx B_z(a)$. Recall that an RFP has much more compression leading to $B_z(0) \gg B_z(a)$ and in fact $B_z(a)$ actually reverses sign.

In current high β tokamak experiments the plasma always passes through an ohmic phase before the auxiliary heating is applied. Ohmic heating is a simple way to produce a relatively high-temperature, high-density, high-quality target plasma into which auxiliary power can be efficiently injected and absorbed.

The analysis below focuses on radial pressure balance in a low β tokamak. The pressure and toroidal current profiles are determined by a simple ohmic heating power balance relation. The goal is to show that for typical experimental parameters, β will always be very low and the toroidal field will be slightly paramagnetic.

Simple transport profiles for p and B$_\theta$

For a low β tokamak it is convenient to specify the $p(r)$ and $B_\theta(r)$ profiles and then use the general screw pinch pressure balance relation to determine $B_z(r)$. To specify reasonable profiles a short transport calculation is first required. Assume the torus has been straightened into a cylinder of length $2\pi R_0$ and minor radius a. The $p(r)$ and $B_\theta(r)$ profiles are determined by noting that during flat top operation power balance requires $P_\Omega = P_\kappa$: the input ohmic heating power balances the thermal conduction losses. This power balance can be expressed as

$$\frac{1}{r}\frac{d}{dr}\left(r n \chi \frac{dT}{dr}\right) + \eta J_z^2 = 0 \tag{5.53}$$

Here, it is assumed that $T_e = T_i \equiv T$. The resistivity is given by the usual Spitzer formula: $\eta = C_\eta/T^{3/2}\Omega$-m, where $C_\eta = 3.3 \times 10^{-8}$ when $T = T_k$ is measured in keV or $C_\eta = 6.7 \times 10^{-32}$ for T in joules.

The quantity χ is the thermal diffusivity which is in general anomalous and not well understood theoretically. A qualitatively correct model for the thermal diffusivity that greatly simplifies the analysis is given by $\chi = \chi_0 (T_0/T)^{1/2}$, where χ_0 and T_0 are the on axis values for the diffusivity and temperature respectively. This expression takes into account that χ, as determined experimentally, is an increasing function of radius. Also, experimental data shows that $\chi_0 \gtrsim 1\,\mathrm{m}^2/\mathrm{sec}$.

The density profile is qualitatively similar to that of the temperature and is therefore chosen as $n = n_0(T/T_0)$, where n_0 is the density on axis. Lastly, the current density profile can be expressed in terms of the temperature by noting that Ohm's law requires $E_z = \eta J_z$. Furthermore, in steady state flat top operation Faraday's law implies that $\nabla \times \mathbf{E} = 0$ leading to $E_z = E_{z0} = $ constant. Thus, $J_z = E_{z0}/\eta$. In practice, the electric field is adjusted to produce a given desired current so that J_{z0} rather than E_{z0} is the parameter of experimental interest. The end result is that one can write the current density as $J_z = J_{z0}(T/T_0)^{3/2}$.

Combining these results leads to the following simplified differential equation for the quantity $U = (T/T_0)^{3/2}$ in terms of normalized radius $\rho = r/a$:

$$\frac{1}{\rho}\frac{d}{d\rho}\left(\rho\frac{dU}{d\rho}\right) + k^2 U = 0$$

$$k^2 = \frac{3C_\eta a^2 J_{z0}^2}{2n_0\chi_0 T_0^{5/2}} = 0.31\left(\frac{I_M}{a\langle U\rangle}\right)^2 \frac{1}{n_{20}\chi_0 T_{k0}^{5/2}} \qquad (5.54)$$

where in the practical formula $I_M = I(\mathrm{MA})$, $n_{20} = n_0(10^{20}\,\mathrm{m}^{-3})$ and use has been made of the relations

$$\langle Q\rangle = 2\int_0^1 Q\rho\,d\rho$$

$$I = 2\pi\int Jr\,dr = \pi a^2 J_{z0}\langle U\rangle \qquad (5.55)$$

The boundary conditions on U for a confined plasma with no internal sources are given by $U'(0) = 0$ and $U(1) = 0$. Also, U has been normalized so that $U(0) = 1$. The solution is easily found and is given by

$$U = \mathrm{J}_0(k\rho) \qquad (5.56)$$

where $\mathrm{J}_0(k\rho)$ is the usual Bessel function and $k = 2.405$ corresponds to the first zero. Also, it immediately follows that $\langle U\rangle = 2J_1(k)/k \approx 0.43$. Using this value for k one finds from Eq. (5.54) that the temperature on axis has the value

$$T_{k0} = 0.61\left(\frac{I_M^2}{n_{20}\chi_0 a^2}\right)^{2/5} \mathrm{keV} \qquad (5.57)$$

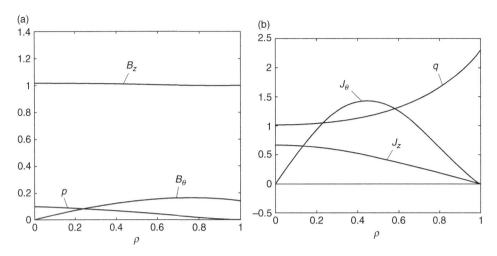

Figure 5.14   Equilibrium profiles for an ohmically heated tokamak for the case $\varepsilon = 1/3$, $q_* = 2.3$, $q_0 = 1$, and $\beta_0 = 0.002$: (a) B_θ/B_0, B_z/B_0, $(q_*/\varepsilon)^2 (2\mu_0 p/B_0^2)$ and (b) $a\mu_0 J_z/B_0$, $a\mu_0 J_\theta/B_0$, q.

It is now straightforward to evaluate the desired profiles of J_z, B_θ, and p:

$$
\frac{a\mu_0 J_z(\rho)}{B_0} = \frac{a\mu_0 J_{z0} U}{B_0} = \frac{\varepsilon}{q_*} \frac{k J_0(k\rho)}{J_1(k)}
$$

$$
\frac{B_\theta(\rho)}{B_0} = \frac{\mu_0 a}{B_0 \rho} \int_0^\rho J_z \rho \, d\rho = \frac{\varepsilon}{q_*} \frac{J_1(k\rho)}{J_1(k)} \tag{5.58}
$$

$$
\frac{2\mu_0 p(\rho)}{B_0^2} = \frac{4\mu_0 n T}{B_0^2} = \beta_0 [J_0(k\rho)]^{4/3}
$$

Here, $B_{\theta a} = \mu_0 I/2\pi a$, $\varepsilon = a/R_0$, $q_a = q_* = \varepsilon B_0/B_{\theta a}$, $\beta_0 = 4\mu_0 n_0 T_0/B_0^2$, and B_0 is the applied toroidal field at $r = a$.

These profiles are illustrated in Fig. 5.14 for typical values $\varepsilon = 1/3$, $q_* = 2.3$, and $\beta_0 = 0.002$. Observe that both pressure and current density are monotonically decreasing functions that vanish at the plasma edge. The pressure gradient also vanishes at the edge.

The B_z profile

The remaining quantity of interest is the B_z profile which is determined by integrating the general screw pinch pressure balance relation. A short calculation yields

$$
B_z^2(\rho) = B_0^2 - 2\mu_0 p + \mu_0 a \int_\rho^1 J_z B_\theta \, d\rho
$$

$$
= B_0^2 - 4\mu_0 n_0 T_0 J_0^{4/3}(k\rho) + \frac{B_{\theta a}^2}{2 J_1^2(k)} J_0^2(k\rho) \tag{5.59}
$$

A good approximation to B_z is obtained by noting that in a large aspect ratio tokamak (i.e. $\varepsilon \ll 1$) the toroidal field pressure is large compared to both the poloidal field pressure and plasma pressure. In this limit

$$\frac{B_z(\rho)}{B_0} \approx 1 - \frac{\beta_0}{2} J_0^{4/3}(k\rho) + \frac{\varepsilon^2}{k^2 q_0^2} J_0^2(k\rho) \qquad (5.60)$$

where $q_0 = 2B_0/\mu_0 R_0 J_{z0}$ is the safety factor on axis. The corresponding current density profile is given by

$$\frac{a\mu_0 J_\theta(\rho)}{B_0} = -\frac{1}{B_0} \frac{dB_z}{d\rho}$$
$$\approx \left[-\frac{2}{3} k\beta_0 J_0^{1/3}(k\rho) + \frac{2\varepsilon^2}{k q_0^2} J_0(k\rho) \right] J_1(\rho) \qquad (5.61)$$

The B_z and J_θ profiles are illustrated in Fig. 5.14. Note that $B_{z0}/B_0 > 1$ implying that for the parameters chosen, the toroidal field is paramagnetic. This, as shown below, is actually a general feature of ohmic tokamaks.

Figures of merit

Having calculated the equilibrium profiles one is now in a position to evaluate the safety factor. For the ohmic tokamak model under consideration one finds

$$q(\rho) = \frac{\varepsilon \rho B_z}{B_\theta} \approx q_0 \frac{k\rho}{2J_1(k\rho)} \qquad (5.62)$$

The safety factor is illustrated in Fig. 5.14. Observe that q is a monotonically increasing function of the radius.

A crucial feature of tokamaks is that the toroidal magnetic field must be large enough for a given toroidal current to ensure that $q_0 \approx 1$ for favorable MHD stability. Another property of interest predicted by the model is the ratio of edge to central safety factor. Using the value $k = 2.405$ leads to

$$\frac{q_a}{q_0} = \frac{q_*}{q_0} = \frac{k}{2J_1(k)} \approx 2.3 \qquad (5.63)$$

This is reasonably close to typical experimental values for high-current ohmic tokamaks.

Global pressure balance

The final topic of interest is the global pressure balance. There are two points to discuss. First it is shown that in practical situations the poloidal beta in an ohmic tokamak is always less than unity, implying that the toroidal field is paramagnetic. Second, it is shown that the corresponding value of beta is always very small, thereby motivating the need for additional auxiliary heating.

The poloidal beta is defined as

$$\beta_p = \frac{2\mu_0 \langle p \rangle}{B_{\theta a}^2} \tag{5.64}$$

The expression for β_p can be written in a simple and convenient form which is a function of the actual thermal diffusivity χ_0 and the ideal classical collisional thermal diffusivity $\chi_\perp$ as follows. First, one notes that $B_{\theta a} = \mu_0 I/2\pi a$ and $\langle p \rangle = 2n_0 T_0 \langle U^{4/3} \rangle$. This yields

$$\beta_p = 16\pi^2 \langle U^{4/3} \rangle \frac{a^2 n_0 T_0}{\mu_0 I^2} = 56.8 \frac{a^2 n_0 T_0}{\mu_0 I^2} \tag{5.65}$$

The quantity $n_0 T_0$ can be rewritten by making use of the definition of k^2 given by Eq. (5.54) plus several classical transport results from Braginskii. Specifically, Eq. (5.54) can be rearranged as

$$n_0 T_0 = \left(\frac{3}{2\pi^2 k^2 \langle U \rangle^2} \right) \left(\frac{C_\eta}{T_0^{3/2}} \right) \left(\frac{I^2}{a^2 \chi_0} \right) = 0.142 \left(\frac{C_\eta}{T_0^{3/2}} \right) \left(\frac{I^2}{a^2 \chi_0} \right) \tag{5.66}$$

while Braginskii has shown that

$$\eta_\| = \frac{C_\eta}{T_0^{3/2}} = 0.51 \frac{m_e}{n_0 e^2 \tau_{e0}}$$

$$\chi_\perp \approx \chi_i = 2^{1/2} \left(\frac{m_i}{m_e} \right)^{1/2} \frac{m_e T_0}{e^2 B_0^2 \tau_{e0}} \tag{5.67}$$

Here $\eta_\|$ is the parallel resistivity and $\chi_\perp$ is the perpendicular thermal diffusivity (essentially due to the ions), both evaluated on axis. The electron collision time on axis τ_{e0} is now eliminated from these two quantities yielding an expression for $C_\eta/T_0^{3/2}$ as a function of $\chi_\perp$. This result is then substituted into Eq. (5.66). A short calculation leads to

$$(n_0 T_0)^2 = 0.0512 \left(\frac{m_e}{m_i} \right)^{1/2} \left(\frac{I^2 B_0^2}{a^2} \right) \left(\frac{\chi_\perp}{\chi_0} \right) \tag{5.68}$$

The desired expression for β_p is obtained by substituting Eq. (5.68) into Eq. (5.65) and making use of the definition $q_* = 2\pi a^2 B_0/\mu_0 R_0 I$:

$$\beta_p \approx 2.0 \left(\frac{q_*}{\varepsilon} \right) \left(\frac{m_e}{m_i} \right)^{1/4} \left(\frac{\chi_\perp}{\chi_0} \right)^{1/2} \tag{5.69}$$

For a typical ohmic tokamak $q_* \approx 2.3$ and $\varepsilon \approx 1/3$. Assuming a deuterium mass ratio one then finds

$$\beta_p \approx 1.82 \left(\frac{\chi_\perp}{\chi_0}\right)^{1/2} \tag{5.70}$$

For classical transport (i.e., $\chi_0 = \chi_\perp$) then $\beta_p \approx 2$. The plasma is slightly diamagnetic. However, the actual plasma transport is highly anomalous with the smallest, most optimistic, measured value of diffusivity given by $\chi_0 \approx 1 \, \text{m}^2/\text{sec}$. On the other hand, Braginskii gives the classical value of thermal diffusivity in Eq. (5.67) as $\chi_\perp = 0.085 \, n_{20}/B_0^2 T_k^{1/2} = 1.06 \times 10^{-2} \, \text{m}^2/\text{sec}$, with the numerical value corresponding to $n_{20} = 2$, $B_0 = 4$, $T_k = 1$, and $\ln \Lambda = 19$. This yields a very small value of poloidal beta: $\beta_p \approx 0.18$. The conclusion is that without a large improvement in the thermal diffusivity an ohmic tokamak will always have $\beta_p \ll 1$. This in turn implies that the toroidal field will always be paramagnetic, which can be seen by rewriting the general screw pinch pressure balance relation given by Eq. (5.21) as follows:

$$\text{paramagnetism} \equiv \frac{1}{2B_{\theta a}^2} \int_0^1 \left(B_z^2 - B_0^2\right) \rho \, dr = 1 - \beta_p > 0 \tag{5.71}$$

Clearly for the right-hand side to be positive for a monotonic $B_z(\rho)$ profile requires $B_z^2(\rho) > B_0^2$. The condition $B_z^2(\rho) > B_0^2$ is the definition of toroidal paramagnetism.

The second topic of interest for global pressure balance is the value of beta. Since $B_z \gg B_\theta$ it follows that

$$\beta = \frac{2\mu_0 \langle p \rangle}{B_0^2 + B_{\theta a}^2} \approx \frac{2\mu_0 \langle p \rangle}{B_0^2} = \beta_t \tag{5.72}$$

The total beta is approximately equal to the toroidal beta. This quantity is always small as can be seen by eliminating $\langle p \rangle$ using the definition of β_p. One finds

$$\beta \approx \beta_t = \frac{B_{\theta a}^2}{B_0^2} \beta_p = \frac{\varepsilon^2}{q_a^2} \beta_p \tag{5.73}$$

For the model profile $\varepsilon = 1/3$, $q_a = 2.3$, and $\beta_p = 0.094$ yielding $\beta \approx 0.002$. Such small values of β are typical for ohmically heated tokamaks and are too low to be of practical use in a fusion reactor. The low value of β combined with the practical upper limit on temperature due to ohmic heating provide a strong motivation to add a substantial amount of auxiliary power to the tokamak. This leads to the high β tokamak which is inherently a 2-D configuration, discussed in Chapters 6 and 12.

Summary of the low β tokamak

The low β ohmically heated tokamak is an axisymmetric toroidal device with a large toroidal magnetic field. The plasma pressure and poloidal magnetic field are both small. The toroidal field is paramagnetic and thus does not help maintain the plasma in radial pressure balance. The main purpose of the toroidal field is to provide stability against MHD modes driven by the toroidal plasma current; that is, a large toroidal field is necessary to keep $q \gtrsim 1$. A large safety factor is one main reason why tokamak physics performance has surpassed that of other confinement concepts. However, from a technological point of view the need for a large toroidal field adds complexity and cost to a tokamak fusion reactor. Overall, the practically achievable temperatures and MHD stable beta values, while good, are still too low for a fusion reactor, thereby motivating the addition of auxiliary heating.

5.6 Summary

An examination of MHD equilibria in one-dimensional cylindrically symmetric configurations focuses attention on the problem of radial pressure balance. Toroidal force balance effects do not enter. Both toroidal and poloidal fields can provide radial pressure balance. This has been demonstrated for the limiting cases of the θ-pinch and Z-pinch, as well as for the general case of the screw pinch, the latter configuration containing an arbitrary combination of B_z and B_θ. A summary of the properties of these configurations is as follows:

- **The θ-pinch:** The θ-pinch is the one-dimensional analog of the torus with purely toroidal fields: $B_z(r)$, $J_\theta(r)$, $p(r)$. In this configuration, radial confinement is provided by the externally applied magnetic pressure. The θ-pinch is capable of a wide range of toroidal β, $0 < \beta_t < 1$. It has an infinite safety factor, $q(r) = \infty$, which is favorable for stability, although detailed calculations show that the worst modes are marginally stable.
- **The Z-pinch:** The Z-pinch is the one-dimensional analog of the torus with purely poloidal fields: $B_\theta(r)$, $J_z(r)$, $p(r)$. Here, radial pressure balance is provided by the tension in the magnetic field lines. In a Z-pinch the poloidal β is always unity: $\beta_p = 1$. The configuration has a zero safety factor, $q(r) = 0$, which is a strong indication of instability. One way to potentially avoid these instabilities is by the addition of a hard-core current-carrying wire along the axis of the device.
- **The general screw pinch:** The screw pinch is the one-dimensional analog of toroidal configurations with both toroidal and poloidal magnetic fields; for example, the low β ohmically heated tokamak and the reversed field pinch. Screw pinch configurations are capable of a wide variation in both β_t and β_p. In general, the kink safety factor q_* and the safety factor $q(r)$ are non-zero. The function $q(r)$ can increase or decrease away from the origin. For an RFP the

safety factor is small, $q \sim \varepsilon$. It decreases away from the axis and reverses near the outside of the plasma. For the low β tokamak, the safety factor is an increasing function of radius with $q_0 \sim 1$, $q_a \sim 2.5$. The implications are that RFPs should be susceptible to current driven kink modes thereby requiring some form of feedback stabilization, while tokamaks should have more favorable stability with respect to these modes.

References

Bennett, W.H. (1934). *Phys. Rev.* **45**, 980.
Bodin, H.A.B., and Newton, A.A. (1980). *Nucl. Fusion* **20**, 1255.
Braginskii, S.I. (1965). In *Reviews of Plasma Physics*, Vol. 1, ed. M.A. Leontovich. New York: Consultants Bureau.
Butt, E.P., Curruthers, R., Mitchell, J.T.D., Pease, R.S., Thonemann, P.C., Bird, M.A., Blears, J. and Hartill E.R. (1958). In *Proceedings of the Second United Nations International Conference on the Peaceful Uses of Atomic Energy*. Geneva: United Nations, Vol. 32, p. 42.
Robinson, D.C. (1971). *Plasma Phys.*, **13**, 439.

Further reading

Below are some recent general references that specifically discuss 1-D cylindrical MHD equilibria
Bateman, G. (1978). *MHD Instabilities*. Cambridge, MA: MIT Press.
Boyd, T.J.M. and Sanderson, J.J. (2003). *The Physics of Plasmas*. Cambridge: Cambridge University Press.
Goedbloed, J.P. and Poedts, S. (2004). *Principles of Magnetohydrodynamics*. Cambridge: Cambridge University Press.
Goldston, R.J. and Rutherford, P.H. (1995). *Introduction to Plasma Physics*. Bristol: IOP Publishing Ltd.
Stacey, W.M. (2005). *Fusion Plasma Physics*. Weinheim, Germany: Wiley-VCH.

Problems

5.1 Consider a θ-pinch with a magnetic field $\mathbf{B} = B(r)\mathbf{e}_z$ and pressure $p = p(r)$. Assume $p(r) = p_0 \exp(-r^2/a^2)$, $B(\infty) = B_0$, $n(0) = 3 \times 10^{22} \, \mathrm{m}^{-3}$, $T_i(0) = 2 \, \mathrm{keV}$, $T_e(0) = 0.8 \, \mathrm{keV}$, $B_0 = 8 \, T$, and $a = 0.01 \, \mathrm{m}$. The temperature profiles are related to the density profile as follows: $T_i(r)/T_i(0) = T_e(r)/T_e(0) = n^2(r)/n^2(0)$.

(a) Calculate $\beta_0 \equiv \beta(0)$
(b) Draw a diagram indicating the direction of $\mathbf{B}$ and the plasma current density $\mathbf{J}$.
(c) Calculate the magnitude and sign of the curvature drift of a thermal ion and a thermal electron at $r = a$.
(d) Calculate the magnitude and sign of the ∇B drift of a thermal ion and a thermal electron at $r = a$.
(e) The signs in step (d) appear to be inconsistent with the current direction in step (b)? Explain how this inconsistency is resolved.

5.2 Consider a static ideal MHD Z-pinch equilibrium with

$$J = c_1 \frac{r^2/a^2}{(1 + r^2/a^2)^3}$$

where c_1 is a constant.

(a) Calculate $B_\theta(r)$ and $p(r)$. Express your answers in terms of I, the total current. Sketch the fields and the currents.
(b) Since $J(r)$ vanishes for large r the Z-pinch is apparently confined by its own current. Doesn't this violate the virial theorem? Explain.

5.3 In MHD equilibrium theory one often specifies $p(\psi)$ rather than $p(r)$. This problem demonstrates how the specification of $p = p(\psi)$ leads to a determination of the equilibrium fields.

(a) Consider a one-dimensional θ-pinch. Show that the flux contained within a given radius r is equal to $2\pi r A_\theta$, where A_θ is the vector potential.
(b) Define $\psi = r A_\theta$ and assume p is specified as follows: $p(\psi) = p_0 \exp(-\psi/\psi_0)$ with p_0 and ψ_0 constants. Using the one-dimensional radial pressure balance for a θ-pinch derive a differential equation for $\psi(r)$. This equation should be a first-order, non-linear differential equation. For convenience write the equation in terms of the normalized variables and parameters defined by

$$H = \psi/\psi_0 \qquad \lambda = a^2 B_0/\psi_0$$
$$x = r^2/a^2 \qquad \beta_0 = 2\mu_0 p_0/B_0^2$$

Here, B_0 is the magnetic field far from the plasma and a is a scale factor representing the characteristic plasma radius.

(c) Solve the equation for $H(x)$ subject to the condition $\psi(0) = 0$. Calculate and sketch $p(r)$ and $B_z(r)$.
(d) Calculate $\lambda = \lambda(\beta)$ by defining the scale length a such that

$$\pi p_0 a^2 = \int p(r)\, r dr d\theta$$

5.4 Prove that the safety factor on axis for a screw pinch is given by

$$q(0) = \frac{2B_z(0)}{\mu_0 R_0 J_z(0)}$$

5.5 The magnetic shear in a screw pinch is defined as

$$s(r) = -\frac{r}{q}\frac{dq}{dr}$$

Assume now that the current density vanishes at the edge of the plasma: $J_z(a) = J_\theta(a) = 0$. Prove that the edge shear has the value

$$s(a) = 2$$

5.6 In tokamaks and reversed field pinches the safety factor $q(r) = rB_z/R_0B_\theta$ is an important quantity of physical interest.

(a) Assume a force-free configuration (i.e., $\nabla p = 0$) and derive a differential equation for $B_z(r)$ assuming that $q(r)$ is the known free function.
(b) Consider the $q(r)$ profile given by

$$q(r) = q_0 \frac{r_0}{(r_0^2 - r^2)^{1/2}}$$

where q_0 is the safety factor on axis and r_0 is a parameter. Find the value of r_0 as a function of the plasma edge a that makes $J_z(a) = J_\theta(a) = 0$.
(c) Calculate $\mathbf{B}(r)$ and $\mathbf{J}(r)$ assuming that $B_z(0) = B_{z0}$ and $q_0^2 = (2/3)\,\varepsilon^2$, where $\varepsilon = a/R_0$. Plot your results.

5.7 A rigid rotor θ-pinch is characterized by a current density $\mathbf{J} = J_\theta(r)\mathbf{e}_\theta$ that is related to the number density by $J_\theta(r) = -e\Omega rn(r)$, where $\Omega = \text{const}$. Assume for simplicity that the temperature and density are related by $T(r) = T_0n(r)/n_0$, where T_0 and n_0 are the on-axis values.

(a) Derive the differential equation that determines $n(r)$.
(b) Determine $n(r)$ assuming that $n(a) = 0$.

5.8 Consider a constant pitch screw pinch defined by the requirement that $B_\theta(r) = krB_z(r)$, where $k = \text{constant}$. Assume that the pressure is given by $p = p_0(1 - r^2/a^2)^2$.

(a) Calculate $B_z(r)$
(b) Show that there is no choice of k and p_0 that leads to $J_\theta(a) = 0$. The current can never vanish at the edge of a constant pitch screw pinch with confined pressure.

5.9 This problem investigates the use of "gravity" in a slab geometry to model plasma rotation and magnetic field line curvature.

(a) Derive the one-dimensional radial pressure balance relation for a general screw pinch including the effect of a stationary angular velocity $\Omega(r)\mathbf{e}_\theta$.
(b) Derive the one-dimensional pressure balance relation for a slab model of a plasma including a gravity $g(x)\mathbf{e}_x$. Assume $\rho(x)$, $p(x)$, $B_y(x)$, and $B_z(x)$ are non-zero. Also assume there is no equilibrium flow: $\mathbf{v} = 0$.

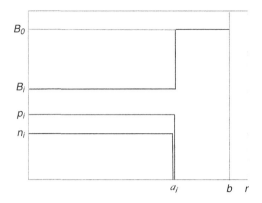

Figure 5.15   Surface current model of a θ-pinch.

(c) By comparing the results in (a) and (b) determine a correspondence between B_θ, Ω, and g so that gravity models the effects of rotation and poloidal field line tension.

5.10 This problem treats the adiabatic compression of a θ-pinch. Consider the simple surface current model of an infinitely long, straight θ-pinch as shown in Fig. 5.15. Here, n_i, p_i, B_i, a_i, and B_0 represent the initial known state of the plasma. At $t = 0$, the applied field at $r = b$ is slowly (i.e., adiabatically) increased until B_0 reaches the value $B_0' = \lambda B_0$. The quantity $\lambda > 1$ is the multiplication factor of the applied field and is assumed known. Also, the plasma obeys an adiabatic equation of state.

(a) Derive, but do not solve, an algebraic equation which relates the final β of the plasma (β_f) to the initial beta of the plasma (β_i). Express this equation in the form $F(\beta_f, \beta_i, \lambda, \gamma) = 0$, where γ is the ratio of specific heats.
(b) Solve this equation analytically for the special case $\gamma = 1$, $\beta_i = 0.5$, and $\lambda = 2$.
(c) Solve this equation numerically for the special case $\gamma = 5/3$, $\beta_i = 0.5$, and $\lambda = 2$.
(d) If you did the problem correctly you should find that $\beta_f < \beta_i$ for both cases. Explain why the final β decreases even though the plasma has been compressed.

5.11 An early description of a low β screw pinch is provided by the "force-free paramagnetic" model. In this model one sets $\nabla p = 0$ and assumes a resistive Ohm's law $\mathbf{E} + \mathbf{v} \times \mathbf{B} = \eta \mathbf{J}$. In steady state $\mathbf{E} = E_0 \mathbf{e}_z$.

(a) Use the pressure balance relation and the parallel component of Ohm's law to derive a coupled set of differential equations for $B_z(r)$ and $B_\theta(r)$.

(b) Solve these equations (numerically) assuming that $\eta = $ constant. Assume the minor radius of the plasma is $a = 0.3\,\mathrm{m}$ and the current density on axis is $J_z(0) = 1\,\mathrm{MA/m^2}$.

5.12 Magnetic profile measurements on a straight screw pinch indicate that the B_θ profile is given by

$$B_\theta(r) = \frac{\mu_0 I}{2\pi} \frac{r}{r^2 + r_0^2}$$

where I is the plasma current and r_0 is the characteristic radius of the plasma. For simplicity assume that p and B_z are related by

$$B_z^2(r) = B_0^2 - 2\mu_0 \lambda p(r)$$

Here, B_0 is the externally applied field and λ is a constant to be determined. Diamagnetic loop measurements indicate that the "excluded flux" has the value

$$\Delta\psi = 0.02 \left(\pi r_0^2 B_0 \right)$$

The "excluded flux" represents the difference between the toroidal flux contained within the conducting wall (of radius $a \gg r_0$) for the case of no plasma present and the case where plasma is present assuming the same value of B_0:

$$\Delta\psi = \psi_{tor}\,(\text{no plasma present}) - \psi_{tor}(\text{plasma present})$$

(a) Calculate λ. To simplify the analysis assume:

$$\mu_0 I / 2\pi r_0 B_0 = 0.2 \ll 1.$$

(b) Calculate the toroidal beta if $a/r_0 = 3$.

5.13 In Section 5.5.2 it was shown that for practical situations an ohmically heated tokamak is paramagnetic with respect to the toroidal field. The goal of this problem, which makes use of the analysis in Section 5.5.2, is to determine the amount of auxiliary power required to transition the plasma from paramagnetic to diamagnetic. To carry out the analysis assume that auxiliary heating is applied to the plasma which is much greater in magnitude than the ohmic power. Specifically, replace the ohmic heating power with the auxiliary power in Eq. (5.53): $\eta J_z^2 \rightarrow S_{aux}(r)$. Next, assume that the auxiliary power deposition profile is peaked on axis, which for mathematical simplicity can be modeled as $S_{aux}(r) = S_0(T/T_0)^{3/2}$.

(a) Make use of the analytic results in Section 5.5.2 to find $U(r) = (T/T_0)^{3/2}$.

(b) Note that S_0 has the units of W/m^2. Derive an expression for S_0 in terms of the total power P_T and the geometry.

(c) Show that the temperature on axis is given by

$$T_0 = \left(\frac{3}{8\pi^2 kJ_1(k)}\right)\left(\frac{P_T}{R_0 n_0 \chi_0}\right)$$

(d) Show that the poloidal beta in the plasma is given by

$$\beta_p = \left(\frac{6\langle U^{4/3}\rangle}{\mu_0 kJ_1(k)}\right)\left(\frac{a^2 P_T}{I^2 R_0 \chi_0}\right) = 0.36\left(\frac{6}{\mu_0 kJ_1(k)}\right)\left(\frac{a^2 P_T}{I^2 R_0 \chi_0}\right)$$

(e) Calculate the value of P_T at the transition point for diamagnetism using typical parameters for TFTR: $R_0 = 2.4$ m, $a = 0.8$ m, $I = 1.2$ MA, $\chi_0 = 1$ m^2/sec, and $n_0 = 0.5 \times 10^{20}$ m^{-3}. Calculate the corresponding value of T_{k0}.

6

Equilibrium: two-dimensional configurations

6.1 Introduction

In Chapter 5 it was shown that a one-dimensional, cylindrically symmetric magnetic geometry accurately describes radial pressure balance in many fusion configurations. The primary goal of Chapter 6 is to address the problem of toroidal force balance in a two-dimensional axisymmetric toroidal geometry. A secondary goal analyzes straight systems with two-dimensional helical symmetry.

The discussion starts with a derivation of the Grad–Shafranov equation, the basic equation describing axisymmetric toroidal equilibrium. For configurations possessing such symmetry, the solutions to this equation provide a complete description of ideal MHD equilibria: radial pressure balance, toroidal force balance, equilibrium β limits, rotational transform, and kink safety factor. A wide number of configurations are well described by the Grad–Shafranov equation. Included among them are all types of tokamaks, the reversed field pinch, the levitated dipole, the spheromak, and the field reversed configuration.

Two strategies are utilized to solve the Grad–Shafranov equation. The first makes use of asymptotic expansions, with the inverse aspect ratio $\varepsilon = a/R_0$ serving as the small expansion parameter. The asymptotic analysis mathematically separates the problems of radial pressure balance and toroidal force balance into two distinct parts. It thereby provides a method for obtaining analytic results as well as developing physical intuition. In a certain sense each different asymptotic expansion (i.e., corresponding to different orderings of β, q, B_p/B_t with respect to ε) serves to mathematically identify a specific MHD configuration. The small ε expansion works well for the ohmically heated tokamak and the reversed field pinch.

The second strategy focuses on configurations in which radial pressure balance and toroidal force balance cannot be separated. This clearly occurs for tight aspect ratio geometries where $\varepsilon \sim 1$, such as the spherical tokamak, spheromak, and field reversed configuration. Interestingly the non-separation of effects can also occur for configurations where ε is small but β/ε is finite, such as the auxiliary heated tokamak. The analysis for this strategy makes use of special exact analytic solutions to the Grad–Shafranov equation where simple choices are made for the two free functions inherent in any MHD equilibrium.

The last section of the chapter focuses on two-dimensional helically (rather than toroidally), symmetric equilibria. The main configuration of interest here is the "straight" stellarator. A helically symmetric version of the Grad–Shafranov equation is derived and then analyzed for stellarators with a single helicity. Of interest are the evaluation of the rotational transform and the calculation of equilibria with zero net current. Bending a straight stellarator into a torus requires a full three-dimensional analysis and is discussed in Chapter 7. Even so, a straight helical system provides a good introduction to some of the basic MHD physics of the stellarator.

6.2 Derivation of the Grad–Shafranov equation

The Grad–Shafranov equation is a two-dimensional, non-linear, partial differential equation obtained from the reduction of the ideal MHD equations (Eq. (4.16)) for the case of toroidal axisymmetry (Grad and Rubin, 1958; Shafranov, 1960; Lust and Schluter, 1957). The geometry of interest is illustrated in Fig. 6.1.

Here, R, ϕ, Z correspond to the familiar right-handed cylindrical coordinate system. The assumption of toroidal axisymmetry implies that $\partial S/\partial \phi = 0$, where S is any scalar. The derivation proceeds by analyzing the equations in the order described in Chapter 4.

6.2.1 The $\nabla \cdot \mathbf{B} = 0$ equation

The axisymmetric assumption implies that $\nabla \cdot \mathbf{B} = 0$ can be written as

$$\frac{1}{R}\frac{\partial}{\partial R}(RB_R) + \frac{\partial B_Z}{\partial Z} = 0 \qquad (6.1)$$

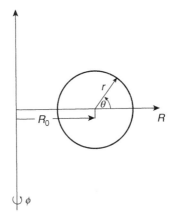

Figure 6.1   Geometry for axisymmetric toroidal equilibrium.

As with many two-dimensional systems it is useful to introduce a stream function ψ for the poloidal magnetic field:

$$B_R = -\frac{1}{R}\frac{\partial \psi}{\partial Z}$$

$$B_Z = \frac{1}{R}\frac{\partial \psi}{\partial R}$$

(6.2)

where $\psi = RA_\phi$ and A_ϕ is the toroidal component of vector potential. Note that B_ϕ is unaffected by Eq. (6.1). In more compact notation, one can write

$$\mathbf{B} = B_\phi\, \mathbf{e}_\phi + \mathbf{B}_p$$

$$\mathbf{B}_p = \frac{1}{R}\nabla\psi \times \mathbf{e}_\phi$$

(6.3)

The stream function ψ is closely related to the poloidal flux in the plasma ψ_p, which is given by

$$\psi_p = \int \mathbf{B}_p \cdot d\mathbf{A}$$

(6.4)

The area of interest in the integral is a washer-shaped surface lying in the $Z=0$ plane as shown in Fig. 6.2. The inner radius of the washer is located at $R = R_a$ corresponding to the magnetic axis. The outer radius extends to an arbitrary ψ contour defined by $\psi = \psi(R,\, 0)$. It then follows that

$$\psi_p = \int_0^{2\pi} d\phi \int_{R_a}^{R} R'B_Z(R',0)\, dR' = 2\pi\psi$$

(6.5)

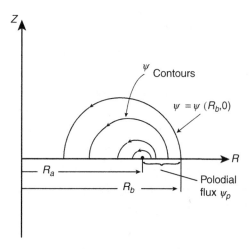

Figure 6.2   Washer-shaped surface through which the poloidal flux ψ_p passes.

Here, the arbitrary integration constant associated with ψ has been chosen so that the poloidal magnetic flux on the axis is zero: $\psi(R_a, 0) = 0$.

As previously stated, it is convenient to label the flux surfaces with the values of ψ rather than p.

6.2.2 Ampere's law

The next step is to substitute the expression for the magnetic field into Ampere's law to obtain an expression for the current density. A straightforward calculation leads to

$$\mu_0 \mathbf{J} = \mu_0 J_\phi \mathbf{e}_\phi + \frac{1}{R} \nabla(R B_\phi) \times \mathbf{e}_\phi$$

$$\mu_0 J_\phi = -\frac{1}{R} \Delta^* \psi$$

(6.6)

where the operator Δ^* is defined by

$$\Delta^* \psi \equiv R^2 \nabla \cdot \left(\frac{\nabla \psi}{R^2} \right) = R \frac{\partial}{\partial R} \left(\frac{1}{R} \frac{\partial \psi}{\partial R} \right) + \frac{\partial^2 \psi}{\partial Z^2} \tag{6.7}$$

6.2.3 Momentum equation

The final step in the derivation is to substitute the expressions for $\mathbf{B}$ and $\mathbf{J}$ (Eqs. (6.3) and (6.6)) into the momentum equation (Eq. (4.1)). An efficient way to carry out this step is to decompose the momentum equation into three components along $\mathbf{B}$, $\mathbf{J}$, and $\nabla \psi$. The $\mathbf{B}$ component yields $\mathbf{B} \cdot \nabla p = 0$, which can be rewritten as

$$\mathbf{e}_\phi \cdot \nabla \psi \times \nabla p = 0 \tag{6.8}$$

The general solution to this equation is

$$p = p(\psi) \tag{6.9}$$

As expected, the pressure is a surface quantity. It is a free function; that is, one can specify $p(\psi)$ arbitrarily. The MHD model does not contain sufficient physics to determine the functional dependence of the pressure on the flux. Transport theory, experimental data, or physical intuition are needed to properly specify $p(\psi)$ as well as $F(\psi)$ introduced below. Also, note that the pressure is a function of a single variable ψ even though the geometry is two dimensional.

Consider next the $\mathbf{J}$ component of the momentum equation which reduces to $\mathbf{J} \cdot \nabla p = 0$. Substituting for $\mathbf{J}$ leads to

$$\mathbf{e}_\phi \cdot \nabla \psi \times \nabla(R B_\phi) = 0 \tag{6.10}$$

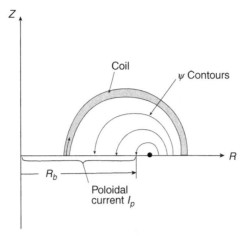

Figure 6.3 Disk-shaped surface through which the total (plasma plus coil) poloidal current I_p flows.

Hence, RB_ϕ is also a free surface function,

$$RB_\phi = F(\psi) \tag{6.11}$$

The quantity $F(\psi)$ is related to the combined poloidal current flowing in the plasma plus toroidal field coils. To see this examine Fig. 6.3 and focus on the poloidal current flowing through a disk-shaped surface lying in the $Z = 0$ plane. The disk extends from the center line $R = 0$ out to an arbitrary flux surface defined by $\psi = \psi(R, 0)$. The value of the total enclosed poloidal current is given by

$$I_p = \int \mathbf{J}_p \cdot d\mathbf{A} = \int_0^{2\pi} d\phi \int_0^R R' J_Z(R', 0) dR' \tag{6.12}$$

which reduces to

$$I_p(\psi) = 2\pi F(\psi) \tag{6.13}$$

Lastly, substituting into the $\nabla \psi$ component of the momentum equation yields the Grad–Shafranov equation. The individual terms appearing simplify as follows:

$$\nabla \psi \cdot \nabla p = (\nabla \psi)^2 \frac{dp}{d\psi}$$

$$\nabla \psi \cdot \mathbf{J}_p \times (B_\phi \, \mathbf{e}_\phi) = -(\nabla \psi)^2 \frac{F}{\mu_0 R^2} \frac{dF}{d\psi}$$

$$\nabla \psi \cdot (J_\phi \, \mathbf{e}_\phi) \times \mathbf{B}_p = -(\nabla \psi)^2 \frac{\Delta^* \psi}{\mu_0 R^2} \tag{6.14}$$

$$\nabla \psi \cdot \mathbf{J}_p \times \mathbf{B}_p = 0$$

$$\nabla \psi \cdot (J_\phi \, \mathbf{e}_\phi) \times (B_\phi \, \mathbf{e}_\phi) = 0$$

The final step is to combine these terms. The result, including a summary of the field relations, can be written as

$$\Delta^*\psi = -\mu_0 R^2 \frac{dp}{d\psi} - \frac{1}{2}\frac{dF^2}{d\psi} \tag{6.15}$$

with

$$\mathbf{B} = \frac{1}{R}\nabla\psi \times \mathbf{e}_\phi + \frac{F}{R}\mathbf{e}_\phi$$

$$\mu_0\mathbf{J} = \frac{1}{R}\frac{dF}{d\psi}\nabla\psi \times \mathbf{e}_\phi - \frac{1}{R}\Delta^*\psi\,\mathbf{e}_\phi \tag{6.16}$$

Equation (6.15) is the Grad–Shafranov equation (Grad and Rubin, 1958; Shafranov, 1960, 1966), a second-order, non-linear partial differential equation describing general axisymmetric toroidal equilibria. The nature of the equilibria (e.g., tokamak, reversed field pinch, etc.) is to a large extent determined by the choice of the free functions $p(\psi)$ and $F(\psi)$ and, of course, the boundary conditions. The properties of the reversed field pinch and straight tokamak are discussed individually in the sections that follow once relations are obtained for the basic plasma parameters and figures of merit.

6.3 Plasma parameters and figures of merit

Consider now the evaluation of the basic plasma parameters and figures of merit. By exploiting the axisymmetry it is possible to obtain general expressions for the quantities of interest in terms of the flux function ψ. In the analysis that follows it is assumed that $p(\psi)$ and $F(\psi)$ have been specified and that a solution has been found that gives ψ and the corresponding magnetic fields and current densities as functions of (R, Z). The calculations below are relatively straightforward. It is, however, convenient for purposes of intuition to first introduce a set of simple "flux coordinates." These coordinates are analogous to polar coordinates which are easier to visualize for circular-like cross sections than the (R, Z) coordinates. The (R, Z) coordinates are more rectangular in nature with respect to the plasma cross section. Once the coordinates are defined it is then an easy task to derive expressions for the plasma parameters and figures of merit.

6.3.1 Simple flux coordinates

The motivation to introduce flux coordinates arises because the evaluation of many of the figures of merit involves averages over the plasma volume or cross section. For example, when calculating the toroidal plasma current one needs to integrate the current density over the cross sectional area $d\mathbf{A} = dR\,dZ\,\mathbf{e}_\phi$. While this is a formally correct expression it is cumbersome to use and visualize because the plasma boundary is a complicated function of (R, Z). If the plasma cross section is circular then cylindrical-like coordinates (r, θ) as shown in Fig. 6.1 would be easier to use. For an arbitrary cross section one would like to introduce generalized radial-like and angular-like coordinates for which the plasma surface and interior flux surfaces have a simple representation.

The choice of flux coordinates made here is as follows. For the radial-like coordinate a good choice is the flux ψ itself. Any flux surface, including the plasma surface, is defined by $\psi = $ constant. For the angular-like coordinate a simple and convenient choice is poloidal arc length l. Other choices are possible and often used when one considers stability, but for present purposes arc length is perhaps the easiest coordinate to visualize.

In the analysis that follows one should view the quantities (ψ, l) as a two-dimensional coordinate transformation replacing the (R, Z) coordinates: $\psi = \psi(R, Z)$, $l = l(R, Z)$ and the inverse transformation $R = R(\psi, l)$, $Z = Z(\psi, l)$. The definition of the flux $\psi(R, Z)$ is given by the solution to the Grad–Shafranov equation and, as stated, is assumed to be known. What remains is to define the arc length coordinate and in particular to evaluate the Jacobian of the transformation J: $dRdZ = Jd\psi dl$.

Now, for any general 2-D coordinate transformation (not necessarily corresponding to arc length) it follows that

$$\frac{1}{J} = \begin{vmatrix} \psi_R & \psi_Z \\ l_R & l_Z \end{vmatrix} = \psi_R l_Z - \psi_Z l_R = R\,\mathbf{B}_p \cdot \nabla l \qquad (6.17)$$

The definition of l equivalent to the special choice of poloidal arc length requires two additional properties: (1) the direction of the arc length vector should be parallel to the poloidal magnetic field and (2) the magnitude of its poloidal gradient should be unity; in other words $\nabla l = \mathbf{B}_p/B_p$. This definition implies that as expected $\mathbf{B}_p \cdot \nabla l = B_p$ and

$$J = 1/RB_p \qquad (6.18)$$

Lastly, it is worth noting that if $\psi(R, Z)$ is known then one practical way to calculate $l(R, Z)$ is by taking the divergence of $\nabla l = \mathbf{B}_p/B_p$ and then solving

$$\nabla^2 l = -\frac{1}{B_p^2}\mathbf{B}_p \cdot \nabla B_p \qquad (6.19)$$

With the coordinate transformation in hand, it now becomes a simple task to calculate the various differential volume and area elements needed to evaluate the plasma parameters and figures of merit. These elements are given by

Volume: $d\mathbf{r} = R\,dR\,dZ\,d\phi = 2\pi\dfrac{d\psi\,dl}{B_p}$

Poloidal surface area: $d\mathbf{S} = dR\,dZ\,\mathbf{e}_\phi = \dfrac{d\psi\,dl}{RB_p}\mathbf{e}_\phi$ (6.20)

Plasma surface area: $d\mathbf{A} = R\,dl\,d\phi\,\dfrac{\nabla\psi}{|\nabla\psi|} = 2\pi R\,dl\,\mathbf{n}$

Here, $\mathbf{n} = \nabla\psi/|\nabla\psi|$ is the outward normal vector on the plasma surface.

6.3.2 The volume of a flux surface

The volume contained within any specified flux surface ψ is given by

$$V(\psi) = \int d\mathbf{r} = 2\pi\int_0^\psi d\psi \oint \frac{dl}{B_p}$$ (6.21)

Here, the poloidal flux on the magnetic axis has been set to zero. Note also that

$$\frac{dV(\psi)}{d\psi} = 2\pi\oint\frac{dl}{B_p}$$ (6.22)

The volume of the entire plasma V_p is obtained by setting $\psi = \psi_a$, the value of flux on the plasma surface,

$$V_p = V(\psi_a) = 2\pi\int_0^{\psi_a}d\psi\oint\frac{dl}{B_p}$$ (6.23)

6.3.3 The plasma beta

The plasma beta is determined from the definitions in Eqs. (4.19) and (4.20). Consider first the toroidal beta:

$$\beta_t = \frac{2\mu_0\langle p\rangle}{B_0^2}$$ (6.24)

Once the solution to the Grad–Shafranov equation is known one can evaluate $\langle p\rangle$ as follows:

$$\langle p\rangle = \frac{1}{V_p}\int p\,d\mathbf{r} = \frac{1}{V_p}\int_0^{\psi_a}p(\psi)\frac{dV}{d\psi}d\psi = \frac{1}{V_p}\int_0^{V_p}p(V)dV$$ (6.25)

Similarly, to evaluate the poloidal beta,

$$\beta_p = \frac{4\pi^2a^2(1+\kappa^2)\langle p\rangle}{\mu_0 I^2}$$ (6.26)

one needs an expression for the toroidal current I, which can be written as

$$\mu_0 I = \oint \mathbf{B}_p \cdot d\mathbf{l} = \oint B_p dl \tag{6.27}$$

where the line integral is carried out on the plasma surface $\psi = \psi_a$. The total beta is given by

$$\beta = \frac{\beta_t \beta_p}{\beta_t + \beta_p} \tag{6.28}$$

6.3.4 The kink safety factor

The kink safety factor is defined as

$$q_* = \frac{2\pi a^2 B_0}{\mu_0 R_0 I} \left(\frac{1 + \kappa^2}{2} \right) \tag{6.29}$$

It is only the current I that needs to be evaluated from the solution to the Grad–Shafranov equation and this task has already been described in Eq. (6.27).

6.3.5 Rotational transform and the MHD safety factor

The concepts of rotational transform and MHD safety factor have been discussed in Section 4.6.4. Recall that the normalized transform $\iota/2\pi$ is defined as the average change in poloidal angle $\Delta\theta$ of a magnetic field line per single transit in the toroidal direction, $\Delta\phi = 2\pi$; that is, $\iota/2\pi \equiv \langle \Delta\theta \rangle / 2\pi$. The average is taken over many toroidal transits. (Here, in terms of the present notation the relation between the (R, ϕ, Z) coordinates and the cylindrical-like (r, θ, z) coordinates is given by $R = R_0 + r\cos\theta$, $Z = r\sin\theta$, and $\phi = -z/R_0$.)

Because of axisymmetry the general expression for the safety factor $q(\psi) = 2\pi/\iota$ given by Eq. (4.28) can be simplified as follows. To begin Eq. (4.28) is repeated here for convenience:

$$q(\psi) = \frac{\displaystyle \lim_{L\to\infty} \int_0^L \frac{B_\phi}{RB} dl_{\text{tot}}}{\displaystyle \lim_{L\to\infty} \int_0^L \frac{B_\theta}{rB} dl_{\text{tot}}} \tag{6.30}$$

where l_{tot} is arc length along the total field. The integrands are converted from line integrals parallel to the total magnetic field to line integrals parallel to the poloidal magnetic field by the geometric relation

$$\frac{dl_{\text{tot}}}{B} = \frac{dl}{B_p} \tag{6.31}$$

The integrals over the infinite length L are now replaced by an infinite sum of integrals, each one over one poloidal period L_p.

$$q(\psi) = \frac{\lim\limits_{m\to\infty} \int_0^{mL_p} \frac{B_\phi}{RB_p} dl}{\lim\limits_{m\to\infty} \int_0^{mL_p} \frac{B_\theta}{rB_p} dl} = \frac{\sum\limits_{m=0}^{\infty} \int_{mL_p}^{(m+1)L_p} \frac{B_\phi}{RB_p} dl}{\sum\limits_{m=0}^{\infty} \int_{mL_p}^{(m+1)L_p} \frac{B_\theta}{rB_p} dl} \qquad (6.32)$$

The critical point is that as a consequence of toroidal axisymmetry each term in the numerator sum is identical. After one complete poloidal transit, the magnetic line trajectory repeats itself because of the toroidal symmetry. A similar conclusion applies to the denominator. Thus, the average over many poloidal periods is the same as the average over one poloidal period:

$$q(\psi) = \frac{\oint \frac{B_\phi}{RB_p} dl}{\oint \frac{B_\theta}{rB_p} dl} \qquad (6.33)$$

Finally, the numerator is simplified by recalling that $F(\psi) = RB_\phi$ with $F(\psi)$ a constant on a flux surface. The denominator is simplified by noting that $dl/B_p = rd\theta/B_\theta$ implying that

$$\oint \frac{B_\theta}{rB_p} dl = \int_0^{2\pi} d\theta = 2\pi \qquad (6.34)$$

Combining these results leads to the desired expression for the safety factor in an axisymmetric torus:

$$q(\psi) = \frac{F(\psi)}{2\pi} \oint \frac{dl}{R^2 B_p} \qquad (6.35)$$

6.3.6 The MHD safety factor on axis

Equation (6.35) is a general expression for the safety factor. In most practical cases it is evaluated numerically. However, the value on axis can be difficult to evaluate because the circumference of the flux surface and the value of B_p simultaneously approach zero, leading to an undefined limit. The limit is finite and in this subsection an explicit practical expression for the safety factor on axis is derived.

The analysis proceeds as follows. On the magnetic axis $R = R_a$, $Z = Z_a$, the value of ψ can be set to zero without loss in generality. Also, by definition, the first

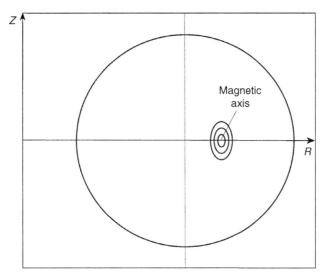

Figure 6.4   Elliptical flux surfaces near the magnetic axis.

derivatives of ψ vanish on the axis: $\psi_R(R_a, Z_a) = \psi_Z(R_a, Z_a) = 0$. Therefore, the Taylor expansion of ψ in the vicinity of the magnetic axis can be written as

$$\psi \approx \frac{\psi_{RR}}{2} (R - R_a)^2 + \frac{\psi_{ZZ}}{2} (Z - Z_a)^2 + \cdots \tag{6.36}$$

Here, the second derivatives are constants evaluated on the magnetic axis: $\psi_{RR} = \psi_{RR}(R_a, Z_a)$ and $\psi_{ZZ} = \psi_{ZZ}(R_a, Z_a)$. As shown in Fig. 6.4 the flux surfaces near the axis are ellipses.

Since the shape of the flux surfaces is known the general expression for the safety factor can now be easily integrated. One introduces new coordinates (ρ, α) defined by

$$
\begin{aligned}
R - R_a &= \left(\frac{2}{\psi_{RR}}\right)^{1/2} \rho \cos \alpha \\[2mm]
Z - Z_a &= \left(\frac{2}{\psi_{ZZ}}\right)^{1/2} \rho \sin \alpha
\end{aligned}
\tag{6.37}
$$

In these coordinates the equation for the flux surfaces become $\psi = \rho^2$.

To evaluate the integral, note that on a $\rho = $ constant flux surface,

$$dl = \left[(dR)^2 + (dZ)^2\right]^{1/2} = \left(\frac{2}{\psi_{RR}\psi_{ZZ}}\right)^{1/2} \left(\psi_{ZZ} \sin^2 \alpha + \psi_{RR} \cos^2 \alpha\right)^{1/2} \rho \, d\alpha$$

$$\tag{6.38}$$

Similarly, the components of the poloidal magnetic field can be written as

$$B_R = -\frac{1}{R}\frac{\partial \psi}{\partial Z} = -\frac{1}{R_a}\psi_{ZZ}(Z - Z_a) = -\frac{1}{R_a}(2\psi_{ZZ})^{1/2}\rho \sin\alpha$$

$$B_Z = \frac{1}{R}\frac{\partial \psi}{\partial R} = \frac{1}{R_a}\psi_{RR}(R - R_a) = \frac{1}{R_a}(2\psi_{RR})^{1/2}\rho \cos\alpha$$

$$(6.39)$$

implying that

$$B_p = \left(B_R^2 + B_Z^2\right)^{1/2} = \frac{2^{1/2}}{R_a}\left(\psi_{ZZ}\sin^2\alpha + \psi_{RR}\cos^2\alpha\right)^{1/2}\rho \qquad (6.40)$$

Observe that in the ratio dl/B_p the ρ factors exactly cancel thereby showing that as one approaches the magnetic axis, $\psi = \rho^2 \rightarrow 0$, the integrand remains finite.

Combining these results leads to the following simple integral for the safety factor on axis

$$q_0 \equiv q(0) = \frac{F(0)}{2\pi R_a}\frac{1}{(\psi_{RR}\psi_{ZZ})^{1/2}}\int_0^{2\pi} d\alpha \qquad (6.41)$$

which reduces to the desired expression

$$q_0 = \frac{B_\phi}{(\psi_{RR}\psi_{ZZ})^{1/2}} = \left(\frac{B_\phi}{\mu_0 R_a J_\phi}\right)\left(\frac{1 + \kappa_0^2}{\kappa_0}\right) \qquad (6.42)$$

where $B_\phi = B_\phi(R_a, Z_a)$, $J_\phi = J_\phi(R_a, Z_a)$, and the elongation κ_0 are the on axis values.[1] The second form of q_0 is obtained from the facts that on the axis $\psi_{RR}/\psi_{ZZ} = \kappa_0^2$ and $\psi_{RR} + \psi_{ZZ} = -\mu_0 R_a J_\phi(R_a, Z_a)$.

6.3.7 Alternate choices for F(ψ)

To solve the Grad–Shafranov equation it is necessary to specify the two free functions $p(\psi)$ and $F(\psi)$. Usually one has reasonable intuition based on experimental data or transport modeling to specify the pressure profile $p(\psi)$. The situation is not as straightforward with respect to the toroidal field profile $F(\psi)$. The reason is that $F(\psi)$ is often nearly a constant. Thus, the evaluation of $dF/d\psi$, the function actually needed in the Grad–Shafranov equation, is entirely dependent on the small deviation of $F(\psi)$ away from constancy. In practice, this small deviation is not easily known from intuition.

---

[1] Unfortunately the symbol κ is widely used in the fusion community for two descriptive purposes, the curvature and the plasma elongation. No attempt is made to counter this long standing tradition but it will be clear in the text which application is under consideration.

To resolve this problem it is sometimes useful to replace $F(\psi)$ with an alternate free function that is easier to specify intuitively. Two common replacements are the safety factor profile $q(\psi)$ and the average toroidal current density profile (which is defined below). For each case one needs to determine the relation between the alternate function and $dF^2/d\psi$ in order to make the proper replacement in the Grad–Shafranov equation.

Consider first the safety factor given by Eq. (6.35). The relation is straightforward in principle but complicated in practice:

$$F(\psi) = q(\psi)\left(\frac{1}{2\pi}\oint\frac{dl}{R^2 B_p}\right)^{-1} \tag{6.43}$$

Thus, if $q(\psi)$ is specified as a free function, one sees that the Grad–Shafranov equation becomes

$$\Delta^*\psi = -\mu_0 R^2\frac{dp}{d\psi} - \frac{1}{2}\frac{d}{d\psi}\left[q^2(\psi)\left(\frac{1}{2\pi}\oint\frac{dl}{R^2 B_p}\right)^{-2}\right] \tag{6.44}$$

Note that the integrand must be viewed as a function of ψ and l. Specifically, $R^2 B_p = f(R, Z) = f[R(\psi, l), Z(\psi, l)] = \hat{f}(\psi, l)$.

Observe that the equation is now more complicated than the original one where $F(\psi)$ is specified. The right-hand side now contains not only functions of ψ, but poloidal arc length integrals over functions of ψ. The Grad–Shafranov equation has become an integro-differential equation whose properties are more difficult to understand analytically. However, in numerical schemes involving an iteration procedure, the more complicated right-hand side has little impact on obtaining a converged solution.

The second choice for an alternate free function is $\langle J_\phi\rangle$, the average toroidal current density. It is defined as follows. With the help of Eq. (6.20) one sees that the total toroidal current flowing within any given flux surface can be written as

$$I_\phi(\psi) = \int J_\phi\,\mathbf{e}_\phi\cdot d\mathbf{S} = \int_0^\psi d\psi\left(\oint J_\phi\frac{dl}{RB_p}\right) \tag{6.45}$$

The quantity $\langle J_\phi\rangle$ is the average current density, defined so that when integrated over the cross sectional area it yields the same value for $I_\phi(\psi)$:

$$I_\phi(\psi) \equiv \int \langle J_\phi\rangle\,\mathbf{e}_\phi\cdot d\mathbf{S} = \int_0^\psi d\psi\left(\langle J_\phi\rangle\oint\frac{dl}{RB_p}\right) \tag{6.46}$$

Since both expressions for $I_\phi(\psi)$ must be equal for any value of ψ it follows that $\langle J_\phi\rangle$, which is a function only of ψ, is given by

$$\langle J_\phi \rangle = \frac{\oint J_\phi \dfrac{dl}{RB_p}}{\oint \dfrac{dl}{RB_p}} \tag{6.47}$$

The relation between $\langle J_\phi \rangle$ and $dF^2/d\psi$ is obtained by recalling from Eqs. (6.15) and (6.16) that $\mu_0 R J_\phi$ is the negative of the right-hand side of the Grad–Shafranov equation. Substituting this information into the integral in the numerator of Eq. (6.47) leads to a relation between $\langle J_\phi \rangle$, $dp/d\psi$, and $dF^2/d\psi$. Assuming that $p(\psi)$ and $\langle J_\phi \rangle$ are the two free functions, one then finds from a short calculation that the Grad–Shafranov equation has the form

$$\Delta^* \psi = -\mu_0 R^2 \frac{dp}{d\psi} + \left(\mu_0 \frac{dp}{d\psi} \oint \frac{dl}{B_p} - \mu_0 \langle J_\phi \rangle \oint \frac{dl}{RB_p} \right) \left(\oint \frac{dl}{R^2 B_p} \right)^{-1} \tag{6.48}$$

The equation has again been transformed into an integro-differential equation.

Other choices for an alternate free function to replace $dF^2/d\psi$ can be made depending upon the application of interest. By and large these choices also lead to integro-differential equations.

6.4 Analytic solution in the limit $\varepsilon \ll 1$ and $\beta_p \sim 1$

The Grad–Shafranov equation has been derived and the figures of merit defined. The next step is to analyze the equation with the goals of obtaining some insight plus quantitative information as to how toroidal effects modify a purely cylindrical MHD equilibrium. The approach used in this section makes use of an asymptotic expansion in which the inverse aspect ratio is assumed to be small: $\varepsilon \ll 1$. Also, the poloidal beta is assumed to be of order unity: $\beta_p \sim 1$. It is shown that these assumptions lead to a sharp separation between the effects of radial pressure balance and toroidal force balance. Specifically, toroidal force balance enters the analysis as an order ε correction to radial pressure balance.

The separation of the two effects provides good insight into plasma behavior as the geometry is bent from a cylinder into a torus. Qualitatively, one consequence of toroidicity is to shift the inner flux surfaces outward (along R) with respect to the location of the plasma surface. A second consequence is that the plasma surface itself can be shifted outward with respect to the geometric center of the confining vacuum vessel. In fact the main goal of much of the analysis is to obtain quantitative, analytic expressions for these shifts.

With respect to the plasma parameters and figures of merit, one might expect that toroidicity simply adds small, relatively unimportant ε corrections to the cylindrical values. This is a correct intuition in the large aspect ratio, $\beta_p \sim 1$ regime.

The analysis is carried out by converting from the general R, ϕ, Z coordinates to a set of cylindrical-like coordinates r, θ, z centered in the plasma chamber. It then becomes a relatively straightforward procedure to solve the equations order by order in ε with the leading order corresponding to radial pressure balance and the first-order correction corresponding to toroidal force balance. Interestingly, the solution for the first-order correction can be obtained essentially analytically for an arbitrary zeroth-order circular cross section equilibrium.

Once derived the asymptotic analysis is then applied to the two cylindrical configurations already discussed in Chapter 5: the reversed field pinch and the low β, circular cross section, ohmically heated tokamak.

6.4.1 The coordinate transformation

To exploit the small ε expansion it is useful to transform the coordinates used in the Grad–Shafranov equation from R, ϕ, Z to a set of cylindrical-like coordinates r, θ, z. This transformation has been previously introduced (see Fig. 6.1) and is repeated here for convenience:

$$
\begin{aligned}
R &= R_0 + r \cos \theta \\
Z &= r \sin \theta \\
\phi &= -z/R_0
\end{aligned}
\tag{6.49}
$$

Note that both R, ϕ, Z and r, θ, z are right-handed coordinate systems. After a straightforward calculation one finds that the basic vector operations are given by

$$
\nabla \psi = \nabla_c \psi
$$

$$
\nabla \cdot \mathbf{B} = \nabla_c \cdot \mathbf{B} + \frac{\mathbf{e}_R \cdot \mathbf{B}}{R} = \frac{1}{rR}\frac{\partial}{\partial r}(rRB_r) + \frac{1}{rR}\frac{\partial}{\partial \theta}(RB_\theta) + \frac{R_0}{R}\frac{\partial B_z}{\partial z}
$$

$$
\nabla \times \mathbf{B} = \nabla_c \times \mathbf{B} - \frac{B_z}{R}\mathbf{e}_z = \left[\frac{1}{rR}\frac{\partial}{\partial \theta}(RB_z) - \frac{R_0}{R}\frac{\partial B_\theta}{\partial z}\right]\mathbf{e}_r
$$
$$
+ \left[\frac{R_0}{R}\frac{\partial B_r}{\partial z} - \frac{1}{R}\frac{\partial}{\partial r}(RB_z)\right]\mathbf{e}_\theta + \left[\frac{1}{r}\frac{\partial}{\partial r}(rB_\theta) - \frac{1}{r}\frac{\partial B_r}{\partial \theta}\right]\mathbf{e}_z
$$

$$
\begin{aligned}
\mathbf{e}_R &= \cos \theta\, \mathbf{e}_r - \sin \theta\, \mathbf{e}_\theta \\
\mathbf{e}_Z &= \sin \theta\, \mathbf{e}_r + \cos \theta\, \mathbf{e}_\theta \\
\mathbf{e}_\phi &= -\mathbf{e}_z
\end{aligned}
\tag{6.50}
$$

Here, ∇_c is the standard gradient operator in a linear r, θ, z cylindrical coordinate system with $\partial/\partial z$ replaced by $(R_0/R)\partial/\partial z$.

$$
\nabla_c = \mathbf{e}_r \frac{\partial}{\partial r} + \mathbf{e}_\theta \frac{1}{r}\frac{\partial}{\partial \theta} + \mathbf{e}_z \frac{R_0}{R}\frac{\partial}{\partial z}
\tag{6.51}
$$

A short calculation that makes use of these results shows that the two-dimensional Δ^* operator appearing in the Grad–Shafranov equation can be expressed as

$$\Delta^* \psi = \frac{1}{r} \frac{\partial}{\partial r} \left(r \frac{\partial \psi}{\partial r} \right) + \frac{1}{r^2} \frac{\partial^2 \psi}{\partial \theta^2} - \frac{1}{R} \left(\cos \theta \frac{\partial \psi}{\partial r} - \frac{\sin \theta}{r} \frac{\partial \psi}{\partial \theta} \right) \tag{6.52}$$

Consequently, in terms of the cylindrical-like coordinates the Grad–Shafranov equation becomes

$$\nabla_c^2 \psi = -\mu_0 R^2 \frac{dp}{d\psi} - \frac{1}{2} \frac{dF^2}{d\psi} + \frac{1}{R} \left(\cos \theta \frac{\partial \psi}{\partial r} - \frac{\sin \theta}{r} \frac{\partial \psi}{\partial \theta} \right) \tag{6.53}$$

where $R = R_0 + r \cos \theta$.

6.4.2 *The asymptotic expansion*

An asymptotic solution to the Grad–Shafranov equation can be obtained by focusing attention on a large aspect ratio, circular cross section plasma whose pressure corresponds to $\beta_p \sim 1$. In order to carry out the analysis a small parameter must be identified and all quantities ordered with respect to this parameter. For the situation of interest the small parameter is the inverse aspect ratio

$$\varepsilon \equiv a/R_0 \ll 1 \tag{6.54}$$

where R_0 and a are the major and minor radii of the plasma respectively. Conceptually, the plasma geometry is more akin to a bicycle tire than a donut.

To define the expansion it is convenient to normalize all quantities with respect to the poloidal magnetic field in the plasma. The assumption $\beta_p \sim 1$ thus requires that the pressure be ordered as

$$\frac{2\mu_0 p}{B_p^2} \sim 1 \tag{6.55}$$

Similarly, the definition of the flux function requires that

$$\frac{\psi}{aR_0 B_p} \sim 1 \tag{6.56}$$

The situation with the free function $F(\psi)$ is slightly more complicated. The reason is that, in general, F^2 can be written as $F^2(\psi) = R_0^2 B_0^2 + G(\psi)$, where B_0 = constant is the applied vacuum toroidal field. The quantity $R_0 B_0$ can be zero, small, medium, or large. Its value does not affect the Grad–Shafranov equation since only the derivative $dF^2/d\psi$ is required. The value of $dF^2/d\psi$ thus depends only on the new free function $G(\psi)$, which represents the diamagnetism or paramagnetism of the toroidal field. The magnitude of $G(\psi)$ cannot be too large. For large positive G (i.e., too much paramagnetism) radial pressure balance is possible only if $\beta_p \ll 1$. For large negative G (i.e., too much diamagnetism) radial pressure balance is possible but only if $\beta_p \gg 1$. Both of these extremes thereby violate the basic ordering assumption $\beta_p \sim 1$. The net result is that the quantity $dF^2/d\psi = dG/d\psi$ can be positive or negative but must be ordered such that it is competitive in size with the pressure gradient term. This requires that

$$\frac{\varepsilon}{B_p}\frac{dF^2}{d\psi} \sim 1 \tag{6.57}$$

Equations (6.55)–(6.57) define the ordering scheme necessary to solve the Grad–Shafranov equation. The asymptotic analysis now proceeds as follows. First, the flux function is expanded as a series in ε:

$$\begin{aligned}\psi(r,\theta) &= \psi_0(r) + \psi_1(r,\theta) + \cdots \\ \psi_0/aR_0B_p &\sim 1 \\ \psi_1/aR_0B_p &\sim \varepsilon\end{aligned} \tag{6.58}$$

Here, $\psi_0(r)$ represents the leading order cylindrically symmetric contribution to ψ while $\psi_1(r,\theta)$ represents the small correction due to toroidal effects. Similarly, the free functions are Taylor expanded as

$$\begin{aligned}\frac{dp(\psi)}{d\psi} &= \frac{dp}{d\psi_0} + \frac{d^2p}{d\psi_0^2}\psi_1 + \cdots \\ \frac{dF^2(\psi)}{d\psi} &= \frac{dF^2}{d\psi_0} + \frac{d^2F^2}{d\psi_0^2}\psi_1 + \cdots\end{aligned} \tag{6.59}$$

One substitutes these expansions into the Grad–Shafranov equation (i.e., Eq. (6.53)), and sets the coefficient of each power of ε to zero. This leads to a sequence of equations, one for each power of ε. For the present analysis only the ε^0 and ε^1 equations are needed and these are discussed below.

6.4.3 The ε^0 equation: radial pressure balance

The zeroth-order contribution to the Grad–Shafranov equation is given by

$$\frac{1}{r}\frac{d}{dr}\left(r\frac{d\psi_0}{dr}\right) = -\mu_0 R_0^2 \frac{dp}{d\psi_0} - \frac{1}{2}\frac{dF^2}{d\psi_0} \tag{6.60}$$

This equation can be put in a more familiar form by noting that the zeroth-order fields can be written as

$$B_{r0}(r) = 0$$

$$B_{\theta 0}(r) = \frac{1}{R_0}\frac{d\psi_0}{dr}$$

$$B_{z0}(r) = -\frac{F(\psi_0)}{R_0} \tag{6.61}$$

$$p_0(r) = p(\psi_0)$$

Then, after making the replacement $(d/d\psi_0) = (1/R_0 B_{\theta 0})(d/dr)$, one obtains

$$\frac{d}{dr}\left(p_0 + \frac{B_{z0}^2}{2\mu_0}\right) + \frac{B_{\theta 0}}{\mu_0 r}\frac{d}{dr}(rB_{\theta 0}) = 0 \tag{6.62}$$

As expected, this is just the radial pressure balance for the general screw pinch.

An important step that further simplifies the analysis is the following. To solve the Grad–Shafranov equation it is necessary to specify two free "one-dimensional" functions, $p(\psi)$ and $F(\psi)$. They are "one dimensional" in the sense that even though the geometry is two dimensional, each free function depends only on a single variable, ψ. Therefore, it is entirely equivalent to specify two alternate "one-dimensional" functions. For the present analysis a convenient choice is $p_0(r)$ and $B_{\theta 0}(r)$. Radial pressure balance then gives $B_{z0}(r)$ and the flux is obtained from $\psi_0(r) = R_0\int B_{\theta 0}dr$. Finally, the original free functions can be evaluated by inverting the flux relation, giving $r = r(\psi_0)$ and then substituting into $p(\psi_0) = p_0[r(\psi_0)]$ and $F(\psi_0) = -R_0 B_{z0}[r(\psi_0)]$.

Hereafter it is assumed that a zeroth-order cylindrical equilibrium solution has been specified that satisfies Eq. (6.62) with $p_0(r)$ and $B_{\theta 0}(r)$ serving as the two free functions.

6.4.4 The ε^1 equation: toroidal force balance

Consider now the first-order toroidal corrections to the Grad–Shafranov equation. A straightforward calculation leads to a complicated looking equation for determining the perturbed flux $\psi_1(r, \theta)$:

$$\nabla^2\psi_1 + \left[\frac{d^2}{d\psi_0^2}\left(\mu_0 R_0^2 p + \frac{F^2}{2}\right)\right]\psi_1 = -\left(2\mu_0 R_0 r\frac{dp}{d\psi_0} - \frac{1}{R_0}\frac{d\psi_0}{dr}\right)\cos\theta \tag{6.63}$$

Here, $\nabla^2 = \nabla_c^2$ is the usual two-dimensional cylindrical Laplacian (with the subscript c suppressed from ∇_c^2 for simplicity) and p, F are free functions of ψ_0 whose dependence is determined by the procedure described in the previous subsection.

Equation (6.63) can actually be solved analytically. To begin, observe that the inhomogeneous forcing terms all have the same dependence on poloidal angle, proportional to $\cos \theta$. This fact, coupled with the assumption that the plasma boundary is circular, implies that $\psi_1(r, \theta)$ can be written as

$$\psi_1(r, \theta) = \overline{\psi}_1(r) \cos \theta \tag{6.64}$$

After some algebra, where multiple use is again made of the replacement $(d/d\psi_0) = (1/R_0 B_{\theta 0})(d/dr)$ and then combined with the zeroth-order radial pressure balance relation, one obtains the following simplified equation for $\overline{\psi}_1(r)$:

$$\frac{d}{dr}\left[rB_\theta^2 \frac{d}{dr}\left(\frac{\overline{\psi}_1}{B_\theta}\right)\right] = rB_\theta^2 - 2\mu_0 r^2 \frac{dp}{dr} \tag{6.65}$$

Here the subscript 0 has been dropped from the zeroth-order quantities to simplify the notation.

This equation can be easily integrated once boundary conditions are specified. For the moment, however, the boundary conditions are left open and the general solution is written in terms of two arbitrary integrations constants. The choices for these constants depend upon the configuration of interest. Specific choices are described in the next two subsections where the theory is applied to the reversed field pinch and the low β tokamak.

To solve for $\overline{\psi}_1(r)$ observe that the first integral of Eq. (6.65) is given by

$$\frac{d}{dr}\left(\frac{\overline{\psi}_1}{B_\theta}\right) = \frac{1}{rB_\theta^2(r)} \int_{r_1}^r \left[yB_\theta^2(y) - 2\mu_0 y^2 \frac{dp(y)}{dy}\right] dy \tag{6.66}$$

Here, r_1 is the first free integration constant. Equation (6.66) can be integrated a second time yielding an expression for $\overline{\psi}_1(r)$:

$$\overline{\psi}_1(r) = B_\theta(r) \int_{r_2}^r \frac{dx}{xB_\theta^2(x)} \int_{r_1}^x \left[yB_\theta^2(y) - 2\mu_0 y^2 \frac{dp(y)}{dy}\right] dy \tag{6.67}$$

with r_2 being the second integration constant.

This is the desired solution for the toroidal correction to the Grad–Shafranov equation. Once the cylindrical profiles $p(r)$, $B_\theta(r)$ plus two boundary conditions are specified, it is then a straightforward task, at least in principle, to evaluate the integrals to determine $\overline{\psi}_1(r)$.

Lastly, observe that qualitatively the toroidal solution shows that bending a cylindrical screw pinch into a torus results in a set of nested circular flux surfaces whose centers are no longer concentric but are now shifted radially outward along R with respect to one another. This can be seen by noting that the equation for a shifted circle of radius r_0 is given by

$$(x - \Delta)^2 + y^2 = r_0^2 \tag{6.68}$$

where Δ is the shift along R and the coordinate system is centered about the major radius $R = R_0$. If one sets $x = r \cos \theta$, $y = r \sin \theta$, and assumes that the shift is small, $\Delta/r_0 \sim r_0/R_0 \ll 1$, it then follows that in (r, θ) coordinates the equation for a shifted circle can be approximated by

$$r \approx r_0 + \Delta \cos \theta \tag{6.69}$$

Now, the equation for the flux surfaces is given by

$$\psi(r, \theta) \approx \psi_0(r) + \overline{\psi}_1(r) \cos \theta = \text{constant} \tag{6.70}$$

For the surface whose minor radius is r_0 this equation can be inverted to give $r = r(r_0, \theta)$ by writing $r \approx r_0 + r_1(r_0, \theta)$ assuming that $r_1 \ll r_0$. The result is

$$r \approx r_0 + r_1(r_0, \theta) \approx r_0 - \frac{\overline{\psi}_1(r_0)}{\psi_0'(r_0)} \cos \theta \tag{6.71}$$

A comparison of Eqs. (6.71) and (6.69) shows that the flux surfaces are indeed shifted circles with the shift given by

$$\Delta(r_0) = -\frac{\overline{\psi}_1(r_0)}{R_0 B_\theta(r_0)} \tag{6.72}$$

The shift is outward since $\overline{\psi}_1 < 0$ for realistic boundary conditions. In the applications that follow attention is focused on the evaluation of the flux surface shift.

6.4.5 Application to early reversed field pinches (RFP)

The first application of the toroidal correction to the Grad–Shafranov equation is to early RFP experiments. Such experiments were relatively short lived, usually due to power supply limitations. Importantly, the plasma was surrounded by a thick conducting wall. Relative to the short plasma lifetime the skin diffusion time of the conducting wall was long, implying that for practical purposes the wall behaved like a perfect conductor. The wall thus defined the edge of the plasma: $p(a) = 0$.

The question of interest that then arises is to determine the shift of the magnetic axis with respect to the center of the conducting wall. The geometry is illustrated in Fig. 6.5. To address this question two boundary conditions are needed to determine $\overline{\psi}_1$ and the corresponding shift of the axis. Since the conducting wall defines the edge of the plasma the first boundary condition requires that $\psi(a, \theta) = \text{constant}$ or equivalently $\overline{\psi}_1(a) = 0$. From Eq. (6.67) it then follows that the integration constant $r_2 = a$.

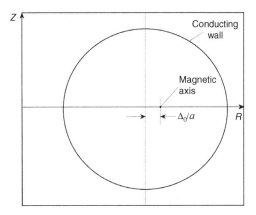

Figure 6.5   Magnetic axis shift with respect to the center of the conducting wall.

The other integration constant is determined from the condition that the solution for $\overline{\psi}_1$ be regular everywhere in the plasma, including $r = 0$. Equation (6.66) shows that regularity requires $r_1 = 0$. The solution for the perturbed flux function reduces to

$$\overline{\psi}_1(r) = -B_\theta(r) \int_r^a \frac{dx}{xB_\theta^2(x)} \int_0^x \left[yB_\theta^2(y) - 2\mu_0 y^2 \frac{dp(y)}{dy} \right] dy \qquad (6.73)$$

The quantity $\overline{\psi}_1(r)$ is determined by using the profiles discussed in Section 5.5.2, repeated here for convenience:

$$\frac{2\mu_0 p(\rho)}{B_{z0}^2} = \frac{140}{81} \alpha_p (1 - \rho^6)^3$$

$$\frac{B_\theta^2(\rho)}{B_{z0}^2} = \frac{\alpha_p}{9} (35\rho^6 - 40\rho^{12} + 14\rho^{18}) + \frac{\alpha_z}{15}[30\rho^2 - 20(2\alpha_z+1)\rho^4 + 45\alpha_z\,\rho^6 - 12\alpha_z\rho^8]$$

$$\alpha_z = 1 - \frac{B_{za}}{B_{z0}}$$

$$\alpha_p = \frac{\alpha_z(10 - 7\alpha_z)}{15} \frac{\beta_p}{1 - \beta_p}$$

$$\rho = \frac{r}{a}$$

$$(6.74)$$

These are substituted into Eq. (6.73). The y integration can be carried out analytically. The resulting expression for $\overline{\psi}_1(r)$ is then used to calculate the shift of the

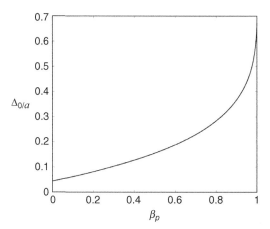

Figure 6.6　Curve of Δ_0/a vs. β_p for the case of $R_0/a = 5$ and $\alpha_z = 1.1$.

axis, defined by Eq. (6.72): $\Delta_0 = -\psi_1(r \to 0)/R_0B_\theta(r \to 0)$. The limit as $r \to 0$ is finite and a short calculation yields

$$\frac{\Delta_0}{a} = \frac{a}{R_0} \int_0^1 dx \frac{x^3}{B_\theta^2/B_{z0}^2}\, g(x, \alpha_p, \alpha_z)$$

$$g = \frac{\alpha_p}{3}(109x^4 - 200x^{10} + 98x^{16}) + \frac{\alpha_z}{2}\left[1 - \frac{4}{9}(2\alpha_z + 1)x^2 + \frac{3}{4}\alpha_z\, x^4 - \frac{4}{25}\alpha_z\, x^6\right]$$

$$(6.75)$$

A numerically obtained curve of Δ_0/a vs. β_p is illustrated in Fig. 6.6 for the case of $R_0/a = 5$ and $\alpha_z = 1.1$ corresponding to a 10% reversal. Observe that as expected the shift of the axis increases with β_p primarily because of the larger tire tube force. The shift increases substantially as β_p increases. Even so the absolute shift remains relatively small for moderate β_p in an RFP because of the small inverse aspect ratio, $a/R_0 \sim 0.2$.

6.4.6 Application to early ohmic tokamaks and modern reversed field pinches

The last application of interest involves early ohmically heated tokamaks and modern RFPs. These devices are characterized by circular cross section plasmas with values of poloidal beta satisfying $\beta_p \lesssim 1$. The plasmas are often surrounded by a thin circular conducting wall of moderate resistivity (e.g. stainless steel), which serves as the vacuum chamber. Such devices therefore satisfy the assumptions made in the derivation of the perturbed flux function.

The key point to recognize is that typical experimental pulse lengths are long compared to the skin diffusion time of the wall. The implication is that during an experimental pulse the fields have time to diffuse through the wall which then loses its ability to provide a toroidal restoring force. A vertical field is thus required to hold the plasma in toroidal force balance.

The fact that the wall is thin has both pros and cons. On the positive side a thin wall is good in order to (1) allow rapid vertical field penetration, resulting in a sufficiently fast response time for the horizontal positioning feedback system, and (2) keep the transient currents in the vacuum chamber to a tolerable level because of the higher wall resistivity. On the negative side the faster wall diffusion time through a thin wall implies a quicker loss of plasma equilibrium. Solving this problem requires a faster response time in the feedback system, which adds to the technological difficulty. In practice, careful engineering tradeoffs are balanced to arrive at an optimum design.

In this subsection the basic MHD question addressed (originally by Shafranov, 1960) is to determine the magnitude and direction of the vertical field required to keep the plasma centered in the vacuum chamber.

Now, from a practical point of view, any metallic wall in contact with the plasma would force the edge temperature of the plasma to become low and the wall temperature to become high. In particular, as particles and energy are transported across the field, they would have to be absorbed by the thin wall which, because of its fragile structure, would rapidly become damaged, perhaps developing burn spots and leaks. Clearly this is unacceptable.

Two methods have been proposed to alleviate this problem: limiters and divertors. The limiter idea is most applicable to early tokamaks and modern RFPs. As such, a temporary digression is made from the toroidal equilibrium calculation to describe the operation of a limiter. This discussion will help formulate the proper boundary conditions required to calculate $\overline{\psi}_1$. The divertor is discussed in Section 6.6.5.

Operation of a limiter

A limiter is a robust piece of material often made of tungsten, molybdenum, or graphite, placed just inside the vacuum chamber. It protects the vacuum chamber from plasma bombardment and serves to define the edge of the plasma. Limiters come in various shapes. Three common examples are the poloidal ring limiter, the toroidal hoop limiter, and the rail limiter, shown in Fig. 6.7.

The basic idea of limiter operation is as follows. To a lowest approximation, within the plasma, particles spiral along the magnetic field with a typical velocity comparable to their thermal speed: $v_{\parallel} \sim V_T$. Both electrons and ions slowly diffuse across the magnetic field at the same collisional transport rate (i.e., due to

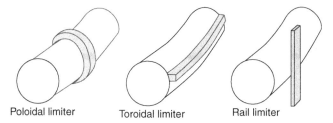

Poloidal limiter   Toroidal limiter   Rail limiter

Figure 6.7   Schematic diagram of a poloidal ring limiter, toroidal hoop limiter, and toroidal rail limiter.

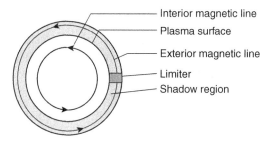

Interior magnetic line
Plasma surface
Exterior magnetic line
Limiter
Shadow region

Figure 6.8   Shadow region of a limiter. Observe that magnetic lines in the shadow region make direct contact with the limiter.

ambipolar diffusion) and enter the shadow region of the limiter as shown in Fig. 6.8. In the shadow region the particles, because of their high parallel velocities, strike the limiter surface in a short time, well before they have a chance to diffuse radially and strike the vacuum chamber wall: $L_\parallel/C_\parallel \ll L_\perp^2/D_\perp$. Here, $L_\parallel$ is the parallel distance traveled by a particle, $C_\parallel = (T_e + T_i)^{1/2}/m_i^{1/2}$ is the parallel sound speed, $L_\perp$ is the perpendicular width of the scrape-off layer, and $D_\perp$ is the cross field diffusion rate, usually anomalous. Note that parallel to the field, electrons and ions must be lost at the same rate (again because of ambipolarity) in a characteristic time $\tau \sim L_\parallel/C_\parallel$. By this operation the limiter protects the vacuum chamber and defines the edge of the plasma.

   Ideally one wants a limiter of sturdy and durable construction that can absorb a large flux of particles and energy while maintaining its structural integrity and not producing any adverse effects on the plasma. In practice plasma bombardment slowly causes damage but limiters are designed for much easier replacement than the vacuum chamber itself. Even so there remains an important problem. The plasma–limiter interaction causes particles to escape from the limiter surface, usually by sputtering. Especially troublesome are limiter neutrals that can penetrate some distance into the plasma before becoming ionized. These high-Z impurities can lead to significant energy loss through radiation. It is clearly important to minimize these losses. This may not be possible in very long pulse experiments, or

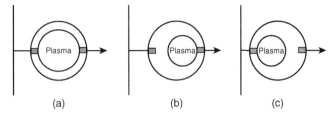

Figure 6.9   Plasma position between the limiters with (a) a correct choice of vertical field, (b) too small a vertical field, and (c) too large a vertical field.

in fusion reactors. The need for impurity shielding in addition to wall protection has led to the development of a second idea, the plasma divertor discussed in Section 6.6.5.

From the ideal MHD point of view the boundary conditions appropriate to a system with axisymmetric toroidal limiters correspond to those of a plasma whose last closed flux surface coincides with the inner surface of each limiter. For a plasma with a minor radius $r = a$ the limiters are located at $R = R_0 \pm a$. The wall radius is denoted by $r = b > a$. Between a and b is a cold, low-density plasma with a small pressure and very little, if any, current. For all practical purposes the magnetic field in the region $a < r < b$ is a vacuum field. The purpose of the vertical field is thus to position the last closed plasma flux surface exactly between the limiters as shown in Fig. 6.9. Also shown are the consequences of too little or too much applied vertical field.

Mathematically, the goal of the MHD analysis is to calculate the shift of the last closed flux surface with respect to the center of the vacuum chamber and then choose the vertical field so that this shift is zero: $\Delta(a) \equiv \Delta_a = 0$. One could also easily calculate the shift of the magnetic axis but this is of less interest.

Calculation of the vertical field

The discussion of the limiter enables one to specify the boundary conditions necessary to complete the solution for $\overline{\psi}_1$ given by Eq. (6.67); that is, to give values for the integration constants r_1 and r_2.

Regularity of the solution in the interior of the plasma again requires that $r_1 = 0$. The constant r_2 is evaluated by the following argument. The end goal is to determine the vertical field necessary to keep the plasma surface, $r = a$, centered in the vacuum chamber which, as stated above, is equivalent to setting $\Delta_a = 0$. Now, recall from Eq. (6.72) that the shift of the last closed flux surface is given by

$$\Delta_a = -\frac{\overline{\psi}_1(a)}{R_0 B_\theta(a)} \tag{6.76}$$

The $\Delta_a = 0$ requirement, therefore, corresponds to the boundary condition $\overline{\psi}_1(a) = 0$ which implies that $r_2 = a$. The perturbed flux function reduces to

$$\overline{\psi}_1(r) = B_\theta(r) \int_a^r \frac{dx}{x B_\theta^2(x)} \int_0^x \left[y B_\theta^2(y) - 2\mu_0 y^2 \frac{dp(y)}{dy} \right] dy \qquad (6.77)$$

The vertical field is determined by extending the solution for $\overline{\psi}_1(r)$ to the region far outside the plasma surface, $a \ll r \ll R_0$, and then matching the solution to the known fields in this region. Specifically, in this region the magnetic field consists of two contributions: the vacuum field due to a "plasma wire" carrying a current I centered at $R = R_0 + \Delta_a \approx R_0$, and an externally applied vacuum vertical field $\mathbf{B}_V = B_V \mathbf{e}_Z$, which is treated as a small, first-order quantity. All plasma diamagnetic effects are small and can be neglected far from the plasma corresponding to $a \ll r \ll R_0$.

The magnetic flux due to the plasma wire is easily evaluated from the standard theory of magnetostatics. For a thin circular wire of major radius R_0 carrying a current I the flux can be expressed as (Stratton, 1941)

$$\psi_I = -\frac{\mu_0 I}{\pi} \frac{(R R_0)^{1/2}}{2k} \left[(2 - k^2) K - 2E \right] + C_0$$

$$k^2 = \frac{4 R R_0}{(R + R_0)^2 + Z^2} \qquad (6.78)$$

where $K(k)$ and $E(k)$ are the complete elliptic integrals and C_0 is an arbitrary, additive constant. In the relevant regime where $a \ll r \ll R_0$ the elliptic functions can be expanded as follows:

$$k \approx 1$$

$$k' = (1 - k^2)^{1/2} \approx \frac{r}{2R_0} \left(1 - \frac{r}{2R_0} \cos \theta \right)$$

$$E \approx 1 \qquad (6.79)$$

$$K \approx \ln \frac{4}{k'} = \ln \frac{8R_0}{r} + \frac{r}{2R_0} \cos \theta$$

Substituting these relations into Eq. (6.78) leads to a simplified form of the wire contribution to the flux far from the plasma:

$$\psi_I(r, \theta) \rightarrow -\frac{\mu_0 I R_0}{2\pi} \left[\ln \frac{8R_0}{r} - 2 + C_0' + \left(\ln \frac{8R_0}{r} - 1 \right) \frac{r}{2R_0} \cos \theta \right] \qquad (6.80)$$

Here, C_0' is a new arbitrary constant replacing C_0. Its value is not important for calculating the flux surface shift; that is, the zeroth-order magnetic field, whose value is necessary to calculate the shift, is given by $R_0 B_\theta = d\psi_0/dr$, which is independent of the additive constant.

Next, note that the flux function for a vacuum vertical field of magnitude B_V is given by $\psi_V = R_0 B_V r \cos \theta$. Thus, combining this relation with Eq. (6.80) shows that the total flux function $\psi = \psi_I + \psi_V$ for $r \gg a$ must approach

$$\psi(r, \theta) \rightarrow -\frac{\mu_0 R_0 I}{2\pi} \left[\ln \frac{8R_0}{r} - 2 + C_0' + \left(\ln \frac{8R_0}{r} - 1 \right) \frac{r}{2R_0} \cos \theta \right] \tag{6.81}$$
$$+ R_0 B_V r \cos \theta$$

The condition on the zeroth-order contribution to ψ is satisfied automatically; that is, for $r \gg a$ one finds from Eq. (6.81) that $B_\theta(r) = (1/R_0)d\psi_0/dr = \mu_0 I / 2\pi r$, which is the correct result. The first-order contribution requires matching the $\cos \theta$ in Eq. (6.81) with $\overline{\psi}_1(r) \cos \theta$ evaluated at $r \gg a$,

$$\overline{\psi}_1(r) \rightarrow -\frac{\mu_0 I}{4\pi} \left(\ln \frac{8R_0}{r} - 1 \right) r + R_0 B_V \, r \tag{6.82}$$

The task now is to evaluate the $\overline{\psi}_1$, given by Eq. (6.77), for large r. Matching in accordance with Eq. (6.82) then yields the desired value of B_V.

Consider first the terms in the square bracket in Eq. (6.77), keeping in mind that the regime of interest corresponds to $x > a$. In this regime the pressure is zero, implying that

$$-\int_0^x 2\mu_0 y^2 \frac{dp(y)}{dy} \, dy = -\int_0^a 2\mu_0 y^2 \frac{dp(y)}{dy} \, dy$$
$$= 4\mu_0 \int_0^a p(y) y \, dy \tag{6.83}$$
$$= a^2 B_{\theta a}^2 \beta_p$$

where $B_{\theta a} = \mu_0 I / 2\pi a$.

Next, in the region $x > a$ there is zero toroidal current, implying that B_θ is a vacuum field: $B_\theta = B_{\theta a}(a/r)$. Therefore, the magnetic energy term can be split into two contributions, one from $0 < y < a$ and the other from $a < y < x$:

$$\int_0^x y B_\theta^2(y) dy = \int_0^a y B_\theta^2(y) dy + \int_a^x y B_\theta^2(y) dy$$
$$= a^2 B_{\theta a}^2 \left(\frac{l_i}{2} + \ln \frac{x}{a} \right) \tag{6.84}$$

Here, l_i is the normalized internal plasma inductance per unit length defined by

$$l_i \equiv \frac{L_i/2\pi R_0}{\mu_0/4\pi}$$
$$\frac{1}{2} L_i I^2 = \int_P \frac{B_\theta^2}{2\mu_0} \, d\mathbf{r} \tag{6.85}$$

which leads to

$$l_i = \frac{2}{a^2 B_{\theta a}^2} \int_0^a B_\theta^2 \, y d\, y \tag{6.86}$$

For the model ohmic tokamak profiles discussed in Section 5.5.2, where $\rho = y/a$ and $B_\theta = B_{\theta a} J_1(k\rho)/J_1(k)$, the value of the normalized internal inductance is $l_i = 1$.

Combining these results leads to the following expression for the square bracket in Eq. (6.77):

$$\int_0^x \left[y B_\theta^2(y) - 2\mu_0 y^2 \frac{dp(y)}{dy} \right] dy = a^2 B_{\theta a}^2 \left(\beta_p + \frac{l_i}{2} + \ln \frac{x}{a} \right) \tag{6.87}$$

The remaining integral over x can be easily evaluated by again recalling that for $x > a$, the magnetic field $B_\theta = B_{\theta a}(a/x)$. This yields

$$\overline{\psi}_1(r) = \frac{\mu_0 I}{4\pi r} \left[\left(\beta_p + \frac{l_i}{2} - \frac{1}{2} \right) (r^2 - a^2) + r^2 \ln \frac{r}{a} \right] \tag{6.88}$$

The limit of this expression for $r \gg a$ is

$$\overline{\psi}_1(r) \to \frac{\mu_0 I}{4\pi} \left(\beta_p + \frac{l_i}{2} - \frac{1}{2} + \ln \frac{r}{a} \right) r \tag{6.89}$$

The final step is to equate Eq. (6.89) to Eq. (6.82) and solve for B_V:

$$B_V = \frac{\mu_0 I}{4\pi R_0} \left(\beta_p + \frac{l_i - 3}{2} + \ln \frac{8 R_0}{a} \right) \tag{6.90}$$

This well-known and often used result was first derived by Shafranov (1966). It provides valuable information about the design requirements for the vertical field circuit, in particular how large is the required vertical field to center the plasma as a function of toroidal current and geometry. The formula also shows that B_V is positive which, with the present sign convention, implies that the vertical field increases the overall B_p on the outside of the torus and decreases it on the inside. As expected the magnetic tension on the outside increases in order to support higher values of pressure in toroidal force balance.

While Eq. (6.90) is simple in appearance, it is somewhat unintuitive in that the individual contributions are not easily recognizable in terms of the forces holding the plasma in toroidal equilibrium. A more intuitive form, valid for an ohmic tokamak, is obtained by multiplying by $2\pi R_0 I$ and then performing some standard manipulations. This enables Eq. (6.90) to be rewritten as

$$2\pi R_0 I B_V = \frac{I^2}{2} \frac{\partial}{\partial R_0} (L_e + L_i) + 4\pi^2 \int_0^a \left(p - \frac{B_0 \delta B_z}{\mu_0} \right) r dr \tag{6.91}$$

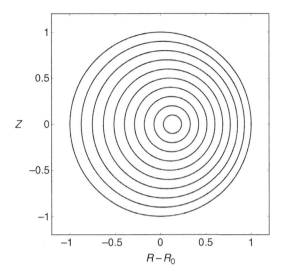

Figure 6.10   Numerically computed equilibrium for the Alcator C tokamak at MIT. Shown are flux surface plots for a typical low β ohmically heated plasma.

where $\delta B_z(r) = B_z(r) - B_0 \ll B_0$. The left-hand side represents the vertical field force. The first term on the right-hand side represents the hoop force[2] and includes the total internal plus external poloidal flux. To obtain this term requires use of the facts that $L_i(R_0) = \mu_0 R_0 l_i/2 \propto R_0$ and for a thin wire $L_e(R_0) = \mu_0 R_0[\ln(8R_0/a)-2]$. The pressure term represents the tire tube force while the last term represents the $1/R$ force due to the toroidal field.

The basic MHD features of an ohmically heated tokamak as described by the large aspect ratio expansion are verified by a full non-linear numerical solution of the Grad–Shafranov equation for the Alcator C tokamak as illustrated in Fig. 6.10. The analytic, shifted circle equilibrium does indeed appear to be a good approximation to the exact numerical results.

6.4.7 Summary

The Grad–Shafranov equation has been solved by means of an asymptotic expansion in the inverse aspect ratio: $\varepsilon \ll 1$ and $\beta_p \sim 1$. The ε^0 contribution describes cylindrical radial pressure balance. The ε^1 contribution is a small correction that describes toroidal force balance. The main consequence of toroidity is to shift the central flux surfaces outward along R with respect to the plasma surface.

---

[2] Conceptually, the hoop force can be calculated as the gradient of the "potential energy" subject to the perfect conductivity constraint of fixed flux within the plasma: $F_{\text{hoop}} = -\mathbf{e}_R \cdot \nabla (LI^2/2)$ subject to $LI = $ constant.

An analytic formula is derived that gives the value of the flux surface shift. This formula is applied to two configurations of interest: (1) the early RFP in which the shift of the magnetic axis with respect to the plasma surface is calculated, and (2) the ohmic tokamak and modern RFP in which the vertical field necessary to center the plasma within the limiter is calculated.

Because of the assumption of small ε the plasma parameters and figures of merit are nearly identical to their simple cylindrical values.

6.5 Analytic solution in the limit $\varepsilon \ll 1$ and $\beta_p \sim 1/\varepsilon$ (the high β tokamak)

The discussion in the previous section has demonstrated several desirable MHD features of the ohmically heated tokamak: good toroidal equilibrium (i.e., small shift), good stability (i.e., $q \sim 1$). Experimentally, ohmically heated tokamaks also exhibit good confinement, and good ohmic heating. Nevertheless, the fact that β is inherently small is an important disadvantage with respect to the efficient production of energy in a fusion reactor. It is thus of interest to investigate modifications to the basic ohmic tokamak which attempt to alleviate this problem. This is the motivation behind the high β tokamak.

Qualitatively, there are two modifications to the ohmic tokamak that lead to higher values of β. The first is the application of external heating sources such as neutral beams or RF power. In addition to raising β, external heating sources are required because ohmic heating by itself, while good, is still not sufficient to raise the temperature of a plasma to the level needed for ignition. To reach ignition the externally applied power in a modern tokamak must be several to many times larger than the ohmic power.

The second modification changes the cross section of the plasma from circular to non-circular. Typical modern tokamaks have vertically elongated cross sections with an outward pointing "D" shape. Cross sectional optimization represents a new degree of freedom in machine design and is shown to lead to higher values of β for the same q_*. This section presents a description of both the circular and elliptical high β tokamak configurations to demonstrate the benefits of external heating and cross sectional optimization.

The basic physics of the high β tokamak is as follows. Application of an independent source of external heating at fixed toroidal current, toroidal field, and number density causes an increase in plasma temperature and hence pressure. The higher β that results is confined by newly generated poloidal diamagnetic currents induced in the plasma. When the external heating becomes sufficiently large, radial pressure balance is provided almost entirely by the toroidal field rather than the poloidal field as in an ohmic tokamak. This is the regime of the high β tokamak: the plasma pressure is decoupled from the ohmic heating current and confinement is more closely related to that of a θ-pinch than a Z-pinch. The critical physics limitation of the high β tokamak is that β can only be externally raised to a certain maximum value that is consistent with toroidal force balance and subject to the constraint that $q_* \gtrsim 1$ for good stability.

6.5.1 The high β tokamak expansion

Taking these requirements into account leads to the following inverse aspect ratio expansion for the high β tokamak:

$$
\begin{aligned}
\frac{B_p}{B_\phi} &\sim \varepsilon \\
\beta_t &\sim \frac{2\mu_0 p}{B_\phi^2} \sim \varepsilon \\
\beta_p &\sim \frac{2\mu_0 p}{B_p^2} \sim \frac{1}{\varepsilon} \\
q &\sim \frac{rB_\phi}{RB_p} \sim 1
\end{aligned}
\tag{6.92}
$$

Note that $\beta_t \sim \varepsilon$ is one order larger in ε than in an ohmic tokamak. Also, the ordering $\beta_p \sim 1/\varepsilon$ indicates that the poloidal field plays only a minor role in radial pressure balance. Indeed, it is shown that radial pressure balance is provided by a small diamagnetic depression in the toroidal field. Even though the depression is small, of order ε, it still confines $1/\varepsilon$ more plasma than the small poloidal pressure $B_p^2 \sim \varepsilon^2 B_\phi^2$. The conclusion is that in a high β tokamak the toroidal field provides both radial pressure balance and stability while the poloidal field is required primarily for toroidal force balance with a small contribution to the heating.

Because of the higher values of β the toroidal shift of the flux surfaces is no longer small but is of order unity: $\Delta/a \sim 1$. Similarly, even with a circular plasma surface, the interior flux surfaces are no longer shifted circles, but develop a finite non-circularity. The mathematical consequence is that while the inverse aspect ratio expansion does somewhat simplify the Grad–Shafranov equation, the leading-order (in ε) contribution remains inherently two dimensional; that is, radial pressure balance and toroidal force balance enter simultaneously.

As a result of this behavior only the 2-D leading order contribution to the Grad–Shafranov equation is required to understand the high β tokamak. The appropriate formal expansion for the high β tokamak analysis is thus given by

$$
\begin{aligned}
\psi(r, \theta) &= \psi_0(r, \theta) + \cdots \\
p(\psi) &= p(\psi_0) + \cdots
\end{aligned}
\tag{6.93}
$$

where

$$\frac{\psi_0}{rRB_\phi} \sim \varepsilon$$

$$\frac{2\mu_0 p(\psi_0)}{B_\phi^2} \sim \varepsilon \tag{6.94}$$

The expansion for $F(\psi)$ is slightly subtle because the Grad–Shafranov equation basically determines the poloidal field whereas radial pressure balance is dominated by the toroidal field. The expansion for $F(\psi)$ must automatically take this into account. The procedure is to introduce a new function $\tilde{B}(\psi)$ in place of $F(\psi)$ as follows:

$$F^2(\psi) = R_0^2 B_0^2 \left[1 - \frac{2\mu_0 p(\psi)}{B_0^2} + \frac{2\tilde{B}(\psi)}{B_0} \right] \tag{6.95}$$

where, as before, B_0 is the vacuum toroidal field at the geometric center of the plasma $R = R_0$. Equation (6.95) implies that the deviation from a θ-pinch pressure balance relation is of order $\tilde{B}/B_\phi \sim \varepsilon^2$, which is one order smaller than the pressure itself.

Substituting the expansion into the Grad–Shafranov equation gives rise to a leading-order contribution which is of the form of a 2-D partial differential equation. This equation describes the behavior of high β tokamak equilibria and is given by

$$\nabla^2 \psi_0 = -R_0^2 B_0^2 \left[\frac{1}{B_0} \frac{d\tilde{B}}{d\psi_0} + \frac{2\mu_0}{B_0^2} \frac{dp}{d\psi_0} \left(\frac{r}{R_0} \right) \cos\theta \right] \tag{6.96}$$

Note that the second term on the right-hand side is often referred to as the large aspect ratio limit of the Pfirsch–Schluter (1962) current,

$$J_{PS} = 2 \frac{dp}{d\psi_0} r \cos\theta \tag{6.97}$$

The quantity J_{PS} represents the return current flowing parallel to the magnetic field that is necessary to short out the vertical charge accumulation caused by the ∇B and curvature drifts.

Equation (6.96) is the desired equation that is used to investigate circular and elliptical high β tokamaks.

6.5.2 The circular high β tokamak

Mathematical solution

Since Eq. (6.96) is a non-linear partial differential equation it must in general be solved numerically. However, for special choices of $p(\psi_0)$ and $\delta B(\psi_0)$ it can be solved analytically, and it is such solutions that provide a great deal of insight into the behavior of high β tokamak equilibria. A well-known example was first proposed by Solov'ev (1967) and later investigated by Haas (1972) as well as many other authors (see for instance Cerfon and Freidberg, 2010). Solov'ev suggested that $p(\psi_0)$ and $\tilde{B}(\psi_0)$ be chosen as linear functions of ψ_0 so that

$$R_0^2 B_0 \frac{d\tilde{B}}{d\psi_0} = -A = \text{constant}$$

$$2\mu_0 R_0 \frac{dp}{d\psi_0} = -C = \text{constant}$$

(6.98)

With this assumption Eq. (6.96) reduces to

$$\nabla^2 \psi_0 = A + C\, r \, \cos\, \theta$$

(6.99)

Observe that the linear dependence of the free functions on ψ_0 implies that the toroidal current density (proportional to the right-hand side) is nearly constant in the plasma and must abruptly jump to zero across the plasma surface into the surrounding vacuum region. This jump in the edge current density is probably the biggest drawback of the Solov'ev model in terms of realistic application to experiment. Still, all other important physical properties are well described by the model, which is why it has been so extensively used by the fusion community.

Consider next the boundary conditions on ψ_0. For simplicity, the first geometry investigated corresponds to a plasma with a circular surface located at $r = a$. Outside the plasma surface is a vacuum region. The free additive constant associated with the flux function is chosen to make $\psi_0 = 0$ on the surface. Also, the flux function must be regular in the plasma interior. These conditions are represented mathematically as follows:

$$\psi_0(a, \theta) = 0$$

$$\psi_0(r, \theta) = \text{regular for } r \leq a$$

(6.100)

The solutions for ψ_0, B_θ, and p are easily found and can be written as

$$\psi_0(r, \theta) = \frac{A}{4}(r^2 - a^2) + \frac{C}{8}(r^2 - a^2)r \, \cos\, \theta$$

$$B_\theta(r, \theta) = \frac{1}{R_0} \frac{\partial \psi_0}{\partial r} = \frac{1}{2R_0}\left[Ar + \frac{C}{4}(3r^2 - a^2)\cos\, \theta\right]$$

$$p(r, \theta) = -\frac{C}{2\mu_0 R_0}\psi_0 = -\frac{C}{2\mu_0 R_0}\left[\frac{A}{4}(r^2 - a^2) + \frac{C}{8}(r^2 - a^2)r \, \cos\, \theta\right]$$

(6.101)

The solutions, while formally correct, are not very intuitive since the two constants A and C do not have simple physical interpretations. It is shown below that one constant is related to the toroidal current and the other to the pressure. It is therefore convenient to replace them with the more physical parameters q_* and β_t. This is a straightforward task and the results are

$$\frac{1}{q_*} = \frac{\mu_0 R_0 I}{2\pi a^2 B_0} = \frac{R_0}{2\pi a^2 B_0} \int_0^{2\pi} B_\theta(a, \theta) a d\theta = \frac{A}{2B_0}$$

$$\beta_t = \frac{2\mu_0 \langle p \rangle}{B_0^2} = \frac{2\mu_0}{\pi a^2 B_0^2} \int_0^{2\pi} \int_0^a p \, r \, dr d\theta = \frac{a^2 A \, C}{8 R_0 B_0^2}$$

(6.102)

Substituting these expressions leads to the following equilibrium solutions for the circular high β tokamak:

$$\frac{\psi_0}{a^2 B_0} = \frac{1}{2q_*} [\rho^2 - 1 + v(\rho^3 - \rho)\cos\theta]$$

$$\frac{B_\theta}{\varepsilon B_0} = \frac{1}{q_*} \left[\rho + \frac{v}{2}(3\rho^2 - 1)\cos\theta \right]$$

$$\frac{B_r}{\varepsilon B_0} = -\frac{v}{2q_*}(\rho^2 - 1)\sin\theta$$

$$\frac{2\mu_0 p}{B_0^2} = 2\beta_t(1 - \rho^2)(1 + v\rho\cos\theta)$$

$$\frac{\mu_0 R_0 J_\phi}{B_0} = -\frac{2}{q_*}(1 + 2v\rho\cos\theta)$$

$$\frac{B_\phi}{B_0} = 1 - \varepsilon\rho\cos\theta - \beta_t(1 - \rho^2)(1 + v\rho\cos\theta)$$

(6.103)

Here, $\varepsilon = a/R_0$, $\rho = r/a$, and v is the crucial high β tokamak parameter defined by

$$v = \frac{\beta_t q_*^2}{\varepsilon}$$

(6.104)

Having determined the analytic solutions one can now investigate the basic equilibrium properties of the high β tokamak.

Plasma parameters and figures of merit

Intuition into the nature of radial pressure balance and toroidal force balance can be developed by plotting the curves of B_ϕ, p, and B_θ along the midplane $Z = 0$. These curves are illustrated in Fig. 6.11. First, observe that there is a small diamagnetic depression of order ε superimposed on the $1/R$ decay of the vacuum toroidal field.

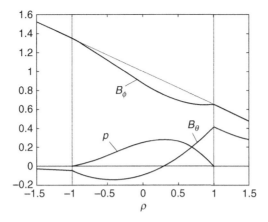

Figure 6.11 Profiles of B_ϕ, B_θ, and p along the $Z = 0$ axis for the slightly exaggerated case corresponding to $\varepsilon = 0.35$, $\nu = 0.8$, $q_* = 1.5$, and $\beta_t = 0.12$. Here, $\rho = (R - R_0)/a$.

It is in this depression that the plasma pressure $p \sim \varepsilon B_0^2/\mu_0$ is held in radial pressure balance. Since $B_\theta/B_\phi \sim \varepsilon$ when $q \sim 1$, the poloidal field can at most confine a pressure of the order $B_\theta^2/\mu_0 \sim \varepsilon^2 B_\phi^2/\mu_0$. As previously stated, this is one order smaller in ε than the actual plasma pressure, implying that the poloidal field has only a very small effect on radial pressure balance.

Next, note that the general equilibrium relation between β_t, β_p, and q_* is the same as for a circular ohmic tokamak

$$\beta_t = \varepsilon^2 \beta_p/q_*^2 \tag{6.105}$$

except that β_t and β_p are both one order larger in ε: $\beta_t \sim \varepsilon$, $\beta_p \sim 1/\varepsilon$, and $q_* \sim 1$. Furthermore, it follows from the definition of ν given by Eq. (6.104) that

$$\nu = \frac{\beta_t q_*^2}{\varepsilon} = \varepsilon \beta_p \sim 1 \tag{6.106}$$

An examination of Eq. (6.103) suggests that it is useful to view ν as the critical parameter that distinguishes different regimes of tokamak operation with $\nu \ll 1$ corresponding to the ohmic tokamak and $\nu \sim 1$, the high β tokamak.

Attention is now focused on how toroidal high β effects modify some of the basic equilibrium properties of the tokamak. To begin observe that the pressure profile is parabolic for small ν but develops a finite shift and a modified shape for $\nu \sim 1$. This is demonstrated in Fig. 6.12, which illustrates the $Z = 0$ mid-plane curves of the pressure p. Also shown is the midplane curve of J_ϕ, which for small ν is nearly constant across the plasma and jumps to zero at the plasma surface. For $\nu \sim 1$ the current accumulates on the outside in order to provide a stronger toroidal restoring force to balance the increased tire tube and

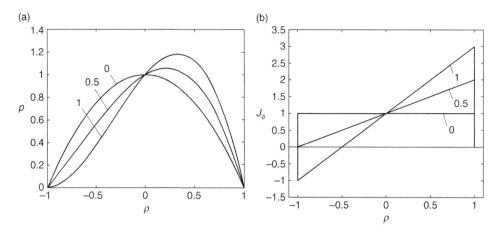

(a)

(b)

Figure 6.12   Midplane profiles of (a) p and (b) J_ϕ for $v = 0$, 0.5, 1. Here, $\rho = (R - R_0)/a$.

$1/R$ expansion forces. In fact for $v > 1/2$ the current reverses direction on the inside of the plasma surface $r = a$, $\theta = \pi$. A current reversal may be difficult to maintain in a plasma with finite resistivity but this issue lies beyond the scope of ideal MHD.

Another interesting feature in the $v \sim 1$ regime is that the flux surfaces $\psi_0 = $ constant, while round, are no longer pure circles except for the surface $r = a$. Equally importantly, the shift of the magnetic axis is now finite and not small as in the ohmic tokamak. These points can be quantified by examining the flux function in the vicinity of the magnetic axis, the point defined by $\partial\psi_0/\partial\theta = \partial\psi_0/\partial\rho = 0$. For the model under consideration the first condition $\partial\psi_0/\partial\theta = 0$ occurs at $\theta = 0$. The second condition $\partial\psi_0/\partial\rho = 0$ occurs at the point $\rho = \Delta_0/a$ whose value is determined by setting $\rho \cos\theta = x + \Delta_0/a$, $\rho \sin\theta = y$ (see Fig. 6.13), substituting into the expression for the flux function, and assuming that near the magnetic axis $\Delta_0/a \gg x, y \to 0$. A short calculation yields

$$\frac{2q_*}{a^2 B_0}\psi_0 = [1 - (x + \delta_0)^2 - y^2][1 + v(x + \delta_0)]$$

$$\approx (1 - \delta_0^2)(1 + v\delta_0) - (3v\delta_0^2 + 2\delta_0 - v)x - (1 + 3v\delta_0)x^2 - (1 + v\delta_0)y^2$$

$$(6.107)$$

where $\delta_0 = \Delta_0/a$ and in the approximate formula terms of the order x^3, xy^2 have been neglected. The magnetic axis condition in x, y coordinates is equivalent to setting $\partial\psi_0/\partial y = \partial\psi_0/\partial x = 0$ at $x = 0$, $y = 0$ and requires that the coefficient of the term linear in x vanishes. Carrying out this task yields the desired expression for the magnetic axis shift with respect to the plasma surface

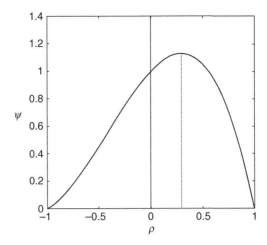

Figure 6.13   Midplane profile of ψ showing the location of the magnetic axis. Here, $\rho = (R - R_0)/a$.

$$\frac{\Delta_0}{a} = \frac{v}{1 + (1 + 3v^2)^{1/2}} \tag{6.108}$$

Clearly, for $v \sim 1$ the shift is finite and not small as in the ohmic tokamak.

With respect to the shape of the flux surfaces, the approximate form of the flux function given in Eq. (6.107) indicates that these surfaces are ellipses centered on the magnetic axis:

$$x^2 + \frac{y^2}{\kappa_0^2} = \text{constant} \tag{6.109}$$

Here, κ_0 is the on-axis elongation of the ellipses and is given by

$$\kappa_0^2 = \frac{1 + 3v\delta_0}{1 + v\delta_0} = 1 + \frac{2v^2}{1 + v^2 + (1 + 3v^2)^{1/2}} \tag{6.110}$$

For $v \ll 1$, then $\kappa_0^2 \approx 1$: the surfaces approach circles in the ohmic limit. However, in the limit $v \sim 1$ Eq. (6.110) shows that the surfaces have a finite ellipticity.

In essence, for a high β tokamak the increased tire tube and $1/R$ forces push the central plasma outward along R towards the outside of the torus. This results in a finite shift and a squashing of the interior flux surfaces causing them to become elliptical.

The final quantity of interest involves the evaluation of the MHD safety factor. This is in general a complicated calculation when the flux surfaces do not have simple shapes. For the circular high β tokamak numerical calculations show that the q profile is an increasing function of flux. Consequently, for present purposes it is sufficient to calculate q at the two extremes, the magnetic axis and the plasma surface, to obtain an idea of the variation. Both of these values can be calculated analytically.

For the magnetic axis safety factor recall that an exact expression has been derived in Section 6.3.6 and is repeated here for convenience:

$$q_0 = \frac{B_\phi}{(\psi_{RR}\psi_{ZZ})^{1/2}} \tag{6.111}$$

All quantities are evaluated on the magnetic axis. This expression simplifies by substituting the high β tokamak expansion and transforming to the x, y coordinates defined above. The result is

$$q_0 = \frac{a^2 B_0}{(\psi_{0xx}\psi_{0yy})^{1/2}} \tag{6.112}$$

Eliminating the ψ_0 derivatives by means of Eq. (6.107) yields the desired result

$$q_0 = \frac{q_*}{(1 + 3\nu\delta_0)^{1/2}(1 + \nu\delta_0)^{1/2}} = q_* \left[\frac{3}{\eta(2 + \eta)}\right]^{1/2} \tag{6.113}$$

where $\eta = (1 + 3\nu^2)^{1/2}$. Since η is always greater than unity then $q_0 < q_*$. In the low β limit, $\nu \ll 1$ and $q_0 \rightarrow q_*$. As ν is increased the ratio q_0/q_* becomes progressively smaller.

Next consider the safety factor on the plasma surface. In the high β ordering the variation of the poloidal field B_θ around the cross section is of order unity. The value of B_θ is larger on the outside of the torus in order to support the higher pressure in toroidal force balance. This variation averages to zero when calculating the total plasma current, implying that $q_* \propto 1/I$ as required by the definition of q_*. However, when calculating q_a as defined by Eq. (6.35), the poloidal field appears in the denominator of the integrand, which is thus more heavily weighted on the inside of the torus where $B_\theta(a, \theta)$ is smaller. This leads to the result that $q_a > q_*$. Specifically, on the plasma surface $r = a$, q_a can be easily evaluated from Eq. (6.35) by using the high β expansion

$$\begin{aligned} q_a &= \frac{F(\psi)}{2\pi} \oint \frac{dl}{R^2 B_p} \\ &\approx \frac{aB_0}{2\pi R_0} \int_0^{2\pi} \frac{d\theta}{B_\theta(a, \theta)} \\ &= \frac{q_*}{(1 - \nu^2)^{1/2}} \end{aligned} \tag{6.114}$$

In the low β limit, $q_a \to q_*$. In contrast in the high β regime it is important to distinguish between q_a and q_*, particularly since $1/q_a$ is no longer linearly proportional to I. The most striking feature of Eq. (6.114) is the breakdown of the solution when $v > 1$. The explanation is a very important aspect of high β tokamak behavior and is the next topic discussed.

The high β tokamak equilibrium limit

The breakdown of the analytic solution for $v > 1$ actually corresponds to an equilibrium β limit. Specifically, there is an upper limit on the value of β that can be confined assuming the plasma current is held fixed. This can be quantified by requiring that $v \leq 1$ in the basic plasma parameters and figures of merit assuming that $q_* \propto 1/I$ is held fixed as one attempts to raise β by external heating. The values of these quantities at the $v = 1$ limit are given by

$$\begin{aligned}
\beta_t &= \varepsilon/q_*^2 \\
\beta_p &= 1/\varepsilon \\
\Delta_0/a &= 1/3 \\
\kappa_0 &= (3/2)^{1/2} \\
q_0 &= (3/8)^{1/2} q_* \\
q_a &\to \infty
\end{aligned}$$
(6.115)

For a given inverse aspect ratio ε the actual maximum numerical value of β_t is obtained by setting q_* to its minimum allowable value (i.e., maximum I) as set by MHD stability limits.

The origin of the equilibrium limit can be understood by examining the magnetic field in the vacuum region. The solution $\hat{\psi}_0$ that matches onto ψ_0 at the plasma surface must satisfy the following equation and jump conditions:

$$\begin{aligned}
\nabla^2 \hat{\psi}_0 &= 0 \\
\hat{\psi}_0(a, \theta) &= \psi_0(a, \theta) = 0 \\
\hat{B}_\theta(a, \theta) &= B_\theta(a, \theta) = (\varepsilon B_0/q_*)(1 + v \cos \theta)
\end{aligned}$$
(6.116)

Note that the boundary conditions require matching the flux and its normal derivative across the boundary, which guarantees that no surface currents are allowed to flow. Equation (6.116) represents an ill-conditioned problem (i.e., Cauchy boundary conditions for an elliptic differential operator) but the difficulties can be overcome because the solutions are known analytically.

The solution for $\hat{\psi}_0$ satisfying Eq. (6.116) is found to be

$$\frac{\hat{\psi}_0}{a^2 B_0} = \frac{1}{q_*} \left[\ln \rho + \frac{v}{2} \left(\rho - \frac{1}{\rho} \right) \cos \theta \right]$$
(6.117)

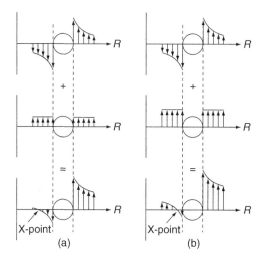

Figure 6.14   Schematic diagram illustrating the appearance of a separatrix in a high β tokamak: (a) low β case and (b) high β case. Note the increased vertical field in the high β case to provide toroidal force balance.

The first term represents the equivalent "cylindrical" contribution from the toroidal current. The last two terms represent the applied vertical field and corresponding plasma diamagnetic response, respectively. A critical feature of the solution is that the vacuum field has a particular flux surface known as the separatrix which is defined by the X-point condition $\hat{B}_r(r_s, \theta_s) = \hat{B}_\theta(r_s, \theta_s) = 0$. The separatrix is shown graphically in the bottom frame of Fig. 6.14.

A simple calculation shows that the separatrix is located at

$$\theta_s = \pi$$
$$\rho_s = \frac{r_s}{a} = \frac{1 + (1 - v^2)^{1/2}}{v} \tag{6.118}$$

For low β, $v \ll 1$ and the separatrix is located at $\rho_s \approx 2/v$, far from the plasma surface. As β increases at fixed current, the vertical field must increase to provide toroidal force balance. An increasing vertical field moves the separatrix closer to the plasma surface as shown in Fig. 6.14, which shows qualitatively how toroidal force balance is achieved by summing the contributions of the vertical field and the $1/r$ field due to the toroidal current.

Now, as $v \rightarrow 1$ then $\rho_s \rightarrow 1$. The X-point moves onto the plasma surface. It can move no further as long as the plasma surface is required to remain circular. It is this behavior that corresponds to the equilibrium β limit.

From the discussion above one can reasonably conclude that all high β tokamaks, regardless of profiles, have an equilibrium β limit with a similar scaling in ε.

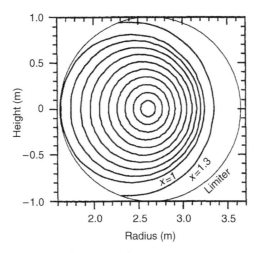

Figure 6.15 Numerically computed equilibrium for the TFTR tokamak at PPPL. Shown is a flux surface plot corresponding to auxiliary heated high β tokamak operation. From Budny, 1994. Reproduced with permission.

This is a correct conclusion but only under the condition that increasing values of β are maintained in toroidal force balance by *increasing the vertical field, while keeping the toroidal current fixed.* Experimentally, avoiding the equilibrium β limit turns out to not be a serious concern. The reasons for this are discussed in Section 6.5.3.

Numerical results for a high β tokamak

One of the large international experiments with circular cross section and substantial external heating was Tokamak Fusion Test Reactor TFTR at Princeton Plasma Physics Laboratory. This experiment had substantial neutral beam power (up to 40 MW) and used carbon limiters to protect the first wall. In other words, from the MHD point of view, the experiment closely satisfied the assumptions made in analyzing the high β tokamak. One disappointing result found on TFTR as well as all other tokamaks is that the energy confinement time degrades with increasing external heating power.

Many numerical simulations have been carried out to model the overall TFTR behavior. With respect to ideal MHD, the agreement between experiment and theory is quite good. One such numerical simulation is illustrated in Fig. 6.15. Observe the large shift of the peak plasma pressure on the outside of the torus.

All in all, in spite of its simplicity, the large aspect ratio Solov'ev equilibrium is a reasonably good model for realistic high β tokamaks.

Summary

A high β tokamak is a configuration in which radial pressure balance is provided almost entirely by a small diamagnetic depression in the large, nearly vacuum toroidal field. The B_ϕ field is also primarily responsible for MHD stability as in an ohmic tokamak. The toroidal current provides toroidal force balance but is limited in magnitude by the stability constraint $q > 1$ plus the maximum technologically allowable toroidal field. Qualitatively, by means of external heating, β_t is raised to the largest value consistent with the $q > 1$ constraint. This leads to the scaling $\beta_t \sim \varepsilon$, $\beta_p \sim 1/\varepsilon$, and $q \sim 1$. The value of β_t is one order larger in ε than in an ohmic tokamak.

An interesting feature of the high β tokamak is the existence of an apparent equilibrium β limit. This occurs by raising β by means of external heating while holding the toroidal current fixed. The increased outward tire tube and $1/R$ forces on the plasma are compensated by raising the vertical field, a strategy that causes the X-point of the vacuum separatrix to move onto the plasma surface, corresponding to the equilibrium limit.

At the limit the value of β_t is reasonably large. For example, for typical experimental values of $\varepsilon = 1/3$ and $q_* = 2$ one finds that $\beta_t = 8\%$, which is adequate, although without much safety margin, for reactor purposes.

6.5.3 The flux conserving tokamak – avoiding the equilibrium β limit

As has been stated, the existence of an equilibrium β limit for high β tokamaks is strongly coupled to the constraint that requires $I \propto 1/q_*$ be held fixed as β_t increases. This constraint does not necessarily represent the evolution of a typical tokamak discharge. An alternate constraint that in fact eliminates the equilibrium β limit was proposed by Clarke and Sigmar (1977) and leads to the flux conserving tokamak concept.

The basic idea behind flux conservation

The basic idea of the flux conserving tokamak is to model the actual evolution of a given plasma discharge by a sequence of plausibly chosen free functions $p(\psi, t)$ and $F(\psi, t)$. The rationale for prescribing this set of functions is as follows. Consider a plasma that is operating initially as an ohmically heated tokamak: $\beta_t \sim \varepsilon^2$, $\beta_p \sim 1$. The plasma is then heated by a high-power external source, for instance neutral beams or RF waves. It is a good approximation to assume that the time scale for heating τ_H is long compared to a typical MHD time τ_M:

$$\tau_H \gg \tau_M \tag{6.119}$$

Here, $\tau_M = a/V_{Ti}$ is the MHD time and $\tau_H = T/(\partial T/\partial t)$ is the time for the temperature to change by a finite amount. Under this assumption the MHD inertial effects are negligible (i.e., the $\rho\, d\mathbf{v}/dt$ term in the MHD momentum equation can be neglected). The plasma evolution can thus be viewed as a series of quasistatic equilibria, each one satisfying the Grad–Shafranov equation with slowly varying free functions $p(\psi, t)$ and $F(\psi, t)$.

In contrast, the time scale for heating is assumed to be short compared to the magnetic diffusion time τ_D,

$$\tau_H \ll \tau_D \tag{6.120}$$

where $\tau_D = \mu_0 a^2/\eta$ and η is the resistivity which can be classical or anomalous. This too is a well-satisfied approximation experimentally. The long diffusion time implies that the plasma behaves like a perfect conductor on the heating time scale.

Choosing p(ψ, t) *and* F(ψ, t)

These two assumptions lead to the following prescription for determining the time evolution of the free functions $p(\psi, t)$ and $F(\psi, t)$ corresponding to the flux conserving tokamak. The function $p(\psi, t)$ is treated as a known function. Initially $p(\psi, 0) = p_\Omega(\psi)$ where $p_\Omega(\psi)$ is the ohmic profile. As time progresses $p(\psi, t)$ increases due to the external heating. Since the theory and computation of heat deposition in a plasma is reasonably well understood, it is not difficult to generate a plausible model for the time evolution of the pressure. A simple model that is adequate for present purposes is to assume that

$$p(\psi, t) = K(t)p_\Omega(\psi) \tag{6.121}$$

with $K(0) = 1$. The function $K(t)$ is assumed to be known and increases with time. This model directly specifies the evolution of $p(\psi, t)$ in the Grad–Shafranov equation and eliminates the need for solving a separate set of particle and energy transport equations. Keep in mind that the time t enters calculation only as a parameter – there are no time derivatives. This is what is meant by the "quasistatic" assumption.

The interesting new feature of the flux conserving tokamak is the specification of $F(\psi, t)$. Since the plasma acts like a perfect conductor on the heating time scale, the function $F(\psi, t)$ must be chosen such that both the toroidal and poloidal fluxes are conserved. How exactly does one mathematically specify flux conservation? As the plasma is heated the shape of the poloidal flux function $\psi(\mathbf{r}, t)$ slowly evolves in time as the increased plasma pressure shifts the magnetic axis outward. Flux conservation requires that the toroidal flux contained within any given poloidal flux surface must remain constant during this evolution if the plasma is a perfect conductor. No toroidal flux is allowed to diffuse into or out of any poloidal flux surface.

The flux conserving constraint can be expressed mathematically by calculating the toroidal flux $\Phi(\psi, t)$ from the definition

$$\Phi(\psi, t) = \int B_\phi \, dS = \int_0^\psi d\psi F(\psi, t) \oint \frac{dl_p}{R^2 B_p} = 2\pi \int_0^\psi q(\psi, t) d\psi \qquad (6.122)$$

Therefore, the condition to conserve the toroidal flux within any given poloidal flux surface requires that

$$\frac{\partial \Phi(\psi, t)}{\partial t} = 2\pi \int_0^\psi \frac{\partial q(\psi, t)}{\partial t} \, d\psi = 0 \qquad (6.123)$$

In other words, $F(\psi, t)$ must be chosen such that the safety factor $q(\psi, t)$ is identical in each quasistatic state. Since this applies in particular to the initial state, the actual functional dependence of $q(\psi, t)$ must correspond to the initial ohmically heated plasma,

$$q(\psi, t) = q_\Omega(\psi) \qquad (6.124)$$

The desired expression for flux conservation relating $F(\psi, t)$ to $q(\psi, t)$ has already been obtained in Eq. (6.43), and is repeated here for convenience

$$F(\psi, t) = q_\Omega(\psi) \left[\frac{1}{2\pi} \oint \frac{dl}{R^2 B_p} \right]^{-1} \qquad (6.125)$$

Note that in this equation the term in the square bracket is a slowly varying function of time since $B_p = B_p(\psi, l, t)$.

Implementing the flux conserving constraint

The Grad–Shafranov equation that incorporates the flux conserving p and F (i.e., Eqs. (6.121) and (6.125)) is equivalent to the form derived in Eq. (6.44), also repeated here for convenience,

$$\Delta^* \psi = -\mu_0 R^2 K \frac{dp_\Omega}{d\psi} - \frac{1}{2} \frac{d}{d\psi} \left[q_\Omega^2(\psi) \left(\frac{1}{2\pi} \oint \frac{dl_p}{R^2 B_p} \right)^{-2} \right] \qquad (6.126)$$

This complicated equation must in general be solved numerically at each instant of time during the heating evolution. One can avoid this difficulty and obtain some insight by means of a simple idea as follows. As a reasonable approximation to flux conservation assume that as the plasma evolves only q_0 and q_a are exactly conserved. Since the $q(\psi)$ profile for most tokamaks is a monotonically increasing function, it should be sufficient to ascertain the main features of the evolution by conserving only the end points, even if the interior profiles are slightly modified.

One can attempt to implement this idea using the simple circular high β Solov'ev equilibrium discussed in Section 6.5.2. There is, however, a problem. The Solov'ev equilibrium is characterized by only two free parameters, A and C. Qualitatively, one parameter can be used to model the plasma heating, proportional to β_t, while the other can be used to model some form of the safety factor. The problem is that there is not enough freedom to fix two values of the safety factor and still have a free parameter available to allow the plasma β to increase as the heating increases. This would require three free parameters in the Solov'ev solution. The problem is overcome by making the gross approximation that flux conservation requires that only q_a, and not q_0, be conserved. Even with this gross approximation it is shown below that the model no longer exhibits an equilibrium limit; that is, well-behaved solutions exist for arbitrarily large values of β_t.

The flux conserving Solov'ev equilibrium

The flux conserving constraint can be applied to the Solov'ev equilibrium as follows. The key step is to introduce a heating parameter defined by

$$H = \beta_t q_a^2 / \varepsilon \tag{6.127}$$

This is a useful measure of heating for a flux conserving evolution in which ε, B_0, and by definition q_a are held fixed. Under these conditions the heating parameter $H \propto p$ indicating that increases in plasma energy due to external heating correspond to increases in the value of H. For the high β tokamak operating at fixed q_* rather than fixed q_a the equivalent heating parameter would be defined as

$$H = \beta_t q_*^2 / \varepsilon \tag{6.128}$$

with equilibria existing only in the range $0 < H < 1$.

The next step in the analysis is to evaluate the important plasma parameters q_*, I, β_p, ρ_s, and B_V as a function of the flux conserving heating parameter H. These quantities can be easily calculated from the high β tokamak results already derived in Section 6.5.2. To begin a simple relation can be obtained relating q_* to q_a and H by using Eqs. (6.106) and (6.114):

$$\left(\frac{q_a}{q_*}\right)^4 - \left(\frac{q_a}{q_*}\right)^2 - H^2 = 0 \tag{6.129}$$

The solution is given by

$$q_* = q_a \left[\frac{2}{(1 + 4H^2)^{1/2} + 1}\right]^{1/2} \tag{6.130}$$

It immediately follows that the relation between the plasma current I and the initial ohmic current I_Ω (corresponding to $H = 0$ and $q_* = q_a$) is given by

$$\frac{I}{I_\Omega} = \frac{q_a}{q_*} = \left[\frac{(1 + 4H^2)^{1/2} + 1}{2}\right]^{1/2} \tag{6.131}$$

The expression for $\varepsilon\beta_p$ is now obtained from Eq. (6.106),

$$\varepsilon\beta_p = \frac{2H}{(1 + 4H^2)^{1/2} + 1} \tag{6.132}$$

The separatrix X-point location ρ_s is calculated by direct substitution into Eq. (6.118),

$$\frac{r_s}{a} = \rho_s = \left[\frac{(1 + 4H^2)^{1/2} + 1}{2H}\right]\left\{1 + \left[\frac{2}{(1 + 4H^2)^{1/2} + 1}\right]^{1/2}\right\} \tag{6.133}$$

The last quantity of interest is B_V which is obtained from Eq. (6.117) after noting that the flux function for a vacuum vertical field is given by $\hat\psi_V = aR_0B_V\rho \cos\theta$,

$$B_V = \frac{\varepsilon B_0}{2q_a}\left[\frac{2H^2}{(1 + 4H^2)^{1/2} + 1}\right]^{1/2} \tag{6.134}$$

These results are illustrated in Fig. 6.16. Also illustrated are the results for the high β tokamak with fixed q_* and H defined by Eq. (6.128). There are several points to be made. Each of the plasma parameters is finite and well behaved for arbitrarily large β_t (i.e., H) when q_a is held fixed; there is no equilibrium β limit for the flux conserving tokamak. This is in contrast to the high β tokamak for which q_* is held fixed. In this case there is an equilibrium limit given by $H = \beta_t q_*^2/\varepsilon = 1$. The explanation is that in the flux conserving case the current I must increase as β_t increases in order to keep q_a constant. A larger I requires a smaller increase in B_V to hold a given β_t in toroidal force balance. Thus, the ratio of B_V/I is smaller in a flux conserving tokamak than in a high β tokamak with fixed q_*. As one can see intuitively from Fig. 6.16, this behavior slows the motion of the separatrix X-point toward the plasma surface.

Another important aspect of flux conserving operation involves q_*. Since I increases with H, then $q_* \propto 1/I$ decreases. Consequently, a plasma that satisfies the kink stability condition $q_* > q_{*\mathrm{min}}$ in the ohmic regime may violate the condition as β_t is increased. One must start with a sufficient safety margin initially to prevent this from happening as the plasma is heated.

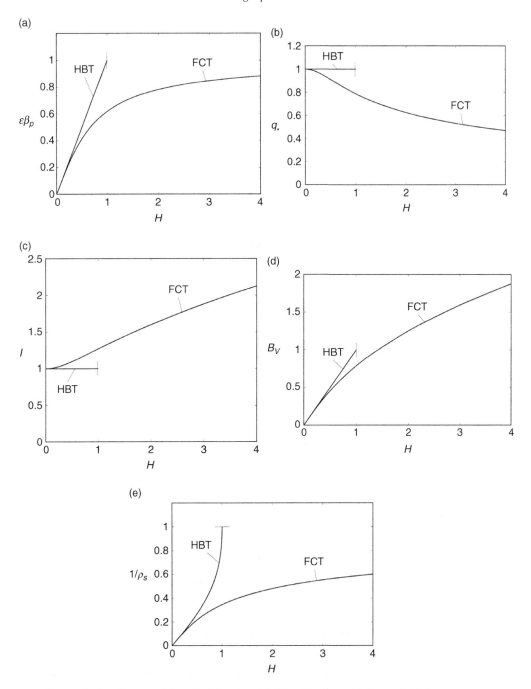

Figure 6.16   Curves of (a) $\varepsilon\beta_p$, (b) q_*/q_a, (c) I/I_Ω, (d) $(2q_a/\varepsilon B_0)\,B_V$, and (e) $1/\rho_s$ vs. the heating parameter H for a flux conserving equilibrium sequence. Also shown are the corresponding high β tokamak curves for $I = $ constant. (HBT = high β tokamak, FCT = flux conserving tokamak.)

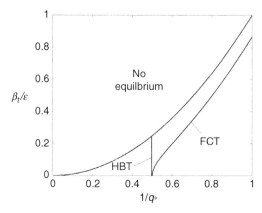

Figure 6.17   Flux conserving trajectory in β_t/ε vs. $1/q_*$ space. Also shown is the trajectory corresponding to the high β tokamak at fixed current.

These results can be conveniently summarized by plotting the trajectory of a flux conserving evolution in a β_t/ε vs. $1/q_*$ space as illustrated in Fig. 6.17 for the case $q_a = 2$. This relation is obtained by eliminating H between Eqs. (127) and (6.130),

$$\frac{\beta_t}{\varepsilon} = \left[\frac{1}{q_*^2} \left(\frac{1}{q_*^2} - \frac{1}{q_a^2} \right) \right]^{1/2} \tag{6.135}$$

Also plotted for comparison is the trajectory for the high β tokamak with fixed q_*: $\beta_t/\varepsilon = 1/q_*^2$. One sees that for a flux conserving equilibrium, as β_t increases the value of q_* decreases in such a way that the equilibrium limit is never crossed. However, for the high β tokamak the trajectory is a vertical line which will always intersect the equilibrium limit at a sufficiently large β_t/ε.

Summary and discussion

A flux conserving tokamak is an externally heated high β tokamak in which the heating rate is slow compared to the MHD time and fast compared to the magnetic diffusion time. In this regime the plasma behaves like a perfect conductor, evolving through a series of quasistatic equilibrium states as the heating increases. The requirements of flux conservation combined with a simple model of plasma heating serve to determine the free functions $p(\psi, t)$ and $F(\psi, t)$ for each quasistatic state. The most interesting feature of the flux conserving tokamak is that the high β tokamak equilibrium limit no longer exists. The reason is directly related to a reduction in the ratio B_V/I implied by the requirements of flux conservation which slows the motion of the separatrix X-point towards the plasma surface.

The relation between the high β tokamak and the flux conserving tokamak is very useful in developing physical intuition about the operation of externally heated tokamaks. In practice, however, flux conservation does not play a major role in the experimental operation of tokamaks. The reason is a negative one – well before the plasma can reach its potential equilibrium limit, its achievable β is limited by MHD instabilities. In fact, many tokamaks are programmed to operate at fixed current as heating occurs so as not to decrease q_*, which could excite current-driven instabilities. There is little concern about reaching the equilibrium limit because pressure-driven MHD instabilities will occur first.

6.5.4 The elliptic high β tokamak

Statement of the problem

While the achievable MHD stable values of β_t in an externally heated high β tokamak are much larger than for an ohmic tokamak, they are marginal for reactor applications if the plasma has a circular cross section. An important way to further increase the achievable β is by shaping the cross section, in particular, making it vertically elongated. This extra degree of optimization freedom is now investigated for the elliptically shaped high β tokamak, whose simple cross section allows for an analytic solution when used in conjunction with the Solov'ev profiles.

The strategy is as follows. As a simple measure of the potential gains in β, the equilibrium limit for an elliptical plasma is calculated as a function of elongation, and then compared with that of a circular plasma. Although recognizing that the actual β limit is set by stability and not equilibrium, one can still invoke the intuition (not yet proven) that a higher equilibrium β limit will lead to a higher stability β limit. It is also of interest to calculate and compare the total current I and average current density $\bar{J}_\phi$. The current is related to the energy confinement time and the current density to the ohmic heating capabilities.

In carrying out the cross sectional optimization and making the corresponding comparisons it is essential to identify the constraints that must be applied. In particular, which quantities should be held fixed as the plasma elongation is varied? Four constraints are applied in the analysis. First, the inverse aspect ratio ε is held fixed. Since $\beta_t \propto \varepsilon$, a fair comparison between different devices suggests that ε be held fixed as elongation is varied. Second, for the same reason the toroidal field B_0 is held fixed. Third, as an attempt to keep MHD stability constant, the value of q_* is fixed as the elongation changes. Numerical simulations of MHD stability indicate that the marginal stability value of q_* as defined by Eq. (6.29) is indeed almost independent of elongation. Fourth, when calculating I and $\bar{J}_\phi \equiv I/\text{area}$, which are un-normalized quantities, an additional constraint is required. The choice is to fix the device volume $V = 2\pi^2 R_0 a^2 \kappa$, which is a rudimentary attempt to keep the cost constant. Here, κ (without a subscript) is the surface elongation.

To summarize, the goal of the analysis is to derive the condition corresponding to the equilibrium limit and then evaluate β_t, I, and $\bar{J}_\phi$ as functions of κ subject to the constraints of fixed ε, q_*, B_0, and V.

Mathematical solution

The task now is to solve the high β tokamak Grad–Shafranov equation with Solov'ev profiles for an elliptically shaped plasma. This is a straightforward task if one introduces rectangular coordinates $x = r \cos\theta$, $y = r \sin\theta$ into Eq. (6.99) leading to the following formulation of the equilibrium problem:

$$\frac{\partial^2 \psi_0}{\partial x^2} + \frac{\partial^2 \psi_0}{\partial y^2} = A + Cx$$

$$\psi_0(S) = 0$$

$$\psi_0(x, y) = \text{regular in the plasma}$$

(6.136)

Here, the surface S corresponds to the ellipse

$$\frac{x^2}{a^2} + \frac{y^2}{\kappa^2 a^2} = 1 \tag{6.137}$$

Because of the simple Solov'ev profiles the solution can be easily found:

$$\psi_0 = K_1 \left(\frac{x^2}{a^2} + \frac{y^2}{\kappa^2 a^2} - 1 \right)(1 + K_2 x)$$

$$K_1 = \frac{a^2}{2} \left(\frac{\kappa^2}{1 + \kappa^2} \right) A$$

$$K_2 = \left(\frac{1 + \kappa^2}{1 + 3\kappa^2} \right) \frac{C}{A}$$

(6.138)

The next step is to rewrite the new constants K_1, K_2 in terms of more physical quantities and then evaluate the plasma parameters and figures of merit.

Plasma parameters and figures of merit

The constants K_1, K_2 can be readily evaluated by introducing yet another set of coordinates ρ, α defined by $x = \alpha\rho \cos\alpha$, $y = \kappa\alpha\rho \sin\alpha$. Note that the plasma surface corresponds to $\rho = 1$. In terms of these coordinates the flux function plus the surface values of the differential poloidal arc length and poloidal magnetic field are given by

$$\psi_0 = -K_1(1 - \rho^2)(1 + K_2 a\rho \cos \alpha)$$

$$dl = a(\kappa^2 \cos^2\alpha + \sin^2\alpha)^{1/2} da \tag{6.139}$$

$$B_p = (2K_1/R_0 a\kappa)(\kappa^2 \cos^2\alpha + \sin^2\alpha)^{1/2}(1 + K_2 a \cos \alpha)$$

Using these relations one can show after a short calculation that K_1, K_2 are related to q_*, β_t (defined in Eqs. (6.29) and (6.24)) by the following relations:

$$K_1 = \frac{B_0 a^2 \kappa}{2q_*}$$

$$K_2 = \frac{4\beta_t q_*^2}{\varepsilon a(1 + 3\kappa^2)} \tag{6.140}$$

Sufficient information is now available to calculate the equilibrium β limit. The shortest path to this goal is to calculate the safety factor on the surface (i.e., q_a) and determine the condition under which $q_a \to \infty$. This is the condition for the separatrix to move onto the plasma surface. The value of q_a is found as follows:

$$q_a = \frac{F(\psi)}{2\pi} \oint \frac{dl}{R^2 B_p}$$

$$\approx \frac{B_0}{2\pi R_0} \oint \frac{dl}{B_p} \tag{6.141}$$

$$= \frac{q_*}{(1 - v^2)^{1/2}}$$

where

$$v = K_2 a = \frac{\beta_t q_*^2}{\varepsilon} \frac{4}{1 + 3\kappa^2} \tag{6.142}$$

Observe that the equilibrium β_t limit again occurs when $v \to 1$ and has the value

$$\beta_t = \left(\frac{\varepsilon}{q_*^2}\right) \frac{1 + 3\kappa^2}{4} \tag{6.143}$$

Here and below the quantity in the parenthesis corresponds to the circular value. Equation (6.143) implies that the β_t limit increases rapidly with elongation. For example, when $\kappa = 2$ the gain in β_t from a circle to an ellipse is about a factor of 3.3.

The other parameters and figures of merit can be calculated in a similar manner as for the circular case. At the equilibrium limit their values are

$$\beta_p = \left(\frac{1}{\varepsilon}\right)\frac{1+3\kappa^2}{2(1+\kappa^2)}$$

$$\frac{\Delta_0}{a} = \left(\frac{1}{3}\right)$$

$$\kappa_0 = \left(\frac{3}{2}\right)^{1/2}\kappa$$ (6.144)

$$q_0 = \left(\frac{3}{8}q_*^2\right)^{1/2}$$

Equation (6.144) implies that as κ increases, β_p increases and then saturates, Δ_0/a and q_0 remain unchanged, and κ_0 scales linearly with κ.

Next, it is of interest to examine the behavior of I and $\bar{J}_\phi$ for arbitrary v as κ increases subject to the constraints described above. Consider first the current, which is determined from

$$\mu_0 I = \oint B_p\, dl \tag{6.145}$$

A straightforward calculation yields

$$\mu_0 I = \left(2\pi\varepsilon\bar{a}\,\frac{B_0}{q_*}\right)\frac{1+\kappa^2}{2\kappa^{1/3}}$$

$$\bar{a} = \left(\frac{\varepsilon V}{2\pi^2}\right)^{1/3} \tag{6.146}$$

where $\bar{a} = \bar{a}(\varepsilon, V)$ is the average plasma radius, which is held fixed as the elongation is varied. Note that $\bar{a} = a$ for a circular plasma. There is an important conclusion to be drawn from this relation based on experimental observations which show that the energy confinement τ_E is approximately linearly proportional to I. Equation (6.146) implies that when comparing different machines with the same aspect ratio and plasma volume, increasing the elongation increases the energy confinement time. Elongation is good for energy confinement as well as MHD.

The second quantity of interest is the average current density defined by

$$\bar{J}_\phi = \frac{I}{\pi a^2 \kappa} \tag{6.147}$$

The radius a is eliminated by means of the aspect ratio and volume constraints: $a = \bar{a}/\kappa^{1/3}$. The expression for the current density can thus be written as

$$\mu_0 \bar{J}_\phi = \left(\frac{2\varepsilon B_0}{\bar{a} q_*} \right) \frac{1 + \kappa^2}{2\kappa^{2/3}} \tag{6.148}$$

The implication here is that the average current density also increases with elongation for a fixed aspect ratio and fixed volume plasma. A higher average current density is favorable for the ohmic heating phase of a tokamak. Once again elongation is good.

Numerical results

Since several important plasma properties – MHD, energy confinement, and ohmic heating – have been shown to improve with increasing κ, one might expect tokamak experiments to be designed with very elongated cross sections. This is true to a degree but there is a serious problem that arises which prevents this strategy from being implemented to the fullest desirable extent. Specifically, as a tokamak is elongated it becomes MHD unstable to axisymmetric instabilities which are only weakly affected by the value of q_*. These modes essentially correspond to a rigid vertical motion of the plasma causing it to strike the top or bottom of the vacuum chamber. When this occurs it is very bad for both the plasma and the chamber. To avoid these instabilities most tokamaks make use of an active feedback system. However, the technological constraints of the feedback circuits (e.g., amplitude and response time) become more severe as the plasma becomes more elongated. In practice, elongations are usually restricted to the range $\kappa \lesssim 2$ in order for the feedback to prevent these modes from occurring. Vertical instabilities are discussed in more detail in Chapter 12.

In terms of numerical studies, many investigations have shown that the overall stability of non-circular tokamaks can be further improved by adding an outward pointing triangularity to the elongated cross section. This is particularly useful in suppressing pressure gradient driven modes and is also discussed in Chapter 12. All modern tokamaks are designed to have this elongated, outward pointing "D" shaped cross section. One example is the DIII-D experiment located at General Atomics (GA) in San Diego. The GA group has produced long-lived plasmas with values of $\beta_t \sim 8\%$, indeed an impressive achievement. A numerically computed equilibrium of a high β_t DIII-D equilibrium is illustrated in Fig. 6.18. Observe that the plasma has an elongation of $\kappa = 1.8$ and that the magnetic axis has shifted out considerably, corresponding to $\Delta_0/a \approx 0.3$.

Summary

A non-circular cross section represents an additional degree of freedom in the design of tokamaks, allowing a further optimization in performance compared to the circular cross section. A comparison of tokamaks with fixed ε, B_0, q_*, and V shows that as κ increases the maximum attainable β_t, I, and $\bar{J}_\phi$ also increase. This

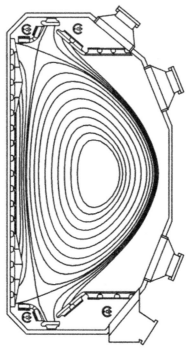

Figure 6.18   Numerically computed equilibrium for the DIII-D tokamak at General Atomics. Shown are flux surface plots corresponding to auxiliary heated high β tokamak operation. From DIII-D Team, 1998. Reproduced with permission from Elsevier.

conclusion has been accepted by the international fusion community and all modern tokamaks are designed to have elongated cross sections (with outward pointing triangularity). There is a practical limit on the maximum achievable elongation set by the onset of fast growing vertical MHD instabilities.

6.6 Exact solutions to the Grad–Shafranov equation (standard and spherical tokamaks)

The use of asymptotic expansions in ε has proven useful for developing intuition about radial pressure balance and toroidal force balance in tokamaks. However, since $\varepsilon \sim 1/3$ for a standard tokamak, this is not that small a value that one might expect the results to be quantitatively accurate for experimental applications. The situation is even more difficult for the ultra-tight aspect ratio spherical tokamak which has $\varepsilon \sim 3/4$. The implication is that for high-accuracy MHD applications to experiment, one needs "exact" solutions to the Grad–Shafranov equation.

In general exact solutions are obtained by numerically solving the Grad–Shafranov partial differential equation. Many such codes are in existence today.

Although the numerical solutions are essential for detailed experimental comparisons, they are not as convenient for developing simple scaling laws describing the effects of aspect ratio, elongation, and triangularity on tokamak equilibria. Stated differently, it would be desirable if in addition to the numerical codes there existed exact analytical solutions to the Grad–Shafranov equation.

Obtaining such exact analytic solutions is the main goal of this Section. It is shown how the exact problem for an up–down symmetric plasma with a smooth surface can be formulated and solved for the Solov'ev profiles. The results are first applied to a large circular tokamak (TFTR at Princeton Plasma Physics Laboratory, USA) and a large non-circular tokamak (JET in Abingdon, UK). A second application is the ultra-tight aspect ratio spherical tokamak (ST). A brief description of the ST, including its scientific motivation, is also presented. The analytic equilibrium solutions are then applied to an ST experiment (MAST at the Culham Laboratory, UK). A further related question addressed by means of the exact analytic solutions is the scaling of the equilibrium β limit with inverse aspect ratio.

An up–down symmetric configuration with a smooth plasma surface is the simplest case to analyze, although often not the most relevant experimentally because of the presence of a divertor. A brief discussion is presented describing the operation of a divertor which essentially consists of a set of coils and collector plates placed in close proximity to the plasma surface. It is the most common way that a plasma makes first contact with a material surface. The divertor has the dual goals of vacuum wall protection and impurity control. Modern tokamaks utilize divertors rather than limiters for these purposes. From the MHD point of view divertors are typically up–down asymmetric configurations with an X-point separatrix in either the upper or lower portion of the plasma. It is shown how the analytic solutions can be modified to treat the divertor geometry. The solutions are then applied to a large standard tokamak (the proposed ITER experiment being built in Cadarache, France) and a spherical tokamak (the NSTX experiment at Princeton Plasma Physics Laboratory, USA).

6.6.1 Mathematical formulation

Equations

The goal is to solve the exact Grad–Shafranov equation with the free functions again chosen as Solov'ev profiles: $F(dF/d\psi) = -A$ and $\mu_0(dp/d\psi) = -C$. With these choices the exact Grad–Shafranov equation becomes

$$R\frac{\partial}{\partial R}\left(\frac{1}{R}\frac{\partial\psi}{\partial R}\right) + \frac{\partial^2\psi}{\partial Z^2} = A + CR^2 \tag{6.149}$$

At this point it is useful to normalize the flux function and the coordinates as follows: $R = R_0 X$, $Z = R_0 Y$, $\psi = \psi_0 U$ where $\psi_0 = R_0^2(A + CR_0^2)$. Also the signs are chosen so that $\psi_0 > 0$ and $U(X, Y) < 0$. The Grad–Shafranov equation reduces to

$$X \frac{\partial}{\partial X}\left(\frac{1}{X}\frac{\partial U}{\partial X}\right) + \frac{\partial^2 U}{\partial Y^2} = \alpha + (1 - \alpha)X^2 \qquad (6.150)$$

Here, $\alpha = A/(A + CR_0^2)$. Effectively α and ψ_0 have replaced A and C as the free parameters. The advantage of this normalization is that the differential equation itself is now a function of only a single parameter, α.

In terms of the formulation, there is a crucial difference in philosophy in the way that the exact equation is solved as compared to the previous asymptotic equations. For the asymptotic equations a standard approach is used. A surface is specified, for instance a circle or ellipse, and the equations are solved subject to the boundary conditions of regularity and $\psi = 0$ on the surface.

This approach does not readily work for the exact problem because simple analytic solutions cannot be found that exactly satisfy the boundary conditions on simple surfaces such as a circle or ellipse. Instead, the approach used is to find an exact analytic solution to Eq. (6.149) consisting of the superposition of a finite number of terms each with an undetermined amplitude. A series of boundary constraints is then applied, forcing the analytic solutions to a match a finite number of specified conditions on a known desired surface – one boundary constraint for each unknown amplitude. The resulting solution is now uniquely defined. One then simply plots the contours of $U =$ constant including the plasma surface $U = 0$. Obviously, matching properties at a finite number of points does not guarantee that the resulting $U = 0$ surface will be close to the desired matching surface at all points. One must just "take whatever the solution produces."

The mathematical challenges to make this approach work involve choosing (1) the right set of analytic basis functions, (2) the right number of terms in the finite sum, and (3) the right set of matching constraints, so that the resulting $U = 0$ surface closely matches the desired surface for a very wide range of plasma parameters. The procedure described in this Section does just that and follows the analysis presented in Cerfon and Freidberg (2010).

Mathematical solutions

The solution to Eq. (6.149) consists of particular and homogeneous contributions. A convenient way to write the particular solution is as follows:

$$U_P(X, Y) = \frac{\alpha}{2}X^2 \ln X + \frac{1 - \alpha}{8} X^4 \qquad (6.151)$$

The homogeneous solution satisfies

$$X\frac{\partial}{\partial X}\left(\frac{1}{X}\frac{\partial U_H}{\partial X}\right)+\frac{\partial^2 U_H}{\partial Y^2}=0 \tag{6.152}$$

Simple basis functions that exactly satisfy Eq. (6.149) consist of combinations of polynomials in X and Y. The exact form of each polynomial solution can be easily found by direct substitution.

The approach used here assumes that the overall solution can be written as a Taylor series in these polynomial solutions, starting from a constant and increasing up to and including sixth-order terms. Truncating the series at sixth-order polynomials is not an obvious choice and indeed was arrived at by trial and error. In any event, the exact analytic solution to the Grad–Shafranov equation used to model tokamaks is given by

$$U(X,Y)=\frac{\alpha}{2}X^2\ln X+\frac{1-\alpha}{8}X^4+\sum_0^6 c_j U_j$$

$$U_0=1$$
$$U_1=X^2$$
$$U_2=Y^2-X^2\ln X$$
$$U_3=X^4-4X^2Y^2$$
$$U_4=2Y^4-9Y^2X^2-(12Y^2X^2-3X^4)\ln X$$
$$U_5=X^6-12X^4Y^2+8X^2Y^4$$
$$U_6=8Y^6-140Y^4X^2+75Y^2X^4-(120Y^4X^2-180Y^2X^4+15X^6)\ln X \tag{6.153}$$

Observe that all the polynomials are even in Y indicating that at this point attention is focused on up–down symmetric configurations.

Boundary constraints

The task now is to define seven boundary constraints that will serve to determine the seven unknown coefficients c_j. These constraints are chosen to match seven properties on a known desired plasma surface. A good choice for this reference surface, which is often used in the fusion community, is given parametrically in terms of τ as follows:

$$X=1+\varepsilon\,\cos\left(\tau+\delta_0\,\sin\,\tau\right)$$
$$Y=\varepsilon\kappa\,\sin\,\tau \tag{6.154}$$

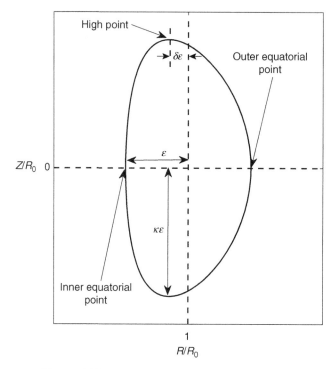

Figure 6.19   Geometry of the reference surface.

The surface and corresponding geometry, are illustrated in Fig. 6.19. Observe that there are three dimensionless parameters that define the geometry: the inverse aspect ratio $\varepsilon = a/R_0$, the elongation κ, and the triangularity $\delta = \sin \delta_0$.

The boundary constraints, again found by some trial and error, require matching the analytic flux function and its first and second derivatives at three separate points on the reference surface: the outer equatorial point, the inner equatorial point, and the high point maximum. This might appear to be nine constraints but two are automatically satisfied – the first derivative conditions at the outer and inner equatorial points by virtue of up–down symmetry. The seven boundary constraints can thus be written as

1. $U(1 + \varepsilon, 0) = 0$ outer point flux
2. $U_{YY}(1 + \varepsilon, 0) = -N_1 U_X(1 + \varepsilon, 0)$ outer point curvature
3. $U(1 - \varepsilon, 0) = 0$ inner point flux
4. $U_{YY}(1 - \varepsilon, 0) = -N_2 U_X(1 - \varepsilon, 0)$ inner point curvature   (6.155)
5. $U(1 - \delta\varepsilon, \kappa\varepsilon) = 0$ high point flux
6. $U_X(1 - \delta\varepsilon, \kappa\varepsilon) = 0$ high point slope
7. $U_{XX}(1 - \delta\varepsilon, \kappa\varepsilon) = -N_3 U_Y(1 - \delta\varepsilon, \kappa\varepsilon)$ high point curvature

Here, the curvature coefficients N_j can be easily calculated from the model surface

$$N_1 = \left[\frac{d^2X}{dY^2}\right]_{\tau=0} = \frac{(1+\delta_0)^2}{\varepsilon\kappa^2}$$

$$N_2 = \left[\frac{d^2X}{dY^2}\right]_{\tau=\pi} = -\frac{(1-\delta_0)^2}{\varepsilon\kappa^2} \tag{6.156}$$

$$N_3 = \left[\frac{d^2Y}{dX^2}\right]_{\tau=\pi/2} = -\frac{\kappa}{\varepsilon\,\cos^2\delta_0}$$

Once the free Solov'ev constants A and C, or equivalently α and ψ_0, are specified, the constraint conditions given by Eq. (6.155) translate into a set of seven linear inhomogeneous algebraic equations for the seven unknown coefficients c_j, a trivial numerical problem.

In analogy with the high β tokamak expansion, the last step in the formulation is to derive relationships between α and ψ_0 and the more physical parameters q_*, β_p, and β_t. From their definitions it can easily be shown that

$$\frac{1}{q_*} = \frac{1}{2\pi}\left(\frac{2}{1+\kappa^2}\right)\left(\frac{\psi_0}{a^2 B_0}\right)K_1$$

$$\beta_p = 8\pi^2\varepsilon^2(1-\alpha)\left(\frac{1+\kappa^2}{2}\right)\frac{K_2}{K_1^2 K_3} \tag{6.157}$$

$$\beta_t = \left[\frac{8\pi^2\varepsilon^4(1-\alpha)}{q_*^2}\right]\left(\frac{1+\kappa^2}{2}\right)^2\frac{K_2}{K_1^2 K_3}$$

where

$$K_1 = \int dXdY\left[\frac{\alpha + (1-\alpha)X^2}{X}\right]$$

$$K_2 = \int XdX\,dY(-U) \tag{6.158}$$

$$K_3 = \int XdXdY$$

and the integrals are carried out over the plasma cross section.

To summarize, exact Grad–Shafranov equilibria are found by specifying the geometric parameters ε, κ, and δ and the plasma parameters q_* and α. (Practically it is more convenient to specify α rather than β_t.) The equation for $U(X,Y)$ is now fully specified and the expansion coefficients c_j are easily found by applying the boundary constraints and solving the resulting set of linear algebraic equations.

Table 6.1 *MHD parameters for TFTR and JET.*

Parameters	Symbol	Units	TFTR	JET
Major radius	R_0	m	2.5	3
Minor radius	a	m	0.87	1
Aspect ratio	R_0/a	-	2.9	3
Elongation	κ	-	1	1.7
Triangularity	δ	-	0	0.25
Toroidal magnetic field	$B_0 = B(R_0)$	T	5.6	3.6
Plasma current	I	MA	2.7	4
Kink safety factor	q_*	-	3.1	2.7
Average temperature	$(\overline{T}_e + \overline{T}_i)/2 \equiv \overline{T}_k$	keV	23	10
Average density	$\overline{n}_{20}$	$10^{20}\,\mathrm{m}^{-3}$	1.0	0.5
Toroidal beta	β_t	-	2.9%	3%

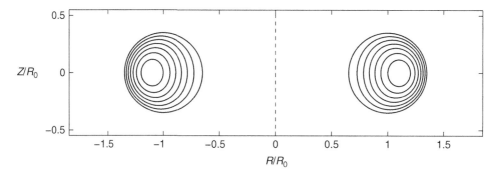

Figure 6.20   Exact Solov'ev equilibrium for TFTR.

Once $U(X, Y)$ is known then $\psi_0/a^2 B_0$, β_p, and β_t are easily numerically evaluated from Eq. (6.157).

6.6.2 Examples: TFTR and JET

The simplest initial test of the analytic solution procedure is the previously discussed circular cross section TFTR experiment at Princeton Plasma Physics Laboratory, whose MHD parameters are given in Table 6.1.

The analytic solution procedure is applied and the resulting flux surfaces are illustrated in Fig. 6.20. One sees that the procedure reliably reproduces shifted circular flux surface equilibria.

Consider now MHD equilibria of the Joint European Torus (JET), a large tokamak built as a collaborative European project. It is located near Oxford, UK, adjacent to the Culham Laboratory site. JET has an elongated, outward pointing "D" cross section. It is a "standard" tokamak in the sense of having a typical aspect ratio of about $R_0/a \approx 3$. Compared to other tokamaks it is unique in that it has an

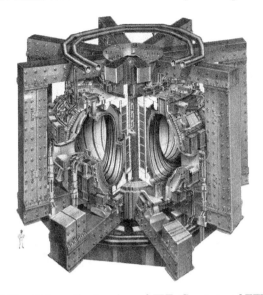

Figure 6.21   Schematic drawing of JET. Courtesy of EFDA-JET.

iron (rather than air) core transformer to produce the ohmic heating current. The iron core improves the efficiency of the primary of the ohmic transformer, but the experiment must live with the maximum field limitations imposed by saturation of the iron. JET, along with TFTR, are the only two tokamaks at this point in time that have operated with tritium. One of the most impressive experimental campaigns was carried out in the late 1990s. Here, 16 MW of fusion power was produced for a short period of time in a 50–50 D-T plasma. About 25 MW of external heating power was required to accomplish this goal.

A drawing of JET is shown in Fig. 6.21. It is indeed a large facility. The experimental MHD parameters characterizing high-performance D-T plasmas are also given in Table 6.1. Although JET normally operates with a divertor, the analytic procedure assumes that the surface has been smoothed out. JET, in fact, is a relatively easy test for the equilibrium procedure. Using the values given above, the equilibrium flux surfaces for the Solov'ev profiles have been calculated and are illustrated in Fig. 6.22. Observe the elongated "D" shaped plasma and the finite shift of the magnetic axis. The flux surfaces are smooth and nested showing that the analytic solutions provide a credible representation of high-performance JET equilibria.

6.6.3 Example: the spherical tokamak (ST)

Description of the spherical tokamak

A more challenging test for the analytic equilibrium procedure is the spherical tokamak (Peng *et al.*, 1985). As stated, a "spherical tokamak" is a tokamak with a

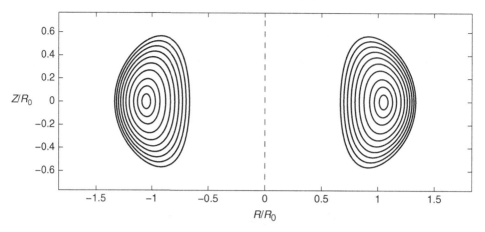

Figure 6.22   Exact Solov'ev equilibrium for JET.

very tight aspect ratio, typically $R_0/a \sim 1.2$–1.4. It is this tight aspect ratio that is challenging to model with the analytic solutions.

Before proceeding with the MHD equilibrium analysis it is useful to briefly review the basic physical properties of an ST, in particular, the motivation behind the concept. One of the primary motivations behind the ST is based on the fact that the maximum achievable β in a tokamak from both MHD equilibrium and stability limits scales as $\beta \sim \varepsilon$. Clearly as the aspect ratio becomes tighter, then the maximum β increases. Higher values of β could potentially add to the attractiveness of a fusion reactor either by allowing the use of lower magnetic fields or alternatively by resulting in a more compact design. Both options would lead to a lower cost.

However, the tight aspect ratio strategy has both pros and cons with respect to reactor desirability. In fact, the discussion below suggests that when engineering considerations are taken into account the ST is marginal at best in terms of leading to improved reactor designs as compared to a standard tokamak. Nevertheless, there is a different application for which the ST may be better suited – a volume neutron source for materials testing.

The basis for these statements is as follows. The gains in β and compactness must be balanced against several problems that are more serious than for the standard tokamak. First, achieving high power density in a fusion reactor requires high pressure which is related to but not the same as high β. Specifically, engineering constraints limit the maximum allowable magnetic field on the inner leg of the toroidal field magnet to a value $B_\phi(R_0 - a, 0) = B_{max}$. Thus, tight aspect ratio leads to a much larger reduction in B_0 at the plasma center where β is defined, because of the strong $1/R$ dependence as $\varepsilon \to 1$: $B_0/B_{max} = 1 - \varepsilon$. The implication is that $\langle p \rangle \propto \beta_t B_{max}^2 (1 - \varepsilon)^2$. One sees that even if the maximum beta scales as $\beta_t \sim \varepsilon$ the gains in pressure ultimately diminish for a very tight aspect ratio.

Second, to achieve a very tight aspect ratio the blanket and almost all the shield must be removed from the inboard side of the plasma. With only a small shield it is no longer possible to use superconducting magnets for the toroidal field coils. They must be made of copper. One consequence is that a copper central leg dissipates a substantial amount of ohmic power. Detailed reactor studies show that favorable power balance in a reactor requires that B_{max} in an ST be less than about 7.5 T. This should be compared to $B_{max} \approx 13$ T for superconducting magnets which are limited by stress and not ohmic losses. A lower B_{max} again leads to a lower plasma pressure.

The overall conclusion from this discussion, as well as from detailed systems studies, is that the ST does not lead to the large gains in reactor attractiveness over the standard tokamak that might have been originally anticipated. However, when viewed as a volume neutron source, the ST may be more desirable. A source of 14 MeV fusion neutrons is essential for developing and testing the advanced materials that are required in a reactor. No such source currently exists in the world fusion program. A key feature of such a source is that its main goal is the production of a high-intensity 14 MeV neutron flux in a relatively small volume compared to a power reactor. The goal is not economical fusion electricity. Therefore, even if the cost per ST neutron is somewhat higher than for a standard tokamak, the compact ST volume may lead to a lower absolute cost because the total number of neutrons produced is smaller, although still more than adequate for a volume neutron source. This would be highly desirable.

Example: the Mega Amp Spherical Tokamak (MAST)

Having motivated the ST concept, one can now focus on calculating a corresponding MHD equilibrium from the analytic solutions. Towards this goal note that several ST experiments are currently in operation in the world's fusion program with the two largest devices being the Mega Amp Spherical Tokamak (MAST) at the Culham Laboratory in the UK and the National Spherical Torus Experiment (NSTX) at the Princeton Plasma Physics Laboratory in the USA.

The MAST experiment serves as a good model to test the equilibrium procedure. A drawing of MAST is shown in Fig. 6.23. Observe the tight aspect ratio. MAST also has a large plasma–wall separation and poloidal field coils located inside the vacuum chamber. This latter feature allows more flexibility and easier experimental control of the plasma shape. On the other hand, with the large plasma–wall separation, smaller image currents flow as the plasma moves, thereby reducing the effects of external flux conservation on MHD equilibrium and stability.

Table 6.2 *MHD parameters for MAST.*

Parameters	Symbol	Units	MAST
Major radius	R_0	m	0.85
Minor radius	a	m	0.65
Aspect ratio	R_0/a	-	1.3
Elongation	κ	-	2.45
Triangularity	δ	-	0.5
Toroidal magnetic field	$B_0 = B(R_0)$	T	0.52
Plasma current	I	MA	1.35
Kink safety factor	q_*	-	3.35
Average temperature	$(\overline{T}_e + \overline{T}_i)/2 \equiv \overline{T}_k$	keV	0.94
Average electron density	$\overline{n}_{20}$	$10^{20}\,\mathrm{m}^{-3}$	0.5
Toroidal beta	β_t	-	0.14

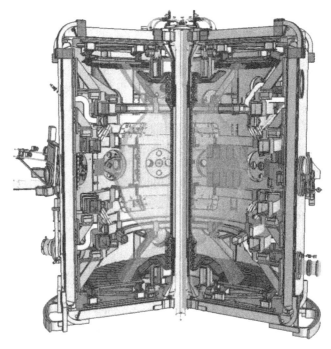

Figure 6.23   Schematic drawing of MAST. Courtesy of William Morris.

High-performance discharges on MAST are characterized by the MHD parameters listed in Table 6.2. The higher values of β_t anticipated for a spherical tokamak have indeed been achieved experimentally.

With all the input parameters specified it is now a straightforward matter to calculate an analytic MAST equilibrium. The resulting flux surface plots are

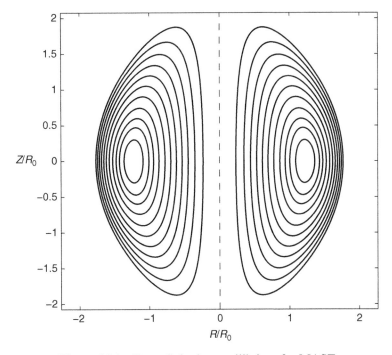

Figure 6.24   Exact Solov'ev equilibrium for MAST

illustrated in Fig. 6.24. Again, the procedure has no difficulty producing a set of realistic, nested flux surfaces, even with a very tight aspect ratio.

6.6.4 The equilibrium β limit

The JET and MAST examples demonstrate that it is possible to calculate exact analytic equilibria using realistic experimental parameters. It is also of theoretical interest to examine the equilibrium β limit using the exact solutions. The issue is to determine whether the basic scaling $\beta_t \propto \varepsilon$ extends into the regime of tight aspect ratio or is only valid for small ε where the asymptotic theories are accurate.

The question can be addressed in a straightforward manner as follows. Keep in mind that the free inputs to the exact solutions are the shape factors ε, κ, and δ and the plasma parameters q_* and α (with α simpler to specify than β_t). Now, the equilibrium β limit corresponds to the situation where the separatrix X-point moves onto the inside of the plasma surface. Mathematically, this is equivalent to setting $B_Z(R_0 - a, 0) = 0$, a condition that puts a constraint on the value of α; that is, α is no longer a free input parameter. In terms of the normalized equations the X-point constraint can be written as

$$U_X(1 - \varepsilon, 0) = 0 \qquad (6.159)$$

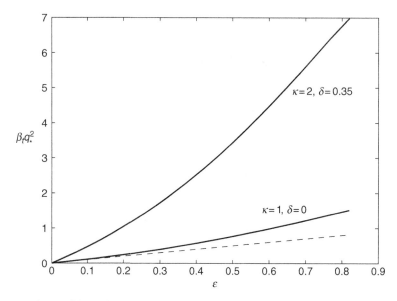

Figure 6.25   Exact equilibrium limit using the Solov'ev profiles.

With this extra condition the analytic problem has eight, not seven, unknown coefficients: the seven c_j plus α.

An examination of Eqs. (6.150) and (6.153) shows that the new constraint equation is also a linear algebraic equation with α being a new unknown. Thus instead of solving seven linear algebraic equations one now has to solve eight linear algebraic equations, a negligible issue computationally. This procedure has been carried out and the results are illustrated in Fig. 6.25. Plotted here are two curves of $\beta_t q_*^2$ vs. ε at the equilibrium limit. The first curve fixes $\kappa = 1$ and $\delta = 0$, approximately corresponding to circular surfaces. Also shown as a dashed line is the previously derived asymptotic result $\beta_t q_*^2 = \varepsilon$. The second curve fixes $\kappa = 2$ and $\delta = 0.35$, which is more typical of present day standard and spherical tokamaks.

There are two conclusions to note. First, a tokamak with an elongated cross section has a higher equilibrium β_t limit than a circular tokamak assuming the same aspect ratio and the same critical value of q_* for stability. This is in agreement with the simple analytic large aspect ratio limit for the Solov'ev profiles previously discussed.

Second, the actual scaling of critical β_t with ε is more optimistic than the simple linear relation predicted by the large aspect ratio analysis. In other words, tight aspect ratio may be even more desirable than originally thought in terms of increasing β_t. However, keep in mind that this conclusion is based on the simple definition of β_t, which is normalized to the central toroidal field. In a tight aspect ratio tokamak one should realistically include the poloidal magnetic energy as well as the toroidal magnetic energy in the evaluation of β_t, which would reduce the improvements over the linear scaling of β_t with ε.

6.6.5 Up–down asymmetric solutions

The discussion related to exact analytic solutions has thus far been focused on up–down symmetric configurations with smooth plasma surfaces. However, as stated, most modern tokamaks operate with an up–down asymmetric geometry because of the presence of a single null divertor. This section shows how the analytical procedure can be extended to include the two new effects associated with a divertor: (1) up–down asymmetry and (2) the presence of an X-point separatrix.

The discussion begins with a brief description of the operation of a divertor. Next, it is shown how the analytic solutions must be modified to allow for up–down asymmetry. Lastly, it is shown how the boundary constraints must be modified to permit an X-point. The sections that follow demonstrate the analytic procedure by means of two examples. A standard tokamak (ITER) and a spherical tokamak (NSTX).

Operation of a divertor

In a qualitative sense a divertor and a limiter protect the vacuum chamber in the same way. Particles which diffuse across the edge of the plasma are lost by rapid parallel motion to a robust target plate before they can diffuse radially across the magnetic field and strike the vacuum wall. Recall though that a basic problem with the limiter is that neutral impurities from the target can diffuse back into the plasma and strongly degrade performance. This is difficult to prevent because the edge of the plasma is by construction in contact with the limiter.

The divertor has the attractive feature of substantially reducing impurity build-up while continuing to protect the vacuum wall. It does this by guiding a narrow layer of magnetic lines away from the edge of the plasma as shown in Fig. 6.26. Physically, the layer is produced by a set of poloidal coils in close proximity to the plasma surface. The currents in these coils are chosen to generate a magnetic field which opposes the poloidal field in the plasma. The result is the formation of a field null corresponding to an X-point on the separatrix flux surface. The magnetic lines in the region just outside the separatrix, known as the scrape-off layer, now must travel a substantial distance away from the plasma edge before they ultimately make contact with the target plate. This is the main benefit of the divertor. The first contact between the plasma and a material surface is a relatively large distance away from the plasma edge, and not the short distance as for a limiter.

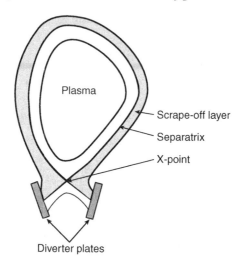

Figure 6.26   Schematic diagram of a divertor.

The divertor thus helps prevent impurity build-up as follows. Neutral impurities emitted from the target surface must now travel a substantial distance before crossing the plasma surface. There is a reasonably high probability that they will become ionized before doing so. Once ionized the charged particles flow parallel to the magnetic field where they are once again absorbed by the target plate – they do not cross the plasma surface.

Another advantage of the divertor over the limiter is that the target plates can be tilted at an angle with respect to the magnetic field as shown in Fig. 6.26. This allows the field lines to be spread over a larger surface area. These features cause a given plasma heat load to be deposited out over a larger area than is usually possible for a limiter. Lowering the heat load is a very important issue because typical values in a fusion plasma are very high – high enough in fact to cause melting if the field lines are concentrated and perpendicular to the target plate. Lower heat load has the added benefit of reducing the sputtering of neutral atoms from the target plate.

A final important advantage of the divertor occurs because the plasma edge is defined by the magnetic separatrix and not a solid surface such as a limiter. This allows the plasma to have a much higher edge temperature, on the order of several keV, which has a very favorable effect on overall plasma confinement and fusion power production.

Of course all of these advantages do not come without a price. One main disadvantage is that additional coils are necessary and extra volume is needed to make room for the divertor target plates. This disadvantage translates into higher costs. The general consensus at present is that the improved performance provided by a divertor is worth the extra cost.

Most modern tokamaks have a single null divertor. Clearly though it is possible to have an up–down symmetric double null divertor which could potentially halve the heat load on each target plate. This is often not the primary choice for two reasons. First, an additional divertor requires even more volume for coils and hence a further increase in cost. Second, the divertors must be very accurately balanced or else the entire heat load will be deposited on the closer X-point target plate, effectively "wasting" the second divertor.

Based on this discussion one sees that to model an MHD equilibrium containing a single null divertor the solutions must allow for an up–down asymmetric geometry with a separatrix containing an X-point.

Up–down asymmetric equilibria

The analytic solution to the Grad–Shafranov equation with Solov'ev profiles for up–down symmetric equilibria has been given by Eq. (6.153). It is straightforward to generalize the solution to include up–down asymmetric equilibria by adding polynomial solutions which are odd in Y. A short calculation shows that the generalized equilibrium can be written as

$$U(X, Y) = \frac{a}{2}X^2 \ln X + \frac{1 - a}{8}X^4 + \sum_0^{11} c_j U_j \qquad (6.160)$$

where five new terms have been added to the sum. The terms U_0–U_6 have already been defined in Eq. (6.153). The additional terms U_7–U_{11} are given by

$$
\begin{aligned}
U_7 &= Y \\
U_8 &= YX^2 \\
U_9 &= Y^3 - 3YX^2 \ln X \\
U_{10} &= 4Y^3X^2 - 3YX^4 \\
U_{11} &= 8Y^5 - 45YX^4 - (80Y^3X^2 - 60YX^4)\ln X
\end{aligned}
\qquad (6.161)
$$

Observe that polynomials up to fifth order have been maintained. This is the number, again found by trial and error, that makes the equilibrium procedure robust.

X-point boundary constraints

Consider now the boundary constraints that determine the 12 c_j coefficients. The original seven constraints specified in Eq. (6.155) are still assumed to be valid and apply to the upper half of the plasma surface. Five new terms have been added to the solution thereby requiring five new boundary constraints for closure. These are as follows. The slope conditions at the outer and inner equatorial points are no longer automatically satisfied since the solution contains polynomials that are odd in Y. Satisfying the slope conditions corresponds to two new constraints. The remaining three constraints are used to define the single null X-point, which is assumed to be located at $X = X_{sep}$, $Y = Y_{sep}$ with $Y_{sep} < 0$. At the X-point the flux and both of its derivatives must vanish. The vanishing of the flux puts the X-point on the plasma surface while the vanishing of both derivatives corresponds to a null in the poloidal field, the definition of an X-point.

The new conditions can be written mathematically as follows:

$$
\begin{aligned}
&8. \quad U_Y(1 + \varepsilon, 0) = 0 && \text{outer point slope} \\
&9. \quad U_Y(1 - \varepsilon, 0) = 0 && \text{inner point slope} \\
&10. \quad U(X_{sep}, Y_{sep}) = 0 && \text{X-point flux} && (6.162) \\
&11. \quad U_X(X_{sep}, Y_{sep}) = 0 && \text{X-point } B_Y = 0 \\
&12. \quad U_Y(X_{sep}, Y_{sep}) = 0 && \text{X-point } B_X = 0
\end{aligned}
$$

The additional constraints also lead to linear algebraic equations. The total set of constraints now consists of 12 simultaneous linear algebraic equations, still a trivial numerical problem.

6.6.6 Example: the International Thermonuclear Experimental Reactor (ITER)

The overall goals of ITER are to (1) investigate burning plasma physics in a long pulse, high-temperature, D-T experiment and (2) address and solve a number of the fusion technology issues that will arise in a fusion reactor. ITER thus has the crucial role of being the flagship facility for the world's fusion program for the next two decades. The project is enormously important in that future progress towards a fusion reactor will be directly tied to the physics and technological performance of ITER. A brief technical description of ITER is presented below.

The primary physics mission of ITER is to produce a stable, well-confined, $Q = 10$ plasma lasting for a sufficiently long duration to reach quasi steady state operation. Here, $Q = $ fusion power out/external heating power in. A second physics mission is to achieve steady state operation using non-inductive (i.e., no transformer) current drive at $Q \gtrsim 5$. With respect to technology, construction of ITER would demonstrate the viability of large superconducting magnets, various plasma facing materials, and large-scale remote handling. It would also test the effectiveness of the divertor design and begin to explore tritium breeding.

The ITER design is illustrated in Fig. 6.27. Note that ITER has a single null divertor and superconducting magnets constructed of niobium–tin. The magnetic field at the center of the plasma is $B_0 = 5.3\,\text{T}$. To keep the cost as low as possible, the size of the machine has been minimized subject to the constraint of achieving $Q = 10$ operation assuming high-performance energy confinement scaling (i.e., H-mode scaling) and MHD stability without a perfectly conducting wall. This

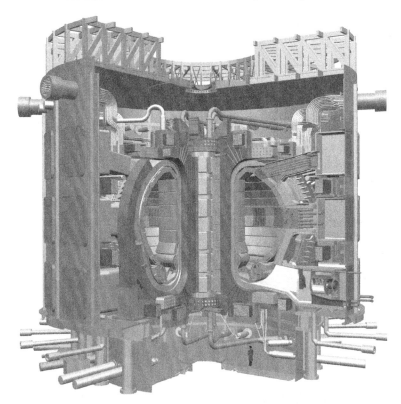

Figure 6.27   Schematic drawing of ITER. Courtesy of ITER.

leads to a machine design whose MHD related parameters are summarized in Table 6.3. Many parameters are quite comparable to those of a full power producing fusion reactor. The main difference is that ITER is still largely an experimental facility. Therefore it has not been designed to have the very high duty factor associated with a steady state power producing reactor. Also, the thermal power output is still about an order of magnitude lower than that required in a reactor.

For base operation, ITER will have about 70 MW of external heating power, divided among negative ion driven neutral beams, ion cyclotron heating, and electron cyclotron heating. ITER will operate for pulse durations of about $\tau_{\text{pulse}} \approx$ 400 sec, driven by the ohmic transformer. If successful, ITER should produce a $Q = 10$ plasma corresponding to a fusion performance factor of $\bar{p}\tau_E = 6.4$ atm-sec.

The parameters in the table have been used as inputs to the analytic solution procedure. The resulting flux surfaces have been calculated and are illustrated in Fig. 6.28. The flux surfaces are quite reasonable in appearance exhibiting smooth nested contours bounded by an up–down asymmetric single null separatrix. The analytic solution has thus just been shown to be capable of modeling a fusion relevant plasma with a sophisticated plasma shape.

Table 6.3 *MHD parameters for base operation of ITER.*

Parameters	Symbol	Units	ITER
Major radius	R_0	m	6.2
Minor radius	a	m	2.0
Aspect ratio	R_0/a	-	3.2
Elongation (95% flux surface)	κ	-	1.7
Triangularity (95% flux surface)	δ	-	0.23
Horizontal X-point location	$X_{sep} = R_{sep}/R_0$	-	0.93
Vertical X-point location	$Y_{sep} = Z_{sep}/R_0$	-	−0.65
Toroidal magnetic field	$B_0 = B(R_0)$	T	5.3
Plasma current	I	MA	15
Kink safety factor	q_*	-	1.9
Safety factor (95% flux surface)	q_{95}	-	3.0
Average temperature	$\overline{T}_e \approx \overline{T}_i \equiv \overline{T}_k$	keV	11.2
Average electron density	$\bar{n}_{20}$	$10^{20}\mathrm{m}^{-3}$	0.91
Toroidal beta	β_t	-	0.026

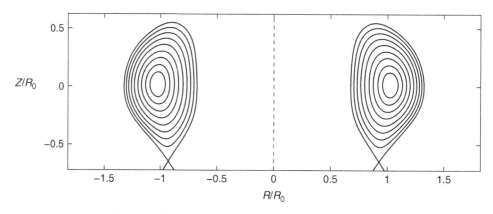

Figure 6.28   Exact Solov'ev equilibrium for ITER.

6.6.7 Example: the National Spherical Torus Experiment (NSTX)

The spherical tokamak (ST) is the most challenging test for the analytic solution procedure. The configuration is characterized by finite aspect ratio, up–down asymmetry, a single null divertor, and high β. The motivation and potential role for the ST in fusion research has been discussed in Section 6.6.3. The MHD parameters for a high-performance NSTX discharge are summarized in Table 6.4.

These parameters are supplied as input to the analytic solution procedure. The corresponding flux surfaces are illustrated in Fig. 6.29. The surfaces are smooth and bounded by a single X-point separatrix. Even for this relatively difficult test the analytic procedure produces a very reasonable solution.

Table 6.4 *MHD parameters for NSTX.*

Parameters	Symbol	Units	NSTX
Major radius	R_0	m	0.85
Minor radius	a	m	0.68
Aspect ratio	R_0/a	-	1.25
Elongation (95% flux surface)	κ	-	2
Triangularity (95% flux surface)	δ	-	0.4
Horizontal X-point location	$X_{\text{sep}} = R_{\text{sep}}/R_0$	-	0.59
Vertical X-point location	$Y_{\text{sep}} = Z_{\text{sep}}/R_0$	-	−1.76
Toroidal magnetic field	$B_0 = B(R_0)$	T	0.3
Plasma current	I	MA	1
Kink safety factor	q_*	-	2.0
Average temperature	$(\overline{T}_e + \overline{T}_i)/2 \equiv \overline{T}_k$	keV	1.1
Average electron density	$\overline{n}_{20}$	10^{20} m^{-3}	0.2
Toroidal beta	β_t	-	0.2

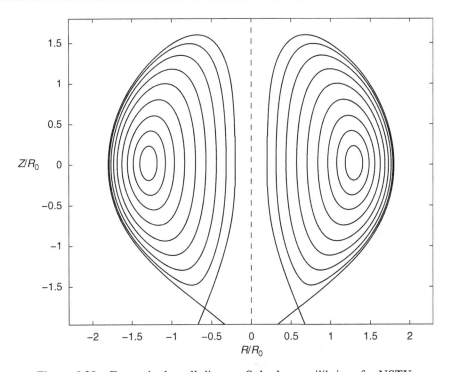

Figure 6.29   Exact single null divertor Solov'ev equilibrium for NSTX.

6.6.8 Summary

An exact analytic solution to the Grad–Shafranov equation using the Solov'ev profiles for $p(\psi)$ and $F(\psi)$ has been derived consisting of a finite number of polynomial solutions. A procedure involving the application of boundary constraints is discussed which serves to determine the unknown amplitudes multiplying each polynomial. Solving for these coefficients requires the solution of a set of simultaneous linear algebraic equations, a trivial task numerically.

The procedure is robust and is capable of calculating a wide range of exact tokamak equilibria of interest to fusion research. The range of equilibria varies from the easiest case, an up–down symmetric circular plasma with an aspect ratio of $R_0/a = 3$, to the most challenging case, an up–down asymmetric non-circular plasma with a single null divertor and a tight aspect ratio, $R_0/a = 1.4$. The specific equilibria considered correspond to the following tokamaks: TFTR, JET, MAST, ITER, and NSTX.

Although "exact" fully numerical Grad–Shafranov solvers are readily available in the fusion community, the analytic solutions are useful for providing insight, deriving scaling laws, and benchmarking numerical codes.

6.7 The helical Grad–Shafranov equation (the straight stellarator)

6.7.1 Overview

The equilibrium analysis in Section 6.6 focused on 2-D systems with toroidal axisymmetry. It is a fact that the MHD equilibrium of systems with one degree of symmetry can always be described by a single partial differential equation – for example the Grad–Shafranov equation when $\partial/\partial\phi = 0$.

A second useful type of 2-D symmetry that also leads to a Grad–Shafranov-like equation is helical symmetry, a good model for a "straight stellarator." Stellarators are inherently 3-D helical–toroidal configurations but some insight to their behavior can be obtained by un-bending the torus into a straight helix. It is of particular interest to demonstrate one of the major attractive features of a stellarator, the ability to achieve good MHD equilibrium and stability without the need for a net toroidal current. A configuration with this property does not require an ohmic transformer or a means of external non-inductive current drive, both of which are key requirements for a tokamak. Also, a system with no net current is expected to be more stable against current-driven kinks (which often lead to major disruptions) than a tokamak.

Mathematically, helical symmetry can be understood by first recalling that toroidal axisymmetry assumes that all physical quantities have the following 2-D geometrical dependence: $\psi(R,\phi,Z) \rightarrow \psi(R,Z)$. Similarly, helical symmetry expressed in a standard r,θ,z straight cylindrical system implies that $\psi(r,\theta,z) \rightarrow \psi(r,l\theta + hz)$. All quantities are only a function of radius r and helical angle $\alpha = l\theta + hz$. The integer l refers to the poloidal periodicity number while $2\pi/h$ is the axial wavelength of the helix. A drawing of a straight helical system is shown in Fig. 6.30. If one wants to think in terms of an equivalent torus then r is the minor radius measured from the center of the vacuum chamber, $z = -R_0\phi$, where R_0 is the major radius, and $h = N / R_0$, where N is the number of helical periods around the torus.

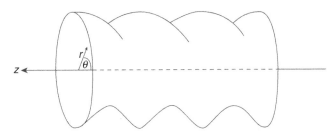

Figure 6.30   A straight helical model of a stellarator

There are three goals in this section. The first is to derive an analogous Grad–Shafranov equation for a system with helical symmetry. The second goal is to solve this equation in the limit of low β. A key question is to see whether or not it is possible to construct equilibria with zero net z current on every flux surface. The third goal is to calculate the rotational transform in order to address the following question. Can a system with no net current have a non-zero rotational transform? This is a non-trivial question since the magnetic lines in a helical system tend to oscillate back and forth on a flux surface. It is not obvious that in addition to oscillating, the lines also wrap around the plasma.

These "introduction to the stellarator" topics are now discussed in detail in the remainder of the section. A more detailed discussion of stellarators is presented in Chapter 7.

6.7.2 The helical Grad–Shafranov equation

The derivation of the helical Grad–Shafranov equation (Solov'ev, 1967) is similar to the toroidally symmetric version presented in Section 6.2. One modification though is the introduction of a new set of unit vectors which are convenient for carrying out the calculation. The new unit vectors and their inverse relations are defined as follows:

$$
\begin{array}{cc}
\mathbf{e}_r & \mathbf{e}_r \\
\mathbf{e}_\alpha = (l\,\mathbf{e}_\theta + hr\,\mathbf{e}_z)/l_0 & \mathbf{e}_\theta = (l\,\mathbf{e}_\alpha - hr\,\mathbf{e}_s)/l_0 \\
\mathbf{e}_s = (l\,\mathbf{e}_z - hr\,\mathbf{e}_\theta)/l_0 & \mathbf{e}_z = (l\,\mathbf{e}_s + hr\,\mathbf{e}_\alpha)/l_0
\end{array}
\tag{6.163}
$$

where $l_0(r) = (l^2 + h^2 r^2)^{1/2}$. Helical symmetry implies that $\mathbf{e}_s \cdot \nabla(\text{scalar}) = 0$. The basic steps in the derivation are now outlined below.

The first equation of interest is $\nabla \cdot \mathbf{B} = 0$. The 2-D symmetry again makes it possible to introduce a stream function ψ as follows:

$$
B_r = -\frac{1}{r}\frac{\partial \psi}{\partial \alpha}
$$

$$
B_\alpha = \frac{1}{l_0}(lB_\theta + hrB_z) = \frac{1}{l_0}\frac{\partial \psi}{\partial r}
\tag{6.164}
$$

The total magnetic field can thus be written in a compact form as

$$\mathbf{B} = \frac{1}{l_0}\mathbf{e}_s \times \nabla\psi + B_s\mathbf{e}_s \tag{6.165}$$

Here, $B_s = (1/l_0)(lB_z - hrB_\theta)$ is the third component of magnetic field and is unknown at this point.

The next step is to substitute into Ampere's law $\nabla \times \mathbf{B} = \mu_0\mathbf{J}$. A straightforward calculation yields an expression for $\mu_0\mathbf{J}$ which can be written as

$$\mu_0\mathbf{J} = -\frac{1}{l_0}\mathbf{e}_s \times \nabla(l_0 B_s) + \left(l_0\Delta^*\psi - \frac{2hl}{l_0^2}B_s\right)\mathbf{e}_s \tag{6.166}$$

where for the case of helical symmetry

$$\Delta^*\psi \equiv \frac{1}{r}\frac{\partial}{\partial r}\left(\frac{r}{l_0^2}\frac{\partial\psi}{\partial r}\right) + \frac{1}{r^2}\frac{\partial^2\psi}{\partial\alpha^2} \tag{6.167}$$

It is also of interest to single out the z component of $\mathbf{J}$ in view of the end goal of designing stellarators with no net current on every flux surface:

$$\mu_0 J_z = -\frac{hr}{l_0^2}\frac{\partial}{\partial r}(l_0 B_s) + l\left(\Delta^*\psi - \frac{2hl}{l_0^2}B_s\right) \tag{6.168}$$

The calculation now proceeds with the momentum equation $\mathbf{J} \times \mathbf{B} - \nabla p = 0$. Forming the dot product of this equation with $\mathbf{B}$ yields

$$\mathbf{B} \cdot \nabla p = \left(\frac{1}{l_0}\mathbf{e}_s \times \nabla\psi + B_s\mathbf{e}_s\right) \cdot \nabla p = \frac{1}{l_0}\mathbf{e}_s \times \nabla\psi \cdot \nabla p = 0 \tag{6.169}$$

which implies that

$$p = p(\psi) \tag{6.170}$$

As expected, the pressure is a flux function.

A similar relation results from forming the dot product of the momentum equation with $\mathbf{J}$:

$$\mu_0\mathbf{J} \cdot \nabla p = -\frac{1}{l_0}\mathbf{e}_s \times \nabla(l_0 B_s) \cdot \nabla\psi\frac{dp}{d\psi} = 0 \tag{6.171}$$

In analogy with the toroidally axisymmetric case one finds

$$l_0 B_s = F(\psi) \tag{6.172}$$

Here $F(\psi)$ is a free function of flux. Note that an ideal stellarator, defined as one with zero net current on every flux surface, can only be realized by a special choice of $F(\psi)$. This is discussed further in the next subsection.

The final step in the derivation involves forming the dot product of the momentum equation with $\nabla\psi$. A short calculation leads to

$$\Delta^*\psi = -\mu_0 \frac{dp}{d\psi} - \frac{F}{l_0^2}\frac{dF}{d\psi} + \frac{2hl}{l_0^4}F \tag{6.173}$$

This is the desired helical Grad–Shafranov equation. As for the axisymmetric torus it is a non-linear partial differential equation with two free functions. Equation (6.173) reduces to the axisymmetric Grad–Shafranov equation for the special case of $l = 0$.

6.7.3 Low β analytic solution

The expansion

An analytic solution to the helical Grad–Shafranov equation can be obtained by means of an asymptotic expansion where the small parameter is the ratio of the helical magnetic field to the axial (i.e., toroidal) magnetic field. The analysis is qualitatively similar to the low β, large aspect ratio expansion for the tokamak presented in Section 6.4.

The basic idea is to expand about a vacuum magnetic field consisting of two components: a large axial field and a smaller helical field. As the expansion progresses, the free function $F(\psi)$ is chosen so that order by order the net current on every flux surface is zero. The end goal is to solve for ψ and see how a non-zero pressure modifies the vacuum flux surfaces.

To begin the analysis one must define the mathematical ordering scheme. The basic expansion parameter is denoted by δ, which measures the ratio of helical field B_h to the axial field B_0. The appropriate ordering scheme for present purposes is defined as follows:

$$\begin{aligned} B_h/B_0 &\sim \delta & ha &\sim 1 \\ \psi/aB_0 &\sim 1 & \varepsilon = a/R_0 &\sim \delta^2 \\ F/B_0 &\sim 1 & \beta = 2\mu_0 p/B_0^2 &\sim \delta^4 \end{aligned} \tag{6.174}$$

The ordering $ha \sim 1$ implies that the helical wavelength is comparable to the minor radius. The ordering $\varepsilon \sim \delta^2$ implies that there are many helical wavelengths around the torus. The ordering $\beta \sim \delta^4$ is equivalent to $\beta \sim \varepsilon^2$, which is the same as for the low β tokamak expansion. To determine the effect of β on the equilibrium therefore requires carrying out the analysis to include four orders in δ, not as daunting a task as one might expect since two orders correspond to simple vacuum solutions.

Now, the detailed expansion for the flux function consistent with this ordering can be written as

$$\psi(r, \alpha) = \psi_0(r) + \psi_1(r, \alpha) + \psi_2(r, \alpha) + \psi_3(r, \alpha) + \cdots \tag{6.175}$$

For simplicity the vacuum helical field $\psi_1(r, \alpha) = \overline{\psi}_1(r) \cos \alpha$ is assumed to have only a single harmonic.

The last point to consider is the boundary conditions. Clearly ψ must be regular in the plasma. With regard to the plasma surface there are two different ways in which to express the boundary condition: (1) specify a helical shape for the boundary and require that $\psi = 0$ on this surface or (2) specify the axial plus helical field on a simple surface, $r = a$. The second choice actually reduces the amount of analysis required and is the one used here. It can be conveniently expressed as

$$\frac{1}{l_0} \frac{\partial \psi}{\partial r} \bigg|_{r=a} = B_\alpha(a, \alpha) = B_0 \left(\frac{ha}{l_a} + \delta \cos \alpha \right) \tag{6.176}$$

where $l_a = (l^2 + h^2 a^2)^{1/2}$, B_0 is the vacuum axial field, and δB_0 explicitly defines the amplitude of the helical field. Note that the constant term in the boundary condition (i.e., $B_0 ha/l_a$) has been chosen to make $B_\theta(a) = 0$, which is the requirement for no net z current in the plasma. In terms of the expansion, Eq. (6.176) translates into

$$
\begin{aligned}
\frac{\partial \psi_0}{\partial r} \bigg|_{r=a} &= ha B_0 \\[2mm]
\frac{\partial \psi_1}{\partial r} \bigg|_{r=a} &= \delta l_a B_0 \cos \alpha \\[2mm]
\frac{\partial \psi_2}{\partial r} \bigg|_{r=a} &= 0 \\[2mm]
\frac{\partial \psi_3}{\partial r} \bigg|_{r=a} &= 0
\end{aligned}
\tag{6.177}
$$

The problem has been properly formulated and the next task is to solve the helical Grad–Shafranov equation order by order to obtain the desired solution.

Zeroth-order solution

The zeroth-order equation and boundary condition determining $\psi_0(r)$ are given by

$$\frac{1}{r} \frac{d}{dr} \left(\frac{r}{l_0^2} \frac{d\psi_0}{dr} \right) = -\frac{F_0}{l_0^2} \frac{dF_0}{d\psi_0} + \frac{2hl}{l_0^4} F_0$$

$$\frac{d\psi_0}{dr} \bigg|_a = ha B_0 \tag{6.178}$$

A short calculation shows that for no net current one must choose

$$F_0(\psi_0) = lB_0 = \text{constant} \tag{6.179}$$

which leads to the solution

$$\psi_0(r) = -\frac{1}{2}hB_0(a^2 - r^2) \tag{6.180}$$

The free constant associated with the flux function has been chosen to make $\psi_0 = 0$ on the surface. The resulting magnetic field and current density are easily evaluated

$$\begin{aligned} \mathbf{B}_0 &= B_0 \mathbf{e}_z \\ \mathbf{J}_0 &= 0 \end{aligned} \tag{6.181}$$

which are the expected results.

First-order solution

The first-order equation and boundary condition determine $\psi_1(r, \alpha)$ and can be written as

$$\frac{1}{r}\frac{\partial}{\partial r}\left(\frac{r}{l_0^2}\frac{\partial \psi_1}{\partial r}\right) + \frac{1}{r^2}\frac{\partial \psi_1}{\partial \alpha^2} = -\frac{F_0}{l_0^2}\frac{dF_1}{d\psi_0} + \frac{2hl}{l_0^4}F_1$$

$$\left.\frac{\partial \psi_1}{\partial r}\right|_a = \delta l_a B_0 \cos \alpha \tag{6.182}$$

Observe that the helical terms on the left-hand side of the equation are all proportional to $\cos \alpha$ while the terms on the right-hand side are only functions of r since $\psi_0 = \psi_0(r)$. The only way to resolve this incompatibility is to choose

$$F_1(\psi_0) = 0 \tag{6.183}$$

If one now writes $\psi_1(r, \alpha) = \overline{\psi}_1(r) \cos \alpha$ then $\overline{\psi}_1$ satisfies the following equation and boundary condition

$$\frac{1}{r}\frac{d}{dr}\left(\frac{r}{l_0^2}\frac{d\overline{\psi}_1}{dr}\right) - \frac{\overline{\psi}_1}{r^2} = 0$$

$$\left.\frac{d\overline{\psi}_1}{dr}\right|_a = \delta l_a B_0 \tag{6.184}$$

The solution for $\overline{\psi}_1$ is easily found and can be written as

$$\overline{\psi}_1(r) = \frac{\delta a B_0}{l_a I_a}\left[r\frac{d}{dr}I_l(hr)\right] \tag{6.185}$$

where $I_l(hr)$ is the modified Bessel function and $I_a = I_l(ha)$. The first-order fields and current density are easily evaluated:

$$B_{r1} = \frac{\delta a B_0}{l_a I_a} \left[\frac{d}{dr} I_l(hr) \right] \sin \alpha$$

$$B_{\theta 1} = \frac{l}{hr} B_{z1} = \frac{\delta a B_0}{l_a I_a} \left[\frac{l}{r} I_l(hr) \right] \cos \alpha \tag{6.186}$$

$$\mathbf{J}_1 = 0$$

As expected, the solution corresponds to a vacuum helical field.

Second-order solution

The second-order equations are the first place where the effects of plasma pressure enter the equilibrium problem. The second-order equation and boundary condition are given by

$$\frac{1}{r} \frac{\partial}{\partial r} \left(\frac{r}{l_0^2} \frac{\partial \psi_2}{\partial r} \right) + \frac{1}{r^2} \frac{\partial^2 \psi_2}{\partial \alpha^2} = -\mu_0 \frac{dp}{d\psi_0} - \frac{F_0 \, dF_2}{l_0^2 \, d\psi_0} + \frac{2hl}{l_0^4} F_2$$

$$\left. \frac{\partial \psi_2}{\partial r} \right|_a = 0 \tag{6.187}$$

Since the right-hand side is only a function of r, this implies that $\psi_2 (r, \alpha) \rightarrow \psi_2(r)$.

Consider next the choice for $F_2(\psi_0)$. In order for the net current on every flux surface to be zero one must set $B_{\theta 2}(r) = 0$. From the definitions of the B_α and B_s in terms of ψ and $F(\psi)$ this can be accomplished by setting

$$F_2[\psi_0(r)] = F_2(r) = lB_{z2}(r) = \frac{l}{hr} \frac{d\psi_2}{dr} \tag{6.188}$$

Equation (6.188) is substituted into Eq. (6.187). After a short calculation one finds

$$\frac{d}{dr} (B_0 B_{z2} + \mu_0 p) = 0 \tag{6.189}$$

which is just the equation for radial pressure balance in a θ-pinch. With zero average axial current it is only the diamagnetism in the axial magnetic field that can provide radial pressure balance.

A good way to think about Eq. (6.189) is to view $p[\psi_0(r)] = p(r)$ as a free function. The resulting second-order physical quantities can then be expressed in terms of $p(r)$ as follows:

$$\psi_2 = \frac{\mu_0 h}{B_0} \int_r^a p \, r \, dr$$

$$\mathbf{B}_2 = -\frac{\mu_0 p}{B_0} \mathbf{e}_z$$

$$\mathbf{J}_2 = \frac{1}{B_0} \frac{dp}{dr} \mathbf{e}_\theta$$

$$F_2 = -\frac{\mu_0 l \, p}{B_0}$$

(6.190)

This completes the second-order solution.

Third-order solution

It is in third-order that the plasma pressure first has an impact on the structure of the helical fields. The third-order equation and boundary condition are given by

$$\frac{1}{r} \frac{\partial}{\partial r} \left(\frac{r}{l_0^2} \frac{\partial \psi_3}{\partial r} \right) + \frac{1}{r^2} \frac{\partial \psi_3}{\partial \alpha^2} = -\left(\mu_0 \frac{d^2 p}{d\psi_0^2} + \frac{F_0}{l_0^2} \frac{d^2 F_2}{d\psi_0^2} \right) \psi_1 - \frac{F_0}{l_0^2} \frac{dF_3}{d\psi_0} + \frac{2hl}{l_0^4} F_3$$

$$\left. \frac{\partial \psi_3}{\partial r} \right|_a = 0$$

(6.191)

Note that the terms on the right proportional to $\psi_1(r, \alpha)$ all vary as $\cos \alpha$ while those containing $F_3(\psi_0)$ are functions only of r. As shown below the correct choice for F_3 leading to no net current on every flux surface is

$$F_3(\psi_0) = 0 \qquad (6.192)$$

Using this condition implies that $\psi_3(r, \alpha) = \overline{\psi}_3(r) \cos \alpha$ with $\overline{\psi}_3$ satisfying

$$\frac{1}{r} \frac{d}{dr} \left(\frac{r}{l_0^2} \frac{d\overline{\psi}_3}{dr} \right) - \frac{\overline{\psi}_3}{r^2} = -\left(\mu_0 \frac{d^2 p}{d\psi_0^2} + \frac{F_0}{l_0^2} \frac{d^2 F_2}{d\psi_0^2} \right) \overline{\psi}_1$$

$$\left. \frac{d\overline{\psi}_3}{dr} \right|_a = 0$$

(6.193)

After a slightly lengthy calculation one can write down an exact but complicated solution to Eq. (6.193) as follows:

$$\overline{\psi}_3 = \overline{\psi}_1 U$$

$$U = \frac{\mu_0}{B_0^2} \left[\int_r^a dx \frac{xl_0^2}{\overline{\psi}_1^2} \int_0^x \frac{\overline{\psi}_1^2}{y^2} \left(\frac{yp'}{l_0^2} \right)' dy + C \right]$$

$$C = \frac{al_a^2}{\overline{\psi}_1(a)\overline{\psi}_1'(a)} \int_0^a \frac{\overline{\psi}_1^2}{y^2} \left(\frac{yp'}{l_0^2} \right)' dy$$

(6.194)

In principle one can substitute profiles and carry out the integrals leading to an expression for $\overline{\psi}_3$ but this will require a numerical calculation. A simpler procedure, providing more intuition, is described below.

The loosely wound helix

The integrals in Eq. (6.194) can be evaluated analytically by introducing two assumptions. First one assumes a simple profile for the pressure:

$$p(r) = 2\langle p \rangle \left(1 - \rho^2 \right)$$

(6.195)

where $\rho = r/a$. The second assumption is that the helix is loosely, but not too loosely, wound: $\delta \ll ha \ll 1$. This leads to the simplifying approximations

$$l_0^2(r) \approx l^2 = \text{constant}$$

$$\overline{\psi}_1(r) \approx \delta a B_0 \rho^l$$

(6.196)

Under these plausible assumptions the integrals can be easily evaluated:

$$U(r) = -\frac{\beta}{l^2} \left[2 + l - l\rho^2 \right]$$

(6.197)

where $\beta = \beta_t = 2\mu_0 \langle p \rangle / B_0^2$.

The contributions from the various orders can now be combined and simplified using the loose helix approximations. This results in the following expression for the total flux

$$\psi(\rho, \alpha) = \overline{\psi}(\rho) + \tilde{\psi}(\rho) \cos \alpha$$

$$\overline{\psi} = -\frac{B_0 ha^2}{2} (1 - \rho^2) \left[1 - \frac{\beta}{2} (1 - \rho^2) \right]$$

$$\tilde{\psi} = \delta a B_0 \rho^l \left[1 - \frac{\beta}{l^2} (2 + l - l\rho^2) \right]$$

(6.198)

where $\tilde{\psi}/\overline{\psi} \sim \delta/ha \ll 1$.

Two basic conclusions can be drawn from this expression. First, since $\tilde{\psi}/\overline{\psi} \ll 1$ the shape of the flux surfaces are readily obtained by letting $\rho(\rho_0, \alpha) = \rho_0 + \rho_1(\rho_0, \alpha)$ with ρ_0 serving as the flux surface label and $\rho_1 \ll \rho_0$. One finds $\rho_1 = -\left[\tilde{\psi}(\rho_0)/\overline{\psi}'(\rho_0)\right]\cos\alpha$. For the specific cases of $l = 1$, $l = 2$, and $l = 3$ the values of ρ_1 are given by

$$l = 1 \qquad \rho_1 = -\frac{\delta}{ha}(1 - 2\beta)\cos(\theta + hz)$$

$$l = 2 \qquad \rho_1 = -\frac{\delta}{ha}\left(1 - \frac{\beta\rho_0^2}{2}\right)\rho_0 \cos(2\theta + hz) \qquad (6.199)$$

$$l = 3 \qquad \rho_1 = -\frac{\delta}{ha}\left[1 - \frac{\beta}{9}(6\rho_0^2 - 4)\right]\rho_0^2 \cos(3\theta + hz)$$

These flux surfaces are illustrated in Fig. 6.31. Observe that an $l = 1$ system corresponds to a rotating shifted circle similar to a corkscrew. An $l = 2$ system has the form of a rotating ellipse while $l = 3$ corresponds to a rotating triangular shape. Non-zero β effects make small modifications to the location of a given flux surface but do not alter its shape.

The second basic conclusion concerns a fundamental property of stellarators and should not get lost in the details of the analysis. Even as $\beta \to 0$ a stellarator magnetic field exhibits closed nested flux surfaces. There is a unique center to the field which automatically defines the location of the peak pressure. Contrast this with the tokamak. When no plasma is present and the poloidal field circuit is energized the resulting "flux surfaces" are a series of vertical lines, also shown in Fig. 6.31. Only when the plasma draws current and the vertical field is properly adjusted will a tokamak plasma be centered in the vacuum chamber. Stated differently, a stellarator provides a robust, immovable, magnetic cage to confine the plasma while a tokamak requires help from the plasma itself in the form of a toroidal current to provide equilibrium. This is indeed an advantage for stellarators, but as discussed in the next chapter other problems related to single-particle confinement arise.

The net z current

A final point to discuss in this subsection is the net z current flowing in a stellarator. In carrying out the analysis a choice for $F(\psi)$ has been made at each order in the expansion which is assumed to eliminate the net z current flowing on every flux surface. The task now is to verify this assumption. The required analysis is straightforward. Starting with the flux function given by Eq. (6.198) one can easily calculate B_r and B_θ. Taking the z component of $\nabla \times \mathbf{B}$ then leads to

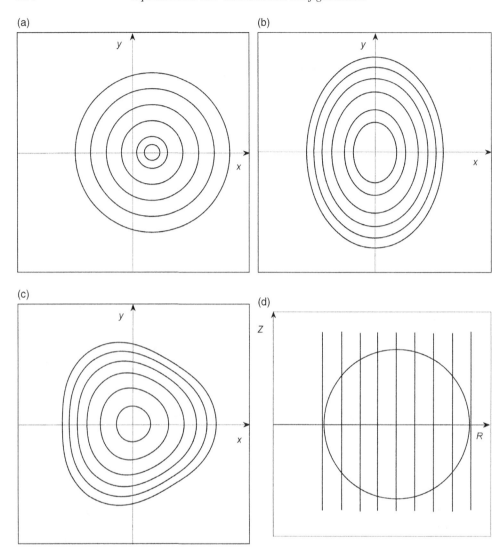

Figure 6.31   Vacuum flux surfaces for (a) $l = 1$, (b) $l = 2$, and (c) $l = 3$ stellarators. Also shown in (d) are the vacuum flux surfaces for a tokamak.

$$\mu_0 a J_z = \beta \delta B_0 \frac{(4 - 2l^2)}{l^2} \rho^l \cos \alpha \qquad (6.200)$$

Note that this contribution corresponds to a third-order term in the expansion. All lower-order contributions have vanished identically because of the choices for F (ψ). Since $J_z \sim \cos \alpha$ the integral over poloidal angle on any flux surface $\rho \approx \rho_0$ averages to zero. Stellarator equilibria are possible with no net current on any flux surface.

6.7.4 The rotational transform

The last topic of interest concerning the straight stellarator is the evaluation of the rotational transform. There is a key point to be made. Even without a plasma, a vacuum helical field produces a non-zero rotational transform. In the limit of small helical field amplitude the transform can be evaluated analytically. The existence of a vacuum rotational transform is not an obvious fact. The helical fields oscillate back and forth on a flux surface. One might think that the oscillating contribution to the transform would average to zero. It is shown that to leading order this is indeed true. The magnitude of the transform is a higher-order effect resulting from the fact that the flux surfaces have a small helical deformation.

Derivation of the rotational transform

The calculation starts with the general definition of transform given in Chapter 4 and then proceeds by exploiting the helical symmetry. The final result is then applied to several common stellarator fields corresponding to $l = 1, 2,$ and 3. The details are as follows.

The general definition of rotational transform derived in Chapter 4, repeated here for convenience, is given by

$$\frac{\iota}{2\pi} = \frac{\lim_{L\to\infty} \int_0^L \frac{B_\theta}{rB}\, dl}{\lim_{L\to\infty} \int_0^L \frac{B_\phi}{RB}\, dl} \tag{6.201}$$

where now l and B represent the total (rather than poloidal) arc length and magnetic field, respectively. One now uses the geometric relations

$$\frac{dl}{B} = \frac{r\,d\theta}{B_\theta} = \frac{dz}{B_z} \tag{6.202}$$

and the definition of the helical angle $\alpha = l\theta + hz$ to obtain

$$\frac{dl}{B} = \frac{r\,d\alpha}{\partial\psi/\partial r} \tag{6.203}$$

Since the magnetic line geometry repeats itself every time α increases by 2π it is only necessary to integrate Eq. (6.201) over the range $0 \le \alpha \le 2\pi$ to evaluate the transform. Recall that a similar conclusion was derived for the tokamak where periodicity corresponds to one full poloidal transit $0 \le \theta \le 2\pi$. For the straight stellarator, helical periodicity thus implies that

$$\frac{\iota}{2\pi} = \frac{\int_0^{2\pi} B_\theta \dfrac{d\alpha}{\partial \psi/\partial r}}{\int_0^{2\pi} \dfrac{rB_\phi}{R} \dfrac{d\alpha}{\partial \psi/\partial r}} \tag{6.204}$$

The quantities appearing in the integrand need to be evaluated on the flux surface $r = r_0 + \bar{r}_1 \cos \alpha$ and must include leading- and first-order contributions. These quantities are obtained from the previous section in the vacuum limit $\beta \to 0$. A short calculation leads to

$$\frac{\partial \psi}{\partial r} = \frac{d\psi_0}{dr_0} + \left(\frac{d^2 \psi_0}{dr_0^2} \bar{r}_1 + \frac{d\bar{\psi}_1}{dr_0} \right) \cos \alpha$$

$$B_\theta = \frac{l}{l_0^2} \frac{d\bar{\psi}_1}{dr_0} \cos \alpha + \frac{d}{dr_0} \left(\frac{l}{l_0^2} \frac{d\bar{\psi}_1}{dr_0} \right) \bar{r}_1 \cos^2 \alpha \tag{6.205}$$

$$B_\phi = -\frac{R}{R_0} B_z \approx -B_0 - \frac{hr}{l_0^2} \bar{\psi}_1 \cos \alpha \approx -B_0$$

where $r_1(r_0, \alpha) = \bar{r}_1(r_0) \cos \alpha$, $\bar{r}_1 = -\bar{\psi}_1/\psi'_0$, and $\psi_0, \bar{\psi}_1$ are given by Eqs. (6.180) and (6.185) respectively. Note that only the leading-order contribution to B_ϕ enters the calculation.

One now substitutes this information into Eq. (6.204). The leading-order contribution averages to zero as expected. The first-order correction is obtained after a straightforward but slightly tedious calculation. The desired relation for the rotational transform is given by

$$\frac{\iota}{2\pi} = \frac{\delta^2 a^2 l R_0}{2h l_a^2 I_a^2} \left[\frac{1}{r} \frac{d}{dr} \left(\frac{I_l}{r} \frac{dI_l}{dr} \right) \right] \tag{6.206}$$

Here, for simplicity the subscript "0" has be suppressed from r_0.

The loose helix approximation

Equation (6.206), while exact in the context of the straight stellarator expansion, is still a relatively complicated expression. More intuition about the transform can be obtained by again introducing the loose helix approximation $\delta \ll ha \ll 1$ and using the small argument expansion of the Bessel functions

$$I_l(u) = \frac{1}{l!} \left(\frac{u}{2} \right)^l \left[1 + \frac{1}{l+1} \left(\frac{u}{2} \right)^2 + \cdots \right] \tag{6.207}$$

For $l \geq 3$ only the leading-order contribution to the Bessel functions is required. The expression for the transform reduces to

$$\frac{\imath(r)}{2\pi} \approx \frac{(l-1)\delta^2}{ha\varepsilon} \left(\frac{r}{a}\right)^{2l-4} \qquad l \geq 3 \qquad\qquad (6.208)$$

Observe that when ha is reconsidered to be of order unity the transform is also of order unity: $\imath/2\pi \sim 1$. It is zero on axis and rises rapidly towards the edge, the rate of rise increasing as l increases. The profile has magnetic shear (i.e., $(r/\imath)(d\imath/dr) \neq 0$), which will be shown to be favorable for stability. Qualitatively, the profile shape is the opposite of that of a tokamak. For a stellarator $\imath/2\pi$ increases with r while for a tokamak $\imath/2\pi = 1/q$ decreases with r.

For an $l = 2$ system it is useful to include the first-order correction to the Bessel functions since, as shown in Eq. (6.208), the transform is constant to leading order. The first-order correction will show whether the profile increases or decreases with r although in either case the rate of change is small for $ha \ll 1$. A short calculation shows that for an $l = 2$ system the transform is given by

$$\frac{\imath(r)}{2\pi} \approx \frac{\imath_H}{2\pi} \left(\frac{1 + h^2 r^2/2}{1 + h^2 a^2/2}\right) \qquad l = 2$$

$$\frac{\imath_H}{2\pi} = \frac{\imath(a)}{2\pi} = \frac{\delta^2}{ha\varepsilon} \left(\frac{1 + h^2 a^2/2}{1 + 5h^2 a^2/12}\right) \approx \frac{\delta^2}{ha\varepsilon} \qquad\qquad (6.209)$$

Observe that the transform increases with radius.

The last example of interest corresponds to $l = 1$. Equation (6.208) shows that to leading order in ha the rotational transform in an $l = 1$ system vanishes. To find the transform one must therefore keep the first-order correction to the Bessel functions. By definition the resulting transform is therefore smaller by a factor $h^2 a^2$. Another short calculation shows that for an $l = 1$ system the transform can be written as

$$\frac{\imath(r)}{2\pi} \approx \frac{\delta^2 ha}{\varepsilon} \qquad l = 1 \qquad\qquad (6.210)$$

The transform is constant in radius.

To summarize, a vacuum helical field produces a non-zero rotational transform. For $l \geq 3$ the transform is zero on axis but has shear across the entire profile. For $l = 2$ the transform is finite on axis and has weak shear across the profile. An $l = 1$ system has weak shear and a small transform.

6.7.5 Summary

The equilibrium of a plasma with helical symmetry (in a linear geometry) represents a simple model of a "straight stellarator." The system obeys a "Grad–Shafranov"-like partial differential equation with two free functions, one being the pressure and the other related to the net toroidal current flowing on a flux surface. This latter function can be chosen so that the net current is zero on each flux surface. Even with zero plasma current, the vacuum helical field still possesses a non-zero rotational transform. As a consequence a helical system has a unique center determined by the external helical coil locations and forms a magnetic cage to confine the plasma.

While many features of a stellarator can be understood from the straight helical model the crucial problem of toroidal force balance has not yet been addressed. This is the topic of the next chapter where it is shown that even using a large aspect ratio expansion the analysis becomes considerably more complex.

6.8 Overall summary

Chapter 6 demonstrates how the ideal MHD model can be used to investigate the equilibrium properties of two-dimensional systems. Two types of systems are considered: a toroidally axisymmetric system and a linear helical system. For each system the equilibrium is described by a single non-linear partial differential equation.

For the toroidally axisymmetric case the equation, known as the Grad–Shafranov equation, provides a complete description of radial pressure balance and toroidal force balance. It can also be used to evaluate important plasma parameters and figures of merit related to equilibrium β limits and MHD stability. A number of concepts are discussed including the reversed field pinch, and various forms of tokamaks.

The helical Grad–Shafranov equation describes the equilibrium properties of a 2-D "straight stellarator" thereby providing a good introduction to the actual stellarator concept which is an inherently 3-D configuration. Certain important properties of stellarators are obtained from the 2-D model, although the critical problem of toroidal force balance is postponed until Chapter 7.

A summary of each of these systems is presented below.

The reversed field pinch (RFP)

The RFP is a large aspect ratio circular cross section configuration with $\beta_p \sim \beta_t \sim 1$ and $q_* \sim \varepsilon \ll 1$. Radial pressure balance is provided primarily by the toroidal current as in a Z-pinch. In fact the toroidal field is highly paramagnetic. Early RFPs were short pulsed devices in which toroidal force balance was provided by flux compression against a conducting wall. Modern RFPs have much longer pulses and are held in toroidal force balance by a vertical field. To achieve good MHD stability against modes driven by the pressure gradient the pressure profile must be very flat near the magnetic axis and the toroidal field must reverse sign near the edge of the plasma.

However, the low value of $q_* \sim \varepsilon$ is of concern with respect to current-driven kink modes and some form of feedback will likely be needed for good MHD stability. RFPs have been operated for many decades and experimental performance has been improving, getting close to, but still not as good as a tokamak.

The ohmically heated tokamak

Ohmically heated tokamaks are early fusion devices, usually having a circular cross section, and with characteristic parameters that scale as $\beta_t \sim \varepsilon^2$, $\beta_p \sim 1$, and $q_* \sim 1$. Radial pressure balance, toroidal force balance, and plasma heating are all provided by the toroidal plasma current. Even early ohmic tokamaks had long pulses requiring a vertical field rather than a conducting wall for toroidal equilibrium. The primary function of the large, costly, and slightly paramagnetic toroidal field is to provide MHD stability. The resulting high $q_* \sim 1$ provides stability against kink modes without a conducting wall or feedback. These early devices achieved quite good performance but were still limited in temperature and pressure since the only source of external power was the ohmic heating.

The high β tokamak

The high β tokamak is a large aspect ratio configuration whose goal is to achieve higher values of β than are possible in an ohmically heated tokamak. Initial high β tokamaks had circular cross sections. Higher values of β are accomplished by confining the plasma in a shallow diamagnetic well in the toroidal field produced by the application of a large amount of auxiliary power. Radial pressure balance is thus similar to that in a θ-pinch. The scaling relations in the high β regime are as follows: $\beta \sim \beta_t \sim \varepsilon$, $\beta_p \sim 1/\varepsilon$, and $q_* \sim 1$. This value of β is typically $1/\varepsilon$ times larger than in an ohmically heated tokamak. The scaling relations imply that the toroidal field is required for both radial pressure balance and MHD stability. The combination of poloidal field and vertical field provides toroidal force balance. The toroidal current also produces ohmic heating but this is usually a small effect compared to that due to the auxiliary power. Stability against current-driven kink modes is again provided by high q_*. Stability against pressure gradient driven modes is accomplished by profile shaping and keeping the value of β below a critical value. High β tokamaks have the interesting property of exhibiting an equilibrium β limit when the current is held constant and the pressure is increased, resulting from a separatrix moving onto the plasma surface. The equilibrium limit scales as $\beta_t \lesssim \varepsilon/q_*^2$.

Non-circular tokamaks

The experimentally achievable values of β_t in both ohmically heated and high β tokamaks can be substantially increased by allowing the plasma to have a non-circular cross section. The improvements are based on the observation that the stability boundary against current-driven kink modes depends only on the value of q_*, essentially independent of elongation. Since elongated cross sections are characterized by higher values of I and J_ϕ for the same q_*, this leads to a higher equilibrium β_t limit, larger ohmic heating, and improved confinement. However, there is a limit to the allowable elongation of the cross section, typically $\kappa \lesssim 2$, set by axisymmetric vertical instabilities. Virtually all modern tokamaks have elongated cross sections with an outward pointing "D" shape.

The equilibrium β_t limit

In practice the equilibrium β_t limit is not a serious experimental problem. It can be avoided by relaxing the assumption that the plasma current be held fixed as the pressure is increased. If the plasma current instead increases in parallel with the pressure then less burden is placed on the vertical field to hold the plasma in toroidal force balance. A smaller increase in the vertical field slows the motion of the separatrix X-point onto the plasma surface thereby delaying and preventing the onset of the equilibrium limit. In actual experiments MHD instabilities occur at lower values of β_t than the equilibrium limit and are therefore more important in setting practical operational boundaries.

The spherical tokamak (ST)

The spherical tokamak attempts to achieve the highest possible values of β_t by exploiting the fact that both MHD equilibrium and stability limits scale as $\beta_t \sim \varepsilon$. Consequently, a compact tokamak with an ultra-tight aspect ratio, $R_0/a \sim 1.2$–1.6, should be able reach high values of β_t while maintaining $q_* \sim 1$. This has indeed been demonstrated experimentally. In practice, however, the pressure in an ST is similar to or even lower than in a standard tokamak in spite of the gains in β_t. The reason is that the toroidal field in the center of an ST is small because of the rapid $1/R$ fall-off of B_ϕ in a tight aspect ratio device. A possible attractive application of the ST is as a component test facility to test materials for a fusion reactor.

Limiters and divertors

An important practical problem in the operation of tokamak and stellarator experiments is the interface between the edge of the plasma and the first solid material with which it comes into contact. In high-performance experiments the plasma boundary is defined by either a mechanical limiter or magnetic divertor. The divertor offers a more flexible method of removing particles from the plasma, protecting the vacuum chamber, and preventing impurities from entering the plasma. However, divertors are more expensive and more costly. Both limiter and divertor tokamaks can be modeled by the Grad–Shafranov equation. Modern tokamaks almost always choose divertors as their first option.

The straight stellarator

A stellarator is a 3-D helical–toroidal fusion configuration. It receives much attention in the fusion community because of the possibility of operating with no net plasma current. This feature eliminates the need for external current drive (i.e., the plasma is inherently a steady state device) and reduces the possibility of major disruptions due to current-driven kink modes. As an introduction to stellarators a Grad–Shafranov equation for a straight cylindrical system with helical symmetry has been derived. Solutions are obtained using the low β, small helical amplitude expansion which demonstrate the following properties: (1) closed nested flux surfaces with a unique center exist even without plasma; (2) radial pressure balance similar to that in a θ-pinch; (3) zero net current on every flux surface; and (4) finite rotational transform. The important effects of toroidicity are addressed in the next chapter.

From the wide variety of realistic equilibria described by the two Grad–Shafranov equations one can see why the ideal MHD model is so widely used in the fusion program.

References

Budny, R.V. (1994). *Nucl. Fusion* **34**, 1247.
Cerfon, A.J. and Freidberg, J.P. (2010). *Phys. Plasmas* **17**, 032502.
Clarke, J.F. and Sigmar, D.J. (1977). *Phys. Rev. Lett.* **38**, 70.
DIII-D Team and Simonen, T.C. (1998). *Fusion Engineering and Design* **39**, 83.
Grad, H. and Rubin, H. (1958). In *Proceedings of the Second United Nations International Conference on the Peaceful Uses of Atomic Energy.* Geneva: United Nations, 31, p. 190.
Haas, F.A. (1972). *Phys. Fluids* **15**, 141.
Lust, R. and Schluter, A. (1957). *Z. Naturforsch* **12a**, 850.
Peng, M., Boroski, S.K., Dalton, B.R., *et al.* (1985). In *Proceedings of the Topical Conference on the Technology of Fusion Energy*, San Francisco, CA, p. 19.
Pfirsch, D. and Schluter, A. (1962). Max-Planck-Institut Report MPI/PA/7/62.
Shafranov, V.D. (1960). *Sov. Phys.-JETP* **26**, 682.
Shafranov, V.D. (1966). In *Reviews of Plasma Physics*, Vol. 2, ed. M.A. Leontovich. New York: Consultants Bureau.
Solev'ev, L.S. (1967). In *Reviews of Plasma Physics*, Vol. 3, ed. M.A. Leontovich. New York: Consultants Bureau.
Stratton, J.A. (1941). *Electromagnetic Theory*. New York: McGraw-Hill.

Further reading

Reversed field pinch

Bodin, H.A.B. and Newton, A.A. (1980). *Nucl. Fusion* **20**, 1255.
Butt, E.P., Curruthers, R., Mitchell, J.T.D. *et al.* (1958). In *Proceedings of the Second United Nations International Conference on the Peaceful Uses of Atomic Energy.* Geneva: United Nations, Vol. 32, p. 42.
Kikuchi, M., Lackner, K., and Tran, M.Q., eds. (2012). *Fusion Physics.* Vienna: International Atomic Energy Agency.
Martin, P., Adamek, J., Agostinetti, P. *et al.* (2011). *Nuclear Fusion* **51**, 1.
Robinson, D.C. (1971). *Plasma Phys.* **13**, 439.

Tokamaks (general)

Bateman, G. (1978). *MHD Instabilities*. Cambridge, MA: MIT Press.
Furth, H.P. (1975). *Nucl. Fusion* **15**, 487.
Kikuchi, M., Lackner, K., and Tran, M.Q., eds. (2012). *Fusion Physics*. Vienna: International Atomic Energy Agency.
Shafranov, V.D. (1966). In *Reviews of Plasma Physics*, Vol 2, ed. M.A. Leontovich. New York: Consultants Bureau.
Wesson, J.A. (1978). *Nucl. Fusion* **18**, 87.
Wesson, J. (2011). *Tokamaks*, 4th edn. Oxford: Oxford University Press.
White, R.B (2006). *Theory of Toroidally Confined Plasmas*, 2nd edn. London: Imperial College Press.

High β tokamak

Callan, J.D. and Dory, R.A. (1972). *Phys. Fluids* **15**, 1523.
Dory, R.A. and Peng, Y.-K.M. (1977). *Nucl. Fusion* **17**, 21.
Freidberg, J.P. and Haas, F.A. (1973). *Phys. Fluids* **16**, 1909.
Shafranov, V.D. (1971). *Plasma Phys.* **13**, 757.
Wesson, J. (2011). *Tokamaks*, 4th edn. Oxford: Oxford University Press.

Non-circular tokamaks

Laval, G. and Pellat, R. (1973). In *Controlled Fusion and Plasma Physics*, Proceedings of the Sixth European Conference, Moscow, Vol. II, p. 640.
Laval, G., Pellat, R. and Soule, J.L. (1972). In *Controlled Fusion and Plasma Physics*, Proceedings of the Fifth European Conference, Euratom CEA, Grenoble, Vol. I, p. 25.
Laval, G., Pellat, R. and Soule, J.L. (1974). *Phys. Fluids* **17**, 835.
Solov'ev, L.S., Shafranov, V.D., and Yurchenko, E.I. (1969). In *Plasma Physics and Controlled Nuclear Fusion Research* (1968), IAEA, Vienna, Vol. I, p. 173.
Wesson, J. (2011). *Tokamaks*, 4th edn. Oxford: Oxford University Press.

Flux-conserving tokamaks

Clarke, J.F. and Sigmar, D.J. (1977). *Phys. Rev. Lett.* **38**, 70.
Dory, R.A. and Peng, Y.-K.M. (1977). *Nucl. Fusion* **17**, 21.

Straight stellarator

Morozov, A.I. and Solov'ev, L.S. (1963). In *Reviews of Plasma Physics*, Vol. 2, ed. M.A. Leontovich. New York: Consultants Bureau.
Solov'ev, L.S. (1967). In *Reviews of Plasma Physics*, Vol. 3, ed. M.A. Leontovich. New York: Consultants Bureau.
Wakatani, M. (1998). *Stellarators and Heliotron Devices*. New York: Oxford University Press.

Numerical MHD equilibria

Goedbloed, J.P., Keppens, R., and Poedts, S. (2010). *Advanced Magnetohydrodynamics*. Cambridge: Cambridge University Press.
Jardin, S. (2010). *Computational Methods in Plasma Physics*. New York: CRC Press.

Problems

6.1 Consider the two-dimensional Grad–Shafranov equation for a straight stellarator (Eq. (6.173)). Show that in the limit where the helical field amplitude approaches zero (i.e., $\psi(r, \alpha) \to \psi(r)$), the general one-dimensional radial pressure balance relation for the screw pinch is obtained.

6.2 Using the ideal MHD model derive a Grad–Shafranov equation for stationary axisymmetric toroidal equilibrium in which there is a steady state velocity $V_\phi(R, Z)$ flowing in the toroidal direction. For simplicity assume $\rho = \rho(\psi)$. Hints: Show that

$$\mathbf{B} = B_\phi \mathbf{e}_\phi + \frac{1}{R} \nabla \psi \times \mathbf{e}_\Phi$$

$$\mu_0 \mathbf{J} = -\frac{1}{R} \Delta^* \psi \mathbf{e}_\phi + \frac{1}{R} \nabla (R B_\phi) \times \mathbf{e}_\phi$$

$$V_\phi / R = \Omega(\psi)$$

$$p - \frac{1}{2} \rho \Omega^2 R^2 = \Pi(\psi)$$

$$R B_\phi = F(\psi)$$

$$\Delta^* \psi = -FF' - \mu_0 R^2 \Pi' - \frac{\mu_0 R^4}{2} \left(\rho \Omega^2 \right)'$$

6.3 Consider the transformation of coordinates (R,Z,ϕ') to (r,θ,ϕ) defined by

$$R = R_0 + r \cos \theta$$

$$Z = r \sin \theta$$

$$\phi' = -\phi$$

where (R,Z,ϕ') represents a standard cylindrical coordinate system. Show that the following relations hold in the (r,θ,ϕ) system:

$$\nabla \psi = \nabla_c \psi$$

$$\nabla \cdot \mathbf{B} = \nabla_c \cdot \mathbf{B} + \frac{\mathbf{e}_R \cdot \mathbf{B}}{R}$$

$$\nabla \times \mathbf{B} = \nabla_c \times \mathbf{B} - \frac{B_\phi}{R} \mathbf{e}_Z$$

$$\Delta^* \psi = \nabla_c^2 \psi - \frac{\mathbf{e}_R \cdot \nabla \psi}{R}$$

Here

$$\nabla_c = \mathbf{e}_r \frac{\partial}{\partial r} + \frac{\mathbf{e}_\theta}{r} \frac{\partial}{\partial \theta} + \frac{\mathbf{e}_\phi}{R} \frac{\partial}{\partial \phi}$$

and

$$\mathbf{e}_R = \cos \theta \mathbf{e}_r - \sin \theta \mathbf{e}_\theta$$

$$\mathbf{e}_Z = \sin \theta \mathbf{e}_r + \cos \theta \mathbf{e}_\theta$$

$$\mathbf{e}_{\phi'} = -\mathbf{e}_\phi$$

6.4 Consider a static cylindrically symmetric reversed field pinch characterized by $p(r)$, $B_\theta(r)$, $B_z(r)$. In realistic experimental situations Ohm's law should include an anisotropic resistivity: $\mathbf{E} = \eta_\| \mathbf{J}_\| + \eta_\perp \mathbf{J}_\perp$ with $\eta_\|$ and $\eta_\perp$ assumed constant for simplicity.

(a) In steady state show that $E_\theta = 0$ and $E_z = E_0 = $ constant.
(b) Show that B_z satisfies the differential equation

$$\frac{dB_z}{dr} = \frac{\eta_\| - \eta_\perp}{\eta_\| \eta_\perp} \frac{B_\theta B_z}{B_\theta^2 + B_z^2} \mu_0 E_0$$

(c) Show that under the assumptions made, *no* solutions exist with B_z reversed for any positive values of $\eta_\|$, $\eta_\perp$. What does this suggest to you about the steady state operation of an RFP?

6.5 Show that in an ohmically heated tokamak the normal curvature, correct to order ε^2, is given by

$$\kappa_n(r, \theta) = -\frac{B_{\theta 1}^2}{r B_0^2} - \left(1 - \frac{r \cos \theta}{R_0}\right) \frac{\cos \theta}{R_0} + \frac{\Delta \sin^2 \theta}{r R_0}$$

Here, $\Delta(r) = -\psi_1(r)/\psi_0'(r)$ is the shift of the flux surfaces.
6.6 Show that Shafranov's formula for the vertical field required to hold a tokamak in toroidal force balance

$$B_V = \frac{\mu_0 I}{4\pi R_0} \left(\beta_p + \frac{l_i - 3}{2} + \ln \frac{8R_0}{a}\right)$$

can be written in the following more intuitive form:

$$2\pi R_0 I_0 B_V = \frac{I_0^2}{2} \frac{\partial}{\partial R_0} (L_e + L_i) + 4\pi^2 \int_0^a \left(p - \frac{B_0 B_{\phi 2}}{\mu_0}\right) r \, dr$$

where L_e and L_i are the external and internal inductances associated with the plasma current. Hint: Make use of the fact that L_i is linearly dependent on R_0.

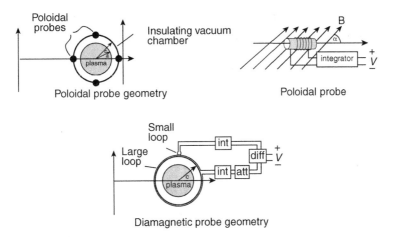

Figure 6.32   Diagram for Problem 6.7.

6.7 Consider an ohmically heated tokamak plasma held in equilibrium by a vertical field B_V. It is desired to measure β_p, l_i, and $\Delta(a)$ by placing various magnetic probes just outside the vacuum chamber, which for simplicity is assumed electrically insulating, and whose radius is c (see Fig. 6.32).

(a) Consider an individual probe as illustrated. Assume the probe has N turns and cross sectional area A. The **B** field makes an angle α with the axis of the probe. What signal appears on the output of the integrator?

(b) Show that the toroidal correction to the flux function ψ_1 (r, θ), evaluated at a radius r in the vicinity of the vacuum chamber, is given by

$$\psi_1(r,\theta) = \frac{I_0}{2\pi}\left[\frac{r^2 - a^2}{2r}\left(\beta_p + \frac{l_i - 1}{2}\right) + \frac{r}{2}\ln\frac{r}{a} - \Delta(a)\frac{R_0}{r}\right]\cos\theta$$

(c) Calculate the difference in signals from two "B_θ" probes located at $r = c, \theta = 0$ and $r = c, \theta = \pi$

(d) Calculate the signal on a "B_r" probe located at $r = c$, $\theta = \pi/2$.

(e) Assuming $c = a(1 + \delta)$ with $\delta \ll 1$, show how the "B_r" and "B_θ" signals can be combined to give separate measurements of $\Delta(a)$ and $\beta_p + l_i/2$.

(f) Consider now a set of "diamagnetic" probes as shown in the diagram. This system consists of a small B_ϕ probe located at $r = c, \theta = \pi/2$ and a large single-turn loop of radius c, surrounding the vacuum chamber. By what factor must the large loop signal be attenuated in order to produce a null signal at the output of the "difference" circuit when no plasma is present? Calculate the difference signal V when plasma is present and show how it can be used to determine β_p.

(g) Which of these measurements do you think might be difficult in a high-field experiment like Alcator C? Explain.

6.8 Consider a low β ohmically heated tokamak. Probe measurements on the outside of the discharge indicate that

$$\beta_p + \frac{l_i}{2} = 0.75$$

Profile measurements indicate peaked densities such that

$$n(r) = CT^2(r), \quad C = \text{const}$$

In addition the plasma diamagnetism is non-zero and for simplicity is modeled as

$$\frac{B_{\phi 2}}{B_0} = \lambda \frac{\mu_0 p}{B_0^2}$$

with λ a constant, yet to be determined. The above data correspond to the "steady state" flat top portion of the current in an ohmic discharge. Assume Spitzer resistivity: $\eta = KT^{-3/2}$ with $T_e = T_i = T$ and $K = 3.0 \times 10^{-8}$, T in keV, η in Ωm.

(a) Calculate the profile $B_\theta(r)/B_\theta(a)$.
(b) If the discharge is operating just at the onset of sawtooth oscillations (i.e., $q_0 = 1$) calculate $q(a)$ assuming $J_\phi(a) = 0$.
(c) Calculate β_p and the diamagnetic constant λ. Is the discharge diamagnetic or paramagnetic?
(d) If $B_0 = 10$ T, $R_0 = 0.64$ m, and $a = 0.16$ m, calculate the electron temperature if the toroidal loop voltage is 1.6 V. See Table 6.5 for helpful relations involving Bessel functions $J_0(x)$ and $J_1(x)$.

Table 6.5 *Helpful relations for Bessel functions $J_0(x)$, $J_1(x)$.*

$$xJ_0'' + J_0' + xJ_0 = 0$$
$$x^2 J_1'' + xJ_1' - (1 - x^2)J_1 = 0$$
$$J_1 = -J_0'$$
$$J_0 \approx 1 \text{ as } x \to 0$$
$$J_1 \approx x/2 \text{ as } x \to 0$$
$$\int xJ_0^2 dx = x^2(J_0^2 + J_1^2)/2$$
$$\int xJ_1^2 dx = x^2(J_0^2 + J_1^2)/2 - xJ_0J_1$$
$$\int \frac{J_0}{J_1}dx = \ln(xJ_1)$$
$$\int \frac{xJ_0^2}{J_1^2}dx = -\frac{xJ_0}{J_1} + 2\ln(xJ_1) - x^2/2$$
$$J_0(x_0) = 0 \to x_0 = 2.4 \text{ (first zero)}$$
$$J_1(x_0) = 0.52$$

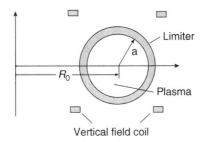

Figure 6.33   Diagram for Problem 6.9.

6.9 This problem demonstrates a practical application of MHD theory to the diagnosing of tokamak experiments. As a particular example, consider the geometry in Fig. 6.33. Assume that $B_0 = 8$ T, $R_0 = 0.64$ m, and $a = 0.16$ m and that the discharge is operating just at the onset of sawtooth oscillations (i.e., $q_0 = 1$). As a simple model of transport assume $n(r) \sim T(r) \sim J_\phi(r)$ and that $J_p(r) = 0$. Using the ohmically heated tokamak expansion calculate:

(a) $B_\theta (r)$
(b) $J_\theta (r)$
(c) $p (r)$
(d) I
(e) $q_a \approx q_*$
(f) β_t
(g) $B_t (0)$
(h) β_p
(i) l_i
(j) B_V (for zero shift of the outer surface)
(k) Δ_0 (shift of the magnetic axis relative to the outer surface).

See Table 6.5 for helpful relations involving Bessel functions $J_0 (x)$ and $J_1 (x)$.

6.10 The MHD equations describing the equilibrium of a plasma with an external force derivable from a potential are given by

$$\mathbf{J} \times \mathbf{B} - \nabla p - \rho\nabla\phi = 0$$

$$\nabla \times \mathbf{B} = \mu_0\mathbf{J}$$

$$\nabla \cdot \mathbf{B} = 0$$

Assume (1) two-dimensional slab symmetry (all quantities are functions only of x and y), (2) $p = (T/m_i)\rho$ with T a constant, (3) all three components of $\mathbf{B}$ are nonzero, and (4) $\phi (x, y)$ is a known prescribed function.

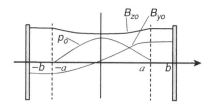

Figure 6.34   Diagram for Problem 6.10.

(a) Derive a two-dimensional Grad–Shafranov equation for this configuration. Hints: (1) Use A (the z component of vector potential) as the flux function; (2) show that $p \exp(m\phi/T) = \Pi(A)$, where $\Pi(A)$ is arbitrary; (3) show that $B_z = F(A)$, where $F(A)$ is arbitrary; and (4) the final equation is given by

$$\nabla^2 A = -FF' - \Pi' \exp(m_i\phi/T)$$

(b) Show that in the one-dimensional gravitational case ($\phi = -gx$ with $g = $ const), the Grad–Shafranov equation reduces to the appropriate one-dimensional limit.

(c) Consider now a two-dimensional case

$$m_i\phi/T = \varepsilon f(x)y/L$$

with $\varepsilon \ll 1, y \geq 0, -b \leq x \leq b$. Assume the currents and pressure are confined to the region $-a < x < a$ and that p and B_z are even about $x = 0$ while B_y is odd (see Fig. 6.34). The region between a and b is a vacuum region and at $x = \pm b$ there is a perfectly conducting boundary. Expand $A(x, y) = A_0(x) + A_1(x, y)$. Find a solution for A_1 expressed in terms of integrals of the zeroth-order quantities.

(d) Derive an expression for the shift of the last flux surface carrying current for the case $f(x) = -x/a$. Sketch the contour describing this surface, including the correction, and give a physical description of the surface distortion resulting from $f(x)$.

6.11 It has been known since the time of Newton that the Earth's gravitational field causes objects to "fall" towards the center of the Earth. Thus, if unsupported, a tokamak plasma in a 200-msec discharge would fall almost 0.2 m during its lifetime. The purpose of this problem is to calculate the equilibrium of an axisymmetric toroidal tokamak under the influence of a downward gravitational force and to determine how effective the magnetic forces are in supporting the mass of the plasma against gravity.

(a) Consider the axisymmetric geometry shown in Fig. 6.35. Assume there is an external potential $\phi(R, Z)$ and calculate the Grad–Shafranov equation for this system. For simplicity, assume $p = (T/m_i)\rho$ with $T = $ const.

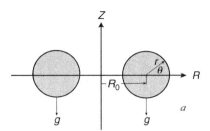

Figure 6.35   Diagram for Problem 6.11.

(b) What choice of ϕ (R, Z) corresponds to a downward gravitational force?
(c) Use this choice and treat ϕ as a small quantity; that is, assume gravitational effects enter in the same order as toroidal corrections. Expand the flux function as $\psi = \psi_0\ (r) + \psi_1\ (r,\ \theta)$ and find the contribution to $\psi_1\ (r,\ \theta)$ due to gravity. For simplicity, assume there is a conducting wall at $r = b$.
(d) Calculate the influence of gravity on the shape and position of the last flux surface that carries current. Estimate the size of these corrections for typical fusion plasmas. Is gravity an important effect?

6.12 Consider the simple model of the high β tokamak discussed in Section 6.5.

Assume that as the plasma evolves it is constrained by stability considerations to operate at $q_0 = 1$, rather than at fixed I or in a flux-conserving mode. Show that when $q_0 = 1$, the maximum achievable plasma β_t does not occur at the equilibrium limit $v = 1$, but at some lower value. Calculate the value of v and the corresponding value of $\beta_t q_0^2/\varepsilon$ at the optimum condition (see Fig. 6.36).

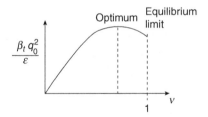

Figure 6.36   Diagram for Problem 6.12.

6.13 In Section 6.5 it was shown, using a simple elliptic model of a tokamak, that the highest equilibrium β_t occurs when $v = 1$ and $\kappa \to \infty$ assuming a fixed ε, q_*. Repeat the analyses assuming that the volume and major radius are held fixed rather than ε, q_*. Show that in the case there is an optimum elongation corresponding to the maximum β_t. Find the optimum κ and the corresponding values of β_t, J_ϕ, and I_0 and compare them to the circular case.

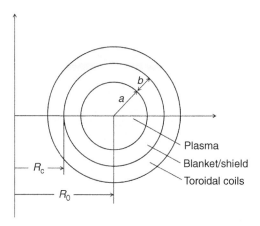

Figure 6.37   Diagram for Problem 6.14.

6.14 Consider a simple model of a superconducting tokamak fusion reactor as illustrated in Fig. 6.37.

(a) Derive an expression for the total thermal power P_T associated with the 14-MeV neutrons.

(b) Assume the wall loading $P_W \equiv P_T/S$ is fixed at the largest value consistent with the properties of the first wall material. Here, S is the plasma surface area. Also, assume the toroidal field coils operate at the maximum permissible value B_c which maintains superconductivity. Note that $B_\phi = B_c$ at $R = R_c$. Derive an expression for the required value of β_t/ε in terms of P_T, P_W, T_i, ε, ε_T, and B_c. Here, $\varepsilon = a/R_0$, $\varepsilon_T = (a + b)/R_0$, $T_i = T_e = T/2$, and β_t is defined as $\beta_t = 4\mu_0 n T_i/B_0^2$, with B_0 the field at the center of the plasma, $R = R_0$.

(c) Assume P_T, P_W, T_i, B_c, and b are fixed. Show that the required β_t/ε exhibits a minimum as a function of ε. Derive an expression for the optimum aspect ratio.

(d) At the optimum point calculate the values of ε, β_t/ε, β_t, a, and R_0 if $T_i = 15$ keV, $P_T = 3500$ MW, $P_W = 4$ MW/m^2, $b = 1$ m, and $B_c = 11$ T. Note: At $T_i = 15$ keV, $\langle\sigma v\rangle = 3 \times 10^{-22}$ m^{-3}/sec for D-T fusion reactions.

7

Equilibrium: three-dimensional configurations

7.1 Introduction

Two-dimensional configurations with toroidal axisymmetry have been investigated in Chapter 6. Many fusion concepts fall into this class – tokamaks of all types, the reversed field pinch, the levitated dipole, the spheromak, and the field reversed configuration. One common feature in each of these concepts is the need for a toroidal current to provide toroidal force balance, either using a perfectly conducting shell or a vertical field.

The need for a toroidal current is of particular importance to the tokamak and RFP, the most advanced of the axisymmetric configurations. The reason is that it is not possible to drive a DC toroidal current indefinitely with a transformer, the method now used in pulsed versions of these configurations. This conflicts with the general consensus that a magnetic fusion reactor must operate as a steady state device for engineering reasons to avoid cyclical thermal and mechanical stresses inherent in a pulsed device. In other words, some form of non-inductive current drive is required. This is an active area of research and while a scientifically sound and technologically viable technique may be possible theoretically, success still depends on current and future experimental development. Overall, non-inductive current drive represents a difficult challenge for the tokamak and RFP concepts.

The difficulty of non-inductive current drive is one of the primary motivations for investigating 3-D MHD concepts. Three-dimensional systems, although obviously more complicated geometrically, offer the potential of providing radial pressure balance and toroidal force balance without the need of a non-inductive current drive system.

Over the years a number of 3-D toroidal concepts have been proposed as possible fusion reactors. The only one that has survived the rigorous challenges of plasma physics is the stellarator. Consequently, Chapter 7 is focused on the MHD equilibrium of stellarators.

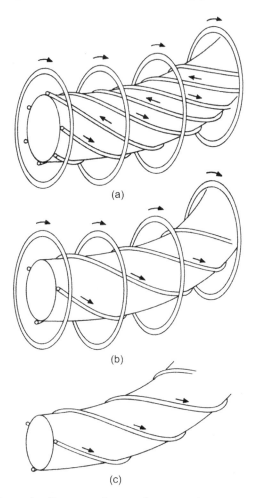

Figure 7.1   Schematic diagram of several types of stellarator windings: (a) original stellarator, (b) heliotron, and (c) torsatron.

Conceptually, a stellarator can be viewed as a straight helix bent into a torus. It is fundamentally a 3-D configuration. For purposes of intuition, several types of stellarator windings are illustrated schematically in Fig. 7.1. The plasma physics advantages of a stellarator can be summarized as follows. First, since stellarators can be held in equilibrium without a net current, or more accurately with only the naturally occurring bootstrap current, there is no need for an external non-inductive current drive system, a major advantage indeed. The stellarator is inherently a steady state device. Second, with zero, or at most a relatively small net bootstrap current, the threat of current-driven major disruptions is greatly if not completely alleviated. Third, even without plasma a vacuum stellarator magnetic field has well-defined closed nested flux surfaces with finite rotational transform and a unique magnetic axis located in the center of the vacuum chamber. The magnetic field acts like a cage surrounding and confining the plasma. No cooperation from the plasma in the form of a large toroidal current is required for toroidal force balance.

These are considerable advantages. There are two counterbalancing disadvantages, one involving plasma physics and the other, engineering. The plasma physics problem is concerned with single-particle confinement and neoclassical transport. Since a stellarator is a three-dimensional configuration there are no exact conserved constants of the motion governing the trajectory of a particle except for the total energy. As shown in Section 7.7 this absence of conserved constants can lead to a rapid loss of particles through helical "loss cones." The resulting neoclassical losses can even exceed anomalous transport and are potentially a very serious problem. The resolution of this problem lies with the invention of two ideas, the quasi-symmetric stellarator and the quasi-isodynamic stellarator, which are special examples of the more general quasi-omnigenous stellarator. These are also explained in Section 7.7. An important geometric feature of these configurations is the necessity of a helical (rather than a circular toroidally symmetric planar) magnetic axis. Whether these new ideas will increase the energy confinement time in a stellarator so that it is equal to or better than that of a tokamak is an area of current interest with the results not yet in.

The second disadvantage involves the coil design. As might be expected the magnet system needed to produce a 3-D stellarator magnetic field is substantially more complicated than that of a 2-D toroidally symmetric configuration. This leads to increased costs and perhaps even to a lower limit on the maximum achievable magnetic field in the plasma because of the need to limit the magnetic stresses in the tight curvature sections of the 3-D coils. Examples of modern stellarator magnet systems are presented in Section 7.8 which describes two of the major stellarators in the world fusion program, the Large Helical Device (LHD) in Japan and Wendelstein 7-X (W7-X) in Germany.

In the end these disadvantages must be weighed against the advantages associated with not requiring an external non-inductive current drive system and the possibility of disruption-free operation. With respect to the overall "competition" between stellarators and tokamaks, it is fair to say that at the time this book is being written, similar size devices perform comparably well, but with the tokamak having an overall edge, primarily because of better energy transport.

Turning to specifics it is perhaps not surprising that a great deal of analysis is required to understand the stellarator concept. To accomplish this task the material in Chapter 7 is organized as follows. The discussion begins with the derivation of a recently developed asymptotic model applicable to modern stellarators. Qualitatively the new model describes a high $\beta \sim \varepsilon$, large aspect ratio stellarator with a loosely wound helical field. The model is analytically solved for a single helicity stellarator including the cases of a low $\beta_t \ll \varepsilon$ system with arbitrary profiles and a high $\beta_t \sim \varepsilon$ system with the Solov'ev profiles. The goal here is to learn the answers to such basic questions as

- If a stellarator has no, or very small, net current how is a force induced to balance the tire tube and $1/R$ outward forces along R?
- If a stellarator has zero, or very small, net current how can changing the vertical field move the plasma inward or outward along R, as is observed experimentally?
- Does a stellarator have an analogous β_t/ε equilibrium limit similar to the one that occurs in a high beta tokamak?

The stellarator model is also solved numerically for a multiple helicity magnetic field for $\beta_t = 0$ and $\beta_t \sim \varepsilon$. It is shown how such systems, even with only two helicities, lead to very complicated sets of flux surfaces.

Lastly, to understand the problem of single-particle confinement in a 3-D geometry, the basic principles of neoclassical transport theory are described. This discussion provides the motivation for the concept of quasi-omnigenity and the corresponding inventions of quasi-symmetric and quasi-isodynamic stellarators. These ideas are then applied to the world's two largest stellarators, LHD and W7-X.

7.2 The high β stellarator expansion

7.2.1 Introduction

To put the present model in context we note that early in the fusion program, before the advent of large high-speed computers, much of MHD stellarator research was carried out using various approximate models, often based on a large aspect ratio or magnetic axis expansion (Greene and Johnson, 1961). These models provided valuable physical insight but, with respect to present day stellarators, are neither accurate enough for precision experimental design nor flexible enough to take into account neoclassical transport requirements. In fact the neoclassical transport requirements had not yet been elucidated during the period when these early theories were developed.

Since then, stellarator research has made great advances with the development of large high-speed computers. Sophisticated 3-D MHD codes have been developed that provide accurate information for experimental design and data analysis including neoclassical constraints resulting in configurations that are quasi-omnigenous, quasi-symmetric, or quasi-isodynamic (see for instance Boozer, 2004, and Helander *et al.*, 2012). However, with such a strong reliance on numerics, some of the physical insight and intuition obtained from simple analytic models has been substantially reduced if not lost. Bridging the gap between the modern 3-D MHD numerical codes and the simpler earlier analytic theories is the focus of the present model.

The basic goal is the development of a relatively simple analytic model that describes the MHD cquilibria of present day stellarators. It is important that, despite its simplicity, the model should allow for a non-planar helical magnetic axis, one the basic constraints imposed by favorable neoclassical transport.

The present work meets this challenge by means of a new high β, large aspect ratio expansion. The end result is a simplified model with the following properties:

- The model treats $\beta \sim \varepsilon$, which is the interesting regime of reactor interest.
- The model allows a finite $l = 1$ non-planar helical magnetic axis.
- The model allows a finite $l = 0$ mirror field modulation (such as in W7-X).
- The model allows finite flux surface modulations due to all $l \geq 2$ helical field components.
- The model treats $\iota/2\pi \sim 1$, which is the interesting regime of reactor interest.

In spite of the "simplicity" of the final model substantial analysis is still required because of the complexities associated with a general 3-D geometry. For ease of presentation the derivation is first carried out for the special case where there is no $l = 0$ mirror field. At the end of the analysis it is shown how the calculation can be generalized to include an $l = 0$ field. The derivation of the stellarator expansion follows below.

7.2.2 The basic equations

The stellarator equilibria of interest are described by the usual ideal MHD equations

$$
\begin{aligned}
\mathbf{J} \times \mathbf{B} &= \nabla p \\
\nabla \times \mathbf{B} &= \mu_0 \mathbf{J} \\
\nabla \cdot \mathbf{B} &= 0
\end{aligned}
\tag{7.1}
$$

The goal is to investigate stellarator equilibria by means of a large aspect ratio expansion. To exploit the large aspect ratio assumption it is convenient to rewrite the MHD equations in an equivalent but alternate form. Also, to help follow the analysis the new equations are listed in the order in which they are to be solved. The alternate form can be written as

$$
\begin{aligned}
\nabla \cdot \mathbf{B} &= 0 \\
\nabla \times \mathbf{B} &= \mu_0 \mathbf{J} \\
\mathbf{J}_\perp &= (\mathbf{B} \times \nabla p)/B^2 \\
\mathbf{B} \cdot \nabla p &= 0 \\
\nabla \cdot \mathbf{J} &= 0
\end{aligned}
\tag{7.2}
$$

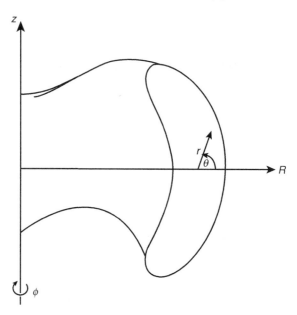

Figure 7.2   Geometry for the stellarator analysis.

7.2.3 The high β stellarator expansion

A stellarator equilibrium consists of a large toroidal magnetic field, small helical and axisymmetric poloidal magnetic fields, and a small pressure.[1] In carrying out the derivation one shall see that five independent dimensionless parameters appear in the analysis:

Inverse aspect ratio	$\varepsilon = a/R_0$
Normalized helical field amplitude	$\delta = |\mathbf{B}_p|/B_\phi$
Normalized plasma pressure	$\beta = 2\mu_0 p/B_\phi^2$
Poloidal periodicity mode number	l
Toroidal periodicity mode number	N

$$(7.3)$$

Here, B_ϕ and $\mathbf{B}_p$ are the toroidal and poloidal magnetic fields respectively. Note also that $\mathbf{B}_p$ contains a sum of helical and axisymmetric harmonics each of which depends on the minor radius, poloidal angle, and toroidal angle as follows:

$$\mathbf{B}_p(\mathbf{r}) = \sum_{l,N} \mathbf{B}_p^{(N,l)}(r, l\theta + N\hat{\phi}) \qquad (7.4)$$

See Fig. 7.2 for the geometry. The axisymmetric component of $\mathbf{B}_p$ (e.g., the vertical field) corresponds to $N = 0$.

---

[1] A "small" pressure corresponds to $\beta_t \sim \varepsilon$ which for historical reasons is referred to as "high beta" because it is much larger than a "low beta" system corresponding to $\beta_t \sim \varepsilon^2$.

Table 7.1 *The high β stellarator expansion.*

Quantity	High β stellarator expansion
δ	ε
β	ε
N	1
l	1

The high β stellarator expansion is summarized in Table 7.1. The inverse aspect ratio $\varepsilon \ll 1$ is used as the basic expansion parameter and all other quantities are ordered with respect to ε.

A primary motivation for the expansion is that the ordering for the rotational transform (except for $l = 1$) must be chosen such that $\iota/2\pi \sim 1$. The ordering requirements to achieve this property can be ascertained by recalling that the rotational transform for a loosely wound, $l \geq 2$, small amplitude single helicity stellarator, derived in Section 6.7.4, is, in the present notation, given by

$$\frac{\iota}{2\pi} \approx \frac{(l-1)\delta^2}{N\varepsilon^2} \left(\frac{r}{a}\right)^{2(l-2)} \tag{7.5}$$

where r is the local plasma radius, a is the average radius of the plasma surface and $N = hR_0$ with $2\pi/h$ the helical wavelength.

Now, analysis shows that in order to produce a stellarator with a non-planar magnetic axis it is essential to assume $N \sim 1$ and indeed this is the ordering used in the expansion presented here. Physically, the implication of $N \sim 1$ is that there are a finite number of helical periods around the torus. Then, Eq. (7.5) implies that in order to make the rotational transform finite ($\iota/2\pi \sim 1$) one must order $\delta \sim \varepsilon$. Lastly, the value of beta is chosen to scale as $\beta \sim \varepsilon$ which is similar to the high beta tokamak ordering described in Section 6.5.

In order to carry out the expansion it is also necessary to distinguish the ordering of the poloidal vs. toroidal coordinates. Based on the orderings given in Table 7.1 one introduces the usual toroidal coordinates leading to

$$R = R_0 + r\cos\theta \qquad \nabla = \nabla_\perp + \nabla_\parallel$$

$$Z = r\sin\theta \qquad a\nabla_\perp = a\left(\mathbf{e}_r\frac{\partial}{\partial r} + \mathbf{e}_\theta\frac{1}{r}\frac{\partial}{\partial\theta}\right) \sim 1 \tag{7.6}$$

$$\hat{\phi} = -\phi/N \qquad a\nabla_\parallel = \mathbf{e}_\phi\frac{aN}{R}\left(\frac{\partial}{\partial\phi}\right) = \mathbf{e}_\phi\frac{aN}{R_0 + r\cos\theta}\left(\frac{\partial}{\partial\phi}\right) \sim \varepsilon$$

where R_0 is the toroidally averaged major radius of the magnetic axis, a is the toroidally averaged minor radius of the plasma, and R, $\hat{\phi}$, Z refer to the actual laboratory coordinate system. Note that normalizing ϕ by N implies that one helical period corresponds to $0 \leq \phi \leq 2\pi$ so that the toroidal angle ϕ is related to the straight stellarator coordinates by $\phi = hz$. Also, $\mathbf{e}_R = \mathbf{e}_r \cos\theta - \mathbf{e}_\theta \sin\theta$, $\mathbf{e}_Z = \mathbf{e}_r \sin\theta + \mathbf{e}_\theta \cos\theta$, and $\mathbf{e}_{\hat\phi} = -\mathbf{e}_\phi$. With these definitions it follows that R, $\hat\phi$, Z and r, θ, ϕ are right-handed coordinate systems.

Here and below note that the characteristic length and magnetic field used to scale the variables appearing in the formulation are the minor radius a and the vacuum toroidal field at the midplane B_0 respectively.

Next, for consistency with the high β stellarator ordering, the expansions for the magnetic field, current density, and pressure must be chosen as

$$\mathbf{B} = \left(B_{\phi 0} \frac{R_0}{R} + B_{\phi 1} \right) \mathbf{e}_\phi + \mathbf{B}_{p1} \approx \left(B_{\phi 0} - B_{\phi 0} \frac{r}{R_0} \cos\theta + B_{\phi 1} \right) \mathbf{e}_\phi + \mathbf{B}_{p1}$$
$$\mathbf{J} = J_{\phi 1}\mathbf{e}_\phi + \mathbf{J}_{p1} \tag{7.7}$$
$$p \approx p_1$$

where the numerical subscripts refer to the order in ε and the terms listed represent the minimum number required to obtain a closed set of equations. Also, the $(r/R_0)\cos\theta$ term is the large aspect ratio expansion of the vacuum toroidal field.

7.2.4 Reduction of the equations

The $\nabla \cdot \mathbf{B} = 0$ equation

The analysis begins with the application of the expansion to the $\nabla \cdot \mathbf{B} = 0$ equation. It is shown that the poloidal magnetic field $\mathbf{B}_{p1}$ can be written in terms of a scalar stream function A_1 equivalent to the toroidal component of vector potential. The analysis proceeds by first noting that to leading order the toroidal field is a vacuum field, a consequence of the assumption that the current density is a first-order quantity. Therefore the leading-order toroidal field is written as

$$B_{\phi 0}(r, \theta, \phi) = B_0 = \text{constant} \tag{7.8}$$

It is here that the $l = 0$ mirror field has been set to zero. If such a field existed then one would have to replace B_0 with $B_0 M(\phi)$ where $M(\phi) = 1 + \Delta_0 \cos\phi$ with the Δ_0 representing the $l = 0$ mirror modulation. As stated however, the derivation below sets $M = 1$. At the end of the section it is shown how to add in the effects of a finite $\Delta_0 \sim 1$ mirror field.

Consider now the $\nabla \cdot \mathbf{B} = 0$ equation. The first non-vanishing contribution expressed in normalized coordinates reduces to

$$\nabla \cdot \mathbf{B} \approx \nabla_\perp \cdot \mathbf{B}_{p1} = 0 \tag{7.9}$$

The solution is easily obtained by exploiting the fact that $\nabla_\perp$ is a two-dimensional operator. This allows the introduction of a stream function $A_1(r,\theta,\phi)$ for the poloidal magnetic field as follows

$$\mathbf{B}_{p1} = \nabla_\perp A_1 \times \mathbf{e}_\phi \qquad (7.10)$$

In this equation ϕ appears only as a parameter.

The $\nabla \cdot \mathbf{B} = 0$ equation yields no information about $B_{\phi 1}$. The first non-vanishing appearance of $B_{\phi 1}$ in the $\nabla \cdot \mathbf{B} = 0$ equation scales as $\nabla_\parallel \cdot (B_{\phi 1} \mathbf{e}_\phi) \sim \varepsilon^2$ and is therefore not required for the analysis. To summarize, the total magnetic field used in the analysis has the form

$$\mathbf{B} = B_0 \left(1 - \frac{r}{R_0} \cos\theta + \frac{B_{\phi 1}}{B_0} \right) \mathbf{e}_\phi + \nabla_\perp A_1 \times \mathbf{e}_\phi \qquad (7.11)$$

where $A_1(r,\theta,\phi)$ and $B_{\phi 1}(r,\theta,\phi)$ are unknowns.

The $\mu_0 \mathbf{J} = \nabla \times \mathbf{B}$ equation

Ampere's law is given by $\mu_0 \mathbf{J} = \nabla \times \mathbf{B}$. A simple calculation leads to the following expression for the first-order current density used in the analysis:

$$\mu_0 \mathbf{J}_1 = -\mathbf{e}_\phi \nabla_\perp^2 A_1 - \mathbf{e}_\phi \times \nabla_\perp B_{\phi 1} \qquad (7.12)$$

The $\mathbf{J}_\perp = (\mathbf{B} \times \nabla p)/B^2$ equation

Perpendicular pressure balance yields an alternate form for $\mathbf{J}_\perp$ which ultimately leads to a relation between $B_{\phi 1}$ and p_1. A simple calculation shows that $\mathbf{J}_\perp \approx \mathbf{J}_{p1}$ where, from perpendicular pressure balance

$$\mathbf{J}_{p1} = \frac{1}{B_0} \mathbf{e}_\phi \times \nabla_\perp p_1 \qquad (7.13)$$

Now, Eq. (7.13) is equated with the poloidal component of Eq. (7.12) leading to

$$\mathbf{e}_\phi \times \nabla_\perp \left(p_1 + \frac{B_0 B_{\phi 1}}{\mu_0} \right) = 0 \qquad (7.14)$$

The solution consistent with the boundary condition that the pressure vanish at the edge of the plasma is

$$p_1 + \frac{B_0 B_{\phi 1}}{\mu_0} = 0 \qquad (7.15)$$

This is just the usual θ-pinch radial pressure balance relation locally valid at every value of ϕ.

Summary of the fields

At this point it is convenient to summarize the expressions for the magnetic field and current density which can be written in terms of two scalar quantities A_1 and p_1:

$$\mathbf{B} = B_0 \left(1 - \frac{r}{R_0} \cos \theta - \frac{\mu_0 p_1}{B_0^2} \right) \mathbf{e}_\phi + \nabla_\perp A_1 \times \mathbf{e}_\phi$$

$$\mu_0 \mathbf{J}_1 = - \mathbf{e}_\phi \nabla_\perp^2 A_1 - \nabla_\perp \left(\frac{\mu_0 p_1}{B_0} \right) \times \mathbf{e}_\phi \tag{7.16}$$

What remains now is to derive the basic equations that determine A_1 and p_1.

The $\mathbf{B} \cdot \nabla$p = 0 equation

The first non-vanishing contribution to parallel pressure balance occurs in second order and is given by

$$\mathbf{B} \cdot \nabla p_1 = 0 \tag{7.17}$$

where

$$\mathbf{B} \cdot \nabla = \frac{B_0 N}{R_0} \frac{\partial}{\partial \phi} + \mathbf{B}_{p1} \cdot \nabla_\perp = \frac{B_0 N}{R_0} \frac{\partial}{\partial \phi} - \mathbf{e}_\phi \times \nabla_\perp A_1 \cdot \nabla_\perp \tag{7.18}$$

This is the first equation relating A_1 and p_1.

The $\nabla \cdot \mathbf{J} = 0$ equation

The final equation that closes the system is $\nabla \cdot \mathbf{J} = 0$. This equation can be written as

$$\mathbf{B} \cdot \nabla \frac{J_\parallel}{B} + \nabla \cdot \mathbf{J}_\perp = 0 \tag{7.19}$$

The first non-vanishing contribution also occurs in second order. The $J_\parallel$ term simplifies by noting that $J_\parallel \approx J_{\phi 1}$ and then substituting from Eq. (7.16).

$$\mathbf{B} \cdot \nabla \frac{J_\parallel}{B} = -\mathbf{B} \cdot \nabla \left(\frac{\nabla_\perp^2 A_1}{\mu_0 B_0} \right) \tag{7.20}$$

The $\mathbf{J}_\perp$ term in Eq. (7.19) simplifies by using perpendicular pressure balance

$$\nabla \cdot \mathbf{J}_\perp = \nabla \cdot \left(\frac{\mathbf{B} \times \nabla p}{B^2} \right)$$

$$= \frac{1}{B^4} \nabla_\perp p \times \mathbf{B} \cdot \nabla_\perp B^2$$

$$= \frac{2}{B^2} \nabla_\perp p \times \mathbf{B} \cdot \boldsymbol{\kappa} \tag{7.21}$$

where $\kappa = \mathbf{b} \cdot \nabla \mathbf{b}$ is the curvature vector, $\mathbf{b} = \mathbf{B}/B$, and use has been made of the relation $\mathbf{J} \cdot \nabla p = 0$. The first non-vanishing contribution to the curvature vector is due to toroidicity; that is if one writes $\mathbf{b} \approx \mathbf{e}_\phi + \mathbf{B}_{p1}/B_0$ then

$$\kappa \approx \mathbf{e}_\phi \cdot \nabla \mathbf{e}_\phi \approx -\frac{\mathbf{e}_R}{R_0} = -\frac{1}{R_0}(\mathbf{e}_r \cos\theta - \mathbf{e}_\theta \sin\theta) \qquad (7.22)$$

This expression is substituted into Eq. (7.21) yielding

$$\nabla \cdot \mathbf{J}_\perp = -\frac{2}{R_0 B_0} \nabla_\perp p_1 \cdot \mathbf{e}_z \qquad (7.23)$$

The two contributions to $\nabla \cdot \mathbf{J} = 0$ can now be combined leading to

$$\mathbf{B} \cdot \nabla(\nabla_\perp^2 A_1) = -\frac{2\mu_0}{R_0} \nabla_\perp p_1 \cdot \mathbf{e}_z \qquad (7.24)$$

where $\mathbf{B} \cdot \nabla$ is again given by Eq. (7.18). This slightly more complicated equation describes the second relation between A_1 and p_1.

Equations (7.17) and (7.24) describe the basic high β stellarator expansion in the absence of an $l = 0$ mirror field.

Including a finite l = 0 mirror field

The inclusion of a finite $l = 0$ mirror field requires a generalization of the form for the background toroidal field and the introduction of a set of simple flux coordinates. Specifically, the equilibrium toroidal field is replaced by $B_{\phi 0} = B_0 \to B_0 M(\phi)$ where[2]

$$M(\phi) = 1 + \Delta_0 \cos\phi \qquad (7.25)$$

Here the ordering for the modulation amplitude is $\Delta_0 \sim 1$. Note that for M to represent a vacuum field (i.e., $\nabla \times \mathbf{B}_{\phi 0} = 0$) consistent with the high β stellarator expansion one requires that $M(r,\theta,\phi) \to M(\phi)$.

The derivation including an $l = 0$ field closely parallels the one just presented without such a field. The details are as follows. The $\nabla \cdot \mathbf{B} = 0$ equation reduces to

$$\nabla \cdot \mathbf{B} \approx \nabla_\perp \cdot \mathbf{B}_{p1} + \frac{B_0 N}{R_0}\dot{M} = 0 \qquad (7.26)$$

with an over dot denoting $d/d\phi$. For the case above $\dot{M} = dM/d\phi = -\Delta_0 \sin\phi$. The solution is easily found and is given by

---

[2] The derivation described here can be further generalized to include multiple toroidal harmonics in ϕ (e.g. terms proportional to $\cos 2\phi$, $\cos 3\phi$, etc.). However, this generalization is not needed for present purposes.

$$\mathbf{B} = B_{\phi 0}\left(1 - \frac{r}{R_0}\cos\theta + \frac{B_{\phi 1}}{B_{\phi 0}}\right)\mathbf{e}_\phi + \nabla_\perp A_1 \times \mathbf{e}_\phi - \frac{N}{2R_0}\dot{B}_{\phi 0}r\mathbf{e}_r \qquad (7.27)$$

Next, Amperes Law ($\nabla \times \mathbf{B} = \mu_0 \mathbf{J}$) and perpendicular pressure balance ($\mathbf{J}_\perp = \mathbf{B} \times \nabla_p/B^2$) lead to a simple generalization of the θ-pinch radial pressure balance relation,

$$p_1 + \frac{B_{\phi 0}B_{\phi 1}}{\mu_0} = 0 \qquad (7.28)$$

These results are substituted into the parallel pressure balance relation ($\mathbf{B} \cdot \nabla p = 0$) and the parallel current relation ($\mathbf{B} \cdot \nabla(J_\parallel/B) = -\nabla \cdot \mathbf{J}_\perp$) leading to two equations for the basic unknowns $p_1(r,\theta,\phi)$ and $A_1(r,\theta,\phi)$,

$$\mathbf{B} \cdot \nabla p_1 = 0$$

$$\mathbf{B} \cdot \nabla\left(\frac{\nabla_\perp^2 A_1}{B_{\phi 0}}\right) = -\frac{2\mu_0}{R_0 B_{\phi 0}}\nabla_\perp p_1 \cdot \mathbf{e}_z \qquad (7.29)$$

$$\mathbf{B} \cdot \nabla = \left(\frac{N}{R_0}B_{\phi 0}\frac{\partial}{\partial\phi} - \frac{N}{2R_0}\dot{B}_{\phi 0}r\frac{\partial}{\partial r} - \mathbf{e}_\phi \times \nabla_\perp A_1 \cdot \nabla_\perp\right)$$

The final step in the formulation is to define a set of flux coordinates which further simplifies the equations. The coordinates are given by

$$u = [M(\phi)]^{1/2}\,r$$
$$\theta = \theta \qquad (7.30)$$
$$\phi = \phi$$

In the new coordinates

$$\frac{\partial}{\partial r} \rightarrow M^{1/2}\frac{\partial}{\partial u}$$

$$\frac{\partial}{\partial\theta} \rightarrow \frac{\partial}{\partial\theta}$$

$$\frac{\partial}{\partial\phi} \rightarrow \frac{\partial}{\partial\phi} + \frac{\dot{M}}{2M}u\frac{\partial}{\partial u}$$

$$\nabla_\perp \rightarrow M^{1/2}\nabla_\perp = M^{1/2}\left(\mathbf{e}_r\frac{\partial}{\partial u} + \frac{\mathbf{e}_\theta}{u}\frac{\partial}{\partial\theta}\right) \qquad (7.31)$$

$$\nabla_\perp^2 \rightarrow M\,\nabla_\perp^2 = M\left(\frac{\partial^2}{\partial u^2} + \frac{1}{u}\frac{\partial}{\partial u} + \frac{1}{u^2}\frac{\partial^2}{\partial\theta^2}\right)$$

$$\mathbf{B} \cdot \nabla \rightarrow M\mathbf{B} \cdot \nabla = M\left(\frac{B_0 N}{R_0}\frac{\partial}{\partial\phi} - \mathbf{e}_\phi \times \nabla_\perp A_1 \cdot \nabla_\perp\right)$$

Here and below $\nabla_\perp$ becomes the standard cylindrical operator applied to the poloidal coordinates u, θ (rather than r, θ) in Eq. (7.31).

Observe that u^2 is proportional to the local (in ϕ) vacuum toroidal flux: $u^2 \propto B_{\phi 0}(\phi) r^2$. However, unlike the usual transformation to flux coordinates, in the present analysis the flux (i.e., u) is not an unknown quantity but is explicitly known because $B_{\phi 0}(\phi) = M(\phi) B_0$ is assumed to have been specified.

The transformation given by Eq. (7.30) is substituted into Eq. (7.29) leading to a simplified form of the new stellarator model:

$$\left(\frac{B_0 N}{R_0} \frac{\partial}{\partial \phi} - \mathbf{e}_\phi \times \nabla_\perp A_1 \cdot \nabla_\perp \right) p_1 = 0$$

$$\left(\frac{B_0 N}{R_0} \frac{\partial}{\partial \phi} - \mathbf{e}_\phi \times \nabla_\perp A_1 \cdot \nabla_\perp \right) \nabla_\perp^2 A_1 = -\frac{2\mu_0}{R_0 M^{3/2}} \nabla_\perp p_1 \cdot \mathbf{e}_z$$

(7.32)

With flux coordinates the only difference between including and not including an $l = 0$ field is the appearance of $M^{-3/2}$ in the right-hand side of the second equation. However, the corresponding magnetic field and current density including an $l = 0$ field are more complicated. Expressed in the new coordinates, these quantities are given by

$$\mathbf{B} = B_0 M \left[1 - \frac{u}{R_0 M^{1/2}} \cos\theta - \frac{\mu_0 p_1}{B_0^2 M^2} \right] \mathbf{e}_\phi + M^{1/2} \nabla_\perp A_1 \times \mathbf{e}_\phi - \frac{B_0 N}{2R_0} \frac{\dot{M}}{M^{1/2}} u \, \mathbf{e}_r$$

$$\mu_0 \mathbf{J}_1 = - \mathbf{e}_\phi M \nabla_\perp^2 A_1 - \frac{\mu_0}{B_0 M^{1/2}} \nabla_\perp p_1 \times \mathbf{e}_\phi$$

(7.33)

Equations (7.32) and (7.33) describe the desired high β stellarator model in terms of un-normalized variables. One final simplifying step is to introduce normalized variables as follows:

$$u = a\rho$$
$$A_1 = -(\varepsilon N a B_0) A$$
$$p_1 = (\varepsilon N^2 B_0^2 / 2\mu_0) \beta$$

(7.34)

with $A \sim \beta \sim 1$ and as before $ha = \varepsilon N$. The high β stellarator model reduces to

$$\left(\frac{\partial}{\partial \phi} + \mathbf{e}_\phi \times \nabla_\perp A \cdot \nabla_\perp \right) \beta = 0$$

$$\left(\frac{\partial}{\partial \phi} + \mathbf{e}_\phi \times \nabla_\perp A \cdot \nabla_\perp \right) \nabla_\perp^2 A = \frac{1}{M^{3/2}} \nabla_\perp \beta \cdot \mathbf{e}_z$$

(7.35)

where $A(\rho, \theta, \phi)$ and $\beta(\rho, \theta, \phi)$ are the primary unknowns and

$$\nabla_\perp = \mathbf{e}_r \frac{\partial}{\partial \rho} + \frac{\mathbf{e}_\theta}{\rho} \frac{\partial}{\partial \theta} \tag{7.36}$$

is the standard cylindrical gradient operator written in ρ, θ (rather than u, θ) coordinates. The corresponding fields and current densities are given by

$$\frac{\mathbf{B}}{B_0} = \left[M \left(1 - \frac{\varepsilon}{M^{1/2}} \rho \cos \theta \right) - \frac{\varepsilon N^2}{2M} \beta \right] \mathbf{e}_\phi - \varepsilon N \left(M^{1/2} \nabla_\perp A \times \mathbf{e}_\phi + \frac{\dot{M}}{2M^{1/2}} \rho\, \mathbf{e}_r \right)$$

$$\frac{\mu_0 a\, \mathbf{J}_1}{B_0} = (\varepsilon N M \nabla_\perp^2 A)\, \mathbf{e}_\phi - \frac{\varepsilon N^2}{2M^{1/2}} \nabla_\perp \beta \times \mathbf{e}_\phi$$

$$\tag{7.37}$$

The examples discussed in the remainder of Chapter 7 are all based on Eq. (7.35) or its equivalent un-normalized form given by Eq. (7.32).

Islands

Before closing the discussion it is worth mentioning magnetic islands. Such islands can form near a separatrix or near rational flux surfaces within the plasma in multiple helicity systems. The basic high β stellarator model includes the effect of islands, which can be important in determining plasma transport. However, for reasons of mathematical simplicity these islands are always ignored (actually averaged out) whenever they arise in the examples that follow. In principle, one can assume that in a well-designed stellarator, the islands are very narrow in width and therefore would not have too large an effect on transport. Even so one must calculate the island width to verify this assumption, a task which only a few numerical codes can carry out. The point is that readers should be aware of the possibly important impact of islands even though they are ignored in the rest of the chapter.

7.3 Relation of the high β stellarator expansion to other models

At first glance, the high β stellarator model given by Eq. (7.32) is sufficiently different in appearance that it has no obvious connection or overlap with the previously derived two-dimensional models. In this section it is shown that such connections do indeed exist. Specifically, the high β stellarator model is shown to overlap with both the high β tokamak model and the straight stellarator model.

7.3.1 The high β tokamak

The high β tokamak is a special case of the high β stellarator obtained by utilizing the assumption of toroidal axisymmetry. For reference the basic high β tokamak equation (i.e., Eq. (6.96)) derived in Chapter 6 is repeated here:

$$\nabla^2 \psi_0 = -R_0^2 B_0^2 \left[\frac{1}{B_0} \frac{d\tilde{B}}{d\psi_0} + \frac{2\mu_0}{B_0^2} \frac{dp}{d\psi_0} \left(\frac{r}{R_0} \right) \cos \theta \right] \tag{7.38}$$

The goal now is to show that the high β stellarator equations reduce to Eq. (7.38) in the appropriate limit.

To demonstrate this one sets $\partial/\partial\phi = 0$ in Eq. (7.32). Similarly all helical field amplitudes including $l = 0$ must be set to zero ($M = 1$) implying that $u = r$. The stellarator equations reduce to

$$\begin{aligned} \mathbf{e}_\phi \times \nabla_\perp A_1 \cdot \nabla_\perp p_1 &= 0 \\ \mathbf{e}_\phi \times \nabla_\perp A_1 \cdot \nabla_\perp (\nabla_\perp^2 A_1) &= \frac{2\mu_0}{R_0} \nabla_\perp p_1 \cdot \mathbf{e}_Z \end{aligned} \tag{7.39}$$

Next, note that A_1 is related to the poloidal flux ψ_0 appearing in the high β tokamak equations by $\psi_0 = R_0 A_\phi = R_0 A_1$. Substituting into Eq. (7.39) yields

$$\begin{aligned} \mathbf{e}_\phi \times \nabla_\perp \psi_0 \cdot \nabla_\perp p_1 &= 0 \\ \mathbf{e}_\phi \times \nabla_\perp \psi_0 \cdot \nabla_\perp (\nabla_\perp^2 \psi_0) &= 2\mu_0 R_0 \nabla_\perp p_1 \cdot \mathbf{e}_\phi \times \nabla(r \cos \theta) \end{aligned} \tag{7.40}$$

The solution to the first equation is, as expected, $p_1 = p_1(\psi_0)$. When inserted into the second equation this leads to

$$\mathbf{e}_\phi \times \nabla_\perp \psi_0 \cdot \nabla_\perp \left(\nabla_\perp^2 \psi_0 + 2\mu_0 R_0 \frac{dp_1}{d\psi_0} r \cos \theta \right) = 0 \tag{7.41}$$

The solution to this equation is

$$\nabla_\perp^2 \psi_0 = -2\mu_0 R_0 \frac{dp}{d\psi_0} r \cos \theta + G(\psi_0) \tag{7.42}$$

which, with a different but equivalent definition of $G(\psi)$, is identical to the high β tokamak equation given by Eq. (7.38).

7.3.2 The straight stellarator

The second comparison of interest involves the straight stellarator. Two steps are required to show that the high β and straight stellarator models coincide in the region of physics overlap. In the straight stellarator it is necessary to take the limit of a low β loosely wound helix. In the high β stellarator one must consider a single helicity system in the limit of very large aspect ratio. In these limits the physics overlaps and each model should reduce to the same basic equation.

Consider first the straight stellarator whose behavior is governed by Eq. (6.173), repeated here for convenience:

$$\Delta^* \psi = -\mu_0 \frac{dp}{d\psi} - \frac{F}{l_0^2} \frac{dF}{d\psi} + \frac{2hl}{l_0^4} F \qquad (7.43)$$

The loosely wound assumption corresponds to approximating $l_0^2 = l^2 + h^2 r^2 \approx l^2$ where h is the pitch number of the helix. In this limit

$$\Delta^* \psi \equiv \frac{1}{r} \frac{\partial}{\partial r} \left(\frac{r}{l_0^2} \frac{\partial \psi}{\partial r} \right) + \frac{1}{r^2} \frac{\partial^2 \psi}{\partial \alpha^2} \approx \frac{1}{l^2} \left[\frac{1}{r} \frac{\partial}{\partial r} \left(r \frac{\partial \psi}{\partial r} \right) + \frac{l^2}{r^2} \frac{\partial^2 \psi}{\partial \alpha^2} \right] = \frac{1}{l^2} \nabla^2 \psi \quad (7.44)$$

and Eq. (7.43) reduces to

$$\nabla^2 \psi = -\frac{d}{d\psi} \left(l^2 \mu_0 \, p + \frac{F^2}{2} \right) + \frac{2h}{l} F \qquad (7.45)$$

The next step is to utilize the low β assumption. If one introduces δ as a small mathematical expansion parameter then the low β limit corresponds to the following ordering: $\beta \sim ha \sim B_p/B_0 \sim \delta \ll 1$. Then, in analogy with the derivation of the high β tokamak equations, the free function $F(\psi)$ is written as

$$F^2(\psi) = l^2 B_0^2 \left[1 - \frac{2\mu_0 p(\psi)}{B_0^2} + 2G_2(\psi) \right] \qquad (7.46)$$

Here, $F(\psi)$ is replaced by $G_2(\psi)$, a new dimensionless free function that scales as $G_2 \sim \delta^2$. This form for $F(\psi)$ guarantees that radial pressure balance is identical to that in a θ-pinch which is the assumption made in the high β stellarator expansion. Substituting into Eq. (7.45) leads to the desired form of the straight stellarator equations in the overlap region,

$$\nabla^2 \psi = -l^2 B_0^2 \frac{dG_2}{d\psi} + 2hB_0 \qquad (7.47)$$

The next task is to take the large aspect ratio, single helicity limit of the high β stellarator equations and see if they reduce to the same form as Eq. (7.47). One begins by setting $\phi = hz$, $M = 1$, and $u = r$. The proper way to take the limit is to assume that a is finite, ha is small but finite, and $R_0 \rightarrow \infty$. Now, the only appearance of R_0 in Eq. (7.32) is in the denominator of the right-hand side of the second equation. In the overlap region this "toroidal" effect can be neglected and Eq. (7.32) reduces to

$$
\left(B_0 \frac{\partial}{\partial z} - \mathbf{e}_\phi \times \nabla_\perp A_1 \cdot \nabla_\perp \right) p_1 = 0
$$

$$
\left(B_0 \frac{\partial}{\partial z} - \mathbf{e}_\phi \times \nabla_\perp A_1 \cdot \nabla_\perp \right) \nabla_\perp^2 A_1 = 0
$$

(7.48)

The last simplification is to apply the single helicity requirement. This requirement is a consequence of the fact that a straight stellarator has, by definition, helical symmetry; that is all quantities are functions only of r and $\alpha = l\theta + hz$, corresponding to a single helicity system. In addition, p_1 is written as $p_1(\psi)$ with ψ now serving as the flux surface label. The first of the high β stellarator equations reduces to

$$
\left(hB_0 r - l\frac{\partial A_1}{\partial r} \right) \frac{\partial \psi}{\partial \alpha} + l\frac{\partial A_1}{\partial \alpha}\frac{\partial \psi}{\partial r} = 0
$$

(7.49)

The solution is

$$
hB_0 r^2 - 2lA_1 = H(\psi)
$$

(7.50)

The free function $H(\psi)$ is determined by identifying ψ with the helical flux in the straight stellarator and then equating the magnetic fields from each model. A short calculation yields

$$
\psi = \frac{1}{2}(hB_0 r^2 - 2lA_1)
$$

(7.51)

The second high β stellarator equation simplifies to

$$
\left(\frac{\partial \psi}{\partial r}\frac{\partial}{\partial \alpha} - \frac{\partial \psi}{\partial \alpha}\frac{\partial}{\partial r} \right) \nabla^2 A_1 = 0
$$

(7.52)

The solution is $\nabla^2 A_1 = \hat{G}(\psi)$ which after eliminating A_1 by means of Eq. (7.51) reduces to

$$
\nabla^2 \psi = G(\psi) + 2hB_0
$$

(7.53)

where $G(\psi) = -l\hat{G}(\psi)$. Observe that Eq. (7.53), with a different but equivalent definition of $G(\psi)$, is identical in form to Eq. (7.47) thereby demonstrating that both models do indeed overlap.

The discussion above has shown that the high β stellarator model reduces to known models in the appropriate regions of physics overlap. These models include the high β tokamak and the straight stellarator. The conclusion is that the high β stellarator model is based on a sound physical foundation and describes a wide range of multidimensional fusion concepts.

7.4 The Greene–Johnson limit

The purpose of this section is to show the relationship between the high β and Greene–Johnson stellarator models. The Greene–Johnson (GJ) model is an early and elegant expansion describing the MHD behavior of stellarators including radial pressure balance and toroidal force balance in a multi-helicity system (Greene and Johnson, 1961). In the context of modern stellarators the only important physics excluded from the model is the existence of a finite helical magnetic axis, which is a requirement to suppress neoclassical particle losses. The nature of the GJ expansion is such that the magnetic axis is forced to be circular with only small helical modulations. In fairness to Greene and Johnson it should be emphasized that at the time of their analysis neoclassical theory had not yet been developed and therefore there was no strong motivation to allow for a finite helical magnetic axis.

A substantial amount of analysis is required to (1) derive the GJ model and (2) reduce the high β stellarator (HBS) model to the GJ model in the overlap region. The critical steps are outlined here and the HBS details can be found in Appendix E.

7.4.1 Comparison of expansions

The discussion begins with a comparison of the different ordering assumptions used in both models. The inverse aspect ratio $\varepsilon \ll 1$ is again used as the basic expansion parameter and all other parameters are ordered with respect to ε. The ordering comparison is shown in the middle two columns in Table 7.2.

Table 7.2 *Comparison of stellarator expansions.*

Quantity	High β stellarator	Greene–Johnson	Overlap model
δ	ε	$\varepsilon^{1/2}$	$\varepsilon^{3/4}$
β	ε	ε	ε
N	1	$1/\varepsilon$	$1/\varepsilon^{1/2}$
l	1	1	1

A primary feature motivating both expansions is that the rotational transform (except for $l = 1$) must be of order unity: $\iota/2\pi \sim 1$. This property can be verified explicitly by recalling that the rotational transform for a loosely wound, $l \geq 2$, single helicity stellarator is given by

$$\frac{\iota}{2\pi} \approx \frac{(l-1)\delta^2}{N\varepsilon^2} \left(\frac{r}{a}\right)^{2(l-2)} \tag{7.54}$$

where r is the local plasma radius and a is the average radius of the plasma surface.

Recall that in order to produce a stellarator with a helical magnetic axis it is necessary to assume $N \sim 1$, which is the ordering used in the HBS expansion. This is in contrast to the GJ ordering, which assumes that $N \sim 1/\varepsilon$ is large. Even so, the GJ equilibrium equations (in the limit of a loosely wound helical field) can be obtained from the HBS model (in the limit of a tightly wound helical field) by the introduction of a subsidiary expansion where N is assumed to be large but not too large.

Physically, the implication of $N \sim 1$ in an HBS is that there are a finite number of helical periods around the torus as compared to a large number in the GJ expansion. Equally important, the assumption $\delta \sim \varepsilon$ in the HBS analysis is required in order to make the rotational transform finite: $\iota/2\pi \sim 1$ as implied by Eq. (7.54). In comparison, the GJ expansion requires $\delta \sim \varepsilon^{1/2}$ for $\iota/2\pi \sim 1$. Lastly, the value of beta for both expansions scales as $\beta \sim \varepsilon$, which is similar to the high β tokamak ordering.

At first glance reducing N from order $1/\varepsilon$ to order 1 might seem to be more restrictive and lead to a simpler analysis. Just the opposite is true. The HBS analysis is far more complicated and much less restrictive. It allows for a finite helical magnetic axis, finite helical modulations of the flux surfaces, and brings toroidal effects into the calculation earlier, in the same order as helical effects.

7.4.2 The Greene–Johnson limit of the HBS model

This subsection outlines the reduction of the HBS model to the GJ model in the appropriate region of overlap. To carry out this task it is necessary to introduce a common subsidiary expansion into each model so that the physics does indeed overlap. An intuitive way to view the physics is as follows. The HBS model will now allow N to become large (i.e., there are many helical periods around the torus) but not too large, subject to the constraints of $\iota/2\pi \sim 1$ and $\beta \sim \varepsilon$. In contrast, the GJ model will now allow N to become small (i.e., there are fewer helical periods around the torus) but not too small, also subject to the constraints of $\iota/2\pi \sim 1$ and $\beta \sim \varepsilon$. This corresponds to the region of overlap. Specifically, it is assumed that

$1 \ll N \ll 1/\varepsilon$ which is satisfied if $N \sim 1/\varepsilon^{1/2} \gg 1$. The complete ordering for the overlap model, consistent with the constraints is shown in the last column of Table 7.2.

The analysis starts with the un-normalized version of the HBS model given by Eq. (7.32). New normalized variables are introduced that are consistent with the overlap ordering. The new definitions are

$$\rho = \frac{r}{a} \sim 1$$

$$A = -\frac{A_1}{aB_0 N^{1/2}\varepsilon} \sim 1 \qquad (7.55)$$

$$\beta = \frac{2\mu_0 p_1}{\varepsilon B_0^2} \sim 1$$

Since the physics overlap is closely tied to the ordering for the number of helical periods the definitions in Eq. (7.55) have been chosen so that $1/N$ rather than ε appears as the expansion parameter in the normalized equations which are given by[3]

$$\frac{\partial \beta}{\partial \phi} = -\frac{1}{N^{1/2}} \mathbf{e}_\phi \times \nabla_\perp A \cdot \nabla_\perp \beta$$

$$\frac{\partial(\nabla_\perp^2 A)}{\partial \phi} = -\frac{1}{N^{1/2}} \mathbf{e}_\phi \times \nabla_\perp A \cdot \nabla_\perp(\nabla_\perp^2 A) + \frac{1}{N^{3/2}} \mathbf{e}_z \cdot \nabla_\perp \beta \qquad (7.56)$$

where $\nabla_\perp$ is the gradient in ρ, θ coordinates.

The subsidiary expansion can now be applied. One sees that in both equations the right-hand side is small when $N \sim 1/\varepsilon^{1/2}$ is large, suggesting a solution by a straightforward expansion technique. The details of the expansion are lengthy and are presented in Appendix E. The end result is that the overlap limit of the HBS model has the form of a single Grad–Shafranov-like partial differential equation for the normalized poloidal flux $\psi(\rho, \theta)$. This equation has two free functions of ψ: the pressure $p(\psi) \propto \beta(\psi)$ and the average toroidal current density $\langle J_\phi \rangle \propto J(\psi)$. There is also a contribution from the vacuum helical field. The final Grad–Shafranov-like equation is given by

$$\nabla_\perp^2 \psi = -J - \frac{d\beta}{d\psi}(x - \langle x \rangle) + \nabla_\perp^2 \left(\sum_1^\infty \frac{i}{n} \mathbf{e}_\phi \cdot \nabla_\perp A_n^* \times \nabla_\perp A_n \right) \qquad (7.57)$$

---

[3] Note that it can easily be shown that the GJ ordering implies that $M = 1 + O(\delta)$. This is the reason why M does not explicitly appear in Eq. (7.56).

Here, $x = \rho \cos\theta$ and

$$\psi(\rho, \theta) = \frac{1}{2\pi a^2 B_0} \psi_p \qquad \text{Normalized poloidal flux } \psi_p$$

$$\beta(\psi) = \frac{2\mu_0}{\varepsilon B_0^2} p \qquad \text{Normalized plasma pressure } p \qquad (7.58)$$

$$J(\psi) = -\frac{\mu_0 R_0}{B_0} \langle J_\phi \rangle \qquad \text{Normalized toroidal current density } J_\phi$$

and $\langle\ \rangle$ denotes the average over poloidal angle (written in terms of poloidal arc length $l_p = l_p(\theta)$)

$$\langle Q \rangle = \frac{\oint Q(r,\theta) \frac{dl_p}{|\nabla_\perp \psi|}}{\oint \frac{dl_p}{|\nabla_\perp \psi|}} \qquad (7.59)$$

Lastly, the vector potential is written as $A(\rho,\theta,\phi) = \overline{A}(\rho,\theta) + \tilde{A}(\rho,\theta,\phi)$. The functions $A_n(r, \theta)$ in Eq. (7.57) are the amplitudes of the vector potential $\tilde{A}$ corresponding to the vacuum helical fields,

$$\tilde{A}(\rho,\theta,\phi) = \sum_{n \neq 0} A_n(\rho,\theta)\, e^{in\phi}$$

$$A_n(\rho,\theta) = \sum_{l \neq 0} \frac{\delta_{nl}}{2lN^{1/2}\varepsilon} \rho^{|l|} e^{il\theta} \qquad (7.60)$$

with $\delta_{nl}^2 = (B_{\rho,nl}^2 + B_{\theta,nl}^2)_{\rho=1}/B_0^2$. The quantity $\overline{A}(\rho,\theta)$ is the toroidally averaged component of the vector potential and is related to the poloidal flux function by

$$\overline{A}(\rho,\theta) = \psi - \sum_{1}^{\infty} \frac{i}{n} \mathbf{e}_\phi \cdot \nabla_\perp A_n^* \times \nabla_\perp A_n \qquad (7.61)$$

Equation (7.57) is the desired GJ limit of the HBS model.

7.4.3 The Greene–Johnson model

The derivation of the Greene–Johnson model also requires a substantial amount of analysis, similar in spirit to the HBS calculation in Appendix E. Derivations of the GJ model can be found in the references (Greene and Johnson, 1961; Freidberg, 1987). For the present purposes of comparison it is sufficient to simply state the original results. After converting to the notation used in Eq. (7.57) one finds that the GJ model can be written as

$$\nabla_\perp^2 \psi = -J - \frac{d\beta}{d\psi}(x - \langle x \rangle) + \nabla_\perp^2 \left(\sum_1^\infty \frac{i}{n} \mathbf{e}_\phi \cdot \nabla_\perp A_n^* \times \nabla_\perp A_n \right)$$
$$+ \frac{\varepsilon N}{2} \frac{d\beta}{d\psi} \sum_1^\infty \left[\nabla_\perp^2 |A_n|^2 - \langle \nabla_\perp^2 |A_n|^2 \rangle \right] \tag{7.62}$$

In the overlap region $\varepsilon N \sim \varepsilon^{1/2} \ll 1$ and the last term can be neglected. The two models, given by Eqs. (7.57) and (7.62), then coincide showing that the physics does indeed overlap. In the original GJ ordering $\varepsilon N \sim 1$ and the last term must be maintained.

7.4.4 Summary

The discussion above has shown that the high β stellarator model overlaps with the Greene–Johnson model in a regime of physical interest corresponding to helical fields that are wound neither too loosely nor too tightly. This adds further confirmation about the reliability of the high β stellarator model.

7.5 Vacuum flux surfaces

In this section and the ones that follow the high β stellarator model is solved for several special cases in order to acquire some intuition about the MHD equilibrium properties of stellarators. The present section focuses on the calculation of vacuum flux surfaces. As previously stated a stellarator has well-defined flux surfaces with a unique magnetic axis even in the vacuum limit $\beta \to 0$.

The problem is surprisingly complex in a 3-D geometry. For example, even when a simple set of vacuum magnetic fields is specified, such as a combined $l = 2, l = 3$ system, there is no simple closed form solution for the flux surfaces. In the examples examined below some analytic and numerical solutions are presented that describe different configurations. A primary goal is to show that there is a maximum limit to the helical field amplitude to avoid the appearance of a separatrix on the surface of the $\beta \to 0$ plasma.

7.5.1 Single helicity – the limiting helical field amplitude

The first example of interest involves a single helicity stellarator. In this case, because of the helical symmetry, an exact solution to the high β stellarator model can be found. The starting point is the second of the normalized high β stellarator equations evaluated in the limit $\beta \to 0$. From Eq. (7.35) it follows that

$$\left(\frac{\partial}{\partial\phi} + \mathbf{e}_\phi \times \nabla_\perp A \cdot \nabla_\perp\right) \nabla_\perp^2 A = 0 \tag{7.63}$$

For a vacuum field the solution is

$$\nabla_\perp^2 A = 0 \tag{7.64}$$

Assume now that the helical component of the magnetic field consists only of a single harmonic with poloidal periodicity l. For this case A is given by

$$A(\rho, \theta, \phi) = \frac{\Delta_l}{l}\rho^l \cos(l\theta + \phi)$$

$$\Delta_l = \frac{\delta_l}{N\varepsilon} \tag{7.65}$$

where $l > 0$ and δ_l is defined by setting $(|B_p|/B_0)_{\rho\,=\,1} = \delta_l$.

Consider next the first of the high β stellarator equations. A flux function ψ is introduced so that $\beta = \beta(\psi)$. Precisely which flux (e.g., poloidal, helical, toroidal) ψ corresponds to is discussed shortly. Also, the single helicity assumption is invoked implying that $\psi(\rho, \theta, \phi) = \psi(\rho, \alpha)$ with $\alpha = l\theta + \phi$. The first high β stellarator equation reduces to

$$\left(1 + \frac{l}{\rho}\frac{\partial A}{\partial\rho}\right)\frac{\partial\psi}{\partial\alpha} - \frac{l}{\rho}\frac{\partial A}{\partial\alpha}\frac{\partial\psi}{\partial\rho} = 0 \tag{7.66}$$

which, after multiplying by 2ρ, is further simplified by substituting A from Eq. (7.65),

$$\frac{\partial\psi}{\partial\alpha}\frac{\partial}{\partial\rho}(\rho^2 + 2\Delta_l\rho^l \cos\alpha) - \frac{\partial\psi}{\partial\rho}\frac{\partial}{\partial\alpha}(\rho^2 + 2\Delta_l\rho^l \cos\alpha) = 0 \tag{7.67}$$

The general solution to this equation is

$$\rho^2 + 2\Delta_l\rho^l \cos\alpha = G(\psi) \tag{7.68}$$

The function $G(\psi)$ is shown here to be proportional to the helical flux ψ_h first defined for the straight stellarator in Chapter 6. This definition, applied to the large aspect ratio vacuum surfaces in the high β stellarator model, is given by

$$\frac{\partial\psi_h}{\partial r} = lB_\theta + hrB_z \approx -l\frac{\partial A_1}{\partial r} + \frac{NB_0}{R_0}r \tag{7.69}$$

This equation can be easily integrated and converted to normalized coordinates. One finds

$$\psi = \rho^2 + 2\Delta_l\rho^l \cos\alpha \tag{7.70}$$

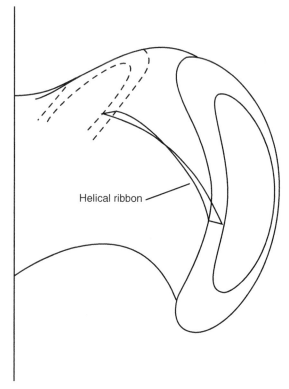

Figure 7.3   Helical flux passing through a rotating helical ribbon whose surface is defined by $\alpha = l\theta + \phi = $ constant.

where $\psi = (2R_0/NB_0a^2)\,\psi_h$ is the normalized helical flux implying that in Eq. (7.68) $G(\psi) = \psi$. Physically the helical flux is the flux passing through a rotating helical ribbon whose surface is defined by $l\theta - N\hat{\phi} = $ constant in un-normalized coordinates or $\alpha = l\theta + \phi = $ constant in normalized coordinates (see Fig. 7.3).

Equation (7.70) is the desired solution for the flux surfaces in a single helicity stellarator. The basic properties of this configuration can now be easily ascertained. The first point to notice is that the solution given by Eq. (7.70) coincides with the vacuum solution for a loosely wound straight stellarator as evidenced by a comparison with Eq. (7.51). This is perhaps not surprising since toroidal effects enter the high β stellarator only through the $\nabla_\perp \beta \cdot \mathbf{e}_z$ term which vanishes in the limit of a vacuum field.

Next it is of interest to plot the flux contours for several values of l as illustrated in Fig. 7.4. These contours rotate helically along ϕ, the basic signature of a stellarator magnetic geometry. As might be expected $l = 2$ leads to elliptical surfaces, $l = 3$ to triangular surfaces, and $l = 4$ to square surfaces.

It is also worth noting from Eq. (7.35) that in the limit $\beta \to 0$ the appearance of the mirror ratio M vanishes from the high β stellarator equations for ψ and

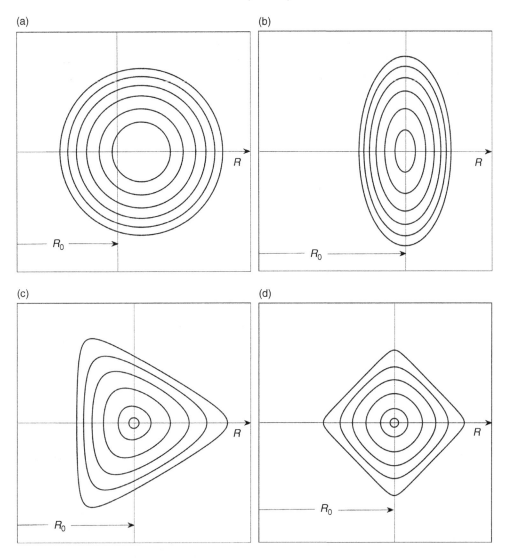

Figure 7.4 Vacuum helical flux surfaces for (a) $l = 1$, (b) $l = 2$, (c) $l = 3$, and (d) $l = 4$ stellarators.

A. Therefore the solution given by Eq. (7.70) still applies. The only difference when $M \neq 1$ is that the transformation back to the un-normalized laboratory coordinates is now given by $\rho = M^{1/2}(r/a)$.

Lastly, a critical feature of the single helicity vacuum surfaces is the existence of a maximum allowable helical amplitude. This limit arises because of the appearance of a separatrix on the plasma surface when the helical field becomes too large. The limiting amplitude can be easily calculated analytically by setting both components of $\nabla\psi$ to zero on the plasma surface.

The only slightly subtle point is defining the value of ψ that corresponds to the $\beta \to 0$ plasma surface. The reason is that the plasma surface has a complicated shape and may be situated in a comparably complicated 3-D vacuum chamber. Where exactly does the plasma make first contact with the surrounding wall, limiter, or diverter which defines the edge of the plasma? For present purposes this difficulty is overcome by defining the $\beta \to 0$ plasma surface to correspond to $r = a$ in the limit of a vanishing small helical field. The plasma surface is thus defined as

$$\psi = 1 \tag{7.71}$$

In this limit the location of the separatrix X-point is given by the simultaneous solution of

$$\left. \frac{\partial \psi}{\partial \alpha} \right|_{\psi=1} = 0$$

$$\left. \frac{\partial \psi}{\partial \rho} \right|_{\psi=1} = 0 \tag{7.72}$$

A short calculation shows that the separatrix moves onto the plasma surface when

$$\alpha = \pi$$

$$\rho = \left(\frac{1}{l \Delta_l} \right)^{\frac{1}{l-2}} \tag{7.73}$$

One now sets $\psi = 1$ and substitutes Eq. (7.73) into Eq. (7.70). The result is that in order to avoid a separatrix on the plasma surface the amplitude of the helical field is limited by

$$\Delta_l = \frac{\delta_l}{N \varepsilon} \leq g(l) \equiv \frac{1}{l} \left(\frac{l-2}{l} \right)^{\frac{l-2}{2}} \tag{7.74}$$

For various values of l one finds that $g(2) = 0.5$, $g(3) = 0.192$, and $g(4) = 0.125$. The flux surfaces for an $l = 3$ system at the equilibrium limit are illustrated in Fig. 7.5 to show the appearance of the separatrix.

Physically, the equilibrium limit can be explained as follows. For small radii the flux surfaces are closed since they are dominated by the toroidal field; that is, for $\rho \ll 1$ the flux scales as $\psi \propto \rho^2$. As the radius increases the helical field starts to dominate. The reason is that for any $l \geq 3$, the radial dependence of the helical field, proportional to ρ^l, increases more rapidly than ρ^2. In the region of large ρ the flux surfaces open up because of the oscillatory behavior of the helical field. The separatrix represents the transition surface separating the regions between where the toroidal field or the helical field dominates the flux surface behavior.

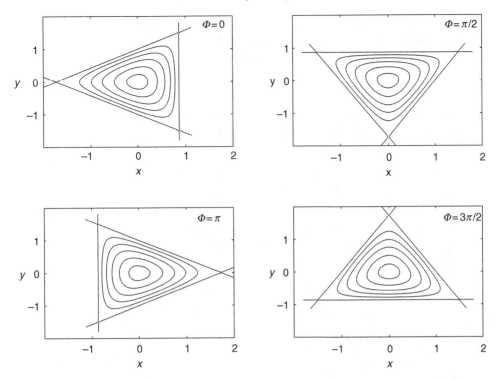

Figure 7.5   Vacuum flux surfaces for an $l = 3$ stellarator at the equilibrium limit showing the separatrix.

In summary, single helicity vacuum surfaces can be calculated analytically. These surfaces are closed out to the edge of the plasma for a sufficiently small helical field amplitude. The surfaces coincide with those calculated for the straight stellarator in Chapter 6.

7.5.2 Multiple helicity stellarators

The next step in building MHD intuition about stellarators involves examining the vacuum flux surfaces in a multiple helicity configuration. The problem is far more difficult than one might imagine. In fact it is not possible to obtain analytic solutions even for a system with only two helical harmonics, for instance $l = 2$ and $l = 3$. The reason is associated with the fact that even in the large aspect ratio, high β stellarator ordering, the problem remains truly three dimensional.

The solution to this problem involves, perhaps not surprisingly, a numerical calculation. There are several numerical procedures that can determine the vacuum surfaces in a fast and accurate manner. These methods are discussed elsewhere allowing the focus here to remain on results (Bauer *et al.*, 1987; Hirshman and Whitson, 1983; Reiman and Greenside, 1986). As an example, results are presented for an $l = 2$, $l = 3$ stellarator. Of interest is the fact that there is now a finite allowable parameter region in Δ_2, Δ_3 space where flux surfaces exist without the appearance of a separatrix within the plasma.

The numerical formulation

The starting point for the analysis is again the second of the normalized high β stellarator equations evaluated in the vacuum limit $\beta \to 0$, given by

$$\nabla_\perp^2 A = 0 \tag{7.75}$$

In general, the vector potential $A(\rho, \theta, \phi)$ consists of a sum of terms, each one of the form $(\Delta_{l,n}/l)\,\rho^l \cos(l\theta + n\phi + n\phi_{l,n})$. It is assumed that the amplitude $\Delta_{l,n}$, phase $\phi_{l,n}$, and toroidal periodicity number, n, are known for each term. In other words, the applied vacuum helical field is assumed to be fully specified. For the $l = 2$, $l = 3$ example under consideration $A(\rho, \theta, \phi)$ is given by

$$A(\rho, \theta, \phi) = \frac{\Delta_2}{2}\rho^2 \,\cos\,(2\theta + \phi) + \frac{\Delta_3}{3}\rho^3 \,\cos\,(3\theta + \phi) \tag{7.76}$$

This expression is substituted into the first of the high β stellarator equations leading to

$$\frac{\partial \psi}{\partial \phi} + \frac{1}{\rho}\frac{\partial A}{\partial \rho}\frac{\partial \psi}{\partial \theta} - \frac{1}{\rho}\frac{\partial A}{\partial \theta}\frac{\partial \psi}{\partial \rho} = 0 \tag{7.77}$$

Equation (7.77) is solved numerically subject to the boundary condition of regularity within the plasma region. Since this is a first-order partial differential equation one is not allowed to also specify the shape of the boundary flux surface. In other words, the solution generates free boundary flux surfaces.

Numerical results

A numerically generated plot of the flux surfaces for the values $\Delta_2 = 0.25$ and $\Delta_3 = 0.03$ is illustrated in Fig. 7.6 for several values of ϕ. Observe that the surfaces have a strong combination of elliptical plus triangular deformations as expected. The deformations are finite in extent showing that simple analytic expansions would have a difficult time accurately approximating the solutions.

A second calculation investigates the transition region of parameter space where surfaces exist with and without the presence of a separatrix in the plasma. This region can be determined as follows. The plasma surface in the limit of low β is again chosen as $\psi = 1$. Next, a value of Δ_3 is chosen that satisfies $\Delta_3 < 0.231$, the limiting value for a single helicity $l = 3$ stellarator. Over a series of computer runs the value of Δ_2 is slowly increased from zero until a separatrix appears on the $\psi = 1$ plasma surface. This is the limiting value of Δ_2 for the given Δ_3. The

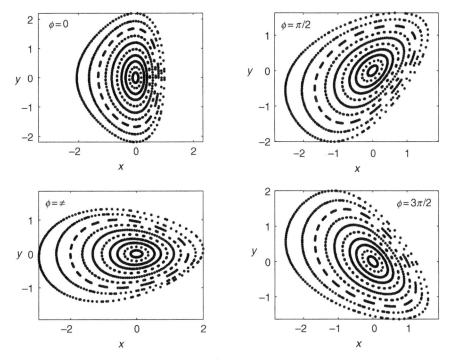

Figure 7.6   Numerically computed vacuum flux surfaces for an $l = 2, 3$ stellarator.

procedure is then repeated for a sequence of different Δ_3 values. The end result is a curve of Δ_2 vs. Δ_3 that defines the separatrix free region of operation and is illustrated in Fig. 7.7. The behavior is qualitatively similar but more difficult to calculate as compared to the pure single helicity stellarator.

Summary

In this subsection it has been shown that a stellarator has closed nested flux surfaces even in the vacuum limit $\beta \to 0$. For a single helicity system, vacuum surfaces can be found analytically. For a multi-helicity system, the vacuum surfaces must be calculated numerically. In general, there is a maximum limit to the size of the helical fields relative to the dominant toroidal field in order to prevent a separatrix from moving onto the plasma surface.

7.6 Effects of finite β

In the previous section the basic properties of vacuum flux surfaces in a stellarator have been derived. The goal of the present section is to determine the effects of

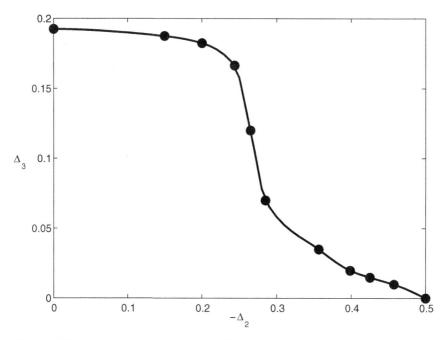

Figure 7.7   Boundary of operation without a separatrix in a Δ_3, Δ_2 space assuming the surface-to-volume ratio of all separatrices is fixed.

finite β on the vacuum surfaces. As might be expected, for reasons of mathematical simplicity, much of the discussion is focused on configurations where the applied vacuum field has only a single helicity, e.g., $\psi = \psi(r, 2\theta + \phi)$. Even so, once finite pressure is included the problem becomes truly three dimensional since toroidal β effects couple through terms proportional to $\cos \theta$.

The results of this section show that finite β stellarator equilibria exist even when there is zero net toroidal current flowing in the plasma. Also, finite β shifts the magnetic axis outward with respect to the vacuum magnetic axis. In this connection a simple physical picture is presented that demonstrates how in a current-free stellarator, a helical field produces an inward toroidal force along R to compensate the outward tire tube and $1/R$ forces.

It is also well known and directly demonstrated that in a stellarator with zero net current a vertical field produces no restoring force. This fact raises an apparent paradox since it is well known experimentally that applying a vertical field can indeed shift the plasma inward or outward along R. An explanation is presented that resolves this paradox, based on a dynamical model that takes into account induced, time-varying surface currents.

Interestingly, it is shown that a stellarator has a high β equilibrium limit similar to that in a high β tokamak. Further calculation shows that a flux conserving stellarator can avoid the equilibrium limit by the natural induction of a net current as β is raised. Lastly numerical solutions are presented that solve the high β stellarator equations for a multi-helicity system.

7.6.1 Low β single helicity solutions

The first problem of interest investigates the equilibrium of a single helicity stellarator with a small but finite β. By assuming that β is small it is possible to derive a simple analytic solution for the flux surfaces from which it is possible to gain some physical insight into the properties of current-free stellarator equilibria.

The basic equations

The analysis is based on the Greene–Johnson overlap model, described by Eq. (7.57). In order to directly relate the solutions to the vacuum flux surfaces previously derived Eq. (7.57) must be renormalized back to the original HBS normalization. The easiest way to do this is to imagine that every term in Eq. (7.57) has a subscript "GJ." The renormalized variables will then have no subscripts and represent the original HBS normalization. The renormalization can be summarized as follows:

$$A_{GJ} = N^{1/2}A = -\frac{A_1}{aB_0 N^{1/2}\varepsilon}$$

$$B_{GJ} = N^2\beta = \frac{2\mu_0 p_1}{\varepsilon B_0^2}$$

$$J_{GJ} = NJ = -\frac{\mu_0 R_0}{B_0}\langle J_\phi \rangle \tag{7.78}$$

$$\psi_{GJ} = N\psi = \frac{\psi_p}{2\pi a^2 B_0}$$

In terms of the original HBS variables the overlap equation for the flux becomes

$$\nabla_\perp^2 \psi = -J - \frac{d\beta}{d\psi}(x - \langle x \rangle) + \nabla_\perp^2 \left(\sum_1^\infty \frac{i}{n} \mathbf{e}_\phi \cdot \nabla_\perp A_n^* \times \nabla_\perp A_n \right) \tag{7.79}$$

which is formally identical to Eq. (7.57) although the variables have different definitions.

The low β expansion

Equation (7.79) is solved by first setting $J = 0$, corresponding to a stellarator with zero net current on every flux surface. Next, it is assumed that the vacuum fields correspond to a single helicity. For this case the vector potential amplitude A_n is easily obtained from Eq. (7.65), repeated here for convenience,

$$A(\rho, \theta, \phi) = A_n(\rho, \theta)\, e^{i\phi} + A_n^*(\rho, \theta)\, e^{-i\phi} = \frac{\Delta_l}{l}\rho^l \cos(l\theta + \phi)$$

$$A_n(\rho, \theta) = \frac{\Delta_l}{2l}\rho^l e^{il\theta} \tag{7.80}$$

Equation (7.79) reduces to

$$\nabla_\perp^2 \psi = -2(l-1)^2 \Delta_l^2 \rho^{2l-4} - \frac{d\beta}{d\psi}(x - \langle x \rangle) \tag{7.81}$$

Note that in the context of the overlap expansion $\Delta_l \sim \delta_l/N\varepsilon \sim \varepsilon^{1/4}$. Mathematically Δ_l must therefore be treated as a small parameter although practically this is not a very good assumption.

Even with these simplifications the 2-D equation is quite difficult to solve – it is still a non-linear partial differential equation. An approximate solution can be obtained by further assuming that β is smaller than assumed in the overlap ordering. The appropriate expansion is to assume that $\psi \sim \Delta_l^2$ and that $\beta/\Delta_l^4 \sim \beta/\varepsilon \ll 1$. Under these assumptions Eq. (7.81) can be solved by expanding

$$\psi(\rho, \theta) = \psi_0(\rho) + \psi_1(\rho, \theta) + \cdots \tag{7.82}$$

where $\psi_1/\psi_0 \sim \beta/\Delta_l^4$. The expansion is qualitatively similar to the low β ohmic tokamak expansion discussed in Chapter 6.

Zeroth-order solution

The zeroth-order equation for ψ_0 is given by

$$\frac{1}{\rho}\frac{d}{d\rho}\left(\rho\frac{d\psi_0}{d\rho}\right) = -2(l-1)^2 \Delta_l^2 \rho^{2l-4} \tag{7.83}$$

The solution is easily found and is given by

$$\psi_0(\rho) = -\frac{\Delta_l^2}{2}\rho^{2l-2} \tag{7.84}$$

To leading order the ϕ averaged flux surfaces are circular.

First-order solution

The first-order equation which determines ψ_1 can be written as

$$\nabla_\perp^2 \psi_1 = -\frac{d\beta}{d\psi_0}(x - \langle x \rangle) \tag{7.85}$$

Note that for circular flux surfaces the poloidally averaged $\langle x \rangle = \langle \rho \cos \theta \rangle = 0$. One now sets $x = \rho \cos \theta$ implying that $\psi_1(\rho, \theta) = \overline{\psi}_1(\rho) \cos \theta$. Also $\beta(\rho)$ replaces $\beta(\psi_0)$ as the free function. This leads to an equation for $\overline{\psi}_1$ given by

$$\frac{1}{\rho}\frac{d}{d\rho}\left(\rho \frac{d\overline{\psi}_1}{d\rho}\right) - \frac{\overline{\psi}_1}{\rho^2} = -\frac{d\beta}{d\rho}\frac{\rho}{d\psi_0/d\rho} \tag{7.86}$$

which simplifies to

$$\frac{1}{\rho^2}\frac{d}{d\rho}\left[\rho^3 \frac{d}{d\rho}\left(\frac{\overline{\psi}_1}{\rho}\right)\right] = \frac{1}{(l-1)\Delta_l^2 \rho^{2l-4}}\frac{d\beta}{d\rho} \tag{7.87}$$

Two boundary conditions must be specified. The first requires regularity on the axis $\rho = 0$. The second follows from the observation that ψ_0 contains the response to the applied vacuum helical fields while ψ_1 describes the diamagnetic modifications due to finite β. For a confined plasma the diamagnetic contributions must vanish far from the plasma which translates into the condition $\overline{\psi}_1(\infty) \to 0$.

The solution for $\overline{\psi}_1$ is easily obtained by integrating Eq. (7.87) twice and applying the boundary conditions. The result is

$$\overline{\psi}_1(\rho) = -\frac{\rho}{(l-1)\Delta_l^2}\int_\rho^\infty \frac{dx}{x^3}\int_0^x \frac{dy}{y^{2l-6}}\frac{d\beta}{dy} \tag{7.88}$$

Observe that for the solution to be well behaved near the origin for $l \geq 3$, the pressure profile must be very flat. The reason is that high l stellarator fields are very small near $\rho = 0$ (i.e., $B_p \propto \rho^{l-1}$) and are thus incapable of providing radial pressure balance. Specifically, for well-behaved solutions β must vary as $c_1 - c_2\rho^{4l-6}$.

The full solution

The solution is completed by adding the helical modulation to the flux surfaces obtained by setting the full $\psi_{tot}(\rho, \theta, \phi) = [\psi_0(\rho) + \overline{\psi}(\rho)\cos\theta] + \psi_{hel}(\rho, \theta, \phi) =$ constant. The quantity ψ_{hel} has been evaluated in Appendix E, Eq. (E.9), and can be written as

$$\psi_{hel}(\rho, \theta, \phi) = \sum_{n\neq 0}\frac{i}{n}(\mathbf{e}_\phi \cdot \nabla_\perp A_n \times \nabla_\perp \psi_0)e^{in\phi}$$
$$= -(l-1)\Delta_l^3 \rho^{3l-4}\cos(l\theta + \phi) \tag{7.89}$$

The full solution is obtained by adding the terms from Eqs. (7.84), (7.88), and (7.89). The final form is simplified by assuming that $\beta = \beta_0(1 - \rho^{4l-6})$ for $0 \le \rho \le 1$. The end result for $l = 2$ is given by

$$\psi_{tot} = -\frac{\Delta_2^2}{2}\left[\rho^2 + 2\Delta_2\rho^2 \, \cos\left(2\theta + \phi\right) - \frac{\beta_0}{2\Delta_2^4}(2\rho - \rho^3) \, \cos\theta\right] \quad \rho \le 1$$

$$\psi_{tot} = -\frac{\Delta_2^2}{2}\left[\rho^2 + 2\Delta_2\rho^2 \, \cos\left(2\theta + \phi\right) - \frac{\beta_0}{2\Delta_2^4}\frac{\cos\theta}{\rho}\right] \qquad \rho \ge 1$$

$$(7.90)$$

Discussion

There are several important conclusions that can be drawn from Eq. (7.90). First and foremost, the solution given by Eq. (7.90) shows that closed nested flux surfaces exist in a small but finite β stellarator having zero net current on every flux surface. In the context of the low β expansion used in the analysis these flux surfaces are nearly circular with a small toroidal shift along R (the cos θ term) and a small rotating helical ellipticity (the $\cos(2\theta + \phi)$ term). Even in this simple model the flux surfaces are three dimensional.

The perturbation to the circular flux surfaces, denoted by σ, can be determined by expanding $\rho = \bar{\rho} + \sigma(\bar{\rho}, \theta, \phi)$ with $\sigma/\bar{\rho} \ll 1$. Substituting into Eq. (7.90) yields

$$\sigma = -\Delta_2\bar{\rho} \, \cos\left(2\theta + \phi\right) + \frac{\beta_0}{4\Delta_2^4}(2 - \bar{\rho}^2)\cos\theta \qquad \bar{\rho} \le 1$$

$$\sigma = -\Delta_2\bar{\rho} \, \cos\left(2\theta + \phi\right) + \frac{\beta_0}{4\Delta_2^4}\frac{1}{\bar{\rho}^2} \cos\theta \qquad \bar{\rho} \ge 1$$

$$(7.91)$$

As stated the elliptical modulation due to the helical fields rotates as one moves along ϕ around the torus.

The toroidal shift is outward along R since the coefficient of the cos θ term is positive. The shift continuously decreases for large $\bar{\rho} \gg 1$ since, as one expects, diamagnetic effects become smaller away from the plasma. Within the plasma (i.e., $\rho \le 1$), it is of interest to calculate the toroidal shift $\bar{\sigma}$ (which is the actual shift normalized to the plasma radius a) near the magnetic axis and at the plasma edge. These shifts are defined as follows:

$$\bar{\sigma}(\text{axis}) \equiv \bar{\sigma}_0 = \frac{1}{2\pi}\int \sigma(0, 0, \phi)d\phi = \frac{\beta_0}{2\Delta_2^4}$$

$$\bar{\sigma}(\text{edge}) \equiv \bar{\sigma}_a = \frac{1}{2\pi}\int \bar{\sigma}(1, 0, \phi)d\phi = \frac{\beta_0}{4\Delta_2^4}$$

$$(7.92)$$

Observe that $\bar{\sigma}(\text{axis}) > \bar{\sigma}(\text{edge})$. The magnetic axis is shifted outward with respect to the edge of the plasma.

Now that the existence of MHD equilibria in a current-free stellarator has been established the next task is to understand physically how a helical magnetic field can produce a restoring force along R to balance the outward tire tube and $1/R$ forces. This is the topic of the next subsection.

7.6.2 Toroidal force balance in a current-free stellarator

In Chapter 6 it was shown that toroidal force balance in a tokamak is achieved by the application of a vertical field which interacts with the net toroidal current in the plasma. This interaction generates an inward force along R of magnitude $2\pi R_0 B_V I$ that balances the outward tire tube, $1/R$, and hoop forces.

In a current-free stellarator the hoop force and most importantly the vertical field force both vanish. How then does a stellarator compensate for the remaining tire tube and $1/R$ forces? Answering this question is the goal of the present subsection. Specifically, one wants to learn which magnetic field $\mathbf{B}$ and current $\mathbf{J}$ interact to produce the inward restoring force required to achieve toroidal force balance. It is shown below that it is the interaction of the vacuum helical magnetic field with the helical modulation of the Pfirsch–Schluter current that produces the restoring force.

Formulation of the toroidal force balance problem

The basic idea used to explain toroidal force balance is straightforward. One starts with the ideal MHD momentum equation, calculates the local toroidal force (i.e., the component along R), and then integrates over the entire plasma volume to obtain the total toroidal force. In this way the various global forces entering toroidal balance can be understood. Mathematically, this procedure can be stated as follows:

$$\int (\mathbf{J} \times \mathbf{B} - \nabla p) \cdot \mathbf{e}_R \, d\mathbf{r} = 0 \tag{7.93}$$

To help obtain physical intuition it is helpful to decompose $\mathbf{B}$ and $\mathbf{J}$ into poloidal and toroidal components,

$$\begin{aligned} \mathbf{B} &= \mathbf{B}_p + B_\phi \mathbf{e}_\phi \\ \mathbf{J} &= \mathbf{J}_p + J_\phi \mathbf{e}_\phi \end{aligned} \tag{7.94}$$

Equation (7.93) then reduces to

$$\int (J_\phi \mathbf{e}_\phi \times \mathbf{B}_p + B_\phi \mathbf{J}_p \times \mathbf{e}_\phi - \nabla p) \cdot \mathbf{e}_R \, d\mathbf{r} = 0 \tag{7.95}$$

Note that there is no $\mathbf{J}_p \times \mathbf{B}_p \cdot \mathbf{e}_R$ contribution since $\mathbf{J}_p \times \mathbf{B}_p$ points in the $\mathbf{e}_\phi$ direction. Next, each of the terms in Eq. (7.95) is evaluated. The first term corresponds to the helical restoring force, the second to the $1/R$ force, and the last to the tire tube force.

The tire tube force

The tire tube force exists in any toroidal geometry and can be evaluated without taking into account the detailed properties of the configuration under investigation. The calculation proceeds by using some simple vector identities

$$
\begin{aligned}
\nabla p \cdot \mathbf{e}_R &= \nabla p \cdot \nabla R \\
&= \nabla \cdot (p \nabla R) - p \nabla^2 R \\
&= \nabla \cdot (p \nabla R) - \frac{p}{R}
\end{aligned}
\tag{7.96}
$$

The divergence term integrates to zero by Gauss's theorem assuming that p vanishes at large r.

In the large aspect ratio limit the tire tube force can thus be written as

$$
F_{TT} = -\int \nabla p \cdot \mathbf{e}_R \, d\mathbf{r} = \int \frac{p}{R} \, d\mathbf{r} \approx \frac{1}{R_0} \int p \, d\mathbf{r}
\tag{7.97}
$$

Since $F_{TT} > 0$ the tire tube force points outward as expected.

The 1/R force

The $1/R$ force is also quite general, depending only on toroidicity but not the details of the configuration under consideration. The force is again easily calculated using some simple vector identities,

$$
\begin{aligned}
B_\phi \mathbf{J}_p \times \mathbf{e}_\phi \cdot \mathbf{e}_R &= -\frac{B_\phi}{\mu_0 R} \nabla (R B_\phi) \cdot \nabla R \\
&= \frac{1}{2\mu_0} \nabla (R^2 B_\phi^2 - R_0^2 B_0^2) \cdot \nabla \frac{1}{R} \\
&= \frac{1}{2\mu_0} \nabla \cdot \left[(R^2 B_\phi^2 - R_0^2 B_0^2) \nabla \frac{1}{R} \right] - \frac{1}{2\mu_0} (R^2 B_\phi^2 - R_0^2 B_0^2) \nabla^2 \frac{1}{R}
\end{aligned}
\tag{7.98}
$$

The divergence term again integrates to zero by Gauss's theorem since $R B_\phi \to R_0 B_0$ for large r. The remaining term is evaluated by (1) recalling from Eq. (7.11) that in a low β stellarator $R B_\phi \approx R_0 B_0 + R_0 B_{\phi 1}$ and (2) that $\nabla^2 (1/R) = 1/R^3$. This leads to

$$F_{1/R} = \int B_\phi \mathbf{J}_p \times \mathbf{e}_\phi \cdot \mathbf{e}_R \, d\mathbf{r} \approx -\frac{1}{\mu_0} \int \frac{R_0^2 B_0 B_{\phi 1}}{R^3} \, d\mathbf{r} \approx -\frac{1}{\mu_0 R_0} \int B_0 B_{\phi 1} d\mathbf{r} \quad (7.99)$$

The last step makes use of the fact that radial pressure balance in a stellarator is similar to that in a θ-pinch. Specifically, Eq. (7.15) shows that $p_1 + B_0 B_{\phi 1}/\mu_0 = 0$. This leads to the final form of the $1/R$ force which can be written as

$$F_{1/R} \approx \frac{1}{R_0} \int p \, d\mathbf{r} \quad (7.100)$$

Observe that the $1/R$ force is equal in sign and magnitude to the tire tube force.

The helical restoring force

The helical restoring force depends on the details of the stellarator magnetic field, and thus requires some additional analysis. The first step is to calculate the quantities $\mathbf{B}_p$ and J_ϕ that appear in the remaining term of toroidal force balance.

In a low β system the only term in $\mathbf{B}_p$ that enters force balance corresponds to the vacuum helical field, denoted by $\hat{\mathbf{B}}_p$. This term is easily evaluated and for $l = 2$ is given by

$$\begin{aligned}
\hat{\mathbf{B}}_p &= \nabla_\perp \hat{A}_1 \times \mathbf{e}_\phi \\
&= -haB_0 \nabla_\perp \left[\frac{\Delta_l}{l} \rho^l \cos (l\theta + \phi) \right] \times \mathbf{e}_\phi \quad (7.101) \\
&= hB_0 \Delta_2 r [\sin \alpha \, \mathbf{e}_r + \cos \alpha \, \mathbf{e}_\theta]
\end{aligned}$$

where $\alpha = 2\theta + \phi$. It then follows that the combination $\mathbf{e}_\phi \times \mathbf{B}_p \cdot \mathbf{e}_R$ can be written as

$$\begin{aligned}
\mathbf{e}_\phi \times \mathbf{B}_p \cdot \mathbf{e}_R &= hB_0 \Delta_2 r \, \mathbf{e}_\phi \times (\sin \alpha \, \mathbf{e}_r + \cos \alpha \, \mathbf{e}_\theta) \cdot (\cos \theta \, \mathbf{e}_r - \sin \theta \, \mathbf{e}_\theta) \\
&= -hB_0 \Delta_2 r (\cos \alpha \, \cos \theta + \sin \alpha \, \sin \theta)
\end{aligned}$$

$$(7.102)$$

The next step describes the evaluation of J_ϕ which must be calculated to leading and first order in Δ_2. The reason is that the leading-order term will be shown to average to zero when integrating over ϕ. The basic equation determining J_ϕ is obtained by recalling that $\mu_0 J_\phi = -\nabla_\perp^2 A_1$ where $\nabla_\perp^2 A_1$ satisfies Eq. (7.32) with $M = 1$,

$$\left(hB_0 \frac{\partial}{\partial \phi} - \mathbf{e}_\phi \times \nabla_\perp A_1 \cdot \nabla_\perp \right) \nabla_\perp^2 A_1 = -\frac{2\mu_0}{R_0} \nabla_\perp p_1 \cdot \mathbf{e}_Z \quad (7.103)$$

This equation can be simplified by writing $A_1 = \hat{A}_1 + \tilde{A}_1$ where $\hat{A}_1$ is the vacuum helical field and $\tilde{A}_1$ is the diamagnetic contribution due to the pressure. It then follows that $\nabla^2_\perp \hat{A}_1 = 0$, $\nabla^2_\perp \tilde{A}_1 = -\mu_0 J_\phi$ and, in the limit of small but finite pressure, $\tilde{A}_1/\hat{A}_1 \sim \beta \ll 1$. After again substituting $\hat{A}_1 = -(hB_0\Delta_2/2)\, r^2 \cos\alpha$ one finds that Eq. (7.103) reduces to

$$\left[\frac{\partial}{\partial\phi} + \Delta_2\left(\cos\alpha\frac{\partial}{\partial\theta} + r\,\sin\alpha\frac{\partial}{\partial r}\right)\right]J_\phi = \frac{2}{hR_0B_0}\nabla_\perp p_1 \cdot \mathbf{e}_Z \tag{7.104}$$

The task now is to solve this equation for the current density by expanding $J_\phi = J_{\phi 0} + J_{\phi 1} + J_{\phi 2} + \cdots$ where Δ_2 is the small expansion parameter. Ultimately it is necessary to calculate both $J_{\phi 0}$ and $J_{\phi 1}$. The leading-order equation is given by

$$\frac{\partial J_{\phi 0}}{\partial\phi} = 0 \tag{7.105}$$

which implies that

$$J_{\phi 0}(r,\theta,\phi) = \bar{J}_{\phi 0}(r,\theta) \tag{7.106}$$

This quantity will be calculated from a higher-order integrability constraint.

The first-order equation determines $J_{\phi 1}$ and can be written as

$$\frac{\partial J_{\phi 1}}{\partial\phi} = -\Delta_2\left(\cos\alpha\frac{\partial}{\partial\theta} + r\,\sin\alpha\frac{\partial}{\partial r}\right)\bar{J}_{\phi 0} \tag{7.107}$$

which has as its solution

$$J_{\phi 1} = \tilde{J}_{\phi 1} + \bar{J}_{\phi 1} = -\Delta_2\left(\sin\alpha\frac{\partial}{\partial\theta} - r\,\cos\alpha\frac{\partial}{\partial r}\right)\bar{J}_{\phi 0} + \bar{J}_{\phi 1}(r,\theta) \tag{7.108}$$

The second-order equation contains the first appearance of the pressure and leads to an expression for $\bar{J}_{\phi 0}$ by means of a periodicity constraint. The second-order equation is obtained by noting that within the context of the Δ_2 expansion the right-hand side of Eq. (7.104) reduces to

$$\begin{aligned}
\frac{2}{hR_0B_0}\nabla_\perp p_1 \cdot \mathbf{e}_Z &= \frac{2}{hR_0B_0}\frac{dp_1}{d\psi}\left(\frac{\partial\psi}{\partial r}\sin\theta + \frac{1}{r}\frac{\partial\psi}{\partial\theta}\cos\theta\right) \\
&\approx \frac{2}{hR_0B_\phi}\frac{dp_1}{dr}\left[\sin\theta + O(\Delta_2)\right]
\end{aligned} \tag{7.109}$$

where use has been made of the fact that $\psi \propto r^2[1 + O(\Delta_2)]$ and $p_1(r)$ rather than $p_1(\psi)$ now serves as the free function. The second-order equation is thus given by

$$\frac{\partial J_{\phi 2}}{\partial\phi} = -\Delta_2\left(\cos\alpha\frac{\partial}{\partial\theta} + r\,\sin\alpha\frac{\partial}{\partial r}\right)J_{\phi 1} + \frac{2}{hR_0B_0}\frac{dp_1}{dr}\sin\theta \tag{7.110}$$

For $J_{\phi 2}$ to be periodic the average value of the right-hand side over one period in ϕ (or equivalently α) must vanish. After the expression for $J_{\phi 1}$ is substituted into Eq. (7.110) a straightforward evaluation of the periodicity constraint yields

$$\bar{J}_{\phi 0}(r,\theta) = \frac{2}{hR_0 B_0 \Delta_2^2}\frac{dp_1}{dr}\cos\theta = 2\frac{q_H}{B_0}\frac{dp_1}{dr}\cos\theta \qquad (7.111)$$

This expression is just the Pfirsch–Schluter current first obtained for a tokamak in Eq. (6.97) except that the safety factor corresponds to the vacuum helical transform (i.e., $q_H = 2\pi/\iota = 1/hR_0\Delta_2^2$) rather than the tokamak safety factor (i.e., $q = rB_0/R_0 B_\theta$). Equation (7.111) is next substituted into Eq. (7.108) yielding an explicit expression for $J_{\phi 1}$

$$J_{\phi 1} = \tilde{J}_{\phi 1} + \bar{J}_{\phi 1} = \frac{2}{hR_0 B_0 \Delta_2}\left(\frac{dp_1}{dr}\sin\alpha\,\sin\theta + r\frac{d^2 p_1}{dr^2}\cos\alpha\,\cos\theta\right) + \bar{J}_{\phi 1}(r,\theta)$$
$$(7.112)$$

The quantities required to evaluate the helical restoring force have now been calculated. If one makes use of the fact that $d\mathbf{r} = R_0 r dr d\theta d\phi = R_0 r dr d\theta d\alpha$ then the helical restoring force F_H can be evaluated from the definition

$$F_H = \int J_\phi (\mathbf{e}_\phi \times \mathbf{B}_p \cdot \mathbf{e}_R)\,d\mathbf{r}$$
$$\approx \int (\bar{J}_{\phi 0} + \tilde{J}_{\phi 1} + \bar{J}_{\phi 1})(\mathbf{e}_\phi \times \hat{\mathbf{B}}_p \cdot \mathbf{e}_R)R_0 r dr d\theta d\alpha \qquad (7.113)$$

The $\bar{J}_{\phi 0}$ and $\bar{J}_{\phi 1}$ terms average to zero over α. Only the $\tilde{J}_{\phi 1}$ contributes to the helical force. The expressions for $\tilde{J}_{\phi 1}$ and $\mathbf{e}_\phi \times \hat{\mathbf{B}}_p \cdot \mathbf{e}_R$ are substituted into Eq. (7.113) and the integrals over θ and α are evaluated. The expression for F_H reduces to

$$F_H = -2\pi^2 \int \left(r^2 \frac{dp_1}{dr} + r^3 \frac{d^2 p_1}{dr^2}\right) dr \qquad (7.114)$$

After several integration by parts this expression can be rewritten as

$$F_H = -8\pi^2 \int p r dr = -\frac{2}{R_0}\int p\,d\mathbf{r} \qquad (7.115)$$

where the subscript "1" has been suppressed from p_1. Equation (7.115) is the desired expression for the helical restoring force.

Discussion

After combining the expressions for the tire tube force (Eq. (7.97)), the 1/R force (Eq. (7.100)), and the helical force (Eq. (7.115)) one finds that

$$F_H = F_{TT} + F_{1/R} \tag{7.116}$$

The helical field does indeed provide an inward toroidal restoring force that balances the outward tire tube and $1/R$ forces. Force balance occurs at a location which is slightly shifted from the natural center of the vacuum helical magnetic axis because of the plasma pressure.

The field and current that interact to produce the helical restoring force are $\hat{\mathbf{B}}_p$ and $\tilde{J}_{\phi 1}\mathbf{e}_\phi$. Here, $\hat{\mathbf{B}}_p$ is the vacuum helical field and $\tilde{J}_{\phi 1}$ is a toroidal current arising from the helical modulation of the Pfirsch–Schluter current.

7.6.3 How does a vertical field shift a stellarator with no net current?

As is well known, an applied vertical field does not produce an inward or outward force along R in a current-free stellarator. This can be easily seen by calculating the force due a uniform vertical field $\mathbf{B} = B_V \mathbf{e}_Z$ as follows:

$$
\begin{aligned}
F_V &= \int (\mathbf{J} \times \mathbf{B_V}) \cdot \mathbf{e}_R \, d\mathbf{r} \\
&= B_V \int J_\phi R \, dr \, d\theta \, d\phi \\
&\approx B_V \int R_0 d\phi \int J_\phi \, r dr \, d\theta \\
&= 2\pi R_0 B_V I
\end{aligned}
\tag{7.117}
$$

Equation (7.117) is the expected result which shows that in a current-free stellarator (defined by $I = 0$), the vertical field force $F_V = 0$. The zero force result leads to an apparent paradox. It is well known experimentally that a current-free stellarator can be shifted inward or outward along R by changing the applied vertical field. However, if a vertical field produces no toroidal force how then can changing B_V shift the plasma?

The answer is associated with the fact that during the time when the vertical field is changing there is, in the context of ideal MHD, a transiently induced toroidal surface current

$$\mathbf{K} = K\mathbf{e}_\phi = (K_I + K_D \cos \theta)\mathbf{e}_\phi \tag{7.118}$$

The amplitude K consists of two components: (1) a net current K_I and (2) a dipole current $K_D \cos \theta$. These interact with the applied helical field and the changing vertical field in a somewhat complex manner, the end result nevertheless being a transient toroidal restoring force which shifts the plasma to a new position. Once the vertical field reaches its final steady state value, the restoring force and the corresponding surface current vanish. This is how a vertical field shifts the plasma. A summary of the interactions that lead to the transient restoring force is as follows.

A convenient way to understand the basic physics is to think of the plasma as a perfectly conducting wire which at time $t = 0$ carries no electric current. A vertical field is then applied in the Z direction. If the wire is straight with its cross section lying in the R, Z plane then a dipole current is induced on the plasma surface to cancel the applied vertical field within the wire. The interior of a perfectly conducting wire is completely shielded from the applied magnetic field.

If the wire instead has the shape of a circular loop then, in addition to the dipole current, there is a component of net current. The magnitude and sign of the net current are determined by the requirement that the flux passing through the hole in the loop remains constant at its initial value. The flux passing through a closed perfectly conducting wire cannot change.

Now, the net current flowing on the surface K_I produces both a vertical field force and a hoop force along R. The combination of these two forces can be shown to be always inward causing the ring to try and collapse on itself. Since K_I is proportional to B_V,[4] this inward force is proportional to $K_I B_V \propto B_V^2$. Consequently, for small B_V the K_I force is small compared to the dipole force which is shown to be linearly proportional to B_V. For simplicity of presentation the effects of the net surface current are hereafter neglected.

The restoring force arising from the dipole current is slightly subtle. The situation is as follows. The interaction of the surface dipole current with the vertical field makes no net toroidal force on the wire – the force averages to zero over θ. Similarly, the interaction of the dipole current with the applied helical magnetic field would also average to zero over ϕ if the minor cross section of the wire was circular. But, the actual plasma cross section is circular plus a helical modulation. This modulation causes the average value of the helical field $\langle \mathbf{B}_h \rangle$ on the plasma surface to be non-zero. Indeed, it is $\langle \mathbf{B}_h \rangle$ that generates the rotational transform. The interaction of the surface dipole current with $\langle \mathbf{B}_h \rangle$ on the helically modulated surface produces the dominant toroidal restoring force. This force can be either inward or outward depending upon the sign of B_V.

A simple model is presented below that translates these ideas into a quantitative prediction of the dynamical motion of the plasma as it moves from one state to another.

The flux function in the presence of a vertical field

The dynamical model is derived by modifying the low β solution to the Greene–Johnson overlap equations derived in Section 7.6.1 to include the effect of an

---

[4] The current required to cancel the vertical field flux is given by $LI = \pi R_0^2 B_V$, which implies that $I \propto K_I \propto B_V$.

applied-time varying vertical field. For analytic simplicity attention is again focused on a single helicity $l = 2$ stellarator in which the free pressure function is chosen as $\beta(\rho) = \beta_0(1 - \rho^2)$. The inclusion of an applied vertical field requires that homogeneous solutions be added to the solutions given by Eq. (7.90). The corresponding amplitudes are determined by a modified set of boundary conditions.

Based on the discussion above, the homogeneous solutions include the effect of the dipole surface current but neglect those due to the net surface current. The dipole contribution to the flux appears in both the interior and exterior of the plasma. The interior contribution is not automatically zero but has a value to cancel the contribution arising from the plasma shift. The modified solutions can be written as

$\rho \leq 1$

$$\psi_{tot} = -\frac{\Delta_2^2}{2}\left[\rho^2 + 2\Delta_2\rho^2 \cos(2\theta + \phi) - \frac{\beta_0}{2\Delta_2^4}(2\rho - \rho^3) \cos\theta\right] + c_1\rho \cos\theta$$

$\rho \geq 1$

$$\psi_{tot} = -\frac{\Delta_2^2}{2}\left[\rho^2 + 2\Delta_2\rho^2 \cos(2\theta + \phi) - \frac{\beta_0}{2\Delta_2^4}\frac{\cos\theta}{\rho}\right] + \left(c_2\rho + \frac{c_3}{\rho}\right) \cos\theta$$

$$(7.119)$$

The terms labeled by the coefficients $c_j = c_j(t)$ are the time-varying homogeneous dipole solutions.

The unknown coefficients are determined by a set of boundary conditions that must be satisfied by the toroidally averaged flux function

$$\overline{\psi}_{tot}(\rho,\theta,t) = \frac{1}{2\pi}\int_0^{2\pi} \psi_{tot}d\phi \tag{7.120}$$

From Eq. (7.119) one sees that

$$\overline{\psi}_{tot} = -\frac{\Delta_2^2}{2}\left[\rho^2 - \frac{\beta_0}{2\Delta_2^4}(2\rho - \rho^3) \cos\theta\right] + c_1\rho \cos\theta \qquad \rho \leq 1$$

$$\overline{\psi}_{tot} = -\frac{\Delta_2^2}{2}\left[\rho^2 - \frac{\beta_0}{2\Delta_2^4}\frac{\cos\theta}{\rho}\right] + \left(c_2\rho + \frac{c_3}{\rho}\right) \cos\theta \qquad \rho \geq 1$$

$$(7.121)$$

The boundary conditions on $\overline{\psi}_{tot}$ are as follows. First, note that the plasma–vacuum interface is a slightly shifted circle. The shift, as discussed, arises from the forces generated by the surface current. This is in addition to the initial shift generated by the tire tube and $1/R$ forces. For a small vertical field the shift is also small, implying that the shape of the plasma–vacuum interface can be written as

$$\rho_S(\theta, t) = 1 + \bar{\sigma}_a(t) \cos \theta \tag{7.122}$$

where, from Eq. (7.92), $\bar{\sigma}_a(0) = \beta_0/4\Delta_2^2$. The shift $\sigma_a(t) \ll 1$ is the basic unknown in the problem. The main goal of the model is to calculate $\bar{\sigma}_a(t)$ as a function of $B_V(t)$.

The first boundary condition requires continuity of flux across the interface and is given by

$$[\![\overline{\psi}_{tot}(\rho_S, \theta, t)]\!] = 0 \tag{7.123}$$

Here, $[\![\;]\!]$ denotes the jump across the surface.

The second boundary condition specifies the change in flux far from the plasma due to the vertical field $B_V(t)$. This change from its initial value is denoted by $\delta\overline{\psi}_{tot}(\rho,\theta,t)$ where $\delta\overline{\psi}_{tot}(\rho,\theta,t) = \overline{\psi}_{tot}(\rho,\theta,t) - \overline{\psi}_{tot}(\rho,\theta,0)$. In terms of the present notation the boundary condition has the form

$$\delta\overline{\psi}_{tot}(\rho, \theta, t)|_{\rho \gg 1} \rightarrow \frac{B_V(t)}{haB_0}\rho \cos \theta \tag{7.124}$$

The third and last boundary condition corresponds to conservation of flux in the plasma. A useful way to view the situation is as follows. Initially the plasma is in equilibrium with the helical field force balancing the tire tube and $1/R$ forces. The fields have diffused into the plasma leading to the low β solutions given by Eq. (7.90). The vertical field is now applied. Its rise time is slow compared to the ideal MHD time but fast compared to the resistive diffusion time. During the evolution the plasma, therefore, behaves like a perfect conductor with respect to the vertical field. In other words, the flux in the plasma at any instant of time must always equal its initial value. This is the third boundary condition which can be expressed as

$$\overline{\psi}_{tot}[\rho_S(t), \theta, t] = \overline{\psi}_{tot}[\rho_S(0), \theta, 0] \tag{7.125}$$

The problem has now been fully specified. Next, to determine the desired dynamical equations requires the following steps: (1) apply the boundary conditions to Eq. (7.121) to evaluate the homogeneous solution coefficients $c_j(t)$; (2) use the resulting solution to calculate the jump in the poloidal field across the plasma surface, which then yields the induced surface current; (3) from the surface current it is straightforward to calculate the net body force on the plasma $F_R(t)$; and (4) the dynamical equations follow from balancing the body force with the plasma inertial force.

The final dynamical equation has the form of a differential equation for $\bar{\sigma}_a(t)$ driven by a forcing term proportional to $B_V(t)$. The solution to this equation is easily obtained showing how a vertical field shifts the plasma in a current-free stellarator.

Applying the boundary conditions

It is a straightforward matter to apply the boundary conditions and evaluate the c_j. The jump condition given by Eq. (7.123) requires that

$$c_1 = c_2 + c_3 \tag{7.126}$$

The far field condition defined by Eq. (7.124) implies that

$$c_2 = \frac{B_V(t)}{haB_0} \tag{7.127}$$

Lastly, keeping in mind that $\rho_S = 1 + \bar{\sigma}_a \cos\theta$ must be expanded for small $\bar{\sigma}_a$ on the surface, one sees that conservation of flux within the plasma as given by Eq. (7.125) yields

$$c_1 = \Delta_2^2[\bar{\sigma}_a(t) - \bar{\sigma}_a(0)] \tag{7.128}$$

The resulting solutions for the c_j are substituted into Eq. (7.121) leading to the following expressions for the plasma flux including the effects of the applied vertical field:

$\rho \leq 1$

$$\overline{\psi}_{tot} = -\frac{\Delta_2^2}{2}\left[\rho^2 - \frac{\beta_0}{2\Delta_2^4}(2\rho - \rho^3)\cos\theta\right] + \Delta_2^2(\bar{\sigma}_a - \bar{\sigma}_{a0})\rho\,\cos\theta$$

$\rho \geq 1$

$$\overline{\psi}_{tot} = -\frac{\Delta_2^2}{2}\left[\rho^2 - \frac{\beta_0}{2\Delta_2^4}\frac{\cos\theta}{\rho}\right] + \frac{B_V}{haB_0}\rho\,\cos\theta + \left[\Delta_2^2(\bar{\sigma}_a - \bar{\sigma}_{a0}) - \frac{B_V}{haB_0}\right]\frac{\cos\theta}{\rho}$$

$$\tag{7.129}$$

where $\bar{\sigma}_{a0} = \bar{\sigma}_a(0) = \beta_0/4\Delta_2^2$.

The dipole surface current

From the flux one can easily calculate the jump in poloidal magnetic field on the surface and the corresponding dipole surface current. In terms of the present notation the jump in poloidal magnetic field is given by

$$[\![B_\theta]\!] = haB_0\left[\frac{\partial\psi_{tot}}{\partial\rho}\bigg|_{S_+} - \frac{\partial\psi_{tot}}{\partial\rho}\bigg|_{S_-}\right] \tag{7.130}$$

$$= 2[B_V - haB_0\Delta_2^2(\bar{\sigma}_a - \bar{\sigma}_{a0})]\cos\theta$$

The dipole surface current is obtained directly from Ampere's law and has the form

$$\mu_0 \mathbf{K}_D = \mu_0 \overline{K}_D \, \cos\theta \, \mathbf{e}_\phi = [\![B_\theta]\!] \mathbf{e}_\phi = 2[B_V - ha B_0 \Delta_2^2 (\overline{\sigma}_a - \overline{\sigma}_{a0})] \, \cos\theta \mathbf{e}_\phi$$

$$(7.131)$$

The surface current body force

The next step is to calculate the net toroidal body force arising from the dipole surface current. As stated this force arises from the interaction of the surface current with the applied helical field on the helically modulated surface and can be written as

$$
\begin{aligned}
F_R &= \int \mathbf{J} \times \mathbf{B} \cdot \mathbf{e}_R \, dr \\
&= -\int J_D dr (B_{hr} \, \sin\theta + B_{h\theta} \, \cos\theta) R_0 r \, d\theta \, d\phi \\
&= -R_0 \int \overline{K}_D \, \cos\theta [r B_{hr} \, \sin\theta + r B_{h\theta} \, \cos\theta]_s d\theta d\phi
\end{aligned}
\qquad (7.132)
$$

where $J_D dr = \overline{K}_D \cos\theta$ is the usual surface current relation and $\mathbf{B}_h$ is the applied helical field. Also, since $\overline{K}_D$ is a surface current the term in the square bracket must be evaluated on S defined by $\rho = 1 + \sigma_h \cos\alpha$.

The applied helical fields have been given in Eq. (7.101), and when evaluated on S have the form

$$
\begin{aligned}
r B_{hr}|_S &= h B_0 \Delta_2 r^2 \, \sin\alpha|_S \approx ha^2 B_0 \Delta_2 (1 + 2\sigma_h \, \cos\alpha) \sin\alpha \\
r B_{h\theta}|_S &= h B_0 \Delta_2 r^2 \, \cos\alpha|_S \approx ha^2 B_0 \Delta_2 (1 + 2\sigma_h \, \cos\alpha) \cos\alpha
\end{aligned}
\qquad (7.133)
$$

Here, σ_h is the helical modulation of the plasma surface which has already been calculated in Eq. (7.91). One sees that $\sigma_h = \overline{\sigma}_h \cos\alpha = -\Delta_2 \cos\alpha$.

The quantities $\overline{K}_D$ and $\mathbf{B}_h$ are substituted into Eq. (7.132) to evaluate the body force. Only the term containing the $B_{h\theta}$ modulation survives the averaging over θ and ϕ leading to

$$
\begin{aligned}
F_R &= -2ha^2 R_0 B_0 \Delta_2 \int \overline{\sigma}_h \overline{K}_D \, \cos^2\theta \, \cos^2\alpha \, d\theta d\phi \\
&= -4\pi^2 h^2 a^3 R_0 \Delta_2^4 \frac{B_0^2}{\mu_0} \left(\overline{\sigma}_a - \overline{\sigma}_{a0} - \frac{B_V}{ha B_0 \Delta_2^2} \right)
\end{aligned}
\qquad (7.134)
$$

The dynamical equations of motion

The dynamical equations of motion are now obtained directly from Newton's law

$$M \frac{d^2 R}{dt^2} = F_R \qquad (7.135)$$

For a uniform density one sets $M = (m_i n_0)(2\pi^2 R_0 a^2)$ and notes that $R = R_0 + a(1 + \bar{\sigma}_a)$. Equation (7.135) reduces to

$$\frac{d^2 \bar{\sigma}_a}{dt^2} + 2h^2 V_a^2 \Delta_2^4 \left(\bar{\sigma}_a - \bar{\sigma}_{a0} - \frac{B_V}{haB_0 \Delta_2^2} \right) = 0 \tag{7.136}$$

Observe that the evolution of $\bar{\sigma}_a$ is determined by the solution to a second order differential equation driven by a forcing term proportional to B_V. To obtain some insight into the behavior assume that the vertical field starts at zero and monotonically increases to a finial value B_{Vf}. Specifically, assume that

$$B_V(t) = B_{Vf}(1 - e^{-t/t_V}) \tag{7.137}$$

where t_V is the characteristic rise time of the vertical field. If one now introduces a normalized time $\tau = t/t_V$ then the equation for $\bar{\sigma}_a$ simplifies to

$$\frac{d^2 \bar{\sigma}_a}{d\tau^2} + \Omega_V^2 (\bar{\sigma}_a - \bar{\sigma}_{a0}) = b_V(1 - e^{-\tau}) \tag{7.138}$$

Here, $\Omega_V^2 = 2(\iota_H/2\pi)^2 (V_A t_V/R_0)^2$, $b_V = \Omega_V^2 (2\pi/\iota_H \varepsilon)(B_{Vf}/B_0)$, $\iota_H/2\pi = hR_0 \Delta_2^2$ is helical transform, and $V_A^2 = B_0^2/\mu_0 m_i n_0$ is the square of the Alfven speed. The parameter Ω_V represents the ratio of the vertical field rise time to the MHD time and for typical experimental parameters $\Omega_V^2 \gg 1$. The initial conditions for a plasma that starts at rest are

$$\bar{\sigma}_a(0) = \bar{\sigma}_{a0}$$
$$\frac{d\bar{\sigma}_a(0)}{d\tau} = 0 \tag{7.139}$$

The solution is easily found and is given by

$$\bar{\sigma}_a(\tau) - \bar{\sigma}_{a0} = \frac{b_V}{\Omega_V^2 (1 + \Omega_V^2)} [1 + \Omega_V^2 - \Omega_V^2 e^{-\tau} - \cos(\Omega_V \tau) - \Omega_V \sin(\Omega_V \tau)]$$

$$\approx \frac{b_V}{\Omega_V^2} \left[1 - e^{-\tau} - \frac{1}{\Omega_V} \sin(\Omega_V \tau) \right]$$

$$\tag{7.140}$$

The approximate form corresponds to the limit $\Omega_V^2 \gg 1$.

Discussion

The dynamical behavior can be understood by plotting the solution given by Eq. (7.140) in Fig. 7.8 for the case $\Omega_V^2 = 25$ and $b_V/\Omega_V^2 = 0.75\bar{\sigma}_{a0}$. One can see that within the context of ideal MHD, $\bar{\sigma}_a(\tau) - \bar{\sigma}_{a0}$ essentially tracks the applied vertical

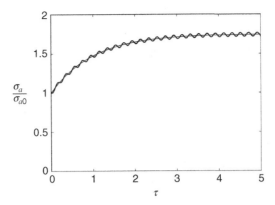

Figure 7.8   Dynamical trajectory of the plasma surface under the action of a time varying vertical field.

field with a small superimposed oscillation of relative amplitude $1/\Omega_V$. A small dissipation would cause these oscillations to damp. Once the vertical field reaches its final value the net toroidal body force on the plasma vanishes and $\bar{\sigma}_a$ reaches a new steady state value given by

$$\bar{\sigma}_a(\infty) = \bar{\sigma}_{a0} + \frac{b_V}{\Omega_V^2} \qquad (7.141)$$

or in real units

$$\bar{\sigma}_a(\infty) = \frac{2\pi}{\iota_H} \left[\frac{\mu_0 p_1(0)}{2NB_0^2} + \frac{B_{Vf}}{\varepsilon B_0} \right] \qquad (7.142)$$

The overall conclusion is that a vertical field produces a net body force only during the transient period when it is changing from its initial value to its final value. In the final steady state the stellarator remains current-free with no toroidal force produced by the vertical field. This analysis resolves the apparent paradox.

7.6.4 The equilibrium β limit in a stellarator

The goal of this subsection is to demonstrate that a current-free stellarator has an equilibrium β limit similar to that in a high β tokamak. The limit may be potentially more important in a stellarator than a tokamak. The reason is that without a net current stellarators are less susceptible to MHD instabilities and thus may be able to operate closer to the equilibrium limit. In other words, the β limit may be ultimately set by equilibrium rather than stability considerations (see for instance Kikuchi *et al.*, 2012, and Helander *et al.*, 2012).

The reason for the existence of an equilibrium β limit is similar to the tokamak. In a current-free stellarator as β increases the plasma surface is shifted outward due to the tire tube and $1/R$ forces. Assuming that the vacuum helical field is held fixed, then in order to keep the plasma centered about the natural helical magnetic axis one must apply a vertical field. Eventually, when β becomes sufficiently large, the resulting vertical field on the plasma surface B_V exceeds the surface averaged helical field $\langle \mathbf{B}_h \rangle$ causing the appearance of a separatrix. When this occurs the stellarator has reached its equilibrium β limit.

The analysis required to calculate the β limit is again based on the Greene–Johnson overlap equations. The problem is more difficult than for the low β equilibrium because the surface shifts become finite; that is, β effects enter in leading order implying that an expansion in β will not be effective. Still, by making several reasonable assumptions it becomes possible to calculate in a simple way the equilibrium β limit in a stellarator.

The assumptions required are as follows. (1) Attention is again focused on a single helicity $l = 2$ configuration. (2) To solve the overlap equations one assumes Solov'ev profiles for the free functions. (3) The plasma is allowed to carry a net current I which is set to zero at the end of the calculation. This will be important in the next subsection that describes the flux conserving stellarator. (4) Most importantly, the normalized pressure is treated as a finite quantity. Specifically, it is assumed that in terms of the overlap equation notation $\beta \sim \Delta_2^4$ which in real units is equivalent to $2\mu_0 p / B_0^2 \sim \varepsilon$.

The calculation of the β limit is similar to the high β tokamak analysis. One minor difference is the choice of sign for the free functions, which are now chosen so that the net current and helical transform both have the same sign for a positive h. The choices are not essential and can be switched, but the present choice makes the algebra a little neater. The analysis is now described below.

Simplification of the overlap equations

The starting model corresponds to the Greene–Johnson overlap equation given by Eq. (7.57). The assumptions described above are now applied and correspond to setting

$$A_n(\rho, \theta) = \frac{\Delta_2}{4}\rho^2 e^{i2\theta}$$

$$\frac{d\beta}{d\psi} = C \tag{7.143}$$

$$J = A + C\langle x \rangle$$

This leads to the following form of the overlap equation,

$$\nabla_\perp^2 \psi = -A - 2\Delta_2^2 - C\rho \, \cos\theta \qquad (7.144)$$

The boundary conditions are based on the idea that the solutions to Eq. (7.144) should describe a sequence of equilibria corresponding to increasing values of β. At low β the shape of the toroidally averaged flux function ψ is a circle. In fact, by symmetry it follows that at low β the surface is a circle for any value of l. Now, as β is increased one must imagine that the vertical field is also increased to keep the plasma surface centered about $\rho = 0$. The surface will remain approximately circular. To keep the calculation simple it is further assumed that small shaping fields are externally applied that keep the surface exactly circular. The formulation is thus transformed from a free boundary to a fixed boundary problem with the boundary condition given by

$$\psi(1, \theta) = 0 \qquad (7.145)$$

Note that without loss in generality, the free additive constant to the flux has been chosen so that $\psi = 0$ on the plasma surface.

The solution to the reduced overlap equation is easily found and is given by

$$\psi = (1 - \rho^2) \left[\frac{1}{4}(A + 2\Delta_2^2) + \frac{C}{8}\rho \, \cos\theta \right] \qquad (7.146)$$

The solution is formally similar to Eq. (6.101) for the high β tokamak. The next task is to express the constants A and C in terms of physical quantities.

Relation of A and C to physical quantities

The constant A is related to the net current I flowing in the plasma. The desired relation is obtained by recalling that

$$I_\phi = \int \langle J_\phi \rangle \, dS = -\frac{NB_0 a^2}{\mu_0 R_0} \int J\rho \, d\rho \, d\theta \qquad (7.147)$$

where the integral is carried out over the cross section of the plasma. Next, the function J is eliminated by means of Eq. (7.143) noting that $\langle x \rangle = \langle \rho \cos\theta \rangle = 0$ on a circular plasma. A short calculation then yields

$$\frac{l_I}{2\pi} \equiv \hat{\imath}_I = \frac{\mu_0 R_0 I_{\hat{\phi}}}{2\pi a^2 B_0} = -\frac{\mu_0 R_0 I_\phi}{2\pi a^2 B_0} = \frac{1}{2}hR_0 A \qquad (7.148)$$

Here, $\hat{\imath}_I = 1/q_*$ is the inverse kink safety factor associated with the net plasma current.

The constant C is related to the plasma beta, which is approximately equal to the toroidal beta β_t defined by

$$\beta_t = \frac{1}{\pi a^2}\int \frac{2\mu_0 p}{B_0^2}\, dS = \frac{\varepsilon N^2}{\pi}\int \beta \rho \, d\rho \, d\theta \qquad (7.149)$$

One now substitutes the relation $\beta(\psi) = C\psi$ with ψ given by Eq. (7.146). Another short calculation leads to

$$\beta_t = \frac{\varepsilon N^2}{8}(A + 2\Delta_2^2)C = \frac{ha}{4}(\hat{\imath}_I + \hat{\imath}_H)C \qquad (7.150)$$

where $(\imath_H/2\pi) \equiv \hat{\imath}_H = hR_0\Delta_2^2$ is the normalized rotational transform due to the vacuum helical field.

The flux function, the rotational transform, and the equilibrium β limit

The expressions for A and C are substituted into the solution for ψ which simplifies to

$$\psi = \frac{\hat{\imath}_I + \hat{\imath}_H}{2hR_0}(1-\rho^2)(1 + v\rho\cos\theta)$$
$$v = \frac{\beta_t}{\varepsilon(\hat{\imath}_I + \hat{\imath}_H)^2} \qquad (7.151)$$

As with the high β tokamak, in order to calculate the equilibrium β limit it is necessary to evaluate the safety factor, or equivalently the rotational transform, on the plasma surface. This task can be readily carried out by using the definition of the safety factor given by Eq. (6.33) and keeping in mind that in the context of the present ordering assumptions the helical component of the flux function is smaller than the toroidally averaged component (see Table 7.2). In other words, the rotational transform is essentially determined by ψ as given by Eq. (7.151). A short calculation using the fact that $B_p = |B_\theta|$ on the plasma surface shows that Eq. (6.33) reduces to

$$q_a \approx -\frac{\varepsilon B_0}{2\pi}\int_0^{2\pi}\frac{d\theta}{B_\theta} = -\frac{1}{2\pi hR_0}\int_0^{2\pi}\frac{d\theta}{[\partial\psi/\partial\rho]_{\rho=1}} \qquad (7.152)$$

The integral can be easily evaluated and is given by

$$\frac{\imath_a}{2\pi} \equiv \hat{\imath}_a = \frac{1}{q_a} = (\hat{\imath}_I + \hat{\imath}_H)(1-v^2)^{1/2} \qquad (7.153)$$

One sees that in order avoid a separatrix moving onto the plasma surface it is necessary that $v^2 \leq 1$. This corresponds to the equilibrium β limit, which can be written as

$$\beta_t \leq \varepsilon(\hat{\imath}_I + \hat{\imath}_H)^2 \qquad \text{Hybrid stellarator/tokamak}$$
$$\beta_t \leq \varepsilon\hat{\imath}_I^2 \qquad \text{Pure tokamak} \qquad\qquad (7.154)$$
$$\beta_t \leq \varepsilon\hat{\imath}_H^2 \qquad \text{Pure stellarator}$$

Equation (7.154) shows that there is a great similarity between the equilibrium β limits in a tokamak and a stellarator with a smooth connection between them. Although the expressions are formally similar there are still some observations that can be made with respect to actual practical values. For both configurations the critical β is proportional to $\varepsilon\imath^2$. Stellarators tend to have a smaller ε but a larger $\imath$. Tokamaks, because of their simpler geometry tend to have a higher maximum magnetic field at the toroidal field coils. However, the $1/R$ fall off of the toroidal field is stronger in a tokamak because of the tighter aspect ratio. Overall, the magnetic field in the center of the plasma, which is the critical location in terms of the plasma physics, can be comparable. Actual numerical comparisons of the β limit depend upon the details of the design but are typically at least as large for a stellarator as a tokamak.

7.6.5 The flux conserving stellarator

As discussed in Chapter 6 a tokamak operating in a flux conserving mode does not actually have an equilibrium limit. The reason is that in order to keep a flux conserving plasma centered in the discharge chamber as β is increased both the plasma current and the vertical field increase. This shared contribution to toroidal force balance delays the appearance of the separatrix on the plasma surface thereby avoiding the equilibrium limit. If the current is held fixed the entire burden falls on the vertical field and in this case a separatrix appears.

A similar situation occurs for the flux conserving stellarator, although in this case the result is a bit more dramatic. To visualize the situation consider starting with a low β plasma in current-free stellarator designed without an ohmic transformer. Auxiliary heating power is supplied on a time scale slow compared to the ideal MHD time but fast compared to the resistive diffusion time, corresponding to conserving operation. As β increases due to the heating a net current must begin to flow in the plasma in order to conserve the flux. Thus, even in the absence of an ohmic transformer and without any effort made to deliberately drive current with microwaves or beams, a net current flows in the plasma – more dramatic than a tokamak which already has a large current and an ohmic transformer. Ultimately the stellarator current will die away on the resistive diffusion time, but even so it will persist for a substantial period of time.

The amount of current that flows in a flux conserving stellarator can be calculated in a straightforward way from the results just derived in subsection 7.6.4. All that is required is to re-interpret the solutions so that they correspond to flux conserving operation. The critical step is to assume that during the entire flux conserving evolution the rotational transform $\hat{\imath}_a$ is held fixed at its initial value. This replaces the condition $\hat{\imath}_I = 0$ which describes a sequence of current-free stellarators.

Now, if the initial low β stellarator starts off current-free then flux conservation requires that $\hat{\imath}_a = \hat{\imath}_H$ during the entire heating evolution. Here, $\hat{\imath}_H$ is the initial helical transform which is assumed to remain constant during the evolution; that is, the helical field amplitude does not change as the heating power is applied. Equation (7.153) subject to the constraint $\hat{\imath}_a = \hat{\imath}_H$ implies that

$$v^2 = 1 - \left(\frac{\hat{\imath}_H}{\hat{\imath}_I + \hat{\imath}_H}\right)^2 \tag{7.155}$$

The parameter v is related to β_t by Eq. (7.151). For flux conservation it makes sense to introduce a heating parameter H consistent with the physics,

$$H \equiv \frac{\beta_t}{\varepsilon \hat{\imath}_H^2} = \frac{(\hat{\imath}_I + \hat{\imath}_H)^2}{\hat{\imath}_H^2} v \tag{7.156}$$

The parameter H is useful since as heat is applied β_t must increase since $\hat{\imath}_H$ remains constant. Thus $\beta_t \propto H$.

The desired relation is obtained by substituting v from Eq. (7.156) into Eq. (7.155). The resulting relation is easily solved for $\hat{\imath}_I/\hat{\imath}_H$ as a function of H

$$\frac{\hat{\imath}_I}{\hat{\imath}_H} = \left\{\frac{1}{2}\left[1 + (1 + 4H^2)^{1/2}\right]\right\}^{1/2} - 1 \tag{7.157}$$

This expression shows how much transform due to net current must be induced relative to the initial helical transform as heat is applied on a flux conserving time scale.

Equation (7.157) is plotted in Fig. 7.9. The main conclusion is that there is no equilibrium β limit. As β_t continually increases a progressively larger net current is induced to hold the plasma in equilibrium. In terms of specific scaling relations observe that for

$$\begin{array}{ll} H \ll 1 & \hat{\imath}_I/\hat{\imath}_H \approx H^2/2 \\ H \gg 1 & \hat{\imath}_I/\hat{\imath}_H \approx H^{1/2} \\ H = 1 & \hat{\imath}_I/\hat{\imath}_H \approx 0.27 \end{array} \tag{7.158}$$

Although the net current transform does not dominate for typical regimes of operation it can become substantial. One must take care to make sure that

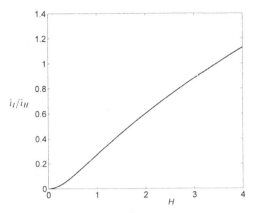

Figure 7.9   Induced normalized net current $\hat{\imath}_I/\hat{\imath}_H$ vs. the heating parameter H in a flux conserving stellarator.

current-driven kink modes are not excited before the net current diffuses away. Likewise, once the current does diffuse away either the helical field and/or the vertical field must be increased to keep the plasma centered while making sure that the equilibrium limit is not violated.

7.6.6 *Multiple helicity, finite* β *stellarators*

The previous analyses of finite β effects in a stellarator have focused on a single helicity, $l = 2$ configuration described by the simplified Greene–Johnson overlap equations. This approach has led to a number of analytic results. The present subsection considers the general HBS model which allows multiple helicities and larger values for the helical field amplitudes and plasma β. This is the more realistic experimental situation and clearly, because of the inherent 3-D nature of the problem, an efficient numerical procedure is required. As with the vacuum case several numerical procedures exist that generate fast and accurate solutions to the HBS equations. These methods are described in the references listed under Further reading at the end of the chapter. For present purposes it is assumed that such a numerical code has been written and used to produce the results described below.

Numerical formulation

To demonstrate the behavior of the flux surfaces in a finite β stellarator attention is focused on two practical examples, specifically the Large Helical Device (LHD) in Japan and the W7-X device in Germany. LHD is predominantly an $l = 2$ heliotron with small additional helical and vertical fields to optimize performance. W7-X is currently under construction and has a more complicated magnetic field configuration aimed at improving transport. It does this by means of a series of linked mirrors with transport optimized "elbows" connecting each mirror.

The form of the HBS equations which have been solved numerically are obtained from Eq. (7.35) and can be written as

$$\left[\frac{\partial}{\partial\phi} + \mathbf{e}_\phi \times \nabla_\perp(\hat{A} + \tilde{A}) \cdot \nabla_\perp\right]\psi = 0$$

$$\left[\frac{\partial}{\partial\phi} + \mathbf{e}_\phi \times \nabla_\perp(\hat{A} + \tilde{A}) \cdot \nabla_\perp\right]\nabla_\perp^2\tilde{A} = \frac{1}{M^{3/2}}\frac{d\beta}{d\psi}\nabla_\perp\psi \cdot \mathbf{e}_z \qquad (7.159)$$

$$\hat{A} = \sum_l \frac{\Delta_l}{l}\rho^{|l|}\cos(l\theta + \phi)$$

where $\hat{A}$ is the known vacuum vector potential and $\tilde{A}$ is the diamagnetic response due to the plasma currents. The boundary conditions require that ψ and $\tilde{A}$ be regular everywhere and that $\tilde{A} \to 0$ far from the plasma since it is a diamagnetic response.

Numerical results

Flux surfaces from the numerical solutions to Eq. (7.159) are illustrated in Figs. 7.10 and 7.11. The first set of flux surfaces represents an LHD-like equilibrium. This is not an especially challenging equilibrium to calculate since it is

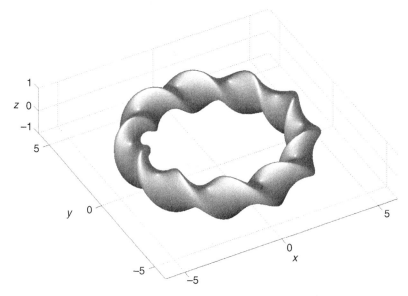

Figure 7.10 Numerically calculated boundary surface for a finite β, LHD-like plasma from the high β stellarator model.

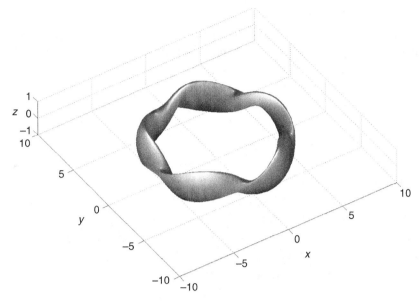

Figure 7.11 Numerically calculated boundary surface for a finite β, W7-X-like plasma from the high β stellarator model.

dominated by a large $l = 2$ field. The flux surfaces have a rotating elliptical structure.

The W7-X-like flux surfaces are more complicated and change substantially at different toroidal locations. At one toroidal location the cross section has the appearance of an elongated "boomerang." In a different toroidal location it looks like a "bullet."

These configurations are discussed in more detail in Section 7.8 following the description of neoclassical transport. The main conclusion is that realistic 3-D stellarator equilibria can be calculated numerically using the HBS expansion. The results are quite similar to numerical solutions obtained from "exact" codes which use no expansions. Detailed comparisons are more difficult than one might imagine, largely due to the differences used in specifying the vacuum magnetic fields.

7.7 Neoclassical transport in stellarators

The previous sections in Chapter 7 have described many of the basic MHD equilibrium properties of stellarators. Even so, one cannot in good conscience complete the discussion without understanding the impact of neoclassical transport on the MHD design of stellarators. The reason is that neoclassical transport in an un-optimized stellarator can substantially exceed anomalous transport. In contrast, neoclassical transport in a tokamak almost always is smaller than anomalous transport. If the situation cannot be improved, the implication is that stellarators will always have poorer confinement than tokamaks, not a good prospect for the future development of the stellarator concept.

Fortunately, the physics of neoclassical transport in stellarators is now reasonably well understood (Helander and Sigmar, 2002; Helander *et al.*, 2012; Kikuchi *et al.*, 2012). This knowledge has led to the development of several very clever ideas to mitigate the problem. These involve the ideas of "omnigenity," "quasi-isodynamics," and "quasi-symmetry." The goal in this subsection is to motivate and explain these basic ideas.

The strategy for accomplishing the goal starts with a review of neoclassical transport in a tokamak in the low collisionality "banana regime," the one of most interest to fusion plasmas. This discussion is also valuable for predicting neoclassical transport in a stellarator. In addition an anomalous "gyro-Bohm" transport estimate is presented which describes the approximate real-world operation of many tokamaks.

With these serving as references it is then possible to describe the problem that arises because of the inherent 3-D nature of a stellarator. Once the problem has been identified, various mitigating ideas (i.e., quasi-omnigenity, quasi-isodynamics, and quasi-symmetry) are presented which substantially improve single-particle confinement and neoclassical transport. Most importantly these ideas are expressed in terms of constraints on the magnetic field which then place the solution in the domain of MHD. In practice MHD design optimization is carried out numerically and several optimized configurations are discussed in Section 7.8.

7.7.1 Review of transport in a tokamak

General neoclassical transport procedure

Readers should keep in mind that the present textbook is focused on MHD and that a self-contained description of transport would require an entire textbook of its own. Several very good ones are listed in the references. Consequently, to keep the discussion in this section relatively brief, simple semi-quantitative ideas are used and in certain places results are simply quoted, the assumption being that readers have some knowledge of neoclassical transport theory in a tokamak (Helander and Sigmar, 2002).

The discussion begins with a summary of the general procedure used to estimate neoclassical transport – the random walk approximation. Attention is focused on the ion thermal diffusivity coefficient χ_i which is responsible for determining the energy confinement time $\tau_E \sim a^2/\chi_i$, one of the critical parameters defining concept desirability.

In general, χ_i is estimated from the random walk approximation given by

$$\chi_i \approx f_{eff} \nu_{eff} \overline{(\Delta r^2)} \tag{7.160}$$

Here, Δr is the step size associated with ion energy transport (i.e., the distance that a single ion's energy is transported after each collision). The quantity $\overline{(\Delta r^2)}$ is the collision averaged step size squared and enters quadratically because of the random nature of collisions resulting in a diffusive rather than a convective transport of energy. The quantity ν_{eff} is the effective rate of collisions. Often, ν_{eff} is just the inverse of the ion–ion Coulomb time. However, in the banana regime where thermal diffusion is dominated by trapped particles, ν_{eff} is higher since fewer collisions are needed to de-trap an ion causing its energy to be transported over a distance Δr. Lastly, f_{eff} is the fraction of the total particle population responsible for energy transport. In many cases all the particles contribute to energy transport and $f_{eff} \approx 1$. In contrast, when transport is dominated by the small fraction of trapped particles, then $f_{eff} \ll 1$.

The random walk approximation is now used to estimate χ_i for the reference case of banana transport in a tokamak.

Neoclassical transport in a tokamak – overview

Energy transport in a tokamak, or any torus for that matter, is substantially more complicated than in a cylinder and is referred to as "neoclassical transport." In a tokamak neoclassical transport, which arises solely from collisions (and not micro-instabilities), has three distinct regimes of behavior, known as the "banana" regime, the "plateau" regime, and the "Pfirsch–Schluter" regime. The regimes are distinguished by the relative size of the ion–ion collision frequency compared to the bounce frequency of trapped particles. Fusion plasmas operate predominantly in the banana regime corresponding to low collisionality.

On top of this complexity is the fact that energy transport over most of the plasma is anomalously large due to turbulence driven by micro-instabilities. The dominant micro-instabilities have been identified although a corresponding analytic expression for χ_i is still to be determined. Even so, basic scaling relations derived from the properties of the micro-instabilities under consideration have been proposed. Often the scaling information is viewed as more important than the hard-to-determine multiplicative numerical coefficients. Nonetheless, a simple expression for the anomalous χ_i is also presented here for the sake of reference. This expression describes the so-called "gyro-Bohm" diffusivity which seems to agree as well as any with experimental measurements.

Neoclassical transport in a tokamak – the banana regime

Banana diffusion corresponds to the low collisionality regime of neoclassical transport. It is dominated by trapped particle collisions. The reason is that the mean step size and effective collision frequency are much larger than for passing particle collisions. Competing with these effects is the fact that trapped particles only constitute a small fraction of the total number of particles. The first two effects win out causing the overall transport to be dominated by trapped particles.

The basic assumption required to estimate banana diffusion is that trapped particles are able to execute many bounce periods before having a collision. The maximum collisionality that defines the transition out of the banana regime into the plateau regime corresponds to the situation where a trapped particle has just one collision during a bounce period. This critical collisionality is estimated as follows.

Trapped particles, because of the condition for mirroring, all have a small parallel velocity $v_\parallel \sim \varepsilon^{1/2} v \sim \varepsilon^{1/2} V_{Ti}$. This leads to a collisionless bounce time between mirror points $\delta t \sim L_\parallel / v_\parallel \sim qR/v_\parallel \sim qR/\varepsilon^{1/2}V_{Ti}$. Next, note that the effective collision time is shorter than τ_{ii} since a trapped particle must only undergo a small angle (not a 90 degree) collision of magnitude $\delta\alpha \sim (v_\parallel/v)(\pi/2) \sim \varepsilon^{1/2}(\pi/2)$ to become de-trapped. This is the change in angle necessary for a particle to jump one banana width because of a collision. Specifically, since angular scattering is a diffusive process the effective collision time is $\delta t \sim [\delta\alpha/(\pi/2)]^2\tau_{ii} \sim \varepsilon\tau_{ii}$. The transition collision frequency where a trapped particle has one collision during a bounce period is obtained by equating the two expressions for δt. The result is $v_{ii} \sim \varepsilon^{8/2}V_{Ti}/qR$. The conclusion is that banana transport occurs when

$$v_{ii} < \varepsilon^{3/2}\frac{V_{Ti}}{qR} \qquad (7.161)$$

The banana thermal diffusivity coefficient can now be easily estimated. To use the random walk approximation one needs to calculate v_{eff}, $(\Delta r)^2$, and f_{eff}. The effective collision frequency has already been estimated. It is the inverse of the time necessary for a particle to diffuse through an angle $\delta\alpha \sim (v_\parallel/v)(\pi/2)$; that is $v_{eff} \sim v_{ii}/\varepsilon$.

To calculate Δr recall that the time it takes a trapped particle to execute one banana orbit scales as $\delta t \sim qR/v_\parallel \sim qR/\varepsilon^{1/2}V_{Ti}$. In the banana regime a particle will not experience a collision during this time, implying that its drift off the flux surface can be estimated as $\Delta r \sim V_D\delta t \sim (q/\varepsilon^{1/2})r_{Li}$. Here, $V_D \sim V_{Ti}(r_{Li}/R)$ is the $1/R$ grad-B guiding center drift off the surface.

Lastly, the fraction of trapped particles can also be easily estimated by calculating the portion of phase space where such particles exist: $f_{eff} \sim \delta\alpha/(\pi/2) \sim (v_\parallel/v) \sim \varepsilon^{1/2}$. Combining these results one finds that the thermal diffusivity coefficient for the banana regime is given by

$$\chi_i \equiv \chi_{NC} \sim f_{eff} v_{eff} \overline{(\Delta r^2)} \sim \varepsilon^{1/2}(v_{ii}/\varepsilon)(qr_{Li}/\varepsilon^{1/2})^2 \sim (q^2/\varepsilon^{3/2}) v_{ii} r_{Li}^2 \quad (7.162)$$

Observe that banana diffusivity is larger by a factor $q^2/\varepsilon^{3/2} \sim 50$ than the pure cylindrical value $v_{ii} r_{Li}^2$.

Anomalous transport in a tokamak – gyro-Bohm diffusion

Even with the neoclassical increase by a factor of about 50, this is still not sufficient to explain experimental observations which are even more pessimistic. The reason is that thermal transport is anomalous, driven by various micro-instabilities such as the ion and electron temperature gradient modes and the trapped electron mode. Progress is being made both analytically and computation-ally to quantify the turbulent transport due to these instabilities but, at the time of this writing, this is still a work in progress.

Many researchers in the fusion community use the "gyro-Bohm" diffusivity as a simple estimate for anomalous transport (see for instance Miyamoto, 2005, and Stacey, 2012). The same estimate is used here. The gyro-Bohm thermal diffusivity coefficient scales as

$$\chi_i \equiv \chi_{GB} \sim \frac{V_{Ti}^2}{\omega_{ci}} \frac{r_{Li}}{a} \sim \frac{V_{Ti}}{a} r_{Li}^2 \quad (7.163)$$

Note that the diffusivity is independent of collision frequency, not a surprise since turbulence replaces collisions as the transport mechanism.

Transport in a tokamak – summary

Thermal transport in a tokamak can be conveniently summarized by plotting thermal diffusivity vs. collision frequency. For plotting purposes the quantities are normalized as follows. The vertical axis illustrates χ_i/χ_{GB}. Thermal diffusivity is normalized to the gyro-Bohm value given by Eq. (7.163). The horizontal axis illustrates the ratio of the trapped particle effective collision frequency to the trapped particle bounce frequency. This ratio is usually called v_* and is given by $v_* \equiv (qR/\varepsilon^{3/2}V_{Ti})v_{ii}$. Banana transport corresponds to $v_* < 1$. Thus, for neoclassical transport corresponding to $\chi_i = \chi_{NC}$, one finds that $\chi_{NC}/\chi_{GB} \approx \varepsilon q v_* \propto 1/T^2$ showing a linear scaling with normalized collision frequency.

The thermal diffusivity curve for a tokamak operating in the banana regime is shown in Fig. 7.12. Also shown is the classical result for a cylinder. Readers should focus on the scaling trends rather than on specific numerical values since all multiplicative constants have been set to unity for simplicity. There are several important conclusions that can be drawn from Fig. 7.12 after having substituted typical experimental values:

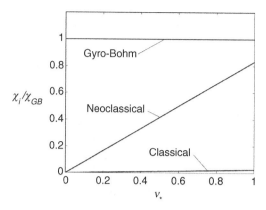

Figure 7.12   Comparison of neoclassical diffusivity in a toroidal and straight tokamak as a function of collisionality for $\varepsilon = 1/3$ and $q = 2.5$.

(1) Neoclassical transport is always substantially higher than cylindrical transport.
(2) Tokamaks operating in the high temperature fusion regime lie almost entirely in the low ν_* banana regime.
(3) In the banana regime anomalous transport dominates over neoclassical transport.
(4) A fusion tokamak can just about reach ignition assuming anomalous transport prevails, but there is not a large margin of safety.

These conclusions along with Fig. 7.12 serve as a useful reference for understanding the problems that arise with neoclassical transport in a stellarator.

7.7.2 The problem with neoclassical transport in a stellarator

Consider now a standard un-optimized stellarator operating in a high temperature, low collisionality regime, equivalent to the banana regime for tokamaks. A major problem arises with the trapped particles because of the helical modulation of B along the parallel motion of a particle's trajectory. This modulation is shown schematically in Fig. 7.13.

The existence of this modulation can be seen explicitly by expanding B to second order using the HBS model,

$$\frac{B}{B_0} \approx 1 + \frac{B_{\phi 1}}{B_0} + \frac{B_{\phi 2}}{B_0} + \frac{1}{2}\frac{B_{r1}^2 + B_{\theta 1}^2}{B_0^2} \tag{7.164}$$

As an example, for a single helicity, vacuum $l = 3$ field this expression reduces to

$$\frac{B}{B_0} \approx 1 - \varepsilon\rho\,\cos\theta + \varepsilon^2\left[\rho^2\,\cos^2\theta + \frac{N^2\Delta_3^2}{2}\rho^4 + \frac{N^2\Delta_3}{3}\rho^2\,\cos\left(3\theta + \phi\right)\right] \tag{7.165}$$

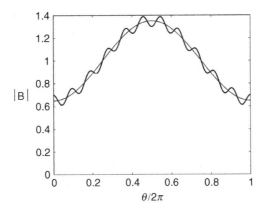

Figure 7.13   Helical modulation of $|B|$ in a toroidal stellarator.

where the helical contribution to $B_{\phi 2}/B_0 \approx (\varepsilon^2 N^2 \Delta_3/3)\rho^2 \cos(3\theta + \phi)$ has been obtained from $\mathbf{e}_r \cdot \nabla \times \mathbf{B} = 0$. The helical part of $B_{\theta 2}/B_0$ appears as the last term in the equation and represents the helical modulation.

As a particle free streams along the field line it sees, in addition to the $1/R$ dependence of the toroidal field, local mirror fields due to the helical modulation. Particles can get trapped in these local mirror fields *virtually anywhere* in the poloidal cross section. A simple physical picture can be obtained by assuming a particle is deeply trapped in such a local mirror in the vicinity of $\rho = \rho_0$, $\theta = \theta_0$ which correspond to the magnetic field minimum. In this case, the dominant variation in B that a particle sees as it tries to stream along the field is due to the ϕ dependence of the helical modulation. Under this assumption Eq. (7.165) reduces to

$$\frac{B(\phi)}{B_0} \approx 1 - \varepsilon\rho_0 \, \cos\theta_0 + \varepsilon^2 \left[\rho_0^2 \, \cos^2\theta_0 + \frac{N^2\Delta_3^2}{2}\rho_0^4 + \frac{N^2\Delta_3}{3}\rho_0^2 \, \cos\left(3\theta_0 + \phi\right) \right]$$

$$\approx 1 + \varepsilon_H \rho_0^2 \, \cos\left(3\theta_0 + \phi\right)$$

$$(7.166)$$

Here, $\varepsilon_H = \varepsilon^2 N^2 \Delta_3/3$ measures the helical mirror ratio $B_{\max}/B_{\min} = (1 + \varepsilon_H)/(1 - \varepsilon_H)$.

It is the fact that particles can get mirror trapped anywhere in the poloidal cross section, i.e., at any value of θ_0, that causes stellarators to have problems. To understand the reason, recall that the grad-B and curvature drifts perpendicular to the flux surface are always uni-directional (either up or down), and are due primarily to the $1/R$ dependence of the toroidal field. The helical fields also produce perpendicular guiding center drifts but these oscillate in sign (are not unidirectional) giving rise to helical bananas which correspond to confined orbits.

What is the consequence of having trapped particles centered about an arbitrary θ_0? For reference, in a tokamak, which has toroidal symmetry, $\theta_0 = 0$ for all trapped particles. Therefore "up" is away from the plasma half of the time and towards the plasma the other half of the time (see Fig. 7.14a). The drift off the flux surface averages to zero over one bounce period thereby implying that trapped particles are confined. Consequently, until a trapped particle has a collision, however long that time may be, its guiding center will continue to execute bounce motion with its maximum excursion off a flux surface equal to its banana width. In other words, when a collision does finally occur, the mean random walk step size is simply $\Delta r \sim V_D \tau_B \sim V_D/v_B$ where $V_D \sim V_{Ti} r_{Li}/R$ is the drift velocity and $\tau_B \sim qR/\varepsilon^{1/2} V_{Ti}$ is the bounce period.

Turning to the stellarator, one sees a much different picture. If a particle is trapped for example, at a θ_0 in the first quadrant of the poloidal plane, it continuously and monotonically drifts upward off the surface because of the unidirectional drifts. There is no opportunity to have its drift canceled since it does not sample the entire poloidal cross section. In fact in the limit of zero collisions most trapped particles would simply drift out of the plasma and strike the first wall. This behavior is illustrated in Fig. 7.14b. When such a particle finally does have a collision its step size will be much, much larger than the banana width characterizing tokamak behavior. A trapped stellarator particle will have drifted a distance equal to $\Delta r \sim V_D \tau_{eff} = V_D/v_{eff}$. In the low collisionality regime $\tau_{eff} \gg \tau_B$ leading to a much larger neoclassical energy loss rate for a stellarator.

A semi-quantitative estimate of neoclassical transport in an un-optimized stellarator can be easily obtained by using a set of analogous approximations as for the tokamak. To begin, note that the trapped particles have a small parallel velocity, related to the helical mirror ratio: $v_\| \sim \varepsilon_H^{1/2} v \sim \varepsilon_H^{1/2} V_{Ti}$. Therefore, a trapped particle must only scatter through a small angle to become de-trapped and jump a distance equal to the mean step size. This angle is given by $\delta\alpha \sim (v_\|/v)(\pi/2) \sim \varepsilon_H^{1/2}(\pi/2)$. Since angular scattering is a diffusive process, the implication is that the effective collision time (and effective collision frequency) are given by $\tau_{eff} = 1/v_{eff} \sim [\delta\alpha/(\pi/2)]^2 \tau_{ii} \sim \varepsilon_H \tau_{ii}$.

Next, the value of f_{eff} is equal to the portion of phase space occupied by trapped particles. A simple calculation yields $f_{eff} \sim \delta\alpha/(\pi/2) \sim (v_\|/v) \sim \varepsilon_H^{1/2}$. Observe that the values of v_{eff} and f_{eff} are the same as for the tokamak with ε replaced by ε_H.

The main difference from the tokamak is in the mean step size. In a tokamak the step size is equal to the banana width: $\Delta r \sim V_D \tau_B$, where V_D is the grad-B drift velocity and τ_B is the bounce period. In a stellarator, where particles drift unidirectionally until they finally have a collision the step size is $\Delta r \sim V_D \tau_{eff}$. In the low collisionality regime $\tau_{eff} \gg \tau_B$ implying a much larger step size for a stellarator. The value of Δr for a stellarator thus scales as $\Delta r \sim (V_{Ti} r_{Li}/R)(\varepsilon_H/v_{ii})$.

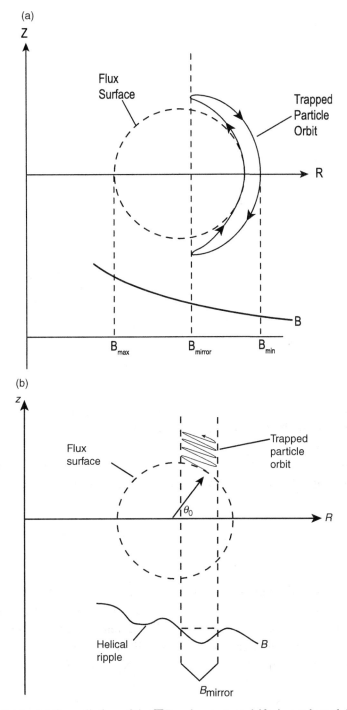

Figure 7.14   (a) Cancellation of the ∇B and curvature drifts in a tokamak leading to trapped particle banana orbits. (b) Direct loss of helically trapped particles in a stellarator because the drift is always away from the surface.

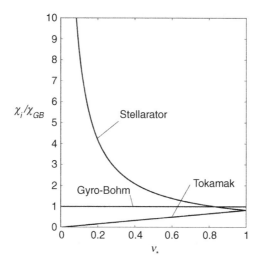

Figure 7.15　Thermal diffusivity in the banana regime of a tokamak and a stellarator vs. collisionality. Note the poorer performance of the stellarator.

These results can be combined leading to the following estimate for neoclassical transport in an un-optimized stellarator,

$$\chi_i \equiv \chi_{NC} \sim \frac{\varepsilon_H^{3/2} V_{Ti}^2 r_{Li}^2}{R^2 \nu_{ii}} \tag{7.167}$$

The critical feature is that $\chi_{NC} \sim 1/\nu_{ii}$ as compared to $\chi_{NC} \sim \nu_{ii}$ for a tokamak. Consequently, in the high temperature, low collisionality regime stellarator transport is inherently worse than in a tokamak.

A more quantitative comparison can be made by introducing the expression for ν_* and normalizing the diffusivity to the tokamak gyro-Bohm value:

$$
\begin{aligned}
\left.\frac{\chi_{NC}}{\chi_{GB}}\right|_{Tok} &\sim q\varepsilon\nu_* \\
\left.\frac{\chi_{NC}}{\chi_{GB}}\right|_{Stel} &\sim (\varepsilon_H/\varepsilon)^{3/2} \frac{q\varepsilon}{\nu_*}
\end{aligned}
\tag{7.168}
$$

Here, q is the tokamak safety factor. For comparable mirror ratios (i.e., $\varepsilon_H \sim \varepsilon$), one sees that neoclassical transport in a stellarator and tokamak are comparable when $\nu_* \approx 1$. As the temperature increases then ν_* decreases, implying that transport improves in a tokamak and gets worse in a stellarator, as illustrated in Fig. 7.15. The specific scalings with temperature for the three thermal diffusivities are given by

$$
\begin{aligned}
\chi_{NC}|_{Tok} &\propto 1/T^{1/2} \\
\chi_{GB} &\propto T^{3/2} \\
\chi_{NC}|_{Stel} &\propto T^{5/2}
\end{aligned}
\tag{7.169}
$$

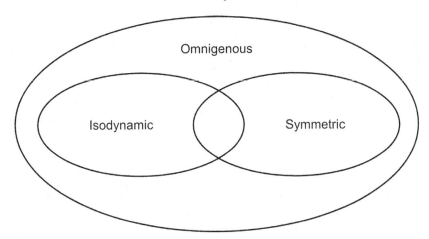

Figure 7.16   Venn diagram relation between ideal omnigenous, ideal isodynamic, and ideal symmetric stellarators.

Since $q\varepsilon \sim 1$ in most tokamaks, the implication is that the neoclassical losses in a low collisionality, un-optimized stellarator are much larger than those produced by anomalous transport. This is the basic problem faced by stellarators.

7.7.3 One solution – the omnigenous stellarator

The previous discussion has shown that greatly enhanced neoclassical transport can occur in un-optimized stellarators because of the unidirectional drift experienced by particles trapped in poloidally localized helical mirror fields. How can one improve this situation? Three related ideas have been proposed: the quasi-omnigenous stellarator, the quasi-isodynamic stellarator, and the quasi-symmetric stellarator. These topics are nicely reviewed in Mynick (2006), Landreman and Catto (2012), and Helander *et al.* (2012).

In terms of an overview it is useful to think of quasi-omnigenous stellarators as a special subclass of general stellarators. Quasi-isodynamic and quasi-symmetric stellarators are then separate subclasses of quasi-omnigenous stellarators. This overview is illustrated schematically in Fig. 7.16. The present subsection is focused on quasi-omnigenous stellarators. The additional subclasses are discussed in the following two subsections.

The goal here is to define and show how the property of omnigenity improves neoclassical transport in a stellarator. The result is the ideal-omnigenity constraint which can be translated into requirements on the magnetic field structure, thereby placing it into the domain of MHD.

The constraint of ideal-omnigenity

The basic idea is to identify the critical geometric requirement on the magnetic field that would prevent particles from being helically trapped in a local section of the poloidal plane. In other words, poloidally localized helical mirror fields must be eliminated. If this could be achieved trapped particles would then behave in a manner similar to a tokamak. In the absence of collisions they would execute confined banana orbits with a low level of neoclassical transport.

The strategy to accomplish this goal is to calculate the net guiding center drift of a banana center from its initial to final flux surface during a single bounce period and then set it equal to zero for all trapped particles. This is equivalent to forcing a particle's upward drift to be away from the plasma half the time and towards the plasma the other half, thereby canceling any net upward motion. A stellarator which exactly satisfies this constraint is defined as having the property of "ideal-omnigenity."

The desired constraint is quantified by calculating the flux surface jump $\Delta\psi$ that is accrued over a single bounce period. This value of $\Delta\psi$ is then easily converted into the mean step size required for the random walk estimate. The quantity $\Delta\psi$ is determined by transforming from r, θ, ϕ to ψ, θ, ϕ coordinates so that ψ now represents the "radial" position of the particle.

The equation of motion for the particle's guiding center ψ position is given by

$$\frac{d\psi}{dt} = \frac{\partial\psi}{\partial t} + \mathbf{V}_D \cdot \nabla\psi = \mathbf{V}_D \cdot \nabla\psi \tag{7.170}$$

Here, $\mathbf{V}_D = d\mathbf{r}_g/dt$ is the guiding center drift velocity and the $\partial\psi/\partial t$ term has been set to zero because of the assumption of equilibrium. The quantity $\mathbf{V}_D$ is dominated by the grad-B drift which is proportional to $v_\perp^2$. The curvature drift is small because it is proportional to $v_\parallel^2$ and the particles of interest are trapped and thus have small $v_\parallel^2$. Single-particle guiding center theory shows that the term $\mathbf{V}_D \cdot \nabla\psi$ for the grad-B drift for ions can be written as

$$\mathbf{V}_D \cdot \nabla\psi = \frac{m_i v_\perp^2}{2eB^3}(\mathbf{B} \times \nabla B \cdot \nabla\psi) = \frac{\mu}{eB^2}(\mathbf{B} \times \nabla B \cdot \nabla\psi) \tag{7.171}$$

Here, $\mu = m_i v_\perp^2/2B$ is the usual adiabatic invariant. (A more complete form of $\mathbf{V}_D$ including the curvature and $\mathbf{E} \times \mathbf{B}$ drifts is given in Section 7.7.6 and leads to the same overall conclusions derived below.)

The flux jump $\Delta\psi$, is obtained by integrating Eq. (7.170) over one bounce period. One finds

$$\Delta\psi = \oint \mathbf{V}_D \cdot \nabla\psi \, dt = \frac{\mu}{e} \oint \frac{1}{B^2}(\mathbf{B} \times \nabla B \cdot \nabla\psi) \, dt \tag{7.172}$$

The constraint of ideal-omnigenity requires that

$$\Delta\psi = 0 \qquad (7.173)$$

for all trapped particles. (The criterion can be shown to be automatically satisfied for all passing particles which, by definition, sample the entire flux surface.)

Equation (7.172) is cumbersome to evaluate because a detailed knowledge of the trapped particle's guiding center trajectory is required. The evaluation can be substantially simplified by assuming that the trapped particle banana width Δr_B is much less than the minor radius of the plasma: $\Delta r_B \sim (q/\varepsilon^{1/2}) r_{Li} \ll r$, usually a very good approximation experimentally. Then, the integrand I, which is a function of the guiding center trajectory $\mathbf{r}_g$ can be expanded as

$$I[\mathbf{r}_g(\mathbf{r},t)] = I[\mathbf{r} + \Delta\mathbf{r}_B(\mathbf{r},t)] \approx I(\mathbf{r}) + O(\Delta r_B/r) \approx I(\mathbf{r}) \qquad (7.174)$$

where $\mathbf{r}$ represents the location of the central magnetic field line that longitudinally bisects the banana orbit.

Similarly, the small Δr_B assumption allows the integration path in Eq. (7.172) to be accurately approximated by

$$dt = \frac{dl}{v_\parallel} = \frac{dl}{(2/m_i)^{1/2}[E - \mu B(\mathbf{r}_g)]^{1/2}} \approx \frac{dl}{(2/m_i)^{1/2}[E - \mu B(\mathbf{r})]^{1/2}} \approx \qquad (7.175)$$

Here, $E = (m_i/2)(v_\perp^2 + v_\parallel^2)$ is the particle's energy, also a constant of the motion and l is arc length along the magnetic field.

The evaluation of $\Delta\psi$ now only requires an integral along the magnetic field line and not the actual particle trajectory, a major reduction in complexity. Another useful simplification is that in the small Δr_B limit the integral along the outer leg of the banana orbit is approximately equal to the return integral along the inner leg.

These results can be combined leading to the following expression for the ideal-omnigenity constraint,

$$\Delta\psi(\psi, \theta_0) \approx 2\frac{\mu}{e}\left(\frac{m_i}{2}\right)^{1/2} \int_{l_1}^{l_2} \frac{(\mathbf{B} \times \nabla B \cdot \nabla\psi)}{B^2(E - \mu B)^{1/2}} \, dl = 0 \qquad (7.176)$$

where l_1, l_2 are the turning points of the banana orbit: $E - \mu B(l_{1,2}) = 0$. Also without loss in generality we can assume that all field lines start their trajectory at $\phi_0 = 0$.

The quasi-omnigenous stellarator

There are several important consequences from Eq. (7.176) that lead to the transformation of the ideal-omnigenous constraint into the more practical quasi-omnigenous constraint. These are as follows.

The ideal-omnigenous constraint has been expressed solely in terms of the magnetic field in physical space coordinates. No detailed information about the guiding center trajectories is required. Consequently, Eq. (7.176) is in a convenient form for MHD analysis. Unfortunately, it is not possible to exactly satisfy Eq. (7.176) experimentally, or even theoretically, for a truly 3-D configuration.

Instead, Eq. (7.176) can only be approximately satisfied by various optimization procedures. The resulting configurations are thus called "quasi-omnigenous." The procedure for generating a quasi-omnigenous equilibrium requires the evaluation of $\Delta\psi = \Delta\psi(\psi, \theta_0, E, \mu)$ for a wide range of trapped particles (i.e., a wide range of E, μ). Actually, since transport is a diffusive process, the quantity that is actually needed is a distribution function weighted average of $(\Delta\psi)^2$ including an additional average over all poloidal angles θ_0:

$$[\Delta\psi_{av}(\psi)]^2 = \frac{\displaystyle\int (\Delta\psi)^2 \frac{f(E,\mu,\psi)}{(E-\mu B)^{1/2}}\, dE d\mu d\theta_0}{\displaystyle\int \frac{f(E,\mu,\psi)}{(E-\mu B)^{1/2}}\, dE d\mu d\theta_0} \tag{7.177}$$

If ψ is now chosen to correspond to the poloidal flux/2π then the equivalent random walk step size is easily related to $(\Delta\psi_{av})^2$ by the expression

$$[\Delta r(\psi)]^2 \sim \left(\frac{\delta r_D}{\tau_B}\tau_{eff}\right)^2 \sim \frac{(\Delta\psi_{av})^2}{\langle(\nabla\psi)^2\rangle}\frac{\tau_{eff}^2}{\tau_B^2} \sim \frac{(\Delta\psi_{av})^2}{\langle R^2 B_p^2\rangle}\frac{\tau_{eff}^2}{\tau_B^2} \tag{7.178}$$

Here, δr_D is the average distance that a banana center moves after one bounce period τ_B and is directly proportional to $\Delta\psi_{av}$. For particles trapped in helical ripples τ_B is shorter than for a tokamak because of the shorter helical wavelength; that is $\tau_B \sim L_H/v_\parallel \sim 2\pi R/N\varepsilon_H^{1/2}V_{Ti}$. The quantity $\langle R^2 B_p^2\rangle$ is the flux surface average of $R^2 B_p^2$. Physically, the ratio $\delta r_D/\tau_B$ is the unidirectional guiding center drift velocity of the banana center leading to enhanced transport.

The step size in Eq. (7.178) can then be used to estimate the neoclassical thermal diffusivity:

$$\chi_{NC}(\psi) \sim f_{eff}(\Delta r)^2 v_{eff} \sim \frac{N^2\varepsilon_H^{5/2}}{4\pi^2 R^2}\frac{(\Delta\psi_{av})^2}{\langle R^2 B_p^2\rangle}\frac{V_{Ti}^2}{v_{ii}} \tag{7.179}$$

Equation (7.179) is a similar but more accurate scaling relation than the simpler estimate given by Eq. (7.167). The value of χ_{NC} does not have to reduce to the neoclassical value for a tokamak to achieve success. It must only be reduced to the size of the gyro-Bohm coefficient, which sets the practical limit for tokamak transport. Qualitatively, stellarator neoclassical transport is improved by designing magnetic geometries which tend to minimize $(\Delta\psi_{av})^2$.

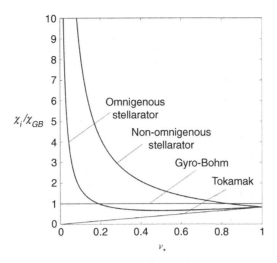

Figure 7.17 Schematic diagram showing the improvement in thermal diffusivity vs. collisionality in an omnigenous stellarator.

Even so, the geometrically optimized value of $\chi_{NC} \propto T^{5/2}$ is still larger for high temperatures. The reasons are that (1) the grad-B drift is proportional to $v_\perp^2 \propto T$ and (2) high energy particles drift a longer time because of their lower collision frequency $v_{eff} \propto 1/T^{3/2}$. Therefore, if omnigenity is not perfect (i.e., $(\Delta \psi_{av})^2 \neq 0$), high energy particles drift further off the flux surface than low energy particles. The implication is that the optimized thermal diffusivity will evolve towards the un-optimized value as the temperature increases. Even so, significant gains are possible over a large portion of the plasma. This behavior is shown schematically in Fig. 7.17.

The procedure just described demonstrates how the thermal diffusivity can be calculated and then compared to gyro-Bohm diffusion. MHD optimization requires that this procedure be repeated many times with different helical field amplitudes, phases and helicities. Optimization corresponds to minimizing $\chi_{NC}(\psi)$, or alternatively maximizing some simpler related parameter such as the global energy confinement time τ_E with respect to the helical field parameters. Often times the optimization leads to configurations with multiple helicities and a non-planar magnetic axis.

Lastly, the fact that no exact 3-D ideal-omnigenous equilibria have been found suggests that a range of widely different stellarator configurations might lead to a comparably optimized configuration in terms of neoclassical transport. To narrow down the range one needs additional constraints. This recognition has led to the ideas of the quasi-isodynamic and quasi-symmetric stellarator and it is here that specific optimized designs appear. The next task is to investigate the ideas behind these additional constraints.

7.7.4 *The isodynamic stellarator*

The idea of the isodynamic stellarator has been developed by German fusion scientists and has served as a guide to their past and future experimental programs (Gori *et al.*, 1996; Helander and Nuhrenberg, 2009; and Nuhrenberg, 2010). Currently the \$1B class W7-X stellarator is under construction at Max Planck Institute for Plasma Physics in Greifswald, Germany and is designed to approximately satisfy the isodynamic constraint. This subsection describes the philosophy behind the constraint. It also provides corresponding mathematical definitions of ideal-isodynamic and quasi-isodynamic stellarators expressed in terms of the properties of the magnetic field. One simple conclusion from the discussion is that an ideal-isodynamic stellarator by definition automatically satisfies the constraint of ideal-omnigenity. The opposite is not true.

Philosophy

There are many additional constraints that can be imposed over and above omnigenity that, if satisfied, would likely lead to improved stellarator performance. In the author's view the philosophy that has led the Germans to the isodynamic constraint is founded on a basic mistrust of plasma behavior as β is increased.

Two potentially undesirable transport effects can happen at high β, even assuming good high β MHD stability. First, the flux surfaces may become more delicate, perhaps forming large islands or stochastic regions of magnetic field, both very detrimental to transport. Also, finite β driven shifts in the flux surface location can impair the performance of the divertor, a more difficult problem in a 3-D than 2-D configuration.

Second, as the pressure increases in a stellarator there is a natural transport-driven net current induced in the plasma, known as the bootstrap current. Thus, even if a stellarator has zero net current at low β a substantial current might flow of its own accord at high β. This current, if large enough, could drive kink instabilities leading to disruptions, thereby eliminating one of the main advantages of the stellarator.

An examination of many years of experimental data would seem to indicate that the German philosophy does indeed have significant merit. How is this philosophy translated into the isodynamic design constraint? Numerical studies have shown that flux surface robustness is related to the size of the bootstrap current. Similarly MHD kink stability is closely coupled to the net toroidal bootstrap current. Therefore eliminating the bootstrap current would alleviate both concerns discussed above.

To summarize: An ideal-isodynamic stellarator is one which satisfies the ideal-omnigenity constraint while simultaneously setting the bootstrap current to zero.

For context, it is worth noting that a stellarator design based on isodynamic optimization usually leads to a large aspect ratio device, $R_0/a \sim 10$.

With the philosophy now defined, the next step is to translate the isodynamic constraint into a mathematical criterion.

The ideal-isodynamic constraint

The ideal-isodynamic constraint requires the derivation of an expression for the bootstrap current in a general stellarator and then setting it equal to zero. Such an expression has been calculated and is presented shortly, but as will be seen it is quite complicated. A simpler constraint is presented here that follows from the basic physics of the bootstrap current. This definition differs from others appearing in the literature, but is used, nevertheless, to keep the concept as simple as possible.

To begin assume that the stellarator of interest satisfies the ideal-omnigenity constraint. Trapped particles now execute helical banana orbits but no longer drift from flux surface to flux surface after every bounce period. Each trapped particle orbit is characterized by a banana width Δr_B. The first point to note is that since Δr_B is finite this leads to a diamagnetic current flowing parallel to the magnetic field.

The diamagnetic effect is illustrated in Fig. 7.18, which shows co ($v_\parallel > 0$) and counter ($v_\parallel < 0$) banana orbits for two trapped particles with the same E, μ that are tangent on the same (dotted) flux surface. If there is a density gradient ($dn/d\psi < 0$) there are more inner than outer bananas resulting in a diamagnetic current flowing parallel to the magnetic field on the flux surface. A similar argument holds if there is a temperature gradient ($dT/d\psi < 0$).

The actual bootstrap current J_B is larger than the trapped particle diamagnetic current J_T and results from friction between trapped and passing particles. Specifically, in order to conserve overall momentum, this friction sets up a differential flow between passing and trapped electrons which generates the actual bootstrap current. The relation between bootstrap current and the trapped particle diamagnetic current is approximately given by

$$J_B \sim \frac{1}{\varepsilon} J_T \qquad (7.180)$$

The conclusion to be drawn from Eq. (7.180) is that in order to completely eliminate the bootstrap current one must set $J_T = 0$. Now, the trapped particle diamagnetic current is clearly proportional to the banana width of the orbits. Therefore, a sufficient condition for setting $J_T = 0$ is to require the banana width of all particles to be zero. This condition actually represents substantial "overkill." It is mostly the case that both positive and negative diamagnetic currents flow within a given flux surface. Minimizing J_B then requires designing a magnetic geometry such that the contributions cancel when averaged over a flux surface.

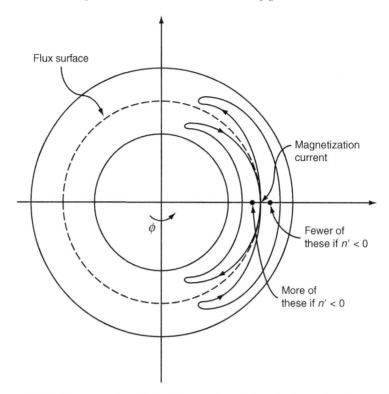

Figure 7.18 Diamagnetic effect of trapped particles due to a density gradient
that ultimately leads to the generation of the bootstrap current.

Even so, the expression for the bootstrap current is sufficiently complicated that it
is convenient to adopt the "overkill" condition as the definition of the ideal-
isodynamic constraint. The condition can never be satisfied theoretically or experimentally but it does point in the right direction. It also helps address the important
issue of alpha-particle confinement in reactor scale devices.

Mathematically, the finite banana width arises from the trapped particle grad-B
drift off the flux surface. If the magnetic field can be designed so that a trapped
particle's grad-B drift is zero along its entire banana orbit, then its banana width
will be zero and there will be no bootstrap current. From Eq. (7.176) one sees that
this requirement, which corresponds to the ideal-isodynamic constraint, can be
written as

$$\mathbf{B} \times \nabla B \cdot \nabla \psi = 0 \qquad (7.181)$$

Equation (7.181) is a very strict condition. It obviously implies that an ideal-
isodynamic stellarator automatically satisfies the ideal-omnigenous constraint; that
is, isodynamic stellarators are a subset of omnigenous stellarators.

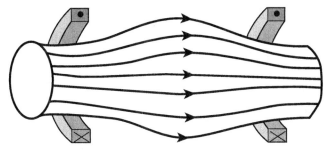

Figure 7.19   A 2-D symmetric mirror that exactly satisfies the ideal isodynamic constraint. Trapped particles never leave the flux surface and thus generate no bootstrap current.

The German approach to satisfy this constraint has been to try and design a magnetic configuration in which **B** and ∇B are parallel. A simple 2-D configuration that exactly satisfies Eq. (7.181) is a linear symmetric mirror machine as shown in Fig. 7.19. In this configuration a particle's grad-B drift is in the θ direction, and by symmetry, its guiding center orbit never drifts off the flux surface. Of course the simple mirror is known to be MHD unstable and all "passing" particles are lost through the ends. Still, in the context of a stellarator, the mirror geometry has the proper features with respect to the trapped particle orbits.

The German design of its new W7-X experiment does indeed have the appearance of a series of linked mirrors, but this is slightly misleading since most of the sophisticated optimization involves the plasma "elbows" which connect one straight section to another. Qualitatively, the W7-X geometry attempts to have the toroidally averaged helical curvature drift cancel the $1/R$ toroidal drift as much as possible. This reduces the bootstrap current and to a certain extent $\Delta\psi_{av}$.

Even with $J_B \to 0$, the $1/v_*$ dependence of χ_{NC} persists since trapped particles still do not sample the entire flux surface, particularly for vacuum fields. However, as β is increased a substantial diamagnetic ∇B drift arises due to the r dependence of $B(r)$ which causes the particles to poloidally drift around the flux surfaces. Therefore, to the extent that the magnetic geometry can be designed such that B does not vary with poloidal angle (i.e., there is no poloidal mirroring), the trapped particles can remain trapped toroidally but still sample the entire flux surface because of the poloidal $\nabla B(r)$ drift. Clearly a sophisticated design effort is required to satisfy these multiple constraints. The W7-X stellarator experiment is discussed in more detail in the next section.

The last point worth reiterating is that the ideal-isodynamic constraint given by Eq. (7.181) cannot be satisfied exactly experimentally or theoretically. Its real purpose is to serve as an idealized design target. Consequently, it is necessary to define a quantitative metric that measures how closely a realistic design comes to satisfying the ideal constraint. This leads to the idea of the quasi-isodynamic stellarator.

The quasi-isodynamic stellarator

The idea behind the quasi-isodynamic stellarator is in principle quite simple. One calculates the bootstrap current as a function of the amplitudes, phases, and helicities of the applied stellarator field. Next, some appropriate form of multi-function optimization is carried out that simultaneously attempts to minimize the thermal diffusivity $\chi_{NC}(\psi)$ and the bootstrap current density $J_B(\psi)$. Alternatively, simpler parameters could be used such as the energy confinement time τ_E and the total bootstrap current I_B. The result of such an optimization is a "quasi-isodynamic" stellarator, a configuration that is as close to being omnigenous as possible while simultaneously minimizing the bootstrap current.

The main technical difficulty in the procedure is the amount of work required to evaluate $J_B(\psi)$. There is an analytic expression for $J_B(\psi)$ for a general stellarator geometry, but it is quite complicated and requires the solution of several magnetic differential equations. Still, the burden of work falls on the computer and not the scientist.

The complexity can be appreciated by examining the actual expression for $J_B(\psi)$ after making the simplifying assumption that $T_i = T_e = T$ and setting the charge number $Z = 1$ (see for instance Landreman and Catto, 2011), and references therein,

$$J_B(\psi) \equiv \frac{\langle J_\parallel B \rangle}{\langle B \rangle} = -6.8 \frac{f_{eff}}{1 - f_{eff}} \frac{\langle G \rangle}{\langle B \rangle} \left(T \frac{\partial n}{\partial \psi} + 0.04 n \frac{\partial T}{\partial \psi} \right) \qquad (7.182)$$

Here, $\langle \ \rangle$ denotes flux surface average and the complicated function $\langle G \rangle$ is given by

$$\langle G \rangle = \frac{1}{f_{eff}} \left(\langle h_1 \rangle - \frac{3}{4} \langle B^2 \rangle \int_0^{1/B_{max}} \frac{\langle h_2 \rangle}{\langle g \rangle} \lambda \, d\lambda \right) \qquad (7.183)$$

with $g = (1 - \lambda B)^{1/2}$ and h_1, h_2 being solutions of the magnetic differential equations

$$\mathbf{B} \cdot \nabla \left(\frac{h_1}{B^2} \right) = \frac{2}{B^3} \mathbf{B} \times \nabla B \cdot \nabla \psi \qquad h_1|_{B=B_{max}} = 0$$

$$\mathbf{B} \cdot \nabla \left(\frac{h_2}{g} \right) = \frac{\lambda}{2g^3} \mathbf{B} \times \nabla B \cdot \nabla \psi \qquad h_2|_{B=B_{max}} = 0 \qquad (7.184)$$

There are two points to note. First, when the ideal-isodynamic constraint is satisfied then $h_1 = h_2 = 0$ and, as expected $J_B = 0$. Second, although the procedure is complicated requiring a numerical evaluation, it is nonetheless expressed solely in terms of the magnetic field; no individual particle orbits have to be calculated.

The quasi-isodynamic stellarator is one of two special classes of stellarator geometries that have been suggested to greatly reduce neoclassical losses in a 3-D system.

7.7.5 The symmetric stellarator

The second class of stellarator geometries that has been suggested to reduce neoclassical losses is the "symmetric stellarator." This configuration has been studied by fusion scientists around the world but has received special attention in the United States. One example is the university scale helically symmetric stellarator experiment, HSX, at the University of Wisconsin (see for instance Canik *et al.*, 2007). Also, there was a major effort to build a $100M class toroidally symmetric stellarator at the Princeton Plasma Physics Laboratory (Zarnstorff *et al.*, 2001). However, to the great frustration of the US fusion community, this experiment, the National Compact Stellarator Experiment (NCSX) was canceled because of budget constraints.

The present subsection describes the basic idea behind the symmetric stellarator. The definition of an ideal-symmetric stellarator is presented as well as a more practical definition for the quasi-symmetric stellarator. Although not obvious, it is also shown that an ideal-symmetric stellarator is omnigenous.

Philosophy

Scientists in the US fusion program have adopted a different strategy than their German colleagues. In the United States, scientists are willing to live with a bootstrap current as long as it is not too large. Instead, the goal is to generate compact stellarator configurations that have both omnigenity and a large macroscopic flow velocity, on the order of the ion thermal speed. This flow velocity, and in particular the corresponding velocity shear, is expected to reduce plasma microturbulence. Thus, omnigenity would reduce neoclassical transport, and large velocity shear would reduce turbulent transport. Overall, the energy confinement time should show significant improvement, indeed a worthy goal.

Generating a large flow velocity in a general 3-D geometry is a difficult task. The reason is that neoclassical viscosity is usually large, and therefore would be expected to damp the flow velocity down to the order of the diamagnetic drift velocity. This velocity is smaller by a factor r_{Li}/a than the thermal velocity. However, by carefully designing the magnetic field, it is possible to dramatically reduce the viscosity, thereby allowing larger flow velocities.

This "careful design" corresponds to the symmetric stellarator configuration. It is important to recognize that stellarator designs optimized with respect to the symmetry constraint can be more compact than isodynamic stellarators. For example, the NCSX design had an aspect ratio $R_0/a \sim 4$. From the US view compactness is an advantage since scaled-up reactor designs would be smaller and presumably more economical. From the German view compactness is apparently not as serious an issue. Instead the large aspect ratio can be viewed as an advantage because of engineering simplicity, lower wall loading, and easier access to the plasma. The best strategy has yet to be determined.

Consider now translating the symmetric stellarator constraint into a mathematical criterion.

The ideal-symmetric stellarator – general velocity constraints

The discussion begins with a description of the general constraints on the flow velocity in any 3-D configuration, temporarily ignoring omnigenity. Once these constraints are established, an additional constraint is introduced that minimizes the viscosity, thereby allowing for large flows. The viscosity constraint can be obtained from a collisionless kinetic MHD model (Helander, 2007), a collisional Braginskii transport model, or even to a large extent from ideal MHD. All lead to the same final result.

The approach adopted here makes use of the ideal MHD equations with flow including a single, but important, contribution from the Braginskii transport model. In addition the model restricts attention to the subclass of flows that are consistent with the assumption that the temperature satisfy $T = T(\psi)$. This constraint arises from hindsight reasoning and, as is shown, is important in showing that symmetric stellarators with large flows are also omnigenous. The constraint is plausible physically since the large parallel thermal conductivity would be expected to quickly equilibrate the temperature on any given flux surface. Still, $T = T(\psi)$ is an added constraint since it cannot be directly deduced from ideal MHD.

The derivation of the general flow constraints starts from the recognition that in steady state $\nabla \times \mathbf{E} = 0$ implying that

$$\mathbf{E} = -\nabla\Phi \tag{7.185}$$

Next, the parallel component of the ideal Ohm's law requires that $\mathbf{E} \cdot \mathbf{B} = -\mathbf{B} \cdot \nabla\Phi = 0$. The consequence is that

$$\Phi = \Phi(\psi) \tag{7.186}$$

where ψ is a flux surface label. The perpendicular component of Ohm's law then leads to an expression for the total velocity that can be written as

$$\mathbf{v} = \frac{\mathbf{E} \times \mathbf{B}}{B^2} + \frac{v_\parallel}{B}\mathbf{B} = -\frac{d\Phi}{d\psi}\frac{\nabla\psi \times \mathbf{B}}{B^2} + \frac{v_\parallel}{B}\mathbf{B} \tag{7.187}$$

The first general constraint on the velocity follows from forming the dot product of Eq. (7.187) with $\nabla \psi$. One obtains

$$\mathbf{v} \cdot \nabla \psi = 0 \tag{7.188}$$

The flow velocity must lie in the flux surface; that is, there is no cross field flow.

The second general constraint arises from the combined use of the ideal MHD energy equation and conservation of mass. The energy equation can be written as

$$\frac{d}{dt}\left(\frac{p}{n^\gamma}\right) = 2\mathbf{v} \cdot \nabla \left(\frac{T}{n^{\gamma-1}}\right) = 2\frac{T}{n^{\gamma-1}}\left[\frac{1}{T}\mathbf{v} \cdot \nabla T - \frac{\gamma-1}{n}\mathbf{v} \cdot \nabla n\right] = 0 \tag{7.189}$$

Since the flow is restricted to be consistent with $T = T(\psi)$ it automatically follows from Eq. (7.188) that $\mathbf{v} \cdot \nabla T = 0$. Thus, the conclusion from Eq. (7.189) is that

$$\mathbf{v} \cdot \nabla n = 0 \tag{7.190}$$

This relation is used in the conservation of mass equation

$$\frac{dn}{dt} + n\nabla \cdot \mathbf{v} = \mathbf{v} \cdot \nabla n + n\nabla \cdot \mathbf{v} = 0 \tag{7.191}$$

Application of Eq. (7.190) leads to the second important constraint on the velocity,

$$\nabla \cdot \mathbf{v} = 0 \tag{7.192}$$

The flow velocity in any 3-D system must be incompressible if one requires that $T = T(\psi)$.

Lastly, in terms of general properties observe that with the form of velocity given by Eq. (7.187) it follows that the steady state Faraday's law, $\nabla \times (\mathbf{v} \times \mathbf{B}) = 0$ is automatically satisfied.

The ideal-symmetric stellarator – the large flow velocity constraint

The large flow velocity constraint is based on the intuition that for such a flow to exist, the viscosity must be small. If not, viscous forces would greatly slow down the flow velocity to a much smaller value. One straightforward way to determine the constraint is to write down the viscosity tensor as, for instance, derived by Braginskii (1965). This is the approach used here. Since the Braginskii viscosity is a very complicated tensor, attention is focused on the largest, leading-order (in $1/\omega_{ci}\tau_{ii}$) contribution. For incompressible flows, the tensor is diagonal and can be written as

$$\mathbf{\Pi} = \eta_0(\mathbf{I} - 3\mathbf{bb})(\mathbf{b} \cdot \nabla \mathbf{v} \cdot \mathbf{b})$$

$$\eta_0 \approx 0.96 n T \tau_{ii} \tag{7.193}$$

The condition on the velocity for the viscosity to vanish for incompressible flows is thus given by

$$\mathbf{b} \cdot \nabla \mathbf{v} \cdot \mathbf{b} = 0 \tag{7.194}$$

Interestingly, the identical condition arises from a low collisionality treatment (Helander, 2007) using the kinetic MHD model. Equation (7.194) can be written in an alternate form by forming the dot product of Faraday's law with $\mathbf{B}$ and using the incompressibility constraint

$$\mathbf{B} \cdot \nabla \times (\mathbf{v} \times \mathbf{B}) = \mathbf{B} \cdot (\mathbf{B} \cdot \nabla \mathbf{v} - \mathbf{v} \cdot \nabla \mathbf{B}) = B^2 \mathbf{b} \cdot \nabla \mathbf{v} \cdot \mathbf{b} - B \mathbf{v} \cdot \nabla B = 0 \tag{7.195}$$

The large flow velocity constraint given by Eq. (7.194) is thus equivalent to

$$\mathbf{v} \cdot \nabla B = 0 \tag{7.196}$$

The constraint requires that the magnitude of $\mathbf{B}$ be constant along the direction of the flow velocity.

With hindsight this condition is perhaps not surprising. Intuitively, Eq. (7.196) implies that a large flow can exist when the plasma flows along a "smooth magnetic channel." A "corrugated magnetic channel" would produce a series of compressions and decompressions of the plasma which, in the presence of any dissipation, would damp the flow. If one had this intuition initially, then Eq. (7.196) could have simply been postulated as the definition of the large velocity constraint without explicitly bringing viscosity into the discussion. Such is the value of hindsight.

The ideal-symmetric stellarator

The constraints that must be satisfied by the velocity in a large flow system have now been determined. The next task is to transform these constraints into a corresponding condition on the magnetic field. This will define the ideal-symmetric stellarator. The basic constraints are summarized below:

$$
\begin{aligned}
\nabla \cdot \mathbf{v} &= 0 &\quad &\text{General flow constraint} \\
\mathbf{v} \cdot \nabla \psi &= 0 &\quad &\text{General flow constraint} \\
\mathbf{v} \cdot \nabla B &= 0 &\quad &\text{Large flow constraint}
\end{aligned} \tag{7.197}
$$

To proceed note that the second and third constraints imply that the velocity can be written as

$$\mathbf{v} = g(\mathbf{r}) \nabla \psi \times \nabla B \tag{7.198}$$

The scalar function $g(\mathbf{r})$ is determined by recalling that the velocity can also be written in the form given by Eq. (7.187). One equates these two expressions and forms the dot product first with ∇B and then with $\mathbf{B}$, leading to the following two relations:

$$\frac{d\Phi}{d\psi} \frac{\nabla\psi \times \mathbf{B} \cdot \nabla B}{B^2} = -v_{\parallel} \frac{\mathbf{B} \cdot \nabla B}{B} \tag{7.199}$$

$$g\nabla\psi \times \mathbf{B} \cdot \nabla B = -v_{\parallel}B$$

Eliminating $v_{\parallel}$ yields

$$g(\mathbf{r}) = \frac{d\Phi}{d\psi} \frac{1}{\mathbf{B} \cdot \nabla B} \tag{7.200}$$

and

$$\mathbf{v} = \frac{d\Phi}{d\psi} \frac{\nabla\psi \times \nabla B}{\mathbf{B} \cdot \nabla B} \tag{7.201}$$

The desired constraint on the magnetic field geometry is now obtained by requiring that the flow be incompressible. Thus, setting $\nabla \cdot \mathbf{v} = 0$ leads to the ideal-symmetric constraint

$$\nabla\psi \times \nabla B \cdot \nabla(\mathbf{B} \cdot \nabla B) = 0 \tag{7.202}$$

This result, while quantitative, is not completely satisfactory in terms of providing physical intuition. There are two reasons. First, Eq. (7.202) is relatively complicated so that the magnetic field strategies required to satisfy the constraint are not immediately obvious. In particular it is not exactly clear what is "symmetric" in the ideal-symmetric stellarator. The second reason is that even if Eq. (7.202) is satisfied over the entire plasma it is not evident whether or not the corresponding configuration is omnigenous. These two issues are discussed below.

What is symmetric in an ideal-symmetric stellarator?

It is shown here that the ideal-symmetric stellarator constraint implies that the three-dimensional function $B = B(\mathbf{r})$ is actually a two-dimensional function of flux and magnetic field arc length: $B = B(\psi, l)$. Assuming that this can be proved then the symmetry is immediately obvious. The quantity $B(\psi, l)$ is symmetric with respect to the third coordinate denoted here as α.

The analysis begins with the introduction of a set of flux coordinates ψ, α, l. The flux satisfies the usual condition $\mathbf{B} \cdot \nabla\psi = 0$. The arc length coordinate also satisfies the familiar condition $\mathbf{B} \cdot \nabla l = B$. The third coordinate can be fairly general. A good choice for α is an angle-like variable which is locally perpendicular to the magnetic field: $\mathbf{B} \cdot \nabla\alpha = 0$. One can always find an α that satisfies this orthogonality property. For unconvinced readers, this is demonstrated explicitly in the next subsection. The property $\mathbf{B} \cdot \nabla \alpha = 0$ is convenient because it leads to the following simplification: $\mathbf{B} \cdot \nabla = B(\partial/\partial l)$.

Introduction of these coordinates into the ideal-symmetric constraint given by Eq. (7.202) leads to

$$B \nabla \psi \times \nabla B \cdot \nabla \left(\frac{\partial B}{\partial l} \right) = 0 \qquad (7.203)$$

The general solution to Eq. (7.203) is

$$\frac{\partial B}{\partial l} = F(\psi, B) \qquad (7.204)$$

where $F(\psi, B)$ is an arbitrary function.

Next, Eq. (7.204) is first differentiated with respect to l and secondly to α resulting in two expressions,

$$\frac{\partial^2 B}{\partial l^2} = \frac{\partial F}{\partial B} \frac{\partial B}{\partial l}$$

$$\frac{\partial^2 B}{\partial l \partial \alpha} = \frac{\partial F}{\partial B} \frac{\partial B}{\partial \alpha} \qquad (7.205)$$

These expressions are divided, yielding

$$\frac{\partial}{\partial l} \ln \left(\frac{\partial B}{\partial l} \right) = \frac{\partial}{\partial l} \ln \left(\frac{\partial B}{\partial \alpha} \right) \qquad (7.206)$$

This equation can be integrated with respect to l, leading to

$$\frac{\partial B}{\partial l} = H(\psi, \alpha) \frac{\partial B}{\partial \alpha} \qquad (7.207)$$

where $H(\psi, \alpha)$ is arbitrary. Now introduce a new set of coordinates with a modified arc length

$$\psi' = \psi$$
$$\alpha' = \alpha \qquad (7.208)$$
$$l' = l + h(\psi, \alpha)$$

with $h(\psi, \alpha)$ to be determined. Note that l' is still a proper arc length coordinate: $\mathbf{B} \cdot \nabla l' = B$. After substituting Eq. (7.208) into Eq. (7.207) one obtains

$$\left(1 - H \frac{\partial h}{\partial \alpha} \right) \frac{\partial B}{\partial l'} = H \frac{\partial B}{\partial \alpha'} \qquad (7.209)$$

The function $h(\psi,\alpha)$ is chosen to satisfy $H(\partial h/\partial \alpha) = 1$ leading to the important conclusion

$$\frac{\partial B}{\partial \alpha'} = 0 \quad \rightarrow \quad B = B(\psi',l') \tag{7.210}$$

This completes the proof, explicitly demonstrating that the ideal-symmetric stellarator constraint corresponds to the symmetry requirement $B(\psi',\alpha',l') \rightarrow B(\psi',l')$.

Physically, the symmetry condition implies that large flows in a flux surface can only occur when B does not depend on the angle-like coordinate α. As the plasma flows around the torus it must, in spite of toroidal effects, encounter a uniform B corresponding to a smooth magnetic channel. If not, then corrugations in B would lead to a strong viscous damping.

Are ideal-symmetric stellarators omnigenous?

Although not obvious, the analysis below shows that an ideal-symmetric stellarator with large flow is also omnigenous. To demonstrate this result one rewrites the ideal quasi-symmetric constraint given by Eq. (7.202) in an alternate form consisting of two terms. One is shown to easily satisfy the omnigenity constraint. The other also satisfies the constraint but the proof requires some effort. The analysis begins by noting that the quantity $\mathbf{B} \times \nabla \psi$ can in general be vector decomposed as

$$\mathbf{B} \times \nabla \psi = g_1 \mathbf{B} + g_2 \nabla \psi \times \nabla B + g_3 \nabla \psi = g_1 \mathbf{B} + g_2 \nabla \psi \times \nabla B \tag{7.211}$$

where $g_3 = 0$ because of the requirement that $\mathbf{B} \cdot \nabla \psi = 0$. Now form the dot product of this relation first with ∇B and then with $\mathbf{B}$ leading to the following two relations:

$$g_1 = -\frac{\mathbf{B} \times \nabla B \cdot \nabla \psi}{\mathbf{B} \cdot \nabla B}$$

$$g_2 = -\frac{B^2}{\mathbf{B} \cdot \nabla B} \tag{7.212}$$

The next step is to take the divergence of Eq. (7.211). The result can be written in terms of the ideal-symmetric stellarator constraint which is then set to zero,

$$\mathbf{B} \cdot \nabla \left(\frac{\mathbf{B} \times \nabla B \cdot \nabla \psi}{\mathbf{B} \cdot \nabla B} \right) + \mu_0 \mathbf{J} \cdot \nabla \psi = \frac{B^2}{(\mathbf{B} \cdot \nabla B)^2} \nabla \psi \times \nabla B \cdot \nabla (\mathbf{B} \cdot \nabla B) = 0 \tag{7.213}$$

The first term on the left-hand side of the constraint can be formally integrated with respect to arc length, yielding

$$\mathbf{B} \times \nabla B \cdot \nabla \psi = G(\psi) B \frac{\partial B}{\partial l} + \mu_0 B \frac{\partial B}{\partial l} \int_0^l \frac{\mathbf{J} \cdot \nabla \psi}{B} dl' \tag{7.214}$$

where $G(\psi)$ is a free integration function. Equation (7.214) is useful because $\mathbf{B} \times \nabla B \cdot \nabla \psi$ is precisely the quantity appearing in the ideal-omnigenous constraint given by Eq. (7.176).

One now observes that the term with $G(\psi)$ automatically satisfies the omnigenity constraint. This can be seen by writing the corresponding contribution to Eq. (7.176) as follows

$$\int_{l_1}^{l_2} \frac{(\mathbf{B} \times \nabla B \cdot \nabla \psi)_G}{B^2 (E - \mu B)^{1/2}} \, dl \rightarrow G(\psi) \int_{l_1}^{l_2} \frac{1}{B(E - \mu B)^{1/2}} \frac{\partial B}{\partial l} \, dl$$

$$= \frac{G(\psi)}{E^{1/2}} \int_{l_1}^{l_2} \frac{\partial}{\partial l} \ln \left[\frac{E - (E - \mu B)^{1/2}}{E + (E - \mu B)^{1/2}} \right] dl \qquad (7.215)$$

The term in the integrand is an exact differential which can be trivially integrated. Furthermore, the end points of the integration path represent the mirror points which by definition satisfy $B_{\max}(l_1) = B_{\max}(l_2)$. Therefore, as stated, this contribution vanishes.

The remaining contribution to the omnigenous constraint involves the quantity $\mathbf{J} \cdot \nabla \psi$. For systems with zero flow $\mathbf{J} \times \mathbf{B} = \nabla p$ implying that $\mathbf{J} \cdot \nabla \psi = 0$. Ideal-symmetric systems without flow are automatically omnigenous. This, however, defeats one main purpose of the ideal-symmetric system which is to achieve good neoclassical transport in the presence of large flow velocities. Therefore, a zero flow ideal-symmetric system, which is more difficult to create than an ideal-omnigenous system, would appear to offer no additional advantages. The increased difficulty is a consequence of the fact that symmetric systems must satisfy a 2-D local criterion in ψ,l, while omnigenous systems average over arc length leading to a 1-D criterion in ψ for many particles.

The quantity $\mathbf{J} \cdot \nabla \psi$ does not obviously vanish in a system with flow because of the inertial terms. However, a nice proof by Sugama *et al.* (2011) shows that $\mathbf{J} \cdot \nabla \psi$ does indeed vanish even in the presence of large flows. An outline of their proof is as follows.

One starts with the momentum equation, assuming that the zero viscosity constraint has been satisfied. Form the dot product of this equation first with $\mathbf{B}$ and then with $\mathbf{v}$. A short calculation that makes use of the previously derived flow constraints leads to two relations given by

$$\mathbf{B} \cdot \nabla \left(\frac{p}{\rho} \ln \rho - \frac{v^2}{2} \right) + \mathbf{v} \cdot \nabla (B v_\parallel) = 0$$

$$\rho \mathbf{v} \cdot \nabla \left(\frac{v^2}{2} \right) - \mathbf{B} \cdot \nabla (B v_\parallel) = 0 \qquad (7.216)$$

The goal now is to eliminate the $Bv_\parallel$ terms from Eq. (7.216). This task is accomplished by some simple differentiation that makes use of the commuting property between $\mathbf{B} \cdot \nabla$ and $\mathbf{v} \cdot \nabla$. Specifically, by making use of Faraday's law it follows that $\mathbf{B} \cdot \nabla(\mathbf{v} \cdot \nabla Q) = \mathbf{v} \cdot \nabla(\mathbf{B} \cdot \nabla Q)$. The resulting equation can be written as

$$(\mathbf{B} \cdot \nabla)^2 \left(\frac{p}{\rho} \ln \rho - \frac{v^2}{2} \right) + \rho (\mathbf{v} \cdot \nabla)^2 \left(\frac{v^2}{2} \right) = 0 \qquad (7.217)$$

Next, note that the quantity of interest $\mathbf{J} \cdot \nabla \psi$ can be expressed in terms of v^2 by rewriting the previously derived $\mathbf{v}$ component of the momentum equation as

$$\mathbf{J} \cdot \nabla \psi = -\frac{\rho}{d\Phi/d\psi} \mathbf{v} \cdot \nabla \left(\frac{v^2}{2} \right) \qquad (7.218)$$

The proof continues by introducing flux coordinates ψ, α, l or more conveniently ψ, α, B. Now, recall that $\mathbf{v} \cdot \nabla \psi = \mathbf{v} \cdot \nabla B = 0$ implying that

$$\mathbf{J} \cdot \nabla \psi = -\frac{\rho}{d\Phi/d\psi} (\mathbf{v} \cdot \nabla \alpha) \frac{\partial}{\partial \alpha} \left(\frac{v^2}{2} \right) = -\rho \frac{\partial}{\partial \alpha} \left(\frac{v^2}{2} \right) \qquad (7.219)$$

where use has been made of the fact that $\mathbf{B} = \nabla \psi \times \nabla \alpha$ (also proven in Section 7.7.6). The final desired equation is obtained by differentiating Eq. (7.217) with respect to α and using the facts that $\rho = \rho(\psi, B)$, $p = p(\psi, B)$ and that for symmetric systems $B = B(\psi, l)$,

$$\left\{ (\mathbf{B} \cdot \nabla B) \frac{\partial}{\partial B} \left[(\mathbf{B} \cdot \nabla B) \frac{\partial}{\partial B} \right] - \rho \Phi'^2 \frac{\partial^2}{\partial \alpha^2} \right\} \frac{\partial v^2}{\partial \alpha} = 0 \qquad (7.220)$$

This is a single partial differential equation for $\partial v^2 / \partial \alpha$. When solved subject to appropriate boundary conditions Sugama *et al.* (2011) pointed out that except for a very narrow class of uninteresting solutions, the only physically acceptable solution is $\partial v^2 / \partial \alpha = 0$ or $v^2 = v^2(\psi, B)$. From Eq. (7.219) it then immediately follows that

$$\mathbf{J} \cdot \nabla \psi = 0 \qquad (7.221)$$

which is the desired result.

The overall conclusion is that the exact ideal-symmetric constraint given by Eq. (7.202) is equivalent to the more convenient form obtained from Eq. (7.214)

$$\mathbf{B} \times \nabla B \cdot \nabla \psi = G(\psi) \mathbf{B} \cdot \nabla B \qquad (7.222)$$

which, as has been shown, proves that ideal-symmetric stellarators automatically satisfy the ideal-omnigenous constraint.

The quasi-symmetric stellarator

As for the other ideal constraints it is not possible to create, either theoretically or experimentally, an exact ideal-symmetric 3-D stellarator. In 2-D, however, a straight cylindrically symmetric mirror with purely poloidal flow satisfies the ideal-symmetric constraint.

To design a practical 3-D large flow stellarator one must define an appropriate set of metrics over which a multi-function optimization can be carried out that simultaneously minimizes the neoclassical transport and the viscous force. The result of this optimization is the "quasi-symmetric stellarator." As before, the thermal diffusivity $\chi_{NC}(\psi)$, given by Eq. (7.179), is a good measure of omnigenity and the associated local neoclassical transport losses. The energy confinement time τ_E is an alternate global parameter.

A useful metric for the viscous forces can be defined as follows. First, form the dot product of the momentum equation (including the Braginskii viscosity) with $\mathbf{v}$. After a short calculation the result can be expressed as

$$\nabla \cdot \left[\left(\frac{\rho v^2}{2} + p \right) \mathbf{v} + \frac{1}{\mu_0} \mathbf{E} \times \mathbf{B} + \Pi_0 (\mathbf{v} - 3v_\parallel \mathbf{b}) \right] = -3 \frac{\eta_0}{B^2} (\mathbf{v} \cdot \nabla B)^2 \quad (7.223)$$

Here, $\Pi_0 = \eta_0 (\mathbf{b} \cdot \nabla \mathbf{v} \cdot \mathbf{b}) = (\eta_0/B)(\mathbf{v} \cdot \nabla B)$, $\eta_0 \approx 0.96 n T \tau_{ii}$, and use has been made of various equilibrium relations and the general velocity constraints. The large flow velocity constraint has not been used.

Next, Eq. (7.223) is integrated over a plasma volume bounded by a flux surface $\psi = $ constant. All the divergence terms are converted to surface integrals by Gauss' theorem. Each surface term vanishes by virtue of the general velocity constraints. What remains is the term on the right-hand side proportional to $(\mathbf{v} \cdot \nabla B)^2$ which would vanish if the large flow constraint was exactly satisfied.

If the constraint is not exactly satisfied the remaining non-zero term represents the viscous power attempting to slow down the velocity. In principle, one needs to solve the time-dependent problem to calculate the actual slowing down of the plasma flow. For design purposes this is not necessary. Instead, it makes sense to use the viscous power as a metric which has to be minimized to allow for the maximum flow. Mathematically, the quantity to be minimized can be written as

$$P_V(\psi) = 3 \int_\psi \frac{\eta_0}{B^2} (\mathbf{v} \cdot \nabla B)^2 d\mathbf{r} \quad (7.224)$$

To utilize Eq. (7.224) one must substitute a desired flow velocity $\mathbf{v}(\mathbf{r})$ and then minimize with respect to the free stellarator harmonic parameters appearing in B. A simpler global parameter is the viscous power integrated over the entire plasma, $P_{Va} = P_V(\psi_a)$.

To summarize a quasi-symmetric stellarator is defined as one that has been optimized to be as close to omnigenous as possible while simultaneously minimizing the viscous force.

7.7.6 Boozer coordinates

Introduction

The previous discussion has led to three possible constraints on the magnetic field, any of which, if satisfied, would greatly improve neoclassical transport in a stellarator. These are summarized below:

$$
\int_{l_1}^{l_2} \frac{(\mathbf{B} \times \nabla B \cdot \nabla \psi)}{B^2 (E - \mu B)^{1/2}}\, dl = 0 \qquad \text{Omnigenous}
$$

$$
\mathbf{B} \times \nabla B \cdot \nabla \psi = 0 \qquad \text{Isodynamic} \tag{7.225}
$$

$$
\mathbf{B} \times \nabla B \cdot \nabla \psi = G(\psi)(\mathbf{B} \cdot \nabla B) \qquad \text{Symmetric}
$$

The constraints listed in Eq. (7.225) have the advantage of being written in a coordinate independent form and are each functions of the same quantity $\mathbf{B} \times \nabla B \cdot \nabla \psi$. Still, it is somewhat difficult to understand intuitively how to design a stellarator magnetic field that approximately satisfies any of these constraints. This is where flux coordinates make an important contribution. By transforming to a clever set of flux coordinates one can show that the constraints reduce to much simpler, easier to understand, forms. These forms are often used to help with the practical design of stellarators. They are also particularly useful for understanding neoclassical transport in a stellarator. Measured against these important benefits is one minor disadvantage. It requires a considerable effort, mainly computational, to actually carry out the transformation from real laboratory coordinates to flux coordinates. The main issues are discussed below.

The first part of the analysis presents a derivation of the Boozer coordinates. In the second part the Boozer coordinates are used to simplify the three neoclassical constraints listed in Eq. (7.225).

Boozer coordinates

The goal is to show that a coordinate transformation always exists that leads to a pair of co-variant and contra-variant representations for the magnetic field which have particularly simple forms. These forms in turn lead to the remarkable result that the perpendicular guiding center drift of a particle off a given flux surface is a function only of $B = |\mathbf{B}|$ and not the individual vector components. The transformed coordinates are known as Boozer coordinates (Boozer, 1982).

The discussion starts with a transformation from the familiar laboratory coordinates r, θ, ϕ to a general set of flux coordinates ψ, χ, ζ. Here, ψ is a radial-like coordinate, χ is a poloidal-like angle, and ζ is a toroidal-like angle. For intuition one can imagine writing the flux coordinates as

$$\psi = \psi_0(r) + \sum_{m,n} \psi_{mn}(r)\, e^{i(m\theta + n\phi)}$$
$$\chi = \theta + \sum_{m,n} \theta_{mn}(r)\, e^{i(m\theta + n\phi)} \tag{7.226}$$
$$\zeta = \phi + \sum_{m,n} \phi_{mn}(r)\, e^{i(m\theta + n\phi)}$$

Observe that χ increases by 2π when θ increases by 2π. Similarly for ζ and ϕ. This form of the transformation is, however, not explicitly needed for the analysis.

To obtain the desired contra-variant representation two pieces of information are required for **B**: $\mathbf{B} \cdot \nabla\psi = 0$, $\nabla \cdot \mathbf{B} = 0$. One can now introduce a set of three independent basis vectors in terms of ψ, χ, ζ:

$$\nabla\psi \times \nabla\chi \qquad \nabla\zeta \times \nabla\psi \qquad \nabla\chi \times \nabla\zeta \tag{7.227}$$

These basis vectors always allow **B** to be written in a contra-variant (i.e., cross-product) form

$$\mathbf{B}(\mathbf{r}) = f_1 \nabla\psi \times \nabla\chi + f_2 \nabla\zeta \times \nabla\psi + f_3 \nabla\chi \times \nabla\zeta \tag{7.228}$$

where $f_j = f_j(\psi, \chi, \zeta)$. Equation (7.228) can be simplified. The condition $\mathbf{B} \cdot \nabla\psi = 0$ requires that $f_3 = 0$. The condition $\nabla \cdot \mathbf{B} = 0$ leads to a constraint between f_1 and f_2,

$$\nabla \cdot \mathbf{B} = J'\left(\frac{\partial f_1}{\partial \zeta} + \frac{\partial f_2}{\partial \chi}\right) = 0 \tag{7.229}$$
$$J' = \nabla\psi \times \nabla\chi \cdot \nabla\zeta$$

where $J' \equiv 1/J$ is the inverse Jacobian of the transformation; that is $d\mathbf{r} = J d\psi d\chi d\zeta$. Equation (7.229) implies that f_1 and f_2 can be written in terms of a stream function. Specifically, the solution to Eq. (7.229) can be expressed as

$$f_1(\psi, \chi, \zeta) = \bar{f}_1(\psi) + \sum_1^\infty \bar{f}_{1n}(\psi)\chi^n + \bar{\bar{f}}(\psi)\zeta + \frac{\partial \tilde{f}(\psi, \chi, \zeta)}{\partial \chi}$$

$$f_2(\psi, \chi, \zeta) = \bar{f}_2(\psi) + \sum_1^\infty \bar{f}_{2n}(\psi)\zeta^n - \bar{\bar{f}}(\psi)\chi - \frac{\partial \tilde{f}(\psi, \chi, \zeta)}{\partial \zeta} \tag{7.230}$$

The non-oscillatory terms have been explicitly displayed so that by definition $\langle \tilde{f} \rangle = 0$ over one period in χ and/or ζ. Periodicity constraints require that

$\bar{f}_{1n} = \bar{f}_{2n} = 0$ and $\bar{\bar{f}} = 0$. Consequently, the magnetic field given by Eq. (7.228) can be rewritten as

$$\mathbf{B}(\mathbf{r}) = \bar{f}_1 \nabla \psi \times \nabla \chi + \bar{f}_2 \nabla \zeta \times \nabla \psi + \nabla \psi \times \nabla \tilde{f} \tag{7.231}$$

An identical relation exists for $\mathbf{J}$ since $\mathbf{J} \cdot \nabla \psi = 0$ and $\nabla \cdot \mathbf{J} = 0$.

$$\mu_0 \mathbf{J}(\mathbf{r}) = \bar{h}_1 \nabla \psi \times \nabla \chi + \bar{h}_2 \nabla \zeta \times \nabla \psi + \nabla \psi \times \nabla \tilde{h} \tag{7.232}$$

Next, the desired co-variant (i.e., gradient) form of the magnetic field can be obtained from Eq. (7.232) by writing

$$\mu_0 \mathbf{J}(\mathbf{r}) = \nabla \times [\bar{g}_1 \nabla \chi - \bar{g}_2 \nabla \zeta - \tilde{h} \nabla \psi]$$
$$\bar{h}_1 = \frac{d\bar{g}_1}{d\psi} \tag{7.233}$$
$$\bar{h}_2 = \frac{d\bar{g}_2}{d\psi}$$

From Ampere's law, $\mu_0 \mathbf{J} = \nabla \times \mathbf{B}$, one sees that

$$\mathbf{B} = \bar{g}_1 \nabla \chi - \bar{g}_2 \nabla \zeta - \tilde{h} \nabla \psi + \nabla \tilde{g} \tag{7.234}$$

where $\tilde{g}(\psi, \chi, \zeta)$ is a free integration function.

Now, the critical step in the transformation to Boozer coordinates is the recognition that introduction of a new set of coordinates where χ and ζ are modified by carefully chosen periodic functions eliminates the appearance of the functions $\tilde{f}$ and $\tilde{g}$ in the two forms for $\mathbf{B}$. The new coordinates are defined as

$$\psi' = \psi$$
$$\chi' = \chi - \tilde{k}_\chi(\psi, \chi, \zeta) \tag{7.235}$$
$$\zeta' = \zeta - \tilde{k}_\zeta(\psi, \chi, \zeta)$$

where $\tilde{k}_\chi$ and $\tilde{k}_\zeta$ are to be determined. After substituting into Eqs. (7.231) and (7.234) one obtains the following representations for $\mathbf{B}$,

$$\mathbf{B}(\mathbf{r}) = \bar{f}_1 \nabla \psi' \times \nabla \chi' + \bar{f}_2 \nabla \zeta' \times \nabla \psi' + \nabla \psi' \times \nabla(\tilde{f} + \bar{f}_1 \tilde{k}_\chi - \bar{f}_2 \tilde{k}_\zeta)$$
$$\mathbf{B}(\mathbf{r}) = \bar{g}_1 \nabla \chi' - \bar{g}_2 \nabla \zeta' + \left(\frac{d\bar{g}_1}{d\psi} \tilde{k}_\chi - \frac{d\bar{g}_2}{d\psi} \tilde{k}_\zeta - \tilde{h} \right) \nabla \psi' + \nabla(\tilde{g} + \bar{g}_1 \tilde{k}_\chi - \bar{g}_2 \tilde{k}_\zeta)$$

$$\tag{7.236}$$

The functions $\tilde{k}_\chi$ and $\tilde{k}_\zeta$ are chosen so that the last terms in each equation are simultaneously set to zero. The appropriate choices are

$$\tilde{k}_\chi = -\left(\frac{\overline{g}_2}{\overline{f}_1\overline{g}_2 - \overline{f}_2\overline{g}_1}\right)\tilde{f} + \left(\frac{\overline{f}_2}{\overline{f}_1\overline{g}_2 - \overline{f}_2\overline{g}_1}\right)\tilde{g}$$

$$\tilde{k}_\zeta = -\left(\frac{\overline{g}_1}{\overline{f}_1\overline{g}_2 - \overline{f}_2\overline{g}_1}\right)\tilde{f} + \left(\frac{\overline{f}_1}{\overline{f}_1\overline{g}_2 - \overline{f}_2\overline{g}_{11}}\right)\tilde{g}$$

$$(7.237)$$

Observe that $\tilde{k}_\chi$ and $\tilde{k}_\zeta$ are periodic with respect to χ and ζ – there are no secular terms. Therefore, as χ, ζ change by 2π, so do χ', ζ'.

With these choices the two forms for **B** reduce to

$$\mathbf{B}(\mathbf{r}) = \nabla\psi \times \nabla(\overline{f}_1\chi - \overline{f}_2\zeta)$$
$$\mathbf{B}(\mathbf{r}) = \overline{g}_1\nabla\chi - \overline{g}_2\nabla\zeta + \tilde{k}\nabla\psi$$

$$(7.238)$$

where for convenience the primes have been suppressed and

$$\tilde{k}(\psi,\chi,\zeta) = \frac{d\overline{g}_1}{d\psi}\tilde{k}_\chi - \frac{d\overline{g}_2}{d\psi}\tilde{k}_\zeta - \tilde{h}$$

$$(7.239)$$

As will be shown the term containing $\tilde{k}$ does not play an important role in the analysis since its contribution is only in the $\nabla\psi$ direction. A very important feature of Eq. (7.238) is that the coefficients $\overline{f}_1$, $\overline{f}_2$, $\overline{g}_1$, $\overline{g}_2$ are functions only of ψ.

Equation (7.239) is almost in the desired final form for the Boozer coordinates. What remains is to relate the functions $\overline{f}_1$, $\overline{f}_2$, $\overline{g}_1, \overline{g}_2$ to more physical quantities.

Relation of $\overline{f}_1, \overline{f}_2, \overline{g}_1, \overline{g}_2$ to physical quantities

It is shown here that the quantities $\overline{f}_1$, $\overline{f}_2$, $\overline{g}_1, \overline{g}_2$ or equivalently $\overline{f}_1$, $\overline{f}_2$, $\overline{h}_1$, $\overline{h}_2$ are closely related to the magnetic fluxes and currents contained within a given pressure contour. The calculations are carried out by introducing the poloidal and toroidal differential surface areas required to calculate the fluxes and currents. These are illustrated in Fig. 7.20. The mathematical expressions for the poloidal and toroidal differential surface areas are given by their usual definitions

$$d\mathbf{A}_p = \frac{\partial\mathbf{r}}{\partial\zeta} \times \frac{\partial\mathbf{r}}{\partial\psi}d\zeta\,d\psi$$

$$d\mathbf{A}_t = \frac{\partial\mathbf{r}}{\partial\psi} \times \frac{\partial\mathbf{r}}{\partial\chi}d\chi\,d\psi$$

$$(7.240)$$

The various derivatives with respect to ψ,χ,ζ can be conveniently carried out in a rectangular coordinate system $x(\psi,\chi,\zeta)$, $y(\psi,\chi,\zeta)$, $z(\psi,\chi,\zeta)$. Thus, writing $\mathbf{r} = x(\psi,\chi,\zeta)\mathbf{e}_x + y(\psi,\chi,\zeta)\,\mathbf{e}_y + z(\psi,\chi,\zeta)\mathbf{e}_z$ one obtains

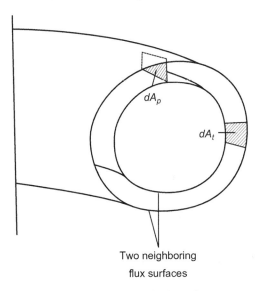

Two neighboring
flux surfaces

Figure 7.20 Differential poloidal and toroidal surface areas used to calculate fluxes and currents.

$$\frac{\partial \mathbf{r}}{\partial \psi} = x_\psi \, \mathbf{e}_x + y_\psi \, \mathbf{e}_y + z_\psi \, \mathbf{e}_z$$

$$\frac{\partial \mathbf{r}}{\partial \chi} = x_\chi \, \mathbf{e}_x + y_\chi \, \mathbf{e}_y + z_\chi \, \mathbf{e}_z \qquad (7.241)$$

$$\frac{\partial \mathbf{r}}{\partial \zeta} = x_\zeta \, \mathbf{e}_x + y_\zeta \, \mathbf{e}_y + z_\zeta \, \mathbf{e}_z$$

Similarly the rectangular unit vectors can be written as

$$\mathbf{e}_x = \nabla x = x_\psi \nabla \psi + x_\chi \nabla \chi + x_\zeta \nabla \zeta$$

$$\mathbf{e}_y = \nabla y = y_\psi \nabla \psi + y_\chi \nabla \chi + y_\zeta \nabla \zeta \qquad (7.242)$$

$$\mathbf{e}_z = \nabla z = z_\psi \nabla \psi + z_\chi \nabla \chi + z_\zeta \nabla \zeta$$

A straightforward but slightly tedious calculation then shows that

$$\frac{\partial \mathbf{r}}{\partial \zeta} \times \frac{\partial \mathbf{r}}{\partial \psi} = J \nabla \chi$$

$$\frac{\partial \mathbf{r}}{\partial \psi} \times \frac{\partial \mathbf{r}}{\partial \chi} = J \nabla \zeta \qquad (7.243)$$

where $J = [x_\psi(y_\chi z_\zeta - z_\chi y_\zeta) + x_\chi (y_\zeta z_\psi - z_\zeta y_\psi) + x_\zeta (y_\psi z_\chi - z_\psi y_\chi)]$. The quantity J is the Jacobian of the coordinate transformation; that is, there are two equivalent ways to write J. The first definition is

$$dx \, dy \, dz = J d\psi \, d\chi \, d\zeta = d\psi \, d\chi \, d\zeta / (\nabla \psi \times \nabla \chi \cdot \nabla \zeta) \qquad (7.244)$$

The second equivalent definition in flux coordinates is

$$
J = \begin{vmatrix} x_\psi & x_\chi & x_\zeta \\ y_\psi & y_\chi & y_\zeta \\ z_\psi & z_\chi & z_\zeta \end{vmatrix} = [x_\psi \, (y_\chi z_\zeta - z_\chi y_\zeta) + x_\chi \, (y_\zeta z_\psi - z_\zeta y_\psi) + x_\zeta \, (y_\psi z_\chi - z_\psi y_\chi)]
$$

$$(7.245)$$

Combining these results leads to the following simple expressions for the differential surface areas

$$
\begin{aligned}
d\mathbf{A}_p &= (J\nabla\chi)d\zeta \, d\psi \\
d\mathbf{A}_t &= (J\nabla\zeta)d\chi \, d\psi
\end{aligned}
$$

$$(7.246)$$

Consider now the poloidal flux defined by

$$
\Psi_p = \int \mathbf{B} \cdot d\mathbf{A}_p = \int_0^{2\pi} \int_0^\psi (\mathbf{B} \cdot \nabla\chi)J \, d\zeta \, d\psi
$$

$$(7.247)$$

Using the contra-variant Boozer representation of $\mathbf{B}$ it follows that

$$
\mathbf{B} \cdot \nabla\chi = \bar{f}_2 \, (\nabla\zeta \times \nabla\psi) \cdot \nabla\chi = \frac{\bar{f}_2}{J}
$$

$$(7.248)$$

The Jacobian factors cancel in the integrand and the ζ integral can be immediately evaluated yielding a factor of 2π. The poloidal flux reduces to

$$
\Psi_p = 2\pi \int_0^\psi \bar{f}_2 \, d\psi
$$

$$(7.249)$$

Recall that ψ is defined as $\psi = \Psi_p/2\pi$. Thus, differentiating Eq. (7.249) with respect to ψ leads to the conclusion that

$$
\bar{f}_2 = 1
$$

$$(7.250)$$

A completely analogous calculation for the toroidal flux shows that

$$
\Psi_t = \int \mathbf{B} \cdot d\mathbf{A}_t = \int_0^{2\pi} \int_0^\psi (\mathbf{B} \cdot \nabla\zeta)J \, d\chi \, d\psi = 2\pi \int_0^\psi \bar{f}_1 \, d\psi
$$

$$(7.251)$$

Again, recalling that $\psi_t = \Psi_t/2\pi$ and then differentiating with respect to ψ leads to the following expression for $\bar{f}_1$.

$$
\bar{f}_1(\psi) = \frac{d\psi_t(\psi)}{d\psi} \equiv q(\psi)
$$

$$(7.252)$$

Here, $q(\psi)$ is an alternate but general definition of the safety factor. It is easy to derive this relation once the straight field line properties of the Boozer coordinates are demonstrated. The derivation is presented in Appendix F.

The next quantities of interest are the currents flowing within a $\psi = $ contour. From the contra-variant Boozer representation of $\mathbf{J}$ one sees that the net toroidal and poloidal currents are given by

$$\mu_0 I_t = \mu_0 \int \mathbf{J} \cdot d\mathbf{A}_t = \mu_0 \int_0^{2\pi} \int_0^{\psi} (\mathbf{J} \cdot \nabla \zeta) J \, d\chi \, d\psi = 2\pi \int_0^{\psi} \overline{h}_1 d\psi$$
$$\mu_0 I_p = \mu_0 \int \mathbf{J} \cdot d\mathbf{A}_p = \mu_0 \int_0^{2\pi} \int_0^{\psi} (\mathbf{J} \cdot \nabla \chi) J \, d\zeta \, d\psi = 2\pi \int_0^{\psi} \overline{h}_2 d\psi$$

(7.253)

Again, differentiating with respect to ψ and defining $i_t = \mu_0 I_t / 2\pi$ and $i_p = \mu_0 I_p / 2\pi$ yields

$$\overline{h}_1 = \frac{d\, i_t}{d\psi} \quad \rightarrow \quad \overline{g}_1 = i_t + i_{t0}$$
$$\overline{h}_2 = \frac{d\, i_p}{d\psi} \quad \rightarrow \quad \overline{g}_2 = i_p + i_{p0}$$

(7.254)

where i_{t0}, i_{p0} are free integration constants chosen to make the magnetic field reduce to the correct limit when no plasma is present. Thus, when there is zero toroidal plasma current (i.e., $i_t = 0$) then there should be zero poloidal field implying that $i_{t0} = 0$. However, when there is zero poloidal plasma current (i.e., $i_p = 0$) there is still a toroidal magnetic field due to the currents in the TF coil. Denoting this current by $\mu_0 I_{coil} = 2\pi i_{coil}$ one sees that $\overline{g}_2 = i_p + i_{coil}$.

The final desired form of Boozer coordinates can now be written as

$$\mathbf{B}(\mathbf{r}) = \nabla \psi \times \nabla (q\chi - \zeta)$$
$$\mathbf{B}(\mathbf{r}) = i_t \nabla \chi - i_p \nabla \zeta + \tilde{k} \nabla \psi$$

(7.255)

where hereafter i_{coil} has been absorbed into i_p; in other words $i_p \rightarrow i_p + i_{coil}$

Properties of Boozer coordinates

There are two important and useful properties of Boozer coordinates that can be easily derived from Eq. (7.255). The first is the fact that Boozer coordinates are a special case of "straight field line" coordinates. This can be seen by using the fact that the contra-variant representation of the magnetic field has the form $\mathbf{B} = \nabla \psi \times \nabla \alpha$. One now plots the two surfaces $\psi = \psi_0 = $ constant and $\alpha = \alpha_B = $ constant as shown in Fig. 7.21. The intersection of these two surfaces defines a magnetic line; that is, since the magnetic field satisfies $\mathbf{B} \cdot \nabla \psi = \mathbf{B} \cdot \nabla \alpha = 0$ then the field lines must lie simultaneously in both surfaces which can only occur where the surfaces intersect. The conclusion is that on the flux surface $\psi = \psi_0$ the equation for the field line trajectory in χ, ζ coordinates is given by $\alpha = \alpha_B$, or

$$\zeta = q(\psi)\chi - \alpha_B$$

(7.256)

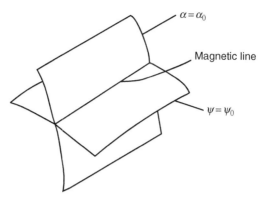

Figure 7.21   Intersection of $\psi =$ constant, $\alpha =$ constant surfaces defining a magnetic line.

This is the equation of a straight line, hence, the name "straight field line coordinates."

The second useful property of Boozer coordinates is that there is a simple representation of the Jacobian. This representation is obtained by evaluating the dot product of the two forms for the magnetic field given in Eq. (7.255). The result is

$$\nabla\psi \times \nabla\chi \cdot \nabla\zeta \equiv \frac{1}{J} = \frac{B^2}{(i_t - q\, i_p)} \qquad (7.257)$$

Observe that $J = J(\psi, B)$ a very important result for the application of Boozer coordinates to neoclassical transport.

Applications of Boozer coordinates

The required relationships have now been derived that allow application of Boozer coordinates to neoclassical transport in a stellarator. The approach is to evaluate the generalized perpendicular guiding center drift of the particles off the flux surface, demonstrate that this drift is proportional to $\mathbf{B} \times \nabla B \cdot \nabla\psi$, and then show how Boozer coordinates simplify this expression.

The analysis begins with the recognition that the guiding center drift normal to the flux surface is responsible for neoclassical transport and is comprised of three contributions: the $\mathbf{E} \times \mathbf{B}$ drift, the grad-B drift, and the curvature drift. The change in flux surface location along the guiding center trajectory satisfies the equation

$$\frac{d\psi}{dt} = \frac{\partial\psi}{\partial t} + \mathbf{V}_D \cdot \nabla\psi = \mathbf{V}_D \cdot \nabla\psi$$

$$= \left(\frac{\mathbf{E} \times \mathbf{B}}{B^2} + \frac{mv_\perp^2}{2e} \frac{\mathbf{B} \times \nabla B}{B^3} + \frac{mv_\parallel^2}{e} \frac{\mathbf{B} \times \boldsymbol{\kappa}}{B^2} \right) \cdot \nabla\psi \qquad (7.258)$$

Since $\mathbf{E} = -\nabla\Phi(\psi)$ then the $\mathbf{E} \times \mathbf{B}$ term automatically vanishes. The grad-B drift is already in the correct form. The curvature term is simplified by noting that

$$\boldsymbol{\kappa} = \mathbf{b} \cdot \nabla\mathbf{b} = \frac{\mathbf{B}}{B} \cdot \nabla\frac{\mathbf{B}}{B} = \frac{1}{B^2}\mathbf{B} \cdot \nabla\mathbf{B} - \frac{\mathbf{B} \cdot \nabla B}{B^3}\mathbf{B}$$

$$= \frac{1}{B^2}\nabla\left(\mu_0 p + \frac{B^2}{2}\right) - \frac{\mathbf{B} \cdot \nabla B}{B^3}\mathbf{B}$$

$$= \frac{1}{B^2}\frac{dp}{d\psi}\nabla\psi + \frac{\nabla B}{B} - \frac{\mathbf{B} \cdot \nabla B}{B^3}\mathbf{B} \tag{7.259}$$

The first and third terms make no contribution to $\mathbf{V}_D \cdot \nabla\psi$. Only the second term enters the analysis. After introducing the constants of the motion E, μ one sees that the perpendicular guiding center drift is given by

$$\mathbf{V}_D \cdot \nabla\psi = \frac{1}{e}(2E - \mu B)\frac{\mathbf{B} \times \nabla B \cdot \nabla\psi}{B^3} \tag{7.260}$$

The critical step is to now show how Boozer coordinates simplify Eq. (7.260). Start with a general co-variant form for the magnetic field: $\mathbf{B} = B_\psi \nabla\psi + B_\chi \nabla\chi + B_\zeta \nabla\zeta$. A short calculation shows that

$$\mathbf{V}_D \cdot \nabla\psi = \frac{1}{e}(2E - \mu B)\frac{1}{JB^3}\left(B_\chi\frac{\partial B}{\partial\zeta} - B_\zeta\frac{\partial B}{\partial\chi}\right) \tag{7.261}$$

Substituting the co-variant Boozer form of $\mathbf{B}$ leads to

$$\mathbf{V}_D \cdot \nabla\psi = \frac{(2E - \mu B)}{eB}\left(\frac{i_t}{i_t - qi_p}\frac{\partial B}{\partial\zeta} + \frac{i_p}{i_t - qi_p}\frac{\partial B}{\partial\chi}\right) \tag{7.262}$$

The remarkable conclusion from Eq. (7.262) is that the normal guiding center drift expressed in Boozer coordinates is only a function of the scalars ψ, B, and derivatives of B. It is not a function of the individual vector components of $\mathbf{B}$. This is not an obvious conclusion since, in general, the individual vector components of $\mathbf{B}$ (i.e., B_χ and B_ζ) appear explicitly in Eq. (7.261).

Equation (7.262) offers the possibility of generating a 2-D symmetric B from a 3-D vector $\mathbf{B}$. As an example illustrating this basic point in a simpler geometry consider the 2-D helical vector field

$$\mathbf{B}(r, \alpha) = B_\phi(r)\,\mathbf{e}_\phi + B_h(r/a)^{l-1}(\mathbf{e}_r \sin\alpha + \mathbf{e}_\theta \cos\alpha) \qquad \alpha = l\theta + n\phi \tag{7.263}$$

It immediately follows that B is a 1-D function,

$$B^2(r) = B_\phi^2(r) + B_h^2\left(\frac{r}{a}\right)^{2l-2} \tag{7.264}$$

In general, it has not been possible to make an exact transition from a 3-D **B** to a 2-D *B*, but one can get very close, giving rise to the concepts of the quasi-isodynamic and quasi-symmetric stellarators.

From a single-particle confinement point of view, symmetry can be very important. It is well known from classical mechanics that an equilibrium system with one degree of geometric symmetry always has two constants of the motion, the energy *E* and a canonical angular momentum *L* associated with the symmetry direction. The existence of these two constants guarantees that single-particle orbits are confined. Without the geometric symmetry only the energy is a constant of the motion and single-particle orbits are in general not confined.

One can immediately see how the Boozer form of $\mathbf{V}_D \cdot \nabla \psi$ helps to design omnigenous systems. One can choose a reasonably general 3-D helical vector magnetic field $\mathbf{B}(r,\theta,\phi)$, convert to Boozer coordinates, calculate $B(\psi,\chi,\zeta)$, and evaluate the omnigenous constraint in Eq. (7.225). This process is then repeated with the aim of optimizing over the harmonic amplitudes and phases to satisfy the constraint as closely as possible. Since the constraint only involves averages over the field lines one intuitively expects that many different stellarator configurations might lead to a similar level of omnigenity. Consequently, there should still be a fair amount of freedom available to further optimize with respect to other goals such as MHD stability, reduced transport, efficient heating and current drive, or engineering simplicity.

This extra freedom can in particular be used to produce a quasi-isodynamic configuration. The corresponding constraint given in Eq. (7.225) translates to

$$ i_t \frac{\partial B}{\partial \zeta} + i_p \frac{\partial B}{\partial \chi} = 0 \tag{7.265} $$

Keeping in mind the 2π periodicity requirements one sees that there are two ways to satisfy Eq. (7.265): (1) $i_t = 0$ and $B = B(\psi, \zeta)$ or (2) $i_p = 0$ and $B = B(\psi, \chi)$. The first choice corresponding to zero toroidal current is the strategy used by the Germans in their W7-X experiment. The result of trying to approach to the ideal goal of $B = B(\psi, \zeta)$ leads to a quasi-isodynamic configuration which can be equally well described as a quasi-poloidal configuration; that is, the dependence on the poloidal angle χ is minimized.

The other option where $i_p = 0$ leads to a configuration which in analogy might be called quasi-toroidal since ideally there is no dependence of *B* on ζ. However, configurations with zero poloidal current and a large toroidal field invariably have low β which is not desirable. As is shown next there is a better way to exploit the idea of quasi-toroidal symmetry.

The last configuration of interest is the symmetric system described in Eq. (7.225). The corresponding constraint is easily evaluated by noting that the non-vanishing contribution of $\mathbf{B} \cdot \nabla B$ is given by

$$
\mathbf{B} \cdot \nabla B = (q\nabla\psi \times \nabla\chi - \nabla\psi \times \nabla\zeta) \cdot \left(\frac{\partial B}{\partial\chi}\nabla\chi + \frac{\partial B}{\partial\zeta}\nabla\zeta \right)
$$

$$
= \frac{B^2}{\mu_0(\mathrm{i}_t - q\,\mathrm{i}_p)} \left(q\frac{\partial B}{\partial\zeta} + \frac{\partial B}{\partial\chi} \right)
$$

(7.266)

The ideal symmetric constraint reduces to

$$
(\mathrm{i}_t - gq)\frac{\partial B}{\partial\zeta} + (\mathrm{i}_p - g)\frac{\partial B}{\partial\chi} = 0
\tag{7.267}
$$

where $g(\psi) = G(\psi)/\mu_0$.

Several conclusions can be drawn from Eq. (7.267). First ideal poloidal symmetry, defined as $B = B(\psi, \zeta)$, is achieved by setting $g = \mathrm{i}_t/q$. In general, one does not have to set $\mathrm{i}_t = 0$ to achieve poloidal symmetry. The special choice $\mathrm{i}_t = 0$ was made by the Germans to minimize the bootstrap current in addition to achieving omnigenity. Thus, based on the definitions presented here, an ideal-isodynamic stellarator is an ideal poloidally symmetric stellarator with the additional requirement that $\mathrm{i}_t = 0$.

Similarly, ideal toroidal symmetry, defined as $B = B(\psi, \chi)$ can be realized by choosing $g = \mathrm{i}_p$. Again, it is not necessary to set $\mathrm{i}_p = 0$ in stellarators with toroidal symmetry. This is important because the presence of poloidal currents leads to configurations with higher values of β. The toroidally symmetric stellarator with $\mathrm{i}_P \neq 0$ was the approach used in the design of the canceled NCSX experiment at PPPL. The NCSX design aimed for omnigenity at reasonably high β values in the presence of substantial flows in a compact geometry.

The last geometry of interest corresponds to helical symmetry: $B = B(\psi, m\chi + n\zeta)$ where m, n are integers. This symmetry is achieved by setting

$$
g = \frac{n\mathrm{i}_t + m\mathrm{i}_p}{m + nq}
\tag{7.268}
$$

A small helically symmetric stellarator experiment, HSX, has been built at the University of Wisconsin with $m = 2, n = 4$. The experiment does indeed observe a marked improvement in confinement when the coil currents are adjusted to approach helical symmetry.

The approximately achievable property $B \approx B(\psi, m\chi + n\zeta)$ is usually referred to in the literature as quasi-symmetry. The special cases are $m = 0$ poloidal quasi-symmetry, $n = 0$ toroidal quasi-symmetry, and $m \neq 0$, $n \neq 0$ quasi-helical symmetry.

7.7.7 Summary of neoclassical transport in a stellarator

A major challenge facing the stellarator concept as it attempts to achieve perform-ance comparable to a tokamak is reducing neoclassical transport losses. These losses are potentially much larger in a stellarator because of the inherent 3-D nature of the magnetic geometry. Specifically, unlike a tokamak, trapped particle orbits are in general not confined in a 3-D system and lead to large energy and particle losses.

Once this situation was recognized strategies were developed to reduce neoclas-sical losses in a stellarator. The basic strategy in modern stellarators is the concept of ideal-omnigenity in which all trapped particles are confined in the absence of collisions. Ideal-omnigenity involves an integral constraint that depends upon the geometric dependence of B and which must be satisfied by all particles, i.e., all E, μ. It has not been possible to exactly satisfy this constraint either theoretically or experimentally which has led to the approximate goal of quasi-omnigenity. How-ever, since the constraint involves averaged and not local requirements on B there are many different types of stellarator configurations that can achieve a similar closeness to omnigenity.

This freedom in the design of quasi-omnigenous stellarators has led to two quite different optimizations. With the advantage of hindsight these can be put in context with the aid of the ideal stellarator constraints discussed above plus the concept of quasi-symmetry as expressed in Boozer coordinates. The key feature is that neoclassical transport depends only on B and not on the vector nature of $\mathbf{B}$ when formulated in terms of Boozer coordinates.

Optimized stellarators try to approximate some form of quasi-symmetry defined as $B(\psi, \chi, \zeta) \rightarrow B(\psi, m\chi + n\zeta)$. If a configuration could exactly satisfy the symmetry constraint it would automatically be ideally omnigenous. The case $m = 0$ corresponds to quasi-poloidal symmetry. In general, a net toroidal current i_t can flow in a quasi-poloidal stellarator. The special case in which the design attempts to make $i_t = 0$ represents the quasi-isodynamic stellarator with zero bootstrap current. This is the approach used by the German's in their W7-X experiment. The W7-X design actually works harder at setting the bootstrap current to zero than achieving quasi-poloidal symmetry and the resulting optimiza-tion has led to a large aspect device. Even so, since quasi-symmetry is sufficient but not necessary for omnigenity, a considerable deviation can occur but still result in good confinement of trapped particles.

A second major design effort has been devoted to the idea of the quasi-toroidal stellarator corresponding to $n = 0$. Here, a toroidal bootstrap current is allowed which is finite but not too large. The advantages of toroidal quasi-symmetry are the possibility of large macroscopic flow velocities and machine compactness. The large flow velocities and corresponding large velocity shear are expected to reduce turbulent transport. The quasi-toroidal symmetry is expected to reduce neoclassical transport. This was the strategy used in the design of NCSX at PPPL.

7.8 Modern stellarators

There are many stellarator experiments operating or being built in the world's fusion program. The two largest flagship facilities are the Large Helical Device (LHD) in Japan and the Wendelstein 7-X (W7-X) in Germany. A brief description is now given of each of these large experiments.

7.8.1 The Large Helical Device (LHD)

The LHD is a billion dollar class experiment which achieved first plasma in 1998 and continues to successfully operate at the National Institute for Fusion Science in Toki, Japan (Motojima *et al.*, 2000). Its design corresponds to one of the classic stellarator configurations known as a "heliotron." In a heliotron the main magnetic field is produced by a pair of continuous superconducting helical-toroidal coils each carrying a current in the same direction. These coils produce both a toroidal magnetic field and a predominantly $l = 2$ helical magnetic field. The coils also produce a substantial vertical field. This field is largely canceled by an additional set of superconducting toroidally symmetric vertical field coils. The current in these coils can be adjusted to leave a small net vertical field which is used to optimize performance. A schematic drawing of LHD is given in Fig. 7.22.

There is a general consensus in the world's fusion community that LHD is an engineering marvel. One main feature of the design results from the fact that at present there does *not* exist a high quality, low loss superconducting joint that could be used to connect different sections of a toroidal coil. As a consequence, the LHD coils are continuous and were actually wound in place. They have worked very successfully with high reliability although there is a belief that this type of technology will not extrapolate well into the reactor regime. The main difficulty is associated with the large cost in time and money that would be required if a fault occurred in the coil. The reactor would have to be essentially completely disassembled to make repairs. This difficulty could perhaps be overcome with the development of low loss superconducting joints which would allow multiple

Table 7.3 *Optimized parameters for LHD discharges.*

	T_e (keV)	T_i (keV)	τ_E (sec)	P_{aux} (MW)	n_e (m^{-3})
T_e maximum	**10.0**	2.0	0.06	1.2	5.0×10^{18}
T_i maximum	4.2	**13.6**	0.06	3.1	3.5×10^{18}
τ_E maximum	1.3	1.3	**0.36**	1.5	4.8×10^{18}
$p_i\tau_E$ maximum			**0.07** atm-sec		
β maximum			**0.05** at $B_\phi = 0.425T$		
n_e maximum			$\mathbf{1.0 \times 10^{21}}$ m^{-3}		
τ_{pulse} maximum			**3900** sec by ECH		

Figure 7.22   Schematic diagram of the LHD stellarator in Japan. Courtesy of Akio Komori.

section helical coils, but there is not a lot of effort, at least in the United States, devoted to achieving this high leverage technological advance. An alternate method to avoid this problem is with stand-alone modular coils as in W7-X but more on this in the next subsection.

In terms of the physics motivation behind LHD it is worth noting that at the time of its design many of the important ideas involving neoclassical transport in 3-D systems had not yet been fully developed. Thus, minimizing neoclassical transport was not a primary consideration. Instead, the design was largely based on optimizing performance with respect to certain classes of MHD instabilities.

One can now ask about the actual experimental performance of LHD. Some general trends can be noted. First, reasonably high values of β have been achieved (see Table 7.3). Furthermore, even when the β limit is exceeded, unlike a tokamak, no major disruptions are observed. Instead there is a gradual degradation of confinement which is detrimental to the physics but is not a threat to machine integrity. This absence of disruptions is a feature of virtually all stellarators and is a major advantage of the concept. At this point it is not clear whether the confinement degradation in LHD is due to MHD instabilities or to β driven destruction of flux surfaces leading to stochastic regions of magnetic field.

With respect to transport at lower β, initial operation produced an adequate energy confinement time but not as high as comparable tokamaks. However, by reprogramming the currents in the vertical field coils to shift the plasma further inward confinement improved considerably, comparable to that in an L-mode tokamak. Presumably this optimization corresponded to a substantial increase in neoclassical confinement of trapped particles – LHD became approximately quasi-omnigenous.

The actual parameters achieved in LHD are listed in Table 7.3. The values given represent the individual optimized peak performance (as of 2006) of each critical plasma parameter; that is, the parameters are not achieved simultaneously. Each optimized parameter is given in boldface type. Observe that β values of 5% have been measured. Also, the plasma has operated over very long pulses, more than an hour, by means of electron cyclotron heating.

Overall the parameters demonstrate that stellarators can indeed achieve fusion relevant parameters. Future research and experimental upgrades will be aimed at simultaneously reaching the individual peak parameters observed in individually optimized discharges.

7.8.2 The Wendelstein 7-X (W7-X) stellarator

The W7-X stellarator is also a billion dollar class experiment currently being built at the Max Planck Institute for Plasma Physics in Greifswald, Germany (Beidler *et al.*, 1990; Grieger *et al.*, 1992). It is scheduled to be completed in 2014. A key feature of the engineering design is the use of modular, non-planar superconducting coils. A schematic diagram of W7-X is shown in Fig. 7.23.

The motivation for such complicated coils is associated with the scale up to fusion reactors. Should a coil failure occur it would be possible to make a repair by removing only one section of the reactor as opposed to disassembling the entire reactor as with continuously wound coils. Clearly this is an advantage. Even so there are two points worth mentioning. Repairing even one section of a modular stellarator might still involve many months of down time which is not desirable from an economic point of view. Also, experience with W7-X and the cancelled NCSX experiment has shown that it is quite difficult technologically to build modular coils – it takes more time and money than for planar, tokamak type coils.

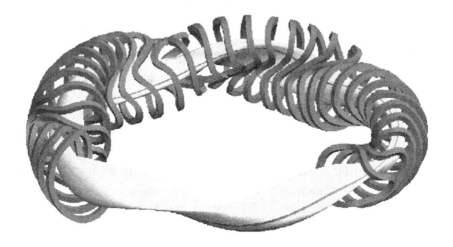

Figure 7.23   Schematic diagram of the W7-X stellarator in Germany. Courtesy of the Max Planck Institute for Plasma Physics.

Hopefully though, the lessons learned from these early modular designs will reduce the cost in future devices.

The physics motivation for W7-X has been previously discussed. Specifically, the design has paid a great deal of attention to neoclassical transport. The final design tries to create a quasi-isodynamic configuration with a minimum bootstrap current, particularly at higher values of β. The idea is that a low bootstrap current will produce robust flux surfaces even as β is increased. This in turn would help alleviate the high β destruction of flux surfaces which would lead to a degradation of confinement. It would also minimize the possibility of current-driven disruptions. The final design of W7-X has substantial $l = 2$ and $l = 0$ helical fields. The $l = 0$ harmonic corresponds to a mirror field which sometimes leads one to view W7-X as a series of linked mirrors. However, this is an oversimplification and a

Table 7.4 *Projected parameters for W7-X.*

Quantity	Value
Major radius R_0	5.5 m
Minor radius a	0.53 m
Magnetic field B_0	3 T
Pulse time τ_{pulse}	~1800 sec with beam power, ∞ with microwaves
Plasma volume V_p	30 m^3
Plasma heating P_h	14 MW
Plasma temperature T	1−6 keV
Plasma β	0.05

considerable design effort has been devoted to producing a quasi-isodynamic configuration with good MHD equilibrium and stability properties at high β.

The projected parameters for W7-X are given in Table 7.4. If W7-X can reach these parameters it, like LHD, will thereby demonstrate that the stellarator can achieve reactor grade performance.

7.9 Overall summary

Chapter 7 describes the basic MHD properties of 3-D magnetic configurations. Attention is focused solely on stellarators which at present represent the main competitor to the tokamak for a fusion reactor, at least in terms of plasma physics performance. There are many different types of configurations that are classified as stellarators. The one common feature is that they all are helical configurations bent into a torus.

In general, stellarators are more complicated and expensive to build than their axisymmetric counterpart, namely the tokamak. There are two primary motivations for investigating these more complex geometries. First, unlike a tokamak, stellarators do not require a net toroidal current to achieve MHD equilibrium. Thus they are inherently steady state devices. Second, while there are MHD β limits associated with both equilibrium and stability, stellarators so far do not observe major disruptions. These are two very important advantages which must be balanced against technology complexity.

An important stellarator concern involves particle and energy transport. Because of the 3-D geometry, stellarators in general have poorer neoclassical transport properties than tokamaks. Large theoretical/computational efforts, backed by experimental observations, have been devoted to reducing neoclassical losses in stellarators resulting in several highly innovative ideas.

The path to good neoclassical confinement is based on the idea of omnigenity. This property requires that the average drift of any plasma particle off a flux surface during one bounce period should be zero. Although it has not been possible to exactly realize this property either theoretically or experimentally there are several configurations that closely approximate this ideal but which also have a sufficient degree of remaining freedom to achieve an additional level of optimization.

One main subset of quasi-omnigenity is the idea of the quasi-isodynamic stellarator which is the basis for W7-X. This configuration is somewhat similar to a series of linked mirror machines with very careful attention paid to the plasma "elbows" connecting each mirror section. Major goals of W7-X are to minimize the bootstrap current in order to produce robust flux surfaces at high β, and to minimize the chance for current driven disruptions.

A second subset of quasi-omnigenity is the quasi-symmetric stellarator. Here, as shown by the introduction of Boozer coordinates, the drift off a given flux surface depends only on ψ and B. Therefore, a carefully designed 3-D vector magnetic field can lead to an approximately 2-D $B = B(\psi,\, m\chi + n\phi)$. Classical mechanics then implies that because of the 2-D symmetry, trapped particles will be confined. An equally important feature of quasi-symmetric configurations is that they allow large macroscopic flow velocities which are believed to be a good method to suppress micro-turbulence. Two examples of quasi-symmetric stellarators are the small HSX at the University of Wisconsin and the unfortunately canceled NCSX at PPPL.

In terms of the MHD equilibrium of stellarators there are several interesting properties to note. First, equilibria are possible without a net current. A toroidal restoring force is produced by the interaction of the vacuum helical magnetic field and the helical modulation of the Pfirsch–Schluter current. Second, while current-free equilibria are possible, the amplitude of the helical magnetic field cannot be too large or else the outer flux surfaces no longer remain closed. Third, a vertical field can shift the position of a stellarator even though the configuration has no net steady state toroidal current. A force is generated only during the transient period when the vertical field is changing but then vanishes once the vertical field has reached its steady state value. It is the interaction of the transiently induced surface dipole current and the average helical field on a flux surface that produces the force that shifts the plasma from one position to another.

Overall, stellarator performance has improved considerably in recent years so that the concept remains as the main competitor to mainline tokamak in terms of achieving fusion relevant plasma physics performance.

References

Bauer, F., Betancourt, O., Garabedian, P., and Wakatani, M. (1987). *The Beta Equilibrium, Stability, and Transport Codes*. Boston, MA: Academic Press.
Beidler, C., Grieger, G., Herrnegger, F., *et al.* (1990). *Fusion Tech.* **17**, 148.
Boozer, A. H. (1982). *Phys. Fluids*, **25**, 520.
Boozer, A. H. (2004). *Rev. Modern Phys.* **76**, 1071.
Braginskii, S. I. (1965). In *Reviews of Plasma Physics*, Vol. 1, ed. M. A. Leontovich. New York: Consultants Bureau.
Canik, J. M., Anderson, D. T., Anderson, F. S. B., *et al.* (2007). *Phys. Rev. Lett.* **98**, 085002.
Freidberg, J. P. (1987). *Ideal Magnetohydrodynamics*. New York: Plenum Press.
Gori, S., Lotz, W., and Nuhrenberg, J. J. (1996). *Theory of Fusion Plasmas*. Bologna: Editrici Compositori.
Greene, J. M. and Johnson, J. L. (1961). *Phys. Fluids* **4**, 875.
Grieger, G., Lotz, W., Merkel, P. *et al.* (1992). *Phys. Fluids B*, **4**, 2081.
Helander, P. and Sigmar, D. J. (2002). *Collisional Transport in Magnetized Plasmas*. Cambridge: Cambridge University Press.
Helander, P. (2007). *Phys. Plasmas*, **14**, 104501.
Helander, P. and Nuhrenberg, J. (2009). *Plasma Phys. Controlled Fusion*, **51**, 055004.
Helander, P., Beidler, C. D., Bird, T. M., *et al.* (2012). *Plasma Phys. Controlled Fusion* **54**, 124009.

Hirshman, S. P. and Whitson, J. C. (1983). *Phys. Fluids* **26**, 3553.

Kikuchi, M., Lackner, K., and Tran, Minh Quang, eds. (2012), *Fusion Physics*. Vienna: International Atomic Energy Agency.

Landreman, M. and Catto, P. J. (2011). *Plasma Phys. Controlled Fusion*, **53**, 035106.

Landreman, M. and Catto, P. J. (2012). *Phys. Plasmas*, **19**, 056103.

Miyamoto, K. (2005). *Plasma Physics and Controlled Nuclear Fusion*. Berlin: Springer-Verlag.

Motojima, O., Akaishi, K., Chikaraishi, H., *et al.* (2000). *Nuclear Fusion* **40**, 599

Mynick, H. E. (2006). *Phys. Plasmas*, **13**, 058102.

Nuhrenberg, J. (2010). *Plasma Phys. Controlled Fusion*, **52**, 124003.

Reiman, A. and Greenside, H. S. (1986). *Comput. Phys. Commun.* **43**, 157.

Stacey, W. M. (2012). *Fusion Plasma Physics*, 2nd edn. Weinheim, Germany: Wiley-VCH.

Sugama, H., Watanabe, T. H., Nunami, M., *et al.* (2011). *Plasma Phys. Controlled Fusion*, **53**, 024004.

Zarnstorff, M. C., Berry, L. A., Brooks, A., *et al.* (2001). *Plasma Phys. Controlled Fusion*, **43**, A237.

Further reading

MHD stellarator theory

Boozer, A. H. (2004). *Rev. Modern Phys.* **76**, 1071.

Kikuchi, M., Lackner, K., and Tran, M. Q., eds. (2012). *Fusion Physics*. Vienna: International Atomic Energy Agency.

Miyamoto, K. (1978). *Nucl. Fusion* **18**, 243.

Shafranov, V. D. (1980). *Nucl. Fusion* **20**, 1075.

Shafranov, V. D. (1983). *Phys. Fluids* **26**, 357.

Solov'ev, L. S. and Shafranov, V. D. (1979). In *Reviews of Plasma Physics*, Vol V, ed. M. A. Leontovich. New York: Consultants Bureau.

MHD stellarator computation

Bauer, F., Betancourt, O., and Garabedian, P. (1980). *J. Comput. Phys.* **39**, 341.

Carraras, B. A., Cantrell, J. L., Charlton, L. A., *et al.* (1984). In *Proceedings of the International Conference on Plasma Physics and Controlled Nuclear Fusion Research*, London, UK.

Hirshman, S. P. and Whitson, J. C. (1983). *Phys. Fluids* **26**, 3553.

Mynick, H. E., Pomphrey, N., and Xanthopoulos, P. (2010). *Phys. Rev. Lett.* **105**, 095004.

Reiman, A. and Greenside, H. S. (1986). *Comput. Phys. Commun.* **43**, 157.

Sovinec, C. R., Glasser, A. H., Gianakon, T. A., *et al.* (2004). *J. Comp. Phys.* **195**, 355.

Strauss, H. R. and Monticello, D. A. (1981). *Phys. Fluids* **24**, 1148.

Problems

7.1 Generalize the formula for the total flux ψ_{tot} given by Eq. (7.90) so that it is valid for arbitrary l.

7.2 Consider a stellarator equilibrium where the applied fields consist of a toroidal field, a helical field, and a vertical field. For simplicity assume $ha \ll 1$. Show that the toroidally averaged shift of the magnetic axis for the vacuum fields is given by

$$\frac{\bar{\sigma}(0)}{a} = \left(\frac{R_0 B_V}{a B_0 \hat{\imath}_H}\right)^{1/(2l-3)}$$

for $l \geq 2$.

7.3 Prove that the expression for the helical restoring force given by Eq. (7.115) is valid for arbitrary l.

7.4 This problem involves the equilibrium β limit for a single helicity, $\beta \sim \varepsilon$ stellarator for arbitrary ℓ. Assume Solov'ev profiles for the free functions as in Section 7.6.4. Also, use the small argument expansion ($ha \ll 1$) of the Bessel functions to simplify the analysis.

(a) Show that the solution for ψ corresponding to a current-free stellarator is given by

$$\psi = \frac{a^2}{2R_0}\left[\frac{\ell\beta_t}{2\hat{\imath}_H\varepsilon}(\rho^3 - \rho)\cos\theta + \frac{\hat{\imath}_H}{\ell - 1}(\rho^{2\ell-2} - 1)\right]$$

where $\rho = r/a$ and $\hat{\imath}_H$ is the vacuum transform/2π at the edge of the plasma $r = a$.
(b) Calculate the MHD safety factor at the edge of the plasma. Show that there is an equilibrium limit, β_t, given by

$$\beta_t < \frac{2\hat{\imath}_H^2\varepsilon}{\ell}$$

7.5 Derive an expression for the rotational transform for a multi-helicity stellarator described by the Greene–Johnson overlap model.

7.6 Consider a stellarator with a toroidal magnetic field $B_\varphi = B_0(R_0/R)$, a vertical field of amplitude B_V, and three helical harmonics $l = 1, 2, 3$, each with the same n. Adjust the amplitudes of the vertical field and helical harmonics so that B satisfies the following constraints as closely as possible as one moves away from the magnetic axis:

(a) $B\,(r,\,\theta,\,\varphi) \rightarrow B(r,\,\theta)$
(b) $B\,(r,\,\theta,\,\varphi) \rightarrow B(r,\,\varphi)$
(c) $B\,(r,\,\theta\,\varphi) \rightarrow B(r,\,2\theta + \varphi)$

Is this the same as quasi-symmetry? Explain.

8

MHD stability – general considerations

8.1 Introduction

In the remainder of the book, it is assumed that an MHD equilibrium has been calculated, either analytically or numerically. The next basic question to ask is whether or not the equilibrium is MHD stable. Qualitatively, the question of stability can be stated as follows. The existence of an MHD equilibrium implies a plasma state in which the sum of all forces acting on the plasma is zero. Assume now that the plasma is perturbed from this state producing a set of corresponding perturbed forces. If the direction of these forces is such as to restore the plasma to its original equilibrium position then the plasma is stable. If, on the other hand, the direction of the forces tends to enhance the initial perturbation then the plasma is unstable.

The question of ideal MHD stability is a crucial one, since plasmas, in general, suffer serious degradation in performance, ranging from enhanced transport to catastrophic termination, as a consequence of such instabilities. Not surprisingly, there is consensus in the international fusion community that a plasma must be MHD stable to be viable in a fusion reactor. Indeed, it is fair to say that MHD stability considerations are a primary driver in the design of virtually all the magnetic geometries that have been proposed as fusion reactors.

The goal of Chapter 8 is to develop a basic understanding of the mechanisms that cause MHD instabilities and to discuss possible ways to avoid them. Several mathematical techniques are available to investigate ideal MHD stability. By far the most common technique is the study of linear stability which is often sufficient for practical situations. Linear theory is also amenable to analytic treatment although in many cases numerical calculations are required to ultimately obtain quantitatively accurate results. Non-linear techniques are more complicated and require numerical calculations from the outset. Often, although not always, non-linear ideal MHD stability analysis is not essential since the details of non-linear plasma degradation are less important than knowing how to avoid such instabilities in the first place. The conditions to excite or avoid MHD instabilities are well described by linear theory.

For this reason the basic discussion in Chapter 8 is focused on the linear theory of MHD instabilities. The treatment is entirely analytic. Important numerical results describing linear stability in realistic geometries are presented in Chapters 11 and 12 in conjunction with practical applications.

The material in Chapter 8 begins with a mathematical definition of stability that is particularly applicable to ideal MHD. Next, as a first application, the dispersion relation for ideal MHD waves in an infinite homogeneous magnetic field is derived. These waves, which are all MHD stable, provide valuable intuition into the dynamical behavior of a plasma. Following this is an extensive discussion of the formulation of the generalized linear stability problem. The discussion starts with the time-dependent equations of motion and culminates with the development of the Energy Principle, a powerful method for testing instability.

Two general results are then derived from the Energy Principle. First, the role of plasma compressibility on MHD stability is analyzed for arbitrary magnetic geometries. There are two types of behavior, the critical distinguishing feature being whether or not the configuration of interest has ergodic field lines or consists entirely of closed field lines. The second issue is subtle and is concerned with the question of whether the region outside the plasma core is best modeled by a vacuum or alternatively by a cold, but still perfectly conducting, force-free plasma. The last topic discussed is a general classification system that describes the different types of MHD instabilities that can be excited in a plasma.

8.2 Definition of MHD stability

To investigate "MHD stability" it is clearly necessary to have a sharp definition of stability. In practice there are several different definitions that might be applicable depending upon the particular physical properties of the system under consideration. Several possible mechanical analogs are illustrated in Fig. 8.1.

In Fig. 8.1a if the ball is moved a small distance away from its initial equilibrium position it simply oscillates indefinitely about this position, assuming an ideal frictionless system. Even though the ball never returns to rest at its equilibrium position the system can be considered to be stable; that is, the ball always remains close to its equilibrium position. This is the best that one can hope for in a system without dissipation. In contrast, Fig. 8.1b shows instability. A small perturbation off the top of the hill sets the ball rolling further and further away from its equilibrium position. Figure 8.1c is a transition point usually referred to as marginal stability or neutral stability. If the ball is placed a short distance away

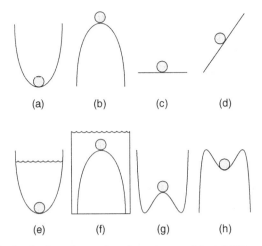

Figure 8.1   Mechanical analogs of various types of instabilities.

from its initial position it will just stay there, neither moving further away nor returning to its starting position. In most cases a change in some plasma parameter such as β or q transforms the system into one that is either stable or unstable. Therefore, marginal stability represents the stability boundary between stability and instability. Figure 8.1d shows a system not in equilbrium.

Some, but certainly not all, of the effects of dissipation can be ascertained by imagining that the ball is immersed in a viscous fluid such as oil. Figure 8.1e shows that a ball moved away from its original position will return and come to rest at its equilibrium position. The dissipation damps out the oscillations. On the other hand, Fig. 8.1f shows that while dissipation can slow down the motion of the ball it cannot convert an unstable system into a stable one. Figures 8.1g and 8.1h are examples of systems that are linearly unstable but non-linearly stable and vice versa, respectively.

To a good approximation ideal MHD is closely analogous to the set of examples illustrated in Figs. 8.1a–8.1c. There is no dissipation in ideal MHD. This recognition suggests the following definition of ideal MHD stability. Assume that all of the MHD quantities of interest are linearized about their equilibrium values

$$Q(\mathbf{r}, t) = Q_0(\mathbf{r}) + \tilde{Q}_1(\mathbf{r}, t) \tag{8.1}$$

Here, $Q_0(\mathbf{r})$ is the zeroth-order equilibrium value while $\tilde{Q}_1(\mathbf{r}, t)$ is a small first-order perturbation satisfying $|\tilde{Q}_1/Q_0| \ll 1$. Now, since the equilibrium is by definition time independent, $\tilde{Q}_1(\mathbf{r}, t)$ can in general be written as

$$\tilde{Q}_1(\mathbf{r}, t) = Q_1(\mathbf{r}) e^{-i\omega t} \tag{8.2}$$

where ω is an eigenvalue to be determined.

With very few exceptions the simplest and most reliable definition of MHD stability corresponds to "exponential stability." Specifically, if any of the eigenvalues of ω correspond to exponential growth the system is MHD unstable. If not the system is MHD stable:

$$\begin{aligned}
\text{Im}(\omega) > 0 \qquad & \text{exponential instability} \\
\text{Im}(\omega) \leq 0 \qquad & \text{exponential stability}
\end{aligned} \tag{8.3}$$

Implicit in this definition is the assumption that the modes are discrete with distinguishable eigenfrequencies. This is not always the case since MHD systems often contain continuous spectra. However, the continua lie in the stable part of the frequency spectrum and thus do not affect the existence of exponential instabilities. The frequency spectrum of ideal MHD is discussed further in Section 8.5.7. The conclusion is that if one is primarily interested in investigating MHD instabilities and the conditions for avoiding them (i.e., marginal stability) then attention should be focused on the unstable part of the spectrum. The definition Im $(\omega) > 0$ then provides a simple and reliable test for instability.

8.3 Waves in an infinite homogeneous plasma

As a first step towards understanding ideal MHD stability consider the simple case of an infinite homogeneous plasma with a uni-directional magnetic field. This configuration is obviously not toroidal and does not confine any plasma. Its "stability" actually corresponds to a determination of the basic waves that can propagate in an MHD plasma and, as such, forms a basic foundation upon which one can develop intuition that can be applied to more realistic magnetic geometries.

The "equilibrium" of the infinite homogeneous system is given by

$$\begin{aligned}
\mathbf{B} &= B_0 \mathbf{e}_z \\
p &= p_0 \\
\rho &= \rho_0 \\
\mathbf{J} &= 0 \\
\mathbf{v} &= 0
\end{aligned} \tag{8.4}$$

where B_0, p_0, and ρ_0 are constants. Since $\nabla p = \nabla \rho = \nabla \times \mathbf{B} = \nabla \cdot \mathbf{B} = 0$, the solution described by Eq. (8.4) automatically satisfies the MHD equilibrium equations.

The stability of this system is determined by the linearization procedure described above. All quantities are expanded as $Q(\mathbf{r}, t) = Q_0(\mathbf{r}) + \tilde{Q}_1(\mathbf{r}, t)$ with $\tilde{Q}_1(\mathbf{r}, t)$ being a small first-order perturbation. Since the equilibrium is independent of both time and space the most general form of the perturbation can be written as

$$\tilde{Q}_1(\mathbf{r},t) = Q_1 \exp[-i(\omega t - \mathbf{k} \cdot \mathbf{r})]$$
$$\mathbf{k} = k_\perp \mathbf{e}_y + k_\parallel \mathbf{e}_z \tag{8.5}$$
$$\mathbf{k} \cdot \mathbf{r} = k_\perp y + k_\parallel z$$

Here, without loss in generality it has been assumed that the coordinate system has been rotated around the z axis so that $\mathbf{k}$ lies in the y, z plane. Also, $\perp$ and $\parallel$ refer to perpendicular and parallel to the equilibrium field, respectively.

The next step is to substitute Eq. (8.5) into the linearized MHD equations, excluding for now the momentum equation. The results show that all perturbed quantities can be expressed in terms of the perturbed velocity $\mathbf{v}_1$,

$$
\begin{array}{ll}
\omega \rho_1 = \rho_0 (\mathbf{k} \cdot \mathbf{v}_1) & \text{conservation of mass} \\
\omega p_1 = \gamma p_0 (\mathbf{k} \cdot \mathbf{v}_1) & \text{conservation of energy} \\
\omega \mathbf{B}_1 = -\mathbf{k} \times (\mathbf{v}_1 \times \mathbf{B}_0) & \text{Faraday's law} \\
\omega \mu_0 \mathbf{J}_1 = -i\mathbf{k} \times [\mathbf{k} \times (\mathbf{v}_1 \times \mathbf{B}_0)] & \text{Ampere's law}
\end{array}
\tag{8.6}
$$

Note that the $\nabla \cdot \mathbf{B} = 0$ equation reduces to $\mathbf{k} \cdot \mathbf{B}_1 = 0$ and is a redundant relation by virtue of Faraday's law. Substituting Eq. (8.6) into the linearized momentum equation yields the following three vector component relations:

$$\left(\omega^2 - k_\parallel^2 V_A^2 \right) v_{1x} = 0$$
$$\left(\omega^2 - k_\perp^2 V_S^2 - k^2 V_A^2 \right) v_{1y} - \left(k_\perp k_\parallel V_S^2 \right) v_{1z} = 0 \tag{8.7}$$
$$- \left(k_\perp k_\parallel V_S^2 \right) v_{1y} + \left(\omega^2 - k_\parallel^2 V_S^2 \right) v_{1z} = 0$$

where $k^2 = k_\perp^2 + k_\parallel^2$, $V_A = (B_0^2/\mu_0\rho_0)^{1/2}$ is the Alfven speed and $V_S = (\gamma p_0/\rho_0)^{1/2}$ is the adiabatic sound speed. Setting the determinant of this system to zero yields the desired dispersion relation

$$\omega^2 = k_\parallel^2 V_A^2$$
$$\omega^2 = \frac{1}{2} k^2 \left(V_A^2 + V_S^2 \right) \left[1 \pm (1 - \alpha^2)^{1/2} \right] \tag{8.8}$$

Here

$$\alpha^2 = 4 \frac{k_\parallel^2}{k^2} \frac{V_S^2 V_A^2}{(V_S^2 + V_A^2)^2} \tag{8.9}$$

Note that there are three branches to the dispersion relation. Since $0 \le \alpha^2 \le 1$ each corresponds to a purely oscillatory solution: $\text{Im}(\omega) = 0$. As stated, the homogeneous magnetic field configuration is exponentially stable. This is not surprising since the system is in thermodynamic equilibrium and there are no sources of free energy available to drive instabilities.

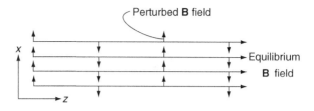

Figure 8.2   Magnetic perturbation for the shear Alfven wave shown as a separate
vector component and combined with the total **B** field.

8.3.1 The shear Alfven wave

Consider now each branch of the dispersion relation. The first branch, $\omega_a^2 = k_\parallel^2 V_A^2$,
is known as the shear Alfven wave and is independent of $k_\perp$ even when $k_\perp^2 \gg k_\parallel^2$.
It is polarized so that the perturbed magnetic field B_{1x} and velocity v_{1x} are aligned
and perpendicular to both $\mathbf{B}_0$ and $\mathbf{k}$; the wave is purely transverse. This causes the
magnetic lines to bend. Plasma is carried with the magnetic perturbation by the $\mathbf{E} \times$
$\mathbf{B}/B^2$ velocity. See Fig. 8.2. Furthermore, the quantities v_{1y}, v_{1z}, ρ_1, p_1, and $\nabla \cdot \mathbf{v}_1$
are all zero for the shear Alfven wave; the mode is incompressible and produces no
density or pressure fluctuations. The shear Alfven wave describes a basic oscillation between perpendicular plasma kinetic energy and perpendicular magnetic
energy; that is, a balance between inertial effects and the magnetic tension due to
field line bending.

8.3.2 The fast magnetosonic wave

The second branch of the dispersion relation corresponding to the $+$ sign in Eq.
(8.8) describes the fast magnetosonic wave, ω_f^2. A simple calculation shows that
$\omega_f^2 \geq \omega_a^2$. This is a wave in which both the magnetic field and the plasma pressure
are compressed so that $\nabla \cdot \mathbf{v}_1$ and p_1 are non-zero. Also, $\mathbf{B}_1$ has both a y and z
component. See Fig. 8.3. In the interesting limit where $\beta \sim V_S^2/V_A^2 \ll 1$, the fast
magnetosonic wave reduces to the compressional Alfven wave

$$\omega_f^2 \approx \left(k_\perp^2 + k_\parallel^2 \right) V_A^2 \tag{8.10}$$

In the low β limit, it can easily be shown that $\mu_0 p_1/B_0 B_{1z} \sim \beta \ll 1$ indicating that
most of the compression involves the magnetic field and not the plasma. The

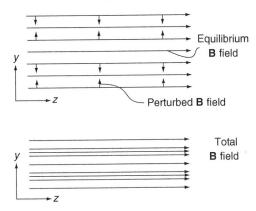

Figure 8.3   Magnetic perturbation for the compressional Alfven wave shown as a separate vector component and combined with the total **B** field.

compressional Alfven wave describes a basic oscillation between perpendicular plasma kinetic energy (plasma inertia) and parallel plus perpendicular magnetic energy. In other words, there is a balance between inertial effects and compression plus tension of the field lines. Also, since $v_{1z}/v_{1y} \sim \beta \ll 1$ the plasma motion is nearly transverse.

8.3.3 The slow magnetosonic wave

The third branch of the dispersion relation corresponds to the slow magnetosonic wave, ω_s^2. This wave always satisfies $\omega_s^2 \leq \omega_a^2$. As in the fast magnetosonic branch the wave is polarized so that both the plasma pressure and the magnetic field are compressed. However, for the slow magneto sonic wave it is the plasma rather than magnetic field that is primarily compressed. In the low β limit, $\beta \sim V_S^2/V_A^2 \ll 1$, the slow magnetosonic wave reduces to a sound wave,

$$\omega_s^2 \approx k_\parallel^2 V_S^2 \tag{8.11}$$

Observe that in this limit the mode is nearly longitudinal since $v_{1y}/v_{1z} \sim \beta \ll 1$ (see Fig. 8.4). Hence, the sound wave describes a basic oscillation between parallel plasma kinetic energy and plasma internal energy; that is, between inertial effects and plasma compression.

A subtle point concerning the sound wave is that the dispersion relation is identical to that of the two-fluid ion acoustic wave. There is an apparent paradox in that the ion acoustic wave is essentially a longitudinal electrostatic oscillation with $E_\parallel = -ik\phi$, while in ideal MHD $E_\parallel = 0$. Both modes are in fact the same mode and the paradox is resolved in Problem 8.2.

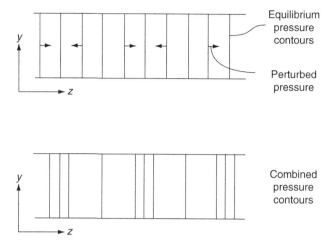

Figure 8.4   Pressure perturbation for the sound wave shown as a separate vector component and combined with the total **B** field.

8.3.4 Summary

To summarize, the three branches of the dispersion relation – the shear Alfven wave, and the fast and slow magnetosonic waves – describe the basic wave propagation characteristics of an ideal MHD plasma. In a homogeneous geometry, all are stable. In more physically interesting inhomogeneous geometries each of these waves is modified and can couple to one another. An important result related to stability is that, for reasons to be discussed later, the most unstable modes almost always couple to the shear Alfven wave.

8.4 General linearized stability equations

There exists an elegant and powerful theoretical formulation, known as the Energy Principle, that can test ideal MHD stability in an arbitrary three-dimensional geometry. To derive this Principle several steps are required. First, the stability problem is formulated as an initial value problem using the general linearized equations of motion. Second, these equations are cast into the form of a normal-mode eigenvalue problem. Next, the eigenmode formulation is transformed into a variational principle. Finally, the variational principle is reduced to the Energy Principle. Each of these steps is discussed in detail in the sections that follow.

8.4.1 Initial value formulation

To begin, assume that a static ideal MHD equilibrium, satisfying

$$\mathbf{J}_0 \times \mathbf{B}_0 = \nabla p_0$$
$$\mu_0 \mathbf{J}_0 = \nabla \times \mathbf{B}_0$$
$$\nabla \cdot \mathbf{B}_0 = 0 \tag{8.12}$$
$$\mathbf{v}_0 = 0$$

is given. All quantities are linearized about this background state: $Q(\mathbf{r}, t) = Q_0(\mathbf{r}) + \tilde{Q}_1(\mathbf{r}, t)$ with $|\tilde{Q}_1/Q_0| \ll 1$. When substituting into the MHD equations it is convenient to express all perturbed quantities in terms of a vector $\tilde{\boldsymbol{\xi}}(\mathbf{r}, t)$ defined by

$$\mathbf{v}_1 = \frac{\partial \tilde{\boldsymbol{\xi}}}{\partial t} \tag{8.13}$$

The vector $\tilde{\boldsymbol{\xi}}$ represents the displacement of the plasma away from its equilibrium position. The goal now is to express all perturbed quantities in terms of $\tilde{\boldsymbol{\xi}}$ and then obtain a single equation describing the time evolution of $\tilde{\boldsymbol{\xi}}$.

In an initial value formulation one needs to specify appropriate initial data. A very convenient choice of initial data for stability problems is as follows:

$$\tilde{\boldsymbol{\xi}}(\mathbf{r}, 0) = \tilde{\mathbf{B}}_1(\mathbf{r}, 0) = \tilde{p}_1(\mathbf{r}, 0) = \tilde{p}_1(\mathbf{r}, 0) = 0$$
$$\frac{\partial \tilde{\boldsymbol{\xi}}(\mathbf{r}, 0)}{\partial t} = \tilde{\mathbf{v}}_1(\mathbf{r}, 0) \neq 0 \tag{8.14}$$

This corresponds to the situation where at $t = 0$ the plasma is in its exact equilibrium position but is moving away with a small velocity $\tilde{\mathbf{v}}_1(\mathbf{r}, 0)$. Under these initial conditions the linearized form of the mass conservation equation, energy relation, and Faraday's law can be integrated with respect to time, yielding

$$\tilde{\rho}_1 = -\nabla \cdot (\rho_0 \tilde{\boldsymbol{\xi}})$$
$$\tilde{p}_1 = -\tilde{\boldsymbol{\xi}} \cdot \nabla p_0 - \gamma p_0 \nabla \cdot \tilde{\boldsymbol{\xi}} \tag{8.15}$$
$$\tilde{\mathbf{B}}_1 = \nabla \times (\tilde{\boldsymbol{\xi}} \times \mathbf{B}_0)$$

Note that the relation $\nabla \cdot \tilde{\mathbf{B}}_1 = 0$ is trivially satisfied by virtue of Faraday's law.

These quantities are substituted into the momentum equation leading to a single vector equation for the displacement $\tilde{\boldsymbol{\xi}}$:

$$\rho \frac{\partial^2 \tilde{\boldsymbol{\xi}}}{\partial t^2} = \mathbf{F}(\tilde{\boldsymbol{\xi}}) \tag{8.16}$$

where the force operator $\mathbf{F}(\tilde{\boldsymbol{\xi}})$ is given by

$$\mathbf{F}(\tilde{\boldsymbol{\xi}}) = \mathbf{J} \times \tilde{\mathbf{B}}_1 + \tilde{\mathbf{J}}_1 \times \mathbf{B} - \nabla \tilde{p}_1 \tag{8.17}$$

In Eq. (8.17) and hereafter, the zero subscript has been dropped from all equilibrium quantities. Equation (8.16), subject to $\tilde{\boldsymbol{\xi}}(\mathbf{r}, 0) = 0$ and $\partial \tilde{\boldsymbol{\xi}}(\mathbf{r}, 0)/\partial t = \tilde{\mathbf{v}}_1(\mathbf{r}, 0)$, plus appropriate boundary conditions (as discussed in Section 3.2) constitute the formulation of the general linearized stability equations as an initial value problem.

The initial value approach has the advantage of directly determining the actual time evolution of a given initial perturbation. The fastest growing mode becomes easily identifiable since it naturally dominates the numerically calculated evolution after a short period of time. The initial value approach is also useful in the numerical formulation of the full non-linear problem.

One drawback of the approach is that it often contains more information than is required to just determine stability boundaries, thereby requiring a substantial implementation effort. In addition, determining the marginal stability boundaries can be difficult since initial value codes have to be run for a long period of time to guarantee that a slow growing mode does not exist.

8.4.2 Normal mode formulation

A more efficient way to investigate linear stability is to reformulate Eq. (8.16) as a normal mode problem. This can easily be done by letting all perturbed quantifies vary as follows: $\tilde{Q}_1(\mathbf{r}, t) = Q_1(\mathbf{r})\exp(-i\omega t)$. Observe that the right-hand sides of Eqs. (8.15) and (8.16) contain no explicit time derivatives. Hence, the conservation of mass, conservation of energy, and Faraday's law reduce to

$$\rho_1 = -\nabla \cdot (\rho_0 \boldsymbol{\xi})$$
$$p_1 = -\boldsymbol{\xi} \cdot \nabla p_0 - \gamma p_0 \nabla \cdot \boldsymbol{\xi} \qquad (8.18)$$
$$\mathbf{B}_1 = \nabla \times (\boldsymbol{\xi} \times \mathbf{B}_0)$$

Upon substituting these relations into the momentum equation, one finds

$$-\omega^2 \rho \boldsymbol{\xi} = \mathbf{F}(\boldsymbol{\xi}) \qquad (8.19)$$

where

$$\mathbf{F}(\boldsymbol{\xi}) = \frac{1}{\mu_0} (\nabla \times \mathbf{B}) \times \mathbf{B}_1 + \frac{1}{\mu_0} (\nabla \times \mathbf{B}_1) \times \mathbf{B} - \nabla p_1 \qquad (8.20)$$

Equation (8.19) represents the normal mode formulation of the linearized MHD stability problem for general three-dimensional equilibria. In this approach only appropriate boundary conditions on $\boldsymbol{\xi}$ are required. Equation (8.19) can then be solved as an eigenvalue problem for the eigenvalue ω^2.

As compared to the initial value approach, the normal mode formulation is more amenable to analysis and more efficient with respect to numerical computations. As such it is often used in the study of ideal MHD stability. Even so, it too often contains more information than is needed to just determine a "yes or no" answer with respect to plasma stability. Thus, this is still not the most efficient way to determine marginal stability.

Another point worth noting is that the usefulness of normal mode approach is coupled to the assumption that for the problems of interest the eigenvalues are discrete and distinguishable so that the concept of exponential stability is valid. This is indeed true for the unstable part of the spectrum, although the full situation is more complicated. To obtain a more complete understanding, additional detailed knowledge of the properties of the force operator $\mathbf{F}(\xi)$ is required. This question is discussed in the next section.

Once this information is established one can proceed to the formulation of the Energy Principle, which is the most efficient way to investigate marginal stability.

8.5 Properties of the force operator $\mathbf{F}(\xi)$

The force operator $\mathbf{F}(\xi)$ possesses an important mathematical property that greatly aids in the analysis of linearized MHD stability. In particular, $\mathbf{F}(\xi)$ is a self-adjoint operator. A major consequence of this property is that each of the discrete eigenvalues ω^2 is purely real. Hence, stability transitions (i.e., $\text{Im}(\omega) = 0$) always occur when ω^2 crosses zero (i.e., $\text{Re}(\omega) = \text{Im}(\omega) = 0$), rather than at some unknown and yet to be determined general point along the real axis (i.e., $\text{Re}(\omega) \neq 0$). This fact ultimately leads to an elegant and efficient formulation for testing linear stability known as the Energy Principle.

A second important consequence of the self-adjoint property is that the discrete normal modes are orthogonal to each other. This property is useful for the following reason. After a short period of time, any arbitrary initial plasma perturbation will always be dominated by the single fastest growing mode as opposed to some combination of modes that would occur if the system were non-orthogonal. Stability is determined solely by learning how to avoid this fastest growing normal mode without regard to the shape of the initial perturbation.

Lastly, a further examination of $\mathbf{F}(\xi)$ shows that the frequency spectrum consists not only of discrete modes, but of continua as well. Even so, the continua lie on the stable side of the ω^2 axis (i.e. $\omega^2 > 0$) or at most reach the origin $\omega^2 = 0$. Thus when attention is focused on exponential instabilities, the difficulties associated with the continuous spectra are avoided.

Each of these issues is now discussed separately.

8.5.1 Self-adjointness of $F(\xi)$

As stated the self-adjointness of $\mathbf{F}(\xi)$ has a major impact on both the analytic and numerical formulation of linearized MHD stability. To demonstrate this property it is necessary to show that for any two arbitrary vectors $\xi(\mathbf{r})$ and $\eta(\mathbf{r})$ both satisfying the same well posed boundary conditions, such as those discussed in Section 3.2, the following relation holds:

$$\int \boldsymbol{\eta} \cdot \mathbf{F}(\boldsymbol{\xi}) d\mathbf{r} = \int \boldsymbol{\xi} \cdot \mathbf{F}(\boldsymbol{\eta}) d\mathbf{r} \tag{8.21}$$

This is the definition of self-adjointness – switching $\boldsymbol{\xi}$ and $\boldsymbol{\eta}$ leaves the integrals unchanged. It is worth emphasizing that $\boldsymbol{\xi}(\mathbf{r})$ and $\boldsymbol{\eta}(\mathbf{r})$ do not in general have to satisfy the actual eigenvalue equation and consequently are usually referred to as "trial functions."

The self-adjoint property, which is by no means obvious because of the complex structure of the force operator, is now directly demonstrated by a series of algebraic manipulations and integrations by parts involving $\mathbf{F}(\boldsymbol{\xi})$. In the process three separate forms of the integrals in Eq. (8.21) are derived. The first corresponds to the "standard form." This form is useful because by appropriate integrations by parts a boundary term is generated that leads to precisely the quantity needed to extend the analysis to include a vacuum region surrounding the plasma. The second form is known as the "intuitive form." Here, additional manipulations, which generate no further boundary terms, lead to a form where the individual contributions appearing in the integral can be given a simple physical interpretation. Note that the first and second forms do not obviously satisfy the self-adjoint property. The third form involves some further manipulations, again generating no new boundary terms. This final form is obviously self-adjoint by construction but still maintains the basic structure of the intuitive form. It is accurately described as the "intuitive self-adjoint form."

8.5.2 The "standard form" of δW

The starting point for the analysis is the left-hand integral in Eq. (8.21). This integral is usually multiplied by a mathematically unimportant factor "$-1/2$" and then defined as $\delta W(\boldsymbol{\eta}, \boldsymbol{\xi})$. The reason for the "$-1/2$" is given in Section 8.6, where it is shown that with this factor $\delta W(\boldsymbol{\xi}^*, \boldsymbol{\xi})$ becomes equal to a physically relevant quantity, the perturbed potential energy of the plasma. The quantity $\delta W(\boldsymbol{\eta}, \boldsymbol{\xi})$ is thus defined as

$$\delta W(\boldsymbol{\eta}, \boldsymbol{\xi}) = -\frac{1}{2} \int \boldsymbol{\eta} \cdot \mathbf{F}(\boldsymbol{\xi}) \, d\mathbf{r}$$

$$= -\frac{1}{2} \int \boldsymbol{\eta} \cdot \left[\frac{1}{\mu_0} (\nabla \times \mathbf{B}_1) \times \mathbf{B} + \frac{1}{\mu_0} (\nabla \times \mathbf{B}) \times \mathbf{B}_1 + \nabla(\boldsymbol{\xi}_\perp \cdot \nabla p + \gamma p \nabla \cdot \boldsymbol{\xi}) \right] d\mathbf{r}$$

$$\tag{8.22}$$

where $\boldsymbol{\xi} = \boldsymbol{\xi}_\perp + \xi_\parallel \mathbf{b}$, $\boldsymbol{\eta} = \boldsymbol{\eta}_\perp + \eta_\parallel \mathbf{b}$, $\mathbf{B}_1(\boldsymbol{\xi}_\perp) = \nabla \times (\boldsymbol{\xi} \times \mathbf{B}) = \nabla \times (\boldsymbol{\xi}_\perp \times \mathbf{B})$ and use has been made of the relation $\boldsymbol{\xi} \cdot \nabla p = \boldsymbol{\xi}_\perp \cdot \nabla p$.

To obtain the standard form of δW two steps are needed. The first requires the integration by parts of the $\boldsymbol{\eta} \cdot (\nabla \times \mathbf{B}_1) \times \mathbf{B}$ and $\boldsymbol{\eta} \cdot \nabla(\gamma p \nabla \cdot \boldsymbol{\xi})$ terms. A short calculation yields

$$\delta W(\boldsymbol{\eta}, \boldsymbol{\xi}) = \delta W_{F1} + BT1$$

$$\delta W_{F1} = \frac{1}{2} \int \left\{ \frac{\mathbf{B}_1(\boldsymbol{\eta}_\perp) \cdot \mathbf{B}_1(\boldsymbol{\xi}_\perp)}{\mu_0} + \gamma p (\nabla \cdot \boldsymbol{\eta})(\nabla \cdot \boldsymbol{\xi}) \right.$$

$$\left. - \boldsymbol{\eta} \cdot [\mathbf{J} \times \mathbf{B}_1(\boldsymbol{\xi}_\perp) + \nabla(\boldsymbol{\xi}_\perp \cdot \nabla p)] \right\} d\mathbf{r}$$

$$BT1 = \frac{1}{2} \int (\mathbf{n} \cdot \boldsymbol{\eta}_\perp) \left[\frac{1}{\mu_0} \mathbf{B} \cdot \mathbf{B}_1(\boldsymbol{\xi}_\perp) - \gamma p \nabla \cdot \boldsymbol{\xi} \right] dS \qquad (8.23)$$

Here the δW_{F1} integral is over the plasma volume. The boundary term $BT1$ arises from the use of Gauss' theorem in the integration by parts. This integral is taken over the plasma surface $dS = \mathbf{n}\, dS$ with $\mathbf{n}$ the outward normal vector.

A short calculation given below shows that $\eta_\parallel \mathbf{b} \cdot [\mathbf{J} \times \mathbf{B}_1(\boldsymbol{\xi}_\perp) + \nabla(\boldsymbol{\xi}_\perp \cdot \nabla p)] = 0$:

(a) $\mathbf{B} \cdot \mathbf{J} \times \mathbf{B}_1 = -\mathbf{B}_1 \cdot \mathbf{J} \times \mathbf{B} = -\mathbf{B}_1 \cdot \nabla p$

$$= \nabla \cdot [\nabla p \times (\boldsymbol{\xi} \times \mathbf{B})] = -\nabla \cdot [(\boldsymbol{\xi} \cdot \nabla p)\mathbf{B}] \qquad (8.24)$$

(b) $\mathbf{B} \cdot \nabla(\boldsymbol{\xi} \cdot \nabla p) = +\nabla \cdot [(\boldsymbol{\xi} \cdot \nabla p)\mathbf{B}]$

The term $\boldsymbol{\eta} \cdot [\mathbf{J} \times \mathbf{B}_1(\boldsymbol{\xi}_\perp) + \nabla(\boldsymbol{\xi}_\perp \cdot \nabla p)]$ thus reduces to $\boldsymbol{\eta}_\perp \cdot [\mathbf{J} \times \mathbf{B}_1(\boldsymbol{\xi}_\perp) + \nabla(\boldsymbol{\xi}_\perp \cdot \nabla p)]$. The $\boldsymbol{\eta}_\perp \cdot \nabla(\boldsymbol{\xi}_\perp \cdot \nabla p)$ term is now integrated by parts. The expression for δW becomes

$$\delta W(\boldsymbol{\eta}, \boldsymbol{\xi}) = \delta W_F + BT \qquad (8.25)$$

where

$$\delta W_F = \frac{1}{2} \int \left[\frac{\mathbf{B}_1(\boldsymbol{\eta}_\perp) \cdot \mathbf{B}_1(\boldsymbol{\xi}_\perp)}{\mu_0} + \gamma p (\nabla \cdot \boldsymbol{\eta})(\nabla \cdot \boldsymbol{\xi}) \right.$$

$$\left. - \boldsymbol{\eta}_\perp \cdot \mathbf{J} \times \mathbf{B}_1(\boldsymbol{\xi}_\perp) + (\boldsymbol{\xi}_\perp \cdot \nabla p)\nabla \cdot \boldsymbol{\eta}_\perp \right] d\mathbf{r}$$

$$BT = \frac{1}{2} \int (\mathbf{n} \cdot \boldsymbol{\eta}_\perp) \left[\frac{1}{\mu_0} \mathbf{B} \cdot \mathbf{B}_1(\boldsymbol{\xi}_\perp) - \gamma p \nabla \cdot \boldsymbol{\xi} - \boldsymbol{\xi}_\perp \cdot \nabla p \right] dS \qquad (8.26)$$

The quantity δW_F is known as the fluid energy and the expression given in Eq. (8.26) corresponds to the "standard form." Observe that the only appearance of $\xi_\parallel$ and $\eta_\parallel$ in δW_F occurs in the $\gamma p(\nabla \cdot \boldsymbol{\eta})(\nabla \cdot \boldsymbol{\xi})$ term. All the other terms are functions only of $\boldsymbol{\xi}_\perp$ and $\boldsymbol{\eta}_\perp$.

The boundary contribution BT contains precisely the right terms to account for a vacuum region surrounding the plasma. However, to temporarily simplify the algebra it is assumed at present that the plasma is surrounded by a perfectly conducting wall which requires setting $\mathbf{n} \cdot \boldsymbol{\xi}_\perp(S) = \mathbf{n} \cdot \boldsymbol{\eta}_\perp(S) = 0$. Under this assumption it follows that $BT = 0$.

The boundary term is re-introduced in Section 8.8 when the vacuum region is considered. It is shown there that BT can also be written in a self-adjoint form thereby generalizing the proof given in the remainder of this section.

Under the $\mathbf{n} \cdot \boldsymbol{\xi}_\perp(S) = \mathbf{n} \cdot \boldsymbol{\eta}_\perp(S) = 0$ assumption, demonstration of the self-adjoint property involves only δW_F plus the requirement that further integration by parts should not produce any additional non-zero boundary terms when $\mathbf{n} \cdot \boldsymbol{\eta}_\perp(S) \neq 0$.

8.5.3 The "intuitive form" of δW

The next step in the proof of the self-adjointness of $\mathbf{F}(\boldsymbol{\xi})$ is to convert δW_F from the standard form to the intuitive form. This form is useful for providing physical insight into the nature of MHD instabilities. The intuitive form is an important result since it is used for most of the applications in the remainder of the book.

The starting point for the analysis is the standard form of δW_F given in Eq. (8.26). The conversion to the intuitive form is straightforward requiring only several algebraic manipulations. In the first step the perturbed magnetic field is separated into its perpendicular and parallel components,

$$\begin{aligned} \mathbf{B}_1(\boldsymbol{\xi}_\perp) &= [\mathbf{b} \times \nabla \times (\boldsymbol{\xi}_\perp \times \mathbf{B})] \times \mathbf{b} + [\mathbf{b} \cdot \nabla \times (\boldsymbol{\xi}_\perp \times \mathbf{B})]\mathbf{b} \\ &= \mathbf{Q}_\perp(\boldsymbol{\xi}_\perp) + Q_\parallel(\boldsymbol{\xi}_\perp)\mathbf{b} \end{aligned} \tag{8.27}$$

The notation using $\mathbf{Q}$ for the perturbed magnetic field is largely historic in origin and fairly common in the literature. Equation (8.27) implies that the first term in the integrand of δW_F can be expressed as

$$\mathbf{B}_1(\boldsymbol{\eta}_\perp) \cdot \mathbf{B}_1(\boldsymbol{\xi}_\perp) = \mathbf{Q}_\perp(\boldsymbol{\eta}_\perp) \cdot \mathbf{Q}_\perp(\boldsymbol{\xi}_\perp) + Q_\parallel(\boldsymbol{\eta}_\perp)Q_\parallel(\boldsymbol{\xi}_\perp) \tag{8.28}$$

Next, the third term in the integrand is rewritten as follows,

$$\boldsymbol{\eta}_\perp \cdot \mathbf{J} \times \mathbf{B}_1(\boldsymbol{\xi}_\perp) = J_\parallel \boldsymbol{\eta}_\perp \times \mathbf{b} \cdot \mathbf{Q}_\perp(\boldsymbol{\xi}_\perp) + Q_\parallel(\boldsymbol{\xi}_\perp)\boldsymbol{\eta}_\perp \cdot \mathbf{J}_\perp \times \mathbf{b} \tag{8.29}$$

This term is simplified by noting that

(a) $$\mathbf{J}_\perp = \frac{\mathbf{b} \times \nabla p}{B}$$

(b) $Q_\parallel(\boldsymbol{\xi}_\perp) = \mathbf{b} \cdot \nabla \times (\boldsymbol{\xi}_\perp \times \mathbf{B})$

$$= \mathbf{b} \cdot (\mathbf{B} \cdot \nabla \boldsymbol{\xi}_\perp - \boldsymbol{\xi}_\perp \cdot \nabla \mathbf{B} - \mathbf{B}\nabla \cdot \boldsymbol{\xi}_\perp)$$

$$= -B(\nabla \cdot \boldsymbol{\xi}_\perp + 2\boldsymbol{\xi}_\perp \cdot \boldsymbol{\kappa}) + \frac{\mu_0}{B}\boldsymbol{\xi}_\perp \cdot \nabla p$$

(8.30)

where as before $\boldsymbol{\kappa} = \mathbf{b} \cdot \nabla \mathbf{b}$ is the curvature vector.

These relations are substituted into Eqs. (8.28) and (8.29) which are then substituted back into Eq. (8.27). A short calculation yields

$$\delta W_F = \frac{1}{2\mu_0} \int \left[\mathbf{Q}_\perp(\boldsymbol{\eta}_\perp) \cdot \mathbf{Q}_\perp(\boldsymbol{\xi}_\perp) + B^2(\nabla \cdot \boldsymbol{\eta}_\perp + 2\boldsymbol{\eta}_\perp \cdot \boldsymbol{\kappa})(\nabla \cdot \boldsymbol{\xi}_\perp + 2\boldsymbol{\xi}_\perp \cdot \boldsymbol{\kappa}) \right.$$

$$\left. + \mu_0 \gamma p(\nabla \cdot \boldsymbol{\eta})(\nabla \cdot \boldsymbol{\xi}) - 2\mu_0(\boldsymbol{\xi}_\perp \cdot \nabla p)(\boldsymbol{\eta}_\perp \cdot \boldsymbol{\kappa}) - \mu_0 J_\parallel \boldsymbol{\eta}_\perp \times \mathbf{b} \cdot \mathbf{Q}_\perp \right]d\mathbf{r}$$

(8.31)

This is the "intuitive form" of δW_F first suggested by Furth *et al.* (1965) and Greene and Johnson (1968). It may not be apparent exactly what is intuitive about this form as presently written. The intuitiveness can be made clearer by noting that since $\boldsymbol{\eta}$ is an arbitrary vector one can always choose $\boldsymbol{\eta} = \boldsymbol{\xi}^*$ as a special case. When this is done Eq. (8.31) can be rewritten as

$$\delta W_F = \frac{1}{2\mu_0} \int \left[|\mathbf{Q}_\perp|^2 \right. \qquad\qquad \text{shear Alfven wave}$$

$$+ B^2|\nabla \cdot \boldsymbol{\xi}_\perp + 2\boldsymbol{\xi}_\perp \cdot \boldsymbol{\kappa}|^2 \qquad \text{compressional Alfven wave}$$

$$+ \mu_0 \gamma p|\nabla \cdot \boldsymbol{\xi}|^2 \qquad\qquad \text{sound wave}$$

$$- 2\mu_0(\boldsymbol{\xi}_\perp \cdot \nabla p)(\boldsymbol{\xi}_\perp^* \cdot \boldsymbol{\kappa}) \qquad \text{pressure-driven modes}$$

$$\left. - \mu_0 J_\parallel \boldsymbol{\xi}_\perp^* \times \mathbf{b} \cdot \mathbf{Q}_\perp(\boldsymbol{\xi}_\perp) \right]d\mathbf{r} \quad \text{current-driven modes}$$

(8.32)

The terms have the following simple interpretation. The $|\mathbf{Q}_\perp|^2$ term represents the magnetic energy required to bend the magnetic field lines. For a homogeneous magnetic field this produces the shear Alfven wave. The second term corresponds to the energy required to compress the magnetic field. For a homogeneous magnetic field this term leads to the fast magnetosonic or compressional Alfven wave. The $\gamma p|\nabla \cdot \boldsymbol{\xi}|^2$ term represents the energy required to compress the plasma. In a homogeneous magnetic field this energy produces the slow magnetosonic or sound wave. Each of these terms is positive, which will be shown to correspond to stability.

The remaining two terms can be positive or negative and can therefore drive instabilities. The first of these is proportional to $\nabla p = \mathbf{J}_\perp \times \mathbf{B}$, while the second is proportional to $J_\parallel$. Thus, while a homogeneous vacuum magnetic field is MHD stable both perpendicular and parallel currents represent potential sources of instability. For obvious reasons instabilities driven by the ∇p term are often referred to as pressure-driven modes. Similarly instabilities driven by the $J_\parallel$ term are known as current-driven modes. A more detailed discussion of the classification of MHD instabilities is given in Section 8.11.

8.5.4 The "intuitive self-adjoint form" of δW

The last step in the analysis transforms the intuitive form of δW_F into the intuitive self-adjoint form. In this final form the self-adjoint property is obvious since each of the individual terms is symmetric in $\boldsymbol{\xi}$ and $\boldsymbol{\eta}$. Consequently, interchanging $\boldsymbol{\xi}$ and $\boldsymbol{\eta}$ leaves each term unchanged which is the definition of self-adjointness. The analysis requires a moderate amount of algebraic manipulations as follows.

First, by adding and subtracting appropriate terms the ∇p contribution is rewritten as a self-adjoint part plus a non-self-adjoint remainder R_1,

$$-2(\boldsymbol{\xi}_\perp \cdot \nabla p)(\boldsymbol{\eta}_\perp \cdot \boldsymbol{\kappa}) = -(\boldsymbol{\xi}_\perp \cdot \nabla p)(\boldsymbol{\eta}_\perp \cdot \boldsymbol{\kappa}) - (\boldsymbol{\eta}_\perp \cdot \nabla p)(\boldsymbol{\xi}_\perp \cdot \boldsymbol{\kappa}) + R_1$$

$$R_1 = -(\boldsymbol{\xi}_\perp \cdot \nabla p)(\boldsymbol{\eta}_\perp \cdot \boldsymbol{\kappa}) + (\boldsymbol{\eta}_\perp \cdot \nabla p)(\boldsymbol{\xi}_\perp \cdot \boldsymbol{\kappa}) \qquad (8.33)$$

Next, the remainder R_1 is simplified by some simple vector manipulations

$$\begin{aligned} R_1 &= -(\boldsymbol{\xi}_\perp \cdot \nabla p)(\boldsymbol{\eta}_\perp \cdot \boldsymbol{\kappa}) + (\boldsymbol{\eta}_\perp \cdot \nabla p)(\boldsymbol{\xi}_\perp \cdot \boldsymbol{\kappa}) \\ &= \nabla p \cdot [\boldsymbol{\kappa} \times (\boldsymbol{\eta}_\perp \times \boldsymbol{\xi}_\perp)] \\ &= (\nabla p \times \boldsymbol{\kappa}) \cdot (\boldsymbol{\eta}_\perp \times \boldsymbol{\xi}_\perp) \end{aligned} \qquad (8.34)$$

Second, a similar but somewhat more complicated analysis is carried out for the $J_\parallel$ contribution,

$$-J_\parallel \boldsymbol{\eta}_\perp \times \mathbf{b} \cdot \mathbf{Q}_\perp(\boldsymbol{\xi}_\perp) = -\frac{1}{2} \left[J_\parallel \boldsymbol{\eta}_\perp \times \mathbf{b} \cdot \mathbf{Q}_\perp(\boldsymbol{\xi}_\perp) + J_\parallel \boldsymbol{\xi}_\perp \times \mathbf{b} \cdot \mathbf{Q}_\perp(\boldsymbol{\eta}_\perp) \right] + R_2$$

$$R_2 = -\frac{1}{2} \left[J_\parallel \boldsymbol{\eta}_\perp \times \mathbf{b} \cdot \mathbf{Q}_\perp(\boldsymbol{\xi}_\perp) - J_\parallel \boldsymbol{\xi}_\perp \times \mathbf{b} \cdot \mathbf{Q}_\perp(\boldsymbol{\eta}_\perp) \right]$$

$$(8.35)$$

The non-self-adjoint contribution R_2 can be rewritten as

$$\begin{aligned} R_2 &= -\frac{1}{2} \left[J_\parallel(\boldsymbol{\eta}_\perp \times \mathbf{b}) \cdot \nabla \times (\boldsymbol{\xi}_\perp \times \mathbf{B}) - J_\parallel(\boldsymbol{\xi}_\perp \times \mathbf{b}) \cdot \nabla \times (\boldsymbol{\eta}_\perp \times \mathbf{B}) \right] \\ &= -\frac{J_\parallel}{2B} \nabla \cdot [(\boldsymbol{\xi}_\perp \times \mathbf{B}) \times (\boldsymbol{\eta}_\perp \times \mathbf{B})] \\ &= -\nabla \cdot \left\{ \frac{J_\parallel}{2B} [(\boldsymbol{\xi}_\perp \times \mathbf{B}) \times (\boldsymbol{\eta}_\perp \times \mathbf{B})] \right\} + \frac{1}{2} [(\boldsymbol{\xi}_\perp \times \mathbf{B}) \times (\boldsymbol{\eta}_\perp \times \mathbf{B})] \cdot \nabla \left(\frac{J_\parallel}{B} \right) \end{aligned}$$

$$(8.36)$$

Now, note that $(\boldsymbol{\xi}_\perp \times \mathbf{B}) \times (\boldsymbol{\eta}_\perp \times \mathbf{B}) = -(\boldsymbol{\eta}_\perp \times \boldsymbol{\xi}_\perp \cdot \mathbf{B})\mathbf{B}$. Using this relation in Eq. (8.36) yields

$$R_2 = \nabla \cdot \left[\frac{J_\parallel}{2B}(\boldsymbol{\eta}_\perp \times \boldsymbol{\xi}_\perp \cdot \mathbf{B})\mathbf{B} \right] - \frac{1}{2}(\boldsymbol{\eta}_\perp \times \boldsymbol{\xi}_\perp \cdot \mathbf{B})\mathbf{B} \cdot \nabla \left(\frac{J_\parallel}{B} \right) \qquad (8.37)$$

Since $\mathbf{n} \cdot \mathbf{B} = 0$ on the plasma surface, the divergence term in Eq. (8.37) integrates to zero by Gauss' theorem (even if $\mathbf{n} \cdot \boldsymbol{\eta}_\perp(S) \neq 0$). Hereafter, this term is suppressed. The remaining term is simplified by using the $\nabla \cdot \mathbf{J} = 0$ relation,

$$
\begin{aligned}
R_2 &= -\frac{1}{2}(\boldsymbol{\eta}_\perp \times \boldsymbol{\xi}_\perp \cdot \mathbf{B})\mathbf{B} \cdot \nabla \left(\frac{J_\parallel}{B} \right) \\
&= \frac{1}{2}(\boldsymbol{\eta}_\perp \times \boldsymbol{\xi}_\perp \cdot \mathbf{B})\nabla \cdot \mathbf{J}_\perp \\
&= \frac{1}{2}(\boldsymbol{\eta}_\perp \times \boldsymbol{\xi}_\perp \cdot \mathbf{B})\nabla \cdot \left(\frac{\mathbf{B} \times \nabla p}{B^2} \right) \\
&= -(\boldsymbol{\eta}_\perp \times \boldsymbol{\xi}_\perp \cdot \mathbf{b})(\mathbf{b} \times \nabla p \cdot \boldsymbol{\kappa})
\end{aligned}
\qquad (8.38)
$$

Next, vector expand $\boldsymbol{\xi}_\perp = f_1\mathbf{n} + f_2\mathbf{n} \times \mathbf{b} + f_3\mathbf{b}$ and $\boldsymbol{\eta}_\perp = h_1\mathbf{n} + h_2\mathbf{n} \times \mathbf{b} + h_3\mathbf{b}$. The coefficients $f_3 = h_3 = 0$ since by definition $\boldsymbol{\xi}_\perp \cdot \mathbf{b} = \boldsymbol{\eta}_\perp \cdot \mathbf{b} = 0$. It then follows that $\boldsymbol{\eta}_\perp \times \boldsymbol{\xi}_\perp = (f_1 h_2 - f_2 h_1)\mathbf{b}$. As might be expected $\boldsymbol{\eta}_\perp \times \boldsymbol{\xi}_\perp$ only has a $\mathbf{b}$ component, which implies that $(\boldsymbol{\eta}_\perp \times \boldsymbol{\xi}_\perp \cdot \mathbf{b})\mathbf{b} = \boldsymbol{\eta}_\perp \times \boldsymbol{\xi}_\perp$. This result is substituted into Eq. (8.38) finally leading to

$$
\begin{aligned}
R_2 &= -(\boldsymbol{\eta}_\perp \times \boldsymbol{\xi}_\perp \cdot \mathbf{b})(\mathbf{b} \times \nabla p \cdot \boldsymbol{\kappa}) \\
&= -[(\boldsymbol{\eta}_\perp \times \boldsymbol{\xi}_\perp) \times \nabla p] \cdot \boldsymbol{\kappa} \\
&= -(\boldsymbol{\eta}_\perp \times \boldsymbol{\xi}_\perp) \cdot (\nabla p \times \boldsymbol{\kappa})
\end{aligned}
\qquad (8.39)
$$

After this slightly lengthy calculation observe that $R_1 + R_2 = 0$. The terms exactly cancel. Collecting all the remaining terms yields the desired expression for "intuitive self-adjoint form" of δW_F

$$
\begin{aligned}
\delta W_F = \frac{1}{2\mu_0} \int \Big\{ & \mathbf{Q}_\perp(\boldsymbol{\eta}_\perp) \cdot \mathbf{Q}_\perp(\boldsymbol{\xi}_\perp) \\
& + B^2(\nabla \cdot \boldsymbol{\eta}_\perp + 2\boldsymbol{\eta}_\perp \cdot \boldsymbol{\kappa})(\nabla \cdot \boldsymbol{\xi}_\perp + 2\boldsymbol{\xi}_\perp \cdot \boldsymbol{\kappa}) \\
& + \mu_0 \gamma p(\nabla \cdot \boldsymbol{\eta})(\nabla \cdot \boldsymbol{\xi}) \\
& - \mu_0[(\boldsymbol{\xi}_\perp \cdot \nabla p)(\boldsymbol{\eta}_\perp \cdot \boldsymbol{\kappa}) + (\boldsymbol{\eta}_\perp \cdot \nabla p)(\boldsymbol{\xi}_\perp \cdot \boldsymbol{\kappa})] \\
& - (\mu_0 J_\parallel / 2)[\boldsymbol{\eta}_\perp \times \mathbf{b} \cdot \mathbf{Q}_\perp(\boldsymbol{\xi}_\perp) + \boldsymbol{\xi}_\perp \times \mathbf{b} \cdot \mathbf{Q}_\perp(\boldsymbol{\eta}_\perp)] \Big\} d\mathbf{r} \qquad (8.40)
\end{aligned}
$$

As previously stated, all the terms are symmetric in ξ and η. Therefore, δW_F is self-adjoint by construction and the proof is completed.

8.5.5 Real ω^2

By making use of the self-adjointness of $\mathbf{F}(\xi)$ it is straightforward to show that the eigenvalue ω^2 for any discrete normal mode is purely real. The proof is obtained by forming the dot product of Eq. (8.19) with $\xi^*(\mathbf{r})$ and integrating over the plasma volume. The result is

$$\omega^2 \int \rho |\xi|^2 \, d\mathbf{r} = - \int \xi^* \cdot \mathbf{F}(\xi) \, d\mathbf{r} \tag{8.41}$$

The procedure is then repeated by forming the dot product of $\xi(\mathbf{r})$ with the complex conjugate of Eq. (8.19),

$$(\omega^2)^* \int \rho |\xi|^2 \, d\mathbf{r} = - \int \xi \cdot \mathbf{F}(\xi^*) \, d\mathbf{r} \tag{8.42}$$

The self-adjoint property of $\mathbf{F}$ implies that the right-hand sides of Eqs. (8.41) and (8.42) are equal. Therefore, subtracting the equations leads to

$$\left[\omega^2 - (\omega^2)^* \right] \int \rho |\xi|^2 \, d\mathbf{r} = 0 \tag{8.43}$$

or

$$\omega^2 = (\omega^2)^* \tag{8.44}$$

that is, ω^2 is purely real.

In terms of the definition of exponential stability a normal mode with $\omega^2 > 0$ corresponds to a pure oscillation and hence is considered stable. Conversely, for a mode with $\omega^2 < 0$, one branch must grow exponentially and thus is unstable. Clearly, the transition from stability to instability occurs when $\omega^2 = 0$. See Fig. 8.5. It is worthwhile emphasizing the importance of this result. In more general plasma models a stability transition occurs when $\text{Im}(\omega) = 0$ but with $\text{Re}(\omega) \neq 0$. The determination of marginal stability boundaries is therefore considerably more complicated to calculate since the value of $\text{Re}(\omega)$ at the stability transition is unknown. It must be calculated separately but in parallel with the rest of the analysis. However, in ideal MHD the self-adjointness of $\mathbf{F}(\xi)$ guarantees that at any marginal stability boundary both $\text{Im}(\omega)$ and the $\text{Re}(\omega)$ must be zero simultaneously.

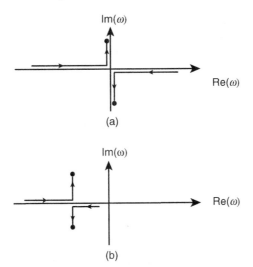

Figure 8.5 Stability transition (a) through $\omega^2 = 0$ in a self-adjoint system and (b) through a complex ω^2 in a non-self-adjoint system.

8.5.6 Orthogonality of the normal modes

A further consequence of the self-adjointness of $\mathbf{F}(\xi)$ is that the discrete normal modes are orthogonal. To show this consider two modes characterized by eigenfunctions and eigenvalues (ξ_m, ω_m^2) and (ξ_n, ω_n^2). These modes satisfy

$$-\omega_m^2 \rho \xi_m = \mathbf{F}(\xi_m)$$
$$-\omega_n^2 \rho \xi_n = \mathbf{F}(\xi_n)$$
(8.45)

One now forms the dot product of the first equation with ξ_n and the second equation with ξ_m. Subtracting the equations and making use of the self-adjointness of $\mathbf{F}(\xi)$ yields

$$(\omega_m^2 - \omega_n^2) \int \rho \xi_m \cdot \xi_n \, d\mathbf{r} = 0$$
(8.46)

Equation (8.46) shows that the non-degenerate discrete normal modes are orthogonal; that is, for two distinct modes with $\omega_m^2 \neq \omega_n^2$ it follows that

$$\int \rho \xi_m \cdot \xi_n \, d\mathbf{r} = 0$$
(8.47)

The modes are orthogonal with weight function ρ.

8.5.7 Spectrum of $\mathbf{F}(\xi)$

Because of the self-adjointness of $\mathbf{F}(\xi)$ one is strongly motivated to choose the normal-mode approach rather than the initial-value approach when considering

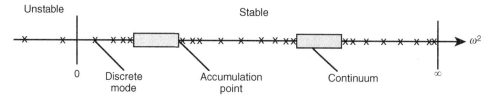

Figure 8.6 Typical ideal MHD spectrum where the continuum does not reach the origin.

the linear stability of ideal MHD plasmas. In fact, if the operator $\mathbf{F}(\xi)$ allowed only discrete eigenvalues, the concept of exponential stability could easily be extended to include both oscillatory and damped motions of the plasma. However, this is not the case. A detailed proof would require a complete spectral analysis of the force operator $\mathbf{F}(\xi)$, a task beyond the scope of the present work but which, nonetheless, has been investigated in the literature (Grad, 1973; Goedbloed, 1975; Goedbloed and Poedts, 2004). Based on these results, it is worthwhile to describe typical spectral properties that can occur and their influence on stability.

The spectral properties of $\mathbf{F}(\xi)$ follow from an examination of the operator $(\mathbf{F}/\rho - \lambda)^{-1}$ for all complex λ. If this operator exists and is bounded for a given λ, then the linearized inhomogeneous MHD equation $(\mathbf{F}/\rho - \lambda)\xi = \mathbf{a}$ (which arises when solving an initial value problem by Laplace transforms) can be inverted yielding $\xi = (\mathbf{F}/\rho - \lambda)^{-1}\mathbf{a}$.

The spectrum of $\mathbf{F}(\xi)$ consists of those values of λ for which the operator $(\mathbf{F}/\rho - \lambda)^{-1}$ cannot be inverted. There are two important cases. First is the familiar situation where λ is such that $(\mathbf{F}/\rho - \lambda)\xi = 0$ possesses a non-trivial solution. These values of λ correspond to the point or discrete spectra of $\mathbf{F}(\xi)$ and represent the normal mode eigenvalues to be examined for exponential stability. In this case it is clear that the quantity $(\mathbf{F}/\rho - \lambda)^{-1}\mathbf{a}$ does not exist.

In the second case of interest λ is such that $(\mathbf{F}/\rho - \lambda)^{-1}\mathbf{a}$ exists, but $(\mathbf{F}/\rho - \lambda)^{-1}$ is itself unbounded over some continuous set of flux surfaces in the plasma. For example, in a cylindrical system one might find $\mathbf{F}/\rho - \lambda = k_{\parallel}^2 V_A^2(r) - \lambda$, where $V_A^2(r)$ is the local Alfven velocity. In these situations there is a continuous range of λ, defined by $\lambda = k_{\parallel}^2 V_A^2(r)$ for $0 < r < a$, over which the operator is ill-behaved. These values of λ constitute the continuous spectra.

A typical ideal MHD spectrum is illustrated schematically in Fig. 8.6. Notice that the continuous spectra would significantly increase the complexity of determining plasma stability if the continuum frequencies were located in that part of the complex plane where $\mathrm{Im}(\omega) > 0$. Fortunately, this appears not to be the case. In all the static ideal MHD equilibria thus far investigated the continuum frequencies, when they exist, are located on the real axis (i.e., $\omega^2 \geq 0$). This important result has

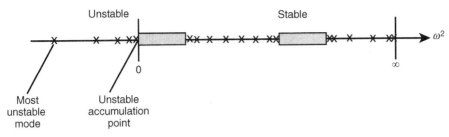

Figure 8.7 Typical ideal MHD spectrum showing a continuum reaching the
origin with an accumulation of eigenvalues on the unstable side of the spectrum.

been explicitly proven for 1-D slab and cylindrical geometries and 2-D axisymmetric toroidal geometries (Goedbloed and Poedts, 2004).

One consequence of the existence of a continuum is that discrete modes can
accumulate at either edge of its boundaries. If one boundary extends to the origin
$\omega^2 = 0$, as it often does, it is possible to have an accumulation of discrete unstable
modes with growth rates approaching zero (see Fig. 8.7). Although such behavior
is perfectly acceptable mathematically, if not accounted for properly it can obscure
the concept of marginal stability. On the other hand, if it is known that a continuum
extends to the origin, it is often possible to derive an analytic criterion that depends
only upon the equilibrium properties of the plasma that determines whether or not
there is an accumulation of unstable modes. This procedure is far simpler than
computing normal modes and gives rise to a necessary condition for stability.

In summary, the spectrum of the force operator $\mathbf{F}(\xi)$ contains both discrete
eigenvalues and continua. However, the continua exist only for $\omega^2 \geq 0$. This result
provides further motivation for examining plasma stability by the normal mode
approach, restricting attention only to the question of whether or not exponentially
growing modes exist. Even so, the situation can become somewhat complicated as
$\omega^2 \to 0$ since unstable modes can either accumulate or make smooth transitions
from instability to stability.

8.6 Variational formulation

Equation (8.19) represents the normal mode formulation of the general three-
dimensional linearized MHD stability problem. These equations have the form
of a set of three coupled homogeneous partial differential equations for the
components of ξ with eigenvalue ω^2.

Because of the self-adjointness of $\mathbf{F}(\xi)$ the stability problem can be easily recast
into the form of a variational principle (Bernstein *et al.*, 1958), which is an
alternate but entirely equivalent integral representation of the linearized MHD
partial differential equations. The variational formulation is an important step on

the path to the Energy Principle. Below the ideal MHD variational integral is presented along with a demonstration of its equivalence to the differential form.

To begin, form the dot product of Eq. (8.19) with ξ^* and then integrate over the plasma volume. This yields

$$\omega^2 = \frac{\delta W(\xi^*, \xi)}{K(\xi^*, \xi)} \tag{8.48}$$

where

$$\delta W(\xi^*, \xi) = -\frac{1}{2} \int \xi^* \cdot \mathbf{F}(\xi) \, d\mathbf{r}$$

$$K(\xi^*, \xi) = \frac{1}{2} \int \rho |\xi|^2 \, d\mathbf{r} \tag{8.49}$$

Here, the unimportant numerical factor of "1/2" has been added to both terms so that $\omega^2 K$ is proportional to the kinetic energy of the plasma $\rho v_1^2/2$. Although $\xi(\mathbf{r})$ and ω^2 have been shown to be real, $\xi(\mathbf{r})$ is treated as complex in Eq. (8.49) in anticipation of cases where, because of symmetry, several spatial coordinates can be Fourier analyzed (e.g., complex quantities appear when $\xi(\mathbf{r}) = \xi(r)\exp(im\theta + ikz)$).

The variational principle states that any allowable trial function $\xi(\mathbf{r})$ that produces an extremum (i.e., maximum, minimum, or saddle point) in the value of ω^2 is an actual eigenfunction of the ideal MHD normal mode equations with eigenvalue $\omega^2 = \delta W(\xi^*, \xi)/K(\xi^*, \xi)$. The proof follows by letting $\xi \rightarrow \xi + \delta\xi$, $\omega^2 \rightarrow \omega^2 + \delta\omega^2$ in Eq. (8.48). Here, $\delta\xi$ is an arbitrary perturbation away from the trial function ξ and $\delta\omega^2$ is the resulting change in ω^2. After substitution one finds that

$$\omega^2 + \delta\omega^2 = \frac{\delta W(\xi^*, \xi) + \delta W(\delta\xi^*, \xi) + \delta W(\xi^*, \delta\xi) + \delta W(\delta\xi^*, \delta\xi)}{K(\xi^*, \xi) + K(\delta\xi^*, \xi) + K(\xi^*, \delta\xi) + K(\delta\xi^*, \delta\xi)} \tag{8.50}$$

For small $\delta\xi$ and $\delta\omega^2$: (1) the leading-order terms yield an identity, (2) the terms quadratic in $\delta\xi$ can be neglected, and (3) the linear terms, including those obtained by Taylor expanding the denominator, lead to an expression for $\delta\omega^2$ that can be written as

$$\delta\omega^2 = \frac{\delta W(\delta\xi^*, \xi) + \delta W(\xi^*, \delta\xi) - \omega^2[K(\delta\xi^*, \xi) + K(\xi^*, \delta\xi)]}{K(\xi^*, \xi)} \tag{8.51}$$

Next, using the self-adjoint property of $\mathbf{F}(\xi)$ in Eq. (8.51), and setting $\delta\omega^2 = 0$, which is the requirement to form an extremum, leads to

$$\int d\mathbf{r}\left\{\delta\boldsymbol{\xi}^* \cdot [\mathbf{F}(\boldsymbol{\xi}) + \omega^2\rho\boldsymbol{\xi}] + \delta\boldsymbol{\xi} \cdot [\mathbf{F}(\boldsymbol{\xi}^*) + \omega^2\rho\boldsymbol{\xi}^*]\right\} = 0 \qquad (8.52)$$

Since Eq. (8.52) must be satisfied for any arbitrary $\delta\boldsymbol{\xi}$, this can only occur if[1]

$$-\omega^2\rho\boldsymbol{\xi} = \mathbf{F}(\boldsymbol{\xi}) \qquad (8.53)$$

In order for the trial function $\boldsymbol{\xi}$ to produce an extremum in ω^2, it must satisfy the actual eigenvalue equation.

This completes the proof and demonstrates that the normal mode eigenvalue equation, Eq. (8.19), and the variational principle, Eq. (8.48), are equivalent formulations of the linearized ideal MHD stability problem.

Physically, the quantity $\delta W(\boldsymbol{\xi}^*, \boldsymbol{\xi})$ represents the change in potential energy associated with the perturbation and is equal to the work done against the force $\mathbf{F}(\boldsymbol{\xi})$ in displacing the plasma by an amount $\boldsymbol{\xi}$.

8.7 The Energy Principle

It is often of primary interest to determine whether a given MHD configuration is stable or unstable without being particularly concerned about the precise value of the growth rate or oscillation frequency for each mode. The reason, which is primarily concerned with instabilities, is that MHD growth times are typically on the order of $\tau < 50\,\mu\text{sec}$. This is much shorter than experimental pulse lengths, which are on the order of $\tau > 1\,\text{sec}$. Thus, it is usually far more important to determine the conditions to avoid instabilities than to calculate precise growth rates (which can easily be estimated). In such cases, the variational formulation just derived can be further simplified, leading to a powerful minimizing principle which determines the exact stability boundaries while providing reasonable estimates of the growth rates. The formulation is known as the Energy Principle and represents the most efficient and often the most intuitive method of determining plasma stability (Bernstein *et al.*, 1958). It receives wide application in the literature and is used extensively throughout the remainder of the book.

The Energy Principle applies to systems in which the plasma is surrounded by either a perfectly conducting wall or by a vacuum region which isolates it from a perfectly conducting wall. The Principle only requires that the force operator $\mathbf{F}(\boldsymbol{\xi})$ be self-adjoint. This property has been explicitly demonstrated for a plasma surrounded by a conducting wall. In Section 8.8 the self-adjoint property is generalized to include a vacuum region leading to the "Extended Energy Principle." For present purposes one does not have to distinguish

---

[1] The reality of $\mathbf{F}$ and ω^2 implies that both contributions in the square brackets yield the same result, namely Eq. (8.53).

between the wall vs. vacuum situations. All that is required is to assume that $\mathbf{F}(\xi)$ is self adjoint.

8.7.1 Statement of the Energy Principle

The physical basis for the Energy Principle is the fact that energy is exactly conserved in the ideal MHD model. As a consequence the particular extremum corresponding to the most negative eigenvalue of ω^2 actually represents the absolute minimum in potential energy δW. This in turn implies that the question of stability or instability can be determined by examining only the sign of $\delta W(\xi^*, \xi)$ and not analyzing the full variational integral or normal-mode equations. Specifically, the Energy Principle states that an equilibrium is stable if and only if

$$\delta W(\xi^*, \xi) \geq 0 \qquad (8.54)$$

for all allowable trial displacements (i.e., ξ bounded in energy and satisfying appropriate boundary conditions); that is, if the value of potential energy is positive for any and all trial displacements, the system is stable. If it is negative for any trial displacement, the system is unstable.

The first part of the statement is intuitively obvious. Since "any and all" trial displacements must include the actual eigenfunctions as special cases, then by assumption $\delta W(\xi^*, \xi) \geq 0$ for every eigenfunction implying that $\omega^2 \geq 0$ for all modes – the system is stable. The converse is not so obvious but is nonetheless true. If a trial displacement can be found that is not an actual eigenfunction but which makes $\delta W(\xi^*, \xi) < 0$ the system is unstable. In other words, the (negative) value of ω^2 obtained by evaluating $\delta W/K$ using the trial ξ is less negative than the actual eigenvalue – the minimum, most unstable eigenvalue is always approached from above (i.e., the stable direction) when using trial functions.

For readers unfamiliar with the concept of an Energy Principle it may not be clear at this point how such a Principle actually simplifies the stability analysis of MHD systems. The simplifications are demonstrated by example in future chapters which examine the stability of various MHD configurations. In general, the most important simplification is that the determination of stability always results in an analysis in which the eigenvalue ω^2 does not explicitly appear. The corresponding equations are far simpler to analyze than the full eigenvalue equations as determined by either the normal-mode or variational formulations. A further discussion of the advantages of the Energy Principle is given in Section 8.7.3.

8.7.2 Proof of the Energy Principle

Proof assuming discrete normal modes

Consider now the proof of the Energy Principle. The proof would be straightforward if $\mathbf{F}(\xi)$ allowed only discrete normal modes which constituted a complete set of basis functions ξ_n, each satisfying

$$-\omega_n^2 \rho \xi_n = \mathbf{F}(\xi_n) \tag{8.55}$$

In this case, any arbitrary trial function $\xi(\mathbf{r})$ could be expanded as

$$\xi(\mathbf{r}) = \sum_0^\infty a_n \xi_n(\mathbf{r}) \tag{8.56}$$

Recall now that the ξ_n have been shown to be orthogonal (Section 8.5.6). Without loss in generality, they can be made orthonormal with respect to the density ρ by choosing their arbitrary amplitudes to satisfy

$$\int \rho \xi_m^* \cdot \xi_n d\mathbf{r} = \delta_{mn} \tag{8.57}$$

Then $\delta W(\xi^*, \xi)$ reduces to

$$\delta W(\xi^*, \xi) = \frac{1}{2} \sum_0^\infty |a_n|^2 \omega_n^2 \tag{8.58}$$

Thus, if a $\xi(\mathbf{r})$ could be found that makes $\delta W < 0$, then at least one $\omega_n^2 < 0$, indicating instability. Conversely, if $\delta W \geq 0$ for all $\xi(\mathbf{r})$ (i.e., any arbitrary choice of a_n) then each $\omega_n^2 \geq 0$ indicating exponential stability. Under the assumption that only discrete modes exist, the above analysis thus proves the validity of the Energy Principle.

A related proof demonstrating convergence of the most unstable eigenvalue from the stable direction follows by noting that any arbitrary trial function ξ can, without loss in generality, always be normalized so that the expansion coefficient $a_0 = 1$:

$$\xi = \xi_0 + \sum_1^\infty a_n \xi_n \tag{8.59}$$

Here, ξ_0 is the actual eigenfunction corresponding to the lowest (i.e., most unstable) eigenvalue ω_0^2. A simple calculation shows that the estimate for ω^2 using the trial function given by Eq. (8.59) has the value

$$\omega^2 = \omega_0^2 + \frac{\sum\limits_1^\infty |a_n|^2 (\omega_n^2 - \omega_0^2)}{1 + \sum\limits_1^\infty |a_n|^2} \geq \omega_0^2 \tag{8.60}$$

Equation (8.60) demonstrates that any estimate for ω^2 is always greater (i.e., more stable) that the actual lowest eigenvalue ω_0^2. Convergence using more terms in the trial function must occur from the stable direction.

General proof

Unfortunately, the proofs just given are not valid for the most general situation because of the existence of continua. An elegant proof of the Energy Principle, outlined below, has been given by Laval *et al.* (1965), which does not invoke the assumption of a complete set of discrete normal modes. The proof is based on the conservation of energy and is carried out in the real time domain. Consider the energy $H(t)$ given by

$$H(t) = K(\dot{\xi}, \dot{\xi}) + \delta W(\xi, \xi)$$
$$= \frac{1}{2} \int d\mathbf{r} \left[\rho \dot{\xi}^2 - \xi \cdot \mathbf{F}(\xi) \right] \tag{8.61}$$

where now $\xi = \xi(\mathbf{r}, t)$ is a real quantity.

A simple calculation that makes use of the self-adjoint property of $\mathbf{F}(\xi)$ yields

$$\frac{dH}{dt} = \int d\mathbf{r}\dot{\xi} \cdot \left[\rho \ddot{\xi} - \mathbf{F}(\xi) \right] = 0 \tag{8.62}$$

Equation (8.62) shows that $H(t) = H_0 = $ constant and corresponds to conservation of energy.

To show sufficiency of the Energy Principle, assume that $\delta W > 0$ for all allowable $\xi(\mathbf{r}, t)$. Energy conservation (i.e., Eq. (8.61)) implies that

$$\delta W(\xi, \xi) = H_0 - K(\dot{\xi}, \dot{\xi}) \tag{8.63}$$

Hence, unbounded growth of the kinetic energy $K(\dot{\xi}, \dot{\xi}) \to +\infty$ (e.g., exponential instability) would violate energy conservation since it has been assumed that $\delta W(\xi, \xi) > 0$: that is, $\delta W(\xi, \xi) > 0$ for all allowable ξ is sufficient for stability.

A simplified proof of necessity of the Energy Principle assumes that a perturbation $\eta(\mathbf{r})$ exists that makes $\delta W(\eta, \eta) < 0$. Now, consider a displacement $\xi(\mathbf{r}, t)$ satisfying initial conditions

$$\xi(\mathbf{r}, 0) = \eta(\mathbf{r})$$
$$\dot{\xi}(\mathbf{r}, 0) = 0 \tag{8.64}$$

Energy conservation implies that

$$H_0 = \delta W(\xi, \xi) + K(\dot{\xi}, \dot{\xi}) = (\delta W + K)_{t=0} = \delta W(\eta, \eta) < 0 \tag{8.65}$$

One now calculates dI^2/dt^2 where

$$I(t) = K(\xi, \xi) = \frac{1}{2} \int d\mathbf{r} \rho \xi^2 \tag{8.66}$$

A simple calculation gives

$$\frac{d^2 I}{dt^2} = \int d\mathbf{r} \left[\rho \dot{\xi}^2 + \xi \cdot \mathbf{F}(\xi) \right] \tag{8.67}$$

$$= 2 \left[K(\dot{\xi}, \dot{\xi}) - \delta W(\xi, \xi) \right]$$

Next, from Eq. (8.61) set $\delta W(\xi, \xi) = H_0 - K(\dot{\xi}, \dot{\xi})$ and substitute into Eq. (8.67). The result is

$$\frac{d^2 I}{dt^2} = 2 \left[2K(\dot{\xi}, \dot{\xi}) - H_0 \right] > -2H_0 > 0 \tag{8.68}$$

where the last inequality is a consequence of $H_0 < 0$ as shown by Eq. (8.65). Equation (8.68) implies that I grows without bound for large t : $I(t) \geq -H_0 t^2$ as $t \to \infty$ indicating that ξ increases as least as fast as t. This proves the necessity of the Energy Principle – any $\delta W(\boldsymbol{\eta}, \boldsymbol{\eta}) < 0$ leads to unbounded growth of $\xi(\mathbf{r}, t)$.

Laval *et al.* (1965) derived a stronger result for necessity than Eq. (8.68) by a more sophisticated analysis. Specifically, they showed that if $\delta W(\boldsymbol{\eta}, \boldsymbol{\eta}) < 0$ then there exists a ξ that grows exponentially (i.e., $\xi \sim \exp(\omega_i t)$) with a growth rate at least as fast as

$$\omega_i = \left[-\frac{\delta W(\boldsymbol{\eta}, \boldsymbol{\eta})}{K(\boldsymbol{\eta}, \boldsymbol{\eta})} \right]^{1/2} \tag{8.69}$$

Their proof, which now follows, starts by considering a perturbation ξ satisfying a modified set of initial conditions given by

$$\xi(\mathbf{r}, 0) = \boldsymbol{\eta}(\mathbf{r})$$
$$\dot{\xi}(\mathbf{r}, 0) = \omega_i \boldsymbol{\eta}(\mathbf{r}) \tag{8.70}$$

which is what one would expect for an exponentially growing mode. Conservation of energy as described by Eq. (8.62) still applies so that for the modified initial conditions the energy constant H_0 has the value

$$H_0 = \delta W(\xi, \xi) + K(\dot{\xi}, \dot{\xi}) = (\delta W + K)_{t=0} = \delta W(\boldsymbol{\eta}, \boldsymbol{\eta}) + \omega_i^2 K(\boldsymbol{\eta}, \boldsymbol{\eta}) = 0 \tag{8.71}$$

The fact that $H_0 = 0$ implies that $\delta W(\xi, \xi) = -K(\dot{\xi}, \dot{\xi})$. Using this result and again introducing the integral $I(t)$ defined in Eq. (8.66) one finds that

$$\frac{d^2 I}{dt^2} = \int d\mathbf{r} \left[\rho \dot{\xi}^2 + \xi \cdot \mathbf{F}(\xi) \right] = 2 \left[K(\dot{\xi}, \dot{\xi}) - \delta W(\xi, \xi) \right] = 4K(\dot{\xi}, \dot{\xi}) \tag{8.72}$$

The next step is to evaluate the quantity $I\ddot{I} - \dot{I}^2$. A short calculation yields

$$I\frac{d^2 I}{dt^2} - \left(\frac{dI}{dt}\right)^2 = \int \rho\xi^2 \, d\mathbf{r} \int \rho\dot{\xi}^2 \, d\mathbf{r} - \left[\int \rho\xi \cdot \dot{\xi} \, d\mathbf{r}\right]^2 \geq 0 \qquad (8.73)$$

The right-hand side is positive because of Schwartz's inequality. Equation (8.73) can be rewritten as

$$\frac{d}{dt}\left(\frac{1}{I}\frac{dI}{dt}\right) \geq 0 \qquad (8.74)$$

Both sides of the equation are integrated from 0 to t. Since integration is a summing operation, it does not change the sign of the inequality. Therefore, one obtains

$$\frac{1}{I(t)}\frac{dI(t)}{dt} - \frac{1}{I(0)}\frac{dI(0)}{dt} \geq 0 \qquad (8.75)$$

Since $\dot{I}(0) = 2\omega_i I(0)$, Eq. (8.75) reduces to

$$\frac{1}{I(t)}\frac{dI(t)}{dt} \geq 2\omega_i \qquad (8.76)$$

One further integration leads to the desired result

$$I(t) \geq I(0)\, e^{2\omega_i t} \qquad (8.77)$$

which implies that ξ grows at least as fast as $\exp(\omega_i t)$.

The conclusion of this analysis is that if any perturbation η exists that makes $\delta W(\eta, \eta) < 0$, then the system is unstable – the displacement grows exponentially without bound. Consequently, $\delta W(\eta, \eta) > 0$ for all allowable trial functions is a necessary condition for stability.

This completes the proof of the Energy Principle.

8.7.3 Advantages of the Energy Principle

Consider now the advantages of using the Energy Principle. First, if one has some intuition about the form of an unstable perturbation, this form can be used as a trial function to evaluate δW. If the value of $\delta W < 0$, the Energy Principle guarantees that the actual eigenvalue must be more negative (i.e., more unstable) than the value $\omega^2 = \delta W/K$ calculated by using the trial function; that is, the existence of an allowable trial function that makes $\delta W < 0$ is sufficient for instability.

Second, when plasma parameters are such that a trial function produces an absolute minimum in δW exactly equal to zero then this trial function satisfies the marginal stability differential equation,

$$\mathbf{F}(\xi) = 0 \tag{8.78}$$

This approach yields the exact marginal stability boundary.

Third, in the integral form of δW the various stabilizing and destabilizing mechanisms appear quite transparently, thus helping in the development of physical intuition.

Finally, there exists a very practical method for numerically testing MHD stability based on the Energy Principle. In this procedure one chooses a suitable set of simple but complete basis functions ξ_n and writes ξ as an arbitrary sum of these functions: $\xi = \sum a_n \xi_n$. Substituting into the expression for δW then yields $\delta W = \sum A_{mn} a_m a_n$, where the A_{mn} are well-defined, computable matrix elements. Once the A_{mn} are known, the expression for δW can be numerically minimized with respect to the coefficients a_n using standard linear algebra techniques. Clearly a clever choice and/or a sufficient number of basis functions gives an increasingly accurate indication of whether or not δW can be made negative. Note that the minimization with respect to the a_n can be carried out for any convenient choice of normalization, not necessarily the orthonormal one given by Eq. (8.57). When a trial function ξ is found that makes $\delta W < 0$ a reasonable estimate of the growth rate is obtained by setting $\omega^2 = \delta W/K$ using the trial ξ. The numerical procedure just described for determining stability is very effective in multidimensional geometries where numerical shooting methods are not easy to implement.

8.8 The Extended Energy Principle

At this point in the discussion it has been shown that (1) the Energy Principle is valid when the force operator $\mathbf{F}(\xi)$ is self-adjoint, and (2) $\mathbf{F}(\xi)$ is self-adjoint for a plasma surrounded by a perfectly conducting wall (see Eq. (8.40)). This section extends the validity of the Energy Principle to plasmas that are separated from a conducting wall by a vacuum region. The resulting stability formulation is known as the "Extended Energy Principle." This is a very important result since the most dangerous MHD instabilities involve perturbations in which the plasma surface moves away from its equilibrium position. Note that such motions are not possible with a perfectly conducting wall surrounding the plasma since in this case, by definition, the surface is not allowed to move: $\mathbf{n} \cdot \xi_\perp (S) = 0$.

The derivation of the Extended Energy Principle requires that $\mathbf{F}(\xi)$ remain self-adjoint when a vacuum region is included in the analysis. Once self-adjointness is demonstrated then, as stated, the analysis in Section 8.7 still applies, thereby extending the validity of the Energy Principle. The analysis proceeds as follows.

8.8.1 Statement of the problem

The potential energy of the plasma plus vacuum has been given by Eq. (8.25) and is repeated here for convenience,

$$\delta W(\boldsymbol{\eta}, \boldsymbol{\xi}) = \delta W_F + BT \qquad (8.79)$$

The fluid contribution δW_F has been shown to be self-adjoint in Eq. (8.40). Furthermore, this result did not require any knowledge of $\mathbf{n} \cdot \boldsymbol{\xi}_\perp$ (S). Thus, δW_F remains self-adjoint when a vacuum region is present. However, the boundary term BT was set to zero because of the simplifying assumption that the plasma was surrounded by a perfectly conducting wall: $\mathbf{n} \cdot \boldsymbol{\xi}_\perp$ (S) = 0. This constraint is now relaxed and the goal is to show that the boundary term given in Eq. (8.26) and repeated here

$$
\begin{aligned}
BT &= \frac{1}{2} \int (\mathbf{n} \cdot \boldsymbol{\eta}_\perp) \left[\frac{1}{\mu_0} \mathbf{B} \cdot \mathbf{B}_1(\boldsymbol{\xi}_\perp) - \gamma p \nabla \cdot \boldsymbol{\xi} - \boldsymbol{\xi}_\perp \cdot \nabla p \right] dS \\
&= \frac{1}{2} \int (\mathbf{n} \cdot \boldsymbol{\eta}_\perp) \left[\frac{1}{\mu_0} \mathbf{B} \cdot \mathbf{B}_1(\boldsymbol{\xi}_\perp) + p_1 \right] dS
\end{aligned}
\qquad (8.80)
$$

is also self-adjoint when the plasma is isolated from the perfectly conducting wall by a vacuum region.

8.8.2 The boundary conditions

The first step in the transformation of Eq. (8.80) to a self-adjoint form is a careful statement of the boundary and jump conditions that must be satisfied by the perturbations. These conditions are required in order to obtain a well-posed problem for the vacuum magnetic field. The resulting solution for the field is expressed in terms of the plasma displacement evaluated on the plasma surface. This relationship is important because it connects the vacuum magnetic energy to the plasma displacement.

The simplest boundary condition is the one that must be satisfied on the rigid conducting wall surrounding the outer edge of the vacuum region. If the surface of the conducting wall is denoted by S_w then the appropriate boundary condition involves only the perturbed magnetic field and is given by

$$\left[\mathbf{n} \cdot \hat{\mathbf{B}}_1 \right]_{S_w} = 0 \qquad (8.81)$$

Here and below a caret signifies a vacuum quantity.

The next conditions apply to the plasma–vacuum interface. These have the form of jump conditions and are obtained by expanding the exact non-linear relations about the perturbed plasma surface $\mathbf{r} = \mathbf{r}_p + \boldsymbol{\xi}$. Two jump conditions are required, one involving $\nabla \cdot \mathbf{B} = 0$ and the other pressure balance across the surface. Consider first the exact $\nabla \cdot \mathbf{B} = 0$ jump condition given by Eq. (3.9). The linearized form can be written as

$$\llbracket \mathbf{n} \cdot \mathbf{B} \rrbracket_S = 0 \quad \rightarrow \quad \left[\mathbf{n}_0 \cdot \hat{\mathbf{B}}_1 + \mathbf{n}_1 \cdot \hat{\mathbf{B}}_0 \right]_{S_p} - \left[\mathbf{n}_0 \cdot \mathbf{B}_1 + \mathbf{n}_1 \cdot \mathbf{B}_0 \right]_{S_p} = 0 \quad (8.82)$$

where S_p denotes the unperturbed plasma surface and $\llbracket \; \rrbracket$ denotes the jump from vacuum to plasma.

The goal now is to reduce this expression to one which relates $\mathbf{n}_0 \cdot \hat{\mathbf{B}}_1$ to $\boldsymbol{\xi}$. This is not as simple task as one might expect. The reason is that the perturbed normal vector $\mathbf{n}_1$ is quite complicated to calculate. There is a simpler approach that makes use of the fact that the plasma surface is a flux surface. This implies that just inside the plasma surface

$$[\mathbf{n} \cdot \mathbf{B}]_{S_-} = 0 \rightarrow [\mathbf{n}_0 \cdot \mathbf{B}_1 + \mathbf{n}_1 \cdot \mathbf{B}_0]_{S_{p-}} = 0 \quad (8.83)$$

Thus, each of the two bracketed terms in the jump condition given by Eq. (8.82) is separately zero.

Within the plasma the relation between $\mathbf{B}$ and $\boldsymbol{\xi}$ is known: $\mathbf{B}_1 = \nabla \times (\boldsymbol{\xi}_\perp \times \mathbf{B})$. Assume for the moment that there are no surface currents. Then all the components of $\mathbf{B}_1$ and $\hat{\mathbf{B}}_1$ are continuous across the interface leading to the condition $[\mathbf{n}_0 \cdot \hat{\mathbf{B}}_1]_{S_p} = [\mathbf{n}_0 \cdot \nabla \times (\boldsymbol{\xi}_\perp \times \mathbf{B}_0)]_{S_p}$. For the general case where surface currents are allowed to flow one can imagine that they do so in a narrow layer just inside the plasma edge as illustrated in Fig. 8.8. In this case there is a jump in the tangential magnetic field but the normal component remains continuous. Therefore, the argument just given concerning continuity still applies. However, one must now use the magnetic field just outside the surface current layer in order to invoke continuity of the normal field across the actual plasma – vacuum interface: $\mathbf{B}_0(S_-) \rightarrow \hat{\mathbf{B}}_0(S_-)$ and $\hat{\mathbf{B}}_0(S_-) = \hat{\mathbf{B}}_0(S_+)$. Also as the layer width shrinks to zero the displacement $\boldsymbol{\xi}$ has the same value on both sides of the current sheet. The result is that the desired boundary condition relating $\mathbf{n}_0 \cdot \hat{\mathbf{B}}_1$ to $\boldsymbol{\xi}$ has the form

$$[\mathbf{n} \cdot \hat{\mathbf{B}}_1]_{S_p} = [\mathbf{n} \cdot \nabla \times (\boldsymbol{\xi}_\perp \times \hat{\mathbf{B}})]_{S_p} \quad (8.84)$$

where for simplicity the zero subscript has been suppressed from the equilibrium quantities. Note that the condition has been derived without having had to calculate $\mathbf{n}_1$, although doing so and substituting directly into Eq. (8.82) would again yield Eq. (8.84).

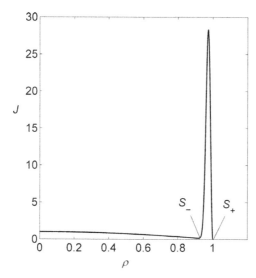

Figure 8.8   Surface currents flowing in a narrow layer at the plasma edge.

The second boundary condition is straightforward to derive but complicated to implement. It corresponds to the pressure balance jump condition also given by Eq. (3.9). The non-linear jump condition is given by

$$\left[\!\!\left[p + \frac{B^2}{2\mu_0} \right]\!\!\right]_S = 0 \tag{8.85}$$

The linearized form can be written as

$$\left[\frac{\hat{\mathbf{B}} \cdot \hat{\mathbf{B}}_1}{\mu_0} + \boldsymbol{\xi} \cdot \nabla\left(\frac{\hat{B}^2}{2\mu_0} \right) \right]_{S_p} - \left[p_1 + \frac{\mathbf{B} \cdot \mathbf{B}_1}{\mu_0} + \boldsymbol{\xi} \cdot \nabla\left(p + \frac{B^2}{2\mu_0} \right) \right]_{S_p} = 0 \tag{8.86}$$

Equations (8.81), (8.84), and (8.86) are the required boundary and jump conditions coupling the field in the vacuum to the plasma displacement.

8.8.3 The natural boundary condition

Consider now the boundary term BT given by Eq. (8.80). There are two main problems to overcome. First, it is not in an obvious self-adjoint form. Second, even if self-adjoint, it is cumbersome to substitute trial functions to estimate stability. The reason is that it is very difficult to generate trial functions that automatically satisfy the complicated jump condition given by Eq. (8.86).

These difficulties are resolved by introducing a natural boundary condition into BT. The boundary term is then separated into two contributions. The first is the surface energy which arises when equilibrium surface currents are present. The second is the magnetic energy in the vacuum volume. Both contributions are self-adjoint by construction.

Most importantly, the principles of variational calculus prove that the introduction of a natural boundary condition eliminates the need to choose trial functions that explicitly satisfy the complicated pressure balance jump condition. In other words, it is perfectly acceptable to substitute a simple trial function consisting of a sum of terms none of which individually or collectively satisfy the boundary condition. As more terms are included in the sum the minimization of δW "naturally" drives the total trial function (i.e., the sum of all the terms) in a direction to satisfy the pressure balance jump condition. As the trial function approaches an actual eigenfunction, the pressure balance jump condition becomes exactly satisfied. For readers unfamiliar with natural boundary conditions, a short discussion is presented in Appendix G.

The introduction of the natural boundary condition into BT is a simple matter. Observe that the term in the square bracket of BT in Eq. (8.80) coincides with some of the terms in the pressure balance jump condition given by Eq. (8.86). These terms are eliminated leading to the following form of BT:

$$BT = \frac{1}{2\mu_0} \int (\mathbf{n} \cdot \boldsymbol{\eta}_\perp) \left[\hat{\mathbf{B}} \cdot \hat{\mathbf{B}}_1(\boldsymbol{\xi}_\perp) + \boldsymbol{\xi} \cdot \nabla \frac{\hat{B}^2}{2} - \boldsymbol{\xi} \cdot \nabla \left(\frac{B^2}{2} + \mu_0 p \right) \right] dS \quad (8.87)$$

This is the natural boundary condition form of BT.

8.8.4 The surface energy

The terms in Eq. (8.87) are now rewritten in a form that is self-adjoint by construction. As stated, there are two contributions – the surface energy and the vacuum energy. The last two terms represent the surface energy δW_S, which can be written as

$$\delta W_S(\boldsymbol{\eta}, \boldsymbol{\xi}) = \frac{1}{2\mu_0} \int (\mathbf{n} \cdot \boldsymbol{\eta}_\perp) \left[\boldsymbol{\xi} \cdot \nabla \frac{\hat{B}^2}{2} - \boldsymbol{\xi} \cdot \nabla \left(\frac{B^2}{2} + \mu_0 p \right) \right] dS$$
$$= \frac{1}{2\mu_0} \int (\mathbf{n} \cdot \boldsymbol{\eta}_\perp) \boldsymbol{\xi} \cdot \left[\!\left[\nabla \left(\frac{B^2}{2} + \mu_0 p \right) \right]\!\right] dS \quad (8.88)$$

Keep in mind that even though $[\![B^2 + 2\mu_0 p]\!] = 0$ across the surface, the jump in its gradient is not necessarily zero. Specifically, $[\![\nabla(B^2 + 2\mu_0 p)]\!]$ is zero if, and only if, there are no equilibrium surface currents. However, even when surface currents are present the fact that $[\![B^2 + 2\mu_0 p]\!] = 0$ remains valid implies that the tangential components of the gradient are continuous; that is, $\mathbf{b} \cdot [\![\nabla(B^2 + 2\mu_0 p)]\!] =$ $\mathbf{b} \times \mathbf{n} \cdot [\![\nabla(B^2 + 2\mu_0 p)]\!] = 0$. The only possible non-zero contribution arises from the normal component of the gradient. Therefore, one can replace $\boldsymbol{\xi}$ in Eq. (8.88) with $(\mathbf{n} \cdot \boldsymbol{\xi})\mathbf{n} = (\mathbf{n} \cdot \boldsymbol{\xi}_\perp) \, \mathbf{n}$. This leads to the usual form of the surface energy,

$$\delta W_S(\boldsymbol{\eta}_\perp, \boldsymbol{\xi}_\perp) = \frac{1}{2\mu_0} \int (\mathbf{n} \cdot \boldsymbol{\eta}_\perp)(\mathbf{n} \cdot \boldsymbol{\xi}_\perp)\mathbf{n} \cdot \left[\!\!\left[\nabla \left(\frac{B^2}{2} + \mu_0 p \right) \right]\!\!\right] dS \qquad (8.89)$$

The surface energy is clearly self-adjoint by construction. It also depends only on the perpendicular components of $\boldsymbol{\xi}$ and $\boldsymbol{\eta}$. If the plasma current is finite or smoothly approaches zero at the plasma edge then $\delta W_S = 0$. When surface currents flow on the plasma boundary then $\delta W_S \neq 0$ and Eq. (8.89) must be used to evaluate its contribution to the total potential energy. In general δW_S can be either positive or negative.

8.8.5 The vacuum energy

As the analysis now stands the natural boundary condition form of BT is given by

$$BT = \delta W_S + \frac{1}{2\mu_0} \int (\mathbf{n} \cdot \boldsymbol{\eta}_\perp)[\hat{\mathbf{B}} \cdot \hat{\mathbf{B}}_1(\boldsymbol{\xi}_\perp)] dS \qquad (8.90)$$

To complete the analysis it is necessary to show that the last term in Eq. (8.90) is equal to the vacuum magnetic energy. Several steps are required.

The starting point is the generalized definition of the vacuum magnetic energy, which is self-adjoint by construction,

$$\delta W_V(\boldsymbol{\eta}, \boldsymbol{\xi}) = \frac{1}{2\mu_0} \int \hat{\mathbf{B}}_1(\boldsymbol{\eta}) \cdot \hat{\mathbf{B}}_1(\boldsymbol{\xi}) d\mathbf{r} \qquad (8.91)$$

where the integration is over the vacuum volume. The derivation proceeds by introducing the vector potential:

$$\begin{aligned} \hat{\mathbf{B}}_1(\boldsymbol{\eta}) &= \nabla \times \hat{\mathbf{A}}_1(\boldsymbol{\eta}) \\ \hat{\mathbf{B}}_1(\boldsymbol{\xi}) &= \nabla \times \hat{\mathbf{A}}_1(\boldsymbol{\xi}) \end{aligned} \qquad (8.92)$$

In the vacuum region the vector potentials satisfy $\nabla \times \nabla \times \hat{\mathbf{A}}_1(\boldsymbol{\eta}) = \nabla \times \nabla \times \hat{\mathbf{A}}_1(\boldsymbol{\xi}) = 0$. A convenient choice of gauge condition assumes that the scalar electric potentials satisfy $\phi_1(\boldsymbol{\eta}) = \phi_1(\boldsymbol{\xi}) = 0$ implying that the perturbed electric fields are given by

$$\begin{aligned} \hat{\mathbf{E}}_1(\boldsymbol{\eta}) &= i\omega \hat{\mathbf{A}}_1(\boldsymbol{\eta}) \\ \hat{\mathbf{E}}_1(\boldsymbol{\xi}) &= i\omega \hat{\mathbf{A}}_1(\boldsymbol{\xi}) \end{aligned} \qquad (8.93)$$

Now, δW_V can be converted into a surface integral by using Gauss' theorem and the vacuum field identity $\nabla \cdot [\hat{\mathbf{A}}_1(\boldsymbol{\eta}) \times \nabla \times \hat{\mathbf{A}}_1(\boldsymbol{\xi})] = \nabla \times \hat{\mathbf{A}}_1(\boldsymbol{\eta}) \cdot \nabla \times \hat{\mathbf{A}}_1(\boldsymbol{\xi})$. The vacuum energy reduces to

$$
\begin{aligned}
\delta W_V(\boldsymbol{\eta}, \boldsymbol{\xi}) &= \frac{1}{2\mu_0} \int \nabla \cdot [\hat{\mathbf{A}}_1(\boldsymbol{\eta}) \times \nabla \times \hat{\mathbf{A}}_1(\boldsymbol{\xi})] d\mathbf{r} \\
&= -\frac{1}{2\mu_0} \int \mathbf{n} \cdot \hat{\mathbf{A}}_1(\boldsymbol{\eta}) \times \hat{\mathbf{B}}_1(\boldsymbol{\xi}) dS \\
&= -\frac{1}{2\mu_0} \int \mathbf{n} \times \hat{\mathbf{A}}_1(\boldsymbol{\eta}) \cdot \hat{\mathbf{B}}_1(\boldsymbol{\xi}) dS
\end{aligned}
\tag{8.94}
$$

The surface integral is over the plasma boundary with the negative sign appearing because $\mathbf{n}$ is defined as the outward pointing normal vector. Also the contribution from the outer perfectly conducting wall is zero since the vanishing of the normal component of magnetic field is equivalent to the vanishing of the tangential component of electric field. In other words, the boundary condition given by Eq. (8.81) is equivalent to $[\mathbf{n} \times \hat{\mathbf{A}}_1(\boldsymbol{\eta})]_{S_w} = 0$.

The next step is to relate the surface value of $\hat{\mathbf{A}}_1$ to the plasma displacement. This is accomplished by considering the region just inside the plasma surface. In this region the ideal Ohm's law still applies: $\mathbf{E} + \mathbf{v} \times \mathbf{B} = 0$. The linearized form is $\mathbf{E}_1 + \mathbf{v}_1 \times \mathbf{B}_0 = 0$. If one introduces the vector potential in the plasma $\mathbf{A}_1(\boldsymbol{\eta})$ and the plasma displacement $\mathbf{v}_1 = -i\omega\boldsymbol{\eta}$ then the Ohm's law reduces to $\mathbf{A}_1(\boldsymbol{\eta}) - \boldsymbol{\eta} \times \mathbf{B}_0 = 0$. Here too the gauge condition $\phi_1(\boldsymbol{\eta}) = 0$ has been assumed. In the general case where surface currents are present the plasma magnetic field must be evaluated at the outer edge of the layer as has been shown in Fig. 8.8. This requires replacing $\mathbf{B}_0 \rightarrow \hat{\mathbf{B}}_0$ in order to exploit the interface continuity properties. Thus the general relation between $\mathbf{A}_1(\boldsymbol{\eta})$ and $\boldsymbol{\eta}$ at the plasma interface is given by $\mathbf{A}_1(\boldsymbol{\eta}) - \boldsymbol{\eta} \times \hat{\mathbf{B}}_0 = 0$. The key step now is to recognize that $\mathbf{A}_1$ is continuous across the interface: $\mathbf{A}_1(\boldsymbol{\eta}) = \hat{\mathbf{A}}_1(\boldsymbol{\eta})$. From this discussion one sees that the relation between $\hat{\mathbf{A}}_1$ and $\boldsymbol{\eta}$ on the unperturbed plasma surface can be written as

$$
\mathbf{n} \times \hat{\mathbf{A}}_1(\boldsymbol{\eta}) = \mathbf{n} \times (\boldsymbol{\eta} \times \hat{\mathbf{B}}_0) = -(\mathbf{n} \cdot \boldsymbol{\eta})\hat{\mathbf{B}}
\tag{8.95}
$$

Equation (8.95) is substituted into Eq. (8.94) leading to

$$
\delta W_V(\boldsymbol{\eta}, \boldsymbol{\xi}) = \frac{1}{2\mu_0} \int \hat{\mathbf{B}}_1(\boldsymbol{\eta}) \cdot \hat{\mathbf{B}}_1(\boldsymbol{\xi}) d\mathbf{r} = \frac{1}{2\mu_0} \int (\mathbf{n} \cdot \boldsymbol{\eta})[\hat{\mathbf{B}} \cdot \hat{\mathbf{B}}_1(\boldsymbol{\xi})] dS
\tag{8.96}
$$

Observe that the last term in Eq. (8.96) is exactly equal to the second term in Eq. (8.90). The conclusion is that the boundary term can be written as

$$
BT = \delta W_S + \delta W_V
\tag{8.97}
$$

where both δW_S and δW_V are self-adjoint by construction. Based on this result it follows that the operator $\mathbf{F}(\xi)$ is self-adjoint with either a conducting wall surrounding the plasma or when a vacuum region isolates the plasma from the conducting wall. The Energy Principle has thereby been generalized to the Extended Energy Principle.

8.8.6 *Summary of the Extended Energy Principle*

Having completed the derivation of the Extended Energy Principle it is convenient to collect and summarize the main results. These are as follows. A plasma is ideal MHD stable if and only if

$$\delta W(\xi^*, \xi) \geq 0 \qquad (8.98)$$

for all allowable displacements (those whose energy is bounded and which satisfy appropriate boundary conditions). If any trial displacement results in $\delta W(\xi^*, \xi) < 0$ the plasma is unstable.

The potential energy comprises three self-adjoint contributions given by

$$\delta W(\xi^*, \xi) = \delta W_F + \delta W_S + \delta W_V \qquad (8.99)$$

After setting $\eta = \xi^*$ one finds that the separate contributions can be written as

The fluid energy

$$\delta W_F(\xi^*, \xi) = \frac{1}{2\mu_0} \int_P \left\{ |\mathbf{Q}_\perp|^2 + B^2 |\nabla \cdot \xi_\perp + 2\xi_\perp \cdot \boldsymbol{\kappa}|^2 + \mu_0 \gamma p |\nabla \cdot \xi|^2 \right. $$
$$- \mu_0 [(\xi_\perp \cdot \nabla p)(\xi_\perp^* \cdot \boldsymbol{\kappa}) + (\xi_\perp^* \cdot \nabla p)(\xi_\perp \cdot \boldsymbol{\kappa})]$$
$$\left. - (\mu_0 J_\parallel / 2)[\xi_\perp^* \times \mathbf{b} \cdot \mathbf{Q}_\perp + \xi_\perp \times \mathbf{b} \cdot \mathbf{Q}_\perp^*] \right\} d\mathbf{r} \qquad (8.100)$$

The surface energy

$$\delta W_S(\xi_\perp^*, \xi_\perp) = \frac{1}{2\mu_0} \int_{S_p} |\mathbf{n} \cdot \xi_\perp|^2 \mathbf{n} \cdot \left[\!\!\left[\nabla \left(\frac{B^2}{2} + \mu_0 p \right) \right]\!\!\right] dS \qquad (8.101)$$

The vacuum energy

$$\delta W_V(\xi_\perp^*, \xi_\perp) = \frac{1}{2\mu_0} \int_V |\hat{\mathbf{B}}_1|^2 d\mathbf{r} \qquad (8.102)$$

In these expressions $\mathbf{Q}(\xi_\perp) = \nabla \times (\xi_\perp \times \mathbf{B})$. The subscripts on the integrals correspond to the following: P = plasma volume, V = vacuum volume, S_p = plasma surface.

The fluid and surface contributions are expressed directly in terms of $\boldsymbol{\xi}$. The perturbed magnetic field in the vacuum $\hat{\mathbf{B}}_1$ is, as shown below, a function only of $\mathbf{n} \cdot \boldsymbol{\xi}_\perp$ on the plasma surface. To evaluate $\hat{\mathbf{B}}_1$ one must solve

$$\nabla \cdot \hat{\mathbf{B}}_1 = \nabla \times \hat{\mathbf{B}}_1 = 0 \tag{8.103}$$

The solutions can be found either by introducing a vector potential or a scalar magnetic potential. In practice the scalar magnetic potential is often the easiest to implement. The boundary conditions coupling $\hat{\mathbf{B}}_1$ to $\boldsymbol{\xi}_\perp$ are given by

$$\begin{aligned} \left[\mathbf{n} \cdot \hat{\mathbf{B}}_1\right]_{S_p} &= \left[\mathbf{n} \cdot \nabla \times (\boldsymbol{\xi}_\perp \times \hat{\mathbf{B}})\right]_{S_p} \\ \left[\mathbf{n} \cdot \hat{\mathbf{B}}_1\right]_{S_w} &= 0 \end{aligned} \tag{8.104}$$

Lastly, it is worth noting that the first boundary condition can be simplified by writing the equilibrium flux surfaces in the vicinity of the plasma surface as $\psi(\mathbf{r}) = \psi_p$ with $\psi_p = $ constant serving as the flux surface label and $\psi_p = 0$ corresponding to the actual plasma–vacuum interface. Next, recall that the normal vector on any surface is, by definition, defined as

$$\mathbf{n}(\mathbf{r}) = \nabla\psi/|\nabla\psi| \tag{8.105}$$

A short calculation then shows that the first boundary condition reduces to

$$\begin{aligned} \left[\mathbf{n} \cdot \hat{\mathbf{B}}_1\right]_{S_p} &= \frac{1}{|\nabla\psi|}\left[\hat{\mathbf{B}} \cdot \nabla(\nabla\psi \cdot \boldsymbol{\xi}_\perp)\right]_{S_p} \\ &= \left[\hat{\mathbf{B}} \cdot \nabla(\mathbf{n} \cdot \boldsymbol{\xi}_\perp) - (\mathbf{n} \cdot \boldsymbol{\xi}_\perp)(\mathbf{n} \cdot \nabla\hat{\mathbf{B}} \cdot \mathbf{n})\right]_{S_p} \end{aligned} \tag{8.106}$$

As stated $\hat{\mathbf{B}}_1$ depends only on the normal component of $\boldsymbol{\xi}_\perp$ evaluated on the surface: $\nabla\psi \cdot \boldsymbol{\xi}_\perp = |\nabla\psi| (\mathbf{n} \cdot \boldsymbol{\xi}_\perp)$.

The summary is now complete and the Extended Energy Principle will be used extensively throughout the remainder of the book.

8.9 Incompressibility

8.9.1 The general minimizing condition

To test for MHD stability one must do as complete a job as possible finding a trial function that minimizes δW. Stability is determined by examining the sign of the resulting minimized δW. In general, the form of the minimizing trial function depends upon the profiles of the equilibrium fields. Also, keep in mind that the minimization must be carried out independently for each of the three vector components of $\boldsymbol{\xi}$.

However, because of the simple way in which $\xi_\parallel$ appears in δW it is possible to minimize the potential energy once and for all with respect to $\xi_\parallel$ for arbitrary geometry. The resulting form of δW is then only a function of $\boldsymbol{\xi}_\perp$ thereby reducing the number of $\boldsymbol{\xi}$ components to vary from three to two.

The general minimizing condition is obtained by recalling that the only appearance of $\xi_\parallel$ in δW occurs in the plasma compressibility term in δW_F. One can then separate δW_F into two components,

$$\delta W_F(\boldsymbol{\xi}^*, \boldsymbol{\xi}) = \delta W_\perp + \delta W_C \qquad (8.107)$$

where

$$\delta W_\perp(\boldsymbol{\xi}_\perp^*, \boldsymbol{\xi}_\perp) = \frac{1}{2\mu_0} \int_P \left\{ |\mathbf{Q}_\perp|^2 + B^2|\nabla \cdot \boldsymbol{\xi}_\perp + 2\boldsymbol{\xi}_\perp \cdot \boldsymbol{\kappa}|^2 \right.$$

$$- \mu_0[(\boldsymbol{\xi}_\perp \cdot \nabla p)(\boldsymbol{\xi}_\perp^* \cdot \boldsymbol{\kappa}) + (\boldsymbol{\xi}_\perp^* \cdot \nabla p)(\boldsymbol{\xi}_\perp \cdot \boldsymbol{\kappa})]$$

$$\left. - (\mu_0 J_\parallel/2)[\boldsymbol{\xi}_\perp^* \times \mathbf{b} \cdot \mathbf{Q}_\perp + \boldsymbol{\xi}_\perp \times \mathbf{b} \cdot \mathbf{Q}_\perp^*] \right\} d\mathbf{r}$$

$$\delta W_C(\boldsymbol{\xi}^*, \boldsymbol{\xi}) = \frac{1}{2\mu_0} \int_P \mu_0 \gamma p |\nabla \cdot \boldsymbol{\xi}|^2 d\mathbf{r} \qquad (8.108)$$

The minimizing $\xi_\parallel$ is found by writing $\boldsymbol{\xi} = \boldsymbol{\xi}_\perp + \xi_\parallel \mathbf{b}$ and letting $\xi_\parallel \rightarrow \xi_\parallel + \delta\xi_\parallel$ in δW_C, the only place where it appears. One then sets the corresponding variation in δW_C to zero:

$$\delta(\delta W_C) = \frac{1}{2} \int_P \gamma p \left[(\nabla \cdot \boldsymbol{\xi}^*)(\nabla \cdot \delta\xi_\parallel \mathbf{b}) + (\nabla \cdot \boldsymbol{\xi})(\nabla \cdot \delta\xi_\parallel^* \mathbf{b}) \right] d\mathbf{r} = 0 \quad (8.109)$$

Integrating the $\delta\xi_\parallel$ terms by parts yields

$$\delta(\delta W_C) = -\frac{1}{2} \int_P \left[\delta\xi_\parallel \mathbf{b} \cdot \nabla(\gamma p \nabla \cdot \boldsymbol{\xi}^*) + \delta\xi_\parallel^* \mathbf{b} \cdot \nabla(\gamma p \nabla \cdot \boldsymbol{\xi}) \right] d\mathbf{r} = 0 \quad (8.110)$$

The boundary term vanishes since $\mathbf{n} \cdot \mathbf{b} = 0$ on the plasma surface. If the variation is set to zero for arbitrary $\delta\xi_\parallel$ and use is made of the fact that $\mathbf{B} \cdot \nabla p = 0$, then one obtains the general minimizing condition

$$\mathbf{B} \cdot \nabla(\nabla \cdot \boldsymbol{\xi}) = 0 \qquad (8.111)$$

8.9.2 Ergodic systems

For most configurations, which are characterized by ergodic field lines, the operator $\mathbf{B} \cdot \nabla$ is non-singular and consequently Eq. (8.111) implies that the general minimizing condition reduces to

$$\nabla \cdot \boldsymbol{\xi} = 0 \qquad (8.112)$$

That is, the most unstable perturbations (in the context of minimizing δW) are incompressible.

In a sense this is obvious from Eq. (8.108) since the only term containing $\xi_\parallel$ is positive. Thus, its smallest value is zero and is obtained by setting $\nabla \cdot \boldsymbol{\xi} = 0$. However, it is the components of $\boldsymbol{\xi}$ that are the physical quantities and not $\nabla \cdot \boldsymbol{\xi}$ itself. Therefore, setting $\nabla \cdot \boldsymbol{\xi} = 0$, requires that a physically allowable $\xi_\parallel$ be found that satisfies

$$\mathbf{B} \cdot \nabla \left(\frac{\xi_\parallel}{B} \right) = -\nabla \cdot \boldsymbol{\xi}_\perp \qquad (8.113)$$

As for Eq. (8.111), physically allowable $\xi_\parallel$ are possible only if the operator $\mathbf{B} \cdot \nabla$ is non-singular. Even if $\mathbf{B} \cdot \nabla$ vanishes on isolated magnetic surfaces, a $\xi_\parallel$ can be constructed that is bounded in the vicinity of these surfaces but that makes a vanishingly small contribution to the plasma compressional energy.

8.9.3 Closed line systems

Although $\mathbf{B} \cdot \nabla$ is non-singular for most configurations there is one class of magnetic geometries for which the operator cannot be inverted. These are closed line configurations, for example a toroidal Z-pinch. For a closed line system the operator $\mathbf{B} \cdot \nabla$ is not automatically non-singular thus leading to $\nabla \cdot \boldsymbol{\xi} \neq 0$.

To demonstrate the problem consider the example of a pure cylindrical Z-pinch with equilibrium magnetic field $\mathbf{B} = B_\theta(r)\mathbf{e}_\theta$. If the perturbations are Fourier analyzed with respect to θ and z so that $\boldsymbol{\xi} = \boldsymbol{\xi}(r)\exp[i(m\theta + kz)]$ then for the $m = 0$ mode the operator $\mathbf{B} \cdot \nabla$ acting on any scalar identically vanishes. In this situation $\nabla \cdot \boldsymbol{\xi} = \nabla \cdot \boldsymbol{\xi}_\perp + \mathbf{B} \cdot \nabla(\xi_\parallel/B) = \nabla \cdot \boldsymbol{\xi}_\perp$; that is, there is no appearance of $\xi_\parallel$ in δW and the compressibility term $\gamma p |\nabla \cdot \boldsymbol{\xi}_\perp|^2$ must be maintained and included in the minimization with respect to $\boldsymbol{\xi}_\perp$.

For a general closed line configuration the operator $\mathbf{B} \cdot \nabla$ does not identically vanish but there is a periodicity constraint requiring $\xi_\parallel(l) = \xi_\parallel(l + L)$ on every magnetic line since the lines are closed. Here l is arc length, $\mathbf{B} \cdot \nabla = B(\partial/\partial l)$, and L is the total length of the field line under consideration. Note that the periodicity constraint only applies to perturbations that maintain the closed line symmetry of the equilbrium. For perturbations that break the closed line symmetry, the most unstable modes are incompressible. For modes which do not break the closed line symmetry it is necessary to add a homogeneous solution to Eq. (8.112) so that

$$\nabla \cdot \boldsymbol{\xi} = f(p) \qquad (8.114)$$

where f is an arbitrary function of the pressure p which satisfies the homogeneous equation $\mathbf{B} \cdot \nabla p = 0$. Solving Eq. (8.114) for $\xi_\parallel$ yields

$$\frac{\xi_\parallel}{B} = -\int_0^l \frac{dl}{B} \nabla \cdot \boldsymbol{\xi}_\perp + f(p) \int_0^l \frac{dl}{B} \tag{8.115}$$

Note that $f(p)$ can be taken outside the integral since p is constant along a field line. The periodicity constraint requires that $f(p)$ be chosen as follows.

$$f(p) = \langle \nabla \cdot \boldsymbol{\xi}_\perp \rangle = \frac{\oint \frac{dl}{B} \nabla \cdot \boldsymbol{\xi}_\perp}{\oint \frac{dl}{B}} \tag{8.116}$$

The end result is that in closed line systems the minimum value of the plasma compressibility term cannot be made zero, but has the value

$$\frac{1}{2} \int_P \gamma p |\nabla \cdot \boldsymbol{\xi}|^2 d\mathbf{r} = \frac{1}{2} \int_P \gamma p |\langle \nabla \cdot \boldsymbol{\xi}_\perp \rangle|^2 d\mathbf{r} \tag{8.117}$$

8.9.4 Summary and discussion

The analysis just presented has shown that δW can be minimized with respect to $\xi_\parallel$ for arbitrary geometry. The displacement $\xi_\parallel$ appears only in δW_F, which, after minimization, has the form

$$\delta W_F(\boldsymbol{\xi}^*, \boldsymbol{\xi}) = \delta W_\perp(\boldsymbol{\xi}_\perp^*, \boldsymbol{\xi}_\perp) + \delta W_C(\boldsymbol{\xi}^*, \boldsymbol{\xi}) \tag{8.118}$$

Here, $\delta W_\perp(\boldsymbol{\xi}_\perp^*, \boldsymbol{\xi}_\perp)$ is given by Eq. (8.108) and

$$\delta W_C(\boldsymbol{\xi}^*, \boldsymbol{\xi}) = \begin{cases} 0 & \text{ergodic systems} \\ \dfrac{1}{2\mu_0} \displaystyle\int_P \mu_0 \gamma p |\langle \nabla \cdot \boldsymbol{\xi}_\perp \rangle|^2 d\mathbf{r} & \text{closed line systems} \end{cases} \tag{8.119}$$

Minimization of δW now only involves the two vector components of $\boldsymbol{\xi}_\perp$.

Based on these results there is an interesting conjecture that one can make concerning the accuracy of ideal MHD marginal stability predictions. Recall that the validity of ideal MHD requires that the plasma be collision dominated, a condition not satisfied in fusion plasmas. A primary consequence of this invalid assumption is that the ideal MHD energy equation has the form of an adiabatic fluid: $d(p/n^\gamma)/dt = 0$. Note the appearance of the ratio of specific heats γ. Therefore, the appearance of γ in any marginal stability condition implies that the adiabatic energy equation has played an important role and the results should therefore be viewed with some suspicion.

In this connection observe that γ does not appear in δW for ergodic systems since the most unstable modes are incompressible – the adiabatic compressibility of the plasma does not play a role. Therefore, one can conjecture that ideal MHD marginal stability predictions may indeed be accurate for ergodic systems since the model is invalid only where it is unimportant. In contrast, for closed line systems γ appears in a stabilizing contribution to the marginal stability predictions and one should be cautious relying too heavily on plasma compressibility to provide stabilization. This last point is discussed further in Chapter 10.

8.10 Vacuum versus force-free plasma

8.10.1 The nature of the problem

An important but somewhat subtle issue is the dependence of stability boundaries on the assumption that the plasma is surrounded by a vacuum rather than a force-free plasma. The issue is as follows. Many confinement configurations consist of a hot core of plasma supporting most of the pressure gradient and carrying nearly all of the current. The edge of the core makes contact with either a limiter or divertor. The space between either of these first points of contact and the vacuum chamber (which for present purposes is assumed to be a perfectly conducting wall), is filled with a cold, low-density, low-pressure, current-free region of plasma. Since the electrical resistivity of this outer region is orders of magnitude larger than that in the core it might appear reasonable to treat this region as a perfect insulator – a vacuum.

However, it is more often the case that the resistivity of the outer region is still sufficiently low that its resistive diffusion time is long compared to the characteristic MHD time. To quantify this statement assume that the width of the region is of order Δr. Now, for the outer plasma the characteristic ideal MHD time is the ion thermal transit time, $\tau_{MHD} \sim \Delta r/V_{Ti}$, while the characteristic resistive time is the magnetic field diffusion time, $\tau_{RES} \sim \mu_0(\Delta r)^2/\eta_\parallel$. Using Spitzer resistivity and typical outer plasma parameters $\Delta r \approx 0.02\,\text{m}$, $T_i \approx 10\,\text{eV} = 0.01\,\text{keV}$, one finds

$$\frac{\tau_{RES}}{\tau_{MHD}} \sim \frac{\mu_0(\Delta r)V_{Ti}}{\eta_\parallel} \sim 1.2 \times 10^7 (\Delta r)T_k^2 \approx 24 \tag{8.120}$$

Since $\tau_{RES} \gg \tau_{MHD}$ the cold plasma still effectively behaves like a perfect conductor. It would thus seem more realistic to treat this region not as a vacuum, but as an ideal MHD plasma with low pressure (i.e., force-free) carrying no equilibrium current.

8.10.2 Vacuum vs. force-free plasma: the same results

The replacement of a perfectly insulating region with a perfectly conducting region might be expected to have a large effect on the overall stability. After all, the electrical resistivity has changed from $\eta = \infty$ to $\eta = 0$. While large effects may occur there are many cases where there is no effect at all. To understand the situation, compare the outer region contribution to δW for each case: (1) the vacuum given by Eq. (8.102) and (2) the force-free plasma with zero current given by the standard form of δW_F in Eq. (8.26) with $\mathbf{J} \to 0$, $p \to 0$,

$$\delta W_V = \frac{1}{2\mu_0} \int_V |\mathbf{B}_1^2| d\mathbf{r} \qquad \text{vacuum}$$

$$\delta W_V = \frac{1}{2\mu_0} \int_V |\mathbf{Q}^2| d\mathbf{r} \qquad \text{force-free plasma}$$

(8.121)

Since $\mathbf{B}_1 = \mathbf{Q} = \nabla \times (\boldsymbol{\xi}_\perp \times \mathbf{B})$ both energy contributions have the identical form. One might then be led to conclude that the magnetic perturbation $\mathbf{B}_1$, which minimizes the vacuum energy, also minimizes the force-free plasma energy. Therefore, both contributions to the overall δW should always be identical leading to the same marginal stability boundaries.

8.10.3 Vacuum vs. force-free plasma: different results

The above conclusion would be true if it were not for an additional constraint on the force-free plasma. Specifically, the plasma displacement resulting from the minimizing magnetic perturbation must correspond to a physically allowable motion, one in which $K(\boldsymbol{\xi}_\perp^*, \boldsymbol{\xi}_\perp)$ is bounded. No such constraint exists in the vacuum since $\boldsymbol{\xi}_\perp$ has no meaning in this region.

For a well-defined, bounded magnetic perturbation $\mathbf{B}_1$, the force-free plasma topological constraint arises when attempting to invert the relationship $\mathbf{Q} = \nabla \times (\boldsymbol{\xi}_\perp \times \mathbf{B})$ to determine $\boldsymbol{\xi}_\perp$. Whether or not the inversion is singular is closely related to the properties of the $\mathbf{B} \cdot \nabla$ operator. For example, in a general one-dimensional screw pinch the relationship between Q_r and ξ_r is given by $Q_r = \mathbf{B} \cdot \nabla \xi_r$. After Fourier analyzing with respect to θ and z one can invert this relationship yielding $\xi_r = -iQ_r/F$, $F = kB_z(r) + mB_\theta(r)/r$. Hence, if $F(r)$ vanishes anywhere in the outer region ξ_r is singular and $K(\boldsymbol{\xi}_\perp^*, \boldsymbol{\xi}_\perp)$ is unbounded.

When this occurs, the stability of the force-free region must be recomputed with $\mathbf{n} \cdot \mathbf{Q}$ set to zero, not on the outer conductor, but on the singular surface where $F(r) = 0$ in order that ξ_r remain bounded. The force-free plasma is more stable than the vacuum since the new boundary condition is equivalent to moving the conducting boundary inward, a more constrained situation.

8.10.4 The real situation: a resistive region

Even with the above distinction between vacuum and force-free plasma there still remains one important physics issue. Each of the two cases discussed represents opposing limits of plasma resistivity. If, however, finite resistivity is included in the Ohm's law, the plasma can diffuse through the magnetic field implying that the ideal MHD frozen-in topological constraint, associated with perfect conductivity, no longer applies. Under this situation, the marginal stability boundaries become identical to the vacuum case (since no additional constraints are necessary) but the growth rates are much lower, depending upon the value of the resistivity at the singular surface.

In summary, the stability behavior of the outer region is as follows. The most pessimistic description corresponds to the vacuum case. Here, no additional constraints need to be imposed and the characteristic growth times are those of ideal MHD. If the vacuum region is replaced by an ideal force-free plasma the stability may or may not change depending upon whether the particular mode under consideration has a singular surface in the outer region. With no singular surface the situation is identical to the vacuum case. If there is a singular surface the force-free plasma is considerably more stable because of the ideal MHD topological constraint which effectively moves the conducting wall inward to the singular surface. Finally, when the outer region is filled with a resistive force-free plasma the topological constraint is eliminated. The marginal stability boundaries are again identical to the vacuum case. The growth rates are the same as for a vacuum when no singular surface is present. They are much smaller, but still finite, when a singular surface exists in the resistive plasma.

The resistive force-free plasma provides the most realistic description of the outer region. Even so, since the remainder of the book is primarily concerned with the determination of stability boundaries, whenever the outer region plays an important role in the stability of a given mode it suffices to treat this region as a vacuum.

8.11 Classification of MHD instabilities

This section describes a method of classifying MHD instabilities in terms of the general structure of the modes and the sources driving the instabilities. Also discussed are those properties of the equilibrium magnetic field that are effective in reducing or eliminating specific types of modes. The classification system divides naturally into two basic categories. First, one must distinguish whether a given instability is predominately an internal or external mode. Second, each instability can be characterized by its dominant driving source. Thus, a given mode can be either pressure driven or current driven. A detailed description of the classification is given below.

8.11.1 Internal/fixed boundary modes

Consider a magnetic configuration in which the plasma is surrounded by a vacuum. Instabilities whose mode structure does not require any motion of the plasma–vacuum interface away from its equilibrium position are called internal or fixed boundary modes. Oftentimes, an internal mode will have a singular surface (i.e., a surface where $\mathbf{B} \cdot \nabla$ vanishes) inside the plasma. The boundary condition applicable to such modes is $[\mathbf{n} \cdot \boldsymbol{\xi}_\perp]_{S_p} = 0$ and is equivalent to moving the conducting wall onto the plasma surface. For internal modes it is only necessary to minimize δW_F since $\delta W_S = \delta W_V = 0$.

8.11.2 External/free boundary modes

If the plasma–vacuum interface moves from its equilibrium position during an unstable MHD perturbation, the corresponding modes are known as external or free-boundary modes. These perturbations often possess a singular surface in the vacuum region. For such instabilities δW_S and δW_V must be evaluated as well as δW_F since $[\mathbf{n} \cdot \boldsymbol{\xi}_\perp]_{S_p} \neq 0$. Although an internal mode can be viewed as the special case of an external mode with $[\mathbf{n} \cdot \boldsymbol{\xi}_\perp]_{S_p} = 0$, the definition assumed here makes each mode mutually exclusive; that is, if $[\mathbf{n} \cdot \boldsymbol{\xi}_\perp]_{S_p} = 0$, the mode is an internal mode. If not, it is an external mode.

8.11.3 Pressure-driven modes

It was shown in Section 8.3 that a plasma immersed in an infinite, homogeneous, unidirectional magnetic field is always MHD stable. It was later shown by Eq. (8.100) that there are two possible sources of MHD instability, one proportional to ∇p and the other to $J_\parallel$. MHD instabilities in which the dominant destabilizing term is the one proportional to ∇p are known as pressure-driven modes. Since $\nabla p \propto \mathbf{J}_\perp$ these modes can exist even if no parallel currents are present in the plasma. Unstable pressure-driven instabilities in which $J_\parallel$ plays no role are usually internal modes with very short wavelengths perpendicular to the magnetic field and long wavelengths parallel to the field. These types of pressure-driven modes are traditionally subdivided into two categories: interchange and ballooning instabilities.

Interchange instabilities

Interchange instabilities are very similar in nature to the Rayleigh–Taylor instability (Chandraseker, 1961). Actually, except in special asymptotic limits the interchange perturbation is not a true mode of the system but represents a special trial function chosen to minimize the line bending contribution in δW_F (i.e., for an

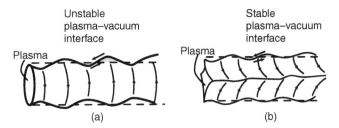

Figure 8.9   Illustration of configurations which are (a) unstable and (b) stable against interchange perturbations.

interchange, $\mathbf{Q}_{\perp} \approx 0$). The interchange instability is important in the RFP, the stellarator, and sometimes in tokamaks.

The interchange perturbation can lead to instability depending upon the relative sign of the magnetic field line curvature with respect to the pressure gradient. If the field lines bend towards the plasma their tension tends to make them shorten and collapse inward. The plasma pressure, on the other hand, has a natural tendency to expand outward. In such cases a perturbation that "interchanges" two flux tubes at different radii leads to a system with lower potential energy and hence instability (see Fig. 8.9). Because of the way the surfaces are fluted, the perturbations are also sometimes known as "flute modes." When the field lines bend away from the plasma, the system is stable to interchange perturbations.

From this description it follows that interchange instabilities represent plasma perturbations that are nearly constant along a field line (i.e., no line bending). Perpendicular to the field the most unstable perturbations have very rapid variation while parallel to the field the variation is much slower (i.e., $k_{\perp}a \gg 1$, $k_{\parallel}/k_{\perp} \ll 1$). Because of the short perpendicular wavelength the modes tend to be highly localized in radius. As such they are often called localized interchanges and are very amenable to analysis. In general, localized interchanges lead to necessary conditions for stability which can be expressed solely in terms of local values of the equilibrium quantities. Examples of such conditions are the Suydam criterion (one dimensional) and the Mercier criterion (two and three dimensional).

There are several properties of the magnetic geometry that can be effective in stabilizing interchanges. First, if there is shear in the magnetic field (i.e., the direction of **B** changes from one flux surface to another), it is not possible to interchange two neighboring flux tubes without at least a little bending of the magnetic lines, which is a stabilizing effect. When the pressure gradient is sufficiently small compared to the shear, the interchange can be stabilized. A second stabilizing method makes use of the fact that on the inside of a torus the local toroidal field line curvature is favorable – the lines bend away from the plasma. Thus, as a given magnetic line encircles the torus it in general passes through

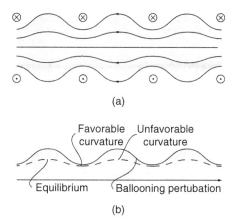

Figure 8.10   (a) Plasma with local regions of favorable and unfavorable curvature. (b) Plasma surface showing a ballooning perturbation.

alternating regions of favorable and unfavorable curvature. By carefully designing the configuration, the "average" curvature can be made favorable thereby stabilizing interchanges. Such configurations are said to possess a "magnetic well" or "average-favorable curvature."

Ballooning modes

Ballooning modes are internal pressure-driven instabilities that occur in toroidal or other multidimensional configurations. These instabilities are important because they determine one set of criteria which limits the maximum achievable value of β.

The designation "ballooning" refers to the fact that in multidimensional geometries the curvature of a magnetic field line often alternates between regions of favorable and unfavorable curvature. Thus, a perturbation that is not constant, but varies slowly along a field line in such a way that the mode is concentrated in the unfavorable curvature region, can lead to more unstable situations than the simple interchange perturbation. However, concentrating the mode also produces some stabilizing line bending (see Fig. 8.10). In effect, the "ballooning" nature of the perturbation in the unfavorable curvature region increases the pressure-driven destabilizing contribution to δW_F. If the localization is not too severe, the accompanying increase in stability from the line bending cannot compensate for this destabilizing effect.

Magnetic shear can be helpful in stabilizing ballooning modes. However, once in the regime where ballooning modes are important the most effective way to stability given magnetic field profiles is to keep β below some critical value.

Like interchanges, ballooning modes are quite amenable to analysis. The most unstable modes also occur for $k_\perp a \gg 1$, $k_\parallel / k_\perp \ll 1$. By exploiting the short perpendicular wavelength nature of the instabilities, the general multidimensional stability problem reduces to the solution of a one-dimensional differential equation along a field line on each flux surface. Substantial progress has been made in the study of ballooning modes in tokamaks and stellarators using this procedure. One property of the ballooning mode equation is that, in a special limit, it leads to the localized interchange criterion (i.e., the Mercier criterion).

8.11.4 Current-driven modes

A current-driven mode is one in which the dominant driving source of instability is proportional to the $J_{\parallel}$ term in Eq. (8.100). These modes are driven by parallel currents and can exist even in a zero-pressure force-free plasma. Current-driven instabilities are often known as "kink" modes. In general the most unstable perturbations have long parallel wavelengths $k_{\parallel}a \ll 1$, and perpendicular wavelengths that are macroscopic in scale $k_{\perp}a \sim 1$. Current-driven modes can have the form of either internal or external perturbations depending upon the location of the singular surface. As such, these modes are usually subdivided into two categories: external kinks and internal kinks.

External kink modes

The external kink mode is a very dangerous instability in tokamak and RFP experiments, often leading to major disruptions. At low β the main destabilizing effect for high m modes (m is the poloidal wave number) is the radial gradient of the parallel current near the plasma edge. For low m numbers, $m = 1$ in particular, the current profile is not as important as the total current itself. The basic form of the perturbation is such that the plasma surface "kinks" into a helix as illustrated in Fig. 8.11. The unstable modes occur for long parallel wavelength in order to minimize line bending. Usually the most dangerous kinks correspond to low perpendicular wave numbers.

At higher values of β the external kink mode develops a distinct ballooning structure. The harmonic content is concentrated at low m values, unlike the short-wavelength internal ballooning mode. At sufficiently high β the external kink mode can become unstable, even if the parallel current is less than the low β stability limit. This is the regime of the "external ballooning-kink" mode. The corresponding β limit is often the most severe stability criterion in the system.

There are several ways in which to improve the stability of a given configuration against external kink. First, for a prescribed geometry there is usually a critical parallel current, which depends on β, indicating the onset of instability. Stability can be achieved by keeping the parallel current below this value. Second, for a fixed parallel current a toroidal device with sufficiently small circumference will prohibit the formation of long-wavelength kinks. In other words, a tight aspect

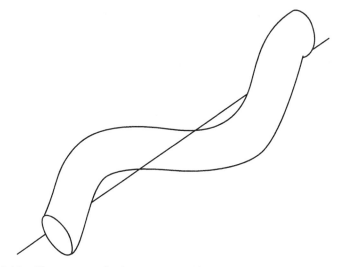

Figure 8.11   Plasma perturbation corresponding to a long wavelength $m = 1$ kink instability.

ratio is beneficial. Third, to the extent that one can experimentally control the current profiles, higher m kink modes can be stabilized by peaking the parallel current profile, keeping the total current fixed. This reduces the edge gradient. Fourth, since low m number kink perturbations have a broad radial extent, they can be stabilized if a perfectly conducting wall is placed sufficiently close to the plasma surface. Finally, one must guarantee that β does not become too large, or else a ballooning-kink instability would be excited.

Overall, the simplest practical advice is "keep the current low, keep the pressure low" if disruptions are to be avoided.

Internal kink modes

The internal kink has many properties similar to those of the external kink, although it is in general a weaker instability. In fact for a tokamak it is usually only $m = 1$ that can become unstable producing what is known as "sawtooth oscillations." As above, a combination of low current and/or tight aspect ratio (i.e., high q) can help to stabilize internal kink modes particularly in a tokamak. Neither of these methods is effective in the RFP since q is very small. However, a broad current profile helps to provide stability; that is, for internal modes, broadening the current profile at fixed total current effectively places a conducting wall closer to the current channel (due to the $\mathbf{n} \cdot \boldsymbol{\xi}_a = 0$ boundary condition), thereby providing stability. In contrast broad profiles are detrimental for external kink stability for systems without a conducting wall. Stabilizing such modes in an RFP requires an actual close-fitting conducting wall plus feedback.

8.12 Summary

In Chapter 8 the general features of MHD stability theory have been described and are summarized below.

- **Exponential stability:** To determine whether or not a given configuration is MHD unstable, the following definition of instability is adopted. A plasma is said to be exponentially unstable if $\text{Im}(\omega) > 0$ for any discrete normal mode of the system.
- **Waves in a homogeneous plasma:** As a simple test of stability, the infinite, homogeneous plasma in a uniform magnetic field is investigated. This system is shown to be stable. The dispersion relation predicts the existence of three distinct propagating waves: the shear Alfven wave, the fast magnetosonic wave (compressional Alfven wave), and the slow magnetosonic wave (sound wave). The shear Alfven wave plays a dominant role in plasma instabilities.
- **General linearized stability equations:** The general three-dimensional MHD stability analysis is formulated first as an initial value problem and then transformed into a normal mode eigenvalue problem. All of the perturbed quantities are expressed in terms of the plasma displacement ξ. The resulting system has the form of three coupled, homogeneous partial differential equations, with eigenvalue ω^2: $-\omega^2 \rho \xi = \mathbf{F}(\xi)$.
- **Properties of the force operator $\mathbf{F}(\xi)$:** The force operator $\mathbf{F}(\xi)$ appearing in the normal mode equation is shown to possess an important mathematical property, self-adjointness. This leads to the conclusion that every eigenvalue ω^2 is purely real. Consequently, normal modes either oscillate, or grow purely exponentially. The self-adjointness also leads to the conclusion that the normal modes are orthogonal. As a general rule, the unstable part of the spectrum associated with $\mathbf{F}(\xi)$ contains only discrete modes. The stable part contains both discrete modes and continua.
- **Variational formulation:** By making use of the self-adjointness of $\mathbf{F}(\xi)$, the normal mode eigenvalue problem is recast into the form of a variational principle. The principle is an equivalent integral representation of the eigenvalue differential equation.
- **Energy Principle:** A further analysis of the variational principle shows that there exists a powerful minimizing principle, known as the Energy Principle, for testing MHD stability. The physical basis for this principle is the exact, non-linear conservation of energy in ideal MHD. The Energy Principle provides an exact test for the stability boundaries although it provides only an estimate of the actual growth rates. Stability is testing by examining the sign of δW for all allowable displacements. The system is stable if and only if $\delta W \geq 0$ for all displacements.
- **Incompressibility:** In attempting to find the most unstable trial functions that minimize δW, it is shown that the minimizing $\xi_\|$ can be calculated once and for

all for arbitrary geometrics. This reduces the number of unknowns to two, corresponding to the components of $\boldsymbol{\xi}_\perp$. For most systems, in which $\mathbf{B} \cdot \nabla$ is non-singular, or at most singular on isolated surfaces, $\xi_\parallel$ must be chosen to make $\nabla \cdot \boldsymbol{\xi} = 0$. For the special case involving a closed line geometry for equilibrium and perturbation there is an additional stabilizing term in δW_F due to the plasma compressibility. This contribution can be expressed explicitly in terms of $\boldsymbol{\xi}_\perp$.

- **Vacuum vs. force-free plasma:** It is shown that it is important to distinguish whether the region surrounding the main plasma core is a vacuum, an ideal MHD plasma, or a resistive MHD plasma. The vacuum case is the most pessimistic with respect to stability boundaries and growth rates. The ideal MHD plasma often predicts considerably improved stability boundaries because the class of allowable perturbations is restricted as a result of the perfect conductivity topological constraint. The most realistic, but more difficult to analyze situation, corresponds to the resistive plasma. This model gives rise to the identical stability boundaries as in the vacuum case, but with much lower growth rates. Since the primary concern is the stability boundaries it is sufficient to treat the outer region as a vacuum in the applications that follows.

- **Classification of MHD instabilities:** A discussion is presented describing the various types of MHD instabilities that can occur in a plasma. In general, instabilities can be driven by currents flowing parallel to the field (current-driven modes) or perpendicular to the field (pressure-driven modes). Modes are also distinguished by whether or not the unstable displacement perturbs the plasma surface: external or internal modes. These classifications are further sub-divided into (1) pressure-driven interchange and ballooning modes and (2) current-driven kinks and ballooning-kink modes. Depending on the mode in question equilibrium properties such as a conducting wall, maximum current, maximum β, magnetic shear, average favorable curvature, or tight aspect ratio may be used to improve stability. As a general feature essentially all MHD instabilities correspond to plasma displacements that are nearly constant along a field line ($k_\parallel a \ll 1$ in order to minimize line bending). However, perpendicular to the field $\boldsymbol{\xi}$ can vary macroscopically for kinks ($k_\perp a \sim 1$) or very rapidly for interchange and ballooning modes ($k_\perp a \gg 1$).

References

Bernstein, I.B., Frieman, E.A., Kruskal, M.D., and Kulsrud, R.M. (1958). *Proc. R. Soc. London, Ser. A* **244**, 17.
Chandraseker, S. (1961). *Hydrodynamic and Hydromagnetic Stability*. Oxford: Clarendon.
Furth, H.P., Killeen, J., Rosenbluth, M.N., and Coppi, B. (1965). In *Plasma Physics and Controlled Nuclear Fusion Research 1964*. Vienna: International Atomic Energy Agency, Vol. I, p. 103.
Goedbloed, J.P. (1975). *Phys. Fluids* **18**, 1258.

Goedbloed, H. and Poedts, S. (2004). *Principles of Magnetohydrodynamics*. Cambridge: Cambridge University Press.

Grad, H. (1973). *Proc. Natl. Acad. Sci.* USA **70**, 3277.

Greene, J.M. and Johnson, J.L. (1968). *Plasma Phys.* **10**, 729.

Laval, G., Mercier, C., and Pellat, R.M. (1965). *Nucl Fusion* **5**, 156.

Further reading

General MHD theory

Bateman, G. (1978). *MHD Instabilities*. Cambridge, MA: MIT Press.

Bernstein, I.B., Frieman, E.A., Kruskal, M.D., and Kulsrud, R.M. (1958). *Proc. R. Soc. London, Ser. A* **244**, 17.

Goedbloed, H. and Poedts, S. (2004). *Principles of Magnetohydrodynamics*. Cambridge: Cambridge University Press.

Kadomtsev, B.B. (1966). In *Reviews of Plasma Physics*, Vol. 2, ed. M.A. Leontovich. New York: Consultants Bureau.

Kikuchi, M., Lackner, K., and Tran, M.Q., eds. (2012). *Fusion Physics*. Vienna: International Atomic Energy Agency.

Miyamoto, K. (2001). *Fundamentals of Plasma Physics and Controlled Fusion*, NIFS-PROC-48. Japan: Toki City.

Variational calculus

Morse, P.M. and Feshbach, H. (1953). *Methods of Theoretical Physics*, Vol. II. New York: McGraw-Hill.

Problems

8.1

(a) Calculate the dispersion relation for waves propagating in an infinite homogeneous MHD plasma with uniform magnetic field $\mathbf{B} = B_0\mathbf{e}_z$. Include the effect of a small resistivity in Ohm's law: $\mathbf{E} + \mathbf{v} \times \mathbf{B} = \eta\mathbf{J}$, $\eta = \text{const}$. Show that the shear Alfvén branch of the dispersion relation can be written as

$$\omega^2\left(1 + i\frac{\gamma_D}{\omega}\right) - k_\parallel^2 V_a^2 = 0$$

Similarly, show that the magnetosonic branch has the form

$$\omega^2\left(\omega^2 - k^2 V_s^2\right)\left(1 + i\frac{\gamma_D}{\omega}\right) - \left(\omega^2 - k_\parallel^2 V_s^2\right)k^2 V_a^2 = 0$$

where $\gamma_D = \eta k^2/\mu_0$.

(b) Prove that each of the three basic MHD waves becomes slightly damped in the presence of resistivity.

8.2 This problem investigates the apparent paradox associated with the ion acoustic wave in a homogeneous plasma. Specifically, the fact that both ideal MHD and two-fluid theory yield the same dispersion relation although $E_\parallel = 0$ in ideal MHD while $E_\parallel \neq 0$ is crucial in two-fluid theory. Consider the following simple two-fluid model: massless electrons, isothermal electrons and ions, quasineutrality, and negligible resistivity.

(a) Assume a background state $\mathbf{B} = B_0\mathbf{e}_z$, $\mathbf{v}_e = \mathbf{v}_i = 0$. Show that the linearized equations describing parallel propagation are given by

$$m_i \frac{\partial v_{zi}}{\partial t} = eE_z - \frac{T_i}{n_0} \frac{\partial n_i}{\partial z}$$

$$0 = -eE_z - \frac{T_e}{n_0} \frac{\partial n_e}{\partial z}$$

$$\frac{\partial n_i}{\partial t} + \frac{\partial}{\partial z} n_0 v_{zi} = 0$$

$$n_e = n_i$$

$$\nabla \times (E_z \mathbf{e}_z) = 0$$

(b) Calculate the dispersion relation and show that

$$\omega^2 = k_{\parallel}^2 (T_e + T_i)/m_i.$$

(c) To obtain a single-fluid derivation, eliminate E_z from the momentum equations and set $n_i = n_e = n$. Note that the resulting mass and momentum equations are identical to those of ideal MHD for parallel propagation. Show that the mass and momentum equations, by themselves, yield the dispersion relation $\omega^2 = k_{\parallel}^2 (T_e + T_i)/m_i$. One thus concludes that the effects of $E_{\parallel}$ *are included* in the ideal MHD momentum equation. It is in Ohm's law that $E_{\parallel}$ is neglected, but this relation is not required for the ion acoustic mode. Thus, the paradox is resolved.

8.3 Consider the integral

$$I = \int_0^1 \eta F(\xi) dx$$

where $\eta(x)$, $\xi(x)$ are two complex functions and F is a differential operator. Determine which of the following forms of F corresponds to a self-adjoint operator:

(a) $F = 1$

(b) $F = \dfrac{\partial}{\partial x}$, BC_1

(c) $F = \dfrac{\partial^2}{\partial x^2}$, BC_1

(d) $F = \dfrac{\partial}{\partial x}\left(h \dfrac{\partial}{\partial x} \right)$, BC_1

(e) $F = \dfrac{\partial^2}{\partial x^2}$, BC$_2$

Here, $h(x)$ is a known function and BC$_1$, BC$_2$ represent the following boundary conditions:

$$BC_1 : \xi(0) = \eta(0) = \xi(1) = \eta(1) = 0$$

$$BC_2 : \xi(0) = \eta(0) = 0, \xi'(1) = A\xi(1), \eta'(1) = A^*\eta(1)$$

8.4 Consider the Grad–Shafranov equation

$$\Delta^*\psi = -\mu_0 R^2 \frac{dp}{d\psi} - F\frac{dF}{d\psi}$$

Derive a Lagrangian density, $\hat{L} = \hat{L}(\psi, \partial\psi/\partial R, \partial\psi/\partial Z, R, Z)$ such that the total Lagrangian

$$L = \int \hat{L}d\mathbf{r}$$

$$d\mathbf{r} = 2\pi R dR dZ$$

represents a variational principle for the Grad–Shafranov equation (i.e., letting $\psi \to \psi + \delta\psi$ in $\hat{L}$ and setting $\delta L = 0$ gives the Grad–Shafranov equation). State what boundary conditions you have assumed.

8.5 Consider the eigenvalue problem

$$y'' + \frac{y'}{x} + \lambda y = 0$$

$$y'(0) = 0, \quad y(1) = 0$$

(a) Multiply the equation by y and then integrate over the region $0 < x < 1$. Note that the integral relation obtained is correct but not variational. Substitute the trial function $y = 1 - x^2$ and evaluate λ.
(b) Derive a variational form for the eigenvalue problem.
(c) Substitute the trial function $y = 1 - x^2$ into the variational form and evaluate λ.
(d) Substitute the trial function $y = (1 - x^2)^\nu$ into the variational form and determine ν so that the Lagrangian is an extremum:

$$\partial L/\partial \nu = 0$$

(e) Compare these results with the exact answer.

8.6 To demonstrate the advantage of using natural boundary conditions consider the following problem:

$$\frac{d^2y}{dx^2} + (\lambda - x^2)y = 0$$

$$y'(0) = 0$$

$$y'(1) = 3y(1)$$

(a) Derive a Lagrangian with the property that the boundary condition at $x = 1$ appears as a natural boundary condition.
(b) Consider now the one-parameter trial function $y = 1 - \alpha x^2$. Choose α so that the boundary condition at $x = 1$ is satisfied and evaluate the eigenvalue λ.
(c) Recall that with natural boundary conditions one is not forced to choose a trial function such that the boundary condition at $x = 1$ is automatically satisfied. The variational principle will do "as good a job as possible" to ensure that this condition is satisfied. Instead, treat α as a variational parameter determined by setting $\delta L = 0$. Use this value of α and calculate the eigenvalue λ.
(d) Compare (b) and (c) with the exact result.

8.7 The boundary condition given by

$$\mathbf{n} \cdot \hat{\mathbf{B}}_1 \big|_{r_p} = \mathbf{n} \cdot \nabla \times (\boldsymbol{\xi} \times \hat{\mathbf{B}}) \big|_{r_p}$$

is intuitively obvious although there are subtleties involved. In particular, the plasma displacement is not defined in the vacuum region. A more rigorous derivation starts from the basic condition $\mathbf{n} \cdot \hat{\mathbf{B}} = 0$. Linearize this relation about the equilibrium solution $\left(\mathbf{n} = \mathbf{n}_0 + \mathbf{n}_1, \hat{\mathbf{B}} = \hat{\mathbf{B}}_0 + \hat{\mathbf{B}}_1, \mathbf{r} = \mathbf{r}_p + \boldsymbol{\xi}\right)$ and show that the equation (above) is indeed the correct boundary condition. Hint: Show that $\mathbf{n}_1 = -(\nabla\boldsymbol{\xi}) \cdot \mathbf{n}_0 + \mathbf{n}_0 \left[\mathbf{n}_0 \cdot (\nabla\boldsymbol{\xi}) \cdot \mathbf{n}_0\right]$.

8.8 The boundary condition

$$\mathbf{n} \cdot \hat{\mathbf{B}}_1 \big|_{r_p} = \mathbf{n} \cdot \nabla \times (\boldsymbol{\xi} \times \hat{\mathbf{B}}) \big|_{r_p}$$

can be expressed solely in terms of the normal component of the plasma displacement evaluated on the plasma surface: $\xi_n = \mathbf{n} \cdot \boldsymbol{\xi}$. Specifically, show that

$$\mathbf{n} \cdot \nabla \times (\boldsymbol{\xi} \times \mathbf{B}) \big|_{r_p} = \mathbf{B} \cdot \nabla \xi_n - \xi_n [\mathbf{n} \cdot (\mathbf{n} \cdot \nabla)\mathbf{B}] \big|_{r_p}$$

8.9 Show by the use of trial functions that the presence of a perfectly conducting wall is always stabilizing.

9

Alternate MHD models

9.1 Introduction

Ideal MHD is the simplest model that describes the macroscopic equilibrium and stability of high-temperature fusion plasmas. A self-consistent derivation of the model has been presented in Chapter 2. The derivation requires that one restrict attention to the MHD length and time scales. The main assumptions for validity of ideal MHD are (1) small ion gyro radius, (2) high collisionality, and (3) negligible resistive diffusion. As pointed out, the high collisionality assumption is never satisfied in fusion-grade plasmas, which makes it perhaps surprising how accurate and reliable the model is in predicting experimental behavior.

The goal of Chapter 9 is to begin to address this unsettling situation. While there are a number of alternate MHD models in various regimes of collisionality, the strategy here is to focus on two specific models that describe MHD behavior in the collisionless regime. These are the models that are most relevant to fusion plasmas. Ultimately, in Chapter 10 a set of general stability "comparison theorems" is derived that enables one to make quantitative comparisons between the predictions of the collision dominated and collisionless models. The present chapter, however, focuses solely upon the introduction of the two alternate models. Also presented is a review of ideal MHD which serves as a reference. The specific models discussed are as follows:

- Ideal MHD: collision dominated fluid
- Kinetic MHD: collisionless kinetic plasma
- Double adiabatic theory: collisionless fluid.

As a general comment it is worth pointing out that as a species of plasma particles evolves from collision dominated to collisionless, its corresponding model switches from a fluid to a kinetic description. *All* models share the common feature of being based on the first two moments of the Boltzmann equation – mass and momentum. The distinction between collisionless and collision dominated behavior arises in the approximations made to achieve closure. Fluid treatments require a macroscopic conservation of energy relation to close the system of equations. In contrast, the kinetic models require an approximate solution for the distribution function which is then used to directly calculate the pressure tensor, thereby providing closure.

The material in Chapter 9 begins with a more detailed analysis of the collision dominated assumption used in ideal MHD. This analysis suggests the different types of MHD models that need to be developed, in particular the models listed above. Next, the conservation of mass and momentum equations are derived. With this as a foundation, the alternate MHD models are then systematically developed.

The end goal is a closed formulation for each of the two alternate MHD models listed above.

9.2 Transition from collision dominated to collisionless regimes

The discussion begins with a re-examination of the collision dominated assumption used in the derivation of ideal MHD, the one that is always violated for fusion plasmas. Recall that the most stringent collisionality requirement used in the derivation requires temperature equilibration on a time scale fast compared to the MHD time scale.

For present purposes it is useful to separately consider three different collisionality regimes each characterized by an appropriate dimensionless parameter $\omega\tau$, where $\omega = V_{Ti}/a$ is the MHD frequency and τ is the collision time of interest. The three regimes are defined as follows:

$$\begin{aligned}
\omega\tau_{eq} &\ll 1 \qquad \text{temperatures equilibrate} \\
\omega\tau_{ii} &\ll 1 \qquad \text{ions are collision dominated} \\
\omega\tau_{ee} &\ll 1 \qquad \text{electrons are collision dominated}
\end{aligned} \tag{9.1}$$

Here the characteristic collision times are chosen as $\tau_{ee} \approx \tau_e$, $\tau_{ii} \approx \tau_i = (2m_i/m_e)^{1/2}\tau_e$, and $\tau_{eq} = (m_i/2m_e)\tau_e$, where τ_e is given by Braginskii (1965)

$$\tau_e = 3(2\pi)^{3/2}\frac{\varepsilon_0^2 m_e^{1/2}T_e^{3/2}}{ne^4\ln\Lambda} = 1.09 \times 10^{-4}\frac{T_k^{3/2}}{n_{20}\ln\Lambda} \quad \text{sec} \tag{9.2}$$

One now sets the Coulomb logarithm to $\ln\Lambda = 19$ and the ion mass to that of deuterium. In order of decreasing stringency the collisionality boundaries reduce to

$$\begin{aligned}
\omega\tau_{eq} &= 3.3 \times 10^3\left(T_k^2/an_{20}\right) \ll 1 \quad \text{temperatures equilibrate} \\
\omega\tau_{ii} &= 1.5 \times 10^2\left(T_k^2/an_{20}\right) \ll 1 \quad \text{ions are collision dominated} \\
\omega\tau_{ee} &= 1.8\left(T_k^2/an_{20}\right) \ll 1 \qquad\quad \text{electrons are collision dominated}
\end{aligned} \tag{9.3}$$

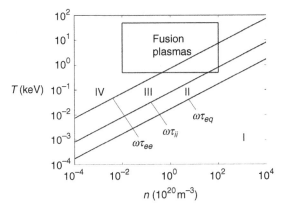

Figure 9.1 Regions of validity for the alternate MHD models in (n, T) space for the case $a = 1\,\mathrm{m}$.

As in Chapter 2, these quantities are plotted (in Fig. 9.1) as curves of T vs. n for the case of $a = 1\,\mathrm{m}$. Also shown is the rectangle corresponding to the region of fusion interest: $10^{18}\,\mathrm{m}^{-3} < n < 10^{22}\,\mathrm{m}^{-3}$ and $0.5\,\mathrm{keV} < T < 50\,\mathrm{keV}$.

Observe that there are four distinct collisionality regimes. Region I corresponds to the highest collisionality regime where the electron and ion temperatures have sufficient time to equilibrate: $\omega\tau_{eq} \ll 1$. This is the reference regime defining the single-fluid ideal MHD model.

In region II both the electrons and ions are collision dominated but there is not sufficient time for temperature equilibration: $\omega\tau_{ii} \ll 1 \ll \omega\tau_{eq}$. In this regime the electron and ion temperatures are decoupled and a separate fluid energy conservation relation is required for each species for closure. This is an intermediate collisionality regime.

As the collisionality continues to decrease one moves into region III where the electrons remain collision dominated but the ions become collisionless: $\omega\tau_{ee} \ll 1 \ll \omega\tau_{ii}$. Higher-density, lower-temperature fusion plasmas, often corresponding to the plasma edge, lie in this regime. For region III a fluid energy conservation relation is needed for closure of the electron model, while a kinetic treatment is required for the ions. This is also an intermediate collisionality regime.

Lastly, fully collisionless plasmas lie in region IV. Here, both the electrons and ions are collisionless: $1 \ll \omega\tau_{ee}$. The core of most fusion plasmas lies in this regime. Since both species are collisionless, each must be described by a kinetic model. A simple kinetic model can be derived by exploiting the small gyro radius assumption. It is known as kinetic MHD and is one of the models discussed below.

As shown, the kinetic MHD model is, in general, much easier to solve than the Vlasov equation but still much more difficult to solve than the ideal MHD model. This fact has led to attempts at fluid closures of the kinetic MHD model. One such result is the Chew, Goldberger, Low (CGL) double adiabatic fluid model also discussed below. The double adiabatic model is far simpler to solve but cannot be justified by a consistent set of physical assumptions. Still, it is a very useful model when considering stability comparison theorems.

With this as background, the task now is to present a mathematical derivation for each of the two alternate models just described.

9.3 General formulation

The MHD models under consideration all have the same starting point consisting of an identical, simplified form of the mass and momentum moments of the Boltzmann–Maxwell kinetic equations. These simplified but accurate moments are coupled to the low-frequency, quasi-neutral form of Maxwell's equations.

The simplified moments are obtained from the exact moments by neglecting electron inertia in the electron momentum equation and neglecting resistivity in both momentum equations. Use is also made of the small gyro radius approximation: $r_{Li}/a \sim k_\perp r_{Li} \ll 1$. Under these conditions the starting equations are easily obtained from the analysis presented in Section 2.3 and can be written as

$$\frac{\partial n}{\partial t} + \nabla \cdot (n \, \mathbf{v}) = 0$$

$$\rho \frac{d \mathbf{v}}{dt} = \mathbf{J} \times \mathbf{B} - \nabla \cdot (\mathbf{P}_e + \mathbf{P}_i)$$

$$\frac{\partial \mathbf{B}}{\partial t} = \nabla \times (\mathbf{v} \times \mathbf{B}) \tag{9.4}$$

$$\nabla \times \mathbf{B} = \mu_0 \mathbf{J}$$

$$\nabla \cdot \mathbf{B} = 0$$

$$n_e = n_i = n$$

Before proceeding, some discussion is required concerning Faraday's law. The simplified form given in Eq. (9.4) can be obtained by recalling that the electron fluid velocity is given by $\mathbf{u}_e = \mathbf{v} - \mathbf{J}/en$. After substituting this into the electron momentum equation one finds

$$\mathbf{E} + \mathbf{v} \times \mathbf{B} = (\mathbf{J} \times \mathbf{B} - \nabla \cdot \mathbf{P}_e)/en \tag{9.5}$$

For the perpendicular component of $\mathbf{E}$, the right-hand side of Eq. (9.5) can be neglected because of the small gyro radius assumption. This implies that

$$\mathbf{E}_\perp + \mathbf{v} \times \mathbf{B} \approx 0 \tag{9.6}$$

The parallel component of electric field is not affected by the $\mathbf{v} \times \mathbf{B}$ term. Its value is given by

$$E_\parallel = -\mathbf{b} \cdot (\nabla \cdot \mathbf{P}_e)/en \tag{9.7}$$

A simple scaling argument shows that $E_\parallel/|\mathbf{E}_\perp| \sim r_{Li}/a \ll 1$.

Therefore, with respect to Faraday's law $\mathbf{E} = \mathbf{E}_\perp + E_\parallel \mathbf{b} \approx \mathbf{E}_\perp$ and the electron momentum equation reduces to the ideal MHD Ohm's law $\mathbf{E} + \mathbf{v} \times \mathbf{B} = 0$. Substituting $\mathbf{E} = -\mathbf{v} \times \mathbf{B}$ into Faraday's law leads to the simple form given in Eq. (9.4). This form of Faraday's law guarantees that the magnetic field lines are frozen into the plasma, the basic condition for ideal MHD behavior. However, one should keep in mind that there is a small non-zero $E_\parallel$ given by Eq. (9.7) which does not enter ideal MHD but will play a role in kinetic MHD. This is the main point of the above discussion.

Return now to Eq. (9.4). One sees that the model, as it now stands, is general, but has more unknowns than equations. Obtaining closure requires a prescription for determining the two pressure tensors $\mathbf{P}_e$ and $\mathbf{P}_i$. There are a variety of models that accomplish this task each corresponding to a different set of assumptions with respect to the collisionality of the electrons and ions. These different closure choices represent various alternate MHD models. In the discussion that follows an explicit mathematical formulation is presented for the MHD models of interest.

9.4 Ideal MHD closure

The closure assumptions for ideal MHD have already been discussed in Section 2.3 and are simply repeated here for completeness. In ideal MHD the ion and electron temperatures are equilibrated, $T_e = T_i \equiv T$, and the pressure is a scalar quantity, $p = p_e + p_i = 2nT$. Furthermore, the pressure satisfies the adiabatic conservation of energy relation leading to the following procedure for closure:

$$\mathbf{P} = \mathbf{P}_e + \mathbf{P}_i = p\mathbf{I}$$
$$\frac{d}{dt}\left(\frac{p}{n^\gamma}\right) = 0 \tag{9.8}$$

where $\gamma = 5/3$.

9.5 Kinetic MHD

9.5.1 Basic assumptions

The kinetic MHD model lies in region IV of the T vs. n diagram (Fig. 9.1) and corresponds to the situation where both the electrons and ions are collisionless.

Many fusion experiments lie in this region. Since both species are collisionless it is necessary to treat each with a kinetic MHD model. To derive the model one needs to exploit both the MHD ordering, $\omega \sim V_{Ti}/a$, $k \sim 1/a$ and the small gyro radius approximation $r_{Li}/a \ll 1$, while ignoring collisions in the starting Boltzmann kinetic equation. That is, the analysis starts with the Vlasov equation which is a function of six phase space variables plus time: $\mathbf{r}$, $\mathbf{v}$, and t. The end result of the analysis is a simplified kinetic equation which is three-dimensional in physical space but only two-dimensional in velocity space: $\mathbf{r}$, ε, μ, and t. Also, the adiabatic invariant μ appears only as a parameter. This represents a substantial saving in complexity.

9.5.2 Derivation of the kinetic MHD model

The derivation of the kinetic MHD model from the Vlasov equation involves a somewhat lengthy analysis. The details are presented below for the ions and closely follow the excellent derivation by Hazeltine and Waelbroeck (1998). The electron equation is obtained by letting $m_i \rightarrow m_e$ and $e \rightarrow -e$.

The starting equations

The derivation of the kinetic MHD equation starts from the Vlasov equation, given by

$$\frac{df}{dt} = \frac{\partial f}{\partial t} + \frac{d\mathbf{r}}{dt} \cdot \nabla f + \frac{d\mathbf{u}}{dt} \cdot \nabla_u f = 0$$

$$\frac{d\mathbf{r}}{dt} = \mathbf{u} \tag{9.9}$$

$$\frac{d\mathbf{u}}{dt} = \frac{e}{m_i}(\mathbf{E} + \mathbf{u} \times \mathbf{B})$$

Coordinate transformations

To obtain the kinetic MHD model new velocity variables are introduced into the Vlasov equation in order to exploit the small gyro radius approximation. Specifically, the single particle velocity vector $\mathbf{u} = u_x \mathbf{e}_x + u_y \mathbf{e}_y + u_z \mathbf{e}_z$ is replaced by

$$\mathbf{u} = \mathbf{w} + \mathbf{v}(\mathbf{r},t) \tag{9.10}$$

where $\mathbf{w}$ is the random component of the particle velocity (i.e., $\langle \mathbf{w} \rangle = 0$). Hopefully the notational confusion is tolerable even though a switch has been made from Chapter 2. Now, $\mathbf{u}$ represents the velocity variable in the Vlasov equation and $\mathbf{v}(\mathbf{r}, t)$ is the traditional MHD macroscopic fluid velocity. Also, recall that the MHD ordering implies that $\mathbf{v}_i = \mathbf{v}$ and $\mathbf{v}_e = \mathbf{v}_i - \mathbf{J}/en \approx \mathbf{v} + O(r_{Li}/a)$. To the order needed in kinetic MHD the ions and electrons have the same macroscopic velocity $\mathbf{v}(\mathbf{r}, t)$.

Next, $\mathbf{w}$ is decomposed as $\mathbf{w} = \mathbf{w}_\perp + w_\parallel \mathbf{b}$, where $\mathbf{b}(\mathbf{r}, t) = \mathbf{B}/B$ and

$$
\begin{aligned}
\mathbf{w}_\perp &= w_1 \mathbf{e}_1 + w_2 \mathbf{e}_2 \\
&= w_\perp \cos \alpha \, \mathbf{e}_1 + w_\perp \sin \alpha \, \mathbf{e}_2
\end{aligned}
\tag{9.11}
$$

Here, $\mathbf{e}_1(\mathbf{r}, t)$, $\mathbf{e}_2(\mathbf{r}, t)$ are a set of local orthogonal unit vectors perpendicular to each other as well as to $\mathbf{b}$: $\mathbf{e}_1 \cdot \mathbf{e}_2 = \mathbf{e}_1 \cdot \mathbf{b} = \mathbf{e}_2 \cdot \mathbf{b} = 0$. The independent velocity variables at this point are considered to be $w_\perp$, $w_\parallel$, and α, where $\alpha = \tan^{-1}(w_2/w_1)$ is the gyro phase angle.

Lastly, the perpendicular and parallel random velocities $w_\perp$ and $w_\parallel$ are replaced with two new variables ε and μ defined as follows:

$$
\varepsilon = \frac{1}{2} m_i \left(w_\perp^2 + w_\parallel^2 \right)
$$

$$
\mu = \frac{m_i w_\perp^2}{2B(\mathbf{r}, t)}
\tag{9.12}
$$

Here, ε is the random particle kinetic energy and μ is the adiabatic invariant. These definitions define the overall velocity variable transformation:

$$
u_x, u_y, u_z \rightarrow w_\perp, w_\parallel, \alpha \rightarrow \varepsilon, \mu, \alpha
\tag{9.13}
$$

The distribution function is now a function of $\mathbf{r}$, ε, μ, α, t and in terms of these variables the Vlasov equation becomes

$$
\frac{df}{dt} = \frac{\partial f}{\partial t} + (\mathbf{v} + \mathbf{w}) \cdot \nabla f + \dot{\varepsilon} \frac{\partial f}{\partial \varepsilon} + \dot{\mu} \frac{\partial f}{\partial \mu} + \dot{\alpha} \frac{\partial f}{\partial \alpha} = 0
\tag{9.14}
$$

Expressions for $\dot{\varepsilon}, \dot{\mu}, \dot{\alpha}$

The next step is to derive expressions for $\dot{\varepsilon}, \dot{\mu}, \dot{\alpha}$ in terms of the new variables. Starting with the equations of motion given in Eq. (9.9) and forming various obvious combinations of the separate vector components one obtains after a straightforward calculation a set of complicated looking relations given by

$$
\dot{\varepsilon} = e w_\parallel E_\parallel + q \, \mathbf{w}_\perp \cdot (\mathbf{E}_\perp + \mathbf{v}_\perp \times \mathbf{B}) - m_i \mathbf{w} \cdot \frac{D\mathbf{v}}{Dt}
$$

$$
\begin{aligned}
\dot{\mu} = &-\frac{\mu}{B} \frac{DB}{Dt} + \frac{e}{B} \mathbf{w}_\perp \cdot (\mathbf{E}_\perp + \mathbf{v}_\perp \times \mathbf{B}) - \frac{m_i}{B} \mathbf{w}_\perp \cdot \frac{D\mathbf{v}}{Dt} \\
&+ m_i \mathbf{w} \cdot \left(\cos \alpha \frac{D\mathbf{e}_1}{Dt} + \sin \alpha \frac{D\mathbf{e}_2}{Dt} \right)
\end{aligned}
\tag{9.15}
$$

$$
\begin{aligned}
\dot{\alpha} = &\,\omega_{ci} - \frac{e}{m_i w_\perp^2} \mathbf{b} \cdot \mathbf{w}_\perp \times (\mathbf{E}_\perp + \mathbf{v}_\perp \times \mathbf{B}) - \frac{1}{w_\perp^2} \mathbf{b} \cdot \mathbf{w}_\perp \times \frac{D\mathbf{v}}{Dt} \\
&- \frac{1}{w_\perp} \mathbf{w} \cdot \left(\cos \alpha \frac{D\mathbf{e}_2}{Dt} - \sin \alpha \frac{D\mathbf{e}_1}{Dt} \right)
\end{aligned}
$$

where $\omega_{ci} = eB/m_i$. In the right-hand side of these equations it is understood that the random velocity $\mathbf{w}$ should be expressed in terms of ε, μ, α. Also the D/Dt operator acting on any macroscopic function of $\mathbf{r}$, t is defined as

$$\frac{D}{Dt} = \frac{\partial}{\partial t} + \mathbf{u} \cdot \nabla = \frac{\partial}{\partial t} + \mathbf{v} \cdot \nabla + (\mathbf{w}_\perp + w_\| \mathbf{b}) \cdot \nabla \qquad (9.16)$$

At this point in the analysis Eqs. (9.14) and (9.15) are still exact. In order to carry out the small gyro radius expansion it is necessary to separate each of the quantities $\dot{\varepsilon}, \dot{\mu}, \dot{\alpha}$ into a gyro averaged contribution plus a contribution that has zero gyro average. That is, each of the quantities $\dot{\varepsilon}, \dot{\mu}, \dot{\alpha}$ is written as

$$\dot{Q}(\mathbf{r}, \varepsilon, \mu, \alpha, t) = \overline{\dot{Q}}(\mathbf{r}, \varepsilon, \mu, t) + \tilde{\dot{Q}}(\mathbf{r}, \varepsilon, \mu, \alpha, t)$$

$$\overline{\dot{Q}} = \frac{1}{2\pi} \int_0^{2\pi} \dot{Q} \, d\alpha$$

$$\tilde{\dot{Q}} = \frac{1}{2\pi} \int_0^{2\pi} \left(\dot{Q} - \overline{\dot{Q}}\right) d\alpha = 0 \qquad (9.17)$$

When the gyro radius expansion is carried out it is shown that only the gyro averaged quantities $\overline{\dot{Q}}$ are needed. Calculating these averages is the next task.

Expressions for $\overline{\dot{\varepsilon}}, \overline{\dot{\mu}}, \overline{\dot{\alpha}}$

The required gyro averages are also straightforward to obtain but require some tedious algebra. The results are

$$\overline{\dot{\varepsilon}} = ew_\| E_\| - m_i w_\| \mathbf{b} \cdot \frac{d\mathbf{v}}{dt} - \frac{m_i w_\perp^2}{2} \nabla \cdot \mathbf{v} - m_i \left(w_\|^2 - \frac{w_\perp^2}{2}\right) \mathbf{b} \cdot (\mathbf{b} \cdot \nabla) \mathbf{v}$$

$$\overline{\dot{\mu}} = 0 + O(r_{Li}/a)$$

$$\overline{\dot{\alpha}} = \omega_{ci} + O(r_{Li}/a) \qquad (9.18)$$

where d/dt operating on a macroscopic function of $\mathbf{r}$, t is defined as

$$\frac{d}{dt} = \frac{\partial}{\partial t} + \mathbf{v} \cdot \nabla \qquad (9.19)$$

The r_{Li}/a corrections to $\overline{\dot{\mu}}$ and $\overline{\dot{\alpha}}$ are complicated expressions which are not needed for the gyro radius expansion.

The gyro radius expansion

The groundwork has now been set for the gyro radius expansion. For the usual MHD ordering the largest term by far in Eq. (9.14) is the one proportional to $\dot{\alpha}$.

This follows because for the MHD ordering the dominant perpendicular fluid motion of both electrons and ions is the $\mathbf{E} \times \mathbf{B}/B^2$ drift: $\mathbf{E}_\perp + \mathbf{v}_\perp \times \mathbf{B} = O(r_i/a)$. Thus, if f is expanded in terms of $f = f_0 + f_1 + \cdots$ then the leading-order contribution to Eq. (9.14) is given by

$$\omega_{ci} \frac{\partial \bar{f}_0}{\partial \alpha} = 0 \tag{9.20}$$

The solution is

$$f_0 = \bar{f}_0(\mathbf{r}, \varepsilon, \mu, t) \tag{9.21}$$

The first-order equation has the form

$$\omega_{ci} \frac{\partial f_1}{\partial \alpha} + \frac{\partial \bar{f}_0}{\partial t} + (\mathbf{v} + \mathbf{w}) \cdot \nabla \bar{f}_0 + \left(\bar{\varepsilon} + \tilde{\varepsilon}\right) \frac{\partial \bar{f}_0}{\partial \varepsilon} = 0 \tag{9.22}$$

For f_1 to be a physical solution it must be periodic in α. Therefore, averaging Eq. (9.22) over one period in α is equivalent to an integrability condition on $\bar{f}_0$. This condition defines the basic kinetic MHD equation. The resulting equations for ions and electrons (after dropping the "0" subscript) are given by

Ions:
$$\frac{\partial \bar{f}_i}{\partial t} + \left(\mathbf{v} + w_\| \mathbf{b}\right) \cdot \nabla \bar{f}_i + \bar{\varepsilon} \frac{\partial \bar{f}_i}{\partial \varepsilon} = 0$$

$$\bar{\varepsilon} = e w_\| E_\| - m_i w_\| \mathbf{b} \cdot \left(\frac{\partial}{\partial t} + \mathbf{v} \cdot \nabla\right) \mathbf{v} - \frac{m_i w_\perp^2}{2} \nabla \cdot \mathbf{v} - m_i \left(w_\|^2 - \frac{w_\perp^2}{2}\right) \mathbf{b} \cdot (\mathbf{b} \cdot \nabla) \mathbf{v} \tag{9.23}$$

Electrons:
$$\frac{\partial \bar{f}_e}{\partial t} + \left(\mathbf{v} + w_\| \mathbf{b}\right) \cdot \nabla \bar{f}_e + \bar{\varepsilon} \frac{\partial \bar{f}_e}{\partial \varepsilon} = 0$$

$$\bar{\varepsilon} = -e w_\| E_\| - m_e w_\| \mathbf{b} \cdot \left(\frac{\partial}{\partial t} + \mathbf{v} \cdot \nabla\right) \mathbf{v} - \frac{m_e w_\perp^2}{2} \nabla \cdot \mathbf{v} - m_e \left(w_\|^2 - \frac{w_\perp^2}{2}\right) \mathbf{b} \cdot (\mathbf{b} \cdot \nabla) \mathbf{v} \tag{9.24}$$

Observe that the same macroscopic fluid velocity $\mathbf{v}$ appears in both the ion and electron equations. These are the desired equations which at this point are still valid non-linearly.

An examination of Eqs. (9.23) and (9.24) shows that as expected the reduced distribution function is independent of the angle α, a consequence of gyro phase averaging. Furthermore, there are no terms containing a derivative with respect to μ since, on the MHD length and time scales, μ is an adiabatic constant of the motion; that is, $\bar{\dot{\mu}} = 0$ and hence the term $\bar{\dot{\mu}}(\partial \bar{f}/\partial \mu) = 0$. The implication is that the independent variable μ appears only as a parameter in the equation.

A further point to note is the explicit appearance of $E_\parallel$ in Eqs. (9.23) and (9.24). While $E_\parallel$ is small in Faraday's law because of the small gyro radius assumption, it is of comparable magnitude to the other terms in $\bar{\varepsilon}$. It cannot be neglected. The appropriate way to determine $E_\parallel$ is to enforce the charge neutrality condition

$$n_e = n_i \rightarrow \int \bar{f}_e d\mathbf{w} = \int \bar{f}_i d\mathbf{w} \qquad (9.25)$$

Final closure

The final step required to complete the discussion of the kinetic MHD model is the closure procedure. Specifically, assuming that a solution for $\bar{f}(\mathbf{r}, \varepsilon, \mu, t)$ has been found, how is closure obtained? The answer is as follows. Knowing $\bar{f}(\mathbf{r}, \varepsilon, \mu, t)$ one can directly calculate the unknown pressure tensors in Eq. (9.4) thereby closing the system of equations. For each species one finds

$$\mathbf{P} = \begin{vmatrix} p_\perp & & \\ & p_\perp & \\ & & p_\parallel \end{vmatrix} = p_\perp \mathbf{I} + \left(p_\parallel - p_\perp \right) \mathbf{bb}$$

$$p_\perp = \int \frac{mw_\perp^2}{2} \bar{f} \, d\mathbf{w} = \int \mu B \bar{f} \, d\mathbf{w} \qquad (9.26)$$

$$p_\parallel = \int mw_\parallel^2 \bar{f} \, d\mathbf{w} = \int 2(\varepsilon - \mu B) \bar{f} \, d\mathbf{w}$$

where (with the $w_\parallel$ dependence explicitly shown)

$$Q(w_\parallel)d\mathbf{w} = \begin{cases} 2\pi Q(w_\parallel)w_\perp dw_\perp dw_\parallel & -\infty < w_\parallel < \infty \\ \dfrac{2^{1/2}\pi}{m^{3/2}} \dfrac{B}{(\varepsilon - \mu B)^{1/2}}[Q(w_\parallel) + Q(-w_\parallel)]d\varepsilon \, d\mu & 0 < w_\parallel < \infty \end{cases}$$

$$(9.27)$$

Observe that the pressure tensor is anisotropic but diagonal in structure. Once $p_\perp = p_{\perp i} + p_{\perp e}$ and $p_\parallel = p_{\parallel i} + p_{\parallel e}$ are known then the pressure tensor appearing in the momentum equation can be evaluated:

$$\nabla \cdot (\mathbf{P}_i + \mathbf{P}_e) = \nabla p_\perp + (p_\parallel - p_\perp)\boldsymbol{\kappa} + \mathbf{b} B \cdot \nabla \left(\frac{p_\parallel - p_\perp}{B} \right) \qquad (9.28)$$

Equation (9.28) is the required form of the pressure tensor, which when combined with the charge neutrality condition given by Eq. (9.25), formally closes the kinetic MHD model.

A last word of caution: one should *not* use the solution for $\bar{f}$ arising from the kinetic MHD equation to evaluate the moments corresponding to either $\mathbf{J}_\perp$ or $\mathbf{v}$. These quantities are determined directly from the fluid moment equations given by Eq. (9.4). They can in principle be derived from higher-order r_L/a corrections to the distribution function, but these are very complicated and unnecessary to calculate. In fact the need *not* to calculate these higher-order corrections is a major advantage of the kinetic MHD model.

From the practical mathematical point of view the unknown distribution function in kinetic MHD is a function of five phase space variables plus time, $\mathbf{r}$, ε, μ, t with μ appearing only as a parameter: $\bar{f} = \bar{f}(\mathbf{r}, \varepsilon, \mu, t)$. Clearly this is a substantial reduction in complexity as compared to the original Vlasov equation. Even so, the simplified kinetic MHD equation is still very complicated to solve in comparison to the ideal MHD fluid model. Nevertheless, it is possible, as shown in Chapter 10, to derive a general "stability comparison theorem" that allows one to compare the kinetic MHD and ideal MHD marginal stability boundaries.

9.6 The Chew, Goldberger, Low (CGL) double adiabatic model

Although the kinetic MHD model represents a substantial simplification of the Vlasov equation it is still quite difficult to solve theoretically or computationally because of the complicated single particle kinetic behavior parallel to the magnetic field. This is unfortunate since kinetic MHD is perhaps the most reliable model in terms of physics content for describing the MHD behavior of fusion plasmas.

These kinetic MHD difficulties have led to attempts at developing collisionless fluid closure models, almost a contradiction in terms. One such early model was developed by Chew, Goldberger, and Low (1956) and is often called the "CGL model" or, for reasons that will become apparent shortly, the "double adiabatic model." The goal in this section is to derive the original model plus a slightly modified form that is very useful when deriving stability comparison theorems.

The approach used consists of calculating the $mw_\perp^2/2$ and $mw_\parallel^2$ moments of the kinetic MHD equation. The result is a set of time evolution equations for $p_\perp$ and $p_\parallel$ in which, however, new unknown heat fluxes $h_\perp$ and $h_\parallel$ also appear. The system is closed by assuming that $h_\perp = h_\parallel = 0$. Neglect of the heat fluxes greatly simplifies the equations but cannot be justified by any rigorous mathematical or physical ordering procedure. The terms neglected are potentially of comparable or even larger size than those that are maintained. The lack of a sound physical basis supporting the model is a primary reason why it is not used very widely in fusion research.

Based on this observation one can ask whether it is actually worth the effort to derive the model. The answer is most definitely "yes" for the following reason. In spite of its lack of physical basis the simplicity of the double adiabatic model leads to a form of δW which together with the ideal MHD δW, bracket the marginal stability boundaries of kinetic MHD. That is, the hard-to-calculate but reliable kinetic MHD predictions are bracketed by two much simpler models, ideal MHD which is a valid model but does not satisfy the collisionless requirement, and double adiabatic MHD which is collisionless but is not a valid model.

The bracketing of kinetic MHD stability predictions has been known for many years (Kruskal and Oberman, 1958; Rosenbluth and Rostoker, 1959). Surprisingly, relatively little work has been carried out to quantitatively determine whether the bracketing gaps are wide or narrow. In many cases, as shown in Chapters 11 and 12, the gaps are indeed narrow, helping to explain why ideal MHD works as well as it does.

For this reason it makes sense to derive the double adiabatic MHD model. In Chapter 10 various forms of δW are derived which explicitly show the bracketing of stability predictions.

9.6.1 Formulation of the problem

The basic strategy to derive double adiabatic MHD involves the following steps: (1) formulate an alternate version of the kinetic MHD equation by transforming the velocity coordinates from ε, μ to $w_\perp$, $w_\parallel$; (2) calculate the $mw_\parallel^2$ moment; (3) calculate the $mw_\perp^2/2$ moment; and (4) close the system of equations.

As is shown, the transformation of velocity coordinates substantially simplifies the derivation. The complete transformation (for ions) is defined by

$$t' = t$$
$$\mathbf{r}' = \mathbf{r}$$
$$w_\perp^2 = \frac{2}{m_i}\mu B \tag{9.29}$$
$$w_\parallel^2 = \frac{2}{m_i}(\varepsilon - \mu B)$$

A straightforward calculation then shows that the derivatives appearing in the kinetic MHD equation are given by

$$\frac{\partial}{\partial t} = \frac{\partial}{\partial t'} + \frac{1}{2B}\frac{\partial B}{\partial t'}\left(w_\perp\frac{\partial}{\partial w_\perp} - \frac{w_\perp^2}{w_\parallel}\frac{\partial}{\partial w_\parallel}\right)$$
$$\nabla = \nabla' + \frac{1}{2B}\nabla'B\left(w_\perp\frac{\partial}{\partial w_\perp} - \frac{w_\perp^2}{w_\parallel}\frac{\partial}{\partial w_\parallel}\right) \tag{9.30}$$
$$\frac{\partial}{\partial\varepsilon} = \frac{1}{m_i w_\parallel}\frac{\partial}{\partial w_\parallel}$$

This transformation is substituted into Eq. (9.23). The result is the desired alternate but still complicated form of the kinetic MHD equation, which can be written as

$$\frac{\partial \bar{f}_i}{\partial t} + (\mathbf{v} + w_{\|}\mathbf{b}) \cdot \nabla \bar{f}_i + C_{\perp}\frac{\partial \bar{f}_i}{\partial w_{\perp}} + C_{\|}\frac{\partial \bar{f}_i}{\partial w_{\|}} = 0$$

$$C_{\perp} = \frac{w_{\perp}}{2B}\left(\frac{dB}{dt} + w_{\|}\mathbf{b}\cdot\nabla B\right) \tag{9.31}$$

$$C_{\|} = \frac{eE_{\|}}{m_i} - \mathbf{b}\cdot\frac{d\mathbf{v}}{dt} - \frac{w_{\perp}^2}{2B}\mathbf{b}\cdot\nabla B - w_{\|}\mathbf{b}\cdot(\mathbf{b}\cdot\nabla)\mathbf{v}$$

Here, for simplicity the primes have been suppressed from t', $\mathbf{r}'$ and use has been made of the following relationship from Faraday's law:

$$\mathbf{b}\cdot\left[\frac{\partial \mathbf{B}}{\partial t} - \nabla\times(\mathbf{v}\times\mathbf{B})\right] = 0 \quad\rightarrow\quad \frac{1}{B}\frac{dB}{dt} + \nabla\cdot\mathbf{v} - \mathbf{b}\cdot(\mathbf{b}\cdot\nabla)\mathbf{v} = 0 \quad (9.32)$$

An equation similar to Eq. (9.31) applies to the electrons.

The kinetic MHD equation is now in a convenient form to calculate the necessary moments.

9.6.2 Derivation of the double adiabatic model

The $m_i w_{\|}^2$ moment

The first required moment equation is obtained by multiplying Eq. (9.31) by $m_i w_{\|}^2$ and integrating over all velocities. To help carry out this calculation note that

$$d\mathbf{w} = 2\pi w_{\perp}dw_{\perp}dw_{\|}$$

$$\int w_{\|}\bar{f}_i\,d\mathbf{w} = 0 \quad (w_{\|} \text{ is a random velocity}) \tag{9.33}$$

Also, the various macroscopic moments that appear in the derivation are defined as

$$n = \int \bar{f}_i\,d\mathbf{w} \qquad\qquad \text{density}$$

$$p_{\perp i} = \int \frac{m_i w_{\perp}^2}{2}\bar{f}_i\,d\mathbf{w} \qquad \text{perpendicular pressure}$$

$$p_{\|i} = \int m_i w_{\|}^2 \bar{f}_i\,d\mathbf{w} \qquad \text{parallel pressure} \tag{9.34}$$

$$h_{\perp i} = \int \frac{m_i w_{\perp}^2}{2} w_{\|}\bar{f}_i\,d\mathbf{w} \quad \text{perpendicular heat flux}$$

$$h_{\|i} = \int m_i w_{\|}^3 \bar{f}_i\,d\mathbf{w} \qquad \text{parallel heat flux}$$

After some standard integration by parts the $m_i w_{\parallel}^2$ moment can be evaluated leading to

$$\frac{dp_{\parallel i}}{dt} + \left(\frac{2}{B}\frac{dB}{dt} + 3\nabla\cdot\mathbf{v}\right)p_{\parallel i} - 2\left(\frac{\mathbf{b}\cdot\nabla B}{B}\right)h_{\perp i} + B\mathbf{b}\cdot\nabla\left(\frac{h_{\parallel i}}{B}\right) = 0 \qquad (9.35)$$

In the last step the ion and electron contributions are combined (e.g., $p_{\perp} = p_{\perp i} + p_{\parallel i}$, etc.) and use is made of the following identity from the conservation of mass equation:

$$\frac{\partial n}{\partial t} + \nabla\cdot(n\mathbf{v}) \quad \rightarrow \quad \nabla\cdot\mathbf{v} = -\frac{1}{n}\frac{dn}{dt} \qquad (9.36)$$

A short calculation then shows that Eq. (9.35) can be rewritten as

$$\frac{n^3}{B^2}\frac{d}{dt}\left(\frac{p_{\parallel}B^2}{n^3}\right) = 2\left(\frac{\mathbf{b}\cdot\nabla B}{B}\right)h_{\perp} - B\mathbf{b}\cdot\nabla\left(\frac{h_{\parallel}}{B}\right) \qquad (9.37)$$

Equation (9.37) is exact in the context of kinetic MHD and represents the first of the required moment relations.

The $m_i w_{\perp}^2/2$ moment

The equation describing the time evolution of $p_{\perp}$ is obtained by multiplying Eq. (9.31) by $m_i w_{\perp}^2/2$ and integrating over all velocities. An analogous calculation to the one just described for $p_{\parallel}$ yields (for the ions)

$$\frac{dp_{\perp i}}{dt} - \left(\frac{1}{B}\frac{dB}{dt} - \nabla\cdot\mathbf{v}\right)p_{\perp i} + B^2\mathbf{b}\cdot\nabla\left(\frac{h_{\perp i}}{B^2}\right) = 0 \qquad (9.38)$$

One again eliminates $\nabla\cdot\mathbf{v}$ by means of Eq. (9.36) and combines the ion plus electron contributions. A short calculation leads to

$$nB\frac{d}{dt}\left(\frac{p_{\perp}}{nB}\right) = -B^2\mathbf{b}\cdot\nabla\left(\frac{h_{\perp}}{B^2}\right) \qquad (9.39)$$

This is the second of the required moment relations.

Closure of the moment equations

As stated previously the closure prescription leading to double adiabatic MHD consists of setting $h_{\perp} = h_{\parallel} = 0$. This assumption cannot be justified on the basis of an asymptotic expansion that exploits the smallness of some physical parameter. In fact from a pure dimensional analysis it follows that after making the usual assumption $m_e w_e^2 \sim m_i w_i^2 \sim T_e \sim T_i$, the electron heat fluxes are large by $(m_i/m_e)^{1/2}$ compared to the terms that are kept. The situation may not be bad as this would suggest since by symmetry the heat fluxes vanish for a pure Maxwellian distribution function. Still, the corrections to the Maxwellian can lead to comparably sized contributions to both the perpendicular and parallel pressure equations.

On the other hand, neglecting the heat fluxes does indeed greatly simplify the model. Plus some, if not all, of the collisionless anisotropic physics is correctly included in the model. In support of Chew, Goldberger, and Low one must appreciate that the double adiabatic model was derived during the very early days of the fusion program – it was important to learn how to walk before trying to run. Furthermore, although more reliable models such as kinetic MHD were also known since the early days of the program, these models were difficult to solve then and remain difficult to solve even now, including the large improvements in computing power.

In any event, following Chew, Goldberger, and Low, one makes the $h_\perp = h_\parallel = 0$ assumption in order to close the kinetic MHD moment equations. The final evolution equations for $p_\perp$ and $p_\parallel$ are thus given by

$$\frac{d}{dt}\left(\frac{p_\parallel B^2}{n^3}\right) = 0$$

$$\frac{d}{dt}\left(\frac{p_\perp}{nB}\right) = 0 \tag{9.40}$$

For obvious reasons the model is referred to as "double adiabatic MHD."

9.6.3 The modified double adiabatic model

As is shown in Chapter 10 the double adiabatic model needs to be slightly modified in order to obtain a sharply defined upper bound for the stability predictions of kinetic MHD. The modification, first suggested by Rosenbluth and Rostoker (1959), requires the replacement of the parallel component of the momentum equation with a simpler relation, namely that the fluid acceleration parallel to the magnetic field be set to zero. Mathematically, the replacement corresponds to

$$\rho\mathbf{b} \cdot \frac{d\mathbf{v}}{dt} + \mathbf{b} \cdot [\nabla \cdot (\mathbf{P}_e + \mathbf{P}_i)] = 0 \quad \rightarrow \quad \mathbf{b} \cdot \frac{d\mathbf{v}}{dt} = 0 \tag{9.41}$$

As with the basic model itself there is no mathematical or physical assumption that can justify this replacement. The justification is purely one of convenience. The model remains simple and leads to a sharply defined upper bound for stability.

Hereafter, whenever the double adiabatic model appears in the discussion or analysis, it is the modified form that is being considered.

9.7 Summary

Two alternate models describing MHD behavior have been derived: kinetic MHD and double adiabatic MHD. Both models are collisionless, their goal being to overcome the unrealistic high collisionality assumption required for the derivation of ideal MHD.

As a general comment kinetic MHD is a mathematically self-consistent, physically reliable model that is difficult to solve because of kinetic effects. In contrast the double adiabatic model cannot be justified on physical grounds but is far easier to solve because of its fluid nature. It is worth re-emphasizing that in spite of its shaky physical underpinnings, the double adiabatic model still serves a very useful purpose because of (1) its simplicity and (2) the fact that in conjunction with ideal MHD the combined stability predictions bracket those of the more difficult-to-calculate kinetic MHD model.

For convenience, the final models are summarized below. All of the models satisfy the same basic conservation of mass and momentum equations and differ only in how closure is achieved. The basic equations are given by

$$\frac{\partial n}{\partial t} + \nabla \cdot (n\mathbf{v}) = 0$$

$$\rho \frac{d\mathbf{v}}{dt} = \mathbf{J} \times \mathbf{B} - \nabla \cdot (\mathbf{P}_i + \mathbf{P}_e)$$

$$\frac{\partial \mathbf{B}}{\partial t} = \nabla \times (\mathbf{v} \times \mathbf{B}) \tag{9.42}$$

$$\nabla \times \mathbf{B} = \mu_0 \mathbf{J}$$

$$\nabla \cdot \mathbf{B} = 0$$

$$n_e = n_i \equiv n$$

The various methods of closure are as follows.

9.7.1 Ideal MHD

In ideal MHD closure is achieved by writing the pressure tensor in terms of a single scalar pressure

$$\mathbf{P} = \mathbf{P}_i + \mathbf{P}_e = p\mathbf{I} \tag{9.43}$$

where p satisfies

$$\frac{d}{dt}\left(\frac{p}{n^\gamma}\right) = 0 \tag{9.44}$$

9.7.2 Kinetic MHD

As its name implies, in kinetic MHD the ions and electrons are treated kinetically. Their corresponding distribution functions satisfy

Ions:
$$\frac{\partial \bar{f}_i}{\partial t} + \left(\mathbf{v} + w_\| \mathbf{b}\right) \cdot \nabla \bar{f}_i + \bar{\varepsilon}\, \frac{\partial \bar{f}_i}{\partial \varepsilon} = 0$$

$$\bar{\varepsilon} = e w_\| E_\| - m_i w_\| \mathbf{b} \cdot \left(\frac{\partial}{\partial t} + \mathbf{v} \cdot \nabla\right)\mathbf{v} - \frac{m_i w_\perp^2}{2} \nabla \cdot \mathbf{v} - m_i \left(w_\|^2 - \frac{w_\perp^2}{2}\right)\mathbf{b} \cdot (\mathbf{b} \cdot \nabla)\mathbf{v}$$

$$(9.45)$$

Electrons:
$$\frac{\partial \bar{f}_e}{\partial t} + \left(\mathbf{v} + w_\| \mathbf{b}\right) \cdot \nabla \bar{f}_e + \bar{\varepsilon}\, \frac{\partial \bar{f}_e}{\partial \varepsilon} = 0$$

$$\bar{\varepsilon} = -e w_\| E_\| - m_e w_\| \mathbf{b} \cdot \left(\frac{\partial}{\partial t} + \mathbf{v} \cdot \nabla\right)\mathbf{v} - \frac{m_e w_\perp^2}{2} \nabla \cdot \mathbf{v} - m_e \left(w_\|^2 - \frac{w_\perp^2}{2}\right)\mathbf{b} \cdot (\mathbf{b} \cdot \nabla)\mathbf{v}$$

$$(9.46)$$

Here, if each species is denoted by $\alpha = e, i$, then $\bar{f}_\alpha = \bar{f}_\alpha(\mathbf{r}, \varepsilon, \mu, t)$ with $\varepsilon = (m_\alpha/2)(w_\perp^2 + w_\|^2)$, $\mu = m_\alpha w_\perp^2/2B$, and $\mathbf{v}_e \approx \mathbf{v}_i \equiv \mathbf{v}$ (the macroscopic fluid velocity).

Closure

Knowing $\bar{f}_i$ and $\bar{f}_e$ one then calculates (for $\alpha = i, e$)

$$\mathbf{P}_\alpha = p_{\perp\alpha}\mathbf{I} + \left(p_{\|\alpha} - p_{\perp\alpha}\right)\mathbf{b}\mathbf{b}$$

$$p_{\perp\alpha} = \int \frac{m_\alpha w_\perp^2}{2} \bar{f}_\alpha \, d\mathbf{w} = \int \mu B \bar{f}_\alpha \, d\mathbf{w}$$

$$p_{\|\alpha} = \int m_\alpha w_\|^2 \bar{f}_\alpha \, d\mathbf{w} = \int 2(\varepsilon - \mu B) \bar{f}_\alpha \, d\mathbf{w} \qquad (9.47)$$

$$n_\alpha = \int \bar{f}_\alpha \, d\mathbf{w}$$

9.7.3 The double adiabatic model

Closure of the modified double adiabatic model requires two separate energy equations, one for $p_\perp$ and the other for $p_\|$. The closure procedure is given by

$$\mathbf{P} = \mathbf{P}_i + \mathbf{P}_e = p_\perp \mathbf{I} + \left(p_\| - p_\perp\right)\mathbf{b}\mathbf{b} \qquad (9.48)$$

where $p_\|$ and $p_\perp$ satisfy

$$\frac{d}{dt}\left(\frac{p_\| B^2}{n^3}\right) = 0$$

$$\frac{d}{dt}\left(\frac{p_\perp}{nB}\right) = 0 \qquad (9.49)$$

The momentum equation is slightly modified from the general form given in Eq. (9.42). The double adiabatic momentum equation is given by

$$\rho \mathbf{b} \times \left(\frac{d\mathbf{v}}{dt} \times \mathbf{b} \right) = \mathbf{J} \times \mathbf{B} - \nabla_\perp p_\perp - (p_\parallel - p_\perp)\boldsymbol{\kappa} \quad \text{perpendicular to } \mathbf{B}$$

$$\mathbf{b} \cdot \frac{d\mathbf{v}}{dt} = 0 \qquad\qquad\qquad \text{parallel to } \mathbf{B}$$

(9.50)

This completes the summary of the models.

References

Braginskii, S.I. (1965). In *Reviews of Plasma Physics*, Vol. 1, ed. M.A. Leontovich. New York: Consultants Bureau.
Chew, G.F., Goldberger, M.L., and Low, F.E. (1956). *Proc. Royal Society (London)* **A236**, 112.
Hazeltine, R.D. and Waelbroeck, F.L. (1998). *The Framework of Plasma Physics*. Reading, MA: Perseus Books.
Kruskal, M.D. and Oberman, C.R. (1958). *Phys. Fluids* **1**, 275.
Rosenbluth, M.N. and Rostoker, N. (1959). *Phys. Fluids* **2**, 23.

Further reading

Hazeltine, R.D. and Meiss, J.D. (2003). *Plasma Confinement*. Redwood City, CA: Addison-Wesley.
Krall, N.A. and Trivelpiece, A.W. (1973). *Principles of Plasma Physics*. New York: McGraw-Hill.
Kulsrud, R.M. (1983). In *Handbook of Plasma Physics*, Vol. 1. Amsterdam: North Holland.

Problems

The first three problems are aimed at understanding the modifications to the dispersion relation for waves in an infinite homogeneous plasma resulting from the new physics included in the alternate MHD models. Towards this goal assume that for each of the models under consideration the plasma equilibrium is given by

$$\mathbf{B} = B_0 \mathbf{e}_z$$
$$n = n_0$$
$$p = p_0$$
$$\mathbf{v} = 0$$

where B_0, n_0, and p_0 are constants. Small amplitude waves are now allowed to propagate in the plasma with all perturbations varying as $Q(\mathbf{r}, t) = \hat{Q}\exp(-i\omega t + \mathbf{k} \cdot \mathbf{r})$. Here, $\hat{Q}$ is a constant and

$$\mathbf{k} = k_\perp \mathbf{e}_y + k_\parallel \mathbf{e}_z$$

Derive the dispersion relation governing wave propagation for the alternate MHD models listed below and compare the results with those of ideal MHD by plotting ω vs. $k_\parallel$ for each of the three branches of the dispersion relation. Are any of the modes damped?

9.1 The original double adiabatic model.

9.2 The modified double adiabatic model.

9.3 The kinetic MHD model.

9.4 Repeat Problems 9.1 and 9.2 for the case of an anisotropic equilibrium using the double adiabatic model. Specifically assume that the equilibrium pressure is characterized by $p_\perp \neq p_\parallel$. Show that with sufficient anisotropy the plasma can become unstable. Derive the condition for stability.

9.5 Repeat Problem 9.4 for the kinetic MHD model.

9.6 Derive the general non-linear conservation of energy relation (analogous to Eq. (3.23)) for the double adiabatic model.

9.7 Derive Eq. (9.15).

9.8 Derive Eq. (9.18).

9.9 Derive Eq. (9.35).

10

MHD stability comparison theorems

10.1 Introduction

Three models have been introduced to investigate the MHD equilibrium and stability properties of a general multidimensional magnetic fusion configuration: ideal MHD, kinetic MHD, and double adiabatic MHD. Ideal MHD is by far the most widely used model although there is concern since the collision dominated assumption used in the derivation is not satisfied in fusion-grade plasmas. The collisionless kinetic MHD model provides the most reliable description of the physics but is difficult to solve in realistic geometries because of the complex kinetic behavior parallel to the magnetic field. Double adiabatic MHD is a collisionless fluid model that is much easier to solve than kinetic MHD but the closure assumptions cannot be justified by any rigorous mathematical or physical arguments.

Based on this assessment one sees that the situation is not very satisfactory from a theoretical point of view. In practice, ideal MHD, because of its mathematical simplicity, is the model that is most widely used to design, predict, and interpret fusion experiments. Many years of experience have shown, perhaps surprisingly, that the model is far more accurate and reliable than one might have anticipated.

Chapter 10 attempts to provide a partial theoretical explanation for this apparent good fortune. The goal is accomplished by examining two basic aspects of MHD behavior. The first issue is MHD equilibrium. The analysis shows that for the special choice of isotropic pressure (i.e., $p_\perp = p_\parallel = p$), both kinetic MHD and double adiabatic MHD lead to the same equilibrium equations as ideal MHD. Therefore, when comparing the stability predictions of each model one is starting with identical equilibria. This is an important property that helps makes the comparisons meaningful.

The second issue is MHD stability. In the analysis, equivalent forms of δW are derived for kinetic MHD and double adiabatic MHD which are then compared to the ideal MHD δW. By focusing on marginal stability boundaries (as opposed to growth rates) it is shown that there is a well-defined hierarchy of stability predictions with ideal MHD being the most pessimistic, double adiabatic MHD being the most optimistic, and kinetic MHD always lying in between. Furthermore, detailed comparisons in Chapters 11 and 12 show that the gaps in stability predictions between ideal and double adiabatic MHD are often not very large. The closeness of the stability predictions, which bracket the intermediate predictions of kinetic MHD, provides a good explanation of why ideal MHD works as well as it does.

The analysis in Chapter 10 is presented as follows. (1) The ideal MHD stability predictions are briefly reviewed in order to establish a point of reference. (2) An Energy Principle is derived for double adiabatic MHD. This is straightforward and closely follows the corresponding derivation for ideal MHD. (3) An Energy Relation is derived for kinetic MHD. This requires a complicated calculation and distinctions must be made between cylindrical and toroidal configurations as well as between ergodic and closed line magnetic geometries. The kinetic MHD stability operator is not self-adjoint (because of kinetic effects) so that only an Energy Relation and not an Energy Principle can be derived. Still, the required stability information can be extracted from the Energy Relation. (4) The results from all three models are collected from which the desired hierarchy of stability predictions can then be easily obtained.

10.2 Ideal MHD equilibrium and stability

Ideal MHD serves as the reference case. The relevant results have been derived in Chapter 8 and are summarized as follows. In ideal MHD any equilibrium must satisfy

$$\mathbf{J} \times \mathbf{B} = \nabla p$$
$$\nabla \times \mathbf{B} = \mu_0 \mathbf{J} \tag{10.1}$$
$$\nabla \cdot \mathbf{B} = 0$$

Stability can be described in terms of the variational integral (Bernstein *et al.*, 1958),

$$\omega^2 = \frac{\delta W_{MHD}}{K_{MHD}} \tag{10.2}$$

The subscript "MHD" denotes expressions corresponding to ideal MHD. Also, for convenience in deriving the comparison theorems, it is useful to separate δW_{MHD} and K_{MHD} into contributions that depend solely on $\xi_\perp$ and those that contain $\xi_\parallel$. Therefore, K_{MHD} is given by

$$K_{MHD}(\xi^*, \xi) = K_\perp + K_\parallel$$

$$K_\perp(\xi_\perp^*, \xi_\perp) = \frac{1}{2} \int \rho |\xi_\perp|^2 \, d\mathbf{r}$$

$$K_\parallel(\xi_\parallel^*, \xi_\parallel) = \frac{1}{2} \int \rho |\xi_\parallel|^2 \, d\mathbf{r}$$

(10.3)

Similarly δW_{MHD} can be written in terms of the "standard" form as follows:

$$\delta W_{MHD}(\xi^*, \xi) = \delta W_\perp + \delta W_C$$

$$\delta W_\perp(\xi_\perp^*, \xi_\perp) = \frac{1}{2\mu_0} \int \left[|\mathbf{Q}|^2 - \mu_0 \xi_\perp^* \cdot \mathbf{J} \times \mathbf{Q} + \mu_0(\xi_\perp \cdot \nabla p) \nabla \cdot \xi_\perp^* \right] d\mathbf{r}$$

$$+ \delta W_S + \delta W_V$$

$$\delta W_S(\xi_\perp^*, \xi_\perp) + \delta W_V(\xi_\perp^*, \xi_\perp) = \frac{1}{2\mu_0} \int (\mathbf{n} \cdot \xi_\perp^*)(\mathbf{B} \cdot \mathbf{Q} + \mu_0 p_1) \, dS$$

(10.4)

$$\delta W_C(\xi^*, \xi) = \frac{1}{2} \int \frac{5}{3} p |\nabla \cdot \xi|^2 \, d\mathbf{r}$$

where δW_C is the contribution due to plasma compressibility and γ, the ratio of specific heats, has been set to $\gamma = 5/3$. As has been shown in Chapter 8 the standard form of δW_{MHD} can, after a series of algebraic manipulations, be rewritten in a self-adjoint form.

The Energy Principle states that a necessary and sufficient condition for MHD stability is that

$$\delta W_{MHD}(\xi^*, \xi) \geq 0$$

(10.5)

for all allowable displacements. A similar set of relations is now derived for double adiabatic MHD and kinetic MHD which then sets the stage to deduce quantitative comparison theorems.

10.3 Double adiabatic MHD equilibrium and stability

As previously stated, in order to make a fair stability comparison with ideal MHD one must, as a special case, choose the double adiabatic MHD equilibrium pressures to be isotropic: $p_\perp = p_\parallel = p$. Under this assumption the double adiabatic pressure tensor reduces to $\mathbf{P} = p_\perp \mathbf{I} + (p_\parallel - p_\perp)\mathbf{bb} = p\mathbf{I}$ and $\nabla \cdot \mathbf{P} = \nabla p$. Clearly, equilibria are then identical in both models. Even so, it is important to keep in mind that the perturbed pressures in double adiabatic MHD are anisotropic.

Consider next the formulation of the stability problem (Kruskal and Oberman, 1958; Rosenbluth and Rostoker, 1959). The analysis closely follows the procedure used for ideal MHD in Chapter 8 with only slightly more algebra. The details are given below.

10.3.1 Relation between the perturbed quantities and ξ

The analysis is again carried out in terms of the perturbed plasma displacement $\boldsymbol{\xi}$ defined as $\mathbf{v}_1 = -i\omega\boldsymbol{\xi}$. First note that for the modified double adiabatic model, the parallel component of the momentum equation, $\mathbf{b} \cdot (d\mathbf{v}/dt) = 0$, reduces to

$$\xi_{\parallel} = 0 \tag{10.6}$$

Second, from Eq. (9.42) it follows that the perturbed density, magnetic field, and current are identical to ideal MHD with only a slight modification in n_1 (i.e., $\nabla \cdot \boldsymbol{\xi} \rightarrow \nabla \cdot \boldsymbol{\xi}_{\perp}$):

$$\begin{aligned}
n_1 &= -\boldsymbol{\xi}_{\perp} \cdot \nabla n - n\nabla \cdot \boldsymbol{\xi}_{\perp} \\
\mathbf{B}_1 &= \nabla \times (\boldsymbol{\xi}_{\perp} \times \mathbf{B}) \\
\mu_0 \mathbf{J}_1 &= \nabla \times [\nabla \times (\boldsymbol{\xi}_{\perp} \times \mathbf{B})]
\end{aligned} \tag{10.7}$$

Also needed for the double adiabatic analysis is an expression for B_1 which is easily obtained from Eq. (10.7),

$$\begin{aligned}
B_1 &= \mathbf{b} \cdot \mathbf{B}_1 = B\,\mathbf{b} \cdot (\mathbf{b} \cdot \nabla)\boldsymbol{\xi}_{\perp} - \mathbf{b} \cdot (\boldsymbol{\xi}_{\perp} \cdot \nabla)\mathbf{B} - B\nabla \cdot \boldsymbol{\xi}_{\perp} \\
&= -B\boldsymbol{\xi}_{\perp} \cdot \boldsymbol{\kappa} - \boldsymbol{\xi}_{\perp} \cdot \nabla B - B\nabla \cdot \boldsymbol{\xi}_{\perp}
\end{aligned} \tag{10.8}$$

The remaining quantities required are the perturbed pressures which are derived from Eq. (9.49). For the parallel pressure the non-linear adiabatic energy equation can be rewritten as

$$\frac{d}{dt}\left(\frac{p_{\parallel}B^2}{n^3}\right) = 0 \quad \rightarrow \quad \frac{1}{p_{\parallel}}\frac{dp_{\parallel}}{dt} + \frac{2}{B}\frac{dB}{dt} - \frac{3}{n}\frac{dn}{dt} = 0 \tag{10.9}$$

Each of these quantities is now linearized: $Q = Q_0 + Q_1$. Using the facts that $dQ_0/dt = 0$ and $(dQ/dt)_1 = -i\omega(Q_1 + \boldsymbol{\xi}_{\perp} \cdot \nabla Q_0)$ one finds that

$$\frac{1}{p}\left(\frac{dp_{\parallel}}{dt}\right)_1 + \frac{2}{B}\left(\frac{dB}{dt}\right)_1 - \frac{3}{n}\left(\frac{dn}{dt}\right)_1 = 0 \tag{10.10}$$

where

$$\begin{aligned}
\left(\frac{dn}{dt}\right)_1 &= -i\omega(n_1 + \boldsymbol{\xi}_{\perp} \cdot \nabla n) = i\omega n\nabla \cdot \boldsymbol{\xi}_{\perp} \\
\left(\frac{dB}{dt}\right)_1 &= -i\omega(B_1 + \boldsymbol{\xi}_{\perp} \cdot \nabla B) = i\omega B(\nabla \cdot \boldsymbol{\xi}_{\perp} + \boldsymbol{\xi}_{\perp} \cdot \boldsymbol{\kappa}) \\
\left(\frac{dp_{\parallel}}{dt}\right)_1 &= -i\omega\left(p_{\parallel 1} + \boldsymbol{\xi}_{\perp} \cdot \nabla p\right)
\end{aligned} \tag{10.11}$$

Straightforward substitution then yields

$$p_{\|1} = -\boldsymbol{\xi}_\perp \cdot \nabla p - p\nabla \cdot \boldsymbol{\xi}_\perp + 2p\,\boldsymbol{\xi}_\perp \cdot \boldsymbol{\kappa} \tag{10.12}$$

Similarly, the non-linear relation for the perpendicular pressure can be rewritten as

$$\frac{d}{dt}\left(\frac{p_\perp}{nB}\right) = 0 \;\longrightarrow\; \frac{1}{p_\perp}\frac{dp_\perp}{dt} - \frac{1}{B}\frac{dB}{dt} - \frac{1}{n}\frac{dn}{dt} = 0 \tag{10.13}$$

Linearization leads to an expression for $p_{\perp 1}$ given by

$$p_{\perp 1} = -\boldsymbol{\xi}_\perp \cdot \nabla p - 2p\nabla \cdot \boldsymbol{\xi}_\perp - p\,\boldsymbol{\xi}_\perp \cdot \boldsymbol{\kappa} \tag{10.14}$$

An examination of Eqs. (10.12) and (10.14) indeed shows that the perturbed pressures are anisotropic.

10.3.2 The double adiabatic MHD Energy Principle

All of the perturbed quantities have now been expressed in terms of $\boldsymbol{\xi}_\perp$. These expressions are to be substituted into the linearized form of the double adiabatic perpendicular momentum equation obtained from Eq. (9.42),

$$-\omega^2\rho\boldsymbol{\xi}_\perp = \mathbf{J}_1 \times \mathbf{B} + \mathbf{J} \times \mathbf{Q} - \nabla_\perp p_{\perp 1} - (p_{\|1} - p_{\perp 1})\boldsymbol{\kappa} \tag{10.15}$$

The next step is to multiply this expression by $\boldsymbol{\xi}_\perp^*$ and integrate over the plasma volume leading to

$$\omega^2 = \frac{\delta W_{CGL}}{K_{CGL}} \tag{10.16}$$

where

$$\delta W_{CGL} = -\frac{1}{2}\int \boldsymbol{\xi}_\perp^* \cdot \left[\mathbf{J}_1 \times \mathbf{B} + \mathbf{J} \times \mathbf{Q} - \nabla_\perp p_{\perp 1} - (p_{\|1} - p_{\perp 1})\boldsymbol{\kappa}\right] d\mathbf{r}$$

$$K_{CGL} = K_\perp = \frac{1}{2}\int \rho\, |\boldsymbol{\xi}_\perp|^2 d\mathbf{r} \tag{10.17}$$

As with ideal MHD, several algebraic manipulations and integrations by parts are required to recast δW_{CGL} in a form suitable for comparisons. The steps are summarized below.

• **The $\boldsymbol{\xi}_\perp^* \cdot \mathbf{J}_1 \times \mathbf{B}$ term**

$$-\int \boldsymbol{\xi}_\perp^* \cdot \mathbf{J}_1 \times \mathbf{B}\, d\mathbf{r} = \frac{1}{\mu_0}\int |\mathbf{Q}|^2 \, d\mathbf{r} + \frac{1}{\mu_0}\int (\mathbf{n} \cdot \boldsymbol{\xi}_\perp^*)(\mathbf{B} \cdot \mathbf{Q})\, dS \tag{10.18}$$

• **The $\xi_\perp^* \cdot \nabla_\perp p_{\perp 1}$ term**

$$\int \xi_\perp^* \cdot \nabla_\perp p_{\perp 1} \, d\mathbf{r} = \int \left[2p|\nabla \cdot \xi_\perp|^2 - \xi_\perp^* \cdot \nabla(\xi_\perp \cdot \nabla p + p\,\xi_\perp \cdot \boldsymbol{\kappa}) \right] d\mathbf{r} \\ - \int (\mathbf{n} \cdot \xi_\perp^*)(2p\nabla \cdot \xi_\perp) \, dS$$

(10.19)

Several terms are now integrated by parts.

• **The $\xi_\perp^* \cdot \nabla(\xi_\perp \cdot \nabla p)$ term**

$$-\int \xi_\perp^* \cdot \nabla(\xi_\perp \cdot \nabla p) \, d\mathbf{r} = \int (\nabla \cdot \xi_\perp^*)(\xi_\perp \cdot \nabla p) \, d\mathbf{r} - \int (\mathbf{n} \cdot \xi_\perp^*)(\xi_\perp \cdot \nabla p) \, dS$$

(10.20)

• **The $\xi_\perp^* \cdot \nabla(p\xi_\perp \cdot \boldsymbol{\kappa})$ term**

$$-\int \xi_\perp^* \cdot \nabla(p\xi_\perp \cdot \boldsymbol{\kappa}) \, d\mathbf{r} = \int (\nabla \cdot \xi_\perp^*)(p\xi_\perp \cdot \boldsymbol{\kappa}) \, d\mathbf{r} - \int (\mathbf{n} \cdot \xi_\perp^*)(p\xi_\perp \cdot \boldsymbol{\kappa}) \, dS$$

(10.21)

The contributions from these steps are combined leading to an expression for δW_{CGL} that can be written as

$$\delta W_{CGL} = \frac{1}{2\mu_0} \int \left[|\mathbf{Q}|^2 + 2\mu_0 p|\nabla \cdot \xi_\perp|^2 - \mu_0 \xi_\perp^* \cdot \mathbf{J} \times \mathbf{Q} + \mu_0 (\nabla \cdot \xi_\perp^*)(\xi_\perp \cdot \nabla p) \right. \\ \left. + \mu_0 p(\nabla \cdot \xi_\perp^*)(\xi_\perp \cdot \boldsymbol{\kappa}) + \mu_0 (p_{\|1} - p_{\perp 1})(\xi_\perp^* \cdot \boldsymbol{\kappa}) \right] d\mathbf{r} + BT$$

(10.22)

Here, the boundary term BT is given by

$$BT = \frac{1}{2\mu_0} \int (\mathbf{n} \cdot \xi_\perp^*)(\mathbf{B} \cdot \mathbf{Q} - \mu_0 \xi_\perp \cdot \nabla p - 2\mu_0 p\nabla \cdot \xi_\perp - \mu_0 p\xi_\perp \cdot \boldsymbol{\kappa}) \, dS \\ = \frac{1}{2\mu_0} \int (\mathbf{n} \cdot \xi_\perp^*)(\mathbf{B} \cdot \mathbf{Q} + \mu_0 p_{\perp 1}) \, dS$$

(10.23)

Observe that BT is in the identical form as for ideal MHD as given by Eq. (8.80) or equivalently Eq. (10.4). Thus, one is again allowed to use the natural boundary condition analysis described in Section 8.8. The result is that

$$BT = \delta W_S + \delta W_V$$

(10.24)

where δW_S and δW_V have the same form as the ideal MHD surface and vacuum energies summarized in Eqs. (8.101) and (8.102).

The last two terms in the fluid contribution to δW_{CGL} are simplified by substituting for $p_{\|1}$, $p_{\perp 1}$ from Eqs. (10.12) and (10.14). A short calculation yields

$$p(\nabla \cdot \boldsymbol{\xi}_\perp^*)(\boldsymbol{\xi}_\perp \cdot \boldsymbol{\kappa}) + (p_{\|1} - p_{\perp 1})(\boldsymbol{\xi}_\perp^* \cdot \boldsymbol{\kappa}) = \frac{p}{3}|\nabla \cdot \boldsymbol{\xi}_\perp + 3\boldsymbol{\xi}_\perp \cdot \boldsymbol{\kappa}|^2 - \frac{p}{3}|\nabla \cdot \boldsymbol{\xi}_\perp|^2$$

(10.25)

Equation (10.25) is substituted into Eq. (10.22) leading to an expression for δW_{CGL} that can be written as

$$\delta W_{CGL} = \frac{1}{2\mu_0} \int \left[|\mathbf{Q}|^2 + \frac{5}{3}\mu_0 p|\nabla \cdot \boldsymbol{\xi}_\perp|^2 - \mu_0 \boldsymbol{\xi}_\perp^* \cdot \mathbf{J} \times \mathbf{Q} + \mu_0(\nabla \cdot \boldsymbol{\xi}_\perp^*)(\boldsymbol{\xi}_\perp \cdot \nabla p) \right.$$

$$\left. + \frac{\mu_0 p}{3}|\nabla \cdot \boldsymbol{\xi}_\perp + 3\boldsymbol{\xi}_\perp \cdot \boldsymbol{\kappa}|^2 \right] d\mathbf{r} + \delta W_S + \delta W_V \qquad (10.26)$$

The final step is to rewrite this equation as

$$\delta W_{CGL} = \delta W_\perp + \delta Q_{CGL} \qquad (10.27)$$

where

$$\delta Q_{CGL} = \frac{1}{2} \int \left[\frac{5}{3}p|\nabla \cdot \boldsymbol{\xi}_\perp|^2 + \frac{p}{3}|\nabla \cdot \boldsymbol{\xi}_\perp + 3\boldsymbol{\xi}_\perp \cdot \boldsymbol{\kappa}|^2 \right] d\mathbf{r} \qquad (10.28)$$

Equation (10.27) represents the desired potential energy relation for double adiabatic MHD.

10.3.3 Summary of CGL stability

To summarize, the variational formulation of the double adiabatic model, originally given by Eq. (10.16), can now be rewritten as

$$\omega^2 = \frac{\delta W_{CGL}}{K_{CGL}} = \frac{\delta W_\perp + \delta Q_{CGL}}{K_\perp} \qquad (10.29)$$

The quantity δQ_{CGL} is the modification to the potential energy due to the plasma compressibility and pressure anisotropy. Since δW_{CGL} is self-adjoint by construction, then an Energy Principle also applies. Specifically $\delta W_{CGL}(\boldsymbol{\xi}^*, \boldsymbol{\xi}) \geq 0$ for all allowable displacements is a necessary and sufficient condition for stability in the modified double adiabatic model.

10.4 Kinetic MHD equilibrium and stability

The goal in this subsection is to derive an Energy Relation describing the stability of the kinetic MHD model (Kruskal and Oberman, 1958; Rosenbluth and Rostoker, 1959). The analysis for general geometry is complicated and lengthy requiring many steps and several clever mathematical tricks. The strategy to reach the final goal has three parts. The first part discusses general equilibrium and is simple and straightforward. The second part focuses on systems with cylindrical symmetry. The symmetry leads to a closed form analytic solution for the perturbed distribution function from which much insight can be obtained. In carrying out this analysis it is necessary to distinguish between ergodic and closed line systems. Overall, the cylindrical part of the analysis is not overly complicated.

The third part of the analysis describes the stability of general toroidal systems where again one must distinguish between ergodic and closed line configurations. The part of the formulation that leads to a general Energy Relation is straightforward although lengthy. It is described in the main text. A considerably more complicated analysis is required to obtain upper and lower bounds on certain terms appearing in the Energy Relation that are needed to make comparisons with the other models. The corresponding details are presented in Appendix H.

It is worth noting at the outset that in order to obtain analytic results for the general toroidal case one must restrict attention to the marginal stability limit $\omega \to 0$. The reason is as follows. Since kinetic MHD is not a self-adjoint model because of the presence of resonant particles it is by no means obvious that $\omega_r = 0$ when $\omega_i \to 0$ at marginal stability; that is, resonant particles lead to Landau damping which in turn leads to a complex dispersion relation. The analysis shows that this damping vanishes when $\omega_r = 0$. Therefore, only by focusing on $\omega \to 0$ does the analysis become sufficiently simplified that it is possible to obtain the necessary upper and lower bounds. The end result is a kinetic MHD Energy Relation from which one can deduce stability comparison theorems.

10.4.1 Equilibrium

To compare kinetic MHD stability with ideal MHD and double adiabatic MHD one again needs to focus on isotropic equilibria. In the kinetic MHD model this is easily accomplished for general geometry by choosing the equilibrium distribution functions for each species to be of the form

$$\bar{f}_0(\mathbf{r}, \varepsilon, \mu) = \bar{f}_0(\varepsilon, \psi) \tag{10.30}$$

where $\psi(\mathbf{r})$ is the equilibrium flux function satisfying $\mathbf{B} \cdot \nabla \psi = 0$. Here and below the species subscript α is suppressed for simplicity except where explicitly needed. This distribution function automatically satisfies the non-linear kinetic MHD equation for $\bar{f}$ and leads to

$$p_{\perp 0}(\mathbf{r}) = p_{\parallel 0}(\mathbf{r}) = p_0(\psi) \tag{10.31}$$

which is identical to ideal MHD. Also, the equilibrium parallel electric field vanishes since

$$E_{\parallel 0} = -\frac{\mathbf{b}_0 \cdot (\nabla \cdot \mathbf{P}_{0e})}{en_0} = -\frac{\mathbf{b}_0 \cdot \nabla p_{0e}(\psi)}{en_0} = -\frac{dp_{0e}}{d\psi}\frac{\mathbf{b}_0 \cdot \nabla \psi}{en_0} = 0 \tag{10.32}$$

The equilibria are isotropic with a one-to-one correspondence to ideal MHD.

10.4.2 Stability of a closed line cylindrical system

The first stability problem considered corresponds to a closed line cylindrically symmetric system subject to perturbations which maintain the closed line symmetry. In practical terms the configuration of interest is a pure Z-pinch subject to an $m = 0$ perturbation where $\boldsymbol{\xi}(\mathbf{r}) = \boldsymbol{\xi}(r)\exp(im\theta + ikz)$. This is the simplest problem to study. The reason is that, as a result of the special symmetries involved, the perturbed kinetic MHD distribution function has only fluid-like contributions – all kinetic effects cancel.

The procedure used to obtain the stability results is to (1) calculate the perturbed distribution function $\bar{f}_1$, (2) from $\bar{f}_1$ calculate the perturbed parallel and perpendicular pressures $p_{\parallel 1}$ and $p_{\perp 1}$, and (3) use $p_{\parallel 1}$ and $p_{\perp 1}$ to obtain the kinetic MHD Energy Relation. The analysis proceeds as follows.

The linearized stability equations

For the Z-pinch it is convenient to solve for the perturbed distribution functions in terms of the general $w_\perp$, $w_\parallel$ formulation of the kinetic MHD equations given by Eq. (9.31) and repeated here for convenience (for a species with mass m and charge q):

$$\frac{\partial \bar{f}}{\partial t} + (\mathbf{v} + w_\parallel \mathbf{b}) \cdot \nabla \bar{f} + C_\perp \frac{\partial \bar{f}}{\partial w_\perp} + C_\parallel \frac{\partial \bar{f}}{\partial w_\parallel} = 0$$

$$C_\perp = \frac{w_\perp}{2B}\left(\frac{dB}{dt} + w_\parallel \mathbf{b} \cdot \nabla B\right) \tag{10.33}$$

$$C_\parallel = \frac{qE_\parallel}{m} - \mathbf{b} \cdot \frac{d\mathbf{v}}{dt} - \frac{w_\perp^2}{2B}\mathbf{b} \cdot \nabla B - w_\parallel \mathbf{b} \cdot (\mathbf{b} \cdot \nabla)\mathbf{v}$$

Here, $d/dt = \partial/\partial t + \mathbf{v} \cdot \nabla$.

Equation (10.33) can be easily linearized for generalized geometry by again using the relations $\mathbf{v}_1 = -i\omega\boldsymbol{\xi}$ and $\mathbf{B}_1 = \nabla \times (\boldsymbol{\xi}_\perp \times \mathbf{B})$. The result is

$$-i\omega \bar{f}_1 + w_\parallel \mathbf{b}_0 \cdot \nabla \bar{f}_1 + (-i\omega \boldsymbol{\xi} + w_\parallel \mathbf{b}_1) \cdot \nabla \bar{f}_0 + C_{\perp 1} \frac{\partial \bar{f}_0}{\partial w_\perp} + C_{\parallel 1} \frac{\partial \bar{f}_0}{\partial w_\parallel} = 0$$

$$C_{\perp 1} = \frac{w_\perp}{2B} \left[\left(\frac{dB}{dt} \right)_1 + w_\parallel (\mathbf{b} \cdot \nabla B)_1 \right] \tag{10.34}$$

$$C_{\parallel 1} = \frac{qE_{\parallel 1}}{m} - \omega^2 \xi_\parallel - \frac{w_\perp^2}{2B} (\mathbf{b} \cdot \nabla B)_1 + i\omega w_\parallel \mathbf{b} \cdot (\mathbf{b} \cdot \nabla) \boldsymbol{\xi}$$

These complicated expressions can be simplified as follows. First, by making use of the fact that $\bar{f}_0 = \bar{f}_0(w_\perp^2 + w_\parallel^2, \psi)$ one finds that

$$(\mathbf{b} \cdot \nabla B)_1 \left(w_\parallel w_\perp \frac{\partial \bar{f}_0}{\partial w_\perp} - w_\perp^2 \frac{\partial \bar{f}_0}{\partial w_\parallel} \right) = 0$$

$$\mathbf{b}_1 \cdot \nabla \bar{f}_0 = \frac{1}{B_0} \nabla \cdot [(\boldsymbol{\xi}_\perp \times \mathbf{B}_0) \times \nabla \bar{f}_0] = \mathbf{b}_0 \cdot \nabla(\boldsymbol{\xi}_\perp \cdot \nabla \bar{f}_0) \tag{10.35}$$

Second, $E_{\parallel 1}$ can be simplified by noting that the total electric field in the plasma can be written in two different but equivalent forms:

$$\begin{aligned} \mathbf{E}_1 &= i\omega \boldsymbol{\xi}_\perp \times \mathbf{B}_0 + E_{\parallel 1} \mathbf{b}_0 \\ &= -\nabla \phi_1 + i\omega \mathbf{A}_{\perp 1} + i\omega A_{\parallel 1} \mathbf{b}_0 \end{aligned} \tag{10.36}$$

Flux coordinates ψ, χ, l are now introduced (as discussed in Chapter 7), where $\mathbf{B}_0 = \nabla \psi \times \nabla \chi$ and $\nabla l = \mathbf{b}_0$. The perpendicular and parallel components of Eq. (10.36) are equated leading to

$$i\omega \boldsymbol{\xi}_\perp \times \mathbf{B}_0 = -\frac{\partial \phi_1}{\partial \psi} \nabla \psi - \frac{\partial \phi_1}{\partial \chi} \nabla \chi + i\omega \mathbf{A}_{\perp 1} \tag{10.37}$$

$$E_{\parallel 1} = -\mathbf{b}_0 \cdot \nabla \phi_1 + i\omega A_{\parallel 1}$$

Equation (10.37) is valid for all ω including $\omega \to 0$. Taking this limit implies that ϕ_1 must be of the form $\phi_1 = \bar{\phi}_1(l) + i\omega \tilde{\phi}_1(\psi, \chi, l)$. Furthermore, $\bar{\phi}_1(l)$ must vanish since it has an incompatible spatial dependence with respect to the mode under consideration. For example, for the $m = 0$ mode in a Z-pinch, $\boldsymbol{\xi} = \boldsymbol{\xi}(r) \exp(ikz)$ while $\bar{\phi}_1 = \bar{\phi}_1(\theta)$. Lastly, the $\tilde{\phi}_1$ contribution can easily be incorporated into $A_{\parallel 1}$ by means of a gauge transformation. The end result is that $E_{\parallel 1}$ can be written as

$$E_{\parallel 1} = i\omega A_{\parallel 1} \tag{10.38}$$

which is the desired simplification.

The last step is to write $\boldsymbol{\xi} = \boldsymbol{\xi}_\perp + \xi_\parallel \mathbf{b}_0$ and to then collect all the $\xi_\parallel$ terms. After substituting all of the simplifications into Eq. (10.34) one obtains a correspondingly reduced form of the kinetic MHD equations,

$$(-i\omega + w_\| \mathbf{b} \cdot \nabla)\left(\bar{f}_1 + \boldsymbol{\xi}_\perp \cdot \nabla \bar{f} + i\omega \xi_\| \frac{\partial \bar{f}}{\partial w_\|}\right) + i\omega\left(\hat{C}_{\perp 1} \frac{\partial \bar{f}}{\partial w_\perp} + \hat{C}_{\|1} \frac{\partial \bar{f}}{\partial w_\|}\right) = 0$$

$$\hat{C}_{\perp 1} = \frac{w_\perp}{2}(\nabla \cdot \boldsymbol{\xi}_\perp + \boldsymbol{\xi}_\perp \cdot \boldsymbol{\kappa})$$

$$\hat{C}_{\|1} = \frac{qA_{\|1}}{m} - w_\| \boldsymbol{\xi}_\perp \cdot \boldsymbol{\kappa}$$

$$(10.39)$$

Here, the "0" subscript has been suppressed from all equilibrium quantities. At this point Eq. (10.39) is still valid for general geometry.

Solution for the m = 0 *mode in a Z-pinch*

In a cylindrical Z-pinch the perturbations can be Fourier analyzed in θ and z: $\boldsymbol{\xi} = \boldsymbol{\xi}(r)\exp(im\theta + ikz)$. For the $m = 0$ mode the closed line symmetry of the equilibrium is preserved since $B_{1r} = B_{1z} = 0$. Both the equilibrium and perturbed magnetic fields point only in the θ direction. The $m = 0$ mode is thus the mode of interest in the present subsection.

Focusing on $m = 0$ in a cylindrical geometry greatly simplifies the analysis. Specifically, for this mode the operator $\mathbf{b} \cdot \nabla S = imS/r = 0$ where S is any scalar quantity. The kinetic MHD equation for each species reduces to a simple algebraic equation for $\bar{f}_{1\alpha}$ whose solution is given by

$$\bar{f}_{1\alpha} = -\boldsymbol{\xi}_\perp \cdot \nabla \bar{f}_\alpha - i\omega\xi_\| \frac{\partial \bar{f}_\alpha}{\partial w_\|} + \frac{w_\perp}{2}(\nabla \cdot \boldsymbol{\xi}_\perp + \boldsymbol{\xi}_\perp \cdot \boldsymbol{\kappa})\frac{\partial \bar{f}_\alpha}{\partial w_\perp} + \left(\frac{q_\alpha A_\|}{m_\alpha} - w_\| \boldsymbol{\xi}_\perp \cdot \boldsymbol{\kappa}\right)\frac{\partial \bar{f}_\alpha}{\partial w_\|}$$

$$(10.40)$$

Knowing $\bar{f}_{1\alpha}$ it is then straightforward to calculate the perturbed pressures. A short calculation yields

$$p_{\|1} = \sum_\alpha \tilde{p}_{\|\alpha} = \sum_\alpha \int m_\alpha w_\|^2 \bar{f}_{1\alpha}\, d\mathbf{w} = -\boldsymbol{\xi}_\perp \cdot \nabla p + 2p\boldsymbol{\xi}_\perp \cdot \boldsymbol{\kappa} - p\nabla \cdot \boldsymbol{\xi}_\perp$$

$$p_{\perp 1} = \sum_\alpha \tilde{p}_{\perp\alpha} = \sum_\alpha \int \frac{m_\alpha w_\perp^2}{2} \bar{f}_{1\alpha}\, d\mathbf{w} = -\boldsymbol{\xi}_\perp \cdot \nabla p - p\boldsymbol{\xi}_\perp \cdot \boldsymbol{\kappa} - 2p\nabla \cdot \boldsymbol{\xi}_\perp$$

$$(10.41)$$

where $\tilde{p}_{\perp\alpha}, \tilde{p}_{\|\alpha}$ are the perturbed pressures for each species and $p = p_i + p_e$. Note that the contributions from $\xi_\|$ and $A_{\|1}$ vanish because they appear in terms with odd symmetry in $w_\|$. The most interesting observation is that the pressures appearing in Eq. (10.41) are identical to those calculated in the modified double adiabatic model as given by Eqs. (10.12) and (10.14).

The Z-pinch Energy Relation

Since the pressures in both models are identical, so then are the stability formulations. The conclusion is that for the $m = 0$ mode in a Z-pinch the Energy Relation for kinetic MHD is actually an Energy Principle given by

$$\omega^2 = \frac{\delta W_{KIN}}{K_{KIN}} = \frac{\delta W_\perp + \delta Q_{KIN}}{K_\perp} \tag{10.42}$$

where

$$\delta Q_{KIN} = \delta Q_{CGL} = \frac{1}{2} \int \left[\frac{5}{3} p |\nabla \cdot \boldsymbol{\xi}_\perp|^2 + \frac{p}{3} |\nabla \cdot \boldsymbol{\xi}_\perp + 3 \boldsymbol{\xi}_\perp \cdot \boldsymbol{\kappa}|^2 \right] d\mathbf{r} \tag{10.43}$$

Since δW_{KIN} is self-adjoint it follows that the stability transition occurs through $\omega^2 = 0$. Detailed comparisons with other models and modes are made in Section 10.5 once the other Energy Relations have been derived.

10.4.3 Stability of an ergodic cylindrical system

The second configuration of interest is a cylindrical system in which all the magnetic lines are ergodic, covering the entire flux surface on which they lie. Isolated rational surfaces are allowed but do not change the final results. The configuration to be studied thus corresponds to a general screw pinch. The analysis also applies to a closed line Z-pinch subject to perturbations that break the closed line symmetry. In other words, modes with $m \geq 1$.

The derivation of the screw pinch Energy Relation is similar to that of the $m = 0$ mode in a Z-pinch in that a purely analytic closed form solution is obtained for the perturbed distribution function. However, the resulting parallel and perpendicular pressures are far more complicated because of the effects of resonant particles, a distinctly kinetic effect. It is the presence of resonant particles that ultimately leads to a close relation between kinetic MHD and ideal MHD.

The analysis presented here is greatly simplified by focusing, at the appropriate point, on the behavior near marginal stability which occurs when $\omega \to 0$. In kinetic MHD, like ideal MHD, the real part of the frequency at marginal stability satisfies $\omega_r = 0$. It is only when $\omega_r = 0$ that the Landau damping due to resonant particles vanishes, which thereby provides an automatic solution to the imaginary part of the dispersion relation.

As for the $m = 0$ mode in a Z-pinch the analysis for the screw pinch consists of (1) calculating the perturbed distribution function, (2) calculating the perturbed pressures, and (3) deriving the corresponding kinetic MHD Energy Relation.

Solution for the perturbed distribution function in a general screw pinch

The perturbed distribution function satisfies the general linearized kinetic MHD equation given by Eq. (10.39). Also, the modes under consideration are described by a displacement vector of the form $\boldsymbol{\xi} = \boldsymbol{\xi}(r) \exp(im\theta + ikz)$. For such modes the operator $\mathbf{b} \cdot \nabla$ acting on any scalar S is given by

$$\mathbf{b} \cdot \nabla S = ik_{\parallel} S \tag{10.44}$$

where

$$k_{\parallel}(r) = \frac{1}{B} \left(\frac{m}{r} B_{\theta} + kB_z \right) \tag{10.45}$$

In general, $k_{\parallel}(r) \neq 0$ except perhaps on isolated surfaces.

Substituting Eq. (10.44) into Eq. (10.39) leads to an analytic expression for the perturbed distribution function,

$$\bar{f}_1 = -\boldsymbol{\xi}_{\perp} \cdot \nabla \bar{f} - i\omega \xi_{\parallel} \frac{\partial \bar{f}}{\partial w_{\parallel}} + \frac{\omega}{\omega - k_{\parallel} w_{\parallel}} \left(\hat{C}_{\perp 1} \frac{\partial \bar{f}}{\partial w_{\perp}} + \hat{C}_{\parallel 1} \frac{\partial \bar{f}}{\partial w_{\parallel}} \right)$$

$$\hat{C}_{\perp 1} = \frac{w_{\perp}}{2} (\nabla \cdot \boldsymbol{\xi}_{\perp} + \boldsymbol{\xi}_{\perp} \cdot \boldsymbol{\kappa}) \tag{10.46}$$

$$\hat{C}_{\parallel 1} = \frac{qA_{\parallel 1}}{m} - w_{\parallel} \boldsymbol{\xi}_{\perp} \cdot \boldsymbol{\kappa}$$

where again the species subscript α has been suppressed. Note the presence of a resonant denominator $\omega - k_{\parallel} w_{\parallel}$ in the last term in $\bar{f}_1$. Specifically, those particles with a parallel velocity $w_{\parallel} = \omega_r/k_{\parallel}$ produce Landau damping thereby generating a complex dispersion relation.

It is at this point that focusing on marginal stability, $\omega \to 0$, greatly simplifies the analysis. The reason is as follows. As stated, the resonant particle effects are contained in the terms multiplied by $\omega/(\omega - k_{\parallel} w_{\parallel})$ in Eq. (10.46). For the $m = 0$ mode in a Z-pinch, $k_{\parallel} = 0$ and $\omega/(\omega - k_{\parallel} w_{\parallel}) \to 1$. The resonance is actually not a resonance, but instead leads to fluid-like contributions. In contrast, for the screw pinch $k_{\parallel}(r) \neq 0$. In this case $\omega/(\omega - k_{\parallel} w_{\parallel}) \to 0$ as $\omega \to 0$. As stated, kinetic effects, Landau damping specifically, vanish when $\omega \to 0$. This automatically guarantees that the imaginary part of the complex dispersion relation is satisfied at marginal stability. The conclusion is that for each species, $\bar{f}_{1\alpha}$ at marginal stability only consists of a simple fluid-like term

$$\bar{f}_{1\alpha} = -\boldsymbol{\xi}_{\perp} \cdot \nabla \bar{f}_{\alpha} \qquad \omega \to 0 \tag{10.47}$$

The perturbed pressures

At marginal stability it is straightforward to calculate the perturbed pressures by taking appropriate moments of $\bar{f}_{1a}$. After summing over species the results are given by

$$p_{\|1} = \sum_\alpha \int m_\alpha w_\|^2 \bar{f}_{1a}\, d\mathbf{w} = -\boldsymbol{\xi}_\perp \cdot \nabla p$$

$$p_{\perp 1} = \sum_\alpha \int \frac{m_\alpha w_\perp^2}{2} \bar{f}_{1a}\, d\mathbf{w} = -\boldsymbol{\xi}_\perp \cdot \nabla p \qquad (10.48)$$

where again $p = p_i + p_e$. The perturbed pressures are quite simple. They are isotropic and describe simple convection with the fluid.

For comparison the exact pressures, valid for arbitrary ω in the limit $m_e \to 0$ are, after a lengthy calculation, given by

$$p_{\|1} = -\boldsymbol{\xi}_\perp \cdot \nabla p - p_i (\nabla \cdot \boldsymbol{\xi}_\perp)(\zeta_i^2 Z_i)' + 2 p_i (\boldsymbol{\xi}_\perp \cdot \boldsymbol{\kappa}) \zeta_i^3 (\zeta_i Z_i)' - (enV_{Ti}A_{\|1})(\zeta_i Z_i')$$

$$p_{\perp 1} = -\boldsymbol{\xi}_\perp \cdot \nabla p + 2 p_i (\nabla \cdot \boldsymbol{\xi}_\perp)(\zeta_i Z_i) + p_i (\boldsymbol{\xi}_\perp \cdot \boldsymbol{\kappa})(\zeta_i^2 Z_i)' + (enV_{Ti}A_{\|1})(\zeta_i^2 Z_i)$$

$$enV_{Ti}A_{\|1} = \frac{2 p_e}{2 T_i - T_e Z_i'} \left[(\nabla \cdot \boldsymbol{\xi}_\perp) Z_i + (\boldsymbol{\xi}_\perp \cdot \boldsymbol{\kappa})(\zeta_i Z_i)' \right]$$

$$(10.49)$$

Here, $\zeta_i = \omega/k_\| V_{Ti}$ and prime denotes $d/d\zeta_i$. Equation (10.49) assumes that the equilibrium distribution functions are Maxwellian,

$$\bar{f}_\alpha = \frac{n}{\pi^{3/2} V_{T\alpha}^3} \exp\left(-\frac{w_\perp^2 + w_\|^2}{V_{T\alpha}^2} \right) \qquad V_{T\alpha}^2 = 2 T_\alpha / m_\alpha \qquad (10.50)$$

which allows one to express the results in terms of the well-known plasma dispersion function $Z = Z(\zeta)$ (Stix, 1992)

$$Z(\zeta) = \frac{1}{\pi^{1/2}} \int_{-\infty}^{\infty} \frac{dz}{z - \zeta} \qquad (10.51)$$

Also Z' is related to Z by $Z' = -2(1 + \zeta Z)$.

The kinetic terms are not derived here since their details are not explicitly needed for the stability comparison theorems. Still there are several points to be made. First and foremost one sees how complicated kinetic MHD is compared to ideal MHD and double adiabatic MHD, even for the simple cylindrical geometry where an analytic solution is known for $\bar{f}_{1a}$. These complications are a primary motivation for attempting to bracket kinetic MHD stability results with simpler fluid predictions. Second, in spite of their complexity one can easily show that the real and imaginary parts of all kinetic terms in the pressures, those that contain Z functions, vanish in the limit $\omega \propto \zeta \to 0$, and the simpler results given by Eq. (10.48) are all that remain.

The cylindrical screw pinch Energy Relation

The kinetic MHD potential energy can essentially be determined by inspection because of the simple relations for $p_{\|1}$ and $p_{\perp 1}$. Recall that the quantity needed to close the system of equations is

$$\nabla p_{\perp 1} + (p_{\|1} - p_{\perp 1})\boldsymbol{\kappa} + \mathbf{bB} \cdot \nabla \left(\frac{p_{\|1} - p_{\perp 1}}{B} \right) \tag{10.52}$$

In the limit of marginal stability $p_{\|1} = p_{\perp 1}$ and $p_{\perp 1} = -\boldsymbol{\xi}_\perp \cdot \nabla p$. Therefore, Eq. (10.52) reduces to

$$\nabla p_{\perp 1} + (p_{\|1} - p_{\perp 1})\boldsymbol{\kappa} + \mathbf{bB} \cdot \nabla \left(\frac{p_{\|1} - p_{\perp 1}}{B} \right) = -\nabla(\boldsymbol{\xi}_\perp \cdot \nabla p) \tag{10.53}$$

This expression should be compared to the corresponding term in ideal MHD

$$\nabla p_1 = -\nabla(\boldsymbol{\xi}_\perp \cdot \nabla p + \gamma p \nabla \cdot \boldsymbol{\xi}) \tag{10.54}$$

Now, for ergodic systems the $\xi_\|$ that minimizes δW for ideal MHD is the one that sets $\nabla \cdot \boldsymbol{\xi} = 0$. Equation (10.54) thus simplifies to

$$\nabla p_1 = -\nabla(\boldsymbol{\xi}_\perp \cdot \nabla p) \tag{10.55}$$

which is identical to the kinetic MHD result given by Eq. (10.53).

The conclusion is that marginal stability for a screw pinch is identical in the ideal MHD and kinetic MHD models:

$$\begin{aligned} \delta W_{MHD} &= \delta W_\perp \\ \delta W_{KIN} &= \delta W_\perp \end{aligned} \tag{10.56}$$

implying that

$$\delta W_{KIN} = \delta W_{MHD} \tag{10.57}$$

This is the desired relation.

10.4.4 Stability of a general toroidal configuration

The insight obtained from the cylindrical analysis sets the stage to investigate stability in general toroidal configurations. The analysis is rather involved, requiring multiple steps and several clever mathematical manipulations. The end result is an Energy Relation valid for arbitrary geometry and arbitrary frequency. From this relation it is then possible to develop stability comparisons between kinetic MHD, ideal MHD, and double adiabatic MHD.

The analysis separates into two parts. The first part is straightforward but lengthy and is presented below. It leads to an Energy Relation that consists of an ideal MHD contribution plus a positive definite contribution resulting from kinetic effects. The kinetic effects result from resonant particles, trapped particles, and particles which do not uniformly sample the entire flux surface because of toroidicity.

The second part of the analysis involves the calculation of upper and lower bounds on the kinetic contribution which are required in order to be able to make stability comparisons with the other models. Here, the details are quite complicated and are presented in Appendix H.

The analysis is carried out by reverting back to the kinetic MHD equations written in terms of ε, μ rather than $w_\parallel, w_\perp$. Although the general Energy Relation is valid for arbitrary ω, it is necessary to take the marginal stability limit $\omega \to 0$ in order to obtain the upper and lower bounds on the kinetic contribution.

The derivation of the Energy Relation presented in this subsection requires two steps. First, an expression is derived for the perturbed distribution function. The expression is formal in nature since it involves an integral along the unperturbed guiding center orbits which in general cannot be analytically evaluated for a 3-D geometry. Fortunately the analytic solutions are not actually required. Second, a quadratic relation is formed from the perpendicular components of the momentum equation. Setting the real and imaginary parts of this relation to zero yields the kinetic MHD Energy Relation.

The analysis proceeds as follows.

Derivation of the perturbed distribution function

The starting point for the derivation is the non-linear kinetic MHD equation given by Eq. (9.23), expressed in ε, μ coordinates. The equation is repeated here for convenience for a species with mass m and charge q:

$$\frac{\partial \bar{f}}{\partial t} + \left(\mathbf{v} + w_\parallel \mathbf{b}\right) \cdot \nabla \bar{f} + \bar{\varepsilon} \frac{\partial \bar{f}}{\partial \varepsilon} = 0$$

$$\bar{\varepsilon} = q w_\parallel E_\parallel - m w_\parallel \mathbf{b} \cdot \left(\frac{\partial}{\partial t} + \mathbf{v} \cdot \nabla\right)\mathbf{v} - \frac{m w_\perp^2}{2} \nabla \cdot \mathbf{v} - m\left(w_\parallel^2 - \frac{w_\perp^2}{2}\right) \mathbf{b} \cdot (\mathbf{b} \cdot \nabla)\mathbf{v}$$

$$(10.58)$$

Readers should keep in mind that in Eq. (10.58), $w_\perp$, $w_\parallel$ must be expressed in terms of ε, μ since these are the independent variables. Specifically, one must set $w_\perp^2 = (2B/m)\mu$ and $w_\parallel^2 = (2/m)(\varepsilon - \mu B)$.

The analysis begins by linearizing Eq. (10.58) using the relations $\mathbf{v}_1 = -i\omega\boldsymbol{\xi}$ and $\mathbf{b}_1 \cdot \nabla \bar{f}_0 = \mathbf{b}_0 \cdot \nabla(\boldsymbol{\xi}_\perp \cdot \nabla \bar{f}_0)$. Also, all the $\xi_\parallel$ terms are collected together. A short calculation yields a complicated differential equation for the perturbed distribution function $\bar{f}_1$,

$$\frac{D}{Dt}\left(\bar{f}_1 + \boldsymbol{\xi}_\perp \cdot \nabla\bar{f}\right) + \bar{\varepsilon}_1 \frac{\partial\bar{f}}{\partial\varepsilon} = 0$$

$$\bar{\varepsilon}_1 = i\omega\left[qw_\parallel A_{\parallel 1} + \frac{mw_\perp^2}{2}\nabla\cdot\boldsymbol{\xi}_\perp + m\left(\frac{w_\perp^2}{2} - w_\parallel^2\right)(\boldsymbol{\xi}_\perp \cdot \boldsymbol{\kappa})\right]$$

$$+ i\omega m\left[-i\omega w_\parallel\xi_\parallel + \frac{w_\perp^2}{2}\mathbf{B}\cdot\nabla\left(\frac{\xi_\parallel}{B}\right) - \left(\frac{w_\perp^2}{2} - w_\parallel^2\right)(\mathbf{b}\cdot\nabla\xi_\parallel)\right] \qquad (10.59)$$

Here, all zero subscripts have been suppressed from equilibrium quantities and

$$\frac{DQ_1}{Dt} = \left(-i\omega + w_\parallel\mathbf{b}\cdot\nabla\right)Q_1 \qquad (10.60)$$

is the derivative moving with the unperturbed parallel velocity of the particle. This operator should be distinguished from

$$\left(\frac{dQ}{dt}\right)_1 = \left(\frac{\partial Q}{\partial t} + \mathbf{v}\cdot\nabla Q\right)_1 = -i\omega(Q_1 + \boldsymbol{\xi}_\perp\cdot\nabla Q_0) \qquad (10.61)$$

which is the linearized derivative moving with the fluid velocity. The $\xi_\parallel$ terms can be simplified as follows:

$$Q_\parallel = i\omega m\left[-i\omega w_\parallel\xi_\parallel + \frac{w_\perp^2}{2}\mathbf{B}\cdot\nabla\left(\frac{\xi_\parallel}{B}\right) - \left(\frac{w_\perp^2}{2} - w_\parallel^2\right)(\mathbf{b}\cdot\nabla\xi_\parallel)\right]$$

$$= i\omega\left[mw_\parallel\frac{D\xi_\parallel}{Dt} - \mu\xi_\parallel\mathbf{b}\cdot\nabla B\right]$$

$$= i\omega\left[\frac{D}{Dt}\left(mw_\parallel\xi_\parallel\right) - m\xi_\parallel\frac{Dw_\parallel}{Dt} - \mu\xi_\parallel\mathbf{b}\cdot\nabla B\right] \qquad (10.62)$$

$$= i\omega\frac{D}{Dt}\left(mw_\parallel\xi_\parallel\right)$$

Substituting Eq. (10.62) into Eq. (10.58) leads to a simplified equation for the perturbed distribution function given by

$$\frac{D}{Dt}\left(\bar{f}_1 + \boldsymbol{\xi}_\perp\cdot\nabla\bar{f} + i\omega mw_\parallel\xi_\parallel\frac{\partial\bar{f}}{\partial\varepsilon}\right) + \bar{\varepsilon}_{\perp1}\frac{\partial\bar{f}}{\partial\varepsilon} = 0$$

$$\bar{\varepsilon}_{\perp1} = i\omega\left[qw_\parallel A_{\parallel 1} + \frac{mw_\perp^2}{2}\nabla\cdot\boldsymbol{\xi}_\perp + m\left(\frac{w_\perp^2}{2} - w_\parallel^2\right)(\boldsymbol{\xi}_\perp \cdot \boldsymbol{\kappa})\right] \qquad (10.63)$$

The formal solution for $\bar{f}_1$ is obtained by integrating Eq. (10.63) along the characteristics (i.e., the unperturbed orbits)

$$\bar{f}_1 = -\boldsymbol{\xi}_\perp \cdot \nabla \bar{f} - i\omega m w_\parallel \xi_\parallel \frac{\partial \bar{f}}{\partial \varepsilon} - i\omega \frac{\partial \bar{f}}{\partial \varepsilon} s \tag{10.64}$$

where the first terms are the fluid-like contributions and

$$s = \int_{-\infty}^{t} \left[q w_\parallel A_{\parallel 1} + (m w_\perp^2/2) \nabla \cdot \boldsymbol{\xi}_\perp + m \left(w_\perp^2/2 - w_\parallel^2 \right)(\boldsymbol{\xi}_\perp \cdot \boldsymbol{\kappa}) \right] dt' \tag{10.65}$$

is the kinetic contribution.

Two points are worth discussing in connection with the evaluation of $\bar{f}_1$. First, the mathematics is simplified by assuming that the spatial variation of the solution is expressed in terms of flux coordinates ψ, χ, l as described in Chapter 7. Recall that in flux coordinates $\mathbf{B} = \nabla \psi \times \nabla \chi$ and $\mathbf{b} \cdot \nabla l = 1$. Then, the integrand in Eq. (10.65) is of the form

$$I' = e^{-i\omega t'} I[\varepsilon, \mu, \psi, \chi, l(t')] \tag{10.66}$$

Here, $l(t') \equiv l'$ represents the parallel motion of a particle along its unperturbed orbit and is found, in principle, by solving

$$\frac{dl'}{dt'} = w_\parallel' = \pm(2/m)^{1/2}[\varepsilon - \mu B(\psi, \chi, l')]^{1/2} \tag{10.67}$$

$$l'(t' = t) = l$$

Note that the "initial" condition corresponds to setting l' equal to its present value l when $t' = t$. The advantage of flux coordinates is that ψ, χ, ε, μ are all constants along the unperturbed orbit which in turn implies that they are constants with respect to the t' integration in Eq. (10.65). The equation for the unperturbed orbits given by Eq. (10.67) is a first order non-linear ordinary differential equation that must be solved for each particle for the given $B(\psi, \chi, l')$. Fortunately, only Eq. (10.67), but not its solution, is required for the analysis.

The second point of interest concerns the lower limit of the time integration in s which has been set to $t' = -\infty$. Observe that there are no contributions to $\bar{f}_1$ or $\boldsymbol{\xi}_\perp$ at $t' = -\infty$ in Eq. (10.64). This requires that $\omega_i = \text{Im}(\omega) > 0$ and corresponds to taking the Laplace transform. Obviously $e^{-i\omega t'} \to 0$ as $t' = -\infty$ when $\omega_i > 0$. An equivalent form for $s = s(\varepsilon, \mu, \psi, \chi, l, t)$ in terms of the usual Laplace transform representation is obtained by letting $t' = t - \tau$ in Eqs. (10.65) and (10.67) where τ is the new integration variable. One finds

$$s = e^{-i\omega t} \int_0^\infty e^{i\omega \tau} I[l'(\tau)] \, d\tau$$

$$= e^{-i\omega t} \int_0^\infty e^{i\omega \tau} \left[q w_\parallel A_{\parallel 1} + (m w_\perp^2/2) \nabla \cdot \boldsymbol{\xi}_\perp + m \left(w_\perp^2/2 - w_\parallel^2 \right)(\boldsymbol{\xi}_\perp \cdot \boldsymbol{\kappa}) \right] d\tau$$

$$\tag{10.68}$$

and

$$\frac{dl'}{d\tau} = -w'_\| = \mp(2/m)^{1/2}[\varepsilon - \mu B(\psi,\chi,l')]^{1/2}$$

$$l'(0) = l \tag{10.69}$$

It is important to keep in mind that $\omega_i > 0$ is assumed in the derivation of the Energy Relation as this will play a major role in interpreting the kinetic MHD stability results.

Derivation of the general Energy Relation

Having obtained a formal expression for the perturbed distribution function one can now proceed to derive the general Energy Relation for kinetic MHD. The starting point is again the perpendicular components of the linearized momentum equation given by

$$-\omega^2 \rho \boldsymbol{\xi}_\perp = \mathbf{J}_1 \times \mathbf{B} + \mathbf{J} \times \mathbf{Q} - \nabla_\perp p_{\perp 1} - (p_{\|1} - p_{\perp 1})\boldsymbol{\kappa} \tag{10.70}$$

Each of the perturbed pressures consists of two contributions, one from the fluid-like contributions in $\bar{f}_1$ and the other from the kinetic contributions. The fluid-like contributions can be easily evaluated by a straightforward integration. For each species this leads to

$$(p_{\perp 1})_\alpha \equiv \tilde{p}_{\perp \alpha} = -\boldsymbol{\xi}_\perp \cdot \nabla p_\alpha + \hat{p}_{\perp \alpha}$$

$$(p_{\|1})_\alpha \equiv \tilde{p}_{\|\alpha} = -\boldsymbol{\xi}_\perp \cdot \nabla p_\alpha + \hat{p}_{\|\alpha} \tag{10.71}$$

where the $\boldsymbol{\xi}_\perp \cdot \nabla p_\alpha$ terms are the fluid contributions and

$$\hat{p}_{\perp \alpha} = -i\omega \int \frac{m_\alpha w_\perp^2}{2} \frac{\partial \bar{f}_\alpha}{\partial \varepsilon} s_\alpha \, d\mathbf{w}$$

$$\hat{p}_{\|\alpha} = -i\omega \int m_\alpha w_\|^2 \frac{\partial \bar{f}_\alpha}{\partial \varepsilon} s_\alpha \, d\mathbf{w} \tag{10.72}$$

are the kinetic contributions. Observe that the contribution from $\xi_\|$ again vanishes because of odd symmetry in $w_\|$.

Next, form the dot product of Eq. (10.70) with $\boldsymbol{\xi}_\perp^*$, integrate over the plasma volume, and sum over species. Some standard integrations by parts yield

$$\omega^2 K_\perp = \delta W_\perp + \delta W_{CK}$$

$$\delta W_\perp = \delta W_S + \delta W_V + \frac{1}{2\mu_0} \int \left[|\mathbf{Q}|^2 - \mu_0 \boldsymbol{\xi}_\perp^* \cdot \mathbf{J} \times \mathbf{Q} + \mu_0 (\boldsymbol{\xi}_\perp \cdot \nabla p)\nabla \cdot \boldsymbol{\xi}_\perp^* \right] d\mathbf{r}$$

$$\delta W_{CK} = -\frac{1}{2} \sum_\alpha \int \left[\hat{p}_{\perp \alpha} \nabla \cdot \boldsymbol{\xi}_\perp^* + (\hat{p}_{\perp \alpha} - \hat{p}_{\|\alpha})(\boldsymbol{\xi}_\perp^* \cdot \boldsymbol{\kappa}) \right] d\mathbf{r}$$

$$\tag{10.73}$$

Here, $\delta W_{\perp}(\xi_{\perp}^{*}, \xi_{\perp})$ is identical to the term in the ideal MHD $\delta W(\xi^{*}, \xi)$ that depends only on $\xi_{\perp}$. See Eq. (10.4). The quantity $\delta W_{CK}(\xi_{\perp}^{*}, \xi_{\perp})$ represents the "compressibility" contribution to the potential energy due to kinetic effects. It ultimately must be compared to the corresponding $\delta W_C(\xi^{*}, \xi)$ term in the ideal MHD $\delta W(\xi^{*}, \xi)$.

The analysis continues by simplifying the expression for $\delta W_{CK}(\xi_{\perp}^{*}, \xi_{\perp})$. Several steps are required. To begin substitute the expressions for $\hat{p}_{\perp\alpha}, \hat{p}_{\parallel\alpha}$,

$$\delta W_{CK} = \frac{i\omega}{2} \sum_{\alpha} \int \frac{\partial \bar{f}_{\alpha}}{\partial \varepsilon} s_{\alpha} \left[(m_{\alpha} w_{\perp}^{2}/2)(\nabla \cdot \xi_{\perp}^{*}) + m_{\alpha}(w_{\perp}^{2}/2 - w_{\parallel}^{2})(\xi_{\perp}^{*} \cdot \kappa) \right] d\mathbf{w}\, d\mathbf{r}$$

$$(10.74)$$

Now, note that from the definition of s_{α} one can write

$$\left(\frac{Ds_{\alpha}}{Dt} \right)^{*} = q_{\alpha} w_{\parallel} A_{\parallel 1}^{*} + (m_{\alpha} w_{\perp}^{2}/2)(\nabla \cdot \xi^{*}) + m_{\alpha}(w_{\perp}^{2}/2 - w_{\parallel}^{2})(\xi_{\perp}^{*} \cdot \kappa) \quad (10.75)$$

which allows Eq. (10.74) to be rewritten as

$$\delta W_{CK} = \frac{i\omega}{2} \sum_{\alpha} \int \frac{\partial \bar{f}_{\alpha}}{\partial \varepsilon} s_{\alpha} \left[\left(\frac{Ds_{\alpha}}{Dt} \right)^{*} - q_{\alpha} w_{\parallel} A_{\parallel 1}^{*} \right] d\mathbf{w}\, d\mathbf{r} \qquad (10.76)$$

The $A_{\parallel 1}^{*}$ term integrates to zero. This can be seen by evaluating the $w_{\parallel}$ moment of $\bar{f}_{1}$ which must be zero because $w_{\parallel}$ is the random velocity. For each species one finds

$$0 = q_{\alpha} \int w_{\parallel} \bar{f}_{1\alpha}\, d\mathbf{w} = i\omega q_{\alpha} n_{\alpha} \xi_{\parallel} - i\omega q_{\alpha} \int \frac{\partial \bar{f}_{\alpha}}{\partial \varepsilon} w_{\parallel} s_{\alpha}\, d\mathbf{w} \qquad (10.77)$$

Now sum over species and make use of the charge neutrality condition

$$0 = i\omega \xi_{\parallel} \sum_{\alpha} q_{\alpha} n_{\alpha}$$

$$= i\omega \sum_{\alpha} q_{\alpha} \int \frac{\partial \bar{f}_{\alpha}}{\partial \varepsilon} w_{\parallel} s_{\alpha} d\mathbf{w} \qquad (10.78)$$

The velocity integral in the second line of Eq. (10.78) is identical to the one multiplying $A_{\parallel 1}^{*}$ in Eq. (10.76). Therefore, as stated, the $A_{\parallel 1}^{*}$ contribution vanishes and Eq. (10.76) reduces to

$$\delta W_{CK} = \frac{i\omega}{2} \sum_{\alpha} \int \frac{\partial \bar{f}_{\alpha}}{\partial \varepsilon} s_{\alpha} \left(\frac{Ds_{\alpha}}{Dt} \right)^{*} d\mathbf{w}\, d\mathbf{r} \qquad (10.79)$$

The next step is to note that

$$\left(\frac{Ds_{\alpha}}{Dt} \right)^{*} = (-i\omega s_{\alpha} + w_{\parallel} \mathbf{b} \cdot \nabla s_{\alpha})^{*} = i\omega^{*} s_{\alpha}^{*} + w_{\parallel} \mathbf{b} \cdot \nabla s_{\alpha}^{*} \qquad (10.80)$$

Substituting into Eq. (10.79) leads to

$$\delta W_{CK} = \delta Q_{KIN} + \frac{i\omega}{2} \sum_\alpha \int \frac{\partial \bar{f}_\alpha}{\partial \varepsilon} s_\alpha \left(w_\parallel \mathbf{b} \cdot \nabla s_\alpha^* \right) d\mathbf{w} d\mathbf{r}$$

$$\delta Q_{KIN} = -\frac{|\omega|^2}{2} \sum_\alpha \int \frac{\partial \bar{f}_\alpha}{\partial \varepsilon} |s_\alpha|^2 d\mathbf{w} d\mathbf{r}$$

(10.81)

The integral in the top relation in Eq. (10.81) can be simplified by writing $s = s_r + i s_i$ for each species. The terms containing s can then be rewritten as

$$s(\mathbf{b} \cdot \nabla s^*) = \frac{1}{2} \mathbf{b} \cdot \nabla |s|^2 + i(s_i \mathbf{b} \cdot \nabla s_r - s_r \mathbf{b} \cdot \nabla s_i)$$

(10.82)

The first term on the right-hand side integrates to zero. This can be seen by switching from ε, μ to $w_\parallel$, $w_\perp$ coordinates. A short calculation shows that

$$
\begin{aligned}
w_\parallel \frac{\partial \bar{f}}{\partial \varepsilon} \mathbf{b} \cdot \nabla |s|^2 \Big|_{\varepsilon,\mu} &= w_\parallel \frac{\partial \bar{f}}{\partial \varepsilon} \mathbf{b} \cdot \nabla |s|^2 \Big|_{w_\perp, w_\parallel} + \frac{w_\parallel w_\perp^2}{B} (\mathbf{b} \cdot \nabla B) \left(\frac{\partial}{\partial w_\perp^2} - \frac{\partial}{\partial w_\parallel^2} \right) \left(|s|^2 \frac{\partial \bar{f}}{\partial \varepsilon} \right) \\
&= w_\parallel \frac{\partial \bar{f}}{\partial \varepsilon} \mathbf{b} \cdot \nabla |s|^2 \Big|_{w_\perp, w_\parallel} - w_\parallel \frac{(\mathbf{b} \cdot \nabla B)}{B} \left(|s|^2 \frac{\partial \bar{f}}{\partial \varepsilon} \right) \\
&= w_\parallel \mathbf{B} \cdot \nabla \left(|s|^2 \frac{1}{B} \frac{\partial \bar{f}}{\partial \varepsilon} \right) \Big|_{w_\perp, w_\parallel}
\end{aligned}
$$

(10.83)

where the terms with a velocity derivative in the top line have been integrated by parts. Therefore, by Gauss' theorem (over physical space) it follows that for each species

$$
\begin{aligned}
I &= \frac{i\omega}{4} \int \frac{\partial \bar{f}}{\partial \varepsilon} w_\parallel \mathbf{b} \cdot \nabla |s|^2 \Big|_{\varepsilon,\mu} d\mathbf{w} d\mathbf{r} \\
&= \frac{i\omega}{4} \int w_\parallel \mathbf{B} \cdot \nabla \left(|s|^2 \frac{1}{B} \frac{\partial \bar{f}}{\partial \varepsilon} \right) \Big|_{w_\perp, w_\parallel} d\mathbf{w} d\mathbf{r} \\
&= \frac{i\omega}{4} \int w_\parallel \nabla \cdot \left[\mathbf{B} \left(|s|^2 \frac{1}{B} \frac{\partial \bar{f}}{\partial \varepsilon} \right) \right] \Big|_{w_\perp, w_\parallel} d\mathbf{w} d\mathbf{r} \\
&= 0
\end{aligned}
$$

(10.84)

The Energy Relation is now obtained by substituting these results into Eq. (10.73),

$$\omega^2 K_\perp = \delta W_\perp + \delta Q_{KIN} - \omega R$$

$$R = \frac{1}{2} \sum_\alpha \int \frac{\partial \bar{f}_\alpha}{\partial \varepsilon} w_\parallel (s_{\alpha i} \mathbf{b} \cdot \nabla s_{\alpha r} - s_{\alpha r} \mathbf{b} \cdot \nabla s_{\alpha i}) d\mathbf{w} d\mathbf{r}$$

(10.85)

where R as well $K_\perp$, $\delta W_\perp$, δQ_{KIN} are all real quantities while ω is complex with $\omega_i > 0$. One now sets the imaginary part of Eq. (10.85) to zero yielding $\omega_i R = -2\omega_i\omega_r K_\perp$ or

$$R = -2\omega_r K_\perp \tag{10.86}$$

Substituting this into the real part of Eq. (10.85) finally leads to the desired form of the kinetic MHD Energy Relation,

$$|\omega|^2 = -\frac{\delta W_\perp + \delta Q_{KIN}}{K_\perp}$$

$$\delta Q_{KIN} = -\frac{|\omega|^2}{2}\sum_\alpha \int \frac{\partial \bar{f}_\alpha}{\partial \varepsilon}|s_\alpha|^2 d\mathbf{w}\, d\mathbf{r} \tag{10.87}$$

This relation is valid for general 3-D geometries with arbitrary complex ω, the only constraint being $\omega_i > 0$

Discussion

After an admittedly lengthy calculation one can now ask how an Energy Relation of the form given by Eq. (10.87) can be used to determine stability. To answer this question several points should be noted. First, observe that $|\omega|^2$ rather than ω^2 appears in the relation. Second, from Eq. (10.81) one sees that $\delta Q_{KIN} \geq 0$ when $\partial \bar{f}_\alpha/\partial \varepsilon < 0$, the usual situation. Third, although δQ_{KIN} would appear to vanish when $|\omega|^2 \to 0$ this is not the case for toroidal configurations. Here, trapped particles and the non-uniform sampling of flux surfaces by passing particles each produce a contribution to $|s_\alpha|^2 \propto 1/|\omega|^2$. The net result is that δQ_{KIN} makes a finite but positive contribution to the energy as $|\omega|^2 \to 0$.

Keeping these points in mind one can deduce information about kinetic MHD stability as follows. If

$$\delta W_{KIN} \equiv \delta W_\perp + \delta Q_{KIN}(\omega = 0) > 0 \tag{10.88}$$

then there is an obvious contradiction because of the negative sign in Eq. (10.87). The only way to resolve this contradiction is to conclude that the assumption $\omega_i > 0$ is violated implying that $\omega_i < 0$. In other words, Eq. (10.88) is a sufficient condition for kinetic MHD stability.

A second stability result can be deduced when

$$\delta W_\perp + \delta Q_{KIN}(\omega = 0) = 0 \tag{10.89}$$

This can only occur if $|\omega|^2 = 0$ (i.e. $\omega_r = \omega_i = 0$) and corresponds to a marginal stability point.

The third stability condition corresponds to

$$\delta W_\perp + \delta Q_{KIN}(\omega = 0) < 0 \qquad (10.90)$$

In this case there is no contradiction in the general Energy Relation. One expects that when Eq. (10.90) is satisfied, the plasma will be unstable since the assumption $\omega_i > 0$ has not been violated.

Physically, the fact that $\omega_r = 0$ when $\omega_i \to 0$ is a consequence of the fact in a kinetic MHD plasma the dispersion relation is in general complex because of the parallel Landau damping generated by resonant particles. The damping vanishes when $\partial \bar{f}/\partial w_\parallel = 0$ for the resonant particles, thereby corresponding to the condition for marginal stability. The requirement $\partial \bar{f}/\partial w_\parallel = 0$ is satisfied when $w_\parallel = \omega_r/k_\parallel = 0$ for a distribution function of the form $\bar{f} = \bar{f}(w_\perp^2 + w_\parallel^2, \psi)$.

The discussion above describes how the Energy Relation can be used to obtain kinetic MHD stability information about a plasma. In order to make stability comparisons with other models such as ideal MHD and double adiabatic MHD it is necessary to obtain upper and lower bounds on the size of δQ_{KIN}. These bounds, which require considerable analysis, are derived in Appendix H. Their values are given as the discussion progresses in the next section.

10.5 Stability comparison theorems

All the necessary derivations for the various contributions to the potential energy have been completed. The task now is to examine these contributions in order to deduce the desired stability comparison theorems (Kruskal and Oberman, 1958; Rosenbluth and Rostoker, 1959). The results are separated into four parts representing cylindrical or toroidal and ergodic or closed line geometries. For all geometries and all MHD models the basic quantity $\delta W_\perp$ defined by

$$\delta W_\perp(\boldsymbol{\xi}_\perp^*, \boldsymbol{\xi}_\perp)$$
$$= \frac{1}{2\mu_0} \int \left[|\mathbf{Q}|^2 - \mu_0 \boldsymbol{\xi}_\perp^* \cdot \mathbf{J} \times \mathbf{Q} + \mu_0(\boldsymbol{\xi}_\perp \cdot \nabla p)\nabla \cdot \boldsymbol{\xi}_\perp^* \right] d\mathbf{r} + \delta W_S + \delta W_V \qquad (10.91)$$

appears as one contribution to the potential energy and serves as a point of reference. The four stability comparison theorems are determined as follows.

10.5.1 Closed line cylindrical geometry

The first theorem applies to a closed line cylindrical system in which the perturbations maintain the closed line symmetry. This corresponds to the $m = 0$ mode in a

Z-pinch. The relevant potential energies can be extracted directly from the text. Specifically, for ideal MHD the potential energy is obtained from Eq. (10.4) after noting that $\nabla \cdot \boldsymbol{\xi} = \nabla \cdot \boldsymbol{\xi}_\perp$ for the $m = 0$ mode in a Z-pinch. The corresponding result for double adiabatic MHD is given in Eqs. (10.27) and (10.28). Similarly the kinetic MHD result follows from Eqs. (10.42) and (10.43). These results are summarized below:

$$\delta W_{MHD} = \delta W_\perp + \frac{1}{2} \int \frac{5}{3} p |\nabla \cdot \boldsymbol{\xi}_\perp|^2 \, d\mathbf{r}$$

$$\delta W_{CGL} = \delta W_\perp + \frac{1}{2} \int \left[\frac{5}{3} p |\nabla \cdot \boldsymbol{\xi}_\perp|^2 + \frac{p}{3} |\nabla \cdot \boldsymbol{\xi}_\perp + 3\boldsymbol{\xi}_\perp \cdot \boldsymbol{\kappa}|^2 \right] d\mathbf{r} \qquad (10.92)$$

$$\delta W_{KIN} = \delta W_\perp + \frac{1}{2} \int \left[\frac{5}{3} p |\nabla \cdot \boldsymbol{\xi}_\perp|^2 + \frac{p}{3} |\nabla \cdot \boldsymbol{\xi}_\perp + 3\boldsymbol{\xi}_\perp \cdot \boldsymbol{\kappa}|^2 \right] d\mathbf{r}$$

It is clear from Eq. (10.92) that the stability comparison theorem for a closed line cylindrical system satisfies

$$\delta W_{MHD} < \delta W_{KIN} = \delta W_{CGL} \qquad (10.93)$$

Kinetic MHD is bracketed by ideal and double adiabatic MHD and is in fact identical to double adiabatic MHD.

10.5.2 Ergodic cylindrical geometry

A similar stability comparison theorem can be derived for the cylindrical screw pinch. For this geometry the ideal MHD potential energy is again obtained from Eq. (10.4) although in this case $\boldsymbol{\xi}_\parallel$ is chosen to make $\nabla \cdot \boldsymbol{\xi} = 0$. The double adiabatic potential energy follows from Eqs. (10.27) and (10.28) while the kinetic MHD potential energy is given by Eq. (10.57). A summary of the potential energies can be written as

$$\delta W_{MHD} = \delta W_\perp$$
$$\delta W_{CGL} = \delta W_\perp + \frac{1}{2} \int \left[\frac{5}{3} p |\nabla \cdot \boldsymbol{\xi}_\perp|^2 + \frac{p}{3} |\nabla \cdot \boldsymbol{\xi}_\perp + 3\boldsymbol{\xi}_\perp \cdot \boldsymbol{\kappa}|^2 \right] d\mathbf{r} \qquad (10.94)$$
$$\delta W_{KIN} = \delta W_\perp$$

An examination of Eq. (10.94) shows that the stability comparison theorem for a cylindrical screw pinch satisfies

$$\delta W_{MHD} = \delta W_{KIN} < \delta W_{CGL} \qquad (10.95)$$

Kinetic MHD is again bracketed by ideal and double adiabatic MHD although in this case it coincides with ideal MHD.

10.5.3 Closed line toroidal geometry

The next configuration of interest is the closed line toroidal geometry where the perturbations maintain the closed line symmetry. Examples of such systems include the toroidal Z-pinch, the levitated dipole, and the field reversed configuration. The potential energy for each model is summarized below.

For ideal MHD the potential energy follows from Eq. (10.4) although one must eliminate $\xi_\parallel$ by the general minimizing condition discussed in Section 8.9 and given by Eq. (8.117). The result is

$$\delta W_{MHD} = \delta W_\perp + \frac{1}{2} \int \frac{5}{3} p \left| \langle \nabla \cdot \xi_\perp \rangle \right|^2 d\mathbf{r}$$

$$\langle \nabla \cdot \xi_\perp \rangle = \frac{\oint \frac{dl}{B} \nabla \cdot \xi_\perp}{\oint \frac{dl}{B}} \tag{10.96}$$

The double adiabatic result is again given by Eqs. (10.27) and (10.28),

$$\delta W_{CGL} = \delta W_\perp + \frac{1}{2} \int \left[\frac{5}{3} p |\nabla \cdot \xi_\perp|^2 + \frac{p}{3} |\nabla \cdot \xi_\perp + 3\xi_\perp \cdot \kappa|^2 \right] d\mathbf{r} \tag{10.97}$$

Observe that ideal MHD and double adiabatic MHD contain compressibility terms proportional to $|\langle \nabla \cdot \xi_\perp \rangle|^2$ and $|\nabla \cdot \xi_\perp|^2$ respectively. One must determine the relative size of these two terms in order to make an accurate comparison between the two models. This comparison can be made by means of Schwartz's inequality after switching to ψ, χ, l flux coordinates and recalling that $d\mathbf{r} = d\psi d\chi dl/B$. Focusing on the l integration one notes that Schwartz's inequality can be written as

$$\int g^2 dl \int |h|^2 dl \geq \left| \int g h dl \right|^2 \tag{10.98}$$

Choosing $g = 1/B^{1/2}$ and $h = (\nabla \cdot \xi_\perp)/B^{1/2}$ yields

$$\int \frac{dl}{B} \int \frac{dl}{B} |\nabla \cdot \xi_\perp|^2 \geq \left| \int \frac{dl}{B} \nabla \cdot \xi_\perp \right|^2 \tag{10.99}$$

which in turn implies that

$$\frac{1}{2} \int \frac{5}{3} p |\nabla \cdot \xi_\perp|^2 d\mathbf{r} \geq \frac{1}{2} \int \frac{5}{3} p |\langle \nabla \cdot \xi_\perp \rangle|^2 d\mathbf{r} \tag{10.100}$$

Using this relation in Eq. (10.97) leads to the conclusion

$$\delta W_{CGL} \geq \delta W_\perp + \frac{1}{2} \int \left[\frac{5}{3} p |\langle \nabla \cdot \boldsymbol{\xi}_\perp \rangle|^2 + \frac{p}{3} |\nabla \cdot \boldsymbol{\xi}_\perp + 3\boldsymbol{\xi}_\perp \cdot \boldsymbol{\kappa}|^2 \right] d\mathbf{r} > \delta W_{MHD}$$

(10.101)

The last result of interest involves the potential energy from the kinetic MHD model, which is written as

$$\delta W_{KIN} = \delta W_\perp + \delta Q_{KIN} \qquad (10.102)$$

Upper and lower bounds on δQ_{KIN} have been derived in Appendix H. The relevant bounds are obtained from Eqs. (H.23) and (H.29) and are given by

$$\delta Q_{\min} < \delta Q_{KIN} < \delta Q_{\max}$$

$$\delta Q_{\min} = \frac{1}{2} \int \frac{5}{3} p |\langle \nabla \cdot \boldsymbol{\xi}_\perp \rangle|^2 d\mathbf{r}$$

(10.103)

$$\delta Q_{\max} = \frac{1}{2} \int \left[\frac{5}{3} p |\nabla \cdot \boldsymbol{\xi}_\perp|^2 + \frac{p}{3} |\nabla \cdot \boldsymbol{\xi}_\perp + 3\boldsymbol{\xi}_\perp \cdot \boldsymbol{\kappa}|^2 \right] d\mathbf{r}$$

The desired comparison theorem is now easily obtained by examining the Energy Relations for each model:

$$\delta W_{MHD} < \delta W_{KIN} < \delta W_{CGL} \qquad (10.104)$$

Kinetic MHD is bracketed by ideal MHD and double adiabatic MHD.

10.5.4 Ergodic toroidal geometry

The final configuration of interest is a general toroidal geometry with ergodic field lines or a closed line system in which the perturbations break the closed line symmetry. The ideal MHD potential energy is given by Eq. (10.4) with $\xi_\parallel$ chosen to make $\nabla \cdot \boldsymbol{\xi} = 0$. In this case

$$\delta W_{MHD} = \delta W_\perp \qquad (10.105)$$

The double adiabatic potential energy remains unchanged and is obtained from Eq. (10.27) and (10.28):

$$\delta W_{CGL} = \delta W_\perp + \frac{1}{2} \int \left[\frac{5}{3} p |\nabla \cdot \boldsymbol{\xi}_\perp|^2 + \frac{p}{3} |\nabla \cdot \boldsymbol{\xi}_\perp + 3\boldsymbol{\xi}_\perp \cdot \boldsymbol{\kappa}|^2 \right] d\mathbf{r} \qquad (10.106)$$

Lastly, the kinetic MHD potential energy also follows from the bounds on δQ_{KIN} derived in Appendix H in Eqs. (H.23) and (H.29). The results are

$$\delta W_{KIN} = \delta W_\perp + \delta Q_{KIN} \qquad (10.107)$$

where

$$\delta Q_{\min} < \delta Q_{KIN} < \delta Q_{\max}$$

$$\delta Q_{\min} = \frac{1}{2} \int \frac{5}{3} p |\langle \nabla \cdot \boldsymbol{\xi}_\perp \rangle|^2 \, d\mathbf{r}$$

$$\delta Q_{\max} = \frac{1}{2} \int \left[\frac{5}{3} p |\nabla \cdot \boldsymbol{\xi}_\perp|^2 + \frac{p}{3} |\nabla \cdot \boldsymbol{\xi}_\perp + 3\boldsymbol{\xi}_\perp \cdot \boldsymbol{\kappa}|^2 \right] d\mathbf{r} \qquad (10.108)$$

Combining these results again yields

$$\delta W_{MHD} < \delta W_{KIN} < \delta W_{CGL} \qquad (10.109)$$

Here too, kinetic MHD is bracketed by ideal MHD and double adiabatic MHD.

10.6 Summary

Marginal stability Energy Relations have been derived for the three models of interest: ideal MHD, double adiabatic MHD, and kinetic MHD. The corresponding potential energy expressions vary somewhat depending on whether the geometry is cylindrical or toroidal and whether the magnetic surfaces are ergodic or closed line. The main result of the analysis is that regardless of which of these situations prevails, there is a hierarchy of stability predictions that is always satisfied,

$$\delta W_{MHD} \le \delta W_{KIN} \le \delta W_{CGL} \qquad (10.110)$$

The ideal MHD model is the most unstable. Since it is a fluid model it is relatively easy to analyze but the invalidity of the collision dominated assumption used in its derivation causes one to be concerned about the reliability of its predictions. The double adiabatic MHD model is the most stable. It too is a fluid model and is thus relatively simple to analyze. Although it is a collisionless model, which makes it relevant for fusion plasmas, the closure assumptions used in its derivation cannot be mathematically justified. This is a cause of concern about the reliability of its stability predictions.

The kinetic MHD stability predictions are bracketed by those of ideal MHD and double adiabatic MHD. Kinetic MHD is a self-consistent collisionless model which, as its name implies, includes kinetic behavior along the field lines. It is the most reliable MHD model in terms of the physics. Its biggest drawback is that it is quite complicated to analyze because of the kinetic effects associated with resonant particles and trapping.

Overall, if the relatively simple to obtain stability prediction gaps between ideal MHD and double adiabatic MHD are found to be small, two main benefits follow: (1) the desired, but complicated to obtain, kinetic MHD stability boundaries are narrowly bracketed; and (2) an explanation is provided of why ideal MHD works as well as it does when comparing to experiments.

In the remainder of the text several examples are analyzed to explicitly determine how close the predictions are between ideal MHD and double adiabatic MHD.

References

Bernstein, I.B., Frieman, E.A., Kruskal, M.D., and Kulsrud, R.M. (1958). *Proc. Royal Society (London)* **A244**, 16.
Kruskal, M.D. and Oberman, C.R. (1958). *Phys. Fluids* **1**, 275.
Rosenbluth, M.N. and Rostoker, N. (1959). *Phys. Fluids* **2**, 23.
Stix, T.H. (1992). *Waves in Plasmas*. New York: American Institute of Physics.

Further reading

Antonsen, T.M. Jr. and Lee, Y.C. (1982). *Phys. Fluids* **25**, 132.
Cerfon, A.J. and Freidberg, J.P. (2011). *Phys. Plasmas* **18**, 012505.
Connor, J.W. and Hastie, R.J. (1974). *Phys. Rev. Lett.* **33**, 202.
Grad, H. (1966). *Phys. Fluids* **9**, 225.
Kulsrud, R.M. (1962). *Phys. Fluids* **5**, 275.

Problems

10.1 Consider the original double adiabatic model. Derive the marginal stability differential equation for an equilibrium with cylindrical symmetry. Assume the equilibrium (but not the perturbed) pressures are isotropic.

10.2 Repeat Problem 10.1 using the modified double adiabatic model.

10.3 Repeat Problem 10.1 assuming the equilibrium pressures are anisotropic.

10.4 Repeat Problem 10.2 assuming the equilibrium pressures are anisotropic.

10.5 Derive an equivalent form of Suydam's criterion for the anisotropic modified double adiabatic model.

10.6 Derive Eq. (10.49).

11

Stability: one-dimensional configurations

11.1 Introduction

Chapter 11 describes the MHD stability of one-dimensional cylindrical configurations, specifically the general screw pinch. The analysis involves both the Energy Principle and in some cases the normal mode eigenvalue equation. The goal is to learn about the properties of a magnetic geometry that lead to favorable or unfavorable MHD stability. Even in a cylindrical geometry a great deal of insight can be obtained regarding MHD stability, although there are important toroidal effects that are described in the next chapter.

The discussion begins with the special case of the θ-pinch. Here, a trivial application of the Energy Principle shows that the θ-pinch has inherently favorable stability properties. Also described is "continuum damping" which has many similarities to the well-known phenomenon of Landau damping of electrostatic plasma oscillations. This analysis is simplified by the introduction of the "incompressible MHD" approximation. It is shown that even though the continuum lies entirely on the real ω axis, an initial perturbation will be exponentially damped.

The next application is to the Z-pinch. This configuration is shown to have unfavorable stability properties. It is potentially unstable to both the $m = 0$ and $m = 1$ modes. However, MHD stability can be achieved with the addition of a solid, current-carrying wire along the axis of the plasma. The resulting configuration is known as a hard-core pinch and is the basis for the levitated dipole concept. Both the Energy Principle and the normal mode equations are used in the analysis. Also, for comparison the $m = 0$ stability criterion is derived for the double adiabatic model. It is shown that in the region of experimental interest the stability boundaries are quite close to those predicted by ideal MHD.

The analysis continues with the stability of the general screw pinch. The discussion begins with a general simplification of the Energy Principle that takes advantage of the cylindrical symmetry of the equilibrium. Although the screw pinch is a relatively basic configuration one finds that the resulting simplified form of δW still exhibits a rather high level of complexity. The condition to minimize δW requires that the radial component of the plasma displacement ξ_r satisfy a second-order ordinary differential equation and this equation is derived in the text. To complete the formulation of the stability problem boundary conditions are presented corresponding to (1) a perfectly conducting wall on the plasma, (2) a plasma isolated from a perfectly conducting wall by a vacuum region, and (3) a plasma isolated from a resistive wall by a vacuum region. For comparison, the general normal mode eigenvalue equation is derived which is shown to be a quite complicated second-order ordinary differential equation.

Three important stability results are derived from the general form of δW for the screw pinch: (1) Suydam's criterion which is a local test for interchange stability; (2) Newcomb's procedure which is a practical general method for testing stability; and (3) the oscillation theorem which shows that in spite of the complexity of the full normal mode equation, the eigenvalues exhibit Sturmian behavior with respect to the number of nodes in the eigenfunction.

Based on the general stability theory of the screw pinch, two applications are discussed: the "straight tokamak" and the reversed field pinch (RFP). The stability of the straight tokamak is analyzed by substituting the large aspect ratio expansion into δW. There is a huge simplification in the minimizing differential equation. Nevertheless, this highly simplified form sheds considerable light on the basic MHD instability drives that give rise to sawtooth oscillations, current- and density-driven disruptions, and edge localized modes (ELMs). These experimentally observed, and often operationally limiting, phenomena are described and then analyzed by means of the simplified form of δW. Toroidal effects also play an important role in some of the phenomena and are discussed in the next chapter. Still, it is useful in terms of intuition to make an initial attempt at understanding the phenomena using the simplified cylindrical model.

The last topic of interest is the stability of the RFP. The ordering of the fields in an RFP is such that the complete form of δW is required. However, somewhat off-setting this complexity is the fact that toroidal effects are in general small, implying that the one-dimensional cylindrical results provide a relatively complete picture of RFP stability. The analysis shows that RFPs can be stable to high values of β against all internal modes but are potentially very unstable to external modes. Thus a conducting wall and feedback are required for stability. The goals of the analysis are to (1) determine which n values are unstable with the wall at infinity and (2) to calculate how close the wall must be to stabilize all external modes. Experimental results are discussed showing how a sophisticated feedback system can dramatically improve MHD stability. In practice once the ideal modes are under control, the RFP is dominated by resistive MHD turbulence. This turbulence reduces β substantially below the wall stabilized ideal limits. However, the resulting βs are still high enough for energy applications. The important question of resistive MHD turbulence is beyond the scope of the present volume.

As a closing comment it is worth noting that the cylindrical screw pinch in general provides a reasonably accurate description of current driven instabilities in a low β plasma. In tokamaks and stellarators (but not RFPs), an accurate description of pressure-driven instabilities requires the inclusion of toroidal effects.

11.2 The basic stability equations

The analysis in this section is based on several formulations of the stability equations, namely the Energy Principle and the normal mode eigenvalue equations. Also, a simplified form of the normal mode equations, known as "incompressible MHD," is used to discuss certain phenomena. For convenience these models are summarized below.

11.2.1 The Energy Principle

Several forms of the Energy Principle have been discussed in Chapter 8. The analysis in Chapter 11 is based on the intuitive form given by

$$\delta W(\boldsymbol{\xi}^*, \boldsymbol{\xi}) = \delta W_F + \delta W_S + \delta W_V$$

$$\delta W_F(\boldsymbol{\xi}^*, \boldsymbol{\xi}) = \frac{1}{2\mu_0} \int_P \left\{ |\mathbf{Q}_\perp|^2 + B^2 |\nabla \cdot \boldsymbol{\xi}_\perp + 2\boldsymbol{\xi}_\perp \cdot \boldsymbol{\kappa}|^2 + \mu_0 \gamma p |\nabla \cdot \boldsymbol{\xi}|^2 \right.$$

$$\left. - 2\mu_0 (\boldsymbol{\xi}_\perp \cdot \nabla p)(\boldsymbol{\xi}_\perp^* \cdot \boldsymbol{\kappa}) - \mu_0 J_{\parallel} \boldsymbol{\xi}_\perp^* \times \mathbf{b} \cdot \mathbf{Q}_\perp \right\} d\mathbf{r}$$

$$\delta W_S(\boldsymbol{\xi}_\perp^*, \boldsymbol{\xi}_\perp) = \frac{1}{2\mu_0} \int_{S_p} |\mathbf{n} \cdot \boldsymbol{\xi}_\perp|^2 \, \mathbf{n} \cdot \left[\kern-0.15em\left[\nabla \left(\frac{B^2}{2} + \mu_0 p \right) \right]\kern-0.15em\right] dS$$

$$\delta W_V(\boldsymbol{\xi}_\perp^*, \boldsymbol{\xi}_\perp) = \frac{1}{2\mu_0} \int_V |\hat{\mathbf{B}}_1|^2 \, d\mathbf{r} \tag{11.1}$$

11.2.2 The normal mode eigenvalue equations

The general normal mode eigenvalue equations have also been derived in Chapter 8 and can be written as

$$-\omega^2 \rho \boldsymbol{\xi} = \mathbf{F}(\boldsymbol{\xi})$$

$$\mathbf{F}(\boldsymbol{\xi}) = \frac{1}{\mu_0} (\nabla \times \mathbf{B}) \times \mathbf{Q} + \frac{1}{\mu_0} (\nabla \times \mathbf{Q}) \times \mathbf{B} + \nabla(\boldsymbol{\xi} \cdot \nabla p + \gamma p \nabla \cdot \boldsymbol{\xi}) \tag{11.2}$$

$$\mathbf{Q}(\boldsymbol{\xi}_\perp) = \nabla \times (\boldsymbol{\xi}_\perp \times \mathbf{B})$$

11.2.3 Incompressible MHD

The incompressible MHD model is a special limit of the general normal mode equations in which the adiabatic equation of state is replaced by the incompressibility condition $\nabla \cdot \xi = 0$. This is a useful limit when the growth time of the MHD instability of interest is much longer than the adiabatic transit time across one wavelength: that is, $\text{Im}(\omega) \ll k_\parallel V_S$. The sound speed is effectively infinite.

There are two ways to obtain this limit. First, if one starts with the full compressible MHD equations one can take the limit $\gamma \to \infty$ (where γ is the ratio of specific heats). In this limit in order for the perturbed pressure $p_1 = -\xi \cdot \nabla p - \gamma p \nabla \cdot \xi$ to remain finite, then $\nabla \cdot \xi \to 0$. However, to take this limit the full compressible normal equations are first required which then defeats the purpose of obtaining a simpler set of equations from the outset.

The second and simpler way to obtain the incompressible MHD equations is to replace the equation for p_1 with the condition $\nabla \cdot \xi = 0$. The system is not over determined since p_1 is now a new unknown in the problem, a consequence of the fact that the product $\gamma p \nabla \cdot \xi$ is indeterminate in the incompressible limit. The basic problem now has four unknowns ξ, p_1, although p_1 can be eliminated from the outset by taking the curl of the momentum equation. The equations for incompressible MHD thus reduce to

$$\nabla \times (\omega^2 \mu_0 \rho \xi + \mathbf{B} \cdot \nabla \mathbf{Q} + \mathbf{Q} \cdot \nabla \mathbf{B}) = 0$$
$$\nabla \cdot \xi = 0 \tag{11.3}$$

This completes the summary of the models used to investigate the MHD stability of cylindrical plasmas.

11.3 Stability of the *θ*-pinch

The first configuration of interest is the *θ*-pinch. Several points are discussed. First, a simple application of the Energy Principle shows that the *θ*-pinch is always MHD stable. Second, a more detailed minimization of δW sheds light and provides intuition about which form of perturbation is the least stable. This is important when one thinks ahead to toroidal configurations. Third, an analysis is presented of the phenomenon known as continuum damping. It is shown here, in analogy with Landau damping of kinetic theory, that MHD perturbations can be exponentially damped by the MHD continua even though the continua lie on the real ω axis. Such damping can occur in cylindrical and toroidal systems, but for mathematical simplicity it is investigated here for the case of a "slab" *θ*-pinch. Continuum damping is in general not very important for determining MHD stability limits but is nonetheless an interesting MHD phenomenon to understand, with applications to plasma heating and space plasma physics.

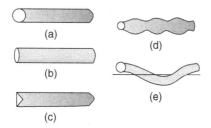

Figure 11.1   Graphical interpretation of the poloidal wave number m and toroidal wave number k : (a) unperturbed column; (b) $m = 2, k = 0$; (c) $m = 3, k = 0$; (d) $m = 0, k \neq 0$; (e) $m = 1, k \neq 0$.

11.3.1 Application of the Energy Principle to the θ-pinch

To begin recall that the basic equilibrium relation for a θ-pinch is given by

$$p + \frac{B^2}{2\mu_0} = \frac{B_0^2}{2\mu_0} \tag{11.4}$$

where $p = p(r)$, $\mathbf{B} = B(r)\mathbf{e}_z$, $\mathbf{J} = J_\theta(r)\mathbf{e}_\theta = -(B'/\mu_0)\mathbf{e}_\theta$, and B_0 is the externally applied magnetic field. Since the equilibrium is symmetric with respect to θ and z the plasma displacement can be Fourier analyzed as follows:

$$\boldsymbol{\xi}(\mathbf{r}) = \boldsymbol{\xi}(r) \exp(im\theta + ikz) \tag{11.5}$$

Here, m and k correspond to the "poloidal" and "toroidal" wave numbers respectively. They are interpreted graphically in Fig. 11.1. Observe that $m = 2, k = 0$ produces an elliptic deformation of the cross section. Similarly, $m = 3$, $k = 0$ produces a triangular deformation. In contrast, when $m = 0, k \neq 0$, the plasma cross section develops a series of bulges along the column. When both $m \neq 0, k \neq 0$ the cross section is helically deformed as illustrated for $m = 1$.

The general stability of a θ-pinch is easily established by the Energy Principle. To see this note that for a θ-pinch, $J_\parallel = 0$ and $\boldsymbol{\kappa} = \mathbf{e}_z \cdot \nabla \mathbf{e}_z = 0$. Then, from Eq. (11.1) one finds that δW_F reduces to

$$\delta W_F(\boldsymbol{\xi}^*, \boldsymbol{\xi}) = \frac{1}{2\mu_0} \int_P \left[|\mathbf{Q}_\perp|^2 + B^2 |\nabla \cdot \boldsymbol{\xi}_\perp|^2 + \mu_0 \gamma p |\nabla \cdot \boldsymbol{\xi}|^2 \right] d\mathbf{r} > 0 \tag{11.6}$$

Each term is positive implying that $\delta W_F > 0$. Furthermore, if the pressure profile vanishes smoothly at the edge of the plasma then $\delta W_S = 0$. Also, it is always true that $\delta W_V > 0$. These results can be combined, yielding

$$\delta W = \delta W_F + \delta W_V > 0 \tag{11.7}$$

Since $\delta W > 0$ for arbitrary perturbations, the conclusion is that a straight θ-pinch is MHD stable.

11.3.2 Minimizing δW_F for a θ-pinch

It is easy to understand why a θ-pinch is stable. Both driving sources of MHD instabilities, parallel current and unfavorable field line curvature, are zero for a θ-pinch. All that remains are the stabilizing terms associated with the three basic MHD waves. Additional insight can be gained by actually minimizing δW_F to show which form of perturbation is the least stable. This is the next task.

The first step in the minimization is to note that the general incompressibility condition $\nabla \cdot \boldsymbol{\xi} = 0$ leads to an expression for $\xi_{\|} = \xi_z$ that can be written as

$$\xi_z = \frac{i}{kr}\left[(r\xi)' + im\xi_\theta\right] \tag{11.8}$$

For simplicity of notation the radial component of the displacement has been defined as $\xi_r \equiv \xi$. Therefore, by choosing ξ_z to satisfy this relation (assuming that $k \neq 0$), it follows that the plasma compressibility term in δW_F vanishes.

The remaining terms in δW_F are evaluated by using the relations

$$\mathbf{Q}_\perp = ikB\boldsymbol{\xi}_\perp = ikB(\xi\mathbf{e}_r + \xi_\theta\mathbf{e}_\theta)$$
$$\nabla \cdot \boldsymbol{\xi}_\perp = \frac{1}{r}(r\xi)' + \frac{im}{r}\xi_\theta \tag{11.9}$$

Substituting in δW_F leads to

$$\frac{\delta W_F}{2\pi R_0} = \frac{\pi}{\mu_0}\int_0^a W(r)\,r\,dr$$

$$W = B^2\left[k^2\left(|\xi|^2 + |\xi_\theta|^2\right) + \frac{1}{r^2}|(r\xi)'|^2 + \frac{m^2}{r^2}|\xi_\theta|^2 + \frac{im}{r}(r\xi^*)'\xi_\theta - \frac{im}{r}(r\xi)'\xi_\theta^*\right]$$

$$\tag{11.10}$$

where a is the outer radius of the plasma and $2\pi R_0$ is the length of the equivalent torus. Observe that ξ_θ appears only algebraically in W. Consequently, the ξ_θ terms can be combined by completing the squares. A short calculation yields

$$W = B^2\left\{\left|k_0\xi_\theta - \frac{im}{k_0 r^2}(r\xi)'\right|^2 + \frac{k^2}{k_0^2 r^2}\left[|(r\xi)'|^2 + k_0^2 r^2|\xi|^2\right]\right\} \tag{11.11}$$

where $k_0^2(r) = k^2 + m^2/r^2$. Now, ξ_θ appears only in the first term, which is positive. Its minimum value is zero, which is obtained by choosing

$$\xi_\theta = \frac{im}{k^2 r^2 + m^2}(r\xi)' \tag{11.12}$$

The value of δW_F thus reduces to

$$\frac{\delta W_F}{2\pi R_0} = \frac{\pi}{\mu_0} \int_0^a \frac{k^2 B^2}{k_0^2 r^2} \left[\left| (r\xi)' \right|^2 + k_0^2 r^2 |\xi|^2 \right] r\,dr \qquad (11.13)$$

The main conclusion to be drawn from Eq. (11.13) is that the least stable modes in a θ-pinch occur for $k^2 \rightarrow 0$ in which case $\delta W_F \rightarrow 0$. In other words, a θ-pinch is always stable but approaches marginal stability for long wavelengths. This corresponds to minimizing the stabilizing effects of line bending. Now, when bending a θ-pinch into a large aspect ratio torus, small additional fields must be added to achieve toroidal force balance. The stability of the plasma against long-wavelength modes is then determined primarily by the properties of these small additional fields since the basic θ-pinch configuration is essentially marginally stable.

11.3.3 *Continuum damping in a "slab" θ-pinch*

The last θ-pinch application concerns continuum damping. As stated this is a phenomenon similar to Landau damping in the kinetic theory of electrostatic waves (Landau, 1946). The basic phenomenon can be described follows. The MHD model in general possesses two continua, one associated with the shear Alfven wave and the other with the sound wave. As their name implies, both exist for a continuous range of frequencies lying on the real axis. Even so, the fact that the frequencies are continuous rather than discrete leads to behavior in which an initial perturbation damps exponentially in time. Alternatively, if an electromagnetic wave with a real frequency is applied, as for heating, it damps exponentially with distance as it deposits energy in the plasma even though the wavelengths defined by the continua are purely real.

Below, a derivation is presented that shows how continuum damping arises when trying to heat a plasma by means of Alfven waves. The analysis assumes that an electromagnetic source is applied at the edge of a plasma with a frequency in the characteristic range of the shear Alfven wave. The resulting boundary value problem is solved allowing one to understand both mathematically and physically how the damping arises. Of specific interest is the calculation of the heating efficiency.

To simplify the mathematics a low β slab model of a θ-pinch is used assuming a density profile that varies linearly with distance. It is shown that under certain conditions an efficiency of 100% can be achieved! It is worth noting that in spite of this, high-efficiency, Alfven wave heating is rarely used in fusion plasmas, but this is associated primarily with practical antenna design problems rather than the physics.

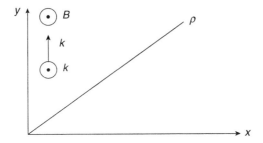

Figure 11.2   Geometry of the "slab" θ-pinch.

The model is deliberately kept as simple as possible to highlight the basic physics. Excellent and far more complete analysis of continuum damping, with applications to fusion and space plasmas, have been given by Tataronis and Grossmann (1973) and Goedbloed and Poedts (2004).

The approach taken here makes use of the incompressible equations of motion given by Eq. (11.3). The incompressibility assumption eliminates the sound wave continuum from the problem so that attention can be focused on the shear Alfven wave. A second-order ordinary differential equation is derived describing the dynamical behavior of the plasma. The assumption of a linearly varying density profile allows one to obtain an analytic solution to this equation from which the desired physical results can be easily extracted. The analysis proceeds as follows.

Derivation of the differential equation

The fields characterizing the equilibrium of a low β slab θ-pinch are given by $\mathbf{B} = B_0 \mathbf{e}_z$, $p = 0$, and $\rho(x) = \rho_0' x$, where B_0, ρ_0' are constants. As stated the density is assumed to vary linearly with distance. The fact that ρ becomes very large at large x is unimportant since, as is shown shortly, deep in the plasma the electromagnetic fields decay exponentially deep in the plasma; that is, the results do not depend on the density profile and the linear density assumption makes the analysis simpler mathematically. The geometry of the problem is illustrated in Fig. 11.2.

The task now is to derive the governing differential equation describing the motion of the plasma, which is assumed to be driven by a source at $x = 0$ with a known frequency ω. Since the equilibrium depends only on x through the density, one can Fourier analyze the spatial dependence with respect to y and z,

$$\boldsymbol{\xi}(\mathbf{r}, t) = \boldsymbol{\xi}(x)\exp\left(-i\omega t + ik_\perp y + ik_\| z\right) \qquad (11.14)$$

where $\boldsymbol{\xi}(x) = \xi(x)\mathbf{e}_x + \xi_y(x)\mathbf{e}_y + \xi_z(x)\mathbf{e}_z$. In the analysis that follows it is assumed that $k_\perp$ and $k_\|$ are set by the structure of the launching source at $x = 0$; they are known quantities.

The analysis begins with the incompressibility condition $\nabla \cdot \boldsymbol{\xi} = 0$ which yields an algebraic relation between ξ_y and ξ_z that can be written as

$$ik_\perp \xi_y + ik_\parallel \xi_z = -\xi' \tag{11.15}$$

A second algebraic relation between ξ_y and ξ_z is obtained from the $\mathbf{e}_x$ component of Eq. (11.3) after using the facts that $\mathbf{Q} \cdot \nabla \mathbf{B} = 0$, $\mathbf{Q} = ik_\parallel B_0 \boldsymbol{\xi}$, and $\mathbf{B} \cdot \nabla \mathbf{Q} = -k_\parallel^2 B^2 \boldsymbol{\xi}$,

$$ik_\perp \xi_z - ik_\parallel \xi_y = 0 \tag{11.16}$$

Equations (11.15) and (11.16) can be solved simultaneously yielding

$$\xi_y = \frac{ik_\perp}{k^2} \xi'$$

$$\xi_z = \frac{ik_\parallel}{k^2} \xi' \tag{11.17}$$

where $k^2 = k_\perp^2 + k_\parallel^2$. From these relations one can easily obtain expressions for the electromagnetic fields in terms of ξ,

$$\mathbf{Q} = ik_\parallel B_0 \boldsymbol{\xi} = ik_\parallel B_0 \left(\xi \, \mathbf{e}_x + i\frac{k_\perp}{k^2} \xi' \mathbf{e}_y + i\frac{k_\parallel}{k^2} \xi' \mathbf{e}_z \right)$$

$$\mathbf{E}_1 = i\omega \boldsymbol{\xi}_\perp \times \mathbf{B} = i\omega B_0 \left(i\frac{k_\perp}{k^2} \xi' \mathbf{e}_x - i\xi \, \mathbf{e}_y \right) \tag{11.18}$$

The final differential equation determining the behavior of ξ is obtained from either the $\mathbf{e}_y$ or $\mathbf{e}_z$ component of Eq. (11.3). It does not matter which is used since one of the equations will always be redundant. A short calculation leads to

$$\frac{d}{dx}\left[\left(\omega^2 \mu_0 \rho - k_\parallel^2 B_0^2 \right) \frac{d\xi}{dx} \right] - k^2 \left(\omega^2 \mu_0 \rho - k_\parallel^2 B_0^2 \right) \xi = 0 \tag{11.19}$$

As it stands, Eq. (11.19) is valid for arbitrary density profiles.

To complete the formulation two boundary conditions are needed. The first requires that $\xi(x \to \infty) = 0$ since there are no sources at $x \to \infty$. The second condition specifies the amplitude of the source at $x = 0$. This amplitude eventually cancels when calculating the efficiency and so can be set to an arbitrary value ξ_0 which is defined shortly. The problem is now completely formulated.

Reference case

As a reference case consider the situation where the plasma density is a constant ρ_0 for $x > 0$ and zero for $x < 0$ (see Fig. 11.3). Assume the applied frequency is not

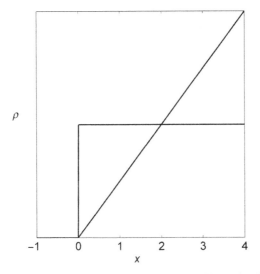

ρ

x

Figure 11.3   Profiles for a step function density profile and a linearly varying density profile.

resonant with the shear Alfven frequency: $\omega^2 \neq k_\parallel^2 B_0^2/\mu_0\rho_0$. Under these assumptions Eq. (11.19) reduces to

$$\frac{d^2\xi}{dx^2} - k^2\xi = 0 \tag{11.20}$$

The solution for ξ is given by

$$\xi = \xi_0 e^{-kx} \tag{11.21}$$

The remaining fields have the form

$$\boldsymbol{\xi} = \left(\mathbf{e}_x - i\frac{k_\perp}{k}\mathbf{e}_y - i\frac{k_\parallel}{k}\mathbf{e}_z \right)\xi$$

$$\mathbf{Q} = ik_\parallel B_0 \left(\mathbf{e}_x - i\frac{k_\perp}{k}\mathbf{e}_y - i\frac{k_\parallel}{k}\mathbf{e}_z \right)\xi \tag{11.22}$$

$$\mathbf{E}_1 = i\omega B_0 \left(-i\frac{k_\perp}{k}\mathbf{e}_x - \mathbf{e}_y \right)\xi$$

Observe that ξ decays exponentially with distance into the plasma as shown in Fig. 11.4. In spite of the exponential decay, the power flow into the plasma, which is given by the real part of the Poynting vector, has the value

$$P_x \equiv \mathrm{Re}(\mathbf{S}\cdot\mathbf{e}_x) = \frac{1}{2}\mathrm{Re}(\mathbf{E}_1 \times \mathbf{Q}\cdot\mathbf{e}_x/\mu_0) = 0 \tag{11.23}$$

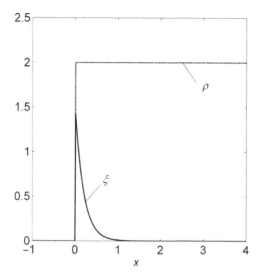

Figure 11.4   Exponential decay of ξ with distance for the step function density profile.

No power flows into the plasma. Electromagnetic energy is stored but not deposited near the surface of the plasma. The reference case therefore corresponds to an evanescent surface wave.

Mathematical solution for a linearly varying density profile

Consider now the more interesting case of a linearly varying density profile whose Alfven frequency is resonant with the applied frequency at some interior point in the plasma (see Fig. 11.3). The governing differential equation given by Eq. (11.19) can be solved by first transforming it to a standard form by introducing a normalized shifted distance for the independent variable. The shift x_0 is determined by the condition that the local Alfven frequency be resonant with the applied frequency: $\omega^2 = \omega_A^2$ where $\omega_A^2 = k_{\parallel}^2 B_0^2/\mu_0 \rho(x_0)$. For the linearly varying density profile, $\rho = \rho_0' x$, the singular surface is located at $x_0 = k_{\parallel}^2 B_0^2/\mu_0 \rho_0' \omega^2$. The normalized distance is now defined as $s = kx - s_0$ with $s_0 = kx_0 > 0$. From this definition it follows that in terms of the new coordinate the resonant surface occurs at $s = 0$ and the plasma exists over the range $-s_0 < s < \infty$.

The coordinate transformation is substituted into Eq. (11.19). A short calculation yields

$$\frac{d^2\xi}{ds^2} + \frac{1}{s}\frac{d\xi}{ds} - \xi = 0 \qquad (11.24)$$

which is just the equation for the modified Bessel functions $I_0(s)$ and $K_0(s)$. The solution that vanishes as $s \to \infty$ (one of the boundary conditions) is given by

$$\xi(s) = \xi_0 K_0(s) \tag{11.25}$$

Here, ξ_0 represents the arbitrary amplitude of the source and scales out of the problem when calculating the efficiency. The source condition represents the second boundary condition.

Equation (11.25) is the desired solution. What are its implications? To answer this question temporarily put aside the physics and focus on the mathematics. The solution is not as innocent as it looks. The reason is that there is a logarithmic singularity at $s = 0$ that arises from the continuum damping. Specifically, from the properties of Bessel functions it follows that near $s = 0$

$$\xi(s) \approx -\xi_0(\ln s + C) \quad s \to 0 \tag{11.26}$$

where $C = $ constant. Therefore there is a jump in the solution across the singular surface which can be determined by noting that

$$
\begin{aligned}
s = 0_+ \quad & \xi(s) = -\xi_0(\ln s + C) \\
s = 0_- \quad & \xi(s) = -\xi_0(\ln s + C) = -\xi_0\{\ln[-(-s)] + C\} = -\xi_0[\ln(-e^{i\pi}s) + C] \\
& = -\xi_0[\ln(-s) + C + i\pi]
\end{aligned}
\tag{11.27}
$$

The jump is thus given by

$$\left[\!\left[\xi(s)\right]\!\right]_{0_-}^{0_+} = i\pi\xi_0 \tag{11.28}$$

This jump can easily be taken into account in the full solution by making use of the following property of Bessel functions: $K_0(s) = K_0(-s) - i\pi I_0(-s)$ for $s < 0$. Specifically, the solution on both sides of the singularity can be rewritten in a more convenient form as follows:

$$
\begin{aligned}
0_+ < s < \infty \quad & \xi(s) = \xi_0 K_0(s) \\
-s_0 < s < 0_- \quad & \xi(s) = \xi_0[K_0(-s) - i\pi I_0(-s)]
\end{aligned}
\tag{11.29}
$$

The real and imaginary parts of ξ are illustrated in Fig. 11.5 for the case $s_0 = 2$. To avoid plotting the singular behavior at $s = 0$ in the illustration an artificial dissipation has been introduced into the solution by letting $\omega \to \omega + i\nu$ with $\nu = 0.005$. It should be emphasized that this replacement is only for the sake of plotting. The power and energy relations derived below are valid for $\nu \to 0$.

Physical implications of the solution

There are two physical implications that can be drawn from the solution given by Eq. (11.29). First energy is absorbed by the plasma and second, there is an optimum frequency that maximizes this absorption. The reasoning is as follows.

To demonstrate that power is absorbed by the plasma one can calculate the real part of the local x-directed Poynting vector. It is actually more instructive to

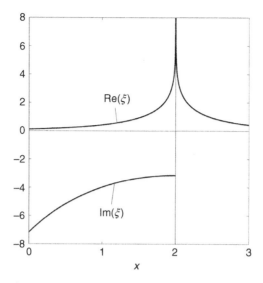

Figure 11.5   Real and imaginary part of ξ for the linearly varying density profile.

calculate the x-directed group velocity which is the ratio of the Poynting vector to the stored energy,

$$V_g(s) = \frac{\mathrm{Re}(\mathbf{S} \cdot \mathbf{e}_x)}{W} \tag{11.30}$$

Here,

$$\mathrm{Re}(\mathbf{S} \cdot \mathbf{e}_x) = \frac{1}{2}\mathrm{Re}\left(\frac{E_{1y}Q_z^*}{\mu_0}\right) = \frac{1}{2}\mathrm{Re}\left(i\omega \frac{k_{\|}^2 B_0^2}{k^2 \mu_0}\xi\,\xi'^*\right)$$

$$W = \frac{1}{2}\left(\frac{\mathbf{Q} \cdot \mathbf{Q}^*}{2\mu_0} + \frac{\rho\,\mathbf{v} \cdot \mathbf{v}^*}{2}\right) = \frac{1}{2}\left(\frac{k_{\|}^2 B_0^2}{2\mu_0} + \frac{\rho\,\omega^2}{2}\right)\left(|\xi|^2 + |\xi'|^2\right) \tag{11.31}$$

with prime denoting d/ds. A straightforward calculation then leads to the following expression for the group velocity

$$V_g(s) = -\frac{\omega}{k}\frac{\pi}{s\,(2 + s/s_0)}\frac{1}{K_0^2 + K_1^2 + \pi^2\left(I_0^2 + I_1^2\right)} \qquad -s_0 < s < 0$$

$$V_g(s) = 0 \qquad\qquad\qquad\qquad\qquad\qquad\qquad\qquad\quad 0 < s < \infty \tag{11.32}$$

where use has been made of the Bessel function relations $K_0'(s) = -K_1(s)$, $I_0'(s) = I_1(s)$, and $I_0(s)K_1(s) + I_1(s)K_0(s) = 1/s$. Also the argument of the Bessel functions in the top equation is $-s$.

Equation (11.32) is illustrated in Fig. 11.6 for the case $s_0 = 0.5$. Observe that power flows into the plasma (i.e., the group velocity is positive) from the

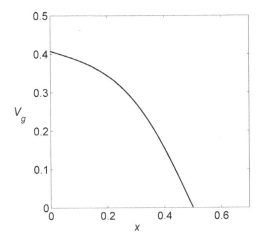

Figure 11.6   Group velocity vs. distance for the case $s_0 = 0.5$.

edge $s = -s_0$ to the singular surface $s = 0$. No power flows into the plasma past the singular surface. Importantly, at the singular surface the group velocity vanishes. This is the classic indicator of a wave resonance. Specifically, the evanescent surface wave resonates with the shear Alfven wave at the resonant surface where $\omega^2 = \omega_A^2$. The group velocity slows down, producing a build-up of energy in the plasma. At the resonant surface the Poynting vector $\mathrm{Re}(\mathbf{S} \cdot \mathbf{e}_x) \propto 1/s$, which explicitly shows this build-up. With even an infinitesimal amount of dissipation the absorbed energy is converted into heat by the plasma.

Consider now the heating efficiency. A useful way to define the efficiency is to examine the power flow from the point of view of the source. A good analog is a simple R-L-C circuit driven by an AC voltage. The source obviously must supply the power dissipated in the resistor. The source also provides reactive power to the inductor and capacitor. However, over one cycle this power averages to zero – half the time the source delivers power and half the time power is returned to the source. Nevertheless, the source must have a high enough volt–amp rating to supply the total power at the time when the reactive power is at its maximum. Supplying the additional power over and above the pure resistive power requires a larger source which translates into additional cost.

The conclusion is that from the point of view of source economics a good definition of efficiency is the ratio of the dissipated power to the total reactive plus dissipated power. This ratio is called the power factor. Also, since the reactive power is always 90° out of phase with the dissipated power they must be added together as the square root of the sum of the squares. Clearly, the highest efficiency occurs at the resonant frequency $\omega = (LC)^{-1/2}$ where the reactive impedances cancel. At this frequency the efficiency is unity.

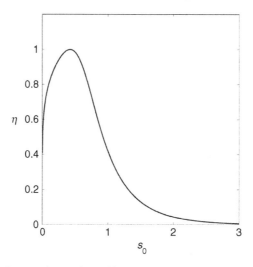

Figure 11.7   Power absorption efficiency vs. resonant frequency location.

In terms of Alfven wave heating this simple circuit picture can be interpreted as follows. The total power absorbed by the plasma P_D is given by the real part of the Poynting vector evaluated at the plasma edge $s = -s_0$ (i.e. $x = 0$): $P_D = \text{Re}[\mathbf{S} \cdot \mathbf{e}_x]_{-s_0}$. Similarly, the peak reactive power supplied to the plasma P_R is given by the imaginary part of the Poynting vector: $P_R = \text{Im}[\mathbf{S} \cdot \mathbf{e}_x]_{-s_0}$. If one takes into account the $90°$ phase shift between absorbed and reactive power, then a good definition of efficiency for Alfven wave heating is

$$\eta(s_0) \equiv \frac{\text{Re}[\mathbf{S} \cdot \mathbf{e}_x]_{-s_0}}{\left|\mathbf{S} \cdot \mathbf{e}_x\right|_{-s_0}} = \frac{\text{Re}[-i\xi\xi'^*]_{-s_0}}{\left|\xi\xi'^*\right|_{-s_0}} \tag{11.33}$$

Substituting for ξ leads to

$$\eta(s_0) = \frac{1}{\left[1 + (s_0/\pi)^2(\pi^2 I_0 I_1 - K_0 K_1)^2\right]^{1/2}} \tag{11.34}$$

where the argument of the Bessel functions is $+s_0$.

The efficiency is plotted in Fig. 11.7. Observe that the efficiency decreases as the resonant surface moves further into the plasma (i.e., as s_0 increases). The reason is that the amplitude of the evanescent surface wave exponentially decays with distance into the plasma so there is less energy available to resonate with the shear Alfven wave. The efficiency is also small when the resonant surface is very close to the edge of the plasma. Here, the surface wave has a large amplitude but there is scarcely any plasma in which to couple energy. There is an optimum location for the resonant layer corresponding to $s_0 \approx 0.42$ for the linear density profile. At this value the efficiency is unity. The reactive powers have exactly canceled. For a given density profile the critical value of $s_0 \propto 1/\omega^2$ defines the optimum frequency at which to operate.

Conclusion

The analysis just presented shows that the existence of the Alfven wave continuum in the MHD spectrum allows efficient heating. Even though the continuum frequencies lie on the real ω axis power is absorbed and dissipated in the plasma because of resonant coupling between the driven surface wave and the continuum.

11.4 Stability of the Z-pinch

The next configuration of interest is the Z-pinch. Several topics are discussed. First, it is shown that the simple Z-pinch has poor MHD stability properties. Specifically, by means of the Energy Principle it is demonstrated that the Z-pinch is always unstable to the $m = 1$ mode and that there are strict requirements on the pressure profile to avoid instability against the $m = 0$ mode. Second, $m = 0$ stability is re-examined using the double adiabatic theory whose stability boundary is identical to that of kinetic MHD. In the regime of physical interest it is shown that the gap in the stability boundaries between ideal and double adiabatic MHD is not very large.

The last topic discussed involves a modified configuration known as the hard-core Z-pinch. This configuration has a rigid, finite radius, current-carrying conductor along the axis. It is shown that the hard core can stabilize the ideal $m = 1$ mode. Also, if the pressure profile decreases sufficiently gently then the $m = 0$ is stabilized. In other words, a hard-core Z-pinch can achieve complete MHD stability.

The stability of the hard-core Z-pinch provides the motivation for a toroidal version of the concept known as the levitated dipole (LDX Group, 1988). The corresponding magnetic geometry is quite simple, an important technological advantage. However, the hard core must be a superconducting coil fully levitated in the vacuum chamber. This is necessary in order to avoid large contact losses with any support hangers that could be used to hold the coil in place. The need to levitate the coil is a technological disadvantage. A small levitated dipole experiment was jointly built by MIT and Columbia University but has been shut down by the Department of Energy for budgetary reasons.

11.4.1 Energy Principle analysis of m $\neq$ 0 *modes*

Evaluation of δW_F

To begin recall that equilibrium pressure balance in a Z-pinch is given by

$$\frac{dp}{dr} + \frac{B_\theta}{\mu_0 r}\frac{d}{dr}(rB_\theta) = 0 \qquad (11.35)$$

Since the equilibrium depends only on r, one can again Fourier analyze the perturbations with respect to θ and z : $\boldsymbol{\xi}(\mathbf{r}) = \boldsymbol{\xi}(r)\exp{(im\theta + ikz)}$.

The stability analysis that follows shows that the most unstable modes are localized within the plasma implying that attention need only be focused on δW_F rather than the full δW. Based on this observation, the first step in the minimization of δW_F is to examine the condition for incompressibility, $\nabla \cdot \boldsymbol{\xi} = 0$. For $m \neq 0$ modes the condition reduces to

$$\xi_\| \equiv \xi_\theta = \frac{i}{m}\left[(r\xi_r)' + ik\xi_z\right] \qquad (11.36)$$

When $m \neq 0$ it is always possible to find a $\xi_\|$ that makes $\nabla \cdot \boldsymbol{\xi} = 0$. The plasma compressibility contribution thus vanishes when minimizing δW_F. The remaining terms in δW_F are easily evaluated using the following relations:

$$\mathbf{Q}_\perp = \frac{imB_\theta}{r}(\xi\mathbf{e}_z + \xi_z\mathbf{e}_z)$$

$$\boldsymbol{\kappa} = \mathbf{e}_\theta \cdot \nabla\mathbf{e}_\theta = -\frac{\mathbf{e}_r}{r} \qquad (11.37)$$

$$\nabla \cdot \boldsymbol{\xi}_\perp + 2\boldsymbol{\xi}_\perp \cdot \boldsymbol{\kappa} = r(\xi/r)' + ik\xi_z$$

$$J_\| = 0$$

where the simplified notation $\xi = \xi_r$ has been introduced. Note that $\mathbf{n} \cdot \boldsymbol{\kappa} = -1/r < 0$ which corresponds to unfavorable curvature.

These expressions are substituted into δW_F (i.e., Eq. (11.1)). A short calculation that makes use of the equivalent torus relation $d\mathbf{r} = 4\pi^2 R_0 r dr$ yields

$$\frac{\delta W_F}{2\pi R_0} = \pi \int_0^a W(r)r dr$$

$$W(r) = \frac{m^2 B_\theta^2}{\mu_0 r^2}\left(|\xi|^2 + |\xi_z|^2\right) + \frac{B_\theta^2}{\mu_0}\left|r(\xi/r)' + ik\xi_z\right|^2 + \frac{2p'}{r}|\xi|^2 \qquad (11.38)$$

Minimization of δW_F

In analogy with the θ-pinch, one sees that ξ_z appears only algebraically. One can then complete the squares resulting in only a single positive term containing ξ_z. This contribution is minimized (i.e., set to zero) by choosing ξ_z as follows:

$$\xi_z = \frac{ikr^3}{m^2 + k^2 r^2} \left(\frac{\xi}{r}\right)'$$

(11.39)

Equation (11.39) is substituted into the expression for δW_F leading to

$$\frac{\delta W_F}{2\pi R_0} = \pi \int_0^a \left[\left(\frac{2p'}{r} + \frac{m^2 B_\theta^2}{\mu_0 r^2}\right)|\xi|^2 + \frac{m^2 B_\theta^2}{\mu_0 (m^2 + k^2 r^2)}\left|r\left(\frac{\xi}{r}\right)'\right|^2\right] r\,dr$$

(11.40)

Observe that k^2 explicitly appears only in the denominator of a positive term. Consequently for any trial function $\xi(r)$, δW_F is minimized by letting $k^2 \to \infty$. The most unstable perturbations correspond to very short wavelengths. With this choice for k^2, δW_F reduces to

$$\frac{\delta W_F}{2\pi R_0} = \pi \int_0^a \left(\frac{2p'}{r} + \frac{m^2 B_\theta^2}{\mu_0 r^2}\right)|\xi|^2 r\,dr$$

(11.41)

Equation (11.41) is the desired expression for δW_F.

Stability criterion for m ≠ 0 modes

An examination of Eq. (11.41) shows that the necessary and sufficient condition for stability against $m \neq 0$ modes in a Z-pinch can be written as (Kadomtsev, 1966)

$$rp' + \frac{m^2 B_\theta^2}{2\mu_0} > 0$$

(11.42)

If this quantity is positive, then $\delta W_F > 0$ obviously showing sufficiency. If, on the other hand, the quantity is negative over any region of the plasma then a trial function localized within the plasma of the form shown in Fig. 11.8 will always make $\delta W_F < 0$, implying instability. This proves necessity.

Two alternate forms of the stability condition can be obtained by eliminating p' using the pressure balance relation. These forms are given by

$$\frac{r^2}{B_\theta}\left(\frac{B_\theta}{r}\right)' < \frac{1}{2}(m^2 - 4)$$

$$\frac{1}{B_\theta^2}(rB_\theta^2)' < m^2 - 1$$

(11.43)

For standard Z-pinch profiles the quantity B_θ/r is a decreasing function of radius: $(B_\theta/r)' < 0$. The first form of Eq. (11.43) thus predicts stability for $m \geq 2$ modes.

Consider now the $m = 1$ mode. At large radii where the current is low, $B_\theta \propto 1/r$ corresponding to a vacuum field. In this region $(rB_\theta^2)' < 0$. The second form of Eq. (11.43) shows that the plasma is stable in this region. However, near the origin

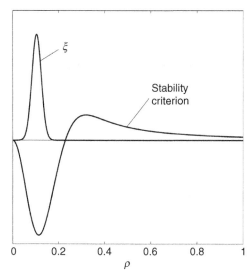

Figure 11.8   Trial function ξ that makes $\delta W_F < 0$ for $m \neq 0$ modes in a pure Z-pinch. Note that $\xi \neq 0$ only in the region where $rp' + m^2 B_\theta^2/2\mu_0 < 0$.

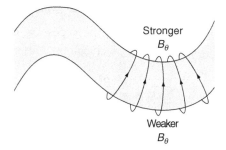

Figure 11.9   Physical mechanism of the $m = 1$ instability in a pure Z-pinch.

$B_\theta \propto r$ and $(rB_\theta^2)' > 0$. The conclusion is that the stability condition is always violated near the core of a standard Z-pinch for the $m = 1$ mode.

The source of the instability can be seen in Fig. 11.9. As the plasma undergoes an $m = 1$ deformation the magnetic lines concentrate in the tighter portion of the column, raising the value of B_θ. The resulting increased magnetic tension produces a force in the direction to further enhance the $m = 1$ deformation; hence instability.

Although the plasma perturbation has the form of a helix the instability does not correspond to a kink mode since $J_{\parallel} = 0$. The minimizing perturbation is best described as a competition between line bending and unfavorable curvature, with magnetic compression making a negligibly small contribution.

The $m = 1$ mode represents one of the basic instabilities present in all standard Z-pinches. It is an important part of the basis for the earlier statement that Z-pinches have unfavorable MHD stability properties.

11.4.2 Energy Principle analysis of the m = 0 mode

Evaluation of δW_F

The remaining mode of interest in a Z-pinch corresponds to $m = 0$. The analysis begins by examining the compressibility contribution to δW_F which is proportional to the square of $\nabla \cdot \xi$. Specifically, for the $m = 0$ mode it follows that

$$\nabla \cdot \boldsymbol{\xi} = \frac{1}{r}(r\xi)' + \frac{im\xi_\theta}{r} + ik\xi_z = \frac{1}{r}(r\xi)' + ik\xi_z = \nabla \cdot \boldsymbol{\xi}_\perp \tag{11.44}$$

Recall now that the only appearance of $\xi_\parallel$ in δW_F occurs in the plasma compressibility term. Since the appearance of $\xi_\parallel \equiv \xi_\theta$ vanishes for $m = 0$ it never appears anywhere else in the calculation. Consequently it is not possible to choose a $\xi_\parallel$ that makes $\nabla \cdot \boldsymbol{\xi} = 0$. One must set $\nabla \cdot \boldsymbol{\xi} = \nabla \cdot \boldsymbol{\xi}_\perp$ and maintain the effects of plasma compressibility when minimizing δW_F.

The evaluation of δW_F is straightforward and again shows that ξ_z appears only algebraically,

$$\frac{\delta W_F}{2\pi R_0} = \pi \int_0^a W(r)rdr$$

$$W(r) = \frac{B_\theta^2}{\mu_0}\left|r(\xi/r)' + ik\xi_z\right|^2 + \gamma p\left|(r\xi)'/r + ik\xi_z\right|^2 + \frac{2p'}{r}\left|\xi\right|^2 \tag{11.45}$$

Minimization of δW_F

The next step is to complete the squares with respect to ξ_z. The result is that ξ_z appears only in a single positive term whose minimum value is set to zero by choosing

$$ik\xi_z = -\frac{B_\theta^2\left[r(\xi/r)'\right] + \mu_0\gamma p\left[(r\xi)'/r\right]}{B_\theta^2 + \mu_0\gamma p} \tag{11.46}$$

Substituting back into δW_F yields

$$\frac{\delta W_F}{2\pi R_0} = \pi \int_0^a \left[\left(\frac{4\gamma B_\theta^2}{B_\theta^2 + \mu_0\gamma p}\right)p + 2rp'\right]\frac{|\xi|^2}{r^2}rdr \tag{11.47}$$

Stability criterion for the m = 0 mode

Following the reasoning associated with the $m \neq 0$ modes (i.e., Eq. (11.41)) one can conclude that the necessary and sufficient condition for stability against the $m = 0$ mode is given by (Kadomstev, 1966)

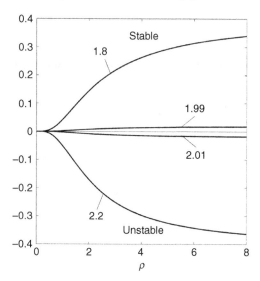

Figure 11.10 Stability criterion given by Eq. (11.48) vs. radius for several values (1.8, 1.99, 2.01, 2.2) of the profile parameter v. The ratio of specific heats has been set to $\gamma = 2$.

$$-\frac{rp'}{p} < \frac{2\gamma B_\theta^2}{B_\theta^2 + \mu_0 \gamma p} \qquad (11.48)$$

This condition can be satisfied near the origin of a standard Z-pinch although the profiles are a little sensitive to the higher-order on-axis curvature terms. However, it can only be satisfied near the outside of the plasma if the pressure decreases sufficiently gradually. Specifically, at large radii in a well-confined plasma, $p \ll B_\theta^2/\mu_0$ and Eq. (11.48) reduces to

$$-\frac{rp'}{p} < 2\gamma \qquad (11.49)$$

For $\gamma = 5/3$ one sees that the pressure must decrease more gradually than $p < C/r^{10/3}$. One important consequence of Eq. (11.49) is that for a Z-pinch to be stable against the $m = 0$ mode it must have a finite, although perhaps small, pressure at the wall. It cannot have $p = 0$ at the wall. These results are summarized in Fig. 11.10 where the stability criterion for the analytically simple case of $\gamma = 2$ is plotted for several pressure and magnetic field profiles given by

$$p(\rho) = \frac{p_0}{(1 + \rho^2)^v}$$

$$B_\theta^2(\rho) = \frac{2p_0}{v - 1} \frac{1}{\rho^2} \left[1 - \frac{(1 + v\rho^2)}{(1 + \rho^2)^v} \right] \qquad (11.50)$$

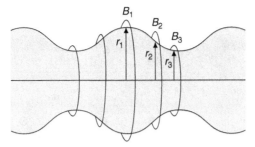

Figure 11.11   Physical mechanism of the $m = 0$ sausage instability in a pure Z-pinch.

where $\rho = r/a$ with a the characteristic scale length of the plasma (but less than the wall radius b). The parameter v determines how gradually the pressure decreases with r at large radii. Observe that for the slowly decreasing profile ($v = 1.8$) the stability criterion is always satisfied. The marginal stability profile corresponds to $v = 2$. For a more rapidly decreasing profile ($v = 2.2$) the stability boundary is violated everywhere.

The $m = 0$ instability is an interchange mode often known as the "sausage instability." The basic force driving the instability is illustrated in Fig. 11.11. When an $m = 0$ sausage perturbation is superimposed on the Z-pinch equilibrium the magnetic field in the throat regions increases since the plasma carries the same current in a smaller cross section. The increased magnetic tension produces a destabilizing force which tends to further constrict the column. However, the plasma pushes back when being compressed, thereby providing a stabilizing force. When the tension force dominates, instability occurs. At marginal stability the two forces are balanced, corresponding to equality of the two terms in Eq. (11.48). The minimizing perturbation for the $m = 0$ instability represents a competition between unfavorable curvature and plasma compression. The line bending is zero.

An alternate physical picture of the sausage instability can be obtained by examining the motion of the plasma from a single particle point of view. This picture shows why the curvature is unfavorable. The physics is illustrated in Fig. 11.12. An $m = 0$ perturbation has been superimposed on the surface of the plasma. The electrons and ions in the bulges (and throats) develop opposing guiding center velocities along z because of the curvature drift. The sign of the drifts, illustrated for a bulge region, corresponds to "unfavorable" negative curvature: $\boldsymbol{\kappa} = -\mathbf{e}_r/r$. Now, because of the sausage perturbation, the opposing drifts set up a charge separation as shown in the figure. This charge separation creates an electric field in the axial direction whose sign also alternates between bulges and

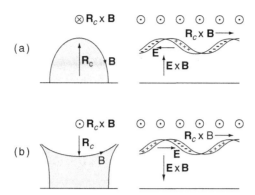

Figure 11.12   Single particle picture of the $m = 0$ interchange mode in a pure
Z-pinch. In (a) an interchange perturbation modulates the plasma surface. The
opposing curvature drifts set up a charge separation leading to a destabilizing
$\mathbf{E} \times \mathbf{B}$ drift. In (b) the curvature direction changes sign leading to stability.

throats. The electric field in turn produces an $\mathbf{E} \times \mathbf{B}$ drift of the plasma. When the
curvature is negative, the direction of the $\mathbf{E} \times \mathbf{B}$ velocity is such as to further
reinforce the perturbation in both the bulges and throats, leading to instability.
When the curvature is positive, as in a cusp geometry, the sign of the drifts reverses
and the perturbation is stabilized.

Validity of ideal MHD stability for the m $= 0$ *mode in a Z-pinch*

The last point to note is that the $m = 0$ stability criterion depends on γ, the ratio
of specific heats. This dependence arises from the ideal MHD adiabatic equation of state, which as previously stated, is highly unreliable for fusion-grade
plasmas. What is needed is an analysis based on the more realistic kinetic
MHD model. In this connection recall from Chapter 10 that $m = 0$ kinetic
MHD stability in a Z-pinch is identical to that of the double adiabatic model.
It is therefore of interest to calculate the corresponding form of the stability
criterion given by Eq. (11.48) to see the differences in the stability boundaries.
This is the next task.

11.4.3 Double adiabatic Energy Principle analysis of the m $= 0$ *mode*

Evaluation of δW_{CGL}

The evaluation of δW_{CGL} from the double adiabatic Energy Principle follows from
the simplification of Eqs. (10.27) and (10.28). One again focuses on the internal
$m = 0$ mode in a Z-pinch. The corresponding form of the Energy Principle can be
written as

$$\delta W_{CGL} = \delta W_F + \frac{1}{2} \int \frac{p}{3} \left| \nabla \cdot \boldsymbol{\xi}_\perp + 3\boldsymbol{\xi}_\perp \cdot \boldsymbol{\kappa} \right|^2 d\mathbf{r} \qquad (11.51)$$

where δW_F is given by Eq. (11.45). Substituting the cylindrical Z-pinch equilibrium into the additional term in Eq. (11.51) leads to

$$\frac{\delta W_{CGL}}{2\pi R_0} = \pi \int_0^a W(r) r \, dr$$

$$W(r) = \frac{B_\theta^2}{\mu_0} \left| r(\xi/r)' + ik\xi_z \right|^2 + \frac{5}{3} p \left| (r\xi)'/r + ik\xi_z \right|^2 \qquad (11.52)$$

$$+ \frac{1}{3} p \left| r^2 (\xi/r^2)' + ik\xi_z \right|^2 + \frac{2p'}{r} \left| \xi \right|^2$$

Minimization of δW_{CGL}

Observe that as with the previous Z-pinch calculations, the quantity $ik\xi_z$ appears only algebraically. Completing the squares and setting the resulting term to zero yields the following relation for the minimizing ξ_z:

$$ik\xi_z = -\frac{B_\theta^2 \left[r(\xi/r)' \right] + (5\mu_0 p/3) \left[(r\xi)'/r \right] + (\mu_0 p/3) \left[r^2(\xi/r^2)' \right]}{B_\theta^2 + 2\mu_0 p} \qquad (11.53)$$

This relation is now substituted into (11.52). A straightforward calculation leads to the desired expression for δW_{CGL}:

$$\frac{\delta W_{CGL}}{2\pi R_0} = \pi \int_0^a \left[\left(\frac{7B_\theta^2 + 5\mu_0 p}{B_\theta^2 + 2\mu_0 p} \right) p + 2rp' \right] \frac{\left| \xi \right|^2}{r^2} r \, dr \qquad (11.54)$$

Stability criterion for the $m = 0$ *mode*

From Eq. (11.54) one can easily obtain the necessary and sufficient condition for double adiabatic stability against the $m = 0$ mode in a Z-pinch. This condition and the corresponding one for ideal MHD (with $\gamma = 5/3$) for comparison are given by

$$-\frac{rp'}{p} < \frac{7B_\theta^2 + 5\mu_0 p}{2B_\theta^2 + 4\mu_0 p} \qquad \text{double adiabatic MHD}$$

$$-\frac{rp'}{p} < \frac{10B_\theta^2}{3B_\theta^2 + 5\mu_0 p} \qquad \text{ideal MHD} \qquad (11.55)$$

Note that near the origin double adiabatic MHD is considerably more stable than ideal MHD: $-rp'/p < 5/4$ for double adiabatic MHD while $-rp'/p < 0$ for ideal

MHD. However, this improved stability is not very important since the ideal MHD model is already stable in this region.

The more interesting region is near the outside of the plasma where a sufficiently weak pressure gradient is required for stability. In this region corresponding to $p \to 0$ the two stability criteria reduce to

$$-\frac{rp'}{p} < \frac{7}{2} = 3.50 \qquad \text{double adiabatic MHD}$$

$$-\frac{rp'}{p} < \frac{10}{3} = 3.33 \qquad \text{ideal MHD} \tag{11.56}$$

There is not a very large difference in the stability boundaries. This analysis thereby provides one example of the small gap between the predictions of ideal MHD and double adiabatic MHD. Consequently, one would not make a very large error using ideal MHD.

Summary of stability in a standard Z-pinch

Based on the analysis just presented, the conclusion is that a standard Z-pinch is always unstable to $m = 1$ perturbations and may likely be unstable to $m = 0$ perturbations. In fact the $m = 0$ mode was often observed experimentally in early, rapidly shock heated Z-pinch experiments leading to a catastrophic termination of the plasma. The theory plus experimental results form the basis for earlier statements attributing very poor ideal MHD stability properties to a standard Z-pinch.

11.4.4 The hard-core Z-pinch

An interesting way to modify the standard Z-pinch that potentially overcomes its poor MHD stability properties is to add a rigid, finite radius, current-carrying conductor along the axis. For obvious reasons the resulting configuration is called a hard-core Z-pinch. A toroidal version of the concept, known as the levitated dipole (LDX Group, 1998), was built several years ago but unfortunately has recently been shut down for budgetary reasons. Even so, it is of help in the development of MHD intuition to demonstrate how the presence of a hard core can stabilize the strong MHD instabilities in a standard Z-pinch. That is the goal of the present subsection.

Hard-core Z-pinch equilibria

The typical equilibrium profiles (specified below) of a hard-core Z-pinch are illustrated in Fig. 11.13. Observe that the pressure vanishes at $r = r_c$, the surface of the hard-core conductor. The pressure rises and then decreases gradually at large

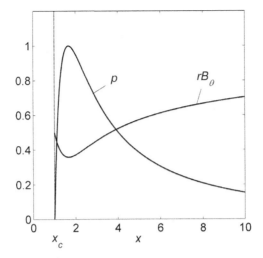

Figure 11.13   Equilibrium profiles for a hard-core Z-pinch. Shown are $p(\rho)/p_{max}$ and $rB_\theta/(rB_\theta)_\infty$ for $\beta = 0.5$. Also $x = r^2/r_c^2$ and $x_c = 1$ corresponds to the coil radius.

radii to avoid the $m = 0$ instability. A gradually decreasing pressure can be generated experimentally by switching from rapid shock heating to slower RF heating with the profile controlled by the location of the resonant frequency.

The presence of the hard core leads to a B_θ profile that is large at the surface of the conductor and then decreases, approximately proportional to $B_\theta \propto 1/r$, although with a superimposed diamagnetic depression due to the plasma pressure. The $1/r$ dependence near the conductor stabilizes the $m = 1$ mode and is in fact the primary MHD motivation for the hard core.

A simple choice for the pressure profile that satisfies the basic requirements of a hard-core Z-pinch is given by

$$p(r) = K\frac{r^2 - r_c^2}{r^5} \tag{11.57}$$

The constant K is a measure of the peak pressure. The exponent 5 has been chosen to guarantee $m = 0$ ideal MHD stability at large radii. This equation can be rewritten in a simpler form by introducing $x = r^2/r_c^2$,

$$p(x) = K_P\frac{x - 1}{x^{5/2}} \tag{11.58}$$

where K_P is a new constant related to the maximum pressure p_{max}, which occurs at $x = 5/3$, and is given by

$$K_P = \frac{3}{2}\left(\frac{5}{3}\right)^{5/2}p_{max} \approx 5.38\,p_{max} \tag{11.59}$$

The magnetic field is easily found by integrating the pressure balance relation and expressing the results in terms of x. The result is

$$xB_\theta^2(x) = K_I - \frac{2\mu_0 K_P}{3} \frac{9x - 5}{x^{3/2}} \qquad (11.60)$$

Here K_I is a free integration constant. Its value is determined by noting that the magnetic field far from the plasma approaches $B_\theta(r \to \infty) = \mu_0(I_C + I_P)/2\pi r$, where I_C is the hard-core current and I_P is the plasma current. Therefore,

$$K_I = \left[\frac{\mu_0(I_C + I_P)}{2\pi r_c}\right]^2 \qquad (11.61)$$

In the analysis that follows it is convenient to introduce normalized variables plus the plasma β. The definitions are as follows:

$$\beta = \frac{16\pi^2}{\mu_0 I^2} \int_{r_c}^{\infty} pr\,dr = \frac{32\,\pi^2 r_c^2}{3\,\mu_0 I^2} K_P \qquad \text{beta}$$

$$P = \frac{2\mu_0 p(x)}{(\mu_0 I/2\pi r_c)^2} = \frac{3\beta}{4}\frac{x - 1}{x^{5/2}} \qquad \text{normalized } p(x) \qquad (11.62)$$

$$b_\theta^2 = \frac{xB_\theta^2(x)}{(\mu_0 I/2\pi r_c)^2} = 1 - \frac{\beta}{4}\frac{9x - 5}{x^{3/2}} \qquad \text{normalized } B_\theta^2(x)$$

with $I = I_C + I_P$. The last equilibrium relation of interest corresponds to the global pressure balance relation. This relation is easily obtained by noting that $B_\theta(r_c) = \mu_0 I_C/2\pi r_c$ and evaluating $b_\theta^2(\infty) - b_\theta^2(1)$. A simple calculation shows that global pressure balance relates β to the coil and plasma currents by

$$\beta = 1 - \left(\frac{I_C}{I_C + I_P}\right)^2 \qquad (11.63)$$

The basic equilibrium relations for the hard-core Z-pinch profiles have now been established. The next goal is to investigate the effectiveness of the hard core in providing high β stabilization of the plasma against MHD instabilities. The analysis that follows shows that there are in fact three limits on β: (1) an equilibrium β limit (which does not exist for the standard Z-pinch); (2) an $m = 1$ stability limit; and (3) an $m = 0$ stability limit. All of these limits occur at high values of β showing that a hard core is a very effective means of stabilizing the standard Z-pinch.

The hard-core Z-pinch equilibrium limit

The hard-core Z-pinch has an equilibrium β limit. This limit can be seen mathematically by examining a plot of b_θ^2 vs. x as shown in Fig. 11.14. Observe that as

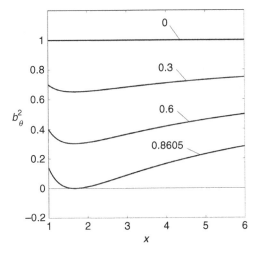

Figure 11.14 Equilibrium β limit in a hard-core Z-pinch as shown in a plot of b_θ^2 vs. x.

β increases the diamagnetic dip due to the plasma pressure becomes increasingly deeper. Eventually a critical value $\beta = \beta_{eq}$ is reached above which the minimum b_θ^2 becomes negative. Since $b_\theta^2 \geq 0$ for physical solutions the value β_{eq} corresponds to the equilibrium β limit.

The value of β_{eq} is easily found by simultaneously setting $b_\theta^2 = db_\theta^2/dx = 0$. A short calculation yields

$$\beta \leq \beta_{eq} = \frac{2}{5}\left(\frac{5}{3}\right)^{3/2} \approx 0.861 \tag{11.64}$$

This clearly is only a mild constraint.

Physically, the limit arises because a minimum amount of coil current is required to push the plasma away from the coil surface $r = r_c$; that is, a coil current is needed to produce a hollow pressure profile. The value of the coil current relative to the plasma current at the equilibrium limit is obtained from Eq. (11.63)

$$\frac{I_C}{I_P} \geq \frac{I_{eq}}{I_P} = \frac{\left(1 - \beta_{eq}\right)^{1/2}}{1 - \left(1 - \beta_{eq}\right)^{1/2}} \approx 0.594 \tag{11.65}$$

The conclusion is that for the simple model under consideration, the hard-core coil current must be about 60% (or greater) of the plasma current to avoid the equilibrium β limit.

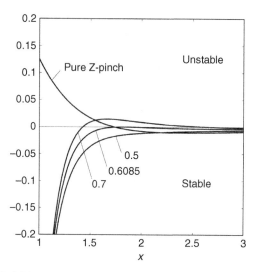

Figure 11.15   Stability criterion given by Eq. (11.66) as a function of radius for the $m = 1$ mode in a pure Z-pinch and a hard-core Z-pinch. The curve labels correspond to different values of β.

The hard-core Z-pinch m = *1 stability limit*

The $m = 1$ stability criterion given by Eq. (11.43) applies to the hard-core Z-pinch as well as the standard Z-pinch since the only assumption used in the derivation is cylindrical symmetry with a purely poloidal equilibrium magnetic field. Recall that the stability criterion is a local criterion that must be satisfied at each value of r. In terms of the normalized variables the $m = 1$ stability criterion has the form

$$\frac{d}{dx}\left(\frac{b_\theta^2}{x^{1/2}}\right) < 0 \qquad (11.66)$$

For the standard Z-pinch the criterion is always violated near the origin since $B_\theta \propto r \propto x^{1/2}$. The hard core dramatically changes this situation. Near the surface of the coil, $B_\theta \propto 1/r \propto 1/x^{1/2}$ implying stability. In fact in the region between the coil radius r_c and the radius of the maximum pressure r_{max} the pressure gradient is positive and the curvature is negative producing an overall favorable curvature term in δW_F. Instability is only possible for $r > r_{max}$ where the curvature contribution becomes negative. Intuitively, it is clear that it becomes more difficult to satisfy the stability criterion as β gets larger since the slope of b_θ^2 becomes increasingly positive. Refer to Fig. 11.15.

A quantitative prediction for the critical β against the $m = 1$ instability can be obtained by substituting the equilibrium profile given by Eq. (11.62) into Eq. (11.66). A short calculation shows that the local stability criterion can be written as

$$\beta \leq \frac{2x^{3/2}}{9x - 10} \qquad (11.67)$$

The function on the right-hand side has a minimum at $x = 10/3 \approx 3.33$ and leads to a β limit given by

$$\beta \leq \beta_{m=1} = \frac{1}{10}\left(\frac{10}{3}\right)^{3/2} \approx 0.609 \tag{11.68}$$

The $m = 1$ stability limit is considerably lower than the equilibrium β limit but nonetheless is still quite high in terms of its absolute value.

Lastly, the value $\beta_{m=1}$ is substituted into the global equilibrium pressure balance relation (i.e., Eq. (11.63)) to determine the minimum coil current to suppress the $m = 1$ instability. One finds

$$\frac{I_C}{I_P} \geq \frac{I_{m=1}}{I_P} = \frac{(1 - \beta_{m=1})^{1/2}}{1 - (1 - \beta_{m=1})^{1/2}} \approx 1.67 \tag{11.69}$$

The coil current must be more than one and a half times larger than the plasma current to avoid the instability.

The hard-core Z-pinch m = 0 stability limit

A similar stability analysis holds for the $m = 0$ sausage instability. In terms of the normalized quantities the stability criterion given by Eq. (11.48) can be rewritten as

$$-\frac{x}{P}\frac{dP}{dx} \leq \frac{10b_\theta^2}{6b_\theta^2 + 5xP} \tag{11.70}$$

where γ has been set to 5/3. The equilibrium profiles are now substituted into Eq. (11.70). A straightforward calculation leads to the following condition on β for stability:

$$\beta \leq \frac{8x^{3/2}(x+5)}{63x^2 - 40x + 25} \tag{11.71}$$

This function has a minimum at $x \approx 6.95$. At this value the β limit and corresponding minimum coil current have the values

$$\beta \leq \beta_{m=0} \approx 0.628$$

$$\frac{I_C}{I_P} \geq \frac{I_{m=0}}{I_P} = \frac{(1 - \beta_{m=0})^{1/2}}{1 - (1 - \beta_{m=0})^{1/2}} \approx 1.56 \tag{11.72}$$

For the model profiles under consideration the $m = 0$ β limit is slightly higher but still comparable to the $m = 1$ limit.

Summary of the hard-core Z-pinch

The standard Z-pinch is ideal MHD unstable to the $m = 0$ and $m = 1$ modes. The addition of a current-carrying, hard-core conductor along the axis greatly improves the stability. For the simple model investigated there is an equilibrium β limit which is quite large: $\beta_{eq} = 0.861$. Both the $m = 0$ and $m = 1$ modes set stricter limits on the maximum allowable β but these limits are high: $\beta_{m=0} \approx 0.628$ and $\beta_{m=1} \approx 0.609$. However, a large wall radius is needed in order for the edge pressure to be small. For example, a normalized wall radius $r_w/r_c = 6$ leads to a pressure ratio $p(r_w)/p_{\max} \approx 0.024$.

A toroidal version of the concept, known as the levitated dipole, must have a levitated superconducting coil in order to avoid losses to any support structure. The fact that the coil has to be superconducting also implies that advanced fuels, such as D-D, must be used in any potential fusion applications. The reason is to avoid neutron heating and quenching of the superconducting magnet as would occur with D-T as the fuel.

11.5 General stability properties of the screw pinch

The screw pinch is the cylindrical analog of the axisymmetric torus. It combines θ-pinch and Z-pinch fields with the goal of discovering configurations that have favorable stability properties at sufficiently high β to be of fusion interest. Both field components are necessary. The θ-pinch field basically provides MHD stability while the Z-pinch field provides equilibrium when the cylinder is bent into a torus.

In this section six general properties of screw pinch stability are discussed:

- A general minimized form of δW plus the corresponding differential equation for the minimizing trial function are derived.
- An analytic criterion, known as "Suydam's criterion" (Suydam, 1958), is derived from the minimized δW which determines MHD stability against localized interchanges.
- A simple and elegant procedure based on the differential equation for the minimizing perturbation is presented that demonstrates how to test for MHD stability in a general screw pinch. This is known as Newcomb's procedure (Newcomb, 1960).
- The analysis of δW is further generalized by the derivation of the full eigenmode differential equation for the general screw pinch, as originally obtained by Hain and Lust (1958).
- The full eigenmode equation is then used to derive Goedbloed and Sakanaka's (1974) "oscillation theorem," which shows how growth rates and the number of radial nodes in the eigenfunction are inversely correlated. This is important in understanding the significance of Suydam's criterion.

• Lastly, the effect of replacing a perfectly conducting wall with a resistive wall is investigated. The result is a general stability criterion that shows how the resistive wall growth rate is related to growth rates with and without a perfectly conducting wall.

All the results in this section are valid for an arbitrary screw pinch. In the following sections the results are applied to two toroidal configurations that are reasonably well represented by their cylindrical analogs: the "straight" tokamak and the reversed field pinch (RFP).

11.5.1 Evaluation of δW for a general screw pinch

The evaluation of δW for a screw pinch is straightforward but involves a considerable amount of analysis. The final result is a one-dimensional integral involving only the radial component of the plasma displacement. The derivation is carried out for the Extended Energy Principle and thus includes the contribution from the vacuum region. The steps are outlined below.

Setting up the evaluation of δW$_F$

To begin recall that the one-dimensional equilibrium pressure balance relation for a screw pinch is given by

$$\frac{d}{dr}\left(p + \frac{B_z^2}{2\mu_0}\right) + \frac{B_\theta}{\mu_0 r}\frac{d}{dr}(rB_\theta) = 0 \qquad (11.73)$$

The cylindrical symmetry again allows one to Fourier analyze the displacement vector with respect to θ and z : $\boldsymbol{\xi}(\mathbf{r}) = \boldsymbol{\xi}(r)\exp(im\theta + ikz)$. It is the symmetry with respect to two coordinates that ultimately leads to the algebraic elimination of two components of $\boldsymbol{\xi}$ when minimizing δW_F.

In carrying out the analysis it is convenient to decompose the displacement vector as follows:

$$\boldsymbol{\xi} = \boldsymbol{\xi}_\perp + \xi_\|\mathbf{b} = \xi\,\mathbf{e}_r + \eta\,\mathbf{e}_\eta + \xi_\|\mathbf{b} \qquad (11.74)$$

where

$$\eta = (\xi_\theta B_z - \xi_z B_\theta)/B \qquad \mathbf{e}_\eta = (B_z\mathbf{e}_\theta - B_\theta\mathbf{e}_z)/B$$
$$\xi_\| = (\xi_\theta B_\theta + \xi_z B_z)/B \qquad \mathbf{b} = (B_\theta\mathbf{e}_\theta + B_z\mathbf{e}_z)/B \qquad (11.75)$$

Also needed are two quantities that frequently appear in the analysis

$$F(r) \equiv \mathbf{k}\cdot\mathbf{B} = kB_z + mB_\theta/r$$
$$G(r) \equiv \mathbf{e}_r\cdot\mathbf{k}\times\mathbf{B} = mB_z/r - kB_\theta \qquad (11.76)$$

These definitions and expressions are substituted into the general form for δW_F given by Eq. (11.1), after which the minimization can proceed.

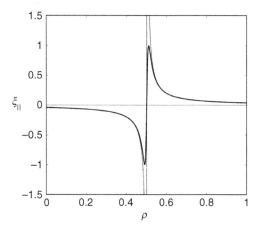

Figure 11.16   Bounded trial function for $\xi_\parallel$ that makes the plasma compressibility contribution to δW_F vanish.

Incompressibility

The first step in the minimization is to examine the incompressibility condition. The question is whether or not a $\xi_\parallel$ can be found that makes $\nabla \cdot \boldsymbol{\xi} = 0$. If so, then this is the $\xi_\parallel$ that minimizes the plasma compressibility term. Setting $\nabla \cdot \boldsymbol{\xi} = 0$ yields an expression for $\xi_\parallel$ given by

$$\xi_\parallel = i\frac{B}{F}\nabla \cdot \boldsymbol{\xi}_\perp \tag{11.77}$$

If the very special case of zero shear (i.e., $(rB_z/B_\theta)' = 0$) is excluded, then the denominator $F(r)$ will in general be non-zero except perhaps at a finite number of discrete radii. When $F(r) \neq 0$ within plasma then a well-behaved $\xi_\parallel$ can always be chosen in accordance with Eq. (11.77), which thereby makes the plasma compressibility term in δW_F vanish.

Even when $F(r) = 0$ at isolated surfaces, the plasma compressibility term can be made negligibly small with a cleverly chosen, well-behaved $\xi_\parallel$. The situation is illustrated in Fig. 11.16 for the case of a single resonant surface at $r = r_s$. The thin curve is a plot of $\xi_\parallel$ as given by Eq. (11.77). Note the singularity at $r = r_s$. The solid curve represents a trial function for $\xi_\parallel$ of the form

$$\xi_\parallel = \frac{iBF}{F^2 + \delta^2}\nabla \cdot \boldsymbol{\xi}_\perp \tag{11.78}$$

where $\delta^2 > 0$ is a small positive constant. This form of $\xi_\parallel$ is bounded and closely approximates the form required to make $\nabla \cdot \boldsymbol{\xi} = 0$ everywhere except near the singular surface. Since $\nabla \cdot \boldsymbol{\xi}$ is not exactly zero everywhere there is a potentially finite contribution to the plasma compressibility contribution from the vicinity of the singular surface whose value will depend upon δ^2.

This contribution can be evaluated by noting that when Eq. (11.78) is substituted into the expression for $\nabla \cdot \boldsymbol{\xi}$ one obtains

$$\nabla \cdot \boldsymbol{\xi} = \frac{\delta^2}{F^2 + \delta^2} \, \nabla \cdot \boldsymbol{\xi}_\perp \tag{11.79}$$

The contribution to δW_F from the plasma compressibility term can now be easily evaluated by expanding the integrand about $r = r_s$: $r = r_s + x$, $F(r) \approx F'(r_s)x$. One finds,

$$\frac{\delta W_C}{2\pi R_0} = \pi \int \gamma p \left| \nabla \cdot \boldsymbol{\xi} \right|^2 r \, dr \approx \pi \left[\gamma p r \left| \nabla \cdot \boldsymbol{\xi}_\perp \right|^2 \right]_{r_s} \int \frac{\delta^4}{\left(F'^2 x^2 + \delta^2 \right)^2} \, dx \tag{11.80}$$

Since the integrand decays rapidly for large $|x|$ the limits of integration can be extended to $-\infty < x < \infty$ with negligible error. Carrying out the integration yields

$$\frac{\delta W_C}{2\pi R_0} = \frac{\pi^2}{2} \left[\frac{\gamma p r \left| \nabla \cdot \boldsymbol{\xi}_\perp \right|^2}{|F'|} \right]_{r_s} |\delta| \propto |\delta| \tag{11.81}$$

Thus, for arbitrarily small but non-zero δ

$$\frac{\delta W_C}{2\pi R_0} \to 0 \tag{11.82}$$

implying that plasma compressibility makes a vanishingly small contribution to δW_F even when isolated singular surfaces exist in the plasma.

Evaluation of δW_F

The remaining terms in δW_F are functions only of $\boldsymbol{\xi}$ and η. The evaluation of δW_F is conceptually similar to that of the θ-pinch and Z-pinch except that the algebra is more cumbersome since the equilibrium contains two non-zero field components. The evaluation of δW_F as given by Eq. (11.1) requires the following quantities.

$$\mathbf{Q}_\perp = \nabla \times (\boldsymbol{\xi}_\perp \times \mathbf{B})_\perp = iF\boldsymbol{\xi}\mathbf{e}_r + \left\{ iF\eta + \frac{\boldsymbol{\xi}}{B} \left[B_\theta B_z' - rB_z(B_\theta/r)' \right] \right\} \mathbf{e}_\eta$$

$$\boldsymbol{\kappa} = \mathbf{b} \cdot \nabla \mathbf{b} = -\frac{B_\theta^2}{rB^2} \mathbf{e}_r$$

$$\nabla \cdot \boldsymbol{\xi}_\perp + 2\boldsymbol{\xi}_\perp \cdot \boldsymbol{\kappa} = \frac{1}{r}(r\boldsymbol{\xi})' - 2\frac{B_\theta^2}{rB^2}\boldsymbol{\xi} + i\frac{G}{B}\eta$$

$$\mu_0 J_\| = \mu_0 \mathbf{J} \cdot \mathbf{B} = \frac{1}{B} \left[B_z(rB_\theta)'/r - B_\theta B_z' \right] \tag{11.83}$$

where F and G are given by Eq. (11.76). These expressions are substituted into δW_F (after setting $\delta W_C = 0$) leading to

$$\frac{\delta W_F}{2\pi R_0} = \frac{\pi}{\mu_0} \int W(r)\, r dr$$

$$W(r) = F^2 |\xi|^2 + \left| iF\eta + \frac{\xi}{B}\left[B_\theta B_z' - rB_z(B_\theta/r)' \right] \right|^2 \qquad \text{shear Alfven}$$

$$+ B^2 \left| \frac{1}{r}(r\xi)' - 2\frac{B_\theta^2}{rB^2}\xi + i\frac{G}{B}\eta \right|^2 \qquad \text{comp. Alfven}$$

$$+ 2\frac{\mu_0 p' B_\theta^2}{rB^2} |\xi|^2 \qquad \text{curvature}$$

$$- \mu_0 J_{\|} \left\{ iF\left(\xi\eta^* - \xi^*\eta\right) - |\xi|^2 \left[\frac{B_\theta B_z'}{B} - \frac{rB_z(B_\theta/r)'}{B} \right] \right\} \qquad \text{kink}$$

$$\tag{11.84}$$

The task now is to simplify this complicated looking expression.

Minimization of δW_F

The minimization of δW_F begins with the observation that η appears only algebraically. After a short calculation the terms in $W(r)$ that explicitly contain η can be grouped together as follows:

$$W_\eta(r) = k_0^2 B^2 |\eta|^2 + 2\frac{kBB_\theta}{r}\left(i\eta\xi^* - i\eta^*\xi\right) + \frac{GB}{r}\left[i\eta\left(r\xi^*\right)' - i\eta^*(r\xi)' \right]$$

$$= \left| ik_0 B\eta + 2\frac{kB_\theta}{rk_0}\xi + \frac{G}{rk_0}(r\xi)' \right|^2 - \left| 2\frac{kB_\theta}{rk_0}\xi + \frac{G}{rk_0}(r\xi)' \right|^2 \qquad (11.85)$$

where $k_0^2 = k^2 + m^2/r^2$. Since η appears only in a positive term, δW_F is minimized by choosing

$$\eta = \frac{i}{rk_0^2 B}\left[2kB_\theta\xi + G(r\xi)' \right] \qquad (11.86)$$

With this choice for η the remaining terms in $W(r)$ can be written as

$$W(r) = A_1 \xi'^2 + 2A_2 \xi\, \xi' + A_3 \xi^2 \qquad (11.87)$$

Note that, without loss in generality, ξ can be considered to be a real quantity since the A_j are real coefficients. A straightforward calculation shows that the A_j are given by

$$A_1 = \frac{F^2}{k_0^2}$$

$$A_2 = \frac{1}{rk_0^2}\left(k^2 B_z^2 - \frac{m^2 B_\theta^2}{r^2}\right)$$

$$A_3 = F^2 + 2\frac{\mu_0 p' B_\theta^2}{rB^2} + \frac{B^2}{r^2}\left(1 - 2\frac{B_\theta^2}{B^2}\right)^2$$

$$- \frac{1}{r^2 k_0^2}(G + 2kB_\theta)^2 + 2\frac{B_\theta B_z}{rB}\left[\frac{B_\theta B_z'}{B} - \frac{rB_z}{B}\left(\frac{B_\theta}{r}\right)'\right]$$

(11.88)

The middle term in Eq. (11.87) is now integrated by parts. The expression for δW_F reduces to

$$\frac{\delta W_F}{2\pi^2 R_0/\mu_0} = \int_0^a \left(f\zeta'^2 + g\zeta^2\right) dr + \left(\frac{FF^\dagger}{k_0^2}\right)_a \zeta^2(a)$$

(11.89)

Here $F^\dagger = kB_z - mB_\theta/r$, $f = rA_1$, and $g = rA_3 - (rA_2)'$. After some tedious algebra that makes use of the equilibrium pressure balance relation, the coefficient g can be greatly simplified. The final form of the coefficients f and g are then given by (Newcomb, 1960)

$$f(r) = \frac{rF^2}{k_0^2}$$

$$g(r) = 2\mu_0 \frac{k^2}{k_0^2}p' + \frac{k_0^2 r^2 - 1}{k_0^2 r^2}rF^2 + 2\frac{k^2}{rk_0^4}FF^\dagger$$

(11.90)

Equations (11.89) and (11.90) are the desired form of δW_F, the fluid contribution to δW.

Evaluation of δW_V and δW_S

To complete the derivation of δW one must evaluate the surface and vacuum contributions. The analysis is simplified by making the realistic practical assumption that no surface currents flow on the plasma–vacuum interface. This implies that $\delta W_S = 0$.

What remains is the vacuum contribution which is evaluated as follows. In the vacuum region the magnetic field can be written as $\hat{\mathbf{B}}_1 = \nabla V_1$ with V_1 satisfying $\nabla^2 V_1 = 0$. The solution for V_1 assuming a perfectly conducting wall boundary condition at $r = b$ (i.e., $(\partial V_1/\partial r)_b = 0$) has the form

$$V_1(r) = A_0\left(K_r - \frac{K_b'}{I_b'}I_r\right)\exp(im\theta + ikz)$$

(11.91)

where $K_\rho = K_m\,(\zeta_\rho)$, $I_\rho = I_m(\zeta_\rho)$, $\zeta_\rho = |k|\,\rho$, and $K_m\,(\zeta_\rho)$, $I_m\,(\zeta_\rho)$ are modified Bessel functions. Also, prime denotes differentiation with respect to the argument and without loss in generality one can assume that $m \geq 0$ and $-\infty < k < \infty$.

The free coefficient A_0 is related to $\xi(a)$ through the boundary condition at the plasma–vacuum interface: $\hat{B}_{1r}(a) = \mathbf{e}_r \cdot \nabla \times \left(\boldsymbol{\xi}_\perp \times \hat{\mathbf{B}}\right)_a$. This condition reduces to

$$\left(\frac{\partial V_1}{\partial r}\right)_a = i(F\xi)_a \tag{11.92}$$

which leads to

$$A_0 = \frac{iF_a}{K_a}\left[1 - \left(\frac{K_b'}{I_b'}\right)\left(\frac{I_a}{K_a}\right)\right]\xi_a \tag{11.93}$$

Here, $F_a = F(a)$ and $\xi_a = \xi(a)$.

The last step is to make use of the fact that $\nabla^2 V_1 = 0$ and transform δW_V from a volume to a surface integral. The result is

$$\delta W_V = \frac{1}{2\mu_0}\int_V\left|\hat{\mathbf{B}}_1\right|^2 d\mathbf{r} = \frac{1}{2\mu_0}\int_V \nabla\cdot\left(V_1^*\nabla V_1\right)d\mathbf{r} = -\frac{1}{2\mu_0}\int_S V_1^*(\mathbf{n}\cdot\nabla V_1)\,dS$$

$$= -\frac{2\pi^2 R_0 a}{\mu_0}\left(V_1^*\frac{\partial V_1}{\partial r}\right)_a$$

$$\tag{11.94}$$

The negative sign appears because $\mathbf{n} = \mathbf{e}_r$ points radially outward. Substituting Eqs. (11.91) and (11.93) into Eq. (11.94) leads to the desired form of δW_V,

$$\frac{\delta W_V}{2\pi^2 R_0/\mu_0} = \left(\frac{r^2 F^2 \Lambda}{m}\right)_a \xi_a^2 \tag{11.95}$$

where Λ, the enhanced stabilization factor due to the conducting wall, can be written as

$$\Lambda = -\frac{mK_a}{|ka|K_a'}\frac{\left[1 - \left(K_b'I_a\right)/\left(I_b'K_a\right)\right]}{\left[1 - \left(K_b'I_a'\right)/\left(I_b'K_a'\right)\right]}$$

$$\approx \frac{1 + (a/b)^{2m}}{1 - (a/b)^{2m}} \qquad kb \sim ka \ll 1 \tag{11.96}$$

$$\approx \frac{m}{ka}\coth[k(b-a)] \qquad kb \sim ka \gg 1$$

Summary

In summary, the Energy Principle for a general screw pinch reduces to a one-dimensional quadratic integral involving only the radial component of the plasma displacement ξ. For internal modes ξ must satisfy $\xi(a) = 0$ with $\delta W = \delta W_F$ given by

$$\frac{\delta W_F}{2\pi^2 R_0/\mu_0} = \int_0^a \left(f \xi'^2 + g\xi^2 \right) dr$$

$$\frac{d}{dr}\left(f \frac{d\xi}{dr} \right) - g\xi = 0 \qquad (11.97)$$

The exact minimizing ξ satisfies the differential equation on the second line, obtained by a simple calculation that sets the variation of δW_F to zero.

For external modes $\xi(a) \neq 0$. In this case the boundary and vacuum terms must be included and δW has the form

$$\frac{\delta W}{2\pi^2 R_0/\mu_0} = \int_0^a (f\xi'^2 + g\xi^2)\, dr + \left(\frac{FF^\dagger}{k_0^2} + \frac{r^2 F^2 \Lambda}{m} \right)_a \xi^2(a) \qquad (11.98)$$

In these expressions f, g are given by Eq. (11.90) and Λ by Eq. (11.96).

The evaluation of δW is exact in that the minimization with respect to $\xi_\parallel$ and η have been carried out without approximation. Note also that $f(r)$ is positive while $g(r)$ can have either sign. Consequently, both terms are competitive and further simplifications as occurred for the θ-pinch and Z-pinch are not possible for the general screw pinch.

11.5.2 Suydam's criterion

Although a general set of stability criteria for the screw pinch cannot be obtained analytically there is one criterion that can be derived that provides an easy-to-test necessary, but not sufficient, condition against a special class of modes – localized interchanges. This criterion is known as "Suydam's criterion" (Suydam, 1958).

The motivation for examining stability against such modes follows from the observation that if $F(r) = 0$ at some radius $r = r_s$ then each term in $f(r_s)$ and $g(r_s)$ also vanishes at this radius except for the one in $g(r_s)$ proportional to $p'(r_s)$. Since $p'(r_s) < 0$ in general, then $g(r_s) < 0$, which is destabilizing. Consequently, a perturbation localized about $r = r_s$ has a reasonable likelihood of causing a pressure-driven instability. Furthermore, the mode has the structure of an interchange since $k_\parallel = \mathbf{k} \cdot \mathbf{B}/B = F/B$ vanishes at $r = r_s$. A perturbation with $k_\parallel = 0$ tends to minimize the bending of the magnetic lines which is the signature of an interchange mode.

It is important to keep in mind that a perturbation localized about $r = r_s$ does not automatically imply instability when $p'(r_s) < 0$. The reason is that when the equilibrium magnetic field has shear then away from the resonant surface $F(r) \neq 0$: $F(r) \approx F'(r_s)(r - r_s)$. Even though this term is small, when multiplied by ξ', which is large because of the localization, the product can become finite. Thus, the $f\xi'^2$ term in δW_F produces a stabilizing contribution.

The end result is that Suydam's criterion describes a competition between unfavorable curvature and localized line bending. The derivation of the criterion proceeds as follows.

Simplification of δW_F

The expression for δW_F can be greatly simplified by exploiting the assumption that the mode is localized about the singular surface. Right at the outset one must recognize that localized eigenfunctions are a special subclass of the most general eigenfunctions and therefore, whatever stability criterion may be derived, will be a necessary but not sufficient condition. With this proviso the functions f and g are expanded about the singular surface, $F(r_s) = 0$ by letting $r = r_s + x$ and assuming the eigenfunction is non-zero only for very small x. The first non-vanishing contributions to f and g are then given by

$$f \approx \left(\frac{rF'^2}{k_0^2} \right)_{r_s} x^2$$

$$g \approx \left(\frac{2\mu_0 k^2 p'}{k_0^2} \right)_{r_s}$$

(11.99)

These expressions lead to a simplified form of δW_F, and its normalized form $\delta \hat{W}_F$, that can be written as

$$\frac{\delta W_F}{2\pi^2 R_0/\mu_0} = \left(\frac{rF'^2}{k_0^2} \right)_{r_s} \delta \hat{W}_F$$

$$\delta \hat{W}_F = \int_{-\Delta}^{\Delta} \left[x^2 \left(\frac{d\xi}{dx} \right)^2 - D_S \xi^2 \right] dx$$

(11.100)

$$D_S = \left(\frac{2\mu_0 k^2 p'}{rF'^2} \right)_{r_s}$$

Here $\Delta \ll a$ is a measure of the localization of the eigenfunction. Also, the parameter D_S can be further simplified by noting that the axial wave number at the resonant surface must be chosen so that

$$k = -\left(\frac{mB_\theta}{rB_z}\right)_{r_s} \tag{11.101}$$

When this is substituted into Eq. (11.100) the result is an expression for D_S that is only a function of the equilibrium fields (and not m),

$$D_S = -\left(\frac{2\mu_0 p' q^2}{rB_z^2 q'^2}\right)_{r_s} \tag{11.102}$$

where $q(r) = rB_z/R_0 B_\theta$ is the safety factor and R_0 is the major radius of the equivalent torus.

Minimization of $\delta \hat{W}_F$

The minimization of δW_F starts off in a straightforward way but subtleties arise shortly thereafter. Even so the resulting difficulties are resolved leading to an easy-to-apply criterion known as "Suydam's criterion." The analysis begins by setting the variation of $\delta \hat{W}_F$ in Eq. (11.100) to zero in order to obtain the equation that determines the minimizing ξ. This equation is given by

$$\frac{d}{dx}\left(x^2 \frac{d\xi}{dx}\right) + D_S \xi = 0 \tag{11.103}$$

which has a simple analytic solution,

$$\xi = c_1 X^{p_1} + c_2 X^{p_2}$$

$$p_{1,2} = -\frac{1}{2} \pm \frac{1}{2}(1 - 4D_S)^{1/2} \tag{11.104}$$

$$X = |x|$$

Now, observe that at least one solution for ξ is always singular. Consequently, one cannot simply use the solution given by Eq. (11.104) as a trial function. The integrals could become infinite depending upon the strength of the singularity. This is the subtlety mentioned above. The resolution is to modify ξ near $x = 0$ so that it becomes a well-behaved, allowable trial function capable of making $\delta \hat{W}_F < 0$. As is shown below, the question of stability or instability depends upon the sign of $1 - 4D_S$.

Consider first the situation where $1 - 4D_S < 0$. In this case the roots $p_{1,2}$ are complex and ξ has the form

$$\xi = \frac{1}{X^{1/2}}[C_1 \sin(k_r \ln X) + C_2 \cos(k_r \ln X)] \qquad X = |x|$$

$$k_r = \frac{1}{2}(4D_S - 1)^{1/2} \tag{11.105}$$

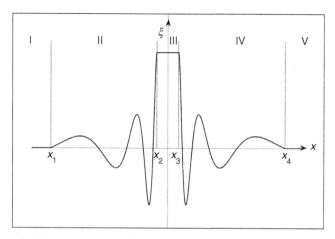

Figure 11.17   Trial function leading to Suydam instability.

The solution as $x \to 0$ oscillates infinitely rapidly with a diverging envelope proportional to $1/X^{1/2}$.

Because of the oscillatory behavior it is always possible to construct a well-behaved ξ that always makes $\delta \hat{W}_F < 0$ (i.e., instability). The modified ξ is illustrated in Fig. 11.17. Observe that $\xi = 0$ in regions I and V so that $\delta \hat{W}_F(I) = \delta \hat{W}_F(V) = 0$. In regions II and IV ξ satisfies the minimizing differential equation with solutions given by Eq. (11.105). The corresponding contributions to $\delta \hat{W}_F$ are obtained by multiplying Eq. (11.103) by ξ and integrating over each region. Since either ξ or ξ' is zero at the endpoints of both regions, one finds

$$\delta \hat{W}_F(II) = \int_{x_1}^{x_2} \left(x^2 \xi'^2 - D_S \xi^2 \right) dx = \left. \left(x^2 \xi \xi' \right) \right|_{x_1}^{x_2} = 0$$

$$\delta \hat{W}_F(IV) = \int_{x_3}^{x_4} \left(x^2 \xi'^2 - D_S \xi^2 \right) dx = \left. \left(x^2 \xi \xi' \right) \right|_{x_3}^{x_4} = 0$$

(11.106)

The total contribution to $\delta \hat{W}_F$ arises from region III where $\xi = \xi_0 = $ constant and is given by

$$\delta \hat{W}_F = \delta \hat{W}_F(III) = \int_{x_2}^{x_3} \left(x^2 \xi'^2 - D_S \xi^2 \right) dx = -D_S \xi_0^2 (x_3 - x_2) \quad (11.107)$$

By assumption $D_S > 1/4$. Therefore, $\delta \hat{W}_F < 0$ and the system is unstable.

When $D_S < 1/4$ both roots $p_{1,2}$ are real. In this case it is not possible to construct a trial function that makes $\delta \hat{W}_F < 0$. Since the solutions are non-oscillatory the solutions in regions II and IV intersect points x_2 and x_3 with a finite slope. Therefore, the contributions to $\delta \hat{W}_F(II)$ and $\delta \hat{W}_F(IV)$ are finite and their net contribution is always positive, dominating the unstable $\delta \hat{W}_F(III)$ contribution.

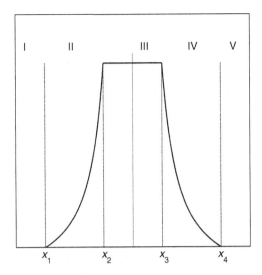

Figure 11.18   Trial function satisfying the Suydam stability criterion.

For instance, for a trial function of the form shown in Fig. 11.18 a straightforward calculation shows that

$$\delta \hat{W}_F = \xi_0^2 (x_3 - x_2) \left(-D_S + \frac{1}{2} \right)$$
$$+ \xi_0^2 (x_3 + x_2) \frac{\alpha}{2} \left[\frac{(x_2/x_1)^\alpha + 1}{(x_2/x_1)^\alpha - 1} + \frac{1 + (x_3/x_4)^\alpha}{1 - (x_3/x_4)^\alpha} \right]$$

(11.108)

where $\alpha = (1 - 4D_S)^{1/2}$. Since $D_S < 1/4$ by assumption it is clear that $\delta \hat{W}_F > 0$ implying stability against this form of perturbation.

The stability criterion $1 - 4D_S > 0$ is known as Suydam's criterion. If it is violated for any r_s in the interval $0 < r_s < a$ the plasma is unstable. The usual form for Suydam's criterion for stability is written as (Suydam, 1958)

$$r B_z^2 \left(\frac{q'}{q} \right)^2 + 8\mu_0 p' > 0$$

(11.109)

The stability boundary defined by this equation represents a balance between two competing effects. The destabilizing term results from the combination of a negative p' and the negative unfavorable curvature of the B_θ field. The stabilizing term, proportional to q'^2 represents the work done bending the magnetic field lines when interchanging two flux tubes in a system with shear.

A final question to address is whether a mode with such a highly localized structure would actually be important in a real experiment or would instead be dominated by other physics. The answer is connected with the oscillation theorem which is derived in Section 11.5.5. Briefly, the theorem shows that violation of Suydam's criterion corresponds to the existence of an accumulation point of discrete unstable eigenvalues collecting at $\omega^2 \to 0$. Equally important the oscillation theorem shows that the eigenvalues and eigenfunctions exhibit Sturmian behavior; that is, if a highly oscillatory localized mode with a certain growth rates exists (as occurs when Suydam's criterion is violated), a gross, non-oscillatory mode with zero nodes must also exist, and this mode will have the maximum growth rate for the given m and k. It is the guaranteed existence of such a large-scale mode that makes violation of Suydam's criterion important.

11.5.3 Newcomb's procedure

A general analysis of the screw pinch Energy Principle has been given by New-comb (1960). This analysis shows how a knowledge of the radial structure of ξ, as determined by the minimizing equation,

$$\frac{d}{dr}\left(f\frac{d\xi}{dr}\right) - g\xi = 0 \tag{11.110}$$

can be used to derive necessary and sufficient conditions for the stability of an arbitrary screw pinch. The end results are not simple analytic criteria, such as Suydam's criterion. Instead, the results provide a simple procedure that can be easily implemented numerically to test stability against all MHD modes. A key feature of the analysis is the use of ideas associated with Sturm's separation theorem, which will be described below.

Overall, there are three steps in Newcomb's procedure, which should be carried out in the following order:

1. Test for Suydam stability.
2. Test for general internal mode stability.
3. Test for external mode stability.

Clearly, if Suydam's criterion is violated the plasma is unstable and no further work is required. Hereafter, it is assume that Suydam's criterion is satisfied and attention is next focused on general internal mode stability. This is the step that involves the most work. Once internal mode stability is established it is then straightforward to test for external mode stability. For a configuration to be MHD stable it must satisfy all three criteria. Steps (2) and (3) are now discussed in detail.

Internal mode stability when $F(r) \neq 0$

To begin consider Eq. (11.110) and assume that m and k are such that $F(r) \neq 0$ for $0 < r < a$. For this case it is shown how Sturm's separation theorem can be used to

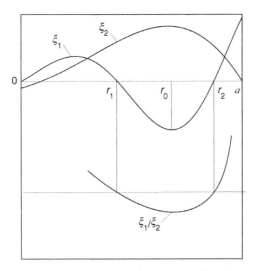

Figure 11.19   Illustration of the separation theorem.

test for internal mode stability. Since $F(r) \neq 0$ then Eq. (11.110) is non-singular (i.e., $f \neq 0$) and has two smooth independent solutions $\xi_1(r)$ and $\xi_2(r)$. Thus, any $\xi(r)$ can be written as

$$\xi(r) = c_1 \xi_1(r) + c_2 \xi_2(r) \tag{11.111}$$

Without loss of generality, $\xi_1(r)$ can be chosen to be regular at $r = 0$ but in general, $\xi_1(a) \neq 0$. Similarly, $\xi_2(r)$ can be chosen so that $\xi_2(a) = 0$ but in general is not regular at $r = 0$.

The first part of the analysis shows that when $\xi_1(r)$ has a zero in the interval $(0, a)$, one can always construct a trial function that makes $\delta W_F < 0$, implying instability. If no zero exists then $\delta W_F > 0$ and the plasma is stable to internal modes for the given m and k. The proof of these statements is based on Sturm's separation theorem, which shows that the zeros of any two independent solutions of Eq. (11.110) (e.g., $\xi_1(r)$ and $\xi_2(r)$) must alternate in radial location.

Sturm's theorem is easily demonstrated by assuming the opposite; that is, assume that $\xi_1(r)$ has two consecutive zeros located at r_1 and r_2 while $\xi_2(r) \neq 0$ in the interval (r_1, r_2) (see Fig. 11.19). Clearly $\xi_1'(r) = 0$ at some intermediate radius r_0 with $r_1 < r_0 < r_2$. Therefore, for $r_1 < r < r_2$, the well-behaved function $Y(r) = \xi_1(r) / \xi_2(r)$ also has zeros at r_1 and r_2 and a zero derivative at some modified internal point. This is also illustrated in Fig. 11.19.

One now computes the derivative $Y'(r)$,

$$Y' = \frac{\xi_2 \xi_1' - \xi_1 \xi_2'}{\xi_2^2} \tag{11.112}$$

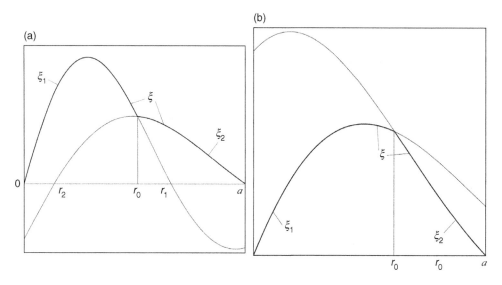

Figure 11.20   Trial function showing (a) instability and (b) stability by means of Newcomb's analysis for the case where $F(r) \neq 0$ for $0 < r < a$.

Since Y' must be zero somewhere in $r_1 < r < r_2$, this can only occur if the Wronskian $\xi_2 \xi_1' - \xi_1 \xi_2' = 0$ in this same interval. However, if this were true then the assumption that $\xi_1(r)$ and $\xi_2(r)$ are independent is contradicted. Specifically, if one solves Eq. (11.111) for arbitrary initial conditions at $r = 0$, the resulting $\xi_{\text{arb}}(r)$ and ξ_{arb}' can always be written as a linear superposition of ξ_1 and ξ_2 assuming they are independent,

$$\xi_{\text{arb}}(r) = c_1 \xi_1(r) + c_2 \xi_2(r)$$
$$\xi_{\text{arb}}'(r) = c_1 \xi_1'(r) + c_2 \xi_2'(r)$$

(11.113)

The condition that finite c_1, c_2 exist for arbitrary initial conditions on $\xi_{\text{arb}}(r)$ requires that the determinant $\xi_2 \xi_1' - \xi_1 \xi_2'$ not vanish anywhere, which contradicts the requirement that Y' vanish somewhere in the interval $r_1 < r < r_2$. The conclusion is that $\xi_2(r)$ must have a zero in this interval. In other words the zeros of $\xi_1(r)$ and $\xi_2(r)$ must alternate. This is Sturm's separation theorem.

Using this theorem, one next constructs a trial function $\xi(r)$ using $\xi_1(r)$ and $\xi_2(r)$ in different parts of the interval $(0, a)$ as shown in Fig. 11.20a. Note that $\xi(r)$ shown as the dark curve is a well-behaved allowable trial function. To show instability, assume that $\xi_1(r)$ has a single zero at $r = r_1$. (The results can be easily generalized to include multiple zeros.) Sturm's separation theorem guarantees that a radius r_2 lying in the range $0 < r_2 < r_1$ must exist for which $\xi_2(r_2) = 0$. Likewise, in the interval $r_2 < r < a$, $\xi_2(r)$ cannot have a second zero. The trial function is constructed so that $\xi(r) = \xi_1(r)$ for $0 < r < r_0$ where r_0 is the intersection point.

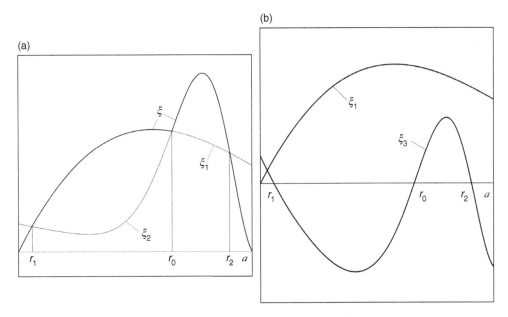

Figure 11.21 Graphical demonstration that an apparently unstable trial function as shown in (a) actually violates the separation theorem as shown in (b).

Similarly $\xi(r) = \xi_2(r)$ for $r_0 < r < a$. The value of $\delta \hat{W}_F$ is obtained by multiplying Eq. (11.110) by $\xi(r)$ and integrating over the separate regions. The result is

$$\delta \hat{W}_F = f\xi_1\xi_1' \Big|_0^{r_0} + f\xi_2\xi_2' \Big|_{r_0}^{a} = f(r_0)\xi(r_0)\left[\xi_1'(r_0) - \xi_2'(r_0)\right] \qquad (11.114)$$

Because of the separation theorem the topology of the eigenfunction must be such that $\xi_1'(r_0) < \xi_2'(r_0)$ implying that $\delta \hat{W}_F < 0$; the system is unstable.

Consider next the situation for stability. In this case assume that $\xi_1(r)$ has no zeros in the interval $(0, a)$. The only type of trial function that can be constructed that is consistent with the separation theorem and the boundary conditions is shown in Fig. 11.20b. In this case Eq. (11.114) still applies but the topology requires that $\xi_1'(r_0) > \xi_2'(r_0)$; the system is stable.

Note that it is not possible to construct a trial function as shown in Fig. 11.21a where neither $\xi_1(r)$ nor $\xi_2(r)$ has a zero, but one function, $\xi_2(r)$ for instance, undulates such that $\xi_1'(r_0) < \xi_2'(r_0)$ implying instability. If neither function has a zero the topology requires that two additional intersections must exist at r_1 and r_2. Consequently, if one replaces $\xi_1(r)$ and $\xi_2(r)$ with an equivalent set of independent functions $\xi_1(r)$ and $\xi_3(r) = \xi_2(r) - \xi_1(r)$ then in the region $r_1 < r < r_2$, $\xi_1(r)$ and $\xi_3(r)$ have the form illustrated in Fig. 11.21b. Clearly the separation theorem is violated and the situation shown in Fig. 11.21a cannot exist.

The first part of Newcomb's analysis can thus be summarized as follows. For values of m and k such that $F(r) \neq 0$ in the interval $(0, a)$ the screw pinch is stable against internal modes if and only if the non-trivial solution to the minimizing equation (i.e., Eq. (11.111)) that is regular at $r = 0$ does not have a zero crossing in $0 < r < a$.

Internal mode stability when $F(r_s) = 0$

The stability procedure is slightly more complicated when $F(r)$ has a single or multiple isolated zeros in $0 < r < a$. In this case the minimizing equation is singular at each radius r_s where $F(r_s) = 0$. A prescription is thus needed to determine how to calculate $\xi(r)$ in the vicinity of each singular surface. The prescription is as follows.

Recall from Eq. (11.104) that in the vicinity of any r_s the behavior of ξ is given by $\xi = c_1 X^{p_1} + c_2 X^{p_2}$, where $X = |x|$ and $p_{1,2} = -(1/2) \pm (1/2)(1 - 4D_S)^{1/2}$. Now, the assumption that Suydam's criterion has been satisfied implies that $1 - 4D_S > 0$. Both roots $p_{1,2}$ are real with $p_1 > -1/2$ and $p_2 < -1/2$. The root p_1 corresponds to a "weak" singularity while p_2 corresponds to a "strong" singularity.

The proper behavior of ξ is then determined by the physical condition that the potential energy remain bounded in the vicinity of r_s. For either root one finds

$$\int \left(x^2 \xi'^2 - D_S \xi^2 \right) dx = px^{2p+1} \Big|_{x \to 0}$$

$$= 0 \qquad\qquad p_1 > -1/2 \qquad\qquad (11.115)$$

$$= \infty \qquad\qquad p_2 < -1/2$$

The conclusion is that the root p_1 is allowable while p_2 leads to a divergent energy. Therefore, in the vicinity of a singular surface the trial function ξ must be chosen to contain only the small solution (i.e., the root p_1). In effect the singularity acts like a regularity condition at the origin in that it singles out one of the two independent solutions as admissible.

Because of this the interval $(0, a)$ must be broken into sub-intervals whose boundaries correspond to the successive zeros of $F(r)$. The first and last sub-intervals are bounded by $r = 0$ and $r = a$ respectively. Each sub-interval must be tested for stability separately. The previous analysis for $F(r) \neq 0$ for $0 < r < a$ can be directly applied to each sub-interval. A plot of the trial function $x^{1/2}\xi$ for the second sub-interval of a one singularity profile is illustrated in Fig. 11.22. The resulting value of $\delta \hat{W}_F$ is given by

$$\delta \hat{W}_F = f \xi_1 \xi_1' \Big|_{r_s}^{r_0} + f \xi_2 \xi_2' \Big|_{r_0}^{a} = f(r_0)\xi(r_0)\left[\xi_1'(r_0) - \xi_2'(r_0)\right] \qquad (11.116)$$

Note that the requirement that $\xi_1(r)$ contain only the small solution guarantees that the contribution at $r = r_s$ vanishes. As before, if $\xi_1(r)$ has a zero in the sub-interval then $\delta \hat{W}_F < 0$. If not, $\delta \hat{W}_F > 0$.

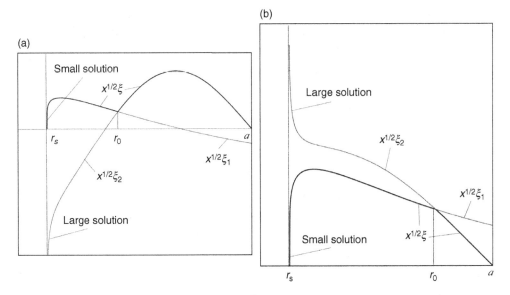

Figure 11.22 Unstable trial function $x^{1/2}\xi$ for a plasma with one singular surface in $0 < r < a$: (a) instability, (b) stability. Note that the trial function contains only the small solution at the resonant surface.

The second part of Newcomb's analysis can thus be summarized as follows for profiles with one singular surface. For values of m and k such that $F(r_s) = 0$ for $0 < r_s < a$ the screw pinch is stable against internal modes if and only if (1) Suydam's criterion is satisfied, (2) the solution to the minimizing equation that is regular at the origin has no zero crossing for $0 < r < r_s$, and (3) the solution containing only the weak singularity solution at $r = r_s$ has no zero crossing for $r_s < r < a$. The criteria are easily generalized to more than one singular surface.

External mode stability

The last step in Newcomb's procedure is to test for external mode stability. To do this assume that the given m and k have been shown to be stable to Suydam's criterion and all internal modes. One then integrates the minimizing equation either from $0 < r < a$ for no singular surfaces or over the last sub-interval from $r_s < r < a$ when one or multiple singular surfaces exist. In either case the fluid contribution to the potential energy reduces to

$$\delta \hat{W}_F = f(a)\xi(a)\xi'(a) \tag{11.117}$$

Note that internal mode stability implies that there will be no zero crossings over the region of integration. Equally important it is not necessary to introduce a separate $\xi_1(r)$ and $\xi_2(r)$ since for an external mode $\xi(a) \neq 0$; in other words, one just sets $\zeta(r) = \zeta_1(r)$, integrates over the region of interest, and evaluates the boundary term given by Eq. (11.117). This is a simple numerical procedure.

Assuming that $\delta\hat{W}_F$ is known then the complete external mode potential energy is obtained directly from Eq. (11.98), which can be rewritten as

$$\delta\hat{W} = \left(\frac{F^2}{k_0^2}\frac{r\zeta'}{\zeta} + \frac{FF^\dagger}{k_0^2} + \frac{r^2F^2\Lambda}{m}\right)_a \zeta^2(a) \tag{11.118}$$

External mode stability requires that $\delta\hat{W} > 0$.

Minimizing the search over m *and* k

A final practical point is related to the fact that a complete stability analysis requires testing all $m \geq 0$ and $-\infty < k < \infty$. Although straightforward in principle this could involve a large amount of work. The actual amount of work can, however, be dramatically reduced by exploiting the m and k dependence of $f(r)$ and $g(r)$.

Consider first internal modes for $m \neq 0$ and treat m and $\bar{k} = k/m$ as the independent wave numbers. From Eq. (11.90) it follows that $f(r)$ is independent of m and $g(r)$ contains only one term with explicit m dependence,

$$g(m, \bar{k}, r) = \bar{g}(\bar{k}, r) + m^2 r(\bar{k}B_z + B_\theta/r)^2 \tag{11.119}$$

Since the m dependent term is always stabilizing the most unstable mode corresponds to $m = 1$. Similarly, for $m = 0$, $f(r)$ is independent of k and $g(r)$ contains only one positive term with explicit k dependence,

$$g(m = 0, k, r) = \bar{g}(r) + rk^2 B_z^2 \tag{11.120}$$

In this case the most unstable mode corresponds to $k^2 \to 0$.

Consequently, when testing for internal mode stability it is necessary and sufficient to examine only $m = 1$, $-\infty < k < \infty$ and $m = 0$, $k^2 \to 0$, a significant reduction in the overall parameter space. With respect to external mode stability it can be easily shown that the same conclusions hold with the wall at infinity ($b \to \infty$). However, when the wall moves sufficiently close to the plasma the situation changes since wall stabilization is more effective for long wavelengths (i.e., small m and /or small k). In this case a more complete search is required.

Summary

Under the assumption that Suydam's criterion is satisfied, Newcomb's analysis, shows how to determine the internal and external stability of a given screw pinch profile by examining the radial structure of $\zeta(r)$ as determined by the solution of the minimizing equation. The required information can be easily obtained numerically. When the internal and external stability criteria are satisfied for the most unstable m and k values, one can then conclude that the screw pinch is MHD stable.

11.5.4 The normal mode eigenvalue equation

All the general analysis thus far presented has been based on the Energy Principle. In this subsection attention is focused on the normal mode formulation of MHD stability. The goal is to derive the normal mode eigenvalue equation and boundary conditions for the general screw pinch.

The motivations for this calculation are as follows. First, it is of interest to demonstrate the procedure for deriving eigenvalue equations starting from the normal mode equations. Second, the actual equation itself can be quite useful. Specifically, while it is difficult to obtain analytic intuition from the equation because of its complexity, it is relatively straightforward to solve the equation numerically. Lastly, there is one important analytic result that can be deduced from the equation, and that is the oscillation theorem of Goedbloed and Sakanaka (1974). This theorem, which is derived in the next subsection, proves that the eigenvalues exhibit Sturmian behavior. It thus demonstrates that the most unstable MHD mode for a given m and k is always the one whose eigenfunction has no radial nodes. The theorem also proves that the existence of highly localized, highly oscillatory modes such as arise from violation of Suydam's criterion, imply the existence of a faster growing macroscopic mode with no nodes, a much more serious instability.

The derivation of the normal mode eigenvalue equation is straightforward but does require extensive algebra. The final form of the equation is a second-order ordinary differential equation in which the eigenvalue ω^2 appears in a quite complicated way (Hain and Lust, 1958).

Formulation of the problem

The equilibrium quantities needed for the formulation are the magnetic field $\mathbf{B} = B_\theta(r)\mathbf{e}_\theta + B_z(r)\mathbf{e}_z$, the pressure $p(r)$, and the density $\rho(r)$. The desired eigenmode equation is obtained from the reduction of the generalized normal mode equations derived in Chapter 8, given by

$$-\omega^2 \rho \boldsymbol{\xi} = \frac{1}{\mu_0}(\nabla \times \mathbf{Q}) \times \mathbf{B} + \frac{1}{\mu_0}(\nabla \times \mathbf{B}) \times \mathbf{Q} - \nabla p_1$$

$$= \frac{1}{\mu_0}(\mathbf{B} \cdot \nabla \mathbf{Q} + \mathbf{Q} \cdot \nabla \mathbf{B}) - \nabla \left(p_1 + \frac{1}{\mu_0}\mathbf{B} \cdot \mathbf{Q} \right)$$

$$(11.121)$$

where

$$\mathbf{Q} = \nabla \times (\boldsymbol{\xi}_\perp \times \mathbf{B})$$
$$p_1 = -\boldsymbol{\xi}_\perp \cdot \nabla p - \gamma p \nabla \cdot \boldsymbol{\xi} \tag{11.122}$$

For cylindrically symmetric systems it is again convenient to decompose the vector components as follows:

$$\boldsymbol{\xi} = \xi \mathbf{e}_r + \eta\, \mathbf{e}_\eta + \xi_\| \mathbf{b}$$
$$\mathbf{Q} = Q_r \mathbf{e}_r + Q_\eta \mathbf{e}_\eta + Q_\| \mathbf{b}$$
$$\mathbf{e}_\eta = (B_z \mathbf{e}_\theta - B_\theta \mathbf{e}_z)/B \tag{11.123}$$
$$\mathbf{b} = (B_\theta \mathbf{e}_\theta + B_z \mathbf{e}_z)/B$$

with

$$\eta = (\xi_\theta B_z - \xi_z B_\theta)/B \qquad \xi_\| = (\xi_\theta B_\theta + \xi_z B_z)/B$$
$$Q_\eta = (Q_\theta B_z - Q_z B_\theta)/B \qquad Q_\| = (Q_\theta B_\theta + Q_z B_z)/B \tag{11.124}$$

As before, all quantities are Fourier analyzed with respect to θ and z. For example $\xi(r,\theta,z) = \xi(r) \exp(im\theta + ikz)$. Also appearing in the analysis are the two functions

$$F(r) = \mathbf{k} \cdot \mathbf{B} = kB_z + mB_\theta/r$$
$$G(r) = \mathbf{e}_r \cdot \mathbf{k} \times \mathbf{B} = mB_z/r - kB_\theta \tag{11.125}$$

Eliminating Q and p₁ in terms of ξ

The expansion of Eq. (11.122) leads to expressions for $\mathbf{Q}$ and p_1 in terms of the components of $\boldsymbol{\xi}$. A short calculation yields

$$Q_r = iF\xi$$

$$Q_\eta = iF\eta + \frac{1}{B}\left[B_\theta B_z' - rB_z(B_\theta/r)'\right]\xi$$

$$Q_\| = -iG\eta - \frac{B}{r}(r\xi)' + \frac{1}{B}\left(\mu_0 p' + 2B_\theta^2/r\right)\xi \tag{11.126}$$

$$p_1 = -p'\xi - \gamma p \left[\frac{(r\xi)'}{r} + i\frac{G}{B}\eta + i\frac{F}{B}\xi_\|\right]$$

These expressions are now to be substituted into Eq. (11.121) leading to a set of three equations and three unknowns (i.e., the three components of $\boldsymbol{\xi}$).

Eliminating η and ξ∥ in terms of ξ

A convenient way to carry out the substitution is to separate Eq. (11.121) into its $(\mathbf{e}_r, \mathbf{e}_\eta, \mathbf{b})$ components. One first forms the dot product of Eq. (11.121) with $\mathbf{b}$. A short calculation leads to

$$\left(\frac{\gamma p}{B^2}F^2 - \omega^2 \rho\right)\xi_{\parallel} + \left(\frac{\gamma p F G}{B^2}\right)\eta = \left(\frac{i\gamma p F}{rB}\right)(r\xi)' \tag{11.127}$$

Next, form the dot product of Eq. (11.121) with $\mathbf{e}_\eta$. Another short calculation yields

$$\left(\frac{\gamma p F G}{B^2}\right)\xi_{\parallel} + \left(\frac{k_0^2 B^2}{\mu_0} + \frac{\gamma p}{B^2}G^2 - \omega^2 \rho\right)\eta = \left[\frac{iG}{rB}\left(\gamma p + \frac{B^2}{\mu_0}\right)\right](r\xi)' + \left(\frac{2ikBB_\theta}{\mu_0 r}\right)\xi \tag{11.128}$$

Equations (11.127) and (11.128) should be viewed as two simultaneous algebraic equations for $\xi_{\parallel}$ and η. The algebraic behavior is a consequence of the cylindrical symmetry of the equilibrium. After solving these equations one obtains

$$\xi_{\parallel} = -\left(\frac{i\gamma p F}{r\rho B D}\right)\left[(\omega^2 - \omega_a^2)(r\xi)' + \left(\frac{2kGB_\theta}{\mu_0 \rho}\right)\xi\right]$$

$$\eta = -\left(\frac{i}{r\rho B D}\right)\left[G\left(\gamma p + \frac{B^2}{\mu_0}\right)(\omega^2 - \omega_h^2)(r\xi)' + \left(\frac{2kB^2 B_\theta}{\mu_0}\right)(\omega^2 - \omega_g^2)\xi\right] \tag{11.129}$$

Here,

$$\omega_a^2 = F^2/\mu_0 \rho$$
$$\omega_h^2 = [V_S^2/(V_S^2 + V_A^2)]\omega_a^2$$
$$\omega_g^2 = (V_S^2/V_A^2)\omega_a^2$$
$$D = (\omega^2 - \omega_f^2)(\omega^2 - \omega_s^2)$$
$$\omega_{f,s}^2 = (1/2)[k_0^2(V_S^2 + V_A^2)]\left[1 \pm \left(1 - \alpha^2\right)^{1/2}\right]$$
$$\alpha^2 = 4V_S^2\omega_a^2/\left[k_0^2\left(V_S^2 + V_A^2\right)^2\right] \tag{11.130}$$

and $V_S^2 = \gamma p/\rho$, $V_A^2 = B^2/\mu_0 \rho$, $k_0^2 = k^2 + m^2/r^2$.

The final equation

The last step in the calculation is to substitute the expressions for $\xi_{\parallel}$ and η into the $\mathbf{e}_r$ component of the normal mode equation which can be written as

$$\rho(\omega^2 - \omega_a^2)\xi = \frac{2B_\theta}{\mu_0 rB}(B_z Q_\eta + B_\theta Q_{\parallel}) + \frac{d}{dr}\left(p_1 + \frac{1}{\mu_0}BQ_{\parallel}\right) \tag{11.131}$$

After a straightforward but lengthy calculation the separate terms appearing in Eq. (11.131) can be evaluated:

$$
p_1 + \frac{1}{\mu_0} B Q_\parallel = - \frac{\rho \left(V_S^2 + V_A^2 \right)}{r} \frac{\left(\omega^2 - \omega_a^2 \right) \left(\omega^2 - \omega_h^2 \right)}{\left(\omega^2 - \omega_f^2 \right) \left(\omega^2 - \omega_s^2 \right)} (r \xi)'
$$

$$
+ \frac{2}{\mu_0 r^2} \left[B_\theta^2 - kGB_\theta \frac{\left(V_S^2 + V_A^2 \right) \left(\omega^2 - \omega_h^2 \right)}{\left(\omega^2 - \omega_f^2 \right) \left(\omega^2 - \omega_s^2 \right)} \right] (r \xi)
$$

$$
\frac{2 B_\theta}{\mu_0 r B} \left(B_z Q_\eta + B_\theta Q_\parallel \right) = - \frac{2}{\mu_0 r^2} \left[B_\theta^2 - kGB_\theta \frac{\left(V_S^2 + V_A^2 \right) \left(\omega^2 - \omega_h^2 \right)}{\left(\omega^2 - \omega_f^2 \right) \left(\omega^2 - \omega_s^2 \right)} \right] (r \xi)'
$$

$$
- \frac{1}{\mu_0 r^3} \left[r^3 \left(B_\theta^2 / r^2 \right)' - 4 k^2 B_\theta^2 V_A^2 \frac{\left(\omega^2 - \omega_g^2 \right)}{\left(\omega^2 - \omega_f^2 \right) \left(\omega^2 - \omega_s^2 \right)} \right] (r \xi)
$$

$$(11.132)$$

These complicated expressions are substituted into Eq. (11.131). A straightforward calculation leads to

$$
\frac{d}{dr} \left(A \frac{dU}{dr} \right) - CU = 0 \tag{11.133}
$$

where $U = r \xi$ and

$$
A(r) = \frac{\rho \left(V_S^2 + V_A^2 \right)}{r} \frac{\left(\omega^2 - \omega_a^2 \right) \left(\omega^2 - \omega_h^2 \right)}{\left(\omega^2 - \omega_f^2 \right) \left(\omega^2 - \omega_s^2 \right)}
$$

$$
C(r) = - \frac{\rho}{r} \left(\omega^2 - \omega_a^2 \right) + \frac{4 k^2 B_\theta^2 V_A^2}{\mu_0 r^3} \frac{\left(\omega^2 - \omega_g^2 \right)}{\left(\omega^2 - \omega_f^2 \right) \left(\omega^2 - \omega_s^2 \right)} \tag{11.134}
$$

$$
+ \frac{d}{dr} \left[\frac{B_\theta^2}{\mu_0 r^2} - \frac{2 k G B_\theta}{\mu_0 r^2} \frac{\left(V_S^2 + V_A^2 \right) \left(\omega^2 - \omega_h^2 \right)}{\left(\omega^2 - \omega_f^2 \right) \left(\omega^2 - \omega_s^2 \right)} \right]
$$

This is the desired form of the eigenvalue equation for a general screw pinch (Hain and Lust, 1958). It is a second-order ordinary differential equation with quite complicated coefficients $A(r)$, $C(r)$. Observe that the eigenvalue ω^2 appears in many terms, also in complicated ways.

Boundary conditions

To complete the formulation of the problem boundary conditions are needed. The first boundary condition, valid for both internal and external modes, requires regularity at the origin. A short calculation shows that near $r = 0$ the solutions behave like $U \propto r^{\pm m}$. Therefore, one practical way to implement regularity

numerically when determining the eigenvalues, for instance by a shooting procedure, is to start integrating the equation a short distance $\delta \ll a$ from the origin,
setting $U(\delta) = \delta^m$ and $U'(\delta) = m\delta^{m-1}$. This automatically selects the regular
branch of the general solution.

A second boundary condition is needed at the plasma surface $r = a$. For internal
modes this is a simple condition. Since the boundary does not move for an internal
mode this corresponds to requiring

$$U(a) = 0 \qquad (11.135)$$

For external modes the boundary condition is more difficult. For this case the two
jump conditions across the interface must be simultaneously satisfied, which translates into a single boundary condition determined as follows. Assume for simplicity
that the current density at the plasma edge satisfies the realistic condition $\mathbf{J}(a) = 0$.
The $\nabla \cdot \mathbf{B} = 0$ and pressure balance jump conditions can then be written as

$$
\begin{aligned}
[\![\mathbf{e}_r \cdot \mathbf{B}_1]\!]_a &= 0 \\
[\![\mathbf{B} \cdot \mathbf{B}_1]\!]_a &= 0
\end{aligned}
\qquad (11.136)
$$

Consider now the fields on the vacuum side of the boundary. From the analysis
presented in connection with Eq. (11.91) one easily can show that

$$
\begin{aligned}
\mathbf{e}_r \cdot \hat{\mathbf{B}}_1 \Big|_a &= A_0 |k| K_a' \left(1 - K_b' I_a' / I_b' K_a' \right) \\
\hat{\mathbf{B}} \cdot \hat{\mathbf{B}}_1 \Big|_a &= i A_0 F(a) K_a \left(1 - K_b' I_a / I_b' K_a \right)
\end{aligned}
\qquad (11.137)
$$

where $F_a = F(a)$, $K_\rho = K_m(\zeta_\rho)$, $I_\rho = I_m(\zeta_\rho)$, $\zeta_\rho = |k|\rho$, and A_0 is an arbitrary (and at
this point) unknown constant.

The same quantities must next be evaluated on the plasma side of the boundary.
The normal component of $\mathbf{B}_1 \equiv \mathbf{Q}$ is directly obtained from Eq. (11.126),

$$\mathbf{e}_r \cdot \mathbf{Q} \Big|_a = i F(a) \xi(a) \qquad (11.138)$$

The parallel component of $\mathbf{Q}$ requires a short calculation. One starts with $Q_\|$ given
by Eq. (11.126), substitutes η from Eq. (11.129), and makes use of the $\mathbf{J} = 0$ edge
relations, $p'(a) = 0$, $B_z'(a) = 0$, $B_\theta'(a) = -B_\theta(a)/a$. A straightforward calculation
yields

$$\mathbf{B} \cdot \mathbf{Q} \Big|_a = - \left[\frac{B^2}{r} \frac{\omega^2 - \omega_a^2}{\omega^2 - k_0^2 V_A^2} \right]_a (r\xi)_a' + \left[\frac{2\omega^2 B_\theta^2 - V_A^2 F(mB_\theta/r)}{\omega^2 - k_0^2 V_A^2} \right]_a \xi(a) \quad (11.139)$$

The required boundary condition is calculated by equating the ratio of the jump
conditions in each region: $\left[\hat{\mathbf{B}} \cdot \hat{\mathbf{B}}_1 / \mathbf{e}_r \cdot \hat{\mathbf{B}}_1 \right]_a = [\mathbf{B} \cdot \mathbf{Q} / \mathbf{e}_r \cdot \mathbf{Q}]_a$. This operation

eliminates the unknown constant A_0 from the vacuum region. After transforming dependent variables from ξ to U one arrives at the desired boundary condition

$$c_1 a U'(a) - c_2 U(a) = 0$$
$$c_1 = \left[B^2 \left(\omega^2 - \omega_a^2 \right) \right]_a \tag{11.140}$$
$$c_2 = \left[2B_\theta^2 - r^2 F^2 \Lambda / m \right]_a \omega^2 - \left[2m B_\theta F / r - k_0^2 r^2 F^2 \Lambda / m \right]_a$$

where Λ is the wall stabilization factor defined in Eq. (11.96). Note that the eigenvalue ω^2 appears in the boundary condition.

Equation (11.133), combined with the regularity condition at the origin and either the internal boundary mode condition Eq. (11.135) or the external boundary mode condition Eq. (11.140) represent the self-consistent formulation of the normal mode eigenvalue problem. As previously stated, the formulation is sufficiently complex that it is difficult to obtain much analytic insight. Still, it is in a very convenient form for numerical implementation.

11.5.5 The oscillation theorem

An important result concerning the stability of a general screw pinch has been derived by Goedbloed and Sakanaka (1974). They demonstrate the existence of an oscillation theorem, analogous to Sturm's well-known oscillation theorem, which relates the number of radial nodes in the eigenfunction to the relative size of the eigenvalue. Specifically, they show that for fixed m and k as the number of internal radial nodes l in a sequence of unstable eigenfunctions monotonically decreases, the corresponding eigenvalue ω_l^2 also decreases (i.e., becomes more negative). In other words, the most unstable eigenfunction is always the one with zero internal radial nodes. Their proof is elegant and non-trivial since the eigenvalue ω^2 appears in a very complicated way in the MHD stability equation (i.e., Eq. (11.133)) as compared to a classic Sturm–Liouville system. The derivation and interpretation of the MHD oscillation theorem is presented below.

The basic reason why an oscillation theorem exists is directly tied to the self-adjoint property of $\mathbf{F}(\xi)$. The analysis proceeds by considering the general 3-D normal mode formulation of MHD stability with attention focused on the unstable part of the spectrum; discrete eigenmodes with $\omega^2 < 0$. After several clever relations are derived that make use of the self-adjointness of $\mathbf{F}(\xi)$, the cylindrical symmetry assumption is introduced, from which one obtains the oscillation theorem. The task that motivates the analysis is the determination of which radial direction (i.e., inward or outward) the node of an unstable trial function moves as the value of ω^2 is changed.

The 3-D analysis

The starting point of the derivation is the general 3-D MHD stability equation: $-\omega^2 \rho \boldsymbol{\xi} = \mathbf{F}(\boldsymbol{\xi})$. Two independent trial functions $\boldsymbol{\xi}$ and $\boldsymbol{\eta}$, each satisfying regularity at the origin, but not $\mathbf{n} \cdot \boldsymbol{\xi} = 0$ or $\mathbf{n} \cdot \boldsymbol{\eta} = 0$ on the plasma surface, are substituted into the following integrals, a step motivated by the analysis in Chapter 8 (i.e., Eq. (8.25)):

$$\frac{1}{2} \int \boldsymbol{\eta} \cdot \mathbf{F}(\boldsymbol{\xi}) \, d\mathbf{r} = -\delta W_F(\boldsymbol{\eta}, \boldsymbol{\xi}) - \frac{1}{2} \int (\mathbf{n} \cdot \boldsymbol{\eta}) \left[\frac{1}{\mu_0} \mathbf{B} \cdot \mathbf{Q}(\boldsymbol{\xi}) + p_1(\boldsymbol{\xi}) \right] dS$$

$$\frac{1}{2} \int \boldsymbol{\xi} \cdot \mathbf{F}(\boldsymbol{\eta}) \, d\mathbf{r} = -\delta W_F(\boldsymbol{\xi}, \boldsymbol{\eta}) - \frac{1}{2} \int (\mathbf{n} \cdot \boldsymbol{\xi}) \left[\frac{1}{\mu_0} \mathbf{B} \cdot \mathbf{Q}(\boldsymbol{\eta}) + p_1(\boldsymbol{\eta}) \right] dS$$

$$(11.141)$$

Here, it is important to keep in mind that the integration volume and surface area correspond to an arbitrary internal flux surface, not in general the plasma boundary. The relations derived in Chapter 8 can easily be shown to apply to this case since all that is required is the fact that $\mathbf{n} \cdot \mathbf{B} = 0$ on the integration surface. Also, the analysis is simplified by treating $\boldsymbol{\xi}$ and $\boldsymbol{\eta}$ as real quantities.

The next step is to subtract these integrals, making use of $\delta W_F(\boldsymbol{\eta}, \boldsymbol{\xi}) = \delta W_F(\boldsymbol{\xi}, \boldsymbol{\eta})$ since they are self-adjoint by construction. The result is

$$\int [\boldsymbol{\eta} \cdot \mathbf{F}(\boldsymbol{\xi}) - \boldsymbol{\xi} \cdot \mathbf{F}(\boldsymbol{\eta})] d\mathbf{r} = - \int [(\mathbf{n} \cdot \boldsymbol{\eta}) \hat{p}_1(\boldsymbol{\xi}) - (\mathbf{n} \cdot \boldsymbol{\xi}) \hat{p}_1(\boldsymbol{\eta})] \, dS \quad (11.142)$$

where $\hat{p}_1 = p_1 + \mathbf{B} \cdot \mathbf{Q}/\mu_0$.

Introducing cylindrical symmetry

At this point it is convenient to introduce the cylindrical symmetry assumption. The volume integral becomes a 1-D integral and the surface integral becomes an evaluation on S. The region of integration is $0 \le r \le r_S$ with $r = r_S$ representing the boundary flux surface of interest. Equation (11.142) reduces to

$$\int_0^{r_S} [\boldsymbol{\eta} \cdot \mathbf{F}(\boldsymbol{\xi}) - \boldsymbol{\xi} \cdot \mathbf{F}(\boldsymbol{\eta})] r dr = -r_S \left[\eta \hat{p}_1(\boldsymbol{\xi}) - \xi \hat{p}_1(\boldsymbol{\eta}) \right]_{r_S} \quad (11.143)$$

Here $\xi = \mathbf{e}_r \cdot \boldsymbol{\xi}$ and $\eta = \mathbf{e}_r \cdot \boldsymbol{\eta}$.

One now has to make clever choices for $\boldsymbol{\xi}$ and $\boldsymbol{\eta}$ to prove the oscillation theorem. To do this, choose an arbitrary negative value of $\omega^2 = \omega_0^2 < 0$, which in general will not be an eigenvalue. The basic differential equation (i.e., Eq. (11.133)) is then integrated from 0 to a assuming that the choice for ω_0^2 leads to a solution with N radial nodes. A typical solution is illustrated in Fig. 11.23. A clever choice for r_S assumes that it coincides with one of the radial nodes, for instance r_n: $r_S = r_n$. This

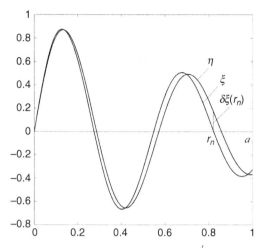

Figure 11.23 Solution for ξ and η used in the derivation of the oscillation theorem.

solution is defined as $\xi(r)$ and satisfies $\xi(r_n) = 0$. Also, since ω_0^2 is not an eigenvalue then in general $\xi(a) \neq 0$.

Consider next the choice for $\eta(r)$. Imagine repeating the previous calculation for a slightly different value of ω^2. Specifically choose $\omega^2 = \omega_0^2 + \delta\omega^2$ with $\delta\omega^2 \ll \omega_0^2$. The resulting solution is defined as $\eta(r)$ and will be only slightly different than $\xi(r)$ when $\delta\omega^2$ is small. In other words, $\eta(r) = \xi(r) + \delta\xi(r)$ with $\delta\xi(r) \ll \xi(r)$. Note that since ω^2 has changed slightly, then the location of the nth node will also have shifted slightly. The direction of this shift can be determined by calculating the value of $\delta\xi(r_n)$. From Fig. 11.23 it follows that, when $\delta\xi(r_n) > 0$, the node has shifted outward. Conversely, when $\delta\xi(r_n) < 0$, the node has shifted inward. The opposite signs apply for node r_{n-1}.

The basic question to be addressed, and whose answer forms the foundation of the oscillation theorem, is whether the nodal shift in is always outwards, always inwards, or can go in either direction, as ω^2 is changed monotonically; that is, mathematically one needs to determine a relation between $\delta\xi(r_n)$ and $\delta\omega^2$.

The relation between $\delta\xi(r_n)$ and $\delta\omega^2$

The answer to the question is obtained by evaluating the various terms in Eq. (11.143). The left-hand side is evaluated by making use of the fact that $\xi(r)$ and $\eta(r) = \xi(r) + \delta\xi(r)$ satisfy

$$-\omega_0^2 \rho \xi = \mathbf{F}(\xi)$$

$$-\left(\omega_0^2 + \delta\omega^2\right)\rho(\xi + \delta\xi) = \mathbf{F}(\eta)$$

(11.144)

Now, form the dot product of the first equation with $\boldsymbol{\eta}$ and the second with $\boldsymbol{\xi}$ and integrate from 0 to r_S. The resulting equations are subtracted with the first non-vanishing term given by

$$\int_0^{r_S} [\boldsymbol{\eta} \cdot \mathbf{F}(\boldsymbol{\xi}) - \boldsymbol{\xi} \cdot \mathbf{F}(\boldsymbol{\eta})] r dr \approx \delta\omega^2 \int_0^{r_n} \rho(\boldsymbol{\xi})^2 r dr \qquad (11.145)$$

The left-hand side, which coincides with the left-hand side of Eq. (11.143), is evaluated by substituting the expressions for $\hat{p}_1$ from Eq. (11.132). A short calculation shows that the first non-vanishing term that survives the cancellation can be written as

$$-r_S\left[\eta\hat{p}_1(\xi) - \xi\hat{p}_1(\eta)\right]_{r_S} = \left[r^2 A\left(\eta\,\xi' - \xi\,\eta'\right)\right]_{r_n}$$

$$\approx \left[r^2 A\xi'\right]_{r_n} \delta\xi(r_n) \qquad (11.146)$$

where $A(r)$ is defined by Eq. (11.134). Equating the terms in Eqs. (11.145) and (11.146) yields the desired relation between $\delta\xi(r_n)$ and $\delta\omega^2$,

$$\delta\xi(r_n) = \frac{\int_0^{r_n} \rho^2(\xi)^2 r dr}{[r^2 A\xi']_{r_n}} \delta\omega^2 \qquad (11.147)$$

The oscillation theorem

The oscillation theorem can now be easily derived by examining Eq. (11.147). Observe that the numerator is positive. Similarly, when $\omega^2 < 0$ the function $A(r)$ is also positive. Lastly for the node r_n illustrated in Fig. 11.23 one sees that $\xi'(r_n)$ is negative. Consequently, when $\delta\omega^2 < 0$ then $\delta\xi(r_n) > 0$; the node always moves outward. The same conclusion applies to the node at r_{n-1}. For this case $\xi'(r_n)$ becomes positive and $\delta\xi(r_n) < 0$, again leading to an outward movement of the node.

Continuous application of this result for a sequence of negative $\delta\omega^2$'s leads to the oscillation theorem. If an unstable eigenfunction exists with n radial nodes then as ω^2 is continuously decreased (corresponding to increased instability), every time $\xi(a)$ happens to cross zero, the result is an unstable eigenfunction with $n - 1$, $n - 2$, $n - 3$, ..., 0 radial nodes. In all cases the zero node mode has the maximum growth rate. This is the main conclusion from the oscillation theorem.

As previously stated the oscillation theorem has important implications with regard to Suydam's criterion. When the criterion is violated the oscillation theorem guarantees the existence of a zero node mode with maximum growth rate. The oscillation theorem also shows that $\omega^2 = 0$ is a point of accumulation from the

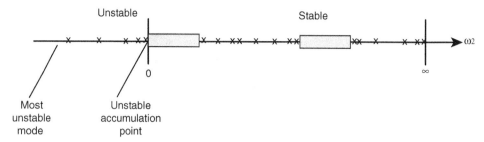

Figure 11.24 Illustration of an accumulation point at $\omega^2 = 0$ from the unstable side of the spectrum when Suydam's criterion is violated.

unstable side of the spectrum when Suydam's criterion is violated as shown in Fig. 11.24. An elegant and complete analysis of the oscillation theorem and its implication on the stable as well as unstable side of the spectrum has been given by Goedbloed and Poedts (2004).

An important practical consequence of the oscillation theorem is that when carrying out numerical studies of MHD stability of a cylindrical screw pinch, one need only focus on the eigenfunction with zero nodes (for the given m and k) to determine the most unstable mode.

11.5.6 The resistive wall mode

The final general theoretical result for the screw pinch concerns the resistive wall mode (Goedbloed *et al.*, 1972). This is a potentially very serious external MHD instability, important in both RFPs and tokamaks. If the mode is excited it often leads to major disruptions. Understanding the resistive wall mode in its simplest form in a cylindrical geometry is the goal of the present subsection.

The mode arises as follows. In fusion-grade experiments the first wall surrounding the plasma, often the vacuum chamber, is metallic. Thus, to a certain extent the wall behaves approximately like a perfect conductor. If so, this leads to a large improvement in MHD stability. The physics is straightforward. As a cylindrical plasma deforms due to an MHD perturbation, parts of the cross section move closer to the wall while other parts move further away. For the closer parts the poloidal field trapped between the surface and the wall increases since the trapped flux must remain constant (i.e., both the plasma and wall are perfect conductors). The resulting increased magnetic tension pushes back on the plasma, attempting to restore it to its original circular shape. This is clearly the wall stabilization effect. A similar argument holds for the parts of the surface that are further away.

Now, in practice the first wall is obviously not a perfect conductor. In fact it is often deliberately designed to have a substantial resistivity (compared to a thick copper shell). This allows externally applied vertical fields to penetrate the chamber sufficiently rapidly to provide effective feedback control of the equilibrium position of the plasma. Also, with a higher resistivity smaller currents are induced in the first wall during transients, alleviating power supply requirements.

The problem that arises with a resistive wall with respect to stability is that over a period of time the compressed and expanded portions of the poloidal field diffuse through the wall. In fact, after several "resistive wall diffusion times" the effect of the wall vanishes and the stability boundaries return to those of the "no-wall" situation. The resulting decrease in MHD stability β limits is important in tokamaks since the no-wall values can be only marginally satisfactory for fusion energy applications. In the RFP the situation is even more serious since external modes are always unstable without a perfectly conducting wall; the no-wall limit is $\beta = 0$.

Avoiding the resistive wall mode requires feedback stabilization. This can be accomplished if the wall is close enough that the plasma would be stable if the wall was perfectly conducting. The reason is that the characteristic resistive wall growth time slows down to the order of the resistive wall diffusion time which is long enough so that practical feedback circuits can be constructed. If the perfectly conducting wall is not close enough to provide stability, then the mode grows with an ideal MHD growth rate which is too fast for practical feedback stabilization.

In the discussion below, the growth rate of the resistive wall mode is calculated and related to the ideal MHD stability boundaries with and without a perfectly conducting wall. This is the basic resistive wall mode in its simplest form. It should be noted that the resistive wall MHD stability limits are favorably affected by equilibrium plasma flows and plasma kinetic effects. While these effects improve the situation, they do not completely and reliably eliminate the problem in all practical situations and feedback will likely still be required. Flow and kinetic effects are advanced topics that will not be covered in this textbook.

The analysis is carried out in four steps: (1) reference cases are established for ideal MHD stability with and without a perfectly conducting wall; (2) magnetic fields are calculated in the vacuum regions and within the resistive wall itself; (3) jump conditions are derived that connect the fields across the resistive wall; and (4) the resulting set of equations is solved yielding the dispersion relation for the resistive wall mode in a general cylindrical screw pinch.

The reference cases

The reference cases needed for the analysis correspond to the values of δW for external modes with and without a perfectly conducting wall. The basic results have already been derived and are given by Eqs. (11.98) and (11.118), repeated here for convenience:

$$\frac{\delta W}{2\pi^2 R_0/\mu_0} = \int_0^a \left(f\xi'^2 + g\xi^2 \right) dr + \left(\frac{FF^\dagger}{k_0^2} + \frac{r^2 F^2 \Lambda}{m} \right) \xi^2(a) \Bigg|_a$$

$$= \left(\frac{F^2}{k_0^2} \frac{r\xi'}{\xi} + \frac{FF^\dagger}{k_0^2} + \frac{r^2 F^2 \Lambda}{m} \right) \xi^2(a) \Bigg|_a \tag{11.148}$$

Recall that the second form is obtained by assuming that a solution to the minim-izing equation for ξ has been obtained and substituted into the integral in the first form. Also, keep in mind that the value of $(r\xi'/\xi)_a$ depends only on the plasma profiles and not the location of the conducting wall.

The two required reference values differ only because of the value of Λ and can be written as

$$\frac{\delta W_\infty}{2\pi^2 R_0/\mu_0} = \left(\frac{F^2}{k_0^2} \frac{r\xi'}{\xi} + \frac{FF^\dagger}{k_0^2} + \frac{r^2 F^2 \Lambda_\infty}{m} \right) \xi^2(a) \Bigg|_a$$

$$\frac{\delta W_b}{2\pi^2 R_0/\mu_0} = \left(\frac{F^2}{k_0^2} \frac{r\xi'}{\xi} + \frac{FF^\dagger}{k_0^2} + \frac{r^2 F^2 \Lambda_b}{m} \right) \xi^2(a) \Bigg|_a \tag{11.149}$$

where

$$\Lambda_\infty = -\frac{mK_a}{|ka|K_a'} \approx 1$$

$$\Lambda_b = \Lambda_\infty \left[\frac{1 - (K_b'I_a)/(I_b'K_a)}{1 - (K_b'I_a')/(I_b'K_a')} \right] \approx \frac{1 + (a/b)^{2m}}{1 - (a/b)^{2m}} \tag{11.150}$$

The approximate formulae are valid when $kb \sim ka \ll 1$. Clearly $\Lambda_b > \Lambda_\infty$. The interesting physical regime occurs when

$$\delta W_\infty < 0 < \delta W_b \tag{11.151}$$

That is, without a wall the plasma is unstable to an external MHD mode while with a wall the mode is stabilized.

Using the Energy Principle δW rather than the full eigenmode solutions is useful for the resistive wall analysis because the expected growth rates are anticipated to be very slow, on the order of the resistive wall diffusion time. Therefore, for the resistive wall analysis one can set $\omega^2 \approx 0$ in the full eigenmode equation which then reduces to the marginal stability minimizing equation $(f\xi')' - g\xi = 0$. The conclusion is that the value of $(r\xi'/\xi)_a$ appearing in the resistive wall analysis is identical to the value in both reference cases.

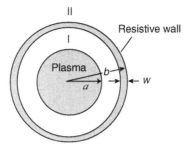

Figure 11.25   Geometry for the resistive wall stability analysis.

What remains is to calculate the exterior magnetic fields which are modified by the presence of the resistive wall.

The vacuum and wall solutions

The perturbed fields exterior to plasma require a three region solution as illustrated in Fig. 11.25. In the two vacuum regions the fields are again expressed in terms of potentials containing modified Bessel functions: $\hat{\mathbf{B}}_I = \nabla V_I$, $\hat{\mathbf{B}}_{II} = \nabla V_{II}$,

$$
\begin{aligned}
V_I &= \left(A_0 K_\rho + A_1 I_\rho\right)\exp[\omega_i t + i(m\theta + kz)] \\
V_{II} &= A_2 K_\rho \exp[\omega_i t + i(m\theta + kz)]
\end{aligned}
\tag{11.152}
$$

As before $K_\rho = K_m(\zeta_\rho)$, $I_\rho = I_m(\zeta_\rho)$, $\zeta_\rho = |k|\rho$. The A_j are unknown coefficients to be calculated from the yet to be determined matching conditions. The time dependence has been explicitly shown with $\omega = i\omega_i$ in anticipation of the fact that the growth rate will turn out to correspond to a purely growing mode. Observe also that V_{II} has only a K_ρ contribution since regularity requires that $V_{II} \to 0$ as $r \to \infty$.

Consider next the magnetic field within the resistive wall which is determined from Faraday's law, after substituting $\hat{\mathbf{E}}_1 = \eta \hat{\mathbf{J}}_1 = (\eta/\mu_0)\nabla \times \hat{\mathbf{B}}_w$ with η the resistivity of the wall. Using the fact that $\nabla \cdot \hat{\mathbf{B}}_w = 0$ one obtains the standard magnetic diffusion equation

$$
\frac{\partial \hat{\mathbf{B}}_w}{\partial t} = \frac{\eta}{\mu_0}\nabla^2 \hat{\mathbf{B}}_w
\tag{11.153}
$$

The solution to Eq. (11.153) is not very complicated but does require some thought. The key point is to assume that $w \ll b$, corresponding to the "thin wall" approximation. This leads to the following simplifications: (1) Only the radial component $\hat{B}_{wr}$ is required for the analysis. (2) The geometry becomes slab-like with the introduction of a local radial coordinate x: $r = b + x$ with $0 \leq x \leq w$. (3) For a thin wall there is rapid radial variation over the narrow distance in x. The variation over θ and z is much slower implying that $\nabla^2 \approx \partial^2/\partial x^2$. (4) The radial magnetic field $\hat{B}_{wr}$ can be Fourier analyzed: $\hat{B}_{wr} = \hat{B}_{wr}(x)\exp[\omega_i t + i(m\theta + kz)]$.

(5) The typical magnitude of ω_i corresponds to the inverse wall diffusion time: $\omega_i \sim \eta/\mu_0 b w$.

Formally the analysis is carried out by introducing the thin wall expansion parameter $\delta = w/b \ll 1$. The various terms appearing in the analysis are ordered as follows:

$$\frac{1}{r^2}\frac{\partial^2}{\partial\theta^2} \sim \frac{\partial^2}{\partial z^2} \sim \delta\frac{\mu_0\omega_i}{\eta} \sim \delta^2\frac{\partial^2}{\partial x^2} \tag{11.154}$$

Under these conditions only the radial component of Eq. (11.153) is required, which reduces to

$$\frac{\partial^2\hat{B}_{wr}}{\partial x^2} - \frac{\mu_0\omega_i}{\eta}\hat{B}_{wr} = 0 \tag{11.155}$$

Furthermore, for this equation to be self-consistent $\hat{B}_{wr}$ must be ordered such that

$$\hat{B}_{wr} = \hat{B}_{r0} + \hat{B}_{r1} + \cdots \tag{11.156}$$

with $\hat{B}_{r1}/\hat{B}_{r0} \sim \delta$ and $\hat{B}_{r0} = $ constant.

The wall solution is obtained by substituting Eq. (11.156) into Eq. (11.155). The leading- and first-order equations are given by

$$\frac{\partial^2\hat{B}_{r0}}{\partial x^2} = 0$$

$$\frac{\partial^2\hat{B}_{r1}}{\partial x^2} = \frac{\mu_0\omega_i}{\eta}\hat{B}_{r0} \tag{11.157}$$

The solutions can be written as

$$\hat{B}_{r0} = B_{r0} = \text{constant}$$

$$\hat{B}_{r1} = B_{r0}\left(\frac{\mu_0\omega_i}{2\eta}x^2\right) + B_{r1}\left(\frac{x}{w}\right) \tag{11.158}$$

where B_{r0} and B_{r1} are unknown constants of zeroth- and first-order respectively. Note that $\mu_0\omega_i x^2/\eta \sim \delta$ in the thin wall approximation. The complete solution for $\hat{B}_{wr}$ has the form

$$\hat{B}_{wr} \approx \hat{B}_{r0} + \hat{B}_{r1} = B_{r0}\left(1 + \frac{\mu_0\omega_i}{2\eta}x^2\right) + B_{r1}\left(\frac{x}{w}\right) \tag{11.159}$$

The solutions in all three regions have now been specified in terms of five unknown constants A_0, A_1, A_2, B_{r0}, B_{r1}. These constants are determined by appropriate matching conditions.

The jump conditions across the wall

There are four boundary conditions across the resistive wall. These conditions can be reduced to a simple set of jump conditions relating the potential functions $\hat{V}_I$ to $\hat{V}_{II}$ at $r = b$.

To begin note that although the wall is thin, it still has a finite thickness. Therefore there are no surface currents on either surface of the wall. The implication is that both the normal and tangential components of the wall magnetic field must be continuous with the corresponding adjacent vacuum fields on each wall surface. Continuity of the normal component of magnetic field requires

$$\frac{\partial \hat{V}_I}{\partial r}\bigg|_{b^-} = \hat{B}_{wr}\bigg|_{x=0} \qquad \frac{\partial \hat{V}_{II}}{\partial r}\bigg|_{b^+} = \hat{B}_{wr}\bigg|_{x=w} \tag{11.160}$$

To evaluate continuity of the tangential magnetic field observe that the condition $\nabla \cdot \mathbf{B} = 0$ within the wall implies that $i\mathbf{k} \cdot \hat{\mathbf{B}}_w = -\partial \hat{B}_{wr}/\partial x$ with $\mathbf{k} = (m/b)\mathbf{e}_\theta + k\mathbf{e}_z$. Similarly, in the vacuum regions $i\mathbf{k} \cdot \hat{\mathbf{B}}_1 = -k_b^2 \hat{V}$, where $k_b^2 = k^2 + m^2/b^2$. Thus continuity of the tangential fields requires

$$\hat{V}_I\bigg|_{b^-} = \frac{1}{k_b^2}\frac{\partial \hat{B}_{wr}}{\partial x}\bigg|_{x=0} \qquad \hat{V}_{II}\bigg|_{b^+} = \frac{1}{k_b^2}\frac{\partial \hat{B}_{wr}}{\partial x}\bigg|_{x=w} \tag{11.161}$$

After a short calculation the constants $\hat{B}_{r0}$ and $\hat{B}_{r1}$ can be eliminated from Eqs. (11.160) and (11.161) resulting in the following set of jump conditions for the vacuum potential functions. Each condition contains the first non-vanishing contribution in the w/b expansion:

$$\frac{\partial \hat{V}_I}{\partial r}\bigg|_{b^-} = \frac{\partial \hat{V}_{II}}{\partial r}\bigg|_{b^+}$$

$$\hat{V}_I\bigg|_{b^-} = \hat{V}_{II}\bigg|_{b^+} - \left(\frac{\mu_0 \omega_i w}{\eta k_b^2}\right)\frac{\partial \hat{V}_{II}}{\partial r}\bigg|_{b^+} \tag{11.162}$$

In terms of the coefficients A_j these conditions can be written as

$$A_0 K_b' + A_1 I_b' - A_2 K_b' = 0$$

$$A_0 K_b + A_1 I_b - A_2\left(K_b - \frac{\mu_0 \omega_i w |k|}{\eta k_0^2}K_b'\right) = 0 \tag{11.163}$$

Equation (11.163) should be viewed as two linear algebraic equations for the two unknown coefficients A_1, A_2 in terms of the third coefficient A_0. Solving these equations yields

$$\Delta A_1 = -\nu K_b' A_0$$
$$\Delta A_2 = \left(I_b K_b' - K_b I_b'\right) A_0$$
$$\nu = \left(\mu_0 |k| w / \eta k_b^2\right) \omega_i \tag{11.164}$$
$$\Delta = I_b K_b' - K_b I_b' + \nu K_b' I_b'$$

Here ν is a normalized form of the growth rate.

The resistive wall dispersion relation

The required information from the exterior region is now available to determine the dispersion relation, which is obtained by matching the two jump conditions across the plasma surface:

$$[\![B_{1r}]\!]_a = 0$$
$$[\![\mathbf{B} \cdot \mathbf{B}_1]\!]_a = 0 \tag{11.165}$$

This calculation can be conveniently carried out by dividing the jump conditions by each other; that is, the dispersion relation is determined by setting

$$\left[\frac{\mathbf{B} \cdot \mathbf{Q}}{Q_r}\right]_a = \left[\frac{\hat{\mathbf{B}} \cdot \hat{\mathbf{B}}_1}{\hat{B}_{1r}}\right]_a \tag{11.166}$$

After eliminating A_1 by means of Eq. (11.164), one finds that the right-hand side can be written as

$$\left[\frac{\hat{\mathbf{B}} \cdot \hat{\mathbf{B}}_1}{\hat{B}_{1r}}\right]_a = \frac{iF_a}{|k|}\left(\frac{\Delta K_a - \nu I_a K_b'^2}{\Delta K_a' - \nu I_a' K_b'^2}\right) \tag{11.167}$$

where $F_a = F(a)$. Similarly, the left-hand side can be evaluated by using the general normal mode relations given by Eqs. (11.126) and (11.132) after setting $\omega^2 = 0$,

$$\left[\frac{\mathbf{B} \cdot \mathbf{Q}}{Q_r}\right]_a = \frac{i}{aF_a}\left[\frac{F^2}{k_0^2}\frac{r\xi'}{\xi} + \frac{FF^\dagger}{k_0^2}\right]_a \tag{11.168}$$

The dispersion relation is now obtained by equating Eq. (11.167) to Eq. (11.168) and solving for ν. After untangling the notation and making use of the reference values of δW one finds

$$\omega_i \tau_w = -\frac{\delta W_\infty}{\delta W_b} \tag{11.169}$$

Here, τ_w is the resistive wall diffusion time

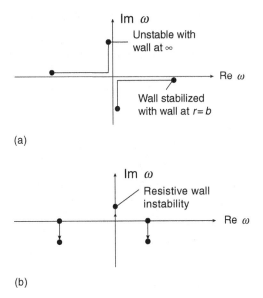

(a)

(b)

Figure 11.26 Spectral behavior of resistive wall instabilities: (a) ideal wall stabilization of a mode that is unstable with the wall at infinity; (b) growth of a resistive wall mode from the origin $\omega^2 = 0$.

$$\tau_w = -\frac{\mu_0 wb}{\eta} g$$

$$g = \frac{|k| K'_b}{k_b^2 b K_b} \left(\frac{1 - I'_a K'_b / I'_b K'_a}{1 - I_b K'_b / I'_b K_b} \right) \approx \frac{1}{2m} \left[1 - \left(\frac{a}{b} \right)^{2m} \right] \tag{11.170}$$

The geometry factor $g \sim 1$ with the approximate value corresponding to $kb \ll 1$.

Discussion

There are a number of points to observe from the remarkably simple dispersion relation given by Eq. (11.169). First, the eigenfrequency is real indicating pure growth or decay depending upon the sign of $\delta W_\infty / \delta W_b$. Second, as assumed, ω_i has a characteristic growth (or decay) time of order $1/\tau_w$. Third, in the interesting case which is unstable with the wall at $r = \infty$ (i.e., $\delta W_\infty < 0$) but is stabilized with a perfectly conducting wall at $r = b$ (i.e., $\delta W_b > 0$) then $\omega_i > 0$; the presence of a resistive wall introduces a new slowly growing mode into the system so that the stability boundary reverts back to the one with the wall at infinity. Both δW_∞ and δW_b must be positive for stability.

The spectral picture is illustrated in Fig. 11.26. In Fig. 11.26a the trajectory of an initially unstable MHD mode is traced in the complex ω plane as a perfectly conducting wall is moved inward from infinity. Since ω^2 is real for an ideal wall, the roots must lie on either the real or imaginary axis. In Fig. 11.26b the roots are traced as the wall at $r = b$ is transformed from perfectly conducting to resistive. Although not proven here, it can be shown that the ideal wall stabilized roots become slightly damped. The resistive wall instability corresponds to the development of a new mode growing out of the origin $\omega^2 = 0$.

To conclude, the resistive wall mode represents a serious MHD instability, often leading to major disruptions. It would be highly desirable for fusion applications to have a configuration that is MHD stable at sufficiently high β with the wall at infinity, thereby avoiding the resistive wall mode. If this is not possible then some form of feedback would be required. Alternatively, one could depend on plasma rotation and/or kinetic effects to stabilize the mode, but this requires highly reliable plasma performance over narrow windows of operation which has not been that easy to achieve experimentally, at least up until the present time.

11.6 The "straight" tokamak

The stage has now been set to apply the general screw pinch analysis to modern fusion concepts. The first such concept is the "straight" tokamak. In this configuration the tokamak is straightened out into a cylindrical column of length $2\pi R_0$ with the fields satisfying the tokamak inverse aspect ratio expansion: $B_\theta/B_z \sim \varepsilon$, $q \sim 1$, and $2\mu_0 p/B_z^2 \sim \varepsilon^2$, where $\varepsilon = a/R_0 \ll 1$. Substituting this expansion into the general screw pinch analysis leads to a remarkably simple form for δW_F. This form, nonetheless, describes at a basic level many practical limits on tokamak operation.

To put the analysis in perspective one should keep in mind that, in general, experimental tokamak performance seriously degrades when either the current or pressure becomes too large. The straight tokamak is surprisingly accurate in describing the effects of large current. It is not very reliable with respect to pressure-driven limitations because the effects of toroidicity tend to dominate and are obviously not included in the straight tokamak model. With this as background, the specific tokamak operational limits discussed in this section include the following:

- Sawtooth oscillations – internal $m = 1$ mode
- Current-driven disruptions – external low m modes
- Density-driven disruptions – external low m modes
- Resistive wall modes – external low m modes
- Edge localized modes (ELMs) – external high m modes.

It is worth emphasizing that each of the above phenomena is macroscopic in nature and readily observed experimentally. A complete description of any of these phenomena requires non-linear simulations plus the inclusion of additional physical effects not contained in the straight, ideal MHD, tokamak model. Such effects include toroidicity, non-circularity, resistivity, two-fluid modeling, kinetic behavior, and turbulent transport. Still, at the initiation, each phenomenon is driven by an ideal MHD instability.

The net result of exceeding any of the above operational limits is highly degraded experimental performance ranging from greatly enhanced transport to catastrophic termination. Ideal MHD achieves the important goal of describing how to avoid such unfavorable behavior from occurring in the first place. The answers can be quite helpful in that the theory provides certain quantitative guidelines that can be implemented experimentally; for example, keep the current or pressure peaked and below a critical value. Other guidelines, while straightforward conceptually, are often not easy to implement experimentally; for example, keep the edge current density gradient below a critical value. Here is where the skill and inventiveness of experimentalists are crucial.

It should also be noted that there are additional operational limits on tokamaks that are not initiated by ideal MHD instabilities. Neo-classical tearing modes leading to disruptions are one example. These require advanced analysis and are not discussed here.

The analysis below starts with a derivation of the reduced form of δW applicable to the straight tokamak. Each of the phenomena listed above is then described physically and investigated by means of the straight tokamak MHD model.

11.6.1 Reduction of δW for the straight tokamak

The large aspect ratio expansion

The straight tokamak potential energy is obtained by substituting the tokamak expansion into the general screw pinch form of δW given by Eqs. (11.98) and 11.90). The result is an asymptotic expansion of the form $\delta W = \delta W_0 + \delta W_2 + \delta W_4 + \cdots$, where each term scales as $\delta W_n \sim \varepsilon^n \delta W_0$ with $\delta W_0 \sim B_0^2 R_0 \xi^2 / \mu_0$. The calculation is straightforward although one must include the toroidal periodicity constraint by setting

$$k = -\frac{n}{R_0} \tag{11.171}$$

Here, n is an integer representing the toroidal mode number. The toroidal periodicity constraint is important in a tokamak since most of the unstable modes have long wavelengths, corresponding to $n \sim 1$; in other words, k cannot realistically be thought of as a continuously varying parameter.

The first non-vanishing contribution to δW is of second order and is given by (Shafranov, 1970)

$$
\delta \hat{W}_2 = \frac{\delta W_2}{\varepsilon^2 W_0} = \int_0^a \left(\frac{n}{m} - \frac{1}{q}\right)^2 \left[r^2 \xi'^2 + \left(m^2 - 1\right)\xi^2\right] r\,dr
$$
$$
+ \left(\frac{n}{m} - \frac{1}{q_a}\right)\left[\left(\frac{n}{m} + \frac{1}{q}\right)_a + m\Lambda\left(\frac{n}{m} - \frac{1}{q_a}\right)\right] a^2 \xi_a^2
$$

$$(11.172)$$

where $W_0 = 2\pi^2 R_0 B_0^2 / \mu_0 a^2$ and the subscript "a" denotes evaluation at $r = a$. Also, keep in mind that without loss in generality it has been assumed that $m \geq 0$ and $-\infty < n < \infty$. One major advantage of the straight tokamak configuration, as demonstrated in the relevant examples below, is that complete MHD stability against all MHD modes can be achieved without the need for a conducting wall.

The absence of pressure

An interesting feature of the straight tokamak form of δW is the absence of the pressure, a consequence of the fact that β is assumed to be small in either the ohmic or high β tokamak expansions. This feature can also be demonstrated by examining Suydam's criterion for localized interchanges which is exact, requiring no expansion in ε. It does, however, simplify when the ε expansion is substituted. Using the definition $\beta(r) = 2\mu_0 p(r)/B_0^2$, one finds that Suydam's criterion for stability reduces to

$$
\left(\frac{rq'}{q}\right)^2 + 4r\beta' > 0
$$

$$(11.173)$$

Since $\beta \sim \varepsilon$ or ϵ^2 and $rq'/q \sim 1$ it follows that Suydam's criterion is easily satisfied over almost the entire profile. The only exception is near the axis where, because of cylindrical geometric effects the shear is small. Expanding near the axis allows Eq. (11.173) to be written as

$$
\left(\frac{q_0''}{q_0}\right)^2 r^2 + 4\beta_0'' > 0
$$

$$(11.174)$$

where the "0" subscript denotes evaluation at $r = 0$. The criterion is always violated near the axis unless the pressure is flattened over a small region,

$$
\Delta r/a = \left(2q_0/aq_0''\right)\left(-\beta_0''\right)^{1/2} \sim \beta^{1/2} \ll 1
$$

$$(11.175)$$

Stability requires only a mild alteration of the profile. It is worth noting, neverthe-less, that this is one example where the pressure effects in the straight tokamak lead to a qualitatively different result than for the toroidal case. The differences are discussed in the next chapter where it is shown that toroidal curvature effects modify the stability criterion such that under reasonable conditions on q_0 no flattening of the pressure profile is required.

11.6.2 Sawtooth oscillations – the internal m = 1 mode

The first important straight tokamak instabilities to investigate are the low m internal kink modes. The analysis shows that only the $m = 1$ mode can lead to an instability whose non-linear evolution results in sawtooth oscillations. The internal kink mode theory and corresponding experimental description of sawtooth oscillations are described below.

Evaluation of δW_F for internal modes

For internal modes one sets $\xi(a) = 0$ and $\delta W = \delta W_F$. The general form of the potential energy, given by Eq. (11.172), reduces to

$$\delta \hat{W}_2 = \int_0^a \left(\frac{n}{m} - \frac{1}{q} \right)^2 \left[r^2 \xi'^2 + (m^2 - 1) \, \xi^2 \right] r \, dr \qquad (11.176)$$

At first glance it would appear that all internal modes are stable since every term is positive. This is certainly true for $m \geq 2$ since both terms in the integrand are positive and non-zero.

For $m = 1$ one needs to be more careful before concluding stability since the second term in the integrand vanishes. The $m = 1$ mode is stable when the $q(r)$ profile increases with r (as it usually does for a tokamak) and $nq_0 > q_0 > 1$ or when $n \leq 0$. For these cases $n/m - q(r)$ never vanishes and only enters the integrand as a positive coefficient.

However, if a $q = 1$ surface exists in the plasma (i.e., $q_0 < 1 < q_a$) then an $n = 1$ trial function can be constructed, as shown in Fig. 11.27, that causes $\delta \hat{W}_2 \to 0$ as the width of the pedestal δ shrinks to zero. Specifically, for the trial function in Fig. 11.27 one can easily show that

$$\delta \hat{W}_2 = \frac{1}{12} \left(r_1^3 q_1'^2 \xi_0^2 \right) \delta \to 0 \qquad (11.177)$$

Here, the subscript "1" denotes evaluation at the $q = 1$ surface.

The implication is that plasma is marginally stable to order ε^2 when a $q = 1$ surface exists in the plasma. To determine stability requires the evaluation of the next non-vanishing contribution to δW which for the straight tokamak occurs

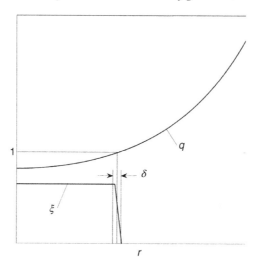

Figure 11.27 Trial function for an $m = 1$ internal kink mode in a straight tokamak.

in order ε^4. This contribution is obtained by a straightforward expansion of Eq. (11.90) and leads to (Rosenbluth *et al.*, 1973)

$$\delta \hat{W}_4 = \frac{\delta W_4}{\varepsilon^4 W_0} = \xi_0^2 \int_0^{r_1} \left[r\beta' + \frac{r^2}{R_0^2} \left(1 - \frac{1}{q} \right) \left(3 + \frac{1}{q} \right) \right] r \, dr \qquad (11.178)$$

Both contributions to the integrand are negative indicating that the plasma is unstable. Physically, when a plasma has an internal $q = 1$ surface, the stabilizing effects of line bending vanish and the destabilizing effects due to the parallel current and unfavorable curvature dominate, leading to instability.

Toroidal effects have a large impact on the form of the correction to δW_2 and are discussed in Chapter 12. The results are more complicated but essentially the same conclusion is reached – stability against the internal $m = 1$, $n = 1$ requires

$$q_0 > 1 \qquad (11.179)$$

The next question to address is "What happens to the plasma when the stability criterion is violated?"

Sawtooth oscillations

Tokamak experiments experience a form of relaxation oscillation known as the "sawtooth oscillation" when $q_0 < 1$. Although the mode structure can be complex in a torus, the dominant harmonic is $m = 1$, $n = 1$. Thus, the mode is closely connected, at least in terms of its initial drive, with the ideal mode just discussed.

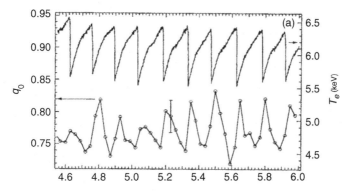

Figure 11.28   Experimental trace of the $T_e(0)$ and q_0 vs. time showing sawtooth oscillations (from Yamada *et al.*, 1994). Reproduced with permission.

Typical experimental traces of the on axis plasma temperature and safety factor as a function of time during a sawtooth oscillation are illustrated in Fig. 11.28 (Yamada *et al.*, 1994). The shape of the time evolution makes it obvious why it is called a sawtooth oscillation. A qualitative picture of the behavior is as follows. The oscillation is basically a two-stage process. In the first stage, corresponding to the long run-up before the collapse, the plasma is heated by classical means, usually ohmic power. Since the current density $J \approx E_0/\eta \propto T^{3/2}$, its central value J_0 increases with time as the heating progresses. Furthermore, since

$$q_0 = \frac{2B_0}{\mu_0 R_0 J_0} \tag{11.180}$$

one sees that an increase in J_0 causes a decrease in q_0.

Eventually, q_0 decreases below the marginal stability value $q_0 = 1$, after which the $m = 1, n = 1$ MHD mode is excited. This second stage corresponds to the rapid "crash" of the sawtooth. During this phase energy and current are expelled from the $q_0 < 1$ region and redistributed to the mid to outer regions of the plasma. Once the crash is completed, another run-up period begins and the process repeats itself. Overall, sawtooth oscillations do not cause a major degradation of plasma performance. Their main effect is to limit the maximum current density on axis to the value

$$J_0 \leq J_{\max} = \frac{2B_0}{\mu_0 R_0} \tag{11.181}$$

Only when the instability criterion is exceeded by a large amount can the sawtooth oscillations negatively affect plasma performance. In this regime the redistributed energy can actually reach the plasma boundary and be lost from the system. This is rarely the regime in which tokamaks are operated.

If one does not look too closely, the above picture would seem to explain sawtooth oscillations and their connection to the internal $m = 1$, $n = 1$ MHD mode. However, a closer comparison of the data with non-linear simulations, including resistive effects, shows that certain important aspects of sawtooth behavior are still not understood from first principles. Specifically, it is difficult to predict (1) the length of the long run-up period, (2) the depth of the crash before the plasma recovers, and (3) the fact that in many tokamaks the central safety factor does not oscillate about the value $q_0 = 1$, but a slightly lower value $q_0 \simeq 0.75$.

At present, the sawtooth oscillation remains an active area of research. Still, stepping back, one sees that from an overall perspective, the ideal MHD stability criterion, $q_0 > 1$ provides a reasonable estimate of the maximum allowable current density on the axis of a tokamak. Also, in the context of the straight tokamak model, internal ideal MHD modes with $m \geq 2$ should be stable.

11.6.3 Current-driven disruptions – external low m modes

The second important class of instabilities described by the straight tokamak model is the low m external kink mode. This is the most dangerous mode experimentally, usually leading to a catastrophic collapse of the plasma current and pressure, known as a major disruption. Mathematically, external kinks are more dangerous energetically than internal kinks since their stability is determined solely by the second-order contribution to δW. The modes can be driven by too much current or a combination of too much current and too much pressure. The straight tokamak model provides a reasonable description of the purely current-driven external kink.

The analysis presented below separates into two parts. The first part involves the $m = 1$ mode and leads to the well-known Kruskal–Shafranov current limit for tokamaks. This is a purely analytic calculation. The second part treats the case of $m \geq 2$ external kinks. These modes produce a more restrictive limit on the allowable current whose value is determined by a combination of analytic and numerical calculations.

The $m \geq 2$ stability boundary depends, not only on the total current, but on the current density profile as well, particularly the steepness of the edge gradient. This behavior is demonstrated by analyzing several profiles with increasing flatness of the edge gradient.

As stated earlier, all external modes can be stabilized for reasonable profiles without the need of a perfectly conducting wall.

Possible ranges of unstable wave numbers

The stability of low m external kinks is determined by an analysis of the full δW given by Eq. (11.172) but with the conducting wall moved to infinity: $b/a \rightarrow \infty$ implying that $\Lambda = 1$. The potential energy is thus given by

$$\delta \hat{W}_2 = \int_0^a \left(\frac{n}{m} - \frac{1}{q} \right)^2 \left[r^2 \xi'^2 + (m^2 - 1) \xi^2 \right] r dr$$
$$+ \left(\frac{n}{m} - \frac{1}{q_a} \right) \left[\left(\frac{n}{m} + \frac{1}{q} \right)_a + m \left(\frac{n}{m} - \frac{1}{q_a} \right) \right] a^2 \xi_a^2$$

(11.182)

One can gain some insight into the possible range of unstable wave numbers by noting that the integral contribution to $\delta \hat{W}_2$ is either positive or zero and again assuming without loss in generality that $q_a > 0$, $m > 0$, and $-\infty < n < \infty$. By examining the boundary term one then sees that sufficient conditions for stability to a given m, n mode can be written as

$$\frac{n}{m} > \frac{1}{q_a}$$
$$\frac{n}{m} < 0$$

(11.183)

Alternatively this implies that a necessary condition for instability is that

$$0 < nq_a < m$$

(11.184)

If $q(r)$ is an increasing function of radius as it is for a tokamak, then Eq. (11.184) implies that the resonant surface for an external mode, located at the radius r_s where $nq(r_s) = m$, must lie in the vacuum region: $q(r_s) > q_a$.

The m = 1 Kruskal–Shafranov limit

Consider now the $m = 1$ mode. For this case the minimizing trial function in the plasma is simply given by $\xi(r) = \xi_a = $ constant. This choice makes the integral contribution to $\delta \hat{W}_2$ vanish. Note that such a trial function can only be chosen for an external mode since the internal mode boundary condition $\xi_a = 0$ no longer need be applied. In addition, since the trial function $\xi(r) = $ constant actually minimizes the integral contribution, it thereby coincides with the true eigenfunction for the mode. A final important point is that the minimizing trial function is independent of the $q(r)$ profile. This leads to the important conclusion that the $m = 1$ stability boundary is a function only of q_a. In other words, the criterion depends only on the total current but not the current profile.

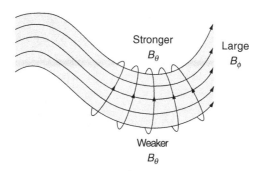

Figure 11.29   Physical driving mechanism for the $m = 1$ Kruskal–Shafranov external kink mode. Note the compression of the field near the inside of the kink perturbation and the bending of the toroidal field lines.

After making use of these results one finds that $\delta \hat{W}_2$ reduces to

$$\delta \hat{W}_2 = \frac{2a^2 \xi_a^2}{q_a} n(nq_a - 1) \qquad (11.185)$$

From Eq. (11.185) it follows that the condition to stabilize the $m = 1$ external kink mode for the most restrictive case, $n = 1$, is given by

$$q_a > 1 \qquad (11.186)$$

This criterion is known as the Kruskal–Shafranov condition (Kruskal and Schwarschild, 1954; Shafranov, 1956). It imposes an important constraint on tokamak operation in that it limits the total toroidal current that can flow to

$$I < I_{KS} = \frac{2\pi a^2 B_0}{\mu_0 R_0} = \frac{5a^2 B_0}{R_0} \, \text{MA} \qquad (11.187)$$

For, $a = 2\,\text{m}$, $R_0 = 6\,\text{m}$, and $B_0 = 6\,\text{T}$, then $I_{KS} = 20\,\text{MA}$. This limit can be increased substantially for tokamaks with an elongated cross section.

The driving mechanism for the instability is similar to that of a Z-pinch as discussed in Section 11.4.1 and illustrated here for the tokamak case in Fig. 11.29. The one main difference is that the modes are restricted to long wavelengths in order to minimize the strong stabilizing effects of bending the B_z field lines. In fact the criterion $q_a > 1$ can also be viewed as a constraint on the geometry that prevents the formation of long wavelength modes by not allowing them to fit in the torus because of toroidal periodicity requirements.

In practical situations the $m = 1$ mode is rarely seen experimentally because of the unfortunate fact that higher m modes produce even stricter requirements on the maximum allowable current. These modes have slightly lower growth rates and are more dependent on the current density profile than the $m = 1$ mode but still lead to major disruptions. These modes are the next topic of discussion.

The m ≥ 2 external kink modes

The stability of $m \geq 2$ external kinks requires an evaluation of both the integral and boundary terms in $\delta \hat{W}_2$. As stated, in this subsection attention is focused on several current density profiles with increasing flatness of the edge gradient. The stability boundaries are determined either analytically or numerically by solving the differential equation for the $\xi(r)$ that minimizes the integral contribution to $\delta \hat{W}_2$. This equation is given by

$$\frac{d}{dr} \left[r^3 \left(\frac{n}{m} - \frac{1}{q} \right)^2 \frac{d\xi}{dr} \right] - (m^2 - 1) r \left(\frac{n}{m} - \frac{1}{q} \right)^2 \xi = 0 \tag{11.188}$$

$$\xi(\delta \to 0) = \delta^{m-1} \qquad \xi'(\delta \to 0) = (m-1)\delta^{m-2}$$

From this solution one evaluates $a\xi_a'/\xi_a$ and substitutes into the following form of $\delta \hat{W}_2$ obtained from Eq. (11.118) after substituting the tokamak expansion,

$$\delta \hat{W}_2 = \frac{1}{m^2 q_a^2} (nq_a - m) \left[(nq_a + m) + \left(m + \frac{a\xi_a'}{\xi_a} \right)(nq_a - m) \right] a^2 \xi_a^2 \tag{11.189}$$

An accurate solution is required because the results are sensitive to the value of $a\xi_a'/\xi_a$, which therefore must be calculated with high precision. When a numerical solution is required it is easy to obtain since for an unstable mode there is no singular surface in the plasma.

A flat J(r) profile with a jump at r = a

The first case of interest corresponds to a flat current density profile which abruptly drops to zero at the plasma edge as illustrated in Fig. 11.30. As shown below this is the maximally unstable profile. The analysis is easy to carry out since the flat profile produces a $B_\theta \propto r$ which in turn leads to a constant safety factor profile: $q_0 = q_a = q(r) = rB_0/R_0 B_\theta(r) = \text{constant}$. Clearly, the factor $n/m - 1/q(r)$ is also constant and factors out of the basic differential equation for ξ (i.e., Eq. (11.188)) which reduces to

$$\frac{d}{dr} \left[r^3 \frac{d\xi}{dr} \right] - (m^2 - 1) r\xi = 0 \tag{11.190}$$

Now, observe that the minimizing $\xi(r)$ that satisfies Eq. (11.190), is given by

$$\xi(r) = \xi_a \left(\frac{r}{a} \right)^{m-1} \tag{11.191}$$

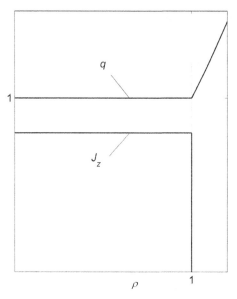

Figure 11.30   A flat $J_z(r)$ profile with a jump at the plasma edge. Also shown is the $q(r)$ profile. Here $\rho = r/a$.

which implies that

$$\frac{a\zeta'_a}{\zeta_a} = m - 1 \qquad\qquad (11.192)$$

Equation (11.192) is substituted into $\delta\hat{W}_2$ (i.e., Eq. (11.189)) yielding

$$\delta\hat{W}_2 = \frac{2}{mq_a^2}(nq_a - m)(nq_a - m + 1) \qquad\qquad (11.193)$$

The conclusion is that, for a given value of m, the condition on q_a for instability can be written as

$$m - 1 < nq_a < m \qquad\qquad (11.194)$$

The unstable values of nq_a are bracketed by adjacent values of m. By setting $n = 1$ and continuously incrementing m by unity it then follows that the plasma is unstable over the entire range $0 \leq q_a \leq \infty$! This is, therefore, the maximally unstable tokamak profile. When it comes to avoiding external kink modes, plasmas do not want to have edge current jumps.

Linear J(r) edge gradient

To show the sensitivity of the stability results to the current profile the calculation just presented is repeated for a $J(r)$ profile which vanishes at $r = a$ but with a linear edge gradient. For this case the model profile is given by

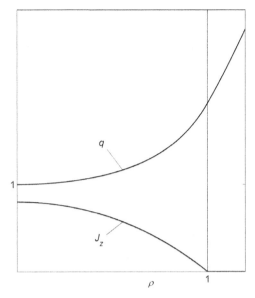

Figure 11.31   Profiles of $J_z(r)$ vs. r with zero edge current but a finite edge current gradient. Also shown is $q(r)$. Here $\rho = r/a$.

$$J(\rho) = J_0 \left(\frac{1 - \rho^2}{1 - \alpha\rho^2} \right) \tag{11.195}$$

Here, $\rho = r/a$, J_0 is the current density on axis, and α is a free profile parameter that must lie in the range $-\infty < \alpha < 1$. The value of α determines the ratio q_a/q_0. For any value of α the edge gradient is a constant: $J'(1) = -2J_0/(1 - \alpha)$. A typical profile is illustrated in Fig. 11.31.

The $J(\rho)$ profile is substituted into Ampere's law yielding $B_\theta(\rho)$ and the safety factor $q(\rho) = (\varepsilon B_0)(\rho/B_\theta)$. The safety factor and its edge value, which are the quantities needed for the evaluation of $\delta\hat{W}_2$, can be written as

$$\begin{aligned}
\frac{q_0}{q(\rho)} &= \frac{1}{\alpha^2\rho^2} \left[\alpha\rho^2 + (1 - \alpha)\ln\left(1 - \alpha\rho^2\right) \right] \\
\frac{q_0}{q_a} &= \frac{1}{\alpha^2} \left[\alpha + (1 - \alpha)\ln(1 - \alpha) \right]
\end{aligned} \tag{11.196}$$

where $q_0 = 2B_0/\mu_0 R_0 J_0$. The $q(\rho)$ profiles is also illustrated in Fig. 11.31.

The variational equation for ξ (i.e., Eq. (11.188)) is now solved numerically in order to evaluate $a\xi_a'/\xi_a$. In all cases the safety factor on axis has been set to $q_0 = 1$. This is a simple way to model the physical effects of sawtooth oscillation while mathematically guaranteeing that the plasma is stable to all internal MHD modes. In addition, since $q(\rho)$ is always an increasing function of ρ for the model profiles, this implies that $q_a \geq q_0 > 1$. The profiles automatically satisfy the Kruskal–Shafranov stability criterion.

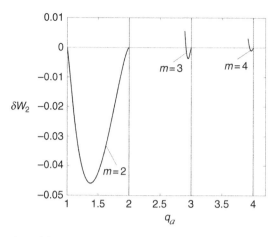

Figure 11.32   Plot of δW vs. q_a assuming $q_0 = 1$ and $n = 1$ for the constant edge gradient profile. Shown here are δW ($m = 2$), $100\delta W$ ($m = 3$), and $100\delta W$ ($m = 4$).

The quantity $a\xi_a'/\xi_a$ is substituted into the expression for $\delta\hat{W}_2$ (i.e., Eq. (11.189)) leading to the stability diagram illustrated in Fig. 11.32. Observe that the $m = 2$, $n = 1$ mode is again unstable for $1 < q_a < 2$. However, unlike the case with a finite current jump, there are narrow, rather than full, bands of instability just below $q_a = 3$ and $q_a = 4$. Note also the expanded vertical scale for $m = 3$ and $m = 4$. In fact it is shown in Section 11.6.6 that for a linear edge gradient these bands exist for all higher m values but become exponentially narrower as m increases. Also, the magnitude of $\delta\hat{W}_2$ corresponding to instability decreases rapidly with increasing m indicating weaker instability.

The overall conclusions are not as definitive as one would like. There are windows of stability between adjacent m values. However, practical operation of an experiment within such a window reduces its robustness against instability as the profiles evolve slowly in time. Often, tokamaks operate reasonably successfully with $q_a > 3$. Operation with $q_a \sim 2.5$ is sometimes possible, but can lead to major disruptions, tolerable in present experiments but not in ITER or a reactor.

Lastly, it is worth noting that the critical q_a for stability is affected by toroidicity and finite pressure. In addition, if the tokamak has a divertor then by definition $q_a = \infty$. By consensus, although not because of any rigorous mathematical analysis, the value of "q_a" in a diverted tokamak is assumed to be $q(\psi)$ evaluated at the ψ surface corresponding to 95% of the total flux and is often abbreviated as q_{95}. These issues are discussed in the next chapter.

For the present, it is reasonable to assume that $q_a > 3$ to safely avoid ideal MHD external kink modes which lead to major disruptions. Other, non-ideal effects can sometimes raise this limit even higher.

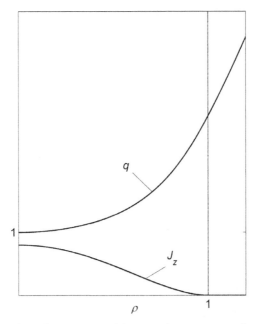

Figure 11.33 Profile of $J_z(r)$ vs. r with zero edge current and zero edge current gradient. Also shown is $q(r)$. Here $\rho = r/a$.

Flat J(r) edge gradient

A convenient choice for the current density profile with a flat edge gradient has the form

$$
J(\rho) = J_0 \left(\frac{1 - \rho^2}{1 - \alpha\rho^2} \right)^2
\tag{11.197}
$$

Again, $\rho = r/a$, J_0 is the current density on axis, and α is a free profile parameter that must lie in the range $-\infty < \alpha < 1$. Observe that both the current density and its gradient vanish at the plasma edge $\rho = 1$ for any value of α. A typical profile is illustrated in Fig. 11.33.

The safety factor and its edge value, obtained from Ampere's law, can be written as

$$
\begin{aligned}
\frac{q_0}{q(\rho)} &= \frac{1}{\alpha^3 \rho^2} \left\{ \alpha\rho^2 \left[1 + \frac{(1-\alpha)^2}{(1 - \alpha\rho^2)} \right] + 2(1-\alpha)\ln(1 - \alpha\rho^2) \right\} \\
\frac{q_0}{q_a} &= \frac{1}{\alpha^3} [\alpha(2-\alpha) + 2(1-\alpha)\ln(1-\alpha)]
\end{aligned}
\tag{11.198}
$$

where $q_0 = 2B_0/\mu_0 R_0 J_0$. The safety factor profile is also plotted in Fig. 11.33.

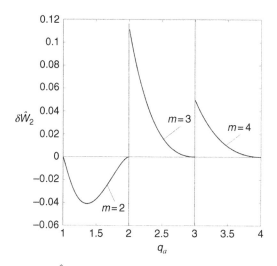

Figure 11.34   Plot of $\delta \hat{W}$ vs. q_a assuming $q_0 = 1$ for several values of m and $n = 1$ corresponding to the zero edge gradient profile.

The stability problem has now been fully formulated. The resulting numerical solutions determine $a \xi'_a / \xi_a$, which is substituted into the expression for $\delta \hat{W}_2$. The stability boundaries are presented in Fig. 11.34 which illustrates curves of $\delta \hat{W}_2$ vs. q_a for $q_0 = 1$, $n = 1$, and various values of m.

From Fig. 11.34 one sees that only $m = 2$ is unstable. The upper boundary is $q_a = 2$ while the lower boundary is $q_a = 1$. Higher $m \geq 3$ bands of instability no longer exist. The overall conclusion is that the condition for a straight tokamak to be stable against external kinks, subject to the constraints that (1) q_0 account for sawtooth oscillations and (2) the edge current density and its gradient are both zero, is given by

$$q_a > 2 \qquad\qquad (11.199)$$

The maximum allowable current is one half of the Kruskal–Shafranov current. The situation has improved over the finite edge gradient case but it is not clear whether such flat edge gradients can be reliably achieved experimentally. Still, complete stability is possible without the need of a conducting wall.

Maximally flat J(r) edge gradient

The last profile of interest corresponds to the maximally flat $J(r)$ edge gradient. This profile, first suggested by Shafranov (1970), is illustrated in Fig. 11.35. Observe that the current density is uniform between $0 < r < r_0$ and zero between $r_0 < r < a$. This implies that the safety factor $q(r) = q_0$ between $0 < r < r_0$ and $q(r) = q_0 (r^2 / r_0^2)$ between $r_0 < r < a$. Clearly, the free radial parameter r_0 sets the edge value of q: $q_a = q_0 (a^2 / r_0^2)$.

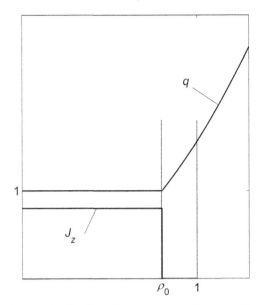

Figure 11.35 The maximally flat edge current gradient profile $J_z(r)$ vs. r. Also shown is the corresponding $q(r)$ profile. Here $\rho_0 = r_0/a$.

There are several reasons for analyzing this idealized profile. First, the calculation is completely analytic. Second, the advantages of a flat edge gradient are explicitly demonstrated. Third, because of its simplicity, it is easy to obtain a global picture of external kink mode stability as a function of q_0 and q_a.

Interestingly, the stability results for this maximally flat $J(r)$ edge gradient model can be directly obtained from the maximally unstable profile with the edge current jump. The reasoning is as follows. The interior region of the plasma has a flat current density which jumps to zero at $r = r_0$. This produces the same contribution to δW_F as the profile with the edge jump if one sets $a \to r_0$ and $q_a \to q_0$.

Next, observe that the contribution to δW_F from the region $r_0 < r < a$ is simply given by

$$\delta W_F = \frac{1}{2\mu_0} \int Q^2 \, d\mathbf{r} = \frac{1}{2\mu_0} \int B_1^2 \, d\mathbf{r} \qquad (11.200)$$

since $J(r) = 0$ in this region and $p(r)$ is negligible because of the tokamak ordering. Equation (11.200) is the same form as the vacuum energy. Therefore, the combined contribution to δW from the region $r_0 < r < a$ plus the actual vacuum region is identical to that obtained from the profile with the edge jump if one again sets $a \to r_0$ and $q_a \to q_0$.

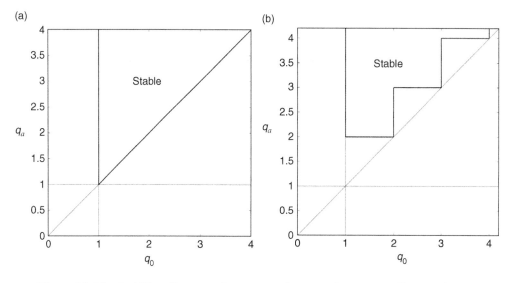

Figure 11.36   Stability diagram of q_a vs. q_0 for $n = 1$ corresponding to the maximum flat edge gradient profile: (a) basic diagram only including the $m = 1$ mode; (b) complete diagram including higher m modes. Note the additional regions of instability. In all cases the region encompassed by the heavier lines is stable.

The end result is that δW for the flat edge gradient model, with one crucial additional constraint, can be directly written down from Eq. (11.193) in terms of the equivalent variables as follows:

$$\delta \hat{W}_2 = \frac{2}{mq_0^2} (nq_0 - m)(nq_0 - m + 1) \qquad (11.201)$$

For a given m the plasma is unstable for

$$m - 1 < nq_0 < m \qquad (11.202)$$

Consider now the additional constraint that follows from the fact that external modes can only be unstable if the resonant surface lies in the vacuum region. Stated differently, the above analysis is only valid if there is no resonant surface in the region $r_0 < r < a$. See the discussion associated with Eq. (11.183) which applies to arbitrary profiles. As has already been shown Eq. (11.183) implies that for the resonant surface to lie in the vacuum region one requires that $nq_a < m$. This condition, combined with the fact that $q_a > q_0$ for a tokamak, leads to an instability constraint on q_a that can be written as

$$nq_0 < nq_a < m \qquad (11.203)$$

Equations (11.202) and (11.203) define the stability boundaries for the flat edge gradient model. They can be conveniently illustrated in Fig. 11.36 as a diagram of

q_a vs. q_0 for the most dangerous case $n = 1$. Figure 11.36a shows the basic diagram consisting of the $q_0 > 1$ boundary for sawtooth stability, $q_a > 1$ for Kruskal–Shafranov stability, and $q_a > q_0$ the basic definition of a tokamak. The area encompassed by the heavier lines is stable under these constraints.

Figure 11.36b shows the additional unstable regions due to higher m external kinks as described by Eqs. (11.202) and (11.203). For a given m Eq. (11.202) corresponds to a vertical band of instabilities. Similarly, Eq. (11.203) produces a triangular wedge. The overlap of these two regions corresponds to an instability. The net result is that the stable region has been reduced in size because of the higher m kink modes. The maximum allowable plasma current (i.e., lowest allowable q_a) corresponds to $q_a = 2$ and occurs for $1 < q_0 < 2$.

The analysis just presented has shown analytically that a flat edge gradient is desirable for good stability to $m > 2$ external kink modes.

Major disruptions

With the theory of external kink modes now established, the final topic of interest is a basic description of a major disruption. The actual physics describing the evolution of a major disruption is quite complicated involving both ideal and resistive MHD. An excellent description has been given by Wesson (2011). Qualitatively what happens during a disruption is as follows. As the total current in the plasma is increased sawtooth oscillations continue to limit the safety factor on axis to $q_0 = 1$, which corresponds to $J_0 = 2B_0/\mu_0 R_0$. Since the current density on axis is fixed and the total current is increasing, the implication is that the current profile must become broader. The edge gradient becomes steeper and the $q = 2$ surface moves closer to the plasma edge.

These are just the conditions necessary to excite internal resistive tearing modes, particularly the $m = 2$ mode. The tearing modes lead to the development of magnetic islands which tend to further flatten the current profile within the islands. Equally important, when islands with different m, n combinations overlap the magnetic field can become stochastic leading to greatly enhanced transport. The enhanced transport in turn first leads to a rapid quenching of the plasma thermal energy. This thermal quench is one of two main dangers due to major disruptions. Specifically, all the plasma energy is rapidly deposited on the first wall which may not be able to withstand the resulting thermal loads, particularly in large devices.

An equally important effect that occurs on a slightly slower time scale is the unfavorable redistribution of the current density. The profile becomes flat over the core because of the turbulence eventually reaching a state where an ideal external MHD mode is excited. In reality, Im(ω) is much smaller than the ideal growth rate because the core is surrounded by a low-temperature resistive plasma and not a

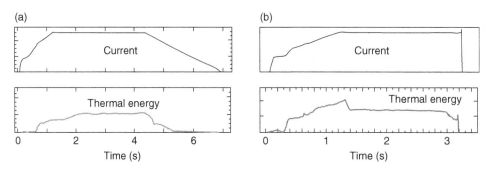

Figure 11.37   Experimental observation of the quenching of pressure and current during a major disruption (reproduced with permission from Kikuchi *et al.*, 2012). (a) No disruption, (b) major disruption at $t = 3.2$ sec.

pure vacuum. Even so the end result is the same in that the entire plasma starts to move away from its equilibrium position.

This motion, which carries the plasma to the wall of the vacuum vessel, is the second main danger due to major disruptions. During the time that the plasma is moving its current quenches because of enhanced magnetic transport and transference of current to the wall by direct contact. In a short time the entire plasma current is quenched and reappears in the wall with a non-uniform distribution approximately corresponding to the m, n mode structure of the MHD instability. This large, non-uniform current produces very large, potentially damaging electromagnetically driven mechanical stresses on the first wall.

Experimental data demonstrating the quenching of the pressure and current during a major disruption are illustrated in Fig. 11.37 (Kikuchi *et al.*, 2012). The rapid increase in wall current leads to enormous forces on the vacuum chamber and this is the main danger of a major disruption. Current experiments are small enough to withstand these forces but in much larger reactor scale devices the forces are also much larger. There is general consensus that major disruptions must be avoided in a tokamak reactor in order to prevent irreversible damage to the vacuum chamber.

11.6.4 Density-driven disruptions – external low **m** *modes*

Overview

High-performance operation of a tokamak requires that the plasma density lie below a critical value, known as the "Greenwald density limit." As the plasma density approaches the limit from below, the edge begins to cool and transport starts to degrade. When the limit is exceeded, a major disruption occurs. The empirical value of the density limit has a surprisingly simple form and is valid over a very wide range of parameters. Its value is given by (Greenwald *et al.*, 1988; Greenwald, 2002)

$$n_{20} < n_G = \frac{I_M}{\pi a^2} \qquad (11.204)$$

There is general agreement that the density limit disruptions are connected to a low m external kink mode. The intuition leading to this connection is as follows. Assume that a tokamak is operating with an edge safety factor sufficiently high that external MHD modes are not excited. As the density is raised, particularly in edge fueled tokamaks, a narrow turbulent boundary layer is formed just inside $r = a$ with high density and low temperature. The resistivity in this region becomes sufficiently high that very little current flows. Since the total current remains constant, it must now flow through a slightly smaller cross section since the edge region no longer carries current. A plasma with a fixed safety factor on axis due to the sawtooth limit, carrying the same current over a slightly smaller cross sectional area, has a correspondingly smaller safety factor at its effective edge.

As the density continues to increase the boundary region grows in width and the safety factor at the effective edge of the plasma core further decreases. Eventually the safety factor at the effective edge of the core falls below the critical value needed to avoid disruptions. Once this occurs the plasma does indeed suffer a major disruption.

While there is basic agreement on this qualitative description of density-driven disruptions, there is no first principles theory that explains why, and how much, the edge region grows as a function of increasing density. At least two possible explanations have been put forth. The first and earliest theory invokes a radiation collapse of the plasma edge. Here, line radiation due to impurities (proportional to $n_e n_Z$) increases rapidly with increasing density shifting the balance of plasma energy loss from thermal conduction to radiation. Eventually, all the energy loss is through radiation. Further increases in density cause the plasma to detach from the wall (or divertor plate). In other words, the plasma core shrinks in radius as it becomes surrounded by an increasingly growing radiation layer. This is just the situation that eventually leads to a major disruption as described above. The theory of the radiation collapse density limit has been derived in a simple but elegant calculation by Wesson (2011) and would seem, at least at first glance, to explain the experimental observations.

However, a closer look at an ever increasing amount of data raises some serious doubts that the radiation collapse is the primary explanation for the density limit. There are two basic problems. First, if line radiation is the dominant loss mechanism then one would expect the density limit to depend directly upon the type of impurity in the edge and its corresponding charge state Z. This dependence is not observed experimentally. Second, the calculation of the density limit depends upon the amount of energy lost from the plasma core which in turn depends on the core thermal conductivity κ. Now, assume that κ is a free quantity chosen so that the resulting density limit coincides with the Greenwald limit. The value and scaling of κ also do not agree with experimental measurements. Therefore, while the radiation collapse mechanism may play a role in the setting the density limit, it would appear to not be the major role.

This leads to a second possible explanation for development and growth of an edge boundary layer (Greenwald, 2002). The explanation again depends on the fact that the edge temperature becomes progressively smaller as the density increases. At low temperatures the edge behaves more like a collision dominated fluid than a collisionless kinetic plasma. Certain collision dominated MHD plasma instabilities driven by the pressure gradient and unfavorable curvature (e.g., resistive ballooning modes) can become noticeably stronger than the collisionless drift wave instabilities that dominate core turbulence, thereby leading to a region of enhanced transport near the plasma edge. It is this enhanced transport that quenches the edge plasma current, forcing it into a smaller core region. One again has a situation where increasing density and lower temperature shrink the plasma core, thus lowering q_a, and eventually leading to a major disruption.

A simple model problem is described below showing how this behavior can occur. The end result is a demonstration of how edge turbulence produces a density limit. However, it is not possible to actually reproduce the Greenwald limit since this would require a detailed knowledge of the anomalous thermal diffusivities in the core and edge regions which at present are not well known from a first principles theory. Still the analysis does provide some intuition into the basic nature of the density limit.

A model problem for a turbulence-driven density limit

The model problem discussed below separates the plasma into two regions, the main core and the turbulent edge boundary layer (see Fig. 11.38). The core is characterized by a thermal conductivity κ_1 which itself may be anomalous. Similarly the edge is characterized by a strongly MHD turbulent conductivity κ_2. Two assumptions are made. First, the edge turbulence is much larger than the core turbulence: $\kappa_2 \gg \kappa_1$. Second, as a result of localized MHD instabilities, the edge turbulence is assumed to be generated when the pressure gradient exceeds a critical value; that is, edge turbulence occurs when $p'(r)/p'_C > 1$. Here, both $p'(r)$ and p'_C are negative and for mathematical simplicity the critical gradient p'_C is assumed to be a constant.

The basic equation to be solved in each region corresponds to steady state conservation of energy, which for the straight tokamak is given by

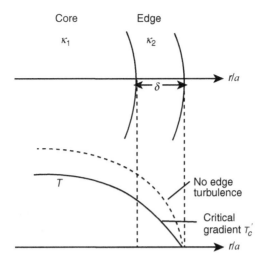

Figure 11.38   Edge geometry for the turbulence-driven density limit model.

$$\frac{1}{r}\frac{d}{dr}\left(\kappa\frac{dT}{dr}\right) + S = 0 \qquad (11.205)$$

For mathematical simplicity the heating source S and plasma density n are both assumed to be constant across the entire plasma. The constant density assumption implies that edge turbulence occurs when $T'(r)/T'_C > 1$ with $T'_C = p'_C/n = $ constant.

In the core and edge regions, denoted by the subscripts "1" and "2" respectively, the thermal conductivities, in accordance with the above discussion, are given by

$$\kappa = \kappa_1 \qquad\qquad \text{core region } T'/T'_C < 1$$
$$\kappa = \kappa_1 + \kappa_2\left(\frac{T'}{T'_C} - 1\right) \qquad \text{edge region } T'/T'_C > 1 \qquad (11.206)$$

Here, κ_1 and κ_2 are both assumed to be constants, also for mathematical simplicity. Observe that there is a linear transition in κ as a function of $T'(r)$ from core transport to edge transport in the edge region.

The boundary conditions assume regularity at the origin and a perfect heat sink at the plasma edge,

$$T'_1(0) = 0$$
$$T_2(a) = 0 \qquad (11.207)$$

Across the core–edge transition both the temperature and heat flux must be continuous. This transition is assumed to occur at a radius $r = r_\delta = a(1 - \delta)$, which leads to

$$[\![T]\!]_{r_\delta} = 0$$
$$[\![T']\!]_{r_\delta} = 0 \tag{11.208}$$

The value of δ is unknown. It is determined by the additional constraint that the temperature gradient at the transition radius r_δ actually be set equal to T'_C:

$$T'(r_\delta) = T'_C \tag{11.209}$$

The main goal of the analysis is in fact to calculate δ and show how it grows as n increases. When δ becomes sufficiently large the edge current forced into the core region lowers $q(r_\delta)$ sufficiently to cause a major disruption.

The problem has now been fully formulated and the solution is obtained as follows. Consider first the core region. The energy equation can be easily integrated and the solution satisfying the boundary condition at the origin can be written as

$$T_1(\rho) = \frac{a^2 S}{4\kappa_1}(1 - c_1 - \rho^2) \tag{11.210}$$

where $\rho = r/a$ and the free integration constant has, for convenience, been defined as $1 - c_1$. Note that when there is no edge turbulence then $c_1 = 0$ and the maximum temperature gradient occurs at the plasma edge. It has the value

$$a\frac{dT_1(a)}{dr} = -\frac{a^2 S}{2\kappa_1} \tag{11.211}$$

Equation (11.210) with $c_1 = 0$ is the correct solution for the entire plasma when the edge gradient is not steep enough to excite MHD edge turbulence. The interesting regime of the density limit occurs when this assumption is violated. In other words, edge turbulence is excited when

$$\zeta = \frac{aS}{2\kappa_1|T'_C|} > 1 \tag{11.212}$$

If $\zeta > 1$ then one must calculate $T_2(r)$ in the edge region and match the solutions across the transition surface. A simple analytic solution can be obtained by assuming that the edge region is narrow. There is only a small region of edge turbulence where $\Delta = \kappa_1/\kappa_2 \ll 1$, which implies that $\delta \ll 1$ and $\zeta - 1 \ll 1$. Under these conditions, to a high degree of accuracy, the edge temperature profile simply tracks the marginally stable critical temperature gradient:

$$\frac{dT_2}{d\rho} \approx -a|T'_C| \tag{11.213}$$

The solution satisfying the edge boundary condition and the temperature gradient constraint at $r = r_\delta$ is given by

$$T_2(\rho) = a|T'_C|(1 - \rho) \qquad (11.214)$$

The solutions now contain two unknown parameters: c_1, δ. These are determined by matching $T_1(\rho)$ and $dT_1(\rho)/d\rho$ across $r = r_\delta$ using the two jump conditions, i.e., Eq. (11.208). One finds

$$c_1 = 2(\zeta - 1)^2 \sim O(\delta^2)$$
$$\delta = (\zeta - 1)/\zeta \qquad (11.215)$$

The last relation is the most important one. It demonstrates that as ζ increases above its marginal value of unity, the edge region becomes wider. The relation can be more conveniently expressed in terms of the safety factor and the density. Specifically, in the region $r_\delta < r < a$ the current density is negligible because of the strong turbulence. In this region $B_\theta \propto 1/r$ implying that the safety factor can be written as $q = q_a\rho^2$, where q_a is assumed to be sufficiently large so that without edge turbulence, external kink modes would not be excited.

However, the actual plasma edge where the current density vanishes occurs at $r_\delta = a(1 - \delta)$ and it is the safety factor at this radius that determines whether or not an external kink mode is excited. If the critical safety factor for external kinks is denoted by q_C, then the condition to avoid a disruption is given by $q(r_\delta) = q_a(1 - \delta)^2 > q_C$. Untangling the notation leads to the following condition to avoid external kink modes: $aS/2\kappa_1|T'_C| < q_a/q_C$. If one now recalls that the onset of edge turbulence arises from MHD modes then, as stated, the threshold gradient is really due to the pressure rather than the temperature. In other words, it is more accurate to write (for a constant density) $T'_C = p'_C/n$. The end result is that the condition to avoid external kink modes, and by inference major disruptions, has the form of a density limit that can be written as

$$n < \frac{2\kappa_1 p'_C}{aS}\left(\frac{q_a}{q_C}\right)^{1/2} \qquad (11.216)$$

Interestingly, when $\kappa_2 \gg \kappa_1$ the density limit is predicted to be independent of the details of the edge turbulence and depends only on the core turbulence.

Finally, it is worth emphasizing that the density limit derived above arises from a simple model calculation. While the scaling information may be qualitatively correct it is still an incomplete calculation. The reason is that p'_C is a function of ε and q_a while κ_1 is at present still only a marginally known function of T, n, B, S, ε, q_a which depends on the nature of the core turbulence. A more accurate knowledge of turbulent core transport is needed before closing the analysis to see whether Eq. (11.216) does indeed reduce to the Greenwald limit. Still, the calculation does show how a major disruption can be initiated when the density becomes too large.

11.6.5 Resistive wall modes – external low **m** modes

The analysis of external low m modes has shown that the current and density must be kept sufficiently small in order to avoid major disruptions. A question then arises as to whether the critical values can be increased (a favorable result) by the presence of a perfectly conducting wall. Of course, if improved stability is possible then the original no-wall instabilities would reappear in the form of resistive wall modes. However, it will be optimistically assumed that resistive wall modes can be stabilized by feedback and/or rotation. The issue then is to determine the effect of a perfectly conducting wall on external kink modes and see how large an improvement can be realized.

The calculation is straightforward but the results are mixed – some favorable and some unfavorable. The required information is easily obtained from the straight tokamak Energy Principle given by Eq. (11.189), generalized to include the effect of a perfectly conducting wall as presented in Eq. (11.172). The generalized relation is given by

$$\delta \hat{W}_2 = \frac{1}{m^2 q_a^2}(nq_a - m)\left[(nq_a + m) + \left(m\Lambda + \frac{a\xi_a'}{\xi_a}\right)(nq_a - m)\right]a^2\xi_a^2$$

$$\Lambda \approx \frac{1 + (a/b)^{2m}}{1 - (a/b)^{2m}} \tag{11.217}$$

The required information is obtained by rewriting Eq. (11.217) as follows:

$$\delta \hat{W}_2 = \frac{m\Lambda + 1 + a\xi_a'/\xi_a}{m^2 q_a^2}(nq_a - m)(nq_a - M_\infty - M_\Lambda) \tag{11.218}$$

Here,

$$M_\infty = m\left(\frac{m - 1 + a\xi_a'/\xi_a}{m + 1 + a\xi_a'/\xi_a}\right) \tag{11.219}$$

is the critical value of nq_a with the wall at infinity and

$$M_\Lambda = 2(\Lambda - 1)\frac{m^2}{\left(m + 1 + a\xi_a'/\xi_a\right)\left(m\Lambda + 1 + a\xi_a'/\xi_a\right)} \tag{11.220}$$

is the modification due to a perfectly conducting wall at $r = b$.

As expected when the wall is at infinity (i.e., $\Lambda = 1$) then $M_\Lambda = 0$ leading to the no-wall stability boundary $nq_a = M_\infty$. Similarly, when the wall is on the plasma (i.e., $\Lambda = \infty$) then $M_\infty + M_\Lambda = m$ and the band of unstable q_a values shrinks to zero – the plasma is stable. Now, keeping in mind that $a\xi_a'/\xi_a$ is always positive, it follows that for the general case a band of unstable q_a values exists between

$$M_\infty + M_\Lambda < nq_a < m \qquad (11.221)$$

What conclusions can be drawn from this result? There are two. First, the upper limit on the unstable band, $nq_a = m$, is unaffected by the presence of a conducting wall. This is an unfavorable result in that the maximum allowable current before the onset of external current-driven kink modes is unchanged by a conducting wall. Second, since $M_\Lambda > 0$ the lower limit on the unstable band, $nq_a = M_\infty + M_\Lambda$ has increased over the no-wall value $nq_a = M_\infty$. That is, the width of the unstable band has narrowed, which is a favorable result.

It is disappointing that a conducting wall does not allow a higher maximum current. However, the wall does have a stronger positive effect on the maximum allowable pressure but this requires the inclusion of toroidal effects in the analysis, which are obviously not included in the straight tokamak model.

11.6.6 Edge localized modes (ELMs) – external high **m** *modes*

The final topic of interest concerns external high m external kink modes. These modes typically occur during H-mode operation of a tokamak. This regime is characterized by improved energy confinement, clearly an advantage. However, an H-mode tokamak also has substantial edge gradients, which can drive high m MHD instabilities, known as "edge localized modes" or "ELMs" for short. Typically the unstable modes have $n \sim 10$, $m \sim 30$. Pressure, density, temperature, and current profiles exhibiting H-mode edge pedestals are illustrated in Fig. 11.39 (Hughes *et al.*, 2013).

ELMs are believed to be driven by a combination of large edge pressure gradients and large edge current gradients; the resulting instability is sometimes referred to as a peeling–ballooning mode (Snyder *et al.*, 2005). A complete description requires a toroidal calculation including pressure effects. The straight tokamak model provides a basic description of the peeling component of the mode which is driven by the edge current gradient. The analysis shows that large edge current gradients can indeed excite high m external kink modes, but it does not predict which particular m values are the most dangerous. This requires the inclusion of pressure-driven ballooning effects which are described in the next chapter.

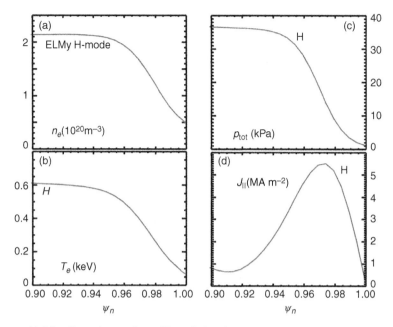

Figure 11.39 Experimental profiles of density, temperature, pressure, and current density profiles in an H-mode discharge showing edge pedestals (reproduced with permission from Hughes *et al.*, 2013).

The specific goal of the present subsection is to derive the stability windows in q_a space for high m modes driven by large edge current gradients neglecting all pressure effects. Following the derivation is an experimental description of the impact of ELMs on present and future tokamaks.

Stability of the constant current density tokamak (reference case)

Stability boundaries for high m external kink modes are again determined by the straight tokamak stability equation given by Eq. (11.189) with the minimizing ξ satisfying Eq. (11.188). These equations (for $\Lambda = 1$) are repeated here for convenience:

$$\delta \hat{W}_2 = \frac{1}{m^2 q_a^2} (nq_a - m) \left[(nq_a + m) + \left(m + \frac{a\xi'_a}{\xi_a} \right) (nq_a - m) \right] a^2 \xi_a^2$$

$$\frac{d}{dr} \left[r^3 \left(\frac{n}{m} - \frac{1}{q_a} \right)^2 \frac{d\xi}{dr} \right] - (m^2 - 1) r \left(\frac{n}{m} - \frac{1}{q_a} \right)^2 \xi = 0 \qquad (11.222)$$

The stability analysis considers a sequence of current density profiles. At one end of the sequence is a profile that is flat across the plasma with a finite jump at the

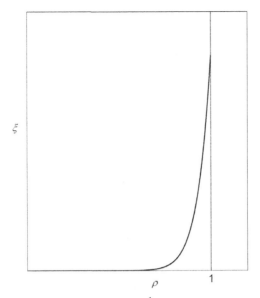

Figure 11.40   Curve of $\xi(r) = \xi(a)(r/a)^{m-1}$ vs. r for $m = 10$. Note the localization of the eigenfunction near the plasma edge. Here, $\rho = r/a$.

plasma edge. This is the reference case already analyzed in Section 11.6.3. As the sequence progresses the current density becomes more peaked but still has a finite jump at the edge. At the end of the sequence, the current density vanishes at the plasma edge but has a finite gradient.

The stability results for the reference case are given by Eq. (11.193), which shows that the condition for instability is given by

$$m - 1 < nq_a < m \tag{11.223}$$

As previously stated, Eq. (11.223) is a very pessimistic result because by setting $n = 1$ and continuously incrementing m by unity, one sees that the plasma is unstable over the entire range $0 \le q_a \le \infty$.

Observe that for high m the eigenfunctions are strongly localized around the plasma surface. This can be seen intuitively by noting that the eigenfunctions behave approximately like

$$\xi(r) \sim r^{m-1} \tag{11.224}$$

over the body of the plasma. A curve of $\xi(r) = \xi(a)(r/a)^{m-1}$ is illustrated in Fig. 11.40 for the case $m = 10$. The conclusion is that because of the localization, a conducting wall some distance away will not have any significant impact, thereby justifying the approximation of setting the wall parameter $\Lambda = 1$ in Eq. (11.222).

The consequence of this behavior is that the integral contribution to $\delta \hat{W}_2$ depends sensitively on the local behavior of $J(r)$ near $r = a$. In particular, as shown below, as $J(a)$ and $J'(a)$ become smaller the unstable regions diminish in size.

The high m stability equation

A quantitative analysis, which determines the stability boundaries in q_a space for large m has been given by Laval and Pellat (1973) and Laval *et al.* (1974). They investigated high m external kink mode stability in circular and non-circular tokamaks. For present purposes it suffices to outline their theory for the circular case.

The analysis begins by introducing a new dependent variable $\psi(r)$ into the equation for $\xi(r)$ contained in Eq. (11.222). The variable $\psi(r)$ is defined by

$$\psi = K\xi$$

$$K(r) = r^{3/2}\left(\frac{n}{m} - \frac{1}{q}\right) \tag{11.225}$$

Substituting into Eq. (11.222) yields

$$\frac{d^2\psi}{dr^2} - \left(\frac{m^2 - 1}{r^2} + \frac{1}{K}\frac{d^2K}{dr^2}\right)\psi = 0 \tag{11.226}$$

The next step is to expand Eq. (11.226) near the plasma surface assuming that $m \gg 1$. To do this, one writes

$$\frac{r}{a} = 1 - \frac{x}{2m} \tag{11.227}$$

where $x > 0$ is the new independent variable. Localization of the mode requires that x be ordered as $x \sim 1$. Similarly if the resonant surface is just outside the edge of the plasma then its location must be ordered so that $1 - nq_a/m \sim 1/m$. Under these assumptions the function $K(r)$ can be expanded as

$$K = K_1 + K_2 + \cdots \tag{11.228}$$

Here, $K_m \sim 1/m$ with

$$K_1 = \frac{a^{3/2}}{2q_a m}\left[2(nq_a - m) - \frac{aq'_a}{q_a}x\right]$$

$$K_2 = \frac{a^{3/2}}{8q_a m^2}\left[6(nq_a - m)x + \left(\frac{a^2 q''_a}{q_a} - 2\frac{a^2 q'^2_a}{q_a^2} + 3\frac{aq'_a}{q_a^2}\right)x^2\right] \tag{11.229}$$

and prime denoting d/dr. These complicated expressions can be simplified by rewriting them in terms of the edge current density J_a and its gradient J_a'. A short calculation yields

$$
\begin{aligned}
K_1 &= \frac{a^{3/2}}{q_a m}\left[nq_a - m - \left(1 - \frac{J_a}{\bar{J}}\right)x\right] \\
K_2 &= \frac{a^{3/2}}{4q_a m^2}\left[3(nq_a - m)x - \frac{aJ_a'}{\bar{J}}x^2\right]
\end{aligned}
\tag{11.230}
$$

The quantity $\bar{J}$ is the average current density defined by $\bar{J} = I/\pi a^2 = 2B_{\theta a}/\mu_0 a$.

Next, these expressions are substituted into the term K''/K. A short calculation leads to the desired equation for ψ valid in the limit of large m,

$$
\frac{d^2\psi}{dx^2} - \left[\frac{1}{4} - \frac{\lambda}{m(x + x_0)}\right]\psi = 0
$$

$$
\lambda = -\frac{1}{2}\frac{aJ_a'}{\bar{J} - J_a}
\tag{11.231}
$$

$$
x_0 = (m - nq_a)\frac{\bar{J}}{\bar{J} - J_a}
$$

Both λ and x_0 are positive and of order unity which might lead one to believe that the $\lambda/m(x + x_0)$ term should be negligible in the limit of large m. This is sometimes true but as is shown shortly subtleties arise when $J_a = 0$.

Assume now that Eq. (11.231) can be solved for $\psi(x)$. Then, the last step in the formulation is to substitute this solution into the expression for $\delta\hat{W}_2$ given by Eq. (11.222). After a short calculation the resulting expression can be written, to lowest order in $1/m$, directly in terms of $\psi(x)$ as follows:

$$
\delta\hat{W}_2 = \frac{1}{mq_a^2}(m - nq_a)\left[(m - nq_a)\left(1 - \frac{2}{\psi}\frac{d\psi}{dx}\right) - 2\frac{J_a}{\bar{J}}\right]a^2\xi_a^2
\tag{11.232}
$$

where ψ and $d\psi/dx$ are evaluated at the plasma surface $x = 0$. Also, for an external mode $m - nq_a > 0$.

Stability for $J_a \neq 0$

The stability analysis separates into two cases depending on whether or not J_a vanishes. Consider first the case where $J(r)$ is a decreasing function of r which, however, still has a finite jump at $r = a$: $J_a \neq 0$. For such profiles it is only necessary to calculate the leading-order contribution to $\psi(x)$ from Eq. (11.231). In other words, the λ term can be neglected. The solution that

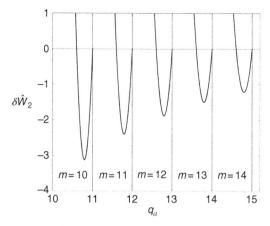

Figure 11.41   Plot of $\delta \hat{W}$ vs. q_a for high m and $n = 1$. Note the bands of instability, which are all of fixed width.

corresponds to a localized eigenfunction (i.e., the one that decays to zero as $x \to \infty$) is given by

$$\psi(x) = e^{-x/2}$$

$$\frac{d\psi(x)}{dx} = -\frac{1}{2}e^{-x/2}$$

$$\left.\frac{1}{\psi}\frac{d\psi}{dx}\right|_0 = -\frac{1}{2}$$
(11.233)

Thus, the value of $\delta \hat{W}_2$ reduces to

$$\delta \hat{W}_2 = \frac{1}{mq_a^2}(m - nq_a)\left(m - \frac{J_a}{\bar{J}} - nq_a\right)a^2\xi_a^2$$
(11.234)

Equation (11.234) predicts that high m external kink modes are unstable for q_a in the range

$$m - \frac{J_a}{\bar{J}} < nq_a < m$$
(11.235)

Any current profile with a finite jump at the plasma surface is unstable. The bands of instability are illustrated in Fig. 11.41. Note that their width remains constant as m increases. The most unstable case corresponds to a uniform current density, $J_a = \bar{J}$, in which case the neighboring bands touch and all q_a values are unstable. As the edge current density decreases the bands become narrower, forming windows of stability. Still, the overall conclusion is that, for a plasma with a finite $J(a)$, there is no minimum value of q_a above which the plasma is stable against high m external kink modes. As stated, stable plasmas do not want to have edge current density jumps.

Stability for $J_a = 0$

Equation (11.235) suggests that when the jump in edge current density vanishes (i.e., $J_a = 0$) the bands of instability shrink to zero and the plasma becomes stable. This is not quite the situation. As $J_a \to 0$ and $nq_a \to m$, the result is $\delta\hat{W}_2 \to 0$ corresponding to marginal stability. To determine the actual stability boundary requires the evaluation of the $1/m$ corrections to $\psi(x)$ resulting from the λ term. In fact, an examination of Eq. (11.232) indicates that these "corrections" must be finite for instabilities to exist. Specifically, one sees that when $J_a = 0$, $\delta\hat{W}_2$ reduces to

$$\delta\hat{W}_2 = \frac{1}{mq_a^2}(m - nq_a)^2\left(1 - \frac{2}{\psi}\frac{d\psi}{dx}\right)a^2\zeta_a^2 \qquad (11.236)$$

implying that the condition determining the lower critical q_a for each band of instabilities is $\psi_x/\psi > 1/2$. The corrections to ψ must somehow change ψ_x/ψ a finite amount, from $-1/2$ to approximately $+1/2$.

The structure of Eq. (11.231) is such that a finite correction to ψ from the λ term is only possible if x_0 is even smaller than assumed; that is, one must alter the assumption $x_0 \sim 1$ to $x_0 \ll 1$. In this case the solution near $x = 0$ is strongly modified, possibly leading to a finite correction in ψ_x/ψ. Physically, the condition $x_0 \ll 1$ corresponds to the requirement that the resonant surface be extremely close to the plasma surface: the original ordering $1 - nq_a/m \sim 1/m$ must change to $1 - nq_a/m \ll 1/m$.

With this mathematical insight one can solve for $\psi(x)$ in Eq. (11.231) as follows. The solution for ψ including the λ term can be written in terms of Whittaker functions. As usual there are two independent solutions since Eq. (11.231) is a second-order ordinary differential equation. The localized solution (i.e., the one that decays to zero as $x \to \infty$) is given by (Gradshteyn and Ryzhik, 2000)

$$\psi(x) = W_{\lambda,\,1/2}(x + x_0) \qquad (11.237)$$

Now, to determine the stability boundary one needs to evaluate ψ_x/ψ at $x = 0$. Since the regime of interest corresponds to $x_0 \ll 1$ one can make use of the small argument expansion of the Whittaker function: $W_{\lambda,\,1/2}(x_0) \approx e^{-x_0/2}[1 - (\lambda/m)x_0 \ln x_0]$. A simple calculation then shows that the condition for instability can be written as

$$1 + \frac{\lambda}{m}\ln x_0 < 0 \qquad (11.238)$$

In terms of the standard notation, this leads to the following band of instabilities:

$$m - \exp\left(\frac{2m\bar{J}}{aJ_a'}\right) < nq_a < m \qquad (11.239)$$

After this rather complicated analysis one sees that instabilities can exist within exponentially narrow bands just below each rational surface. The calculation explicitly demonstrates the sensitivity of the stability boundaries to the behavior of the edge current density. The basic conclusion is that current density profiles which vanish at the plasma edge but have sharp edge gradients can be unstable to high m external kink modes. These modes serve as the peeling component of the peeling–ballooning modes responsible for ELMs. The ballooning component is described in the next chapter.

ELMs

The final topic of interest involves the impact of ELMs on experimental tokamak operation. As stated, ELMs appear in tokamaks operating in the favorable H-mode confinement regime, characterized by high edge pressure and current gradients. Experimentally, ELMS appear as short, periodic, bursts of particles and energy that are ejected from the plasma. In terms of overall performance ELMs are a mixed blessing in present day experiments but whose downside becomes much more serious in reactor scale devices. The issues are as follows.

On the positive side each ELM ejects impurities in addition to plasma particles. Preventing the build-up of impurities improves performance in present experiments and is crucial in reactor plasmas where they can dilute the basic D-T fuel. On the negative side the loss of plasma from too many ELMs reduces the time-averaged particle density and temperature, thereby degrading the overall high confinement properties of H-mode operation.

Equally importantly, while the divertor plates are designed to withstand the time-averaged heat loads resulting from thermal conduction losses, each ELM deposits a short, high-intensity, pulse of heat to the divertor plate. In other words, the heat deposition is not constant in time but has short, high-intensity pulses superimposed over the steady thermal conduction losses. Present experiments can withstand these bursts of heat but reactor scale devices would likely suffer irreversible damage to the divertor plate, clearly an unacceptable situation.

The situation is further complicated by the fact that there are different types of ELMs whose positive and negative contributions to tokamak operation vary in relative size. Type I ELMs are short in duration, well-spaced in time, but very intense in magnitude. These ELMs do not significantly degrade the H-mode confinement properties but would be unacceptable in reactor scale devices because of the high intensity of the pulses.

Type III ELMs lie on the other end of the spectrum. They occur frequently, almost continuously, and have low intensity. Consequently, they do not pose much of a threat to the divertor plates. However, because of the nearly continuous ejection of particles and energy, Type III ELMs substantially degrade confinement performance, thus reducing the hoped-for benefits of H-mode operation.

Type II ELMs perhaps represent the best of both worlds. They are less intense than Type I ELMs but less frequent than Type III ELMs. As a result they do not significantly degrade performance, do not seriously threaten the divertor plates, and still help prevent impurity build-up in the plasma. The biggest difficulty with Type II ELMs is that they exist only over a relatively narrow range of parameters; that is, reliable accessibility to the Type II ELM regime of operation is not an easy task experimentally.

In terms of theory, the causes and properties of ELMs is an important and ongoing topic of research. Substantial progress has been made (Snyder *et al.*, 2005) but further research is still needed. One problem is that while the theory has made excellent progress in the understanding of ELMs, the theory does not in general offer sharply defined experimental techniques to control their behavior. The reason is that the theoretical understanding depends strongly on the properties of the edge region (e.g., turbulent transport, current, and edge gradients), which are difficult to measure and control experimentally.

One suggestion has been to apply resonant magnetic perturbations (RMPs) at the edge of the plasma to afford some level of external control over ELM behavior. This is still a work in progress. Another experimental regime of operation, noted as the I-mode (for "intermediate" mode), has been discovered (Whyte *et al.*, 2010) that is characterized by a high edge temperature gradient but without a high edge density gradient. In the I-mode regime energy confinement is very close to the desirable H-mode confinement, there is very little build-up of impurities, and operation is ELM free so there is no threat to the divertor plates. This appears to be a new and highly desirable regime of operation. Further research is needed to test the reliability and robustness of I-mode operation in different tokamaks operating over longer periods of time.

Clearly, ELMs represent an important operational limit on tokamaks. The straight tokamak model helps to shed some insight on the peeling component of the peeling–ballooning modes which are thought to be the basic driving instabilities for the generation of ELMs.

11.6.7 Summary of the straight tokamak

The straight tokamak model provides a simple but surprisingly reliable description of the basic MHD instabilities that can arise in a large aspect ratio, circular cross section, low-pressure tokamak. The model accurately describes instabilities that arise from too much plasma current or too sharp an edge current density gradient. The tokamak ordering used to define the model leads to a description in which the pressure is negligible and in fact is entirely absent from the analysis. This is actually a benefit to the model in that when pressure effects are included in a straight model they are inaccurate in describing the actual situation. An accurate description of pressure effects requires a toroidal analysis which is presented in the next chapter.

In spite of its simplicity, the straight tokamak model describes MHD instabilities that are the important drives for many practical tokamak operational limits. These are summarized below:

- **Sawtooth oscillations:** The $m = 1$, $n = 1$ internal mode requires $q_0 > 1$ to avoid instability. When $q_0 < 1$ the internal relaxation instability known as the sawtooth oscillation is excited.
- **Current-driven disruptions:** When a plasma has too much current then external current-driven kinks can be excited which lead to major disruptions. The robust $m = 1$, $n = 1$ external Kruskal–Shafranov kink mode requires $q_a > 1$ for stability. Practically, depending on the steepness of the edge current density gradient, one finds that $q_a \sim 2-3$ is needed to avoid external kink instabilities and major disruptions.
- **Density-driven disruptions:** When the plasma density becomes too large a turbulent edge layer is formed, possibly driven by MHD instabilities. This layer effectively shrinks the size of the core plasma lowering the edge q. When the layer becomes wide enough the effective edge q becomes sufficiently small that a current-driven disruption is excited.
- **Resistive wall modes:** When a perfectly conducting wall is moved closer to the plasma surface the bands of unstable q_a values decrease. However, the minimum q_a for complete stability to current-driven modes remains unchanged and it is only the lower values of q_a, which defines the lower edge of the unstable bands, that are increased. Thus, resistive wall stabilization has no impact on the maximum allowable plasma current.
- **ELMs:** ELMs are believed to be caused by peeling–ballooning modes which are driven by a combination of high edge current and pressure gradients. The straight tokamak model provides a simple description of the peeling component of the mode driven by the edge current density gradient. The model shows that bands of instability exist for high $n \sim 10$ modes whose width depends sensitively on the edge current density jump and gradient. The ballooning effect requires a toroidal calculation.

Overall the straight tokamak model provides a good introduction to the basic MHD instabilities that affect the operation of tokamak experiments.

11.7 The reversed field pinch (RFP)

The cylindrical screw pinch analysis provides an accurate description of the MHD stability of the RFP configuration, including internal and external modes driven by the current and/or the pressure gradient. The reason for the good accuracy is that all basic equilibrium quantities in an RFP are of the same order: $B_\theta^2 \sim B_z^2 \sim \mu_0 p$. Thus, when a toroidal RFP is straightened into a cylindrical configuration all of the resulting errors are small, of order ε^2. This is in contrast to the high β tokamak where the ordering $\mu_0 p \sim \varepsilon B_z^2$ leads to finite and important modifications when transforming from a cylinder to a torus.

The basic goal of this section is to learn how much current and pressure can be stably confined in an RFP. The answer actually involves a two-part analysis that treats both ideal and resistive MHD modes. Qualitatively, the first part investigates ways to stabilize ideal modes which, if allowed to exist, would lead to strong turbulence and unacceptably poor performance. Once ideal stability is achieved a second part to the analysis is required that examines ways to minimize the remaining weaker resistive MHD turbulence. This weaker turbulence leads to improved transport compared to that generated by ideal modes. However, if left unchecked resistive MHD turbulence still results in poorer transport than a comparably sized tokamak. Since the present textbook is focused on ideal MHD the important problem of resistive MHD turbulence will not be covered here.

The general conclusions from the ideal MHD analysis are as follows. An RFP can be theoretically completely stable against all internal MHD modes at high values of pressure and current: $\beta \sim 0.2$ and $q_a \sim 0.2$. However, the low edge safety factor causes the plasma to be unstable to a substantial number of external ideal kink modes with typical unstable wave numbers ranging from $1 < |n| < 1/\varepsilon$. A close fitting conducting wall with wall radius/plasma radius ~ 1.5 can in principle stabilize these modes, but as discussed they then transform into resistive wall modes. From a practical point of view stable operation requires feedback, a task once thought by many to be unfeasible because of technological complexity, but which has, nevertheless, actually been successfully demonstrated on several RFP experiments.

The analysis required to demonstrate these points involves the evaluation of the cylindrical δW for ideal MHD modes by means of a combination of trial functions and simple numerical calculations. While the mathematics is not overly complex, obtaining a simple, global view of RFP stability is somewhat complicated because of the large number of parameters that enter the analysis. Specifically the parameters describing the RFP and its stability include q_0, q_a, B_{za}/B_{z0}, β, ε, n, m. The situation is made more complicated by the facts that B_{za}/B_{z0} reverses sign, implying that both positive and negative n values can become unstable.

Indeed, a main goal of the present section is to organize the material in such a way that an overall picture of RFP stability can be obtained. This is accomplished by the following steps:

1. A new set of dimensionless parameters is introduced that is closely tied to experimental operation. These are intuitively simpler to understand and replace the more mathematical parameters characterizing the RFP equilibrium model given in Chapter 5.
2. It is shown at the outset that an RFP with zero current density at the plasma edge and without a B_z reversal is always unstable to an $m = 1$ mode. This result motivates the remainder of the analysis which considers only RFPs with a B_z reversal.
3. Based on the discussion in Section 11.5.3 attention is next focused on the $m = 0$ mode in the limit $k \to 0$. It is shown that stability against this mode sets a limit on β which turns out to be sufficiently large that it does not represent an important operational limit.
4. The main ideal stability limitations arise from the $m = 1$ mode. The first and simplest stability limit is due to Suydam's criterion. High values of critical β are possible but achieving them imposes important constraints on the pressure profile. Stable profiles are flat near the origin and have flat gradients near the plasma edge.
5. Interestingly, when Suydam's criterion is satisfied, then over the usual RFP operating range, ideal internal modes are also shown to be stable. Theoretically, the implication is that RFPs can be completely stable to all internal modes at high values of β.
6. In contrast to these optimistic predictions experimental RFPs usually do not operate at such high values of β. A theory proposed by Taylor helps to explain why practical values of β are much lower. Taylor's theory also sheds insight into the time evolution of the magnetic field profiles.
7. From the ideal MHD point of view the most dangerous instabilities in an RFP are $m = 1$ external modes. These modes are analyzed and shown to be unstable over a wide range of n values, even at $\beta = 0$. A conducting wall is needed for stability and a calculation is presented showing how close the wall must be to stabilize external modes at zero and finite β.

The ambitious set of calculations outlined above proceeds as follows.

11.7.1 Physical parameters describing an RFP

The analysis begins by introducing new, physically intuitive, dimensionless parameters into the model RFP equilibrium discussed in Chapter 5. The model profiles are repeated here for convenience:

$$\frac{2\mu_0 p}{B_{z0}^2} = \frac{2\mu_0 p_{\text{max}}}{B_{z0}^2}(1-\rho^6)^3 = \frac{140\alpha_p}{81}(1-\rho^6)^3$$

$$\frac{B_z}{B_{z0}} = 1 - 2\alpha_z\rho^2 + \alpha_z\rho^4$$

$$\frac{B_\theta^2}{B_{z0}^2} = \frac{\alpha_p}{9}(35\rho^6 - 40\rho^{12} + 14\rho^{18}) + \frac{\alpha_z}{15}\left[30\rho^2 - 20(2\alpha_z + 1)\rho^4 + 45\alpha_z\rho^6 - 12\alpha_z\rho^8\right]$$

$$\frac{a\mu_0 J_\theta}{B_{z0}} = 4\alpha_z\rho\left(1-\rho^2\right)$$

$$\frac{a\mu_0 J_z}{B_{z0}} = \frac{140\alpha_p}{9}\left(\frac{\rho B_{z0}}{B_\theta}\right)\rho^4(1-\rho^6)^2 + 4\alpha_z\left(\frac{\rho B_{z0}}{B_\theta}\right)\left(1-\rho^2\right)\left(1-2\alpha_z\rho^2 + \alpha_z\rho^4\right)$$

$$(11.240)$$

where $\rho = r/a$ and $\langle p \rangle = (81/140)p_{\text{max}}$. Observe that the equilibrium is characterized by two dimensionless parameters α_p, α_z and the scale factor B_{z0}.

Consider first the scale factor B_{z0}. In typical RFP operation, a small uniform B_z bias field initially fills the plasma chamber before the current ramp-up. After the current reaches its peak value the B_z profile has changed substantially and is far from uniform. Thus, the value B_{z0} is not one that is easy to set or measure experimentally. Instead, since the toroidal flux, ψ_t, is reasonably well conserved during the current rise, it makes sense to normalize $B_z(\rho)$ by this quantity. Specifically, one defines $\langle B_z \rangle = \psi_t/\pi a^2$ and normalizes all magnetic fields to $\langle B_z \rangle$. For the model $B_z(\rho)$ in Eq. (11.240) it follows that the relation between $\langle B_z \rangle$ and B_{z0} is given by

$$\langle B_z \rangle = B_{z0}\left(\frac{3 - 2\alpha_z}{3}\right) \qquad (11.241)$$

Using this relation it is possible to define three basic dimensionless parameters that describe the physical operation and performance of an RFP. The first measures how much the B_z field reverses at the edge of the plasma. Its definition is $\mathsf{F} = B_{za}/\langle B_z \rangle$. The second parameter is a measure of the total toroidal current flowing in the plasma and is defined by $\boldsymbol{\theta} = (\mu_0 I/2\pi a)/\langle B_z \rangle = B_{\theta a}/\langle B_z \rangle$. Note the different font that distinguishes F, $\boldsymbol{\theta}$ from F, θ which already appear in the analysis. The third parameter measures the average plasma pressure. Since $\beta \approx \beta_p$ for typical RFP operation a good definition for this parameter is $\beta_p = 2\mu_0\langle p \rangle/B_{\theta a}^2$.

On the basis of this discussion, all of the stability results presented in this section are expressed in terms of the three basic RFP parameters F, $\boldsymbol{\theta}$, β_p. By using the basic definitions $\alpha_z = B_{za}/B_{z0}$, $\alpha_p = 2\mu_0\langle p\rangle/B_{z0}^2$ plus Eq. (11.241) one can show after a short calculation that the physical parameters F, θ, β_p, expressed in terms of the mathematical parameters α_z, α_p, are given by

$$F = \frac{B_{za}}{\langle B_z\rangle} = 3\left(\frac{1 - \alpha_z}{3 - 2\alpha_z}\right)$$

$$\theta = \frac{B_{\theta a}}{\langle B_z\rangle} = \left(\frac{3}{3 - 2\alpha_z}\right)[\alpha_p + \alpha_z(10 - 7\alpha_z)/15]^{1/2}$$

$$\beta_p = \frac{2\mu_0\langle p\rangle}{B_{\theta a}^2} = \frac{\alpha_p}{\alpha_p + \alpha_z(10 - 7\alpha_z)/15} \tag{11.242}$$

The goal of the analysis is to determine the maximum values of θ and β_p, plus the corresponding values of F, that are MHD stable for an RFP.

The last quantity of interest is the expression for the ideal MHD δW written in a form suitable for the RFP. This form is obtained directly from Eq. (11.98) and is written as

$$\frac{\delta W}{2\pi^2 R_0/\mu_0} = \delta\hat{W} = \int_0^1 (f\xi'^2 + g\xi^2)\,d\rho$$

$$+ B_{\theta a}^2\left[\frac{\Lambda}{m}(nq + m)^2 + \frac{1}{\kappa_0^2\rho^2}(n^2q^2 - m^2)\right]\xi_a^2$$

$$f(\rho) = \frac{B_\theta^2}{\rho\kappa_0^2}(nq + m)^2 \tag{11.243}$$

$$g(\rho) = 2\mu_0\frac{\kappa^2}{\kappa_0^2}p' + \frac{\kappa_0^2\rho^2 - 1}{\kappa_0^2\rho^2}\frac{B_\theta^2}{\rho}(nq + m)^2 + 2\frac{\kappa^2}{\kappa_0^4}\frac{B_\theta^2}{\rho^3}(n^2q^2 - m^2)$$

$$= \frac{\kappa^2}{\kappa_0^2}\left[2\mu_0 p' + \frac{3 + \kappa^2\rho^2}{1 + \kappa^2\rho^2}\frac{B_\theta^2}{\rho}(nq + 1)\left(nq - \frac{1 - \kappa^2\rho^2}{3 + \kappa^2\rho^2}\right)\right]$$

Here, $\rho = r/a$, $q(\rho) = \varepsilon\rho B_z/B_\theta$, $\kappa = ka = n\varepsilon$, $\kappa_0^2(\rho) = k_0^2 a^2 = n^2\varepsilon^2 + m^2/\rho^2$, and prime denotes $d/d\rho$. Note that the RFP community defines n with the opposite sign as the tokamak community. Also the second form of $g(\rho)$ corresponds to the important case of $m = 1$.

The stage has now been set to investigate the stability of the RFP.

11.7.2 Instability of an RFP without a B_z reversal

There is a good reason why a reversal in B_z is required for high-performance MHD behavior in an RFP. Without a reversal the RFP is always unstable to an ideal

(a)

(b)

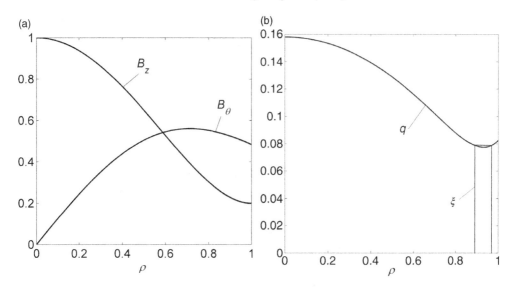

Figure 11.42   Profiles for an RFP without a B_z reversal: (a) $B_\theta(r)$ and $B_z(r)$; (b) $q(r)$ showing the existence of a minimum and the trial function leading to instability.

$m = 1$ instability near the outside of the plasma. The proof of this statement is presented below and is based on the generic shape of the $q(\rho)$ profile for a non-reversed RFP.

To determine the shape of the $q(\rho)$ profile assume that the current density and pressure gradient vanish at the edge of the plasma. The fields in this region are thus approximately vacuum fields (i.e., $B_z \approx B_{za} = $ constant and $B_\theta \approx B_{\theta a}/\rho$). This leads to a radially increasing safety factor profile $q \approx q_a \rho^2$. Now, near the axis of an RFP the $q(\rho)$ profile is a decreasing function of ρ. Without a reversal the only way these two constraints can be made compatible is for $q(\rho)$ to have a minimum as shown in Fig. 11.42.

Because of the minimum it is always possible to choose a trial function that makes $\delta W < 0$. This trial function is superimposed on the $q(\rho)$ profile, also shown in Fig. 11.42. The trial function corresponds to an $m = 1$ mode with n chosen so that $nq = -1$ at two neighboring points about the minimum in $q(\rho)$. Since $q \sim \varepsilon$ then $n \sim -1/\varepsilon$, a high value. Now, $f(\rho) = 0$ at the endpoints of the trial function where $\zeta'(\rho) \to \infty$. Even so, the line bending contributions to $\delta \hat{W}$ vanish at both of these points in exact analogy with the sawtooth analysis of the tokamak (see Section 11.6.2). This trial function thus leads to an upper bound on $\delta \hat{W} = \delta \hat{W}_F$ given by

$$\delta \hat{W}_F = \xi_0^2 \int_{\rho_1}^{\rho_2} g(\rho) \, d\rho \qquad (11.244)$$

From the $m = 1$ expression for $g(\rho)$ given in Eq. (11.243) one sees that $p'(\rho) < 0$ and $-nq(\rho) < 1$ over the range $\rho_1 < \rho < \rho_2$. Similarly, one finds that

$$-nq + \frac{1 - \kappa^2 \rho^2}{3 + \kappa^2 \rho^2} = \frac{4}{3 + \kappa^2 \rho^2} - nq - 1 \approx \frac{4}{3 + \kappa^2 \rho^2} > 0 \qquad (11.245)$$

where the last approximation follows because of the smallness of $nq + 1$ near the minimum of $q(\rho)$.

The conclusion is that $g(\rho) < 0$ over the entire range of integration implying that $\delta \hat{W}_F < 0$; that is, a minimum in the $q(\rho)$ profile in an RFP leads to an $m = 1$ instability driven by both the pressure gradient and the current. This instability is not as "harmless" as the sawtooth instability in a tokamak. For the tokamak the instability redistributes energy and current near the center of the plasma. It does not lead to transport losses across the plasma boundary. However, the RFP instability is located near the edge of the plasma. Consequently, the redistribution of energy and current leads to transport losses across the plasma boundary to the wall, a very undesirable situation. Experimental observations indeed confirm this unfavorable prediction.

It is for this reason that good performance of an RFP requires a reversal in the $B_z(\rho)$ profile in order to eliminate the minimum in $q(\rho)$. The remainder of the analysis in this section focuses on profiles where $B_z(\rho)$ has a reversal. Such profiles motivate the name "reversed field pinch."

11.7.3 The m = 0 instability

An RFP can be unstable to an ideal $m = 0$ internal mode if β_p is too high. As shown below, however, the critical β_p is large and as such does not represent much of a constraint on RFP operation. A good estimate of the β_p limit can be obtained by the use of a simple trial function as illustrated in Fig. 11.43. The trial function ξ increases linearly with ρ near the origin as required by regularity. It continues to increase linearly until it drops abruptly to zero across a narrow layer at the radius ρ_0 where $B_z(\rho)$ reverses. Since $f(\rho_0) = 0$ for $m = 0$, the contribution to $\delta \hat{W}_F$ in the vicinity of $\rho = \rho_0$ from the $f \xi'^2$ term vanishes in analogy with the tokamak sawtooth analysis in Section 11.6.2.

The end result is that for the trial function $\xi = \xi_0(\rho/\rho_0)$ the value of $\delta \hat{W}_F$ reduces to

$$\delta \hat{W}_F = \frac{\xi_0^2}{\rho_0^2} \int_0^{\rho_0} \left(f + \rho^2 g\right) d\rho \qquad (11.246)$$

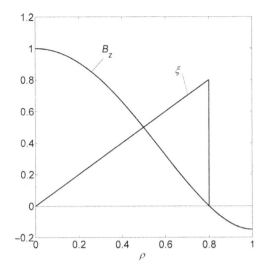

Figure 11.43   Trial function for the $m = 0$ pressure-driven internal mode.

Now, as discussed in Section 11.5.3 the most unstable wave number for $m = 0$ corresponds to $n \propto k \to 0$. In this limit one sees from Eq. (11.243) that

$$f + \rho^2 g = 2\mu_0 \rho^2 p' + 2\rho B_z^2 \tag{11.247}$$

Equation (11.247) is substituted into Eq. (11.246) which is then evaluated for the model RFP profile. The resulting $\delta \hat{W}_F$ is a function of ρ_0 and a numerical calculation shows that $\rho_0 \to 1$ is the most unstable case. In the $\rho_0 \to 1$ limit one finds that the stability boundary for the $m = 0$ mode in an RFP is given by

$$\beta_p < \frac{1}{2} \tag{11.248}$$

Equation (11.248) is only a weak condition on RFP operation as most experiments operate considerably below this value. Physically, the stability limit is a consequence of the fact that there must be a sufficient B_z to "stiffen" the plasma against $m = 0$ sausage perturbations. As shown in Fig. 11.44 $m = 0$ perturbations compress both the plasma and the internal B_z field, each of which is a stabilizing effect against the unfavorable curvature. When B_z is too small, corresponding to $\beta_p > 1/2$, the RFP behaves similarly to a sausage-unstable Z-pinch.

11.7.4 Suydam's criterion

The next class of modes to examine corresponds to localized interchanges whose stability boundary is defined by Suydam's criterion. This involves a very easy-to-implement test, simply substituting the equilibrium profiles into Eq. (11.109) and

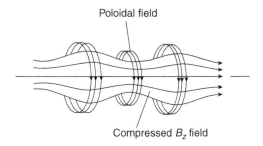

Figure 11.44 "Stiffening" of the plasma against $m = 0$ perturbations by an internal B_z field.

seeing whether or not the criterion is violated. An examination of the criterion shows that stability imposes two constraints on the shape of the pressure profile.

First, as previously discussed in Section 11.6.1, Suydam's stability criterion near the axis can be written for an RFP (in terms of normalized radius) as

$$\left(\frac{q_0''}{q_0}\right)^2 \rho^2 > -\frac{8\mu_0 p_0''}{B_{z0}^2} \tag{11.249}$$

Clearly, the criterion is violated near $\rho = 0$ unless $p_0'' \geq 0$. The implication is that the pressure profile must be very flat or hollow near the axis. Unlike the tokamak, toroidal effects do not improve the situation very much because of the small $q \sim \varepsilon$ in an RFP.

The second constraint involves the pressure gradient near the edge of the plasma for the realistic case where $p' = J_\theta = J_z = 0$ at $\rho = 1$. Suydam's criterion tends to be violated in this region for profiles where B_z is only slightly reversed. This can be seen explicitly by expanding Suydam's criterion about the edge. One sets $\rho = 1 - \delta$ with $\delta \ll 1$, uses the fact that $\rho q'/q = 2$ when $\mathbf{J} = 0$, and assumes that $B_{za}/B_{za}'' \sim \delta^2$. Suydam's criterion (for stability) reduces to

$$8\mu_0 p_a'' < B_{za}''^2 \frac{\left(\delta^2 + 2B_{za}/B_{za}''\right)^2}{\delta} \tag{11.250}$$

For typical RFP profiles, $B_{za} < 0$, $B_{za}'' > 0$, and $p_a'' > 0$. It thus follows that for such profiles the stability criterion is always violated at the radius corresponding to $\delta^2 = -2B_{za}/B_{za}''$. The way to avoid this instability is to require $p_a'' = 0$. In other words, the edge pressure gradient must be very flat.

The two constraints just discussed motivate the choice of pressure profile used in Eq. (11.240) to model an RFP. It is now a simple task to substitute the profiles into Suydam's criterion to determine the marginal stability boundary. A simple numerical evaluation leads to the marginal stability curve of β_p vs. θ as illustrated

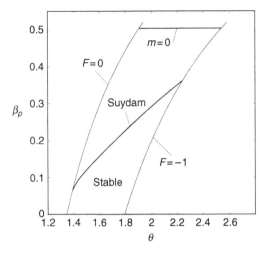

Figure 11.45 Marginal stability diagram of β_p vs. θ as determined by the Suydam criterion. The vertical boundaries correspond to the normal range of RFP operation $F = 0$ and $F = -1$. The regions below the heavier curves are stable.

in Fig. 11.45. In this figure the range of θ values corresponds to the usual RFP experimental operating regime $0 > F > -1$. The conclusion is that over the relevant range of θ the critical β_p can be substantial, ranging from 0.1 to 0.35. The Suydam limit for β_p is usually not the limiting factor in high-performance RFP operation.

11.7.5 Internal modes

Consider now the stability of an RFP against general ideal internal modes. Not much discussion is required. Stability is determined by numerically solving the minimizing equation for ξ given by

$$(f\xi')' - g\xi = 0 \tag{11.251}$$

for the $m = 1$ mode using the model profiles. The numerical solutions are tested for stability against internal mode wave numbers that are resonant within the plasma either inside or outside the reversal radius. Following Newcomb's procedure the solutions between the origin and the resonant surface and between the plasma edge (where $\xi_a = 0$ for an internal mode) and the resonant surface are examined for the existence of a zero crossing. If such a crossing exists in either region the plasma is unstable.

The results are quite simple. For the experimental range of F lying between $0 > F > -1$ the plasma is stable to all internal MHD modes up to the value of β_p

at which the Suydam criterion is violated. In other words the diagram given by Fig. 11.45 represents the complete stability of an RFP against all internal ideal MHD modes and includes the $m = 0$ mode, Suydam's criterion, and global internal modes. It is worth noting that global internal modes can be driven unstable at higher values of current where the plasma is more like a Z-pinch than an RFP. However, these larger currents lie outside the normal operating range of an RFP.

11.7.6 Taylor's theory

The analysis thus far presented has shown that high β RFP profiles exist that are MHD stable to all internal modes. These profiles are characterized by (1) a flat $p(r)$ profile near the origin and the plasma edge and (2) a $B_z(r)$ profile that reverses near the edge of the plasma. One can ask how such profiles were discovered and whether they are generated naturally or require sophisticated external field programming.

Historically, it was observed on the Zeta experiment (Bodin and Newton, 1980) that RFP-like profiles were formed by the natural evolution of ohmically heated discharges. These plasmas, however, were characterized by a high level of turbulence and, most importantly, did not have a $B_z(r)$ reversal. The MHD instabilities inherent in a non-reversed RFP, as discussed in Section 11.7.2, presumably played an important role in producing this poor behavior.

Even so, towards the end of Zeta's experimental operation, it was discovered that under certain conditions, after a turbulent initial phase, the plasma remarkably evolved to a stable quiescent state that included a spontaneous self-reversal of the $B_z(r)$ field. The reduction of turbulence in the quiescent state is illustrated in Fig. 11.46 for a later RFP, HBTX-1A at Culham Laboratory.

The reversal in $B_z(r)$ is quite interesting since no special field programming was utilized. Furthermore, as shown in the analysis below, reversed $B_z(r)$ profiles cannot be generated from a pure ohmically heated evolution. Instead, reversal requires the generation of a dynamo, driven by a relatively low level of turbulence (as compared to the non-reversed state). The low turbulence level helps to explain the improved transport but still causes the β value to be considerably lower than the maximum ideal MHD internal mode limit. An elegant theory explaining the evolution of RFP discharges and the conditions under which self-reversal occurs has been given by Taylor (1974) and is the main topic of this subsection.

The basic idea put forth by Taylor is as follows. Assume the existence of a slightly dissipative plasma surrounded by a perfectly conducting shell which is initially not in a state of stable equilibrium. As the plasma turbulently evolves from its initial state, perhaps violently, it dissipates energy through thermal and particle

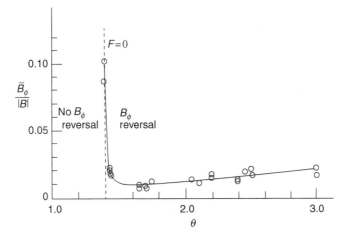

Figure 11.46   Experimental observations of B_θ fluctuations vs. $\theta \propto I$ in the HBTX-1A RFP at Culham Laboratory, UK. Experimentally, field reversal occurs at $\theta = 1.4$. Observe the low-fluctuation, quiescent operation for $\theta > 1.4$. Courtesy of I. Hutchinson.

losses to the wall. It continues to do so until it reaches a state of minimum energy after which it is incapable of further motion. The properties of the final state are determined by the constraints governing the evolution. Clearly the crucial physics issue, resolved by Taylor, is the determination of the appropriate constraints. Once the constraints are known it is relatively straightforward to obtain the final profiles. The presence of dissipation is critical as it allows accessibility to a much wider range of states than those that are accessible under the topological restrictions of ideal MHD.

Taylor's theory requires several steps which are described below after it is demonstrated that reversal cannot occur in a pure ohmically heated discharge.

Impossibility of $B_z(r)$ reversal in a pure ohmic discharge

A pure ohmic discharge is defined as a plasma that maintains exact 1-D cylindrical symmetry during its entire evolution. No instabilities are allowed that can generate multidimensional turbulence. The equations describing the 1-D evolution are given by the standard cylindrical MHD model, the only modification being in the parallel Ohm's law, which changes from $E_\parallel = 0$ to $E_\parallel = \eta J_\parallel$. Even though the resistivity is small and can be neglected in the perpendicular components of Ohm's law it must be kept in the parallel component since $E_\parallel$ is being compared to "zero."

The calculation begins by noting that Ampere's law combined with the definition $J_\parallel = \mathbf{J} \cdot \mathbf{B}/B$ leads to the relation

$$\mu_0 J_{\parallel}(r) = \frac{1}{B}\left[B_z\frac{(rB_\theta)'}{r} - B_\theta B_z'\right] \tag{11.252}$$

Now, assume that the profile is evolving on the slow (compared to ideal MHD) transport time scale implying that the plasma, at any instant of time, satisfies the steady state MHD equations; that is, the plasma is passing through a series of quasistatic equilibria. During this slow evolution it follows that $\nabla \times \mathbf{E} = -\partial\mathbf{B}/\partial t \approx 0$. The solution which is regular at the origin is given by $\mathbf{E} = E_0\mathbf{e}_z$. Here, $E_0(r, t) \approx E_0(t)$ is the toroidal electric field induced by the transformer. The parallel Ohm's law then yields a relation between $J_{\parallel}$ and E_0 given by

$$J_{\parallel}(r) = \frac{E_0}{\eta}\frac{B_z}{B} \tag{11.253}$$

Equation (11.253) is substituted into Eq. (11.252) and combined with the radial pressure balance relation. A short calculation yields

$$B_z' = -\frac{\mu_0}{B^2}\left(\frac{E_0 B_\theta}{\eta} + p'\right)B_z \tag{11.254}$$

This equation can be formally integrated leading to

$$B_z(r) = B_{z0}\exp\left[-\int_0^r\frac{\mu_0}{B^2}\left(\frac{E_0 B_\theta}{\eta} + p'\right)dr\right] \tag{11.255}$$

Clearly, the exponential behavior shows that a 1-D ohmic evolution can never lead to a reversal in $B_z(r)$.

Taylor's minimum energy formulation

Taylor suggested that as an RFP evolves in the presence of fine scale resistive turbulence, the plasma would naturally seek its lowest energy state. Mathematically, this is equivalent to the statement that the profiles must adjust themselves so as to minimize the total potential energy whose exact formula (see Eq. (3.24) with $v = 0$ corresponding to slow evolution) is given by

$$W = \int\left(\frac{B^2}{2\mu_0} + \frac{p}{\gamma - 1}\right)d\mathbf{r} \tag{11.256}$$

As stated, the key point in the analysis is to determine the appropriate constraints under which the minimization is to be carried out. Within the ideal MHD model it is shown below that there are an infinite number of constraints that can be applied. Taylor recognized that almost all of these constraints do not actually apply to a dissipative plasma. This recognition led him to suggest a much more plausible set of constraints. The reasoning behind the final constraints is described below by first focusing on an ideal MHD plasma and then considering the consequences of dissipation.

The ideal MHD helicity constraints

The ideal MHD constraints start with the requirement that any physically acceptable magnetic field must obviously satisfy $\nabla \cdot \mathbf{B} = 0$. Most importantly, if the magnetic field is embedded in a perfectly conducting ideal MHD plasma then any evolution of $\mathbf{B}$ away from its initial state must be consistent with Faraday's law

$$\frac{\partial \mathbf{B}}{\partial t} = \nabla \times (\mathbf{v} \times \mathbf{B}) \tag{11.257}$$

In other words, magnetic field variations must result from physically allowable plasma displacements that exactly preserve the field line topology.

It has been shown (Woltjer, 1958) that the local constraint given by Eq. (11.257) can be replaced by an infinite set of integral constraints involving the helicity H which is defined as

$$H(\psi, t) = \int_\psi \mathbf{A} \cdot \mathbf{B} d\mathbf{r} \tag{11.258}$$

where $\mathbf{A}$ is the vector potential and the integration is carried out over the volume enclosed by an arbitrary flux surface ψ. This intuitive equivalence (i.e., one local constraint $=$ infinite number of integral constraints) follows from the fact that in an ideal MHD plasma

$$\frac{dH}{dt} = 0 \tag{11.259}$$

on every flux surface. The conservation of helicity can be easily derived by noting that on any flux surface, which is allowed to move with a velocity $\mathbf{v}$ as time evolves,

$$\frac{dH}{dt} = \int_\psi \left(\mathbf{A} \cdot \frac{\partial \mathbf{B}}{\partial t} + \mathbf{B} \cdot \frac{\partial \mathbf{A}}{\partial t} \right) d\mathbf{r} + \int_{S_\psi} (\mathbf{A} \cdot \mathbf{B})(\mathbf{n} \cdot \mathbf{v}) dS \tag{11.260}$$

This expression is simplified by substituting $\partial \mathbf{B} / \partial t = -\nabla \times \mathbf{E}$ and $\partial \mathbf{A} / \partial t = -\mathbf{E} - \nabla \phi$. The $\nabla \phi$ term integrates to zero and the remaining terms can be written as

$$\frac{dH}{dt} = -\int_\psi [2\mathbf{E} \cdot \mathbf{B} + \nabla \cdot (\mathbf{E} \times \mathbf{A})] d\mathbf{r} + \int_{S_\psi} (\mathbf{A} \cdot \mathbf{B})(\mathbf{n} \cdot \mathbf{v}) dS \tag{11.261}$$

The divergence theorem is now applied to the $\mathbf{E} \times \mathbf{A}$ term. A short calculation yields

$$\frac{dH}{dt} = -2 \int_\psi \mathbf{E} \cdot \mathbf{B} \, d\mathbf{r} + \int_{S_\psi} (\mathbf{n} \times \mathbf{A}) \cdot (\mathbf{E} + \mathbf{v} \times \mathbf{B}) dS = 0 \qquad (11.262)$$

Clearly, helicity is conserved on every flux surface for a plasma that satisfies the ideal MHD Ohm's law.

Although the energy W can be minimized subject to conservation of H on every flux surface, the result, as pointed out by Taylor, is unrealistic for a plasma with dissipation. The presence of resistivity relaxes the ideal MHD requirement that field line topology be exactly conserved as the plasma evolves thereby providing access to a much wider class of lower energy states. This insight led Taylor to propose a much simpler set of constraints that govern the evolution of RFP plasmas. These are the next topic for discussion.

Taylor's constraints for a dissipative plasma

Since field line topology is not conserved in a dissipative plasma, Taylor reasoned that H would not be conserved on every flux surface. The continual tearing and reconnection of field lines in the presence of fine scale resistive MHD turbulence would destroy the identity of virtually all flux surfaces. He then made the crucial assumption that the only flux surface that would maintain its identity and corresponding helicity is the plasma boundary since it is assumed to be rigid and perfectly conducting; that is, if ψ_a is the total toroidal flux contained within the plasma then $H_a = H(\psi_a)$ is the only helicity that is conserved. Furthermore, because the boundary is perfectly conducting the toroidal flux

$$\psi_a = \int_0^{2\pi} d\theta \int_0^a B_z r dr \qquad (11.263)$$

is itself conserved.

To summarize, Taylor considered the evolution of an RFP plasma and postulated that the profiles at any instant of time would be determined by minimizing the potential energy subject to the constraints of fixed total helicity and fixed total toroidal flux. Mathematically, this is equivalent to minimizing W subject to the constraints of fixed H_a and fixed ψ_a. Implicit in the formulation are the assumptions that the evolution takes place on a time scale slow compared to the ideal MHD time but rapid compared to the time during which the plasma current changes. The plasma thus evolves through a sequence of minimum energy states each one characterized by a different value of the slowly changing current.

The minimum energy equations

The task now is to derive the governing equations obtained by minimizing W subject to the H_a and ψ_a constraints. The procedure is as follows. First, since the flux constraint involves a surface rather than a volume integral it is easily satisfied by a simple normalization procedure as demonstrated shortly. Second, the helicity constraint, which involves a volume integral, can be taken into account by a standard application of the method of Lagrange multipliers. Specifically, the minimum energy state is obtained by setting the variation of

$$W = \frac{1}{2\mu_0} \left[\int_P \left(B^2 + \frac{2\mu_0 p}{\gamma - 1} - \mu \mathbf{A} \cdot \mathbf{B} \right) d\mathbf{r} + \mu H_a \right] \qquad (11.264)$$

to zero where μ is the Lagrange multiplier. In carrying out this step the variation of W with respect to $\mathbf{B}$, p and μ must each be set to zero.

As expected, the variation with respect to μ yields the original helicity constraint:

$$(\Delta W)_\mu = \frac{\delta \mu}{2\mu_0} \left(H_a - \int_P \mathbf{A} \cdot \mathbf{B} \, d\mathbf{r} \right) = 0 \qquad (11.265)$$

The variation with respect to $\mathbf{B}$ and p can be written as

$$(\Delta W)_{\mathbf{B},p} = \frac{1}{2\mu_0} \int_P \left(2\mathbf{B} \cdot \delta\mathbf{B} + \frac{2\mu_0 \delta p}{\gamma - 1} - \mu \delta\mathbf{A} \cdot \mathbf{B} - \mu \mathbf{A} \cdot \delta\mathbf{B} \right) d\mathbf{r} = 0 \quad (11.266)$$

Two steps are required to reduce this expression to a useful simplified form. First, Taylor argues that the term involving δp can be set to zero independent of the variation in $\mathbf{B}$. The reasoning is that the continual tearing and reconnecting of field lines, coupled with an enormous parallel heat conductivity, allows the plasma pressure to redistribute and equalize itself so that $\nabla p = 0$ regardless of the variations in $\mathbf{B}$. This is obviously a pessimistic conclusion but is often closer than not to the realistic experimental situation. In practice the plasma pressure is non-zero since it is being driven by ohmic power. Even so, because of turbulent resistive MHD transport, the resulting ohmic pressure is considerably below the ideal MHD stability limit, the latter corresponding to a tight coupling between δp and $\delta\mathbf{B}$ through the adiabatic equation of state. For simplicity ohmic heating is neglected allowing one to apply Taylor's reasoning and set $\delta p \approx 0$.

Second, the remaining terms in the integral are simplified by noting that $\delta\mathbf{B} = \nabla \times \delta\mathbf{A}$. A short calculation leads to

$$(\Delta W)_{\mathbf{B},p} = \frac{1}{\mu_0} \int_P \delta\mathbf{A} \cdot (\nabla \times \mathbf{B} - \mu\mathbf{B}) \, d\mathbf{r} + \frac{1}{2\mu_0} \int_P \nabla \cdot [\delta\mathbf{A} \times (2\mathbf{B} - \mu\mathbf{A})] \, d\mathbf{r} = 0$$

$$(11.267)$$

The divergence term is converted into a surface integral which then vanishes since $\mathbf{n} \times \delta\mathbf{A} = 0$ on the plasma surface. For the remaining integral to vanish for arbitrary $\delta\mathbf{A}$ requires that at each point in the plasma

$$\nabla \times \mathbf{B} = \mu\mathbf{B} \qquad (11.268)$$

Equation (11.268) with μ a constant, is the desired relation that describes each minimum energy state during the evolution of the RFP plasma.

Mathematical solution for the cylindrically symmetric minimum energy state

The next step in the analysis is to solve for $\mathbf{B}$ and $\mathbf{A}$ assuming that μ is a constant. In general, multiple solutions exist to Eq. (11.268), each satisfying the same boundary conditions – the solution is not unique. For each solution the resulting $\mathbf{B}$ and $\mathbf{A}$ are used to evaluate $H_a(\mu)$ which when inverted determines $\mu = \mu(H_a)$. Substituting this value of μ into the expression for the potential energy allows one to calculate $W(H_a)$ for each solution. Taylor conjectures that the plasma should evolve to that particular state which has the lowest absolute value of W.

Presented below are the mathematical solutions to the minimum energy equation followed by a discussion of their application to experiment. In terms of the mathematics Taylor (1974) and Reiman (1980) have shown that only two of the many non-unique solutions can possibly have an absolute minimum in W. Of these two the one that produces the absolute minimum depends upon the actual value of H_a.

Qualitatively, the two possible minimum energy solutions correspond to (1) an $m = 0$ cylindrically symmetric state and (2) an $m = 0, 1$ mixed helical state. Consider first the $m = 0$ solution which is easily found by eliminating $B_\theta(r)$ from the θ component of Eq. (11.268) and substituting into the z component. The result is a differential equation for $B_z(r)$ given by

$$\frac{1}{r}\frac{d}{dr}\left(r\frac{dB_z}{dr}\right) + \mu^2 B_z = 0 \qquad (11.269)$$

The solution for the fields can then be written as

$$B_z(r) = B_{z0}J_0(\mu r)$$
$$B_\theta(r) = B_{z0}J_1(\mu r) \qquad (11.270)$$

Here, J_0, J_1 are Bessel functions and B_{z0} is the toroidal field on axis. The quantity B_{z0} is directly related to the toroidal flux constraint. Specifically, in accordance with previous notation, B_{z0} is related to the normalized toroidal flux $\langle B_z \rangle \equiv \psi_a/\pi a^2$ by

$$\langle B_z \rangle = \frac{2}{a^2}\int_0^a B_z r\,dr = \left[\frac{2J_1(x)}{x}\right]B_{z0} \qquad (11.271)$$

where $x = \mu a$. Note that B_z has a reversal when $x > 2.4$.

Using the fields given by Eq. (11.270) one can evaluate the potential energy from Eq. (11.256) after invoking the $p \approx 0$ approximation. After a straightforward calculation that makes use of various Bessel function identities (Gradshteyn and Ryzhik, 2000) one obtains

$$\frac{W}{\overline{W}} \equiv W_0(x) = \frac{x^2}{J_1^2}\left(J_0^2 + J_1^2 - \frac{J_0 J_1}{x}\right)$$

$$\overline{W} = \frac{\pi^2 R_0 a^2}{2\mu_0}\langle B_z\rangle^2$$

(11.272)

where the argument of the Bessel functions is x.

The next quantity to calculate is the helicity H_a. To do this requires an expression for $\mathbf{A}$. Since $\nabla \times \mathbf{B} = \mu \mathbf{B} = \mu \nabla \times \mathbf{A}$ it follows that $\mathbf{B} = \mu \mathbf{A} + \nabla \chi$ where χ is a free function. The gauge condition is chosen by the usual relation $\nabla \cdot \mathbf{A} = 0$ implying that $\nabla^2 \chi = 0$. It is important to determine χ since its value impacts the calculation of H_a.

The determination of χ is slightly subtle. A convenient procedure is to assume that the plasma evolves on a slowly varying time scale which then leads to a determination of χ by application of the condition $\mathbf{n} \times \mathbf{E} = 0$ on the perfectly conducting plasma boundary. The electric field can be expressed as

$$\mathbf{E} = -\frac{\partial \mathbf{A}}{\partial t} - \nabla \phi = -\frac{\partial \mathbf{A}}{\partial t}$$

(11.273)

The formulation now has two potential functions, χ and φ in addition to $\mathbf{A}$. Without loss in generality, one potential function can always be absorbed into $\mathbf{A}$. For present purposes it is convenient to absorb φ into $\mathbf{A}$, which is accomplished by introducing a scalar $U(\mathbf{r}, t)$ defined by $\partial U / \partial t = -\varphi$. One then sets $\mathbf{A} = \mathbf{A}' + \nabla U$ and $\chi = \chi' - \mu U$. After supressimg the primes, $\mathbf{B}$ has the same form as initially assumed and $\mathbf{E}$ reduces to the second form in Eq. (11.273).

In general the solutions for $\mathbf{E}, \mathbf{B}, \mathbf{A}$ consist of two contributions, one with cylindrical symmetry $\left(\text{e.g., } \overline{\mathbf{E}}(r,t)\right)$ and the other with helical symmetry that averages to zero over θ or z $\left(\text{e.g., } \tilde{\mathbf{E}}(r, \theta + kz, t)\right)$. The corresponding solutions for χ satisfying $\nabla^2 \chi = 0$ are thus of the form $\chi = c_1(t)\,\theta + c_2(t)\,z + \tilde{\chi}(r, \theta + kz, t)$. The non-single valued contributions (i.e., the c_1, c_2 terms) generate cylindrically symmetric contribution to the fields.

A straightforward calculation that makes use of the boundary conditions $\tilde{B}_r = \tilde{E}_\theta = \tilde{E}_z = 0$ on the plasma boundary leads to a solution for $\tilde{\chi}$. The details are unimportant since application of the divergence theorem shows that the resulting $\tilde{\chi}$ makes no contribution to the value of H_a. Hereafter $\tilde{\chi}$ is ignored in the analysis.

What remains are the values of c_1, c_2. Consider first c_1. The poloidal magnetic field is given by $B_\theta = \mu A_\theta + c_1 / r$. Clearly $c_1 = 0$ for regularity at the origin.

Next, note that $\dot{E}_z = -\dot{A}_z = -(\dot{B}_z - \dot{C}_2/\mu)$. Consequently, for $\overline{E}_z$ to vanish on the plasma boundary requires that

$$\left.\overline{E}_z\right|_a = -\left.\frac{\partial A_z}{\partial t}\right|_a = -\frac{\partial}{\partial t}\left[\frac{1}{\mu}(\overline{B}_z - c_2)\right]_a = 0 \qquad (11.274)$$

This implies that

$$c_2 = \overline{B}_z(a, t) \qquad (11.275)$$

The conclusion from this discussion is that the vector potential can finally be written as

$$\mathbf{A} = \mu[\mathbf{B} - B_{z0}J_0(\mu a)\,\mathbf{e}_z] \qquad (11.276)$$

and is valid for both the cylindrically symmetric and mixed helical states. The helicity is evaluated by substituting Eq. (11.276) into Eq. (11.258) leading to

$$\frac{H_a}{\overline{H}} \equiv H_0(x) = \frac{x}{J_1^2}\left(J_0^2 + J_1^2 - \frac{2J_0J_1}{x}\right) \qquad (11.277)$$
$$\overline{H} = \pi^2 R_0 a^3 \langle B_z\rangle^2$$

Observe that the normalizing factors $\overline{W}$, $\overline{H}$ are functions of $\langle B_z\rangle$ which is a constant during the plasma evolution because of the toroidal flux constraint. This is the reason that these constants are written in terms of $\langle B_z\rangle$ rather than B_{z0}.

Equations (11.272) and (11.277) are two parametric relations in x for $W = W(x)$ and $H_a = H_a(x)$. Eliminating x (numerically) yields the desired relation between energy and helicity: $W = W(H_a)$. This function is plotted shortly after the mixed helicity state is calculated.

Mathematical solution for the mixed helical minimum energy state

The basic equation describing the $m = 0, 1$ mixed helicity state is obtained by operating on Eq. (11.268) with $\mathbf{e}_z \cdot \nabla \times$ and using the relation $\nabla \cdot \mathbf{B} = 0$. The result is a single second-order partial differential equation for B_z given by

$$\nabla^2 B_z + \mu^2 B_z = 0 \qquad (11.278)$$

The solution for the $m = 0, 1$ mixed helicity state is easily calculated and has the form

$$\frac{B_z(r)}{B_{z0}} = J_0(\mu r) + C J_1(ar) \cos(\theta + kz)$$

$$\frac{B_\theta(r)}{B_{z0}} = J_1(\mu r) - \frac{C}{a} \left[\mu J_1'(ar) + \frac{k}{ar} J_1(ar) \right] \cos(\theta + kz) \qquad (11.279)$$

$$\frac{B_r(r)}{B_{z0}} = -\frac{C}{a} \left[k J_1'(ar) + \frac{\mu}{ar} J_1(ar) \right] \sin(\theta + kz)$$

Here, prime denotes differentiation with respect to the argument, k is the helical wave number, $a = (\mu^2 - k^2)^{1/2}$, and C is the helical amplitude. At this point both k and C are arbitrary. Their values are determined shortly by the analysis of the minimum energy state.

The next step is to calculate W and H for the mixed helicity state. The calculation is straightforward but slightly tedious. The results are given by

$$\frac{W}{\overline{W}} = W_0(x) + C^2 W_1(x)$$

$$\frac{H_a}{\overline{H}} = \frac{1}{x} \left[W_0(x) + C^2 W_1(x) - \frac{x J_0(x)}{J_1(x)} \right] \qquad (11.280)$$

where

$$W_1 = \frac{x^2}{2y^2 J_1^2(x)} \left\{ x^2 \left[J_0^2(y) + J_1^2(y) \right] - y J_0(y) J_1(y) - \frac{(x-\kappa)^2}{y^2} J_1^2(y) \right\} \qquad (11.281)$$

and $\kappa = ka$, $y = aa = (\mu^2 - k^2)^{1/2} a = (x^2 - \kappa^2)^{1/2}$.

These admittedly complicated looking expressions are subject to an additional constraint resulting from the boundary condition at the plasma edge which, when applied, greatly simplifies the analysis. Specifically, if the plasma is surrounded by a perfectly conducting wall then $B_r(a, \theta, z) = 0$. This condition is trivially satisfied for the $m = 0$ cylindrical state. However, for the $m = 0, 1$ state one sees after a simple calculation that Eq. (11.279) requires that

$$\kappa y J_0(y) + (x - \kappa) J_1(y) = 0 \qquad (11.282)$$

Equation (11.282) is a transcendental relation which, when solved, determines $x = x(\kappa)$. Now, it can be shown (numerically) that the energy is an increasing function of x in the regime of interest. It thus follows that the lowest energy mixed helicity state is produced by the value of κ that minimizes x. A numerically obtained plot of x vs. κ is illustrated in Fig. 11.47 which shows that the minimum occurs at $\kappa = \kappa_0 \approx 1.2$ corresponding to $x = x_0 \approx 3.11$.

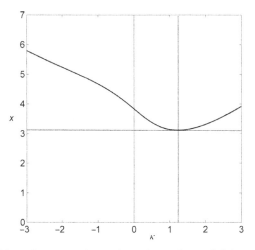

Figure 11.47   Plot of x vs. κ determined from the radial boundary condition. Note the minimum which occurs at $\kappa = \kappa_0 \approx 1.2$ corresponding to $x = x_0 \approx 3.11$.

Observe that all mixed $m = 0, 1$ helical states are characterized by the same value of $x = x_0$. Consequently the values of W and H can change only because the helical amplitude C varies. This is in contrast to the cylindrical state where W and H change because of variations in x. Recognition of this fact allows one to easily calculate the relation between W and H for the $m = 0, 1$ mixed helical state by simply eliminating C in Eq. (11.280). The result is

$$\frac{W}{\overline{W}} = x_0 \left[\frac{H_a}{\overline{H}} + \frac{J_0(x_0)}{J_1(x_0)} \right] \tag{11.283}$$

This is the desired relation.

The minimum energy state

It is now a straightforward matter to determine whether the cylindrical or the $m = 0, 1$ mixed helical state corresponds to the absolute minimum energy. One superimposes plots of $W/\overline{W}$ vs. $H_a/\overline{H}$ for the cylindrical state (Eqs. (11.272) and (11.277)) and the $m = 0, 1$ helical state (Eq. (11.283)) as illustrated in Fig. 11.48. There are several points to be made:

- For low values of H_a only the $m = 0$ solution can exist. Therefore, in this regime the plasma is cylindrical.
- At higher values of H_a both the $m = 0$ and $m = 0, 1$ states can exist simultaneously. The $m = 0, 1$ state has the lowest energy and in this regime the plasma transforms from a cylinder to a helix.

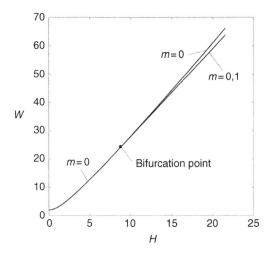

Figure 11.48   Energy vs. helicity for the $m = 0$ cylindrical state and the $m = 0, 1$ mixed helical state.

- The helical state cannot exist for low H_a because the radial boundary condition given by Eq. (11.282) cannot be satisfied for $x \leq x_0 = 3.11$. The condition $x \leq x_0$ corresponds to a minimum allowable value of H_a. This can be seen by considering an $m = 0$ state and an $m = 0, 1$ state, which because of conservation of helicity, have the same H_a. Equating the expressions for H_a given by Eqs. (11.277) and (11.280), with the latter evaluated at $x = x_0$, yields an expression for C^2 given by

$$C^2 = \left[\frac{W_0(x)}{x} - \frac{W_0(x_0)}{x_0} - \frac{J_0(x)}{J_1(x)} + \frac{J_0(x_0)}{J_1(x_0)} \right] \frac{x_0}{W_1(x_0)} \qquad (11.284)$$

For $x \leq x_0$ or equivalently $H_a(x) \leq H_a(x_0)$, one can easily verify (numerically) that $C^2 < 0$, which is unphysical.
- The transition from the $m = 0$ to the $m = 0, 1$ state occurs at the bifurcation point $H_a/\overline{H} \approx 8.21$.
- Recall that B_z reverses when $x \approx 2.4$ in the cylindrical state, which corresponds to $H_a/\overline{H} \approx 2.4$. Therefore, an RFP that evolves from an initial state with zero current to a final state whose helicity lies in the range $2.4 < H_a/\overline{H} < 8.21$ should have the form of a cylindrical plasma with a reversed B_z field.
- The last point concerns the minimizing wave number $ka \approx 1.2$ which is obviously positive. Since the condition for an $m = 1$ resonant surface within the plasma is given by $k = -B_\theta/rB_z$ the implication is that the helix rotation direction aligns with the magnetic lines exterior to the B_z reversal surface (i.e., $B_z < 0$ corresponds to $k > 0$). Furthermore, for $x_0 = 3.11$ the range of resonant k values between the reversal surface and the plasma edge is given by

$-J_1(x_0)/J_0(x_0) = 1.01 < ka < \infty$. This range includes $ka \approx 1.2$. The conclusion is that the wave number of the Taylor helical state corresponds to a resonant ka located between the reversal surface and the plasma edge (reversal occurs at $r/a \approx 0.77$ while resonance occurs at $r/a \approx 0.88$).

This completes the mathematical description of Taylor relaxation theory. The next step is to see how well the theory applies to RFP experiments.

Application to experiment

There are several important applications of Taylor's theory to actual experimental situations. One involves the need for a B_z reversal. The other involves the predicted evolution of an RFP plasma as the current is increased. These are discussed below.

Consider first the importance of the B_z reversal. Strictly speaking Taylor's theory does not suggest that an RFP with a non-reversed B_z should have poorer performance than one with a reversed B_z. In both cases the plasma has relaxed to a minimum energy state. However, actual experiments do exhibit a strong change in performance when reversal occurs and it is interesting to understand the reason.

The answer lies in the fact that a Taylor state is characterized by a current profile with $J_z(a) \neq 0$. In contrast actual experiments almost always have $J_z(a) \approx 0$ because of plasma contact with the cold surrounding wall. This experimental reality is not included in Taylor's theory. The consequence is that the minimum energy profiles must be modified near $r \approx a$ to insure that $J_z(a) \approx 0$. Over the large remaining core of the plasma the Taylor profiles are quite reasonable.

Recall now the discussion in Section 11.7.2 which shows that RFP profiles with $J_z(a) \approx 0$ have a minimum in $q(r)$ when B_z is not reversed. Such profiles are always ideal MHD unstable, being driven by both the parallel current and the pressure gradient. This fact helps to explain the poor performance without a B_z reversal. On the other hand, when B_z has a reversal, the $q(r)$ profile no longer has a minimum and the ideal MHD instabilities are no longer excited. The result is a much improved experimental performance. The conclusion is that the condition for the onset of a B_z reversal as predicted by Taylor's theory plays a far more important role in RFP operation than one might have originally anticipated.

The second application of Taylor's theory concerns the evolution of an RFP plasma as the current is increased. To address this issue from an experimental point of view assume that initially a uniform bias B_z field is produced in the vacuum chamber. This sets the initial the value of the toroidal flux which is then conserved during the plasma evolution. Next, the toroidal current is increased slowly towards its final value. The initially cold plasma begins to heat due to the ohmic power. Ideally one would like to compare the theoretical and experimental magnetic field profiles as the current $I(t)$ increases. However, these are difficult experimental

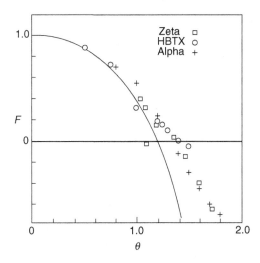

Figure 11.49   Diagram of F vs. θ comparing theoretical and experimental data from several RFP experiments (from Taylor, 1986). Reproduced with permission.

measurements. A simpler test is to compare the theoretical prediction of $B_z(a,t)$ with the easy-to-measure experimental value. If the current rise time is slow (as it is for most RFP plasmas) then Taylor's theory predicts that the RFP should evolve through a sequence of minimum energy states. The desired relation between $B_z(a, t)$ and $I(t)$ can be illustrated in a convenient normalized form by plotting an $F(t)$ vs. $\theta(t)$ diagram. The normalization is useful because it allows a comparison between different experiments on the same diagram.

For the Taylor minimum energy cylindrical profiles, F and θ are related parametrically through x as follows:

$$F(t) \equiv \frac{B_z(a, t)}{\langle B_z \rangle} = \frac{x J_0(x)}{2 J_1(x)}$$

$$\theta(t) \equiv \frac{B_\theta(a, t)}{\langle B_z \rangle} = \frac{x}{2} \tag{11.285}$$

A plot of $F(t)$ vs. $\theta(t)$ is illustrated in Fig. 11.49 (Taylor, 1986). Here, one can view increasing θ as equivalent to increasing time during the plasma evolution. Note that field reversal occurs at $\theta \approx 1.2$ and the $m = 0$, 1 helical bifurcation point corresponds to $\theta \approx 1.56$. Also shown in the diagram is a sequence of data points from early RFP experiments. Observe that there is quite good agreement between theory and experiment. The agreement using the model profiles, although not illustrated, is even further improved because of the more realistic treatment of the edge. However, it must be emphasized that without Taylor's theory there is no justification for believing that the plasma should evolve along the $F(t)$ vs. $\theta(t)$ curve.

To summarize, Taylor's theory provides a simple but elegant explanation of the evolution of ohmically heated RFP plasmas. The theory captures the magnetic field and current density profiles except right near the edge. It predicts zero pressure while in practice the pressure is finite because of ohmic heating. Even so the experimentally measured β_p values are usually small compared to the ideal MHD stability limits. Experimentally, good performance requires raising the current to a sufficiently high value, $\theta > 1.2$, so that B_z reverses.

11.7.7 Ideal external modes

From the point of view of ideal MHD the analysis thus far presented has shown that an RFP can be stable against all internal current-driven and pressure-driven modes at high values of $\beta_p \sim 0.2$. This subsection focuses on ideal external modes, setting aside temporarily Taylor's prediction of $\beta_p \to 0$ which involves resistive MHD. Note that when the plasma is bounded by a perfectly conducting wall the internal mode analysis constitutes a complete description of ideal MHD stability. In this section it is shown that the presence of a conducting wall is crucial. Without such a wall an RFP is always unstable to external modes, even at $\beta_p = 0$. A comprehensive study of ideal and resistive MHD stability for the case $\beta_p = 0$ has been carried out by Antoni *et al.* (1986). The results presented here are consistent with Antoni *et al.* in the appropriate regions of overlap.

Before proceeding, however, a discussion is warranted regarding the way in which external modes can be excited in an RFP. The reason for the discussion is that in practice, RFP plasmas exist out to the wall – there is no well-defined vacuum region separating plasma from wall. Why then worry about external modes? The answer is associated with the fact that the edge of the plasma is cold because of its contact with the wall, or perhaps a limiter, and thus is highly resistive. Since $\eta \propto 1/T^{3/2}$ it follows that the cold edge of an ohmically heated plasma can be treated as a current-free resistive layer. As discussed in Section 8.10, such a layer leads to the same MHD stability boundaries as a vacuum, although with lower growth rates when unstable.

Now, in practice the transition from a high β_p ideal plasma region to a cold current-free resistive plasma region is gradual. Even so, to keep the physics simple but crisp, the analysis presented below adds a narrow resistive plasma layer to the outside of the ideal plasma. The layer is located between $a < r < b$ and can be treated as a vacuum region for the purpose of calculating the marginal stability boundaries. In carrying out these calculations the wall at $r = b$ is assumed to be perfectly conducting.

Here again difficulties arise because in practice the wall is resistive. Thus, modes which are wall stabilized when b/a is sufficiently close to the plasma but are unstable with a wall at infinity become resistive wall modes. Some form of feedback or rotational stabilization is required.

To summarize, the analysis focuses on calculating the marginal stability boundaries against all external modes in an RFP whose core is separated from a perfectly conducting wall by a narrow current-free resistive plasma region. The specific goals are to first determine which k (or n) values are unstable with the wall at infinity and to then determine how close the wall must be (i.e., calculate the critical b/a) to stabilize all such modes. This information is critical to assess the ease or difficulty in providing feedback stabilization. The information is then used to discuss experimental results aimed at avoiding external mode instabilities in an RFP.

Theoretical/computational analysis

The theoretical/computational analysis of external mode stability requires a straightforward numerical calculation. The most dangerous external modes correspond to $m = 1$ and almost always have a resonant surface in the vacuum region. The corresponding values of wave number to be tested are then defined as those values of k for which $F(r) \neq 0$ in the plasma. For RFPs with a field reversal this range of k is given by

$$-\frac{B'_\theta(0)}{B_z(0)} < k < -\frac{B_\theta(a)}{aB_z(a)}$$

$$-\left[\frac{6(1-F)}{3-2F}\right]^{1/2} < n\varepsilon < -\frac{\theta}{F} \tag{11.286}$$

where the second form corresponds to the model profiles and $ka = n\varepsilon$.

The stability boundaries are determined by solving

$$\frac{d}{dr}\left(f\frac{d\xi}{dr}\right) - g\xi = 0 \qquad \xi(0) = 1 \qquad \xi'(0) = 0 \tag{11.287}$$

for $0 < r < a$ and then evaluating $\xi'(a)/\xi(a)$. This quantity is substituted into Eq. (11.118) or equivalently Eq. (11.148), leading to the desired value of $\hat{W} = \hat{W}(n\varepsilon)$. The numerical solution to Eq. (11.287) is easy to obtain since $F(r) \neq 0$ in the plasma implying that there are no singularities in the equation. The procedure just described has been carried out for various values of β_p assuming that $b/a = \infty$, $F = -0.2$, $\theta = 1.45$, and $\varepsilon = 1/5$. Also, for numerical convenience n is treated as a continuous variable but obviously should only be allowed to have integer values when applied to actual experiments. The results are illustrated for two values of β_p (i.e., $\beta_p = 0$ and $\beta_p = 0.1$) in Fig. 11.50, which illustrates $\hat{W}$ vs. n.

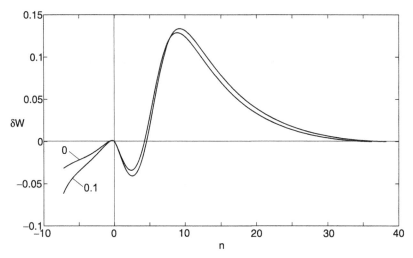

Figure 11.50   Stability diagram of δW vs. n for external mode stability in an RFP with the wall at infinity. The parameters correspond to $F = -0.2$, $\theta = 1.45$. The two curves correspond to $\beta_p = 0$ and $\beta_p = 0.1$.

Observe that the plasma is unstable for both positive and negative n. For $\beta_p = 0$ the unstable positive n values correspond to $1 \leq n \leq 4$. For these modes the helicity has the same sense as the fields outside the reversal surface. The unstable negative n values lie in the range $1 \leq -n \leq 7$. Here, the helicity aligns with the field inside the reversal surface. Note that the curves end abruptly at about $n \approx -7.5$. For more negative values the resonant surface moves into the plasma from the origin. The mode becomes an internal mode which is quickly stablized. All told, the plasma is unstable to 11 external modes, even at $\beta_p = 0$. As expected, finite β_p makes the situation more unstable but only by a small amount. It is not a dominant effect.

It is clear from these results that for the RFP to be a viable fusion concept it is necessary to wall stabilize the external kink modes. The key issue is to determine how close to the plasma must one place a perfectly conducting wall to stabilize all external kink modes. A similar set of numerical calculations as described above has been carried out to answer this question. For this study all the parameters are the same as for Fig. 11.50 except that b/a is allowed to vary. The mode number n is again treated as a continuous variable. The results are given in Fig. 11.51, which illustrates curves of $\hat{W}$ vs. n for various b/a assuming that $\beta_p = 0$.

The main results are as follows. As intuition suggests, moving the wall closer to the plasma raises the curves, producing an increasingly positive value of $\hat{W}$; that is, the plasma is becoming more stable. For the curves shown, corresponding to $\theta = 1.45$, one sees that all external kink modes are stabilized for $b/a < 1.7$. The last mode that is stabilized has a toroidal mode number $n = -7$.

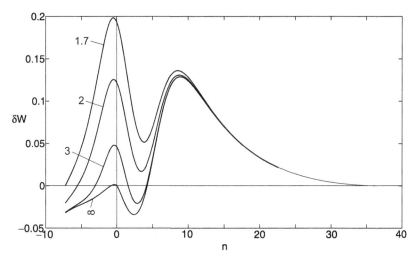

Figure 11.51 Stability diagram of δW vs. n for external mode stability in an RFP for various b/a. The parameters correspond to $F = -0.2$, $\theta = 1.45$, and $\beta_p = 0$. Observe wall stabilization as b/a decreases.

A similar set of curves can be generated for different values of θ. The results are qualitatively similar with one exception. As the current (θ) increases, the modes with positive n become relatively more unstable than the modes with negative n. Eventually, above a critical $\theta = 1.5$, the last mode to become stabilized switches from negative n to positive n. In other words, the helical sense of the most unstable mode switches from being aligned with the field inside the reversal radius to the field on the outside. This point is illustrated in Fig. 11.52, which illustrates the critical b/a vs. θ for the most unstable mode. Also shown are the corresponding values of n for the last mode to be stabilized. The switch from an inside to an outside mode is clearly exhibited at $\theta = 1.5$. The process has been repeated for several values of β_p, also illustrated in Fig. 11.52.

The last point of interest concerns the dependence of stability on the current profiles. The model profiles are fairly broad, approximating the Taylor profiles everywhere except at the edge. More peaked profiles, which are often observed experimentally, effectively reduce the size of the plasma radius to a value denoted by $a_{eff} < a$. Since the unstable toroidal mode numbers scale as $n \sim 1/\varepsilon \rightarrow R_0/a_{eff}$, the implication is that peaked current profiles will be unstable to a wider range of unstable n values than broad profiles. Numerical calculations, not presented here, indeed confirm this intuition.

Experimental stabilization of external kink modes

The above analysis has shown that an RFP will always be unstable to multiple external kink modes, even at $\beta_p = 0$, with the wall at infinity. These modes can be

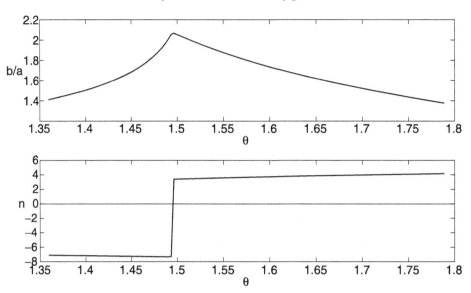

Figure 11.52 Critical b/a, and corresponding n, vs. θ for the most unstable external kink mode. Note the switch from an inside to an outside mode at $\theta = 1.5$ for $\beta_p = 0$. Curves are also illustrated for $\beta_p = 0.05$ and $\beta_p = 0.1$.

stabilized by a close-fitting conducting shell. Of course the stabilized modes are then transformed into resistive wall modes. For the RFP to move forward as a fusion concept it must be possible to simultaneously stabilize all resistive wall modes, at least ten in number.

Many researchers believed it would be *theoretically* possible to simultaneously feedback stabilize all the resistive wall modes. However, these same researchers (the author included) also believed it would be not possible to *experimentally* achieve this goal because of engineering complexity. This is one situation where the author is pleased to admit being wrong. Experiments on two European RFP experiments, EXTRAP T2R (Brunsell *et el.*, 2004) in Sweden and RFX-mod (Paccagnella *et al.*, 2006) in Italy, demonstrated that sophisticated feedback systems could be constructed that would simultaneously stabilize all external kink modes.

For example, the RFX-mod experiment has 192 feedback saddle coils, each one independently controllable. This sophisticated feedback system has been operated experimentally and shown to stabilize all resistive wall modes. A striking example of this stabilization is illustrated in Fig. 11.53. Shown here are experimental traces of plasma current versus time for various progressive states of the feedback control system. The lifetime of the discharge has been extended substantially. The final decay of the plasma is not due to a loss of feedback control, but is the result of having utilized all the volt-seconds available from the transformer.

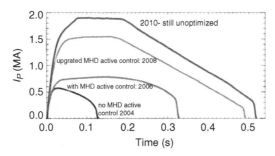

Figure 11.53   RFX experimental traces of plasma current vs. time for progressing states of the feedback control system. Note the extended lifetime with the feedback activated. From Martin et al, 2011. Reproduced with permission.

The conclusion is that feedback control has been able to stabilize all external kink modes in an RFP, a very favorable result. Also internal modes are ideally stable to quite high values of β_p. Still, experimental values of β_p, even in feedback stabilized discharges, are considerably lower than the ideal MHD internal limits, primarily because of anomalous resistive MHD turbulence causing enhanced transport. The resulting energy confinement times are noticeably poorer than those in a tokamak with comparable current. It is this poor confinement, and not MHD, that suggests that for the standard RFP to move forward the transport must be improved. This then is one major goal of RFP research – improve transport. A second major goal is to discover an efficient way to non-inductively drive a large amount of current.

Two innovative methods have been suggested to achieve the transport goal – profile control and single helicity states. These topics involve detailed studies of resistive MHD stability and are thus beyond the scope of the present textbook. Several relevant references are listed at the end of the chapter.

11.7.8 Overview of the RFP

The RFP is an axisymmetric toroidal configuration with $B_\theta^2 \sim B_z^2 \sim \mu_0 p$. It achieves ideal MHD stability with an optimally efficient use of the magnetic field profiles. Specifically, the B_z field has a reversal near the outside of the profile and is much smaller in magnitude than the toroidal field in a tokamak with the same toroidal current. This is a major technological advantage of the RFP.

Good MHD stability at high β is possible against all internal ideal MHD modes. However, external modes with a wall at infinity are potentially very unstable because the value of $q_a \sim \varepsilon$ is small at the plasma edge. A large number of modes, on the order of 10–20, would be simultaneously unstable as resistive wall modes unless they are feedback stabilized. For many years researchers thought this would be too difficult technologically. However, experiments in Sweden and Italy have shown that sophisticated feedback systems could be developed that provide stabilization against all external ideal MHD modes.

The result is longer lived RFP plasmas, whose lifetime is limited by external power supply constraints. This favorable result is somewhat tempered by the fact that transport in these long lived states is still substantially below that in a tokamak of comparable size and current. The reason is the presence of resistive MHD turbulence. There remains a clear need to improve transport and this is one of the major areas of current and future RFP research. A second major area of research requires the development of efficient methods of non-inductive current drive. Some effort has been devoted to this problem but most of RFP research has focused on improved transport.

11.8 Summary

In Chapter 11 the Energy Principle and the eigenvalue formulation of MHD have been used to investigate the stability of 1-D, cylindrically symmetric, equilibria. A summary of these results is given below.

The θ-pinch

The straight θ-pinch is shown to be stable to all m, k modes for arbitrary values of β. The potential energy δW is either positive or approaches zero when $k \rightarrow 0$. This favorable result is a consequence of the fact that a θ-pinch has neither parallel currents nor field line curvature to drive instabilities. Unfortunately the θ-pinch cannot be bent into a torus unless additional fields are added to provide toroidal force balance and it is these fields that ultimately determine stability.

The Z-pinch

The straight Z-pinch is unstable to two potentially dangerous instabilities, each driven by a combination of the pressure gradient and unfavorable curvature. The first instability is the short wavelength $m = 1$ mode while the second corresponds to the $m = 0$ sausage instability. Both of these modes can lead to very undesirable, if not catastrophic behavior of the plasma. Thus, even though a Z-pinch can easily be bent into a torus, its unsatisfactory MHD stability properties make it a poor candidate for fusion energy applications.

Adding a hard-core current-carrying conductor along the axis greatly improves the situation. With sufficient current the $m = 1$ mode is stabilized. Equally importantly, when the pressure gradient decays sufficiently weakly the sausage instability is also stabilized. The toroidal version of the hard-core pinch is known as a levitated dipole. The same favorable stability properties exist although the need to levitate the superconducting hard core is a clear technological disadvantage. In addition, D-D is required as the fuel in order to prevent neutron heating and quenching of the superconductors.

The general screw pinch

The Energy Principle for the general screw pinch reduces to a 1-D quadratic integral in terms of the single unknown ξ, which is the radial component of the plasma displacement. An analysis of the energy integral leads to several general results:

- Suydam's criterion provides a simple necessary condition for stability against localized interchange modes. The criterion requires only a simple substitution of equilibrium profiles to test stability which quantifies the balance between shear and unfavorable profiles.
- Newcomb's analysis demonstrates how the minimizing differential equation associated with the variation of δW can be used to determine necessary and sufficient conditions for stability of arbitrary profiles. Application of Newcomb's procedure requires the (numerical) solution of a 1-D differential equation in r.
- The Goedbloed–Sakanaka oscillation theorem, valid for the general screw pinch, proves that the most unstable (i.e., fastest growing) MHD perturbation always has the structure of a macroscopic zero-node mode. This theorem shows why it is important to satisfy Suydam's criterion – the existence of not very worrisome, highly localized interchange modes always implies the existence of a large-scale macroscopic mode.
- For a screw pinch external modes can be unstable. They can, however, be stabilized by the presence of a perfectly conducting wall sufficiently close to the plasma. Realistically, the wall has a finite resistivity, in which case the mode again becomes unstable although as a resistive wall mode. The stability boundary is the same as when the wall is at infinity, although the growth rates are much slower, comparable to the wall diffusion time, allowing for the possibility of feedback stabilization.

The straight tokamak

The straight tokamak provides reliable information about current-driven kink modes in a circular cross section plasma. It provides unreliable information for pressure-driven modes since these are strongly affected by toroidal curvature which is obviously not included in a straight model. The main conclusions and connections with experiments are as follows:

- With respect to internal modes only the $m = 1$, $n = 1$ internal kink can become unstable. The ideal MHD version is the initial drive for sawtooth oscillations although a full non-linear toroidal calculation, including finite pressure and resistivity, is required to at least partially explain the data. The main consequence of sawtooth oscillations is to clamp the value of safety factor on axis to approximately $q_0 \approx 1$ or slightly lower. This limits the current density, and hence the ohmic heating on axis.

- The $m = 1$, $n = 1$ external kink mode sets a limit on the total current that can flow in the plasma. The corresponding stability limit $q_a > 1$ is known as the Kruskal–Shafranov limit. In practice, external kink modes with slightly higher values of m can also become unstable and these set stricter limits on q_a. The exact limit depends on profiles but typically $q_a \sim 2.5-3$ is needed to avoid an instability. When the external kink mode criterion is violated the result is most likely to be a major disruption in which both the pressure and current quench in a very short time corresponding to a catastrophic termination of the plasma. Too much current, or alternatively too much edge density (which shrinks the size of the plasma lowering the effective q_a), lead to major disruptions.

- Higher m external kinks can also become unstable if the edge current and current gradient are not sufficiently small. High edge gradients often appear in H-mode operation of a tokamak. When higher m external kink modes are excited the usual results are ELMs, periodic bursts of plasma energy and particles. If strong enough, ELMs can produce irreversible damage to the divertor plate. ELMs are thought to be driven by ballooning-kink modes with the kink component of the perturbation being reasonably well treated conceptually by the straight tokamak analysis.

Overall, the straight tokamak analysis suggests that successful operation requires a sufficiently small current with a sufficiently small edge gradient. Small current is readily controllable experimentally. In contrast, control of the edge gradient is a much more difficult task experimentally.

The reversed field pinch

The ideal MHD properties of an RFP are accurately described by the 1-D cylindrical model. There are a variety of modes that can become unstable. A useful way to organize the stability results and develop some intuition is summarized below:

- To begin, as its name implies, an RFP must have a reversal in the B_z field. If not, an internal $m = 1$ is excited localized around the minimum in $q(r)$. This leads to poor transport because of strong MHD turbulence. Therefore, high-performance RFPs always have a field reversal to avoid the problem of very poor transport.

- An RFP always has a limit on β due to the $m = 0$ mode. There must be sufficient toroidal field to prevent the analog of sausage instabilities. The β limit is approximately $\beta < 0.5$, which is a weak requirement in terms of normal experimental operation.

- Suydam's criterion leads to stricter limits on β. To avoid the resulting localized interchange instabilities the pressure profile must be very flat near the origin and have a very small edge gradient. When such profiles are generated, the β limits typically lie in the range $0.1 < \beta < 0.35$, the exact value depending on the amount of B_z reversal.

- Assuming that Suydam's criterion is satisfied, one then finds that all macroscopic internal modes are also stable for the typical operating range of RFP experiments, $0 > F > -1$. This conclusion is valid up until the value of β predicted by Suydam's criterion and is a very favorable conclusion.

- Focusing on internal perturbations, Taylor has provided a theory that explains the typical time evolution of an RFP plasma. His analysis assumes that there is always a background level of resistive MHD turbulence. The theory suggests that an RFP plasma always evolves to a minimum energy state subject to the constraint of conservation of total helicity. The resulting profiles have flat current density profiles out to the wall and zero β. The zero β result does not take into account that an RFP plasma is always driven by ohmic current which keeps the pressure finite. Taylor's theory qualitatively explains the evolution of the magnetic fields in an RFP but clearly underestimates the achievable values of β.

- Next consider external modes. Because of the low value of q_a, an RFP is always unstable to a large number of $m = 1$ modes with the wall at infinity, even when $\beta_p = 0$. Typically 10–20 different n modes can be simultaneously unstable, depending upon profile. While these modes could be theoretically stabilized as resistive wall modes with a sufficiently close wall, many researchers believed this would be an impossibly difficult technological problem. Undeterred by this pessimism, experimentalists in Sweden and Italy developed highly sophisticated feedback systems that demonstrated complete stability against all ideal external modes. This favorable result opens up the opportunity to improve transport in an RFP. However, even after wall stabilization, transport still remains poorer than in a comparable tokamak because of resistive MHD turbulence. Improving transport is a major area of RFP research.

Overall, the RFP has had a more difficult path to follow than the tokamak in terms of plasma physics performance because of its inherently low q. Still, performance has steadily progressed by some very clever innovations but it still remains to be seen if the RFP can overtake the tokamak in terms of performance. If so, the low toroidal field would be a major technological advantage.

References

Antoni, V., Merlin, D., Ortolani, S., and Paccagnella, R. (1986). *Nucl. Fusion* **26**, 1711.
Bodin, H. A. B. and Newton, A. A. (1980). *Nucl. Fusion* **20**, 1255.
Brunsell, P. R., Yadikin, D., Gregoratto, D., *et al.* (2004). *Phys. Rev. Lett.* **93**, 225001.
Goedbloed, J. P., Pfirsch, D., and Tasso, H. (1972). *Nucl. Fusion* **12**, 649.
Goedbloed, J. P. and Sakanaka, P. H. (1974). *Phys. Fluids* **17**, 908.
Goedbloed, J. P. and Poedts, S. (2004). *Principles of Magnetohydrodynamics*. Cambridge: Cambridge University Press.
Gradshteyn, L. S. and Ryzhik, I. M. (2000). *Table of Integrals, Series, and Products*, 6th edn. San Diego, CA: Academic Press.
Greenwald, M., Terry, J. L., Wolfe, S. M. *et al.* (1988). *Nucl. Fusion* **28**, 2199.
Greenwald, M. (2002). *Plasma Phy. Controlled Fusion* **44**, R27.
Hain, K. and Lust, R. (1958). *Z. Naturforsch. Teil A* **13**, 936.
Hughes, J. W., Snyder, P. B., Walk, J. R., *et al.* (2013). *Nucl. Fusion* **53**, 043016.
Kadomtsev, B. B. (1966). In *Reviews of Plasma Physics*, Vol. 2, ed. M. A. Leontovich. New York: Consultants Bureaus.
Kikuchi, M., Lackner, K., and Tran, M. Q., eds. (2012). *Fusion Physics*. Vienna: International Atomic Energy Agency.
Kruskal, M. D. and Schwarzschild, M. (1954). *Proc. R. Soc. London, Ser. A* **223**, 348.
Landau, L. D. (1946). *J. Phys. (USSR)* **10**, 25
Laval, G. and Pellat, R. (1973). In *Controlled Fusion and Plasma Physics*, Proceedings of the Sixth European Conference, Moscow, Vol. II, p. 640.
Laval, G., Pellat, R., and Soule, J. L. (1974). *Phys. Fluids* **17**, 835.
LDX Group (1998). *Innovative Confinement Concepts Workshop*. Princeton, NJ: Princeton Plasma Physics Laboratory.
Martin, P., Adamek, J., Agostinetti, P., Agostini, M. et al. (2011). *Nucl. Fusion* **51** 094023.
Newcomb, W. A. (1960). *Ann. Phys. (NY)* **10**, 232.
Paccagnella, R., Ortolani, S., Zanca, P. *et al.* (2006). *Phys. Rev. Lett.* **97**, 075001.
Reiman, A. (1980). *Phys. Fluids* **23**, 230.
Rosenbluth, M. N., Dagazian, R. Y., and Rutherford, P. H. (1973). *Phys. Fluids* **11**, 1984.
Shafranov, V. D. (1956). *At. Energy* **5**, 38.
Shafranov, V. D. (1970). *Sov. Phys.-Tech.* **15**, 175.
Snyder, P. B., Wilson, H. R., and Xu, X. Q. (2005). *Phys. Plasmas* **12**, 056115.
Suydam, B. R. (1958). In *Proceedings of the Second United Nations International Conference on the Peaceful Uses of Atomic Energy*. Geneva: United Nations, **31**, p. 157.
Tataronis, J. A. and Grossmann, W. (1973). *Z. Physik* **261**, 203.
Taylor, J. B. (1974). *Phys. Rev. Lett.* **33**, 1139.
Taylor, J. B. (1986). *Rev. Modern Phys.* **3**, 741
Wesson, J. A. (2011). *Tokamaks*, 4th edn. Oxford: Oxford University Press.
Whyte, D. G., Hubbard, A. E., Hughes, J. W. *et al.* (2010). *Nucl. Fusion* **50**, 105005.
Woltjer, J. (1958). *Proc. Nat. Acad. Sci. (USA)* **44**, 489.
Yamada, M., Levinton, F. M., Pomphrey, N. *et al.* (1994). *Phys. Plasmas* **1**, 3269.

Further reading

General screw pinch stability theory

Goedbloed, J. P. and Poedts, S. (2004). *Principles of Magnetohydrodynamics*. Cambridge: Cambridge University Press.

The reversed field pinch

Bodin, H. A. B. and Newton, A. A. (1980). *Nucl. Fusion* **20**, 1255.
Cooper, W. A., Graves, J. P., Sauter, O. *et al.* (2011). *Plasma Phys. Controlled Fusion* **53**, 084001.
Escande, D. F., Martin, P., Ortolani, S. *et al.* (2000). *Phys. Rev. Lett.* **85**, 1662.
Kikuchi, M., Lackner, K. and Tran, M. Q., eds. (2012). *Fusion Physics.* Vienna: International Atomic Energy Agency.
Martin, P., Adamek, J., Agostinetti, P. *et al.* (2011). *Nucl. Fusion* **51**, 094023.
Robinson, D. C. (1971). *Plasma Phys.* **13**, 439.
Sarff, J. S., Hokin, S. A., Ji, H. *et al.* (1994). *Phys. Rev. Lett.* **72**, 3670.
Sovinec, C. and Prager, S. C. (1999). *Nucl. Fusion* **39**, 777.

Tokamaks

Bateman, G. (1978). *MHD Instabilities.* Cambridge, MA: MIT Press.
Furth, H. P. (1975). *Nucl. Fusion* **15**, 133.
Furth, H. P. (1985). *Phys. Fluids* **28**, 1595.
Mukhovatov, V. S. and Shafranov, V. D. (1971). *Nucl. Fusion* **11**, 605.
Kikuchi, M., Lackner, K., and Tran, M. Q., eds. (2012). *Fusion Physics.* Vienna: International Atomic Energy Agency.
Wesson, J. A. (1978). *Nucl. Fusion* **18**, 87.
Wesson, J. A. (2011). *Tokamaks*, 4th edn. Oxford: Oxford University Press.
White, R. B. (1986). *Rev. Mod. Phys.* **58**, 183.
White, R. B. (2006). *Theory of Toroidally Confined Plasmas*, 2th edn. London: Imperial College Press.

Problems

11.1 Consider the marginal stability equation for the general screw pinch, given by

$$\mathbf{F}(\xi) = 0$$

By direct reduction of this equation (i.e., do not use a variational minimization procedure) derive the one-dimensional radial equation for the eigenfunction $\xi = \xi_r(r)$. Show that this equation can be written as

$$(f\xi')' - g\xi = 0$$

where $f(r)$ and $g(r)$ are given by Eq. (11.90).

11.2 One term in δW that is always stabilizing for an arbitrary cylindrical screw pinch is given by

$$\delta W_a = \int f\xi'^2 dr$$

where $f = r^3 (kB_z + mB_\theta/r)^2/(k^2r^2 + m^2)$. Show that for an $m = 1$ mode, $\delta W_a \to 0$ as $\Delta \to 0$ for the trial function illustrated in Fig. 11.54. Assume the pinch

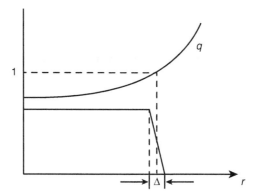

Figure 11.54   Diagram for Problem 11.2.

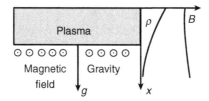

Figure 11.55   Diagram for Problem 11.3.

represents an equivalent torus of length $2\pi R_0$, implying that $k = -n/R_0$. Note that for $\delta W_a \to 0$ there must be a surface in the plasma where $nq(r_s) = 1$

11.3 Consider a slab model of a plasma supported in equilibrium against gravity by a uniform magnetic field $\mathbf{B} = B\mathbf{e}_z$ as shown in Fig. 11.55.

(a) Calculate the one-dimensional radial pressure-balance relation describing the equilibrium.
(b) Derive the one-dimensional eigenvalue equation describing linear stability. Assume the perturbations vary as $\boldsymbol{\xi} = \boldsymbol{\xi}(x)\exp(-i\omega t + ik_y y + ik_z z)$. For simplicity assume incompressible displacements.
(c) Form an Energy Principle from the eigenvalue equation and show that $k_z^2 \to 0$ is the most unstable mode.
(d) Assume that $\rho(x) = \rho_0 \exp(-x/a)$ and show that

$$\omega^2 = -g/a$$

This is the Raleigh–Taylor instability.

11.4 Consider a slab model of a plasma with a constant external gravitational force. Assume a one-dimensional equilibrium with $p(x)$, $B_y(x)$, $B_z(x)$ and $g = $ constant.

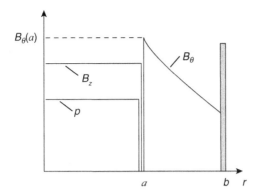

Figure 11.56   Diagram for Problem 11.5.

(a) Show that the Energy Principle for this system reduces to

$$\frac{\delta W}{L_y L_z} = \frac{1}{2} \int dx \left[\frac{F^2}{k^2} \left| \zeta' \right|^2 + (F^2 + g\rho') \left| \zeta \right|^2 \right]$$

where $\mathbf{k} = k_y \mathbf{e}_y + k_z \mathbf{e}_z$, $F(x) = k_y B_y + k_z B_z$ and $\zeta(x) = \zeta_x$.

(b) Derive the equivalent Suydam's criterion for this system. (The final criterion should be a function only of equilibrium quantities.)

(c) By comparing the slab and cylindrical versions of Suydam's criterion, show how gravity can be used to "mock up" field line curvature.

11.5 A standard RFP is always unstable to external modes if the conducting wall is moved to infinity. The purpose of this problem is to estimate how close the conducting wall must be moved in order to obtain stability against all external kinks. Consider the surface current model of an RFP illustrated in Fig. 11.56. Note that all currents flow only on the plasma surface.

(a) Write the equilibrium pressure balance relation across the plasma surface: $\left[p + B^2/2\mu_0 \right] = 0$.

(b) Evaluate δW_F, δW_S, and δW_V. Show that

$$\frac{\delta W}{2\pi R_0} = \frac{\pi B_\theta^2(a) \zeta_a^2}{\mu_0} \left[(1 - \beta) \frac{k a I_a}{I_a'} - 1 - \frac{m^2 K_a}{k a K_a'} \Lambda \right]$$

where $\beta = 2\mu_0 p/B_\theta^2(a)$ and

$$\Lambda = \frac{1 - I_a K_b'/K_a I_b'}{I - I_a' K_b'/I_b' K_a'}$$

Also, $I_x \equiv I_m(kx)$, $K_x \equiv K_m(kx)$ are modified Bessel functions. The notation K_x', I_x' denotes differentiation with respect to the argument. When evaluating

δW_F treat $\mathbf{Q}$ rather than $\boldsymbol{\xi}$ as the quantity to vary. Check that the $\boldsymbol{\xi}$ that results from minimizing $\mathbf{Q}$ is well behaved.

(c) Show that the most unstable mode for $m = 0$ corresponds to $k \to 0$ and requires $\beta < 1/2$ for stability.

(d) Simplify Λ by assuming $kb \ll 1$. (This is only a marginally acceptable approximation, made for mathematical simplicity.)

(e) For $m = 1$ approximate

$$\frac{zI_1(z)}{I'_1(z)} \approx \frac{z^2}{c_1 + c_2 z}$$

$$-\frac{K_1(z)}{zK'_1(z)} \approx \frac{1}{b_1 + b_2 z}$$

Determine c_1, c_2, b_1, b_2 so that each function assumes the correct limiting values as $z \to 0$ and $z \to \infty$.

(f) Calculate the most unstable wave number and the corresponding farthest wall position that leads to stability against $m = 1$ external kinks for (1) $\beta = 0$ and (2) $\beta = 1/2$.

11.6 Show that δW for a general sharp-boundary screw pinch is given by

$$\frac{\delta W}{2\pi R_0} = \frac{\pi \xi_a^2}{\mu_0} \left[(B_0^2 + B_\theta^2 - 2\mu_0 p) \frac{ka I_a}{I'_a} - B_\theta^2 - \Lambda (kaB_0 + mB_\theta)^2 \frac{K_a}{kaK'_a} \right]$$

where B_0 is the external B_z field, $B_\theta = B_\theta(a)$ and Λ, K_a, I_a are defined in Problem 11.5.

11.7 Consider an arbitrary diffuse screw pinch. Prove that when $J_z(r)$ reverses sign at some value $r = r_0$, then Suydam's criterion is automatically satisfied for $r > r_0$ as long as B_θ itself does not reverse. Hint: Assume $rJ_z/B_\theta < 0$ for $r_0 < r < a$ and eliminate p' from Suydam's criterion using the equilibrium pressure balance relation.

11.8 Starting from the resistive MHD equations (i.e., $\mathbf{E} + \mathbf{v} \times \mathbf{B} = \eta \mathbf{J}$) derive a formula for dK/dt where the helicity K is given by

$$K = \int_V \mathbf{A} \cdot \mathbf{B} d\mathbf{r}$$

and the integration volume corresponds to the volume enclosed by a flux surface.

11.9 Estimate the growth rate for the $m = 1$, $n = 1$ internal kink mode in a low β large aspect ratio tokamak as follows. Start with the variational formulation of MHD given by

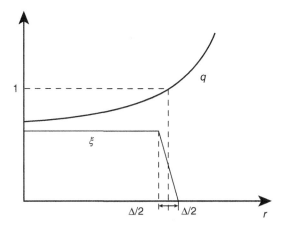

Figure 11.57 Diagram for Problem 11.9.

$$\omega^2 = \frac{\delta W}{K}$$

Use the ohmically heated tokamak expansion and write $\delta W = \delta W_2 + \delta W_4$. Substitute the trial function illustrated in Fig. 11.57 and treat Δ as a variational parameter. Note that for the $m = 1$, $n = 1$ mode, $\delta W_2 \to 0$ as $\Delta \to 0$. Assume Δ is ordered so that $\delta W_2 \sim \delta W_4$ as $\Delta \to 0$ and for simplicity treat $\rho\,(r) \approx \rho_0 = $ const out to the $q = 1$ surface. Show that for the minimizing Δ

$$\omega^2 \approx -\frac{3V_a^2}{R_0^2}\left[\frac{W_4^2}{r_s^2 q'^2(r_s)}\right]$$

where $V_a^2 = B_0^2/\mu_0\rho_0$, r_s is the radius of the $q = 1$ surface and

$$W_4 = -\frac{1}{r_s^2}\int_0^{r_s} rdr\left[r\beta' + \frac{r^2}{R_0^2}\left(1 - \frac{1}{q}\right)\left(3 + \frac{1}{q}\right)\right]$$

11.10 The Kruskal–Shafranov condition for an $m = 1$, $n = 1$ external kink instability in a low β large aspect ratio tokamak is given by

$$q_a \leq 1$$

This condition is independent of the current profile. Show here that the condition is also independent of the conducting wall position as long as the wall is not on the plasma.

11.11 (Due to R. R. Parker.) This problem examines the stability of a large aspect ratio, low β tokamak against external kink modes. Recall that in the uniform current density model, instability exists for values of q_a in the range

$m - 1 < nq_a < m$ for a given m, n mode with the wall at infinity. To simplify the analysis, attention is focused on the $m = 4$, $n = 1$ mode. The q profile is assumed to be

$$q(r) = \frac{4}{1 + 6J_1(\alpha r)/\alpha r}$$

where J_1 is a Bessel function and α is a parameter.

(a) Sketch the $q(r)$ and $J_\phi(r)$ profiles for several values of αa in the range $0 \le \alpha a \le \alpha_0 a$. Here, $\alpha_0 a$ is the value of α for which $J_\phi(a) = 0$. Calculate the value of $\alpha_0 a$ and the corresponding value of q_a.

(b) Find the perturbation which minimizes the plasma contribution to δW. Hint: Make the substitution $b = r^{3/2} \Delta \xi$ where $\Delta(r) = n/m - 1/q(r)$.

(c) Using the solution for ξ found in (b), substitute into δW. Note that the range of unstable q_a values can be written as $q_1 \le q_a \le 4$ where $q_1 = q(\alpha a)$. Sketch q_1 for $0 < \alpha a < \alpha_0 a$. Show that as $\alpha a \to \alpha_0 a$ (corresponding to zero current density at the plasma edge) the range of unstable wave numbers shrinks to zero for the $m = 4$, $n = 1$ mode.

11.12 This problem demonstrates the powerful stabilizing effect that arises for external kink modes when the vacuum region is replaced by an ideal force-free plasma. Consider external kink modes in a low β, large aspect ratio ohmically heated tokamak.

(a) Show that for the external kink mode, the general screw pinch δW reduces to

$$\frac{\delta W}{W_0} = \int_0^a r \, dr \left(\frac{n}{m} - \frac{1}{q} \right)^2 \left[r^2 \xi'^2 + (m^2 - 1)\xi^2 \right] + BT$$

where $W_0 = 2\pi^2 B_0^2/\mu_0 R_0$,

$$BT = a^2 \xi_a^2 \left(\frac{n}{m} - \frac{1}{q_a} \right) \left[\frac{n}{m} + \frac{1}{q_0} + m \left(\frac{n}{m} - \frac{1}{q_0} \right) \Lambda \right]$$

and

$$\Lambda \approx \frac{1 + (a/b)^{2m}}{1 - (a/b)^{2m}}$$

For simplicity assume $m > 0$ and $nb/R \ll 1$ in the expansion of Λ.

(b) Assume now that the vacuum region is replaced by an ideal force-free plasma. In order for ξ to remain bounded, B_{1r}, must equal zero at the resonant surface in the exterior region, rather than at the location of the conducting wall; that is, since $\zeta = -iB_{1r}/F$, then $B_{1r}(r = r_s) = 0$ at the radius where $F(r_s) = 0$. This implies that the value of a/b appearing in Λ should correspond to a/r_s. Show that

$$\frac{a^2}{b^2} \rightarrow \frac{a^2}{r_s^2} = \frac{nq_a}{m}$$

(c) Using this relation show that the boundary term is always positive. All external kinks have been stabilized for arbitrary current profiles in the plasma.

12

Stability: multi-dimensional configurations

12.1 Introduction

Chapter 12 is concerned with the MHD stability of toroidal systems, specifically tokamaks and stellarators. As might be expected the complexities associated with multidimensional systems make analysis quite difficult. Consequently, in practice stability results are often obtained numerically. There are, however, a variety of simple models which do provide physical insight into qualitative geometric features that lead to favorable and unfavorable stability behavior. This chapter describes several of these simple models plus summarizes several of the major numerical studies.

The discussion begins with a general analysis of ballooning modes. These are radially localized modes with short perpendicular wavelengths and long parallel wavelengths. They arise because most toroidal configurations have magnetic geometries with alternating regions of favorable and unfavorable curvature. The most unstable perturbations are those whose amplitude "balloons" out in the region of unfavorable curvature. The modes are important because they set one limit on the maximum stable β in toroidal configurations.

Mathematically, it is possible to exploit the localization of the mode. The result is a general reduction of the Energy Principle to a one-dimensional differential equation that allows testing of stability one magnetic line at a time. This represents an enormous saving in computing time compared to the general fully coupled two- or three-dimensional stability problems. The simplified analysis is another reason why ballooning modes have received so much attention.

Once the general ballooning mode equation is derived, attention is focused on the stability of tokamaks. The analysis shows that toroidicity produces qualitative changes in the stability of pressure-driven modes as compared to the straight tokamak. This is demonstrated explicitly by means of the ballooning mode equation. Two results are derived. First, the condition for the ballooning mode equation to be valid leads to the Mercier criterion, a 2-D generalization of Suydam's criterion. In practice, satisfying the condition requires a minimum q on axis. When the Mercier criterion is satisfied, the ballooning mode equation is valid. When it is not, the plasma is unstable to localized interchange modes.

Assuming the Mercier criterion is satisfied, one then solves the ballooning mode equation for a simple model leading to a prediction for the maximum stable beta which is compared to more extensive numerical results. Elongated, outward pointing triangular shapes are favorable for stability.

Next, a second general reduction of the Energy Principle is derived by utilizing the large aspect ratio, high beta tokamak expansion. The end formulation remains two dimensional but is simplified because the potential energy can be written in terms of a single potential function rather than the two components of $\boldsymbol{\xi}_\perp$. The simplified Energy Principle is applied to the problem of $n = 0$ axisymmetric instability. It is shown that elongated cross sections tend to be unstable. A conducting wall is necessary for stability. This converts the instability into a resistive wall mode which requires feedback stabilization.

The toroidal tokamak is then examined for stability against non-localized internal and external modes. Like for the straight tokamak, internal modes lead to a condition on the minimum allowable q on axis. The external modes are the most dangerous instabilities. Physically the low n unstable current-driven kink modes in a straight tokamak develop a strong ballooning component, transforming them into ballooning-kink modes. These modes set the strictest limits on beta. Furthermore, the highest achievable beta occurs at an optimum value of q_*. The external kink stability results are demonstrated by means of the relatively simple surface current model and compared to extensive numerical results. Again, a combination of elongation and outward pointing triangularity are favorable for stability. Overall, the theoretical maximum stable β_t is probably sufficient for energy applications although there is not a large safety margin.

Still, there remains an important problem facing the tokamak – the need for steady state operation. This requires external non-inductive current drive, which is technologically feasible but inefficient from an energy point of view. The end result is that to overcome current drive inefficiency the toroidal current must be driven primarily by the bootstrap effect. This in turn requires high β_t values, typically exceeding the limits predicted by numerical studies. Again, a conducting wall is required implying conversion of the instability to a resistive wall mode which requires feedback for stabilization. The achievement of steady state operation is encompassed in the advanced tokamak (AT) discussion.

The stability of tokamaks concludes with a discussion of $n = 0$ axisymmetric modes. These modes limit the allowable elongation of the tokamak cross section, a feature that is desirable for both equilibrium and stability. In general, a conducting wall is required to stabilize a finite elongation. As for other instabilities, the presence of a finite conductivity wall transforms the wall stabilized mode into a resistive wall mode. Stabilization is provided by feedback which for practical engineering reasons limits the elongation to about $\kappa_{\max} \simeq 1.8$. Violation of the elongation criterion leads to a major disruption.

The final topic discussed is the stability of stellarators. Since stellarators in general have small or zero toroidal current, kink modes are not very important. Instead, stability is determined primarily by pressure-driven modes, both local and global. Although no simple but rigorous stability criteria exist a useful guideline is introduced, known as the average curvature. This quantity is shown to be related to an alternate figure of merit known as the magnetic well. By using this guideline it is possible to develop some intuition as to how to design a stellarator with favorable MHD stability properties.

Several simple models are analyzed. It is shown that a straight stellarator always has unfavorable average curvature. Shear can provide stability against unfavorable average curvature but only at very low values of beta. Toroidal effects are shown to help stability. Specifically, a vertical field can create a favorable shift of the flux surfaces to provide stability at higher values of β_t. Full 3-D stability results have been obtained numerically. These studies predict complete stability at values of β_t sufficiently high for energy applications without the need for a conducting wall.

12.2 Ballooning and interchange instabilities

12.2.1 Introduction

Recall that interchange stability in a cylinder is determined by Suydam's criterion. The modes of interest correspond to a special subclass of trial functions which are radially localized about a resonant surface $r = r_s : [\mathbf{k} \cdot \mathbf{B}]_{r_s} = 0$. Application of Suydam's criterion is simple – one simply substitutes the equilibrium profiles into the analytic formula for Suydam's criterion and examines whether the resulting expression is positive or negative for all r. Also, since the analysis focuses on a special subclass of trial functions the resulting criterion is a necessary but not sufficient condition for stability.

An analogous situation exists for toroidal plasmas. However, the analysis is substantially more complicated because of the multi-dimensional geometric effects arising from toroidicity. The aim of this section is to investigate the toroidal analogs of Suydam's criterion. This leads to a modified form of Suydam's criterion known as the Mercier criterion. The analysis also leads to a generalized form of the interchange instability known as the ballooning mode.

The difference between interchange and ballooning modes is as follows. Both modes are driven by the interaction of the pressure gradient and the curvature.

Interchange modes are approximately constant along the field lines (i.e., $\mathbf{k} \cdot \mathbf{B} \approx 0$) to minimize the stabilizing effects of field line bending. This form of trial function implies that stability is essentially determined by the "average" field line curvature resulting from the $(\boldsymbol{\xi}_\perp \cdot \nabla p)(\boldsymbol{\xi}_\perp^* \cdot \boldsymbol{\kappa})$ term in δW. In contrast, a ballooning perturbation recognizes that the field line curvature can oscillate between being favorable and unfavorable along a field line. Therefore, by allowing a small amount of line bending the perturbation has the potential to localize in regions of unfavorable curvature, thereby having a net effect which is more unstable than the nearly constant interchange perturbation. In effect, the trial function "balloons out" in the region of unfavorable curvature.

In terms of application the Mercier criterion is qualitatively similar to Suydam's criterion. One substitutes equilibrium profiles into an analytic expression and examines whether or not the resulting value is positive or negative over all flux surfaces in the plasma. The actual substitution procedure is more complicated since flux surface averages must be performed, but this is a technical rather than conceptual distinction. For tokamaks, the Mercier criterion usually results in a limit on the minimum allowable $q(\psi)$ at the magnetic axis.

Application of the ballooning mode stability criterion requires the solution of a one-dimensional differential equation where the independent variable is basically distance along a field line. Newcomb's analysis then implies that stability or instability is determined by whether or not the minimizing solution has a zero crossing at some point along the field line. The procedure must be applied to all flux surfaces in the plasma. Although many solutions to the 1-D differential equation are thus required to test overall ballooning stability this is still a far, far simpler procedure than solving the fully coupled two- or three-dimensional stability problems for tokamaks or stellarators. Ballooning modes are important because they set upper limits on the allowable β values. Excitation of ballooning modes does not in general lead to major disruptions but instead to enhanced MHD turbulent transport.

As stated, the analysis of interchange and ballooning modes carried out in this section is somewhat complicated. Several steps are required as follows:

1 A discussion is presented that focuses on the basic incompatibility that arises when one attempts to create a trial function that maintains poloidal and toroidal periodicity in a multidimensional configuration with magnetic shear. The situation is further complicated by the need for the trial function to be localized around a flux surface. The end result is the introduction of a very clever transformation, known as the "ballooning mode formalism."

2 This transformation is then applied to δW, leading to the general 1-D ballooning mode differential equation. The derivation is somewhat elegant in that the details are coordinate independent.

3 Specific coordinate systems are next introduced resulting in more practical forms of the ballooning mode differential equation for tokamaks and stellarators.

4 Attention is next focused on tokamaks. It is shown that the validity of the ballooning mode differential equation is closely connected to the Mercier criterion. The actual Mercier criterion is derived and in simple limits, applied to tokamaks.

5 Assuming the plasma is Mercier stable one must then test for ballooning mode stability by means of the 1-D differential equation. A simple semi-analytic model for the tokamak is introduced and it is shown how ballooning modes lead to β limits. These results are compared to extensive numerical results for more realistic configurations which provide a good guide to the practical β limit in tokamaks due to ballooning modes.

12.2.2 *Shear, periodicity, and localization*

The stability analysis of interchange and ballooning modes requires the construction of a trial function which is localized around a flux surface and satisfies the periodicity requirements in θ and ϕ. The trial function that meets these requirements results from the ballooning mode formalism and is not at all obvious. The analysis begins with a discussion of the difficulty of constructing such a trial function. Once the difficulties are understood it is then possible to motivate the choice of the ballooning mode trial function.

Straightforward Fourier analysis

Consider a two- or three-dimensional toroidal configuration whose outer flux surface is chosen for simplicity to be circular. The interior flux surfaces are non-circular. A straightforward approach is to write each component of the plasma displacement as a Fourier series. For example, if one focuses on the normal component of plasma displacement $\mathbf{n} \cdot \boldsymbol{\xi}_\perp = \xi(r,\theta,\phi)$, then

$$\xi(r,\theta,\phi) = \sum_{m,n} \xi_{mn}(r)\, e^{im\theta - in\phi} \tag{12.1}$$

While this is a mathematically correct form of the displacement it is not very useful. The reason is that it is not possible to easily exploit the localization requirement since the interior flux surfaces are not circular. In other words, since $r = $ constant is not a flux surface, a local trial function would require many terms in the Fourier series. The resulting analysis is thus a multidimensional coupled harmonic problem with no mathematical advantage due to localization.

A simple attempt at localization

A cleverer approach to exploit localization is to introduce the flux ψ as the radial variable: $r,\theta,\phi \rightarrow \psi,\theta,\phi$. For simplicity focus on axisymmetric configurations (i.e., tokamaks) so that $\psi(r,\theta,\phi) \rightarrow \psi(r,\theta)$. A trial function that is easily localized around a flux surface can then be written as $\xi = \xi_0 \exp[-in\phi + iS_p(\psi,\theta)]$. To localize, one simply expands $S_p(\psi,\theta)$ about the flux surface of interest, $\psi = \psi_0 + \delta\psi$. A valid local trial function vanishes rapidly as $\delta\psi$ increases.

Now, the function $S_p(\psi, \theta)$ must be chosen to correspond to an interchange-like perturbation in order to minimize line bending. This requires setting $\mathbf{B} \cdot \nabla\xi = \xi_0 \mathbf{B} \cdot \nabla(-in\phi + iS_p) = 0$. A short calculation shows that $S_p(\psi,\theta)$ satisfies

$$\frac{\partial S_p}{\partial \theta} = n \frac{rB_\phi}{RB_\theta} \tag{12.2}$$

which has as its solution

$$S_p(\psi,\theta) = n \int_{\theta_0}^{\theta} \frac{rB_\phi}{RB_\theta} d\theta \tag{12.3}$$

where θ_0 is a free integration constant and it is understood that the integrand is evaluated on the flux surface. Note that the radial wave number, which is a measure of localization, is approximately given by $k_\psi \approx (\partial S_p/\partial\psi)|\nabla\psi|$. The implication is that a local trial function requires $k_\psi a \gg 1$ which is equivalent to $n \gg 1$.

The last requirement to be satisfied by $S_p(\psi,\theta)$ is periodicity and corresponds to $S_p(\psi,\theta + 2m\pi) = S_p(\psi,\theta) + 2m\pi$. On $\psi = \psi_0$, the flux surface of interest, Eq. (12.3) implies that periodicity requires

$$n \int_0^{2\pi} \left(\frac{rB_\phi}{RB_\theta}\right)_{\psi_0} d\theta = 2\pi m \tag{12.4}$$

Recalling the definition of the safety factor allows one to rewrite Eq. (12.4) as

$$\frac{1}{2\pi} \int_0^{2\pi} \left(\frac{rB_\phi}{RB_\theta}\right)_{\psi_0} d\theta \equiv q(\psi_0) = \frac{m}{n} \tag{12.5}$$

The localization must take place around a rational surface.

It is at this point that the fundamental incompatibility between shear and periodicity arises. While periodicity is satisfied on a rational surface, a short distance away (in flux) the function $S_p(\psi,\theta)$ has the value

$$S_p(\psi_0 + \delta\psi,\theta) \approx n \left[\int_{\theta_0}^{\theta} \left(\frac{rB_\phi}{RB_\theta}\right)_{\psi_0} d\theta + \delta\psi \int_{\theta_0}^{\theta} \frac{\partial}{\partial\psi_0}\left(\frac{rB_\phi}{RB_\theta}\right) d\theta \right] \tag{12.6}$$

The periodicity constraint reduces to

$$S_p(\psi_0 + \delta\psi, \theta + 2m\pi) - S_p(\psi_0 + \delta\psi, \theta) \approx 2\pi m \left(1 + \frac{1}{q} \frac{\partial q}{\partial \psi_0} \delta\psi \right) \qquad (12.7)$$

In a system with shear (i.e., $\partial q/\partial \psi_0 \neq 0$), a trial function with $m \sim n \gg 1$ is no longer periodic away from the rational surface even if $\delta\psi$ is small!

The ballooning mode formalism

The incompatibility between shear and periodicity was resolved by Connor, Hastie, and Taylor (1979) by the introduction of the ballooning mode formalism. The basic idea is as follows. Since the equilibrium and all perturbed quantities must be periodic in θ, ϕ in a physical system, no information is lost or gained if one extends the angular domain of validity from $0 \leq \theta \leq 2\pi$, $0 \leq \phi \leq 2\pi$ to $-\infty < \theta < \infty$, $-\infty < \phi < \infty$. Actually, as is shown below, only a single combination of angles must be extended, the other combination maintaining a finite range of integration. For present purposes, however, it is convenient to assume both angles are extended as this helps determine the proper combination to be extended.

The motivation for extension is that it allows one, perhaps unexpectedly, to construct a perturbation that is periodic in θ, ϕ by means of an infinite sum of terms, none of which is periodic. This is the ballooning mode formalism. Specifically, the displacement vector is written as

$$\xi(r,\theta,\phi) = \sum_{m,n} \overline{\xi}(r, \theta + 2m\pi, \phi + 2n\pi) \qquad (12.8)$$

The function $\overline{\xi}(r,\theta,\phi)$, often called a "quasi-mode," extends over the doubly infinite domain and is not periodic. Typically it has small amplitude periodic oscillations superimposed on a non-periodic envelope. Note that each term in the summation corresponds to the same quasi-mode, just shifted in angle. The actual eigenfunction $\xi(r,\theta,\phi)$ is periodic since shifting θ and/or ϕ by $2l\pi$ just corresponds to a relabeling of the summation indices: for example, $m = m' - l$. The only possible problem is the accumulation or loss of contributions from the ends of the summation. This problem is resolved by requiring that the quasi-mode decay sufficiently rapidly for large $|\theta|$ and $|\phi|$ so that terms from the ends of the summation are small and thus make no contribution.

As is shown shortly, there is a sharp criterion that distinguishes "sufficiently rapidly" and "not sufficiently rapidly." The transition corresponds to the Mercier criterion. The proper procedure to test stability therefore requires first testing the Mercier criterion. If the Mercier criterion is violated, the ballooning mode formulation is not valid. However, violation of the criterion implies that the plasma is already unstable so one need not proceed any further. If the Mercier criterion is satisfied, then the ballooning mode formulation is valid and one can proceed to test stability by analyzing the ballooning mode differential equation.

An important simplifying property of the quasi-mode expansion is that $\overline{\xi}(r, \theta, \phi)$ satisfies the same equation as $\xi(r,\theta,\phi)$. To see this recall that at marginal stability $\xi(r,\theta,\phi)$ satisfies

$$\mathbf{F}(\xi) = 0 \tag{12.9}$$

where $\mathbf{F}$ is the MHD force operator. Since the equilibria of interest are periodic this implies that $\mathbf{F}(r,\theta + 2m\pi,\phi + 2n\pi) = \mathbf{F}(r,\theta,\phi)$. As a consequence, application of the force operator to Eq. (12.8) shows that

$$
\begin{aligned}
\mathbf{F}(\xi) &= \sum_{m,n} \mathbf{F}(r,\theta,\phi)[\overline{\xi}(r,\theta + 2m\pi,\phi + 2n\pi)] \\
&= \sum_{m,n} \mathbf{F}(r,\theta + 2m\pi,\phi + 2n\pi)[\overline{\xi}(r,\theta + 2m\pi,\phi + 2n\pi)]
\end{aligned}
\tag{12.10}
$$

In other words, each term in the summation satisfies the same equation

$$\mathbf{F}(\overline{\xi}) = 0 \tag{12.11}$$

There is an important conclusion that can be drawn from Eq. (12.11). If a marginally stable solution can be found for a single quasi-mode, for instance leading to a critical β, then this critical β will be identical for all quasi-modes in the summation. The implication is that marginal stability can be investigated by analyzing a single quasi-mode, thereby eliminating the need to consider the entire summation.

To summarize, the ballooning mode formalism allows one to investigate marginal stability by (1) analyzing a single quasi-mode, (2) focusing on a mode structure localized around a flux surface, (3) choosing the mode structure to be interchange-like (i.e., $\mathbf{B} \cdot \nabla \overline{\xi} \approx 0$), but (4) *not* requiring the quasi-mode to be periodic in θ, ϕ. The ballooning mode formalism resolves the incompatibility between shear and periodicity.

Localizing the quasi-mode

One last step is required to finalize the form of the quasi-mode in order to account for the localization around a flux surface. To exploit localization mathematically one needs to introduce a small parameter that measures the width in radius of the quasi-mode compared to the characteristic minor radius of the plasma. This can be conveniently accomplished by introducing a local perpendicular wavelength $\mathbf{k}_{\perp}(r)$ and assuming that $|\mathbf{k}_{\perp}a| \gg 1$. Mathematically, the stability problem now has two scale lengths, the slowly varying equilibrium length a and the rapidly varying perturbation wavelength $\mathbf{k}_\perp^{-1}$. This is the ideal situation to represent the quasi-mode by means of a WKB expansion.

For the ballooning mode analysis under consideration the appropriate form for the quasi-mode can thus be written as

$$\overline{\boldsymbol{\xi}}(\mathbf{r}) = \overline{\boldsymbol{\eta}}(\mathbf{r})\, e^{iS(\mathbf{r})} \qquad (12.12)$$

Here, the envelope $\overline{\boldsymbol{\eta}}(\mathbf{r})$ is assumed to vary slowly on the equilibrium scale. The local wave number of the mode is defined as $\mathbf{k} = \nabla S$ and must correspond to rapid variation. This implies that $S(\mathbf{r})$, known as the eikonal, must have the following properties:

$$\begin{aligned} \mathbf{B} \cdot \nabla S &= 0 && \text{interchange-like perturbation} \\ \nabla S &= \mathbf{k}_\perp \gg 1/a && \text{localized mode structure} \end{aligned} \qquad (12.13)$$

For practical applications to tokamaks or stellarators it will be convenient, particularly for the localization requirement, to introduce some form of flux coordinates. However, this is not necessary for the next task which consists of a general, coordinate free reduction of δW obtained by exploiting the properties of the quasi-mode trial function.

12.2.3 General reduction of δW for ballooning modes

Because ballooning modes are localized in radial extent they naturally correspond to internal MHD modes. Thus, one need only consider δW_F. Furthermore, since marginal stability can be determined by analyzing only a single quasi-mode, the actual quantity of interest is defined as

$$\delta \overline{W}_F = -\frac{1}{2}\int \overline{\boldsymbol{\xi}}^* \cdot \mathbf{F}(\overline{\boldsymbol{\xi}})d\mathbf{r} \qquad (12.14)$$

Here it is understood that the integration is carried out over the extended angular domain.

The reduction of $\delta \overline{W}_F$ makes use of the fact that all of the analysis presented in Chapter 8 involving the actual displacement $\boldsymbol{\xi}$ also applies to the quasi-mode displacement $\overline{\boldsymbol{\xi}}$ since $\boldsymbol{\xi}$ and $\overline{\boldsymbol{\xi}}$ satisfy the same equation, $\mathbf{F}(\boldsymbol{\xi}) = \mathbf{F}(\overline{\boldsymbol{\xi}}) = 0$, and the same boundary condition, $\mathbf{n} \cdot \boldsymbol{\xi} = \mathbf{n} \cdot \overline{\boldsymbol{\xi}} = 0$, on the plasma surface. The starting point of the analysis is chosen as the intuitive form of the potential energy given by Eq. (8.32), which is repeated here for convenience:

$$\delta \overline{W}_F = \frac{1}{2\mu_0}\int \left[|\overline{\mathbf{Q}}_\perp|^2 + B^2 |\nabla \cdot \overline{\boldsymbol{\xi}}_\perp + 2\overline{\boldsymbol{\xi}}_\perp \cdot \boldsymbol{\kappa}|^2 + \mu_0 \gamma p |\nabla \cdot \overline{\boldsymbol{\xi}}|^2 \right.$$

$$\left. -2\mu_0(\overline{\boldsymbol{\xi}}_\perp \cdot \nabla p)(\overline{\boldsymbol{\xi}}_\perp^* \cdot \boldsymbol{\kappa}) - \mu_0 J_\parallel \overline{\boldsymbol{\xi}}_\perp^* \times \mathbf{b} \cdot \overline{\mathbf{Q}}_\perp \right] d\mathbf{r} \qquad (12.15)$$

The first step is to substitute the quasi-mode trial function given by Eq. (12.12) into the expression for $\delta \overline{W}_F$. A key feature of the substitution is that $\overline{\mathbf{Q}}_\perp$ does not contain any explicit gradients in S:

$$\begin{aligned}
\overline{\mathbf{Q}}_\perp &= e^{iS}[\nabla \times (\overline{\boldsymbol{\eta}}_\perp \times \mathbf{B}) + i\nabla S \times (\overline{\boldsymbol{\eta}}_\perp \times \mathbf{B})]_\perp \\
&= e^{iS}[\nabla \times (\overline{\boldsymbol{\eta}}_\perp \times \mathbf{B}) + i(\mathbf{k}_\perp \cdot \mathbf{B})\overline{\boldsymbol{\eta}}_\perp - i(\mathbf{k}_\perp \cdot \overline{\boldsymbol{\eta}}_\perp)\mathbf{B}]_\perp \\
&= e^{iS}\nabla \times (\overline{\boldsymbol{\eta}}_\perp \times \mathbf{B})_\perp
\end{aligned} \tag{12.16}$$

Substituting $\overline{\boldsymbol{\xi}}$ and $\overline{\mathbf{Q}}_\perp$ into the expression for $\delta \overline{W}_F$ leads to

$$\delta \overline{W}_F = \frac{1}{2\mu_0} \int \left[|\nabla \times (\overline{\boldsymbol{\eta}}_\perp \times \mathbf{B})_\perp|^2 + B^2|i\mathbf{k}_\perp \cdot \overline{\boldsymbol{\eta}}_\perp + \nabla \cdot \overline{\boldsymbol{\eta}}_\perp + 2\overline{\boldsymbol{\eta}}_\perp \cdot \boldsymbol{\kappa}|^2 \right.$$

$$\left. - 2\mu_0(\overline{\boldsymbol{\eta}}_\perp \cdot \nabla p)(\overline{\boldsymbol{\eta}}_\perp^* \cdot \boldsymbol{\kappa}) - \mu_0 J_\parallel (\overline{\boldsymbol{\eta}}_\perp^* \times \mathbf{b}) \cdot \nabla \times (\overline{\boldsymbol{\eta}}_\perp \times \mathbf{B})_\perp \right] d\mathbf{r} \quad (12.17)$$

Note that the plasma compressibility term is not included in Eq. (12.17) since in systems with shear the most unstable modes are incompressible; that is, $\overline{\boldsymbol{\xi}}_\parallel$ is chosen to make $\nabla \cdot \overline{\boldsymbol{\xi}} = 0$. Now, observe that the only explicit appearance of S occurs as $\mathbf{k}_\perp$ in the magnetic compressibility term which for $k_\perp a \gg 1$ tends to dominate the integrand. One is thus motivated to systematically minimize $\delta \overline{W}_F$ by expanding

$$\overline{\boldsymbol{\eta}}_\perp = \overline{\boldsymbol{\eta}}_{\perp 0} + \overline{\boldsymbol{\eta}}_{\perp 1} + \cdots \tag{12.18}$$

where $|\overline{\boldsymbol{\eta}}_{\perp 1}|/|\overline{\boldsymbol{\eta}}_{\perp 0}| \sim 1/k_\perp a$.

The zeroth-order contribution to $\delta \overline{W}_F$ reduces to

$$\delta \overline{W}_0 = \frac{1}{2\mu_0} \int B^2 |\mathbf{k}_\perp \cdot \overline{\boldsymbol{\eta}}_{\perp 0}|^2 d\mathbf{r} \tag{12.19}$$

Clearly the minimizing perturbation satisfies $\mathbf{k}_\perp \cdot \overline{\boldsymbol{\eta}}_{\perp 0} = 0$, which implies that $\overline{\boldsymbol{\eta}}_{\perp 0}$ can be written as

$$\overline{\boldsymbol{\eta}}_{\perp 0} = Y \mathbf{b} \times \mathbf{k}_\perp \tag{12.20}$$

Here, $Y(\mathbf{r})$ is a scalar quantity varying on the slow equilibrium length scale.

The first non-vanishing contribution to $\delta \overline{W}_F$ occurs in second order. In this expression the only appearance of the quantity $\overline{\boldsymbol{\eta}}_{\perp 1}$ is in the magnetic compressibility term,

$$\delta \overline{W}_2(\text{comp}) = \frac{1}{2\mu_0} \int B^2 |i\mathbf{k}_\perp \cdot \overline{\boldsymbol{\eta}}_{\perp 1} + \nabla \cdot \overline{\boldsymbol{\eta}}_{\perp 0} + 2\overline{\boldsymbol{\eta}}_{\perp 0} \cdot \boldsymbol{\kappa}|^2 d\mathbf{r} \tag{12.21}$$

Clearly $\delta \overline{W}_2$ is minimized by choosing $i\mathbf{k}_\perp \cdot \overline{\boldsymbol{\eta}}_{\perp 1} = -\nabla \cdot \overline{\boldsymbol{\eta}}_{\perp 0} - 2\overline{\boldsymbol{\eta}}_{\perp 0} \cdot \boldsymbol{\kappa}$. The most unstable perturbations for localized modes characterized by $k_\perp a \gg 1$ do not produce any compression of the magnetic field.

The next step in the evaluation of $\delta \overline{W}_2$ is to simplify the quantity $\nabla \times (\overline{\boldsymbol{\eta}}_\perp \times \mathbf{B})_\perp$ as follows:

$$
\begin{aligned}
\nabla \times (\overline{\boldsymbol{\eta}}_\perp \times \mathbf{B})_\perp &= \nabla \times [YB(\mathbf{b} \times \mathbf{k}_\perp) \times \mathbf{b}]_\perp \\
&= \nabla \times (YB\nabla S)_\perp \\
&= (\nabla X \times \mathbf{k}_\perp)_\perp
\end{aligned}
\tag{12.22}
$$

where $X(\mathbf{r}) = YB$. One now writes $\nabla X = \nabla_\perp X + (\mathbf{b} \cdot \nabla X)\mathbf{b}$. The quantity $\nabla_\perp X \times \mathbf{k}_\perp$ only has a non-vanishing contribution along $\mathbf{b}$ and thus does not make a contribution. The remaining term in $\nabla \times (\overline{\boldsymbol{\eta}}_\perp \times \mathbf{B})_\perp$ can then be expressed as

$$
\nabla \times (\overline{\boldsymbol{\eta}}_\perp \times \mathbf{B})_\perp = (\mathbf{b} \cdot \nabla X)\mathbf{b} \times \mathbf{k}_\perp
\tag{12.23}
$$

This relation is substituted into the kink contribution to $\delta \overline{W}_2$ leading to

$$
\begin{aligned}
\delta \overline{W}_2(\text{kink}) &= -\frac{1}{2} \int J_\parallel (\overline{\boldsymbol{\eta}}_\perp^* \times \mathbf{b}) \cdot \nabla \times (\overline{\boldsymbol{\eta}}_\perp \times \mathbf{B})_\perp d\mathbf{r} \\
&= -\frac{1}{2} \int \frac{J_\parallel}{B} [X^*(\mathbf{b} \cdot \nabla X)][\mathbf{k}_\perp \cdot (\mathbf{b} \times \mathbf{k}_\perp)]d\mathbf{r} \\
&= 0
\end{aligned}
\tag{12.24}
$$

In the limit $k_\perp a \gg 1$ the kink term makes no contribution to ballooning mode instability.

The remaining terms in $\delta \overline{W}_2$ describe a competition between the stabilizing effects of line bending and the destabilizing effects of unfavorable curvature:

$$
\delta \overline{W}_2 = \frac{1}{2\mu_0} \int \left[k_\perp^2 |\mathbf{b} \cdot \nabla X|^2 - \frac{2\mu_0}{B^2}(\mathbf{b} \times \mathbf{k}_\perp \cdot \nabla p)(\mathbf{b} \times \mathbf{k}_\perp \cdot \boldsymbol{\kappa})|X|^2 \right] d\mathbf{r}
\tag{12.25}
$$

This is the desired general form of the potential energy for ballooning modes. The basic unknown is the single scalar variable $X(\mathbf{r})$ which varies on the equilibrium length scale. It is assumed that a function S and correspondingly $\mathbf{k}_\perp = \nabla_\perp S$ can be found since these quantities depend only on the equilibrium fields. Observe that the variation of the integrand leads to a differential equation in only one variable. The reason is that the only derivatives that appear are along a single direction, parallel to $\mathbf{B}$. The one-dimensional nature becomes explicit when specific coordinate systems are introduced for the tokamak and stellarator.

However, before proceeding with the analysis it is of interest to write down the corresponding ballooning mode equation for the double adiabatic model. The derivation closely follows the one given above and is left as an assignment in the Problems at the end of the chapter. The double adiabatic form for $\delta \overline{W}_2$ is derived from Eq. (10.27) and can be expressed as

$$\delta \overline{W}_2 = \frac{1}{2\mu_0} \int \left[k_\perp^2 |\mathbf{b} \cdot \nabla X|^2 + \Omega |X|^2 \right] d\mathbf{r}$$

$$\Omega = \frac{2\mu_0}{B^2} \left[-(\mathbf{b} \times \mathbf{k}_\perp \cdot \nabla p)(\mathbf{b} \times \mathbf{k}_\perp \cdot \boldsymbol{\kappa}) + p(\mathbf{b} \times \mathbf{k}_\perp \cdot \boldsymbol{\kappa})^2 \left(\frac{7B^2 + 5\mu_0 p}{2B^2 + 4\mu_0 p} \right) \right]$$

(12.26)

There is again a stabilizing line bending term plus a modified driving term. As expected the additional driving term in Ω is always stabilizing, regardless of the sign of the curvature vector. In the low β, large aspect ratio limit the additional stabilizing term is nominally smaller by ε than the pressure gradient-curvature term, although the numerical factor of 7/2 may in practice make both terms comparable.

12.3 The ballooning mode equations for tokamaks

Equation (12.25) describes ballooning mode stability for an arbitrary 3-D geometry. To actually apply the equation one needs to specify a specific geometry and introduce a corresponding set of flux coordinates. Once this is done it is then possible to explicitly determine the eikonal function S plus $\mathbf{k}_\perp = \nabla S$. The ultimate result is a 1-D differential equation where the dependent variable is a measure of arc length along the magnetic field. In this subsection the ballooning mode differential equation is derived for tokamaks.

The analysis for the tokamak is somewhat simplified because the equilibrium is toroidally axisymmetric: $\partial/\partial\phi = 0$. Convenient flux coordinates for the calculation are $\psi(R,Z)$, $l(R,Z)$, ϕ, where ψ and ϕ have their usual meaning and l is poloidal arc length. Other forms of poloidal angle rather than l could be used and are equally valid, although the Jacobian J resulting from arc length is particularly simple, leading to a more transparent form of the ballooning mode equation.

The relevant equilibrium relations for the tokamak, derived in Chapter 6, are given by

$$\mathbf{B} = B_\phi \mathbf{e}_\phi + \mathbf{B}_p$$

$$B_\phi = \frac{F(\psi)}{R}$$

(12.27)

$$\mathbf{B}_p = \frac{1}{R} \nabla \psi \times \mathbf{e}_\phi$$

$$\boldsymbol{\kappa} = \mathbf{b} \cdot \nabla \mathbf{b}$$

The magnitude and direction of the poloidal arc length gradient are defined as

$$\nabla l \cdot \nabla \psi = 0$$
$$\mathbf{B}_p \cdot \nabla l = B_p$$

(12.28)

A practical method for calculating the flux coordinates is given in the Problems at the end of the chapter.

Two further equilibrium relations are needed to carry out the analysis. First, the transformation of the toroidal volume element to flux coordinates is given by

$$d\mathbf{r} = R dR \, dZ d\phi = J d\psi \, dl \, d\phi \qquad (12.29)$$

where the Jacobian satisfies

$$\frac{R}{J} = \begin{vmatrix} \partial\psi/\partial R & \partial\psi/\partial Z & 0 \\ \partial l/\partial R & \partial l/\partial Z & 0 \\ 0 & 0 & 1 \end{vmatrix} = \nabla l \cdot \nabla\psi \times \mathbf{e}_\phi \qquad (12.30)$$

From the equilibrium relations one sees that

$$J = \frac{1}{B_p} \qquad (12.31)$$

The second relation of interest defines a set of mutually orthogonal unit vectors,

$$\mathbf{n} = \frac{\nabla\psi}{|\nabla\psi|} = \frac{\nabla\psi}{RB_p}$$

$$\mathbf{b} = \frac{B_p}{B}\mathbf{b}_p + \frac{B_\phi}{B}\mathbf{e}_\phi \qquad (12.32)$$

$$\mathbf{t} = \mathbf{n} \times \mathbf{b} = \frac{B_\phi}{B}\mathbf{b}_p - \frac{B_p}{B}\mathbf{e}_\phi$$

Here, $\mathbf{b}_p = \mathbf{B}_p/B_p$. Note that $\mathbf{n}$ is a normal vector while $\mathbf{b}$ and $\mathbf{t}$ are tangential vectors lying in the flux surface, parallel and perpendicular to the equilibrium magnetic field respectively. All vectors appearing in the analysis are decomposed into these three components.

With this as background one can now start to simplify the ballooning mode equation. The first step is to evaluate the eikonal function S. Recall that $S(\psi,l,\phi)$ must satisfy two requirements: (1) it must produce rapid oscillations perpendicular to the field and (2) it must be constant along a field line, $\mathbf{B} \cdot \nabla S = 0$. For a general tokamak this last condition can be written as

$$\frac{B_\phi}{R}\frac{\partial S}{\partial \phi} + B_p \frac{\partial S}{\partial l} = 0 \qquad (12.33)$$

Now, axisymmetry implies that Fourier analysis can be carried out with respect to ϕ. In other words the solution to Eq. (12.33) has the form $S = -n\phi + S_p(\psi, l)$. Rapid oscillation requires that the toroidal mode number n satisfy $n \gg 1$. With this form of S, Eq. (12.33) can be formally integrated, yielding

$$S = n\left(-\phi + \int_{l_0}^{l} \frac{B_\phi}{RB_p} dl\right) \qquad (12.34)$$

where l_0 is a free integration constant whose determination is discussed shortly. Observe that on a rational surface (i.e. $m'q(\psi_0) = n'$), the function S is periodic in l over an appropriate number of complete poloidal transits, each of length $L(\psi)$:

$$
\begin{aligned}
S(\psi_0, l + m'L, \phi + 2\pi n') - S(\psi_0, l) &= n\left[-2\pi n' + m' \int_0^L \left(\frac{B_\phi}{RB_p}\right)_{\psi_0} dl\right] \\
&= 2\pi n[-n' + m'q(\psi_0)] \\
&= 0
\end{aligned}
\qquad (12.35)
$$

and use has been made of Eq. (6.35), the general definition of $q(\psi)$. Off the rational surface, S is no longer periodic.

Knowing S one can next evaluate the perpendicular wave number appearing in the ballooning mode equation,

$$\mathbf{k}_\perp = k_n \mathbf{n} + k_t \mathbf{t} = \nabla S$$

$$k_n = \mathbf{n} \cdot \nabla S = (\mathbf{n} \cdot \nabla \psi) \frac{\partial S}{\partial \psi} = nRB_p \frac{\partial}{\partial \psi}\left(F \int_{l_0}^{l} \frac{dl}{R^2 B_p}\right) \qquad (12.36)$$

$$k_t = \mathbf{t} \cdot \nabla S = (\mathbf{t} \cdot \nabla \phi)\frac{\partial S}{\partial \phi} + (\mathbf{t} \cdot \nabla l)\frac{\partial S}{\partial l} = n\frac{B}{RB_p}$$

Observe that the free constant l_0 is associated with the normal component of the wave number vector, k_n.

These expressions are substituted into the general ballooning mode equation (i.e., Eq. (12.25)). Two further simplifications should be noted. First, since the ϕ dependence has been Fourier analyzed and included in $S(\psi, l, \phi)$, this implies that $X(\psi, l, \phi) \rightarrow X(\psi, l)$. No further ϕ dependence is required. Second, because the ϕ dependence is automatically periodic by the form chosen for S, there is no need to extend the ϕ domain from $-\infty$ to $+\infty$. One can simply integrate over the single period $0 \le \phi \le 2\pi$. Therefore it follows that

$$\int d\phi = 2\pi \qquad (12.37)$$

However, the integral over θ or equivalently l must be carried out over the extended domain.

The desired form of the ballooning mode potential energy can now be written as

$$\delta \overline{W}_2 = \frac{\pi}{\mu_0} \int_0^{\psi_a} \overline{W}(\psi, l_0) \, d\psi$$

$$\overline{W}(\psi, l_0) = \int_{-\infty}^{\infty} \left[(k_n^2 + k_t^2) \left(\frac{B_p}{B} \frac{\partial X}{\partial l} \right)^2 - \frac{2\mu_0 R B_p}{B^2} \frac{dp}{d\psi} (k_t^2 \kappa_n - k_t k_n \kappa_t) X^2 \right] \frac{dl}{B_p}$$

(12.38)

Here, $\kappa_n = \mathbf{n} \cdot \boldsymbol{\kappa}$ is the normal curvature and $\kappa_t = \mathbf{t} \cdot \boldsymbol{\kappa}$ is the geodesic curvature (which is perpendicular to $\mathbf{B}$ but lies in the flux surface). After a short calculation these quantities can be rewritten in terms of the equilibrium fields and flux coordinates as

$$\kappa_n = \frac{\mu_0 R B_p}{B^2} \frac{\partial}{\partial \psi} \left(p + \frac{B^2}{2\mu_0} \right)$$

$$\kappa_t = \frac{\mu_0 F}{R B^3} \frac{\partial}{\partial l} \left(\frac{B^2}{2\mu_0} \right)$$

(12.39)

The final reasoning used to determine stability is based on the observation that the integrand in $W(\psi, l_0)$ has no ψ derivatives on X, only l derivatives. In other words, ψ, as well as l_0, are parameters in the integrand. Then, for fixed ψ and l_0, if one can find an l dependence of $X(\psi, l_0, l)$ that makes $\overline{W}(\psi, l_0) < 0$ it follows that a suffi-ciently localized trial function about the given ψ will in turn lead to $\delta \overline{W}_2 < 0$. The plasma is unstable. The major benefit of the ballooning mode analysis is that stability can be tested one flux surface at a time, which is a far simpler procedure than solving the coupled 2-D partial differential equations that arise in a general stability analysis.

The standard procedure to construct the $X(\psi, l_0, l)$ with the most unstable l dependence is to simply set the variation of $\overline{W}(\psi, l_0)$ with respect to l equal to zero. The result is the 1-D ballooning mode differential equation given by

$$\frac{\partial}{\partial l} \left[\frac{(k_n^2 + k_t^2) B_p}{B^2} \frac{\partial X}{\partial l} \right] + \frac{2\mu_0 R}{B^2} \frac{dp}{d\psi} (k_t^2 \kappa_n - k_t k_n \kappa_t) X = 0$$

(12.40)

If the solutions decay sufficiently rapidly for $l = \pm\infty$, then Newcomb's procedure can be used to determine stability. Specifically, for a given ψ and l_0, Eq. (12.40) is integrated (usually numerically) from $l = -\infty$ to $l = +\infty$. If X has a zero crossing anywhere in the interval the plasma is unstable. On each flux surface this procedure must be carried out for a range of l_0 defined by $0 < l_0 < L(\psi)$ to find the most unstable l_0. Often, but not always, $l_0 = 0$ is the most unstable case. If there is no zero crossing for all l_0 the plasma is stable to ballooning modes on the given flux surface. Complete stability against ballooning modes requires no zero crossings on any flux surface.

12.4 The ballooning mode equation for stellarators

The derivation of the ballooning mode equation for stellarators is similar to that of the tokamak. The analysis is conveniently carried out by introducing a set of generalized straight line field coordinates. The generalized coordinates become unique when a specific choice is made for the Jacobian. At the end of the derivation two specific examples are illustrated, corresponding to two choices for J: arc length and Boozer coordinates. Although the analysis applies to general 3-D configurations in some ways the details are simpler and more elegant than for the 2-D case. The complexity, however, is hidden in the difficulty of actually calculating and inverting the field line and laboratory coordinates.

The key equilibrium quantities derived in Chapter 7 that are required for the ballooning mode analysis are repeated here for convenience (see Eq. (7.251)). They include the representation of the magnetic field and an expression for the Jacobian in terms of the generalized flux coordinates, ψ, χ, ζ:

$$\mathbf{B} = q\nabla\psi \times \nabla\chi - \nabla\psi \times \nabla\zeta$$

$$\frac{1}{J} = \nabla\psi \times \nabla\chi \cdot \nabla\zeta \qquad (12.41)$$

$$d\mathbf{r} = J\,d\psi\,d\chi\,d\zeta$$

Here, $q = q(\psi)$ is a flux function and the dependence on both χ and ζ is periodic with period 2π.

The derivation below is made simpler by a shift to straight field line coordinates defined by

$$\begin{aligned} \psi' &= \psi \\ \chi' &= \chi \\ \alpha' &= -\zeta + q(\psi)\chi \end{aligned} \qquad (12.42)$$

In terms of the new coordinates the relevant equilibrium relations can be rewritten as

$$\mathbf{B} = \nabla\psi \times \nabla\alpha$$

$$\frac{1}{J} = \nabla\psi \times \nabla\alpha \cdot \nabla\chi \qquad (12.43)$$

$$d\mathbf{r} = J\,d\psi\,d\alpha\,d\chi$$

where the primes have been dropped from ψ', α', χ'. Note that in accordance with the discussion in Section 7.7.6 the quantities ψ, α now serve as field line labels.

The quantity χ has the role of an extended variable analogous to poloidal arc length in the tokamak analysis.

The usefulness of this transformation is that the operator $\mathbf{B} \cdot \nabla$ reduces to the simple form

$$\mathbf{B} \cdot \nabla = \frac{1}{J} \frac{\partial}{\partial \chi} \tag{12.44}$$

Equation (12.44) implies that the eikonal function, which satisfies $\mathbf{B} \cdot \nabla S = 0$, must be of the form $S = S(\psi, \alpha)$. An appropriate choice that produces rapid oscillation while maintaining periodicity in ζ is given by

$$S(\psi, \alpha) = n[\alpha - \alpha_0(\psi)] \tag{12.45}$$

where n is an integer satisfying $n \gg 1$ and α_0 is a free function

Next, the unit vectors $\mathbf{n}$ and $\mathbf{t}$ are defined as

$$\begin{aligned}
\mathbf{n} &= \frac{\nabla \psi}{|\nabla \psi|} \\
\mathbf{t} &= \frac{\nabla \alpha}{|\nabla \alpha|}
\end{aligned} \tag{12.46}$$

Note that both $\mathbf{n}$ and $\mathbf{t}$ are perpendicular to $\mathbf{B}$ but in a 3-D geometry they are not in general orthogonal to each other: $\mathbf{n} \cdot \mathbf{t} \neq 0$. These definitions are used to decompose the perpendicular wave number $\mathbf{k}_\perp$ and curvature vector $\boldsymbol{\kappa}$ as follows:

$$\begin{aligned}
\mathbf{k}_\perp &= \nabla S = \frac{\partial S}{\partial \psi} \nabla \psi + \frac{\partial S}{\partial \alpha} \nabla \alpha = k_\psi \nabla \psi + k_\alpha \nabla \alpha \\
\boldsymbol{\kappa} &= \kappa_\psi \nabla \psi + \kappa_\alpha \nabla \alpha
\end{aligned} \tag{12.47}$$

with

$$\begin{aligned}
k_\psi &= \frac{\partial S}{\partial \psi} = -n\alpha_0' \\
k_\alpha &= \frac{\partial S}{\partial \alpha} = n \\
\kappa_\psi &= \frac{1}{2B^2} \frac{\partial}{\partial \psi} \left(2\mu_0 p + B^2 \right) \\
\kappa_\alpha &= \frac{1}{2B^2} \frac{\partial}{\partial \alpha} \left(B^2 \right)
\end{aligned} \tag{12.48}$$

Here, k_ψ, k_α are wave number-like quantities similar to k_n, k_t in the tokamak, but normalized differently. Similarly for κ_ψ, κ_α and κ_n, κ_t. The last quantity needed for substitution into the ballooning mode potential energy is $\mathbf{b} \times \mathbf{k}_\perp$. A simple calculation leads to

$$\mathbf{b} \times \mathbf{k}_\perp = k_\psi \mathbf{b} \times \nabla\psi + k_\alpha \mathbf{b} \times \nabla\alpha \tag{12.49}$$

The above relations are now substituted into the 3-D expression for the ballooning mode potential energy given by Eq. (12.25). A straightforward calculation results in the desired form of the stellarator ballooning mode potential energy (see for instance Dewar and Glosset (1983) and Hegna and Nakajima (1998))

$$\delta\overline{W}_2 = \frac{1}{2\mu_0} \int_0^{\psi_a} \int_0^{2\pi/q} \overline{W}(\psi, \alpha, \alpha_0) \, d\psi \, d\alpha$$

$$\overline{W}(\psi, \alpha, \alpha_0) = \int_{-\infty}^{\infty} \left[\frac{k_\perp^2}{B^2 J^2} \left| \frac{\partial X}{\partial \chi} \right|^2 - 2\mu_0 \frac{dp}{d\psi} (k_\alpha^2 \kappa_\psi - k_\alpha k_\psi \kappa_\alpha) |X|^2 \right] J \, d\chi \tag{12.50}$$

and the corresponding minimizing ballooning mode differential equation

$$\frac{\partial}{\partial\chi} \left(\frac{k_\perp^2}{B^2 J} \frac{\partial X}{\partial\chi} \right) + 2\mu_0 J \frac{dp}{d\psi} (k_\alpha^2 \kappa_\psi - k_\alpha k_\psi \kappa_\alpha) X = 0 \tag{12.51}$$

Observe that only the χ variable has to be extended over the infinite domain. The α variable has to vary only over one circuit (and not an infinite number of circuits) of the flux surface to ensure that stability is tested on every field line (with each field line identified by ψ, α). Equation (12.51) is quite similar conceptually to the tokamak equation given by Eq. (12.40), allowing stability to be tested one field line at a time.

The potential energy can be further reduced by choosing a specific J to make the field line coordinates unique. One simple choice is arc length corresponding to $J = 1/B$. For this case one relabels $\chi \rightarrow l$ and notes that $\mathbf{B} \cdot \nabla = B(\partial/\partial l)$. The potential energy now has the form

$$\overline{W}(\psi, \alpha, \alpha_0) = \int_{-\infty}^{\infty} \left[k_\perp^2 \left| \frac{\partial X}{\partial l} \right|^2 - 2\mu_0 \frac{dp}{d\psi} (k_\alpha^2 \kappa_\psi - k_\alpha k_\psi \kappa_\alpha) |X|^2 \right] \frac{dl}{B} \tag{12.52}$$

A second choice for J corresponds to Boozer coordinates: $J = (i_t - qi_p)/B^2$. In this case the potential energy can be written as

$$\overline{W}(\psi, \alpha, \alpha_0) = \frac{1}{(i_t - qi_p)} \int_{-\infty}^{\infty} \left[k_\perp^2 \left| \frac{\partial X}{\partial \chi} \right|^2 - \frac{2\mu_0}{B^2} \frac{dp}{d\psi} (i_t - qi_p)^2 (k_\alpha^2 \kappa_\psi - k_\alpha k_\psi \kappa_\alpha) |X|^2 \right] d\chi$$

$$\tag{12.53}$$

Having derived several forms for the ballooning mode potential energy and ballooning mode differential equation, one can now proceed to investigate the criterion for sufficiently rapid convergence at $\chi = \pm\infty$ (or equivalently $l = \pm\infty$) and assuming the condition is satisfied, determine the critical β for stability. Most of the relevant analysis in the literature has focused on the tokamak geometry, and that is the strategy adopted here.

12.5 Stability of tokamaks – the Mercier criterion

12.5.1 Introduction

As has been discussed the ballooning mode formalism is valid only if the envelope function X decays sufficiently rapidly at $l = \pm\infty$. In this section the condition for sufficiently rapid decay is quantified leading to the Mercier criterion (Mercier, 1960), which turns out to be the toroidal analog of Suydam's criterion. Thus, the Mercier criterion examines interchange stability in a torus.

The analysis presented below follows the work of Connor *et al.* (1979) and proceeds in two steps. First, the condition for rapid decay is derived in the large aspect ratio cylindrical limit, thereby reproducing Suydam's criterion. The purpose of this calculation is to demonstrate how the ballooning mode formalism, in which the extended arc length l is the independent variable is related to the standard Suydam analysis where $x = r - r_0$ plays this role. It is easiest to understand this relationship in the familiar r, θ cylindrical geometry.

In the second step the analysis is repeated for the actual ballooning mode equation. For simplicity, the derivation focuses on the 2-D tokamak geometry where most of the practical applications have been carried out. Conceptually, the calculation is quite similar to the large aspect ratio cylinder, although the details are substantially more complicated because of the 2-D nature of the equilibrium. The end result is the Mercier criterion, which is valid for arbitrary β, ε and cross sectional shape. From a practical point of view, the Mercier criterion, when applied to tokamaks, leads to a minimum allowable $q(\psi)$ on axis. Several examples are discussed.

The basic strategy of both calculations is to examine the behavior of X for large l and to see whether the solutions are oscillatory or exponential. Oscillating solutions indicate that the ballooning mode formalism is not valid. However, such solutions violate the Mercier criterion, thus predicting instability. On the other hand, exponential solutions are Mercier stable implying that X decays sufficiently rapidly for large l for the ballooning mode equation to be valid. This equation must then be examined for ballooning mode stability which ultimately leads to a limit on β.

12.5.2 Cylindrical limit: the Suydam criterion

The starting point of the analysis is the axisymmetric ballooning mode potential energy given by Eq. (12.38). In the large aspect ratio cylindrical limit, which assumes that $R_0/a \to \infty$ and $\mu_0 p \sim B_\phi^2 \sim B_p^2$, the various quantities appearing in the integrand simplify to

$$
\begin{aligned}
R &\to R_0 \\
l &\to r\theta \\
\psi(r,\theta) &\to \psi(r) \\
J &\to 1/B_\theta(r)
\end{aligned}
\tag{12.54}
$$

These limiting values in turn imply that

$$
\begin{aligned}
S &\to n[-\phi + q(r)(\theta - \theta_0)] \\
k_n &\to nq'(\theta - \theta_0) \\
k_t &\to nB/R_0 B_\theta \\
\kappa_n &\to -(B_\theta^2/B^2)(1/r) \\
\kappa_t &\to 0
\end{aligned}
\tag{12.55}
$$

where here and below prime denotes d/dr.

Equations (12.54) and (12.55) are substituted into Eq. (12.38) where, because of cylindrical symmetry, one can set $\theta_0 = 0$ without loss in generality. After a short calculation, the resulting expression for $\overline{W}$ reduces to

$$
\frac{\overline{W}(r)}{W_0} = \int_{-\infty}^{\infty} \left[(\theta^2 + f) \left(\frac{\partial X}{\partial \theta} \right)^2 - D_S X^2 \right] d\theta
\tag{12.56}
$$

Here, $W_0 = n^2 q'^2 B_\theta / r B_z^2 > 0$, $f(r) = q^2/r^2 q'^2$, and $D_S = -2\mu_0 p' q^2 / r B_z^2 q'^2$. Note that D_S is Suydam's parameter as defined in Eq. (11.102). Also, with respect to θ, one sees that $f = $ constant.

The variational equation that determines the minimizing $X(\theta)$ for $\theta \to \pm\infty$ is easily obtained and is given by

$$
\frac{d}{d\theta} \left(\theta^2 \frac{dX}{d\theta} \right) + D_S X = 0
\tag{12.57}
$$

The solution can be written as

$$
\begin{aligned}
X &= c_1 |\theta|^{P_1} + c_2 |\theta|^{P_2} \\
P_{1,2} &= -\frac{1}{2} \pm \frac{1}{2}(1 - 4D_S)^{1/2}
\end{aligned}
\tag{12.58}
$$

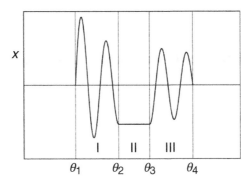

Figure 12.1   Trial function leading to the violation of Suydam's criterion.

Oscillatory solutions exist for $D_S > 1/4$. For this case $|X|^2 \propto 1/|\theta|$ leading to a logarithmic divergence of the potential energy integral. Such behavior is unphysical and implies that the convergence assumption used in the derivation of the ballooning mode equation is not valid. Even so, when $D_S > 1/4$ one can construct a truncated trial function as illustrated in Fig. 12.1, where X satisfies Eq. (12.58) in regions I and III and $X = X_0 = $ constant in region II. A simple calculation then shows that

$$\frac{\overline{W}(r)}{W_0} = -D_S X_0^2 (\theta_3 - \theta_2) \tag{12.59}$$

Since $D_S > 1/4$ by assumption it follows that $\overline{W} < 0$ indicating instability. The condition $D_S > 1/4$ for instability is just Suydam's criterion.

It is interesting to observe that in the ballooning mode formalism Suydam's criterion arises from an analysis of X at large θ while in the standard analysis presented in Chapter 11 attention is focused on the localized behavior of ξ for small $x = r - r_0$. The two analyses are actually equivalent in that one is the Fourier transform of the other. To see this consider the standard analysis where the minimizing equation for ξ is given by (Eq. (11.103))

$$\frac{d}{dx}\left(x^2 \frac{d\xi}{dx}\right) + D_S \xi = 0 \tag{12.60}$$

Upon introducing the Fourier transform

$$X(k) = \int_{-\infty}^{\infty} \xi(x)\, e^{-ikx} dx \tag{12.61}$$

and assuming that ξ converges for large x (as it must for the truncated trial functions used in the derivation) one can easily show that X satisfies

$$\frac{d}{dk}\left(k^2 \frac{dX}{dk}\right) + D_S X = 0 \tag{12.62}$$

Now, when inverting the transform the integrand (including all dependences on angle) has the form

$$\xi(\mathbf{r}) = \frac{1}{2\pi} \int_{-\infty}^{\infty} X(k) \, e^{iS} dk \qquad (12.63)$$

where the eikonal function S is given by

$$S = -n\phi + m\theta + kx \qquad (12.64)$$

A comparison with Eq. (12.55) shows that in the vicinity of a rational surface defined by $r = r_0 + x$, the eikonal (with $\theta_0 = 0$) reduces to

$$S = -n\phi + nq(r_0)\theta + nq'(r_0)x\theta \qquad (12.65)$$

Thus, one can identify $m = nq(r_0)$ and $k = q'(r_0)\theta$. Substituting into Eq. (12.62) then yields

$$\frac{d}{d\theta}\left(\theta^2 \frac{dX}{d\theta}\right) + D_S X = 0 \qquad (12.66)$$

which is identical to Eq. (12.57).

The conclusion is that in the ballooning mode formalism the extended angle θ corresponds to the Fourier transform variable of the radius $x = r - r_0$ in real space. It is a well-known property of Fourier transforms that oscillatory behavior near $x \to 0$ in real space produces oscillatory behavior near $k \propto \theta \to \infty$ in transform space.

12.5.3 Toroidal geometry: the Mercier criterion

The derivation of the criterion for sufficiently rapid decay of X as $l \to \pm\infty$ in toroidal geometry is conceptually similar to the cylindrical case. Specifically, the general solution of the ballooning mode differential equation is examined for large l. Straightforward application of Newcomb's analysis (Section 11.5.3) shows that oscillatory behavior in l leads to instability. Non-oscillatory behavior is a necessary (but not sufficient) condition for stability. The transition point defines the Mercier criterion.

As has been stated, the Mercier criterion can be viewed as a quantification of the convergence requirement for the validity of the ballooning mode formalism. If the criterion is violated the ballooning mode potential energy has a logarithmic divergence at large l which is not allowed physically. Hence the ballooning mode formalism cannot be used. However, since the system is already unstable to Mercier interchange modes one need proceed no further. If the Mercier criterion is satisfied the system is stable to the interchange mode. In this case the convergence requirements on X are satisfied and the ballooning mode equation can be used to test stability.

In carrying out the analysis for the toroidal geometry it will become apparent that the details are more complicated than for the cylindrical case because of the l dependence of the coefficients. Even so the critical feature of the analysis remains the same and involves the separation of the secular algebraic dependence of the coefficients from the periodic behavior. This allows a systematic asymptotic analysis of the solution for large l from which one can distinguish oscillatory from non-oscillatory behavior.

The derivation proceeds as follows. The starting point is the ballooning mode differential equation given by Eq. (12.40), which can be written as

$$\frac{\partial}{\partial l}\left(f\frac{\partial X}{\partial l}\right) - gX = 0$$

$$f = \frac{(k_n^2 + k_t^2)B_p}{B^2} \tag{12.67}$$

$$g = -\frac{2\mu_0 R}{B^2}\frac{dp}{d\psi}(k_t^2\kappa_n - k_t k_n \kappa_t)$$

An examination of f and g indicates that the secular algebraic terms (i.e., those proportional to l^ν as $l \to \pm\infty$) appear only in the coefficient k_n. Specifically, from Eq. (12.36) one sees that secular terms appear in the integral

$$k_n = nRB_p \int_{l_0}^{l} \frac{\partial Q}{\partial \psi}\,dl$$

$$Q(\psi,l) = \frac{F(\psi)}{R^2 B_p} \tag{12.68}$$

The secularity can be seen explicitly by recognizing that Q is periodic in l since it only depends on equilibrium quantities. Thus one can write $Q = \overline{Q}(\psi) + \tilde{Q}(\psi, l)$ where $\tilde{Q}$ has zero average value over $L(\psi)$, one poloidal circuit in l. Next, observe from the definition of $q(\psi)$ given by Eq. (6.35) that the relation between q and Q is given by

$$q(\psi) = \frac{1}{2\pi}\oint \overline{Q}\,dl = \frac{L(\psi)}{2\pi}\overline{Q}(\psi) \tag{12.69}$$

Using this relation in Eq. (12.68) leads to a simple expression for k_n in the limit $l \to \pm\infty$,

$$k_n = 2\pi nRB_p\left(\frac{1}{L}\frac{dq}{d\psi}\right)l \tag{12.70}$$

Equation (12.70) clearly exhibits a linear secular dependence on l.

After a straightforward calculation, the algebraic and periodic behavior in the coefficients f and g can be explicitly displayed by rewriting Eq. (12.67) as follows:

$$\frac{\partial}{\partial l}\left[\left(f_0\hat{l}^2 + f_1\right)\frac{\partial X}{\partial l}\right] + \left(\frac{\partial g_0}{\partial l}\hat{l} + g_1\right)X = 0$$

$$\hat{l}(\psi,l) = \int_{l_0}^{l}\frac{\partial Q}{\partial \psi}\,dl$$

$$f_0(\psi,l) = \frac{R^2 B_p^3}{B^2}$$

$$f_1(\psi,l) = \frac{1}{R^2 B_p}$$ (12.71)

$$g_0(\psi,l) = \frac{\mu_0 F}{B^2}\frac{dp}{d\psi}$$

$$g_1(\psi,l) = \frac{\mu_0}{B^2 B_p}\frac{dp}{d\psi}\frac{\partial}{\partial\psi}\left(2\mu_0 p + B^2\right) = \frac{2\mu_0}{RB_p^2}\frac{dp}{d\psi}\kappa_n$$

In this equation the coefficients f_0, f_1, g_0, g_1 are periodic functions of l. The quantity $\hat{l}(\psi, l)$ is valid for arbitrary l and exhibits secular behavior for large l: $\hat{l} \to (d\overline{Q}/d\psi)\,l$ as $l \to \infty$. In its present form Eq. (12.71) is still exact although obviously quite complicated.

The goal now is to calculate the asymptotic behavior of X as $l \to \infty$ in order to determine the threshold condition for oscillatory behavior. This goal is accomplished by exploiting the fact that $\hat{l}$ becomes large for large l, which suggests an asymptotic expansion for X of the form

$$X = \hat{l}^\nu\left(X_0 + \frac{X_1}{\hat{l}} + \frac{X_2}{\hat{l}^2} + \cdots\right)$$ (12.72)

Here, the $X_j(\psi,l)$ are assumed to be periodic functions of l with the equilibrium period $L(\psi)$. The parameter ν is the indicial coefficient determined in the course of solving Eq. (12.71). When ν is complex, then X oscillates for large l. If ν is real the solutions are non-oscillatory. The stability threshold which defines the Mercier criterion corresponds to the transition value of ν separating oscillatory and non-oscillatory behavior.

To determine ν Eq. (12.72) is substituted into Eq. (12.71). The result is a system of equations that can be written as a polynomial series in $\hat{l}$. By equating the coefficients of descending powers of $\hat{l}$ to zero and making use of appropriate periodicity constraints, one obtains the value of ν and the solutions for the X_j. To determine ν requires carrying out the analysis to second order (i.e., the X_2 level).

The leading-order equation corresponds to the highest power of $\hat{l}$, which is $\hat{l}^{\,\nu+2}$. This equation is given by

$$\frac{\partial}{\partial l}\left(f_0\frac{\partial X_0}{\partial l}\right) = 0 \tag{12.73}$$

The solution that is bounded at $l = \pm\infty$ is simply

$$X_0 = 1 \tag{12.74}$$

The first-order equation corresponding to $\hat{l}^{\,\nu+1}$ can be written as

$$\frac{\partial}{\partial l}\left[f_0\left(\frac{\partial X_1}{\partial l} + v\frac{\partial Q}{\partial \psi}\right)\right] + \frac{\partial g_0}{\partial l} = 0 \tag{12.75}$$

This equation can be integrated yielding

$$\frac{\partial X_1}{\partial l} + v\frac{\partial Q}{\partial \psi} = \frac{G_0 - g_0}{f_0} \tag{12.76}$$

where $G_0(\psi)$ is a free integration function determined by requiring that X_1 be periodic. Thus, integrating Eq. (12.76) over one poloidal circuit yields

$$G_0 = \frac{\oint\left(v\dfrac{\partial Q}{\partial \psi} + \dfrac{g_0}{f_0}\right)dl}{\oint\left(\dfrac{1}{f_0}\right)dl} \tag{12.77}$$

The second-order equation is the last one necessary for the analysis and corresponds to $\hat{l}^{\,\nu}$. A short calculation leads to

$$\frac{\partial}{\partial l}\left[f_0\frac{\partial X_2}{\partial l} + (v-1)f_0 X_1\right]$$

$$+ (v+1)f_0\frac{\partial Q}{\partial \psi}\frac{\partial X_1}{\partial l} + v(v+1)f_0\left(\frac{\partial Q}{\partial \psi}\right)^2 + g_1 + \frac{\partial g_0}{\partial l}X_1 = 0 \tag{12.78}$$

The solution for X_2 is not explicitly needed to determine the indicial coefficient v. All that is required to determine v is to integrate Eq. (12.78) over one poloidal circuit, assuming that X_2 is periodic. The result, after one simple integration by parts, is

$$\oint\left[(v+1)f_0\frac{\partial Q}{\partial \psi}\frac{\partial X_1}{\partial l} + v(v+1)f_0\left(\frac{\partial Q}{\partial \psi}\right)^2 + g_1 - g_0\frac{\partial X_1}{\partial l}\right]dl = 0 \tag{12.79}$$

The expression for $\partial X_1/\partial l$ given by Eq. (12.76) is substituted into Eq. (12.79). After a slightly lengthy calculation one obtains the following equation for v:

$$v^2 + v + D_M = 0 \tag{12.80}$$

where D_M, the toroidal analog of the cylindrical Suydam parameter D_S, is a complicated expression that can be written as

$$
D_M(\psi) = \frac{1}{4\pi^2 q'^2} \left[\left(\oint \frac{dl}{f_0} \right) \left(\oint g_1 dl \right) \right.
$$

$$
+ \left(\oint \frac{dl}{f_0} \right) \left(\oint g_0^2 \frac{dl}{f_0} \right) - \left(\oint g_0 \frac{dl}{f_0} \right)^2
$$

$$
\left. + \left(\oint \frac{\partial Q}{\partial \psi} dl \right) \left(\oint g_0 \frac{dl}{f_0} \right) - \left(\oint \frac{dl}{f_0} \right) \left(\oint g_0 \frac{\partial Q}{\partial \psi} dl \right) \right] \tag{12.81}
$$

The roots of the indicial equation are

$$v = -\frac{1}{2} \pm \frac{1}{2} (1 - 4D_M)^{1/2} \tag{12.82}$$

In analogy with the Suydam criterion one sees that the transition from oscillatory to non-oscillatory behavior occurs for $D_M = 1/4$. Consequently, the Mercier criterion for interchange stability, which coincides with the criterion for validity of the ballooning mode differential equation, is given by

$$D_M < \frac{1}{4} \tag{12.83}$$

The derivation is completed by rewriting D_M in terms of physical variables as follows:

$$
D_M = \frac{\mu_0 p'}{q'^2} \left[2I(B^2)I(RB_p\kappa_n) + I(B^2)I(\Gamma/B^2) - I(1)I(\Gamma) \right]
$$

$$
\Gamma = F \left(\mu_0 F p' - R^2 B_p^3 \frac{\partial Q}{\partial \psi} \right) \tag{12.84}
$$

$$
I(U) = \frac{1}{2\pi} \oint \frac{dl}{R^2 B_p^3} U
$$

Here, prime denotes $d/d\psi$. Although the Mercier criterion involves a complicated expression, it is nonetheless a function only of the equilibrium quantities and can be tested one flux surface at a time. It must be satisfied on each surface to guarantee interchange stability. In general, numerical calculations are required to test the Mercier criterion. However, there are some simple analytic limits that shed more insight into the physics and these are discussed next.

12.5.4 Analytic limits of the Mercier criterion

If one first takes the limit $\varepsilon \ll 1$, $\mu_0 p / B_\phi^2 \sim B_p^2 / B_\phi^2 \sim 1$ then as expected the result is the cylindrical Suydam criterion,

$$\left(\frac{rq'}{q}\right)^2 + \frac{8\mu_0}{B_\phi^2} rp' > 0 \qquad \text{for stability} \tag{12.85}$$

Here, prime now denotes d/dr.

In contrast, if one assumes the tokamak ordering $\varepsilon \ll 1$, $\mu_0 p / B_\phi^2 \sim B_p^2 / B_\phi^2 \sim \varepsilon^2$, and carries out the expansion carefully, then Eq. (12.85) is modified in an important way, leading to the large aspect ratio toroidal limit of the Mercier criterion. The result for a circular cross section plasma with $\beta_p \sim 1$ has been calculated by Ware and Haas (1966) and Shafranov and Yurchenko (1968). The modified stability criterion is given by

$$\left(\frac{rq'}{q}\right)^2 + \frac{8\mu_0}{B_\phi^2} rp'(1 - q^2) > 0 \qquad \text{for stability} \tag{12.86}$$

The expression is similar to Suydam's criterion except for the factor $(1 - q^2)$. The modification shows that for typical configurations with a negative pressure gradient, including the region near the axis, stability can be achieved if $q_0 > 1$. As previously shown in Chapter 11, Suydam's criterion always predicts a small region of instability near the axis.

The additional stability found in the torus is associated with the curvature of the toroidal magnetic field. On the inside of the torus ($\theta = \pi$) the toroidal field bends away from the plasma producing favorable curvature. The opposite is true on the outside ($\theta = 0$). The oscillation in curvature tends to average to zero with an interchange perturbation because such perturbations are nearly constant along a field line; in other words, there is essentially a uniform sampling of the curvature with an interchange perturbation.

However, to second order in ε there is a net favorable contribution to the average toroidal curvature that enters competitively with the poloidal field curvature. This can be seen qualitatively by noting that in the large aspect ratio limit

$$\boldsymbol{\kappa} \approx -\frac{B_\theta^2}{rB_0^2}\mathbf{e}_r - \frac{\mathbf{e}_R}{R} \tag{12.87}$$

Now, the normal component of $\boldsymbol{\kappa}$ is approximately given by $\kappa_n \approx \mathbf{e}_r \cdot \boldsymbol{\kappa}$ which can be written as

$$
\begin{aligned}
\kappa_n &\approx -\frac{B_\theta^2}{r B_0^2} - \frac{\cos\theta}{R_0}\left(1 - \frac{r}{R_0}\cos\theta + \cdots\right) \\
&\approx -\frac{B_\theta^2}{r B_0^2}\left(1 - \frac{q^2}{2}\right) - \frac{\cos\theta}{R_0} + \frac{r\cos 2\theta}{2R_0^2} + \cdots
\end{aligned}
\tag{12.88}
$$

Observe that the toroidal curvature ($\sim 1/R_0$) is actually $1/\varepsilon$ larger than the poloidal curvature ($\sim B_\theta^2/R_0 B_0^2$) under the usual assumption that $B_\theta/B_0 \sim \varepsilon$. Equation (12.88) implies that the last two terms tend to average to zero around one poloidal circuit. The remaining term has an additional stabilizing contribution equal to $(q^2/2)(B_\theta^2/r B_0^2)$.

The correct result, as given by Eq.(12.86), contains an extra factor of $q^2/2$ and requires a surprisingly lengthy calculation. All of the terms in the Mercier criterion must be maintained as well as the small toroidal shift of the flux surfaces. The more careful averaging with the correct weight factors indicates that the net average curvature is favorable when $q > 1$.

For tokamaks, which are usually characterized by low β and finite shear, the rq'/q in Mercier's criterion dominates over most of the profile. However, near the origin the shear is small and it is in this region that the criterion is most easily violated. The conclusion then is that the Mercier criterion basically sets a limit on the minimum q_0.

The value of the minimum q_0 depends on geometry. For circular flux surfaces, as stated above, $q_0 > 1$. The case of non-circular flux surfaces near the axis has been investigated by a number of authors: Solov'ev *et al.* (1969), Laval *et al.* (1971), Lortz and Nuhrenberg (1973), Mikhailovskii and Shafranov (1974), and Mikhailovskii (1974). Their results can be summarized as follows. For non-circular cross sections, the on-axis stability criterion $1 < q_0^2$ is replaced by

$$
1 < q_0^2\left\{1 - \frac{4}{1+3\kappa^2}\left[\frac{3\kappa^2-1}{4\kappa^2+1}\left(\kappa^2 - \frac{2\delta}{\varepsilon}\right) + \frac{(\kappa-1)^2\beta_{p0}}{\kappa(\kappa+1)}\right]\right\}
\tag{12.89}
$$

where κ is the elongation, $\beta_{p0} = -\mu_0 p''(0)/B_\theta'^2(0)$ is the poloidal beta on axis, δ is the triangularity, $\varepsilon = r_0/R_0$ is the inverse aspect ratio of the flux surface of interest, and the axis corresponds to the limit $r_0 \to 0$. Note that in this limit δ/ε remains finite. The geometric interpretation of κ, δ, ε is illustrated in Fig. 12.2.

Observe the following points. For cross sections with zero elongation ($\kappa = 1$), triangularity and finite β_{p0} have no effect on the stability boundary. Elongation by itself ($\kappa > 1$, $\delta = \beta_{p0} = 0$) is destabilizing. However, a combination of elongation

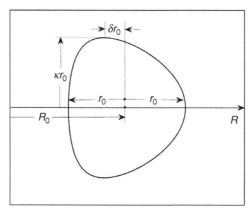

Figure 12.2   Geometry defining κ, δ, and r_0/R_0 in a non-circular tokamak. Note that $\delta > 0$ corresponds to an outward pointing triangle.

and sufficient outward pointing triangularity ($\kappa > 1$, $\delta/\varepsilon > \kappa^2/2$, $\beta_{p0} = 0$) is stabilizing. For any elongation the effect of β_{p0} is destabilizing. Here stabilizing corresponds to a lower value of the minimum q_0.

The improved stability associated with elongated "D" shapes is related to the fact that for such cross sections a magnetic line has a relatively large fraction of its trajectory located in the favorable curvature region on the inside of the torus.

12.5.5 Summary

In summary the Mercier criterion represents one limit on the maximum allowable value of $J_{\phi 0} \propto 1/q_0$ on axis in a toroidal tokamak. For sufficiently high q_0 the average curvature of a field line becomes favorable. Thus, a negative pressure gradient on axis, which is always unstable in the straight cylinder, is stabilized in a torus. As with the Suydam criterion, a Mercier interchange perturbation by itself is unlikely to be important in an actual experiment. However, because of the oscillation theorem, the presence of localized interchange instabilities implies that more dangerous large-scale instabilities are also very likely to exist. Lastly, if a plasma is stable with respect to the Mercier criterion, then the ballooning mode differential equation is valid and can be used to test ballooning mode stability.

12.6 Stability of tokamaks – ballooning modes

12.6.1 Introduction

In this subsection it is assumed that the Mercier criterion is satisfied. The task then is to solve the ballooning mode differential equation with the goal of determining the critical β for stability against ballooning modes. As a general feature, note that ballooning modes can occur almost anywhere in the profile, wherever the pressure gradient is sufficiently steep. Even so, ballooning modes most often do not set the most severe limits on β – these are set by external ballooning-kink modes.

Nevertheless, one area where ballooning modes may be particularly important is near the plasma boundary where there are sharp edge gradients, most often arising in H-mode operation. As described in Chapter 11, MHD instabilities, usually called peeling–ballooning modes, are believed to play an important role in the excitation of edge localized modes (ELMs) (Snyder *et al.*, 2005). The peeling contribution is due to the gradient in the current density and has been qualitatively described by the straight tokamak model. The ballooning contribution, on the other hand, requires a toroidal calculation, namely the solution to the ballooning mode differential equation.

A simple, well-known model is introduced that provides good physical insight into the behavior of ballooning modes near the plasma edge. The quasi-analytic investigation of this model yields a relation between the critical β and the shear. More accurately, the relation involves a limit, not on β, but on $\beta q^2/\varepsilon \sim p/I^2$. In principle, one could then achieve stability against ballooning modes at high pressures by operating at high current (i.e., low q). Ballooning mode stability does not require high q operation since the kink term vanishes in the large n expansion. In addition to the simple analytic model a description is presented of more extensive numerical studies which gives rise to a general ballooning mode stability boundary known as the "Sykes limit."

Mathematically, the marginal stability boundaries can be obtained by Newcomb's analysis. One integrates the ballooning mode differential equation from $-\infty < l < \infty$ assuming that $X(l = -\infty)$ starts off with the small, square integrable branch of the solution. If any zero crossings in X occur before $l = +\infty$, the plasma is unstable. An alternate approach is to introduce a convenient normalization and then compute the corresponding eigenvalue (basically proportional to ω^2). Stability is determined by the sign of the eigenvalue. In both methods, stability must be tested separately for all ψ and l_0.

12.6.2 The s–α model for ballooning modes

Reduction of the general ballooning mode equation

A relatively simple model that provides insight into the behavior of ballooning modes is the large aspect ratio, circular cross section tokamak with a special choice of equilibrium profiles as illustrated in Fig. 12.3 (Connor *et al.*, 1979). The critical feature of the profile is that the average value of the global β is assumed to be small ($\beta \sim \varepsilon^2$) although the local β' in a narrow layer near the plasma edge is larger by $1/\varepsilon$: $r_0\beta' \sim \varepsilon$ near $r \approx r_0$.

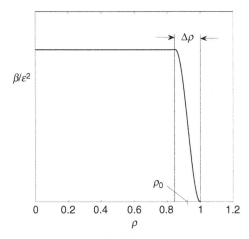

Figure 12.3   Simple $\beta(r)$ profile used in the analytic study of the ballooning mode. Here, $\rho = r/a$.

This equilibrium leads to a great simplification in the analysis. When $\beta \sim \varepsilon^2$ the flux surface shifts are small and most quantities assume their cylindrical values. Even so, there remains a small edge region where $r_0\beta'$ is sufficiently large to drive ballooning mode instabilities. From a practical point of view this is a pessimistic profile in that the global β is small while high $r_0\beta'$ ballooning modes can still be excited. On the positive side the model is quite reliable for developing qualitative insight because of its analytic simplicity.

The analysis begins with the reduction of the general tokamak ballooning mode equation given by Eq. (12.71) which is exact. The coefficients f_0, f_1, g_0, g_1 as well as $\hat{l}$ must all be evaluated using the large aspect ratio expansion. As implied by Eq. (12.71) several of these quantities are easily evaluated using simple cylindrical limits:

$$f_0 = \frac{R_0^2 B_\theta^3}{B_0^2}$$

$$f_1 = \frac{1}{R_0^2 B_\theta} \tag{12.90}$$

$$g_1 = -\frac{2\mu_0}{R_0^3 B_\theta^3} \frac{dp}{dr} \cos\theta$$

Here, $l = r\theta$ and $r \approx r_0 = $ constant describes the flux surface of interest. Also, in g_1 use has been made of the approximation $\kappa_n \approx -(1/R_0)\cos\theta$. The only curvature that enters is due to the toroidal field.

Slightly more care is required to evaluate g_0 and $\hat{l}$. The coefficient g_0 must be calculated to first order since the leading order contribution is only a function of r

while the quantity actually required is $\partial g_0/\partial l = (1/r)(\partial g_0/\partial \theta)$. This evaluation is carried out by noting that, correct to first order in ε, $B \approx B_\phi = F/R$. Thus, from Eq. (12.71) one finds

$$g_0 \approx \left(\frac{\mu_0}{F} \frac{dp}{d\psi} \right) R^2 \approx \left(\frac{\mu_0}{F} \frac{dp}{d\psi} \right) (R_0^2 + 2R_0 r \cos\theta) \qquad (12.91)$$

Since $p(\psi)$ and $F(\psi)$ are constant on a flux surface it follows that

$$\frac{\partial g_0}{\partial l} \approx -\left(\frac{2\mu_0}{R_0 B_0 B_\theta} \frac{dp}{dr} \right) \sin\theta \qquad (12.92)$$

Consider now $\hat{l}$ which involves a poloidal integral of $Q(\psi,l) = F(\psi)/R^2 B_p$. There is a zeroth-order contribution plus a first order contribution arising from the ε expansion. One of the nominally first-order contributions actually produces a zeroth-order effect when the local pressure gradient is large and it is only this contribution that must be maintained. To identify the contribution examine the expansion of the terms that appear in $\hat{l}$:

$$\hat{l}(\psi,l) = \int_{l_0}^{l} \frac{\partial Q}{\partial \psi} \, dl \approx \frac{\partial}{\partial \psi} \left[F(\psi) \int_{\theta_0}^{\theta} \frac{(r_0 + r_1 \cos\theta) d\theta}{(R_0^2 + 2R_0 r_0 \cos\theta)(B_{\theta 0} + B_{\theta 1} \cos\theta)} \right]$$
$$(12.93)$$

The expansion takes place around the flux surface $r \approx r_0 + r_1 \cos\theta$. Most of the correction terms in Eq. (12.93) are small by ε and can be neglected. However, the term with $B_{\theta 1}$ contains a first (radial) derivative of the perturbed flux function: $B_{\theta 1} \propto \psi'_1(r_0)$. While this term by itself is small its derivative with respect to r_0 in the edge layer is large since it results in a contribution to $\hat{l}$ proportional to dp/dr_0. Specifically, the term is larger by $1/\varepsilon$ than the other correction terms because of the large edge gradient assumption. It thus becomes competitive with the zeroth-order term.

The total zeroth-order contribution to $\hat{l}$ reduces to

$$\hat{l}(\psi,l) \approx \frac{\partial}{\partial \psi} \left[\frac{Fr_0}{R_0^2 B_{\theta 0}(r_0)} \int_{\theta_0}^{\theta} \left(1 - \frac{B_{\theta 1}}{B_{\theta 0}} \cos\theta \right) d\theta \right] \qquad (12.94)$$

If one now recalls that $B_\theta = (1/R)(\partial \psi/\partial r)$ then a short calculation shows that Eq. (12.94) simplifies to

$$\hat{l}(\psi,l) \approx \frac{1}{R_0 B_{\theta 0}} \frac{d}{dr_0} \left[q(r_0)(\theta - \theta_0) - \frac{q(r_0)}{R_0 B_{\theta 0}} \frac{d\psi_1}{dr_0} (\sin\theta - \sin\theta_0) \right] \qquad (12.95)$$

The final form is obtained by substituting for ψ''_1 from Eq. (6.65): $\psi''_1 \approx -2\mu_0 r_0 p'/B_{\theta 0}$ plus smaller corrections which are neglected. For the case

$\theta_0 = 0$, which, although not proven, turns out to be the most unstable case, the expression for $\hat{l}$ reduces to

$$\hat{l}(r,\theta) \approx \frac{1}{R_0 B_\theta} \left(\frac{dq}{dr} \theta + \frac{2\mu_0 q}{R_0 B_\theta^2} \frac{r \, dp}{dr} \sin \theta \right) \tag{12.96}$$

where the subscript zero has been suppressed from r_0

The above results are substituted into Eq. (12.71) leading to the desired form of the simplified ballooning mode equation (Connor *et al.*, 1979):

$$\frac{\partial}{\partial \theta} \left[(1 + \Lambda^2) \frac{\partial X}{\partial \theta} \right] + \alpha(\Lambda \sin \theta + \cos \theta)X = 0$$

$$\Lambda = s\theta - \alpha \sin \theta$$

$$s = \frac{r}{q} \frac{dq}{dr} \tag{12.97}$$

$$\alpha = -\frac{2\mu_0 r^2}{R_0 B_\theta^2} \frac{dp}{dr} = -q^2 R_0 \frac{d\beta}{dr}$$

with $\beta(r) = 2\mu_0 p(r)/B_0^2$. Note that $s(r)$ is the magnetic shear while $\alpha(r)$ is a parameter that measures the pressure gradient and scales as $\beta_t q^2/\varepsilon$. The terms with the θ derivatives represent the effects of line bending. The $\alpha \cos \theta$ term represents the effect of the normal curvature, favorable on the inside ($\theta = \pi$) and unfavorable on the outside ($\theta = 0$). The remaining $\alpha \sin \theta$ term represents the effect of the geodesic curvature. The range of θ is $-\infty < \theta < \infty$ and $X(\theta)$ is required to vanish at both $\theta \to \pm\infty$. Although Eq. (12.97) cannot be solved purely analytically it has the advantage of depending only upon two parameters, s and α. As stated previously, there is no separate dependence on the safety factor that would prevent low q operation since the kink term vanishes in the ballooning mode expansion.

Solution

The solution to Eq.(12.97) is easily obtained numerically. A typical marginally stable quasi-mode perturbation (for $-\infty < \theta < \infty$) and the corresponding full mode perturbation (for $0 < \theta < 2\pi$) are illustrated in Fig. 12.4 as well as a sketch of a perturbed plasma flux surface exhibiting a ballooning mode instability. As previously stated the quasi mode has small oscillations superimposed on a decaying envelope. The full eigenfunction, obtained by summing up shifted quasi-modes, is indeed periodic as required.

The main quantitative result of the numerical solutions is a curve of marginal stability in s, α space and is illustrated in Fig. 12.5. Observe that there are two branches in the diagram. The left branch separates the first region of stability from

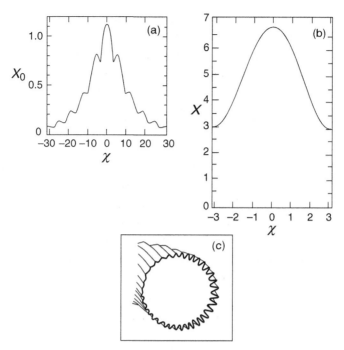

Figure 12.4   Ballooning mode perturbation: (a) quasi-mode, (b) full eigenfunction, and (c) ballooning mode perturbation in physical space.

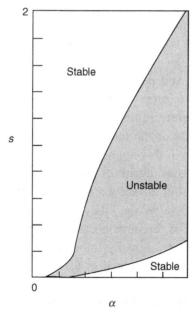

Figure 12.5   The s vs. α marginal stability diagram for the analytic ballooning mode model. From Lortz, 1973. Reproduced with permission from Elsevier.

the unstable region and is consistent with the intuition that has been developed for the behavior of ballooning modes. Specifically, at a fixed value of shear the plasma is stable for sufficiently small pressure gradients. As the pressure gradient (i.e., α) increases the perturbation develops a progressively larger ballooning component. Eventually, at sufficiently large α the destabilizing contribution from the unfavorable curvature region overcomes the shear and the system becomes unstable. As expected, when the shear increases the maximum allowable pressure gradient also increases.

A surprising feature of the s, α diagram is the second curve on the right which separates the unstable region from the so-called second region of stability (Coppi, 1977; Connor *et al.*, 1979; Greene and Chance, 1981). In this region stability is achieved when the pressure gradient becomes sufficiently large! Interestingly, this region is most easily accessed at low values of shear. This behavior is counter-intuitive but can be explained as follows. From Eq. (12.97) it follows that a local (in θ) shear $\hat{s}(r,\theta)$ can defined by

$$\Lambda(r,\theta) \equiv \int_0^\theta \hat{s} \, d\theta \tag{12.98}$$

where

$$\hat{s}(r,\theta) = s(r) - \alpha(r)\cos\theta \tag{12.99}$$

Here $s(r)$ represents the average shear and $\alpha(r)\cos\theta$ is the local pressure driven modulation. Now, as the pressure gradient increases from zero the local shear in the unfavorable curvature region ($\theta = 0$) initially decreases, thereby increasing instability. However, as the pressure gradient further increases so that $\alpha > s$ the shear becomes large and negative in the region of unfavorable curvature. Since shear stabilization is proportional to $\hat{s}^2$ this produces a stabilizing effect in the unfavorable curvature region. Although the local shear becomes zero somewhere near the inside of the torus ($\pi/2 < \theta < \pi$) this is not critical since the curvature is favorable in this region.

For a number of years considerable effort has been devoted to creating plasmas that operate at very high β in the second region of stability. The results have not been completely convincing in that they have mainly been achieved during transient operation. This is because access to the second stability region is sensitive to profile and shape. Equally importantly, other external resistive wall ballooning-kink instabilities are excited at lower values of β and would require successful feedback control before second stability operation could become accessible. Still, second stability is an interesting physical phenomenon that may still play some operational role, particularly near the plasma edge in connection with ELMs.

12.6.3 β *limits due to ballooning modes*

Predictions from the s–α *diagram*

It is worth noting that because of its simplicity the s, α diagram is sometimes used as a local criterion valid at all plasma radii and not just the edge. This clearly violates the original assumptions used in its derivation but, perhaps fortuitously, still gives reasonable scaling predictions for the global critical β for stability. Below a simple calculation based on the analytic model is presented that predicts the global critical β. The results are then compared to a much more comprehensive numerical study that includes finite aspect ratio and plasma shape.

The admittedly poorly justified local application of the s, α diagram can be used to predict global stability as follows. It is assumed that the plasma relaxes to a state that is marginally stable to the first stability boundary in the s, α diagram over the entire pressure profile. For simplicity this boundary can be reasonably well approximated by the simple relation $\alpha = Ks$ where $K \approx 0.6$. The approximation is valid over most of the plasma except near the origin. Therefore, at every radius in the plasma the pressure gradient satisfies

$$-R_0 \frac{d\beta}{dr} = K \frac{r}{q^3} \frac{dq}{dr} \tag{12.100}$$

One now multiplies Eq. (12.100) by r^2 and integrates over the plasma volume. A short calculation yields

$$\beta_t = \frac{K\varepsilon}{2}\left(-\frac{1}{q_a^2} + 3\int_0^1 \frac{\rho^2}{q^2}d\rho\right) \tag{12.101}$$

where $\rho = r/a$ and

$$\beta_t \equiv 2\int_0^1 \beta\rho d\rho \tag{12.102}$$

is the global average beta.

The critical value of β_t can be calculated by choosing a profile for $q(\rho)$ and carrying out the integration. A convenient choice that makes the integration simple is

$$q(\rho) = q_0(1 + \sigma\rho^3) \tag{12.103}$$

Here, the parameter σ is related to q_a/q_0 by $q_a/q_0 = 1 + \sigma$. The $q(\rho)$ profile is flat near the axis (i.e., a broad current profile) and increases monotonically towards the outside of the plasma. The resulting β_t limit is easily evaluated and can be written as

$$\beta_t = 0.3\frac{\varepsilon}{q_0^2}\left[\frac{q_0}{q_a}\left(1 - \frac{q_0}{q_a}\right)\right] = 0.3\varepsilon\left[\frac{1}{q_a}\left(1 - \frac{1}{q_a}\right)\right] \tag{12.104}$$

The second expression corresponds to the interesting case $q_0 = 1$. Observe that the critical β_t increases linearly with the current for large q_a: $\beta_t \approx 0.3\varepsilon/q_a \propto I$. For much larger current the critical β_t starts to decrease. The optimum for arbitrary q_0 occurs at $q_a/q_0 = 2$ and gives $\beta_t = 0.075\varepsilon/q_0^2$.

For typical tokamak parameters, $\varepsilon = 1/3$, $q_0 = 1$, $q_a = 3$, Eq. (12.104) predicts $\beta_t = 2.2\%$. As is shown shortly this value is perhaps fortuitously similar to realistic diffuse profile numerical calculations.

A final point worth noting is that if one substitutes the simple analytic profiles into the double adiabatic model the result is a stabilizing correction whose magnitude can be easily estimated in terms of the potential energies as follows:

$$\delta\overline{W}_{CGL} = \delta\overline{W}_{MHD}\left[1 + O\left(\varepsilon\frac{\Delta r}{a}\right)\right] \tag{12.105}$$

The conclusion is that the corrections are small implying that ideal MHD provides a good estimate of ballooning mode stability.

Numerical studies of ballooning mode stability

Many numerical codes have been developed to study ideal MHD equilibrium and stability. These codes contain an equilibrium Grad–Shafranov solver valid for arbitrary β, ε, and cross section. They can treat either fixed or free boundary configurations. General profiles are allowed and usually $p(\psi)$ and one of $F(\psi)$, $q(\psi)$, or $\langle J_\phi(\psi)\rangle$ are specified. Once an equilibrium is calculated the following stability tests can be made: (1) Mercier criterion; (2) high n ballooning modes; (3) low-n internal modes; and (4) external ballooning-kink modes.

Such studies, while time consuming, provide a complete and detailed picture of tokamak stability. They are very helpful in the design of new experiments and in the interpretation and analysis of existing experimental data. These studies also play a major role in the determination of optimized configurations and are ultimately able to make quantitative predictions for the maximum β_t and q_* that can be stably maintained in MHD equilibrium.

Many of the stability studies have been analyzed and merged, the end result being a surprisingly simple picture of overall tokamak stability. One such study involves ballooning modes. The main results have been assembled, unified, and expanded in the well-known calculations of Sykes and Wesson (1974) and Sykes *et al.* (1983). The problem addressed here is to determine the optimum tokamak profiles that achieve the maximum stable β_t against ballooning modes.

In these studies the safety factor is always set to $q_0 = 1.05$ to ensure stability against low-n internal kinks. Different q profiles are examined by varying the plasma current $I \propto 1/q_*$ For a given q profile the amplitude and shape of the pressure profile is varied iteratively until the maximum β_t is found. In the optimized state the profiles are essentially marginally stable to ballooning modes on every surface and satisfy the Mercier criterion on axis. This optimization procedure is carried out for different shaped cross sections.

The end result has a simple form. The maximized β_t for all cases studied is accurately represented by the Sykes limit,

$$\beta_t = \beta_N \frac{I}{aB_0}$$
$$\beta_N = 0.044 = 4.4\%$$

(12.106)

where the units are I (MA), a (m), and B_0 (T), and β_N is the standard notation for the numerical coefficient multiplying I/aB_0. When β_N is given as a % then β_t also converts from a decimal to a percentage. An alternate form of Eq. (12.106) can be written as

$$\beta_t = 0.22 \left(\frac{1 + \kappa^2}{2} \right) \frac{\varepsilon}{q_*}$$

(12.107)

There are several points to observe. Equation (12.107) implies that high β_t is achieved by operating at high I/B_0. Clearly though, from a physical point of view one would not expect this relation to hold as $I \to \infty$, $B_0 \to 0$. In practice two additional non-ballooning factors limit the maximum ratio of I/B_0. First, as I/B_0 increases the flux surfaces near the axis become highly elongated causing violation of the Mercier criterion. As previously discussed, outward triangularity helps alleviate this problem although not indefinitely. Second, for sufficiently large I/B_0 external kink modes are excited, even at low β_t. Even so, the linear relation between β_t and I is valid over a reasonably wide practical parameter range.

Putting aside the question of external kink stability and assuming Mercier stable profiles, Sykes *et al.* (1983) discovered that the scaling given in Eq. (12.107) persists at moderate triangularity ($\delta = 0.24$) down to the low value of $q_* \approx 1.4$ for $\kappa = 2$. As an example, for $\kappa = 2$, $q_* = 1.5$, and $\varepsilon = 1/3$, one finds $\beta_t \approx 12\%$, a very encouraging value if achievable.

One final slightly puzzling feature of the Sykes limit is the fact that the relation between β_t and $1/q_* \sim I$ is linear and not quadratic as one would expect from the basic scaling relations for the high β tokamak ordering. In fact there is an implied quadratic scaling in the Sykes limit that has been masked since q_0 has been held fixed at a value near unity. Indeed, a comparison of Eq. (12.107) with Eq. (12.104) suggests that the Sykes limit may actually scale as $\beta_t \propto \varepsilon/q_0 q_*$ (after noting that $q_a = q_*$ for the simple model profiles).

12.6.4 Summary

Ballooning modes are high n MHD instabilities that tend to localize near the outside of the torus in the region of unfavorable curvature. They may be particularly important in driving ELMs in H-mode operation because of the high edge pressure and current gradients. In the first region of stability the maximum β_t is limited by a maximum κ to avoid violating the Mercier criterion and by a minimum q_* to avoid external kink modes.

12.7 Stability of tokamaks – low n internal modes

12.7.1 Introduction

Localized high n internal MHD modes set a limit on β due to ballooning instabilities and a limit on q_0 due to interchange instabilities. The q_0 limit is approximately consistent with the current density limit on axis which, if violated, leads to the development of sawtooth oscillations. Still, because of their highly localized structure coupled with the corresponding high m and n mode numbers, interchange instabilities are not believed to be the main drive for sawtooth oscillations.

Instead, there is a large amount of experimental evidence that indicates that sawtooth oscillations have a macroscopic mode structure characterized by a low $m = 1$, $n = 1$ mode structure. These observations motivate an analysis of low n internal modes in a tokamak including the effect of toroidicity. In this connection recall that the corresponding analysis in the straight tokamak shows that $q_0 > 1$ is required for stability.

It is shown below that toroidal effects play an important role in the ideal MHD stability of low n internal modes. There are two qualitatively different regimes of interest which are distinguished by the amount of magnetic shear near the axis. The first corresponds to finite shear while the second corresponds to very small shear. Although a complete picture of the ideal MHD stability of low n is now well in hand, actual theoretical predictions of sawtooth behavior require the inclusion of resistivity, two-fluid effects, and non-linear simulations. Even with these more sophisticated models there is still not an entirely self-consistent explanation of all aspects of sawtooth behavior. This remains an area of active research with more work to be done.

In spite of these uncertainties, if one takes a step back it still remains true that the stability of low n internal ideal MHD modes play an important role in determining the approximate onset condition for sawtooth oscillations. These conditions are described below.

12.7.2 *Low* n *internal modes with finite shear*

Since low n internal modes have a large-scale radial structure the stability bound-aries can often only be determined by numerical calculations. However, an inter-esting analytic calculation has been carried out by Bussac *et al.* (1975) that sheds considerable light on the problem. Their calculation considers the limit of a large aspect ratio, circular cross section tokamak. By applying the low β tokamak expansion, $\beta \sim \varepsilon^2$, $\beta_p \sim 1$, $q \sim 1$ they are able to calculate an expression for δW_F for the toroidal case which shows how toroidicity modifies the stability criterion for low n modes.

To put their analysis into perspective it is useful to briefly review the results of the straight tokamak. The analysis of the $m = 1$ mode in Chapter 11 has shown that δW_F for a straight tokamak, which uses the same low β expansion as Bussac *et al.*, can be written as (see Eqs. (11.176) and (11.178))

$$\frac{\delta W_F}{W_0} = \varepsilon^2 \delta \hat{W}_2 + \varepsilon^4 \delta \hat{W}_4 + \cdots$$

$$W_0 = \frac{2\pi^2 R_0 B_0^2}{\mu_0 a^2}$$

$$\delta \hat{W}_2 = \int_0^a \left(\frac{n}{m} - \frac{1}{q} \right)^2 r^2 \xi'^2 \, rdr \qquad (12.108)$$

$$\delta \hat{W}_4 = \int_0^{r_1} \left[r\beta' + \frac{n^2 r^2}{R_0^2} \left(1 - \frac{1}{nq} \right) \left(3 + \frac{1}{nq} \right) \right] rdr$$

where the resonant surface r_1 is defined by $nq(r_1) = 1$ and without loss in generality the unimportant amplitude scale factor ξ_0 has been set to $\xi_0 = 1$. The reason why the fourth-order correction, $\delta \hat{W}_4$, is required is that $\delta \hat{W}_2$ vanishes for the "top hat" trial function illustrated in Fig. 11.27. Therefore, $\delta \hat{W}_4$ is the first non-vanishing contribution, which is always negative (unstable) if $nq_0 < 1$.

Now, observe that toroidal effects, which could be neglected if $\delta \hat{W}_2 \neq 0$ but not when $\delta \hat{W}_2 = 0$, also enter the analysis in the ε^4 order and are competitive with the cylindrical corrections. Using the same "top hat" trial function Bussac *et al.* again find that $\delta \hat{W}_2 = 0$ but that $\delta \hat{W}_4$ is modified by toroidal effects. After a rather involved calculation too complicated to reproduce here, Bussac *et al.* derive a modified expression for $\delta \hat{W}_4$ that can be schematically written as

$$\frac{\delta W_F}{W_0} = \varepsilon^4 \left[\left(1 - \frac{1}{n^2} \right) \delta \hat{W}_{4C} + \frac{1}{n^2} \delta \hat{W}_{4T} \right] \qquad (12.109)$$

Here, $\delta \hat{W}_{4C}$ is the cylindrical contribution shown in Eq. (12.108) and $\delta \hat{W}_{4T}$ is the toroidal modification. This modification is quite complicated for general profiles.

However, in the sawtooth relevant limit $|q_0 - 1| \ll 1$, $q_0 < 1$, with a parabolic current density profile, a simple expression is obtained, given by

$$\delta \hat{W}_{4T} \approx \frac{3n^2 r_1^4}{R_0^2}(1 - q_0)\left(\frac{13}{144} - \beta_p^2\right)$$

$$\beta_p = -\frac{R_0^2}{n^2 r_1^2}\int_0^{r_1} r^2 \beta' \, dr$$

(12.110)

Note that for $n \gg 1$, $\delta W_F / W_0 \approx \varepsilon^4 \delta \hat{W}_{4C}$ and the stability condition is identical to that in the straight tokamak: $nq_0 > 1$. In the more interesting and restrictive case corresponding to $n = 1$, the cylindrical contribution vanishes and all that remains is the toroidal modification: $\delta W_F / W_0 \approx \varepsilon^4 \delta \hat{W}_{4T}$. Here, in contrast to the cylindrical result, the internal kink mode is stabilized by toroidal effects in the limit $\beta_p \to 0$, $q_0 < 1$. For both the cylindrical and toroidal contributions, increasing β_p is destabilizing. For the $n = 1$ mode the toroidal calculation predicts instability for $\beta_p > (13)^{1/2}/12 \approx 0.3$. Lastly, be aware that one cannot achieve stability at high β_p by setting for $q_0 > 1$ since $q_0 < 1$ is a requirement for the top hat trial function that makes $\delta \hat{W}_2 = 0$.

The physical phenomena at play in Eq. (12.110) involve a competition between the average curvature, which is unfavorable near the axis if $q_0 < 1$, and the toroidal correction to line bending which is a stabilizing effect.

12.7.3 *Low* n *internal modes with small shear*

Wesson and Sykes (1974) have pointed out that the low β_p stabilization of the $m = 1$, $n = 1$ mode is a strong function of the shear near the axis. Quantitatively, in deriving the stability criterion in Eq. (12.110) it has been assumed that near the $q(r_1) = 1$ resonant surface the quantity $1 - q_0$ can be approximated by

$$1 - q(r) \approx -q'(r_1)(r - r_1)$$

(12.111)

Critically, for a system with finite shear it is further assumed that at the radius r_1, the following ordering applies:

$$r_1 q'(r_1) \sim 1$$

(12.112)

This is the definition of "finite" shear.

Wesson and Sykes observed that many tokamaks actually have a very flat $q(r)$ profile within the $q(r_1) = 1$ resonant surface. For sufficient flatness it would then be a better approximation to assume that

$$r q'(r_1) \sim \varepsilon$$

(12.113)

This is the definition of "low" shear.

Under this assumption the previous analysis, which allowed one to write $\delta W_F/W_0 = \varepsilon^2 \delta W_2 + \varepsilon^4 \delta \hat{W}_4$, breaks down. The reason is that $\varepsilon^2 \delta \hat{W}_2$ actually becomes an ε^4 contribution and must be treated competitively with the other cylindrical and toroidal terms. The uniform shift, top hat trial function is no longer the most unstable perturbation. Instead, analysis shows that the most unstable trial function has a much smoother profile having the approximate form of an interchange perturbation. In fact the mode is often referred to as a quasi-interchange instability.

As might be expected, in the regime of low shear, line bending is substantially reduced as compared to the finite shear case. The end result is that one again requires

$$q_0 > 1 \tag{12.114}$$

for stability.

12.7.4 Summary

To summarize, low n internal modes can be excited in a tokamak when $q_0 < 1$. This conclusion, based on the simple model described above, is confirmed in more accurate numerical simulations. Since the instability requires the existence of a $q = 1$ resonant surface in the plasma one does not expect non-circularity to significantly alter the $q_0 = 1$ stability condition, particularly in low-shear systems. The $m = 1$, $n = 1$ internal mode is quite possibly the initial drive for sawtooth oscillations although the actual phenomenon requires more sophisticated physical models plus non-linear simulations for a quantitative explanation of experimental results. Sawtooth oscillations, which limit the current density on axis, are usually not very harmful to tokamak operation. The main concern involves the situation where the q profile is very flat and the $q = 1$ surface occurs a considerable distance out from the axis. Then the non-linear rearrangement of pressure and magnetic flux might have the opportunity to interact with the first wall, an undesirable situation.

12.8 Stability of tokamaks – low n external ballooning-kink modes

12.8.1 Introduction

Low n external ballooning-kink modes are the most dangerous instabilities in a tokamak. If exited, these modes invariably lead to major disruptions. Thus, it is critical to learn how to avoid such modes. As the name implies, the corresponding eigenfunctions are composed of contributions from both the pressure gradient (i.e., ballooning) and parallel current (i.e., kinking) terms in δW_F. For comparison recall that pure high n ballooning modes are driven solely by the pressure gradient term. The kink term vanishes.

The practical consequences of this difference in driving terms are as follows. Ballooning modes actually produce a limit on the ratio $p/I^2 \sim \varepsilon\beta_p$. In principle, to achieve high pressures one could operate at very high currents, equivalent to very low q values, and still be stable against pure ballooning modes. In contrast, low n ballooning-kink modes produce a limit on both p/I^2 and $I/B_0 \sim 1/q_*$, the latter limit arising from the presence of the kink term. In general, there is an optimum I/B_0 at which to operate if the goal is to maximize p/B_0^2.

Another important feature of ballooning-kink modes is the need to distinguish between q_a and q_*. In a low $\beta \sim \varepsilon^2$ circular cross section plasma $q_a \approx q_*$ so there is no need to make this distinction. However, in the high $\beta \sim \varepsilon$ regime the values of q_a and q_* diverge and one must learn which parameter is most relevant for determining ballooning-kink stability. The analysis presented here shows that q_* is the more important parameter.

Lastly, it is worth pointing out that one important advantage of the tokamak configuration is that stability at finite pressures and finite currents is possible without the presence of a perfectly conducting wall. For comparison, recall that an RFP always requires a conducting wall to stabilize external kink modes even when the pressure is zero.

To summarize, the study of low n external ballooning-kink modes shows that there is a maximum stable value of $\beta_t q_*^2/\varepsilon$ which occurs at a specific optimized value of q_*. These limits are finite and reactor relevant even with the conducting wall at infinity.

To help demonstrate the behavior of external ballooning-kink modes described above, three calculations are presented. The first shows how the general expression for δW_F, which is a function of the two components of $\xi_\perp$, can be reduced to a form involving only a single scalar unknown $U(r,\theta)$ by applying the large aspect ratio, high β tokamak expansion. Not surprisingly, the reduced form is simpler than the original δW_F. More importantly it shows that in contrast to the high n pure ballooning analysis, the kink term survives in the expansion. It is this term that sets a separate limit on q_* in addition to the $\beta_t q_*^2/\varepsilon$ limit set by the pressure gradient term.

Even the reduced form of δW_F is still too complicated to obtain a quantitative analytic estimate of the stability limits using smooth simple profiles. This motivates the second calculation which makes use of the ultra-simple surface current model. In most of the analysis thus far presented realistic smooth current and pressure profiles have been assumed without the need for surface currents. Thus, when evaluating the total potential energy for these realistic cases, $\delta W_S = 0$. However, for the problem of interest the surface current model is perhaps the only one that is sufficiently simple to predict an explicit analytic stability boundary in the form of a β_t/ε vs. q_* curve. In addition, the surface current analysis shows in a clear way the presence of both the pressure gradient and kink driving terms. The final results give the correct scaling relations but are slightly optimistic with respect to numerical values obtained from more complete computational studies.

In fact it is these computational studies that represent the third contribution to the study of ballooning-kink modes. Exhaustive numerical studies have been carried out to determine the pressure and current limits in toroidal tokamaks with no assumptions made about the size of β_t, ε, or the shape of the plasma cross section. These studies lead to another remarkably simple stability result known as the "Troyon limit" (Troyon *et al.*, 1984). This and follow-on numerical results are discussed at the end of this subsection.

12.8.2 Simplification of δW$_F$ by the high β tokamak expansion

In general, the evaluation of the potential energy for external modes requires the evaluation of three contributions: $\delta W = \delta W_F + \delta W_S + \delta W_V$. The surface and vacuum terms depend only on the value of the normal component of displacement on the plasma surface: $\mathbf{n} \cdot \boldsymbol{\xi}_\perp(S)$. The fluid contribution is more complicated. Even after setting $\nabla \cdot \boldsymbol{\xi} = 0$ to minimize plasma compressibility, the remaining integrand in δW_F is still a function of the two vector components of $\boldsymbol{\xi}_\perp$. The goal now is to show how the integrand in δW_F can be written in terms of a single scalar $U(r,\theta)$ by introducing the high β tokamak expansion. The starting point is the intuitive form of δW_F for incompressible displacements given by Eq. (8.100) and repeated here for convenience:

$$\delta W_F(\boldsymbol{\xi}^*, \boldsymbol{\xi}) = \frac{1}{2\mu_0} \int_P \left[|\mathbf{Q}_\perp|^2 + B^2 |\nabla \cdot \boldsymbol{\xi}_\perp + 2\boldsymbol{\xi}_\perp \cdot \boldsymbol{\kappa}|^2 \right.$$

$$\left. -2\mu_0(\boldsymbol{\xi}_\perp \cdot \nabla p)(\boldsymbol{\xi}_\perp^* \cdot \boldsymbol{\kappa}) - \mu_0 J_\parallel \boldsymbol{\xi}_\perp^* \times \mathbf{b} \cdot \mathbf{Q}_\perp \right] d\mathbf{r} \quad (12.115)$$

Recall now that the basic ordering assumptions defining the high β tokamak expansion are given by $\beta \sim \varepsilon$, $q \sim 1$, $B_p/B_\phi \sim \varepsilon$, and $\delta B_\phi/B_\phi \sim \varepsilon$. With respect to the terms appearing in the integrand, this ordering implies that

$$\begin{aligned} \boldsymbol{\xi}_\perp &\sim 1 \\ \mathbf{Q}_\perp &\sim \varepsilon \\ p &\sim \varepsilon \\ \boldsymbol{\kappa} &\sim \varepsilon \\ J_\parallel &\sim \varepsilon \end{aligned} \quad (12.116)$$

The crucial difference compared to the ballooning mode analysis is that the toroidal mode number is ordered as

$$n \sim 1 \quad (12.117)$$

rather than $n \gg 1$ and having $1/n$ serve as a small expansion parameter.

However, as with the ballooning mode analysis, the dominant term is still due to magnetic compressibility and is of order unity whereas all other terms are of order ε^2. Since magnetic compressibility is always stabilizing the most unstable perturbations must be chosen to make this term as close to zero as possible. This can be accomplished by expanding the various terms appearing in the integrand as follows:

$$\boldsymbol{\xi}_\perp = (\boldsymbol{\xi}_{p0} + \boldsymbol{\xi}_{p1} + \cdots) + (\xi_{\phi 1} + \cdots)\,\mathbf{e}_\phi$$

$$\mathbf{B} = (B_0 + B_{\phi 1} + \cdots)\,\mathbf{e}_\phi + (\mathbf{B}_{p1} + \cdots)$$

$$\boldsymbol{\kappa} = -\frac{\mathbf{e}_R}{R_0} + \cdots \qquad\qquad (12.118)$$

$$\nabla = \nabla_p + \frac{\mathbf{e}_\phi}{R_0}\frac{\partial}{\partial\phi} + \cdots$$

Here, the subscript p refers to the poloidal (r, θ) plane and the numerical subscripts define the order in ε. Also, the condition that $\boldsymbol{\xi}_\perp$ be perpendicular to $\mathbf{B}$ requires that $\xi_{\phi 1} = -(\mathbf{B}_{p1}\cdot\boldsymbol{\xi}_{p0})/B_0$.

This ordering leads to a simplification of the magnetic compressibility term. Specifically, one can write

$$\nabla\cdot\boldsymbol{\xi}_\perp + 2\boldsymbol{\xi}_\perp\cdot\boldsymbol{\kappa} = \nabla_p\cdot\boldsymbol{\xi}_{p0} + \left(\nabla_p\cdot\boldsymbol{\xi}_{p1} - \frac{2}{R_0}\boldsymbol{\xi}_{p0}\cdot\mathbf{e}_R\right) + \cdots \qquad (12.119)$$

The leading-order contribution to magnetic compressibility can be made to vanish by choosing

$$\boldsymbol{\xi}_{p0} = \mathbf{e}_\phi \times \nabla_p U \qquad\qquad (12.120)$$

which implies that

$$\nabla_p\cdot\boldsymbol{\xi}_{p0} = \frac{1}{R_0}\nabla_p U\cdot\mathbf{e}_Z \sim \varepsilon \qquad\qquad (12.121)$$

Here, $U = U(r,\theta)\exp(-in\phi)$ and except where obviously required the factor $\exp(-in\phi)$ is suppressed. The scalar quantity $U(r,\theta)$ is the basic unknown in the problem and defines the two components of $\boldsymbol{\xi}_{p0}$. All the remaining terms in δW_F can be written as functions of $U(r,\theta)$ to the required order in ε^2.

Note that $\boldsymbol{\xi}_{p1}$ appears only in the residual magnetic compressibility term which is also of order ε^2. This stabilizing contribution can also be made to vanish by choosing $\boldsymbol{\xi}_{p1}$ to satisfy

$$\nabla_p\cdot\boldsymbol{\xi}_{p1} = \frac{2}{R_0}\boldsymbol{\xi}_{p0}\cdot\mathbf{e}_R - \frac{1}{R_0}\nabla_p U\cdot\mathbf{e}_Z = \frac{1}{R_0}\nabla_p U\cdot\mathbf{e}_Z \qquad (12.122)$$

The end result is that magnetic compressibility effects play no role in external ballooning-kink instabilities in a high β tokamak.

The desired form of δW_F for the high β tokamak is obtained by substituting Eq. (12.120) into the remaining terms in Eq. (12.115). A simple calculation yields

$$\delta W_F(U^*, U) = \frac{1}{2\mu_0} \int_P \left[|\mathbf{Q}_p|^2 + 2\mu_0 \frac{dp}{d\psi} (\mathbf{B}_p \cdot \nabla_p U) \frac{\partial U^*}{\partial Z} - \mu_0 J_\parallel \nabla_p U^* \cdot \mathbf{Q}_p \right] d\mathbf{r}$$

(12.123)

where, for convenience, the subscript "1" has been suppressed from $\mathbf{B}_p$. The poloidal component of $\mathbf{Q}_\perp$ requires a short calculation which leads to

$$\mathbf{Q}_p = \mathbf{e}_\phi \times \left[\mathbf{B} \cdot \nabla(\nabla_p U) \right] - \left[(\mathbf{e}_\phi \times \nabla_p U) \cdot \nabla_p \right] \mathbf{B}_p$$

$$\mathbf{B} \cdot \nabla = -in\frac{B_0}{R_0} + \mathbf{B}_p \cdot \nabla_p$$

(12.124)

Observe that the reduced δW_F is considerably simpler than its original form. There are fewer terms and only a single unknown perturbation quantity $U(r,\theta)$. The three contributions to ballooning-kink stability are due to line bending, the pressure gradient, and the parallel current. Even though δW_F is simplified it is still too complicated to analyze for diffuse profiles because the system is essentially two dimensional (r,θ) with no further obvious small parameters about which to expand. In general, 2-D numerical solutions are required for both equilibrium and stability. There is, however, one special equilibrium that can be treated analytically and that is the surface current model which is the next topic of interest.

12.8.3 High β stability of the surface current model

The surface current model describes a particularly simple MHD equilibrium that still contains sufficient physics to describe the stability of external ballooning-kink modes in a high β tokamak (Freidberg and Haas, 1973). The goal of the analysis presented below is to derive the maximum stable value of β/ε as a function of q_*. Five steps are required. First, the surface current equilibrium is calculated including the equilibrium β limit. Second, the surface energy δW_S is evaluated for a ballooning-kink perturbation, demonstrating the effects of both the curvature and kink driving terms. Third, the fluid energy is calculated using the simplified high β tokamak form of δW_F. Fourth, the vacuum energy δW_V is evaluated which, perhaps surprisingly, requires the most amount of work. The total potential energy $\delta W = \delta W_F + \delta W_S + \delta W_V$ is then minimized with respect to the ratio of the ballooning to kink driving terms. This is an easy calculation from which it is then straightforward to calculate the maximum stable β. The analysis proceeds as follows.

Surface current equilibrium

To keep the algebra to a minimum the high β tokamak of interest is assumed to have a circular cross section. In terms of profiles, all the plasma current, as its name implies, flows on the plasma surface. Therefore, in the plasma the poloidal magnetic field is zero, the toroidal magnetic field has a vacuum dependence proportional to $1/R$, and the pressure is a constant. Mathematically this is equivalent to

$$\mathbf{B} = B_i \frac{R_0}{R} \mathbf{e}_\phi$$

$$p = \text{constant}$$

(12.125)

where $B_i = \text{constant}$ represents the amplitude of the internal toroidal magnetic field on the axis $R = R_0$. Equation (12.125) trivially satisfies the MHD equilibrium equations.

In the vacuum region there is both a poloidal and toroidal magnetic field and zero pressure. The toroidal field also has a vacuum $1/R$ dependence. The poloidal magnetic field has a more complicated behavior arising from the non-uniform distribution of surface currents plus external sources, for example vertical field coils. Therefore, in the vacuum the equilibrium quantities can be written as

$$\hat{\mathbf{B}} = B_0 \frac{R_0}{R} \mathbf{e}_\phi + \mathbf{B}_p(r,\theta)$$

$$p = 0$$

(12.126)

where B_0 is the applied toroidal field at $R = R_0$.

The complete poloidal field requires a complicated calculation. Fortunately, for the surface current stability analysis only the poloidal field on the surface is required. This quantity can be easily determined from the pressure balance jump condition across the surface: $2\mu_0 p + B^2 = \hat{B}^2$. Specifically, for a circular plasma the poloidal field on the surface has the form $\mathbf{B}_p(a, \theta) = B_\theta(\theta)\mathbf{e}_\theta$ and satisfies

$$2\mu_0 p + B_i^2 \frac{R_0^2}{R^2} = B_0^2 \frac{R_0^2}{R^2} + B_\theta^2$$

(12.127)

Here, $R(\theta) = R_0(1 + \varepsilon \cos \theta)$ and as usual $\varepsilon = a/R_0$.

In its present form Eq. (12.127) is exact – no aspect ratio expansion has been assumed. The next step is to assume that $\varepsilon \ll 1$ and to introduce the high β tokamak ordering. The appropriate expansion is given by

$$\frac{2\mu_0 p}{B_0^2} \equiv \beta \sim \varepsilon$$

$$\frac{B_\theta^2}{B_0^2} \sim \varepsilon^2$$

$$\frac{B_i^2}{B_0^2} \equiv 1 - \beta + b_2, \quad b_2 \sim \varepsilon^2$$

(12.128)

The quantity b_2 is a new constant replacing B_i and must be chosen as second order in ε to be consistent with the expansion. The first non-trivial contribution to the pressure balance relation occurs in second order and yields a simplified expression for $B_\theta^2(\theta)$,

$$\left(\frac{B_\theta}{\varepsilon B_0}\right)^2 = \frac{b_2}{\varepsilon^2} + 2\frac{\beta}{\varepsilon}\cos\theta = \frac{4\beta}{\varepsilon k^2}\left(1 - k^2\sin^2\frac{\theta}{2}\right) \tag{12.129}$$

where $k^2 = 4\varepsilon\beta/(b_2 + 2\varepsilon\beta)$ is yet another new constant replacing b_2 and ordered as $k^2 \sim 1$. In physical terms the value of k^2 can be related to the plasma current and pressure by means of the familiar definition

$$q_* = \frac{2\pi a^2 B_0}{\mu_0 R_0 I} = \varepsilon B_0\left(\frac{1}{2\pi}\int_0^{2\pi} B_\theta \, d\theta\right)^{-1} \tag{12.130}$$

The integral can be evaluated in terms of elliptic integrals yielding a transcendental relation for k given by

$$\left[\frac{k}{E(k)}\right]^2 = \frac{16}{\pi^2}\frac{\beta q_*^2}{\varepsilon} \tag{12.131}$$

Observe that similar to the diffuse profile case there is an equilibrium β limit which, from Eq. (12.129), corresponds to $k^2 = 1$. A higher value of k^2 leads to negative B_θ^2, which is unphysical. Since $E(1) = 1$ one sees from Eq. (12.131) that the equilibrium limit, which corresponds to the separatrix moving onto the plasma surface at $\theta = \pi$, can be written as

$$\frac{\beta q_*^2}{\varepsilon} \leq \frac{\pi^2}{16} \approx 0.62 \tag{12.132}$$

The required equilibrium information has now been derived, enabling the stability analysis to proceed.

Evaluation of δW_S

The stability analysis begins with the evaluation of the surface energy δW_S whose exact form is given by Eq. (8.101)

$$\delta W_S = \frac{1}{2\mu_0}\int_S |\mathbf{n}\cdot\boldsymbol{\xi}_\perp|^2\mathbf{n}\cdot\left[\!\left[\nabla\left(\mu_0 p + \frac{B^2}{2}\right)\right]\!\right] dS \tag{12.133}$$

The integrand can be cast in a more convenient form for evaluation by using the general equilibrium relation

$$\nabla\left(\mu_0 p + \frac{B^2}{2}\right) = \mathbf{B}\cdot\nabla\mathbf{B} = B^2\boldsymbol{\kappa} + \mathbf{b}(\mathbf{B}\cdot\nabla B) \tag{12.134}$$

Then, a short calculation that makes use of the surface current pressure balance relation, allows one to rewrite the surface energy in terms of the plasma and vacuum curvature vectors, $\boldsymbol{\kappa}$ and $\hat{\boldsymbol{\kappa}}$ respectively,

$$\delta W_S = \frac{1}{2\mu_0} \int_S |\mathbf{n} \cdot \boldsymbol{\xi}_\perp|^2 \left[2\mu_0 p \left(\frac{\hat{\kappa}_n + \kappa_n}{2} \right) + \left(\frac{\hat{B}^2 + B^2}{2} \right) (\hat{\kappa}_n - \kappa_n) \right] dS \quad (12.135)$$

where $\kappa_n = \mathbf{n} \cdot \boldsymbol{\kappa}$ is the normal component of the curvature.

At present this form of δW_S is exact. Note that there are two contributions. The first involves the plasma pressure multiplied by the average curvature across the surface. This is ballooning mode drive for instabilities. The second contribution involves the average magnetic pressure multiplied by the jump in curvature across the surface. The jump in curvature is proportional to the parallel current flowing on the surface. Hence, this is the kink mode drive for instabilities. Both drives are clearly and explicitly exhibited in the surface current model, one of its advantages.

The expression for δW_S can be further simplified by introducing the high β tokamak expansion into the curvature coefficients. The curvatures need to be evaluated to first and second order. A straightforward calculation yields

$$\kappa_n = -\frac{\cos \theta}{R_0} (1 - \varepsilon \cos \theta)$$

$$\hat{\kappa}_n = -\frac{\cos \theta}{R_0} (1 - \varepsilon \cos \theta) - \frac{B_\theta^2}{a B_0^2} \quad (12.136)$$

The surface energy reduces to

$$\delta W_S = -\frac{\pi a^2 B_0^2}{\mu_0 R_0} \int_0^{2\pi} |\mathbf{n} \cdot \boldsymbol{\xi}_\perp|^2 \left[\frac{\beta}{\varepsilon} \cos \theta + \left(\frac{B_\theta}{\varepsilon B_0} \right)^2 \right] d\theta \quad (12.137)$$

As expected the pressure gradient term is destabilizing on the outside of the cross section ($\theta = 0$) and stabilizing on the inside ($\theta = \pi$). The kink term is destabilizing around the entire cross section.

To complete the evaluation of δW_S one must specify $\mathbf{n} \cdot \boldsymbol{\xi}_\perp = \xi(\theta)$ on the plasma surface. A general form can be written in terms of a Fourier series as follows:

$$\xi(\theta) = \sum_{-\infty}^{\infty} \xi_m e^{im\theta} \quad (12.138)$$

With the advantage of hindsight from numerical calculations it is possible to choose a reduced form of Eq. (12.138) that contains only three harmonics, $m = 1, 2, 3$, but provides an accurate approximation to the infinite harmonic solution. The $m = 1$ harmonic ξ_1 is shown to be dominant at low β/ε with the ξ_2, ξ_3 harmonics playing a negligible role. At finite β/ε the $m = 2$ harmonic ξ_2 is shown to be dominant although the ξ_1, ξ_3 sideband harmonics are also finite and are critical to allow the perturbation to balloon in the unfavorable curvature region. The reduced form of ξ is given by

$$\xi(\theta) = \xi_1 e^{i\theta} + \xi_2 e^{2i\theta} + \xi_3 e^{3i\theta} \qquad (12.139)$$

where the relative sizes of the ξ_j are determined at the end of the analysis by minimizing δW.

Equation (12.139) is substituted into Eq. (12.137) leading to the desired expression for δW_S in terms of the ξ_j,

$$\frac{\delta W_S}{W_0} \equiv \delta \hat{W}_S = -\frac{2\beta}{\varepsilon k^2} \left[2(2 - k^2)(\xi_1^2 + \xi_2^2 + \xi_3^2) + 3k^2\xi_2(\xi_1 + \xi_3) \right]$$

$$= -\frac{2}{q_*^2} \left[\frac{\pi}{4E(k)} \right]^2 \left[2(2 - k^2)(\xi_1^2 + \xi_2^2 + \xi_3^2) + 3k^2\xi_2(\xi_1 + \xi_3) \right]$$

$$(12.140)$$

with $W_0 = \pi^2 a^2 B_0^2 / \mu_0 R_0$. The second form is obtained by expressing β/ε in terms of k from Eq. (12.131). Note that the uncoupled quadratic terms correspond to the kink drive while the cross terms represent the ballooning drive.

Evaluation of δW_F

The evaluation of δW_F is obtained from the high β tokamak form given by Eq. (12.123). For the surface current model where $p = $ constant and $J_\parallel = 0$ this form reduces to

$$\delta W_F(U^*, U) = \frac{1}{2\mu_0} \int_P |\mathbf{Q}_p|^2 d\mathbf{r} \qquad (12.141)$$

The expression for $\mathbf{Q}_p$ in Eq. (12.124) is also greatly simplified,

$$\mathbf{Q}_p = -\frac{inB_0}{R_0} \mathbf{e}_\phi \times \nabla U \qquad (12.142)$$

Substituting $\mathbf{Q}_p$ into δW_F yields

$$\delta W_F = \frac{\pi B_0^2 n^2}{\mu_0 R_0} \int_P \left[\left| \frac{\partial U}{\partial r} \right|^2 + \left| \frac{1}{r} \frac{\partial U}{\partial \theta} \right|^2 \right] r\, dr\, d\theta \qquad (12.143)$$

Now, the function $U(r, \theta)$ that is regular at the origin and which minimizes the integral can be written as a general Fourier series

$$U(r,\theta) = \sum_{-\infty}^{\infty} U_m r^{|m|} e^{im\theta} \tag{12.144}$$

The coefficients U_m are chosen to match the trial function on the surface; that is, one chooses the U_m so that

$$\frac{1}{a}\frac{\partial U(a,\theta)}{\partial \theta} = \xi(\theta) = \xi_1 e^{i\theta} + \xi_2 e^{2i\theta} + \xi_3 e^{3i\theta} \tag{12.145}$$

Only three terms are needed for the matching, U_1, U_2, U_3. A short calculation shows that

$$U(r,\theta) = -ia\left(\xi_1\rho\, e^{i\theta} + \frac{\xi_2}{2}\rho^2 e^{2i\theta} + \frac{\xi_3}{3}\rho^3 e^{3i\theta}\right) \tag{12.146}$$

where $\rho = r/a$.

Equation (12.146) is substituted into Eq. (12.143) yielding an expression for δW_F,

$$\frac{\delta W_F}{W_0} \equiv \delta\hat{W}_F = \frac{n^2}{3}\left(6\xi_1^2 + 3\xi_2^2 + 2\xi_3^2\right) \tag{12.147}$$

The simplicity of evaluating δW_F is the primary motivation for using the surface current model.

Evaluation of δW_V

The final term to evaluate is the vacuum energy given by

$$\delta W_V = \frac{1}{2\mu_0}\int_V |\,\hat{\mathbf{B}}_1| d\mathbf{r} \tag{12.148}$$

This is the most complicated term for the following reason. Even though only three harmonics (ξ_1, ξ_2, ξ_3) are needed to define the surface perturbation, each of these couples to an infinite number of magnetic field harmonics in the vacuum because of the non-trivial θ dependence of $B_\theta(\theta)$. Still, with a little bit of work, δW_V can be analytically evaluated for the two most interesting limits, β/ε approaching zero and the equilibrium limit.

The analysis starts by noting that in the vacuum region $\hat{\mathbf{B}}_1 = \nabla V$ with V satisfying $\nabla^2 V \approx \nabla_p^2 V = 0$. The general solution for V corresponding to a wall at infinity can be written as

$$V = V(\rho,\theta)e^{-in\phi} = e^{-in\phi}\sum_m V_m \rho^{-|m|} e^{im\theta} \tag{12.149}$$

where $\rho = r/a$ and it should be understood that the summation includes all m in the range $-\infty < m < \infty$ except $m = 0$. If one now notes that $|\mathbf{B}_1|^2 = |\nabla V|^2 = \nabla \cdot (V^* \nabla V)$ then a simple application of the divergence theorem leads to an expression for δW_V that has the form of a surface integral,

$$\delta W_V = -\frac{1}{2\mu_0} \int_S V^* \frac{\partial V}{\partial r} \, dS = \frac{2\pi^2 R_0}{\mu_0} \sum |m| |V_m|^2 \tag{12.150}$$

The next task is to relate the V_m to the ξ_m, which is accomplished by applying the boundary condition on the plasma surface $\mathbf{n} \cdot \hat{\mathbf{B}}_1(S) = \mathbf{n} \cdot \nabla \times (\boldsymbol{\xi}_\perp \times \hat{\mathbf{B}})$. A short calculation shows that this condition reduces to

$$\frac{\partial V}{\partial \rho}\bigg|_{\rho=1} = -in\varepsilon B_0 \xi + \frac{\partial}{\partial \theta}(B_\theta \xi) \tag{12.151}$$

Straightforward Fourier analysis then leads to an expression for the V_m,

$$V_m = \frac{i\varepsilon B_0}{|m|} \sum_l G_{ml}\xi_l$$

$$G_{ml} = \frac{1}{2\pi} \int_0^{2\pi} \left(-n + \frac{mB_\theta}{\varepsilon B_0}\right) \cos\left[(l-m)\theta\right] d\theta \tag{12.152}$$

For the displacement under consideration l has the values 1,2,3. Even so, because of the θ dependence of $B_\theta(\theta)$ the V_m will be non-zero over the infinite range of m. This represents the coupling of each harmonic of ξ to an infinite number of harmonics in V that was previously mentioned.

The vacuum energy is now evaluated by substituting Eq. (12.152) into Eq. (12.150),

$$\frac{\delta W_V}{W_0} \equiv \delta \hat{W}_V = 2 \sum_{m,l,p} \frac{G_{ml}G_{mp}}{|m|} \xi_l \xi_p \tag{12.153}$$

The coefficients G_{ml} can be evaluated analytically but lead to complicated combinations of elliptic integrals. A simpler approach is to consider the two limiting cases of low and high β/ε corresponding to $k^2 = 0$ and $k^2 = 1$. For these cases simpler analytic expressions can be found resulting in explicit stability boundaries on q_* and β/ε. These calculations are described below.

Stability analysis

The goal of the stability analysis is to determine the marginal stability boundaries resulting from external ballooning-kink modes in the form of a β/ε vs. $1/q_*$ diagram. This important goal is accomplished by evaluating $\delta W = \delta W_F + \delta W_S + \delta W_V$

by summing the three separate contributions just evaluated. As stated above, relatively simple analytic answers are obtained for the two limiting cases $k^2 = 0$ and $k^2 = 1$. Intermediate cases require a minor numerical calculation and these results are discussed as well.

To begin, consider the case of $\beta/\varepsilon \to 0$ corresponding to $k^2 = 0$. This case is easy to analyze because the magnetic field matrix becomes diagonal. Specifically, for $k^2 = 0$ one finds that $B_\theta/\varepsilon B_0 = 1/q_*$ implying that

$$G_{ml} = \left(-n + \frac{m}{q_*}\right)\delta_{l-m} \tag{12.154}$$

where δ_{l-m} is the Kronecker delta function. Using this relation it follows that the vacuum energy reduces to

$$\delta \hat{W}_V = 2\left[\left(1 - \frac{1}{q_*}\right)^2 \xi_1^2 + \left(1 - \frac{2}{q_*}\right)^2 \frac{\xi_2^2}{2} + \left(1 - \frac{3}{q_*}\right)^2 \frac{\xi_3^2}{3}\right] \tag{12.155}$$

Here, the toroidal mode number has been set to $n = 1$ which is the most unstable case. The complete δW can now be evaluated by summing the separate contributions. A short calculation yields

$$\delta \hat{W} = \frac{4}{q_*}(q_* - 1)\xi_1^2 + \frac{2}{q_*^2}(q_* - 2)^2 \xi_2^2 + \frac{1}{3q_*^2}\left[(2q_* - 3)^2 + 3\right]\xi_3^2 \tag{12.156}$$

Observe that there are no cross terms; that is, at low β/ε there is no toroidal coupling and each mode is independent. Stated differently, in the limit of low β/ε the stability result reduces to that of the straight cylinder. Next, Eq. (12.156) shows that the coefficients of the $m = 2$ and $m = 3$ are both positive. Therefore, the most unstable perturbation corresponds to $\xi_2 = \xi_3 = 0$. The only mode that can become unstable is $m = 1$. Stabilization of this mode requires

$$q_* > 1 \tag{12.157}$$

which is just the Kruskal–Shafranov condition. This is the only purely current-driven kink mode that can be unstable in the surface current model.

The second limit of interest corresponds to $k^2 = 1$. This value corresponds to the β/ε equilibrium limit at which $\beta/\varepsilon = (\pi^2/16)(1/q_*^2)$. To calculate the maximum stable β/ε one must determine the smallest value of q_* that keeps δW positive. The procedure to do this is to write $\delta W = \delta W(q_*, \xi_1, \xi_2, \xi_3)$ and then minimize over the relative amplitudes of the harmonics. The resulting δW is then only a function of q_*: $\delta W = \delta W(q_*)$. A simple plot of this function yields the lowest stable q_* and the corresponding maximum stable value of β/ε.

The main mathematical advantage of focusing on $k^2 = 1$ is that it leads to a simple expression for the coefficient G_{ml}, a consequence of the fact that $B_\theta/\varepsilon B_0$ reduces to

$$\frac{B_\theta}{\varepsilon B_0} = \frac{\pi}{2q_*}\left|\cos\frac{\theta}{2}\right| \tag{12.158}$$

From this relation it follows that

$$G_{ml} = -n\delta_{m-l} + \frac{(-1)^{m-l+1}}{q_*}\frac{m}{4(m-l)^2 - 1} \tag{12.159}$$

The next step is to evaluate δW at $k^2 = 1$. The result is an expression of the form

$$\delta \hat{W} = W_{11}\xi_1^2 + W_{22}\xi_2^2 + W_{33}\xi_3^2 + 2W_{12}\xi_1\xi_2 + 2W_{13}\xi_1\xi_3 + 2W_{23}\xi_2\xi_3 \tag{12.160}$$

where use has been made of the symmetry relations $W_{lp} = W_{pl}$. Also each matrix element depends only on q_*: $W_{lp} = W_{lp}(q_*)$. The matrix elements are given shortly.

For the moment focus on the form of δW. Choose one of the amplitudes, for instance ξ_2, as the primary harmonic. The amplitudes ξ_1 and ξ_3 can be expressed in terms of ξ_2 by minimizing δW with respect to ξ_1, ξ_3. A simple calculation leads to two simultaneous equations for ξ_1 and ξ_3,

$$\begin{aligned}
W_{11}\xi_1 + W_{13}\xi_3 &= -W_{12}\xi_2 \\
W_{13}\xi_1 + W_{33}\xi_3 &= -W_{23}\xi_2
\end{aligned} \tag{12.161}$$

whose solution is given by

$$\begin{aligned}
\xi_1 &= \frac{W_{23}W_{13} - W_{12}W_{33}}{W_{11}W_{33} - W_{13}^2}\xi_2 \\
\xi_3 &= \frac{W_{12}W_{13} - W_{23}W_{11}}{W_{11}W_{33} - W_{13}^2}\xi_2
\end{aligned} \tag{12.162}$$

These expressions are substituted in δW. After a short calculation one obtains

$$\delta \hat{W} = W(q_*)\xi_2^2 = \left(W_{22} - \frac{W_{11}W_{23}^2 + W_{33}W_{12}^2 - 2W_{12}W_{13}W_{23}}{W_{11}W_{33} - W_{13}^2}\right)\xi_2^2 \tag{12.163}$$

To test for stability one just has to substitute the expressions for the $W_{lp}(q_*)$ into Eq. (12.163) and plot the resulting function $W(q_*)$ to see when it is positive or negative. The explicit expressions for the $W_{lp}(q_*)$ are obtained by combining the separate contributions to $\delta \hat{W}$. The results, again for $n = 1$, are given by

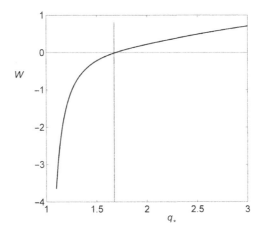

Figure 12.6   Curve of W vs. q_* at the equilibrium β limit.

$$W_{11} = 4 - \frac{4}{q_*} - \left(\frac{\pi^2}{4} - 2M_{11}\right)\frac{1}{q_*^2} = 4 - \frac{4}{q_*} + \frac{0.033}{q_*^2}$$

$$W_{22} = 2 - \frac{4}{q_*} - \left(\frac{\pi^2}{4} - 2M_{22}\right)\frac{1}{q_*^2} = 2 - \frac{4}{q_*} + \frac{2.48}{q_*^2}$$

$$W_{33} = \frac{4}{3} - \frac{4}{q_*} - \left(\frac{\pi^2}{4} - 2M_{33}\right)\frac{1}{q_*^2} = \frac{4}{3} - \frac{4}{q_*} + \frac{4.94}{q_*^2}$$

$$W_{12} = -\frac{4}{3q_*} - \left(\frac{3\pi^2}{16} + 2M_{12}\right)\frac{1}{q_*^2} = -\frac{4}{3q_*} - \frac{0.017}{q_*^2}$$

(12.164)

$$W_{23} = -\frac{4}{3q_*} - \left(\frac{3\pi^2}{16} + 2M_{23}\right)\frac{1}{q_*^2} = -\frac{4}{3q_*} + \frac{1.23}{q_*^2}$$

$$W_{13} = \frac{4}{15q_*} + \frac{2M_{13}}{q_*^2} = \frac{4}{15q_*} + \frac{0.011}{q_*^2}$$

where the M_{lp} are pure numerical coefficients defined by

$$M_{lp} = \sum_{m=-\infty}^{\infty} \frac{|m|}{[4(m-l)^2 - 1][4(m-p)^2 - 1]} \tag{12.165}$$

The coefficients can be evaluated either numerically or, with some effort, analytically. Their values are $M_{11} = 5/4$, $M_{22} = 89/36$, $M_{33} = 1111/300$, $M_{12} = -11/12$, $M_{23} = -277/180$, and $M_{13} = 1/180$. These values have been substituted into Eq. (12.164) to give the simpler numerical forms on the far right.

   The resulting expression for $W(q_*)$ is plotted vs. q_* in Fig. 12.6. As expected the plasma is unstable [$W(q_*) < 0$] for low q_* and stable [$W(q_*) > 0$] for high q_*. The

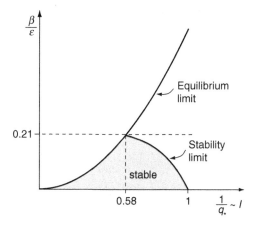

Figure 12.7   Stability diagram for external modes in a high β tokamak described by the surface current model.

transition occurs for $q_* = 1.69$. A more accurate numerical calculation keeping 20 harmonics yields $q_* = 1.71$ and this more accurate value is used hereafter.

The basic conclusion is that the optimum q_* and corresponding highest β/ε that are stable against ballooning-kink modes in a high β tokamak with a circular cross section are given by

$$q_* = q_{crit} = 1.71$$

$$\frac{\beta}{\varepsilon} = \frac{\beta_{crit}}{\varepsilon} = \left(\frac{\pi}{4q_{crit}}\right)^2 = 0.21 \tag{12.166}$$

There are several additional points worth noting. The scaling $\beta_{crit} \sim a/R_0$ suggests that a tight aspect ratio is desirable for favorable MHD stability. Even so one must keep in mind that fusion really requires high p rather than high $\beta \sim p/B_0^2$ although the two are related. For tight aspect ratio tokamaks the strong $1/R$ dependence of the toroidal field reduces B_0 at $R = R_0$ as compared to larger aspect ratio devices assuming the same maximum field on the toroidal field coil. Therefore, while β may be higher in a tight aspect ratio tokamak, the actual pressure may be smaller.

A second point is that the optimized value of stable q_* has been raised from $q_* = 1$ at $\beta/\varepsilon = 0$ to $q_* = 1.71$ at $\beta/\varepsilon = 0.21$. The maximum allowable plasma current has decreased by nearly a factor of 2. Lastly, by comparing with detailed numerical studies using realistic profiles, one finds that the surface current model predicts the correct scaling of $\beta_{crit} \sim \varepsilon$, although the numerical coefficient 0.21 is optimistic. The numerical results are described in the next subsection.

The overall picture of surface current model stability can be summarized by plotting the marginally stable β/ε vs. $1/q_*$ (i.e., essentially p vs. I) as illustrated in Fig. 12.7. Shown here are the equilibrium limit and the marginal stability boundary

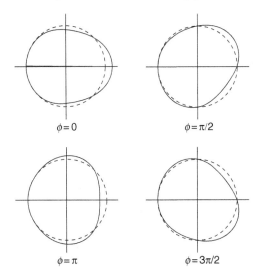

$\phi = 0$ $\phi = \pi/2$

$\phi = \pi$ $\phi = 3\pi/2$

Figure 12.8 Marginally stable eigenfunction for the surface current model at four different ϕ locations. The major axis is to the left. The dashed and solid curves represent the unperturbed and perturbed surfaces respectively.

for arbitrary k^2 obtained using 20 harmonics and numerically evaluating the matrix coefficients. The analytic stability results correspond to the end points on the stability curve. Observe that there is a stable operating regime without a conducting wall for sufficiently low p and sufficiently low I.

The final point of interest concerns the shape of the eigenfunction. One wants to learn whether or not the most unstable mode has the form of a ballooning-kink perturbation of the plasma surface. This is easily demonstrated by substituting $q_* = 1.71$ into Eq. (12.162), yielding

$$\xi_1/\xi_2 = 0.429$$
$$\xi_3/\xi_2 = 0.427$$
(12.167)

Equation (12.167) shows that at the high β/ε limit, the perturbation is dominated by $m = 2$ but that there are substantial $m = 1$, $m = 3$ sidebands. The ballooning-kink nature of the perturbation is shown graphically in Fig. 12.8, where the perturbed surface $r = a + \xi_1 \cos(\theta - \phi) + \xi_2 \cos(2\theta - \phi) + \xi_3 \cos(3\theta - \phi)$ overlays the unperturbed surface $r = a$. The figure shows the poloidal cross section at four different toroidal angles assuming $\xi_2/a = 0.09$. One sees that the surface distortion rotates with the toroidal angle, demonstrating the kink component of the mode. Equally importantly, the perturbation is always much larger on the outside of the torus than the inside: $r(\theta = 0) - a = 1.86\xi_2 \cos \phi$ while $r(\theta = \pi) - a = 0.14\xi_2$ $\cos \phi$. This clearly demonstrates that the perturbation balloons in the region of unfavorable curvature.

This completes the analysis of ballooning-kink instabilities in a high β tokamak as described by the surface current model. It is now of interest to discuss more realistic results using diffuse profiles as obtained from detailed numerical studies.

12.8.4 Numerical studies of ballooning-kink instabilities

Many sophisticated numerical codes have been written that examine the MHD stability of tokamaks against a variety of modes. These codes are "exact" in that they treat finite aspect ratio, finite β, finite shaping, as well as arbitrary diffuse profiles. Several excellent textbooks (e.g., Jardin, 2010; Goedbloed *et al.*, 2010) have been written that describe the numerical aspects of these codes as well as presenting results. The discussion below focuses primarily on the results described in these books as well as many related scientific papers.

In terms of an overview it is accurate to say that standard tokamaks with an increasing $q(\psi)$ profile are dominated by either $n = \infty$ ballooning modes or $n = 1$ ballooning-kink modes. Intermediate n modes lie between these two limiting cases. Invariably, if no conducting wall is present, then the $n = 1$ mode sets the most stringent limit on β. When the wall is very close, then the $n = 1$ mode is wall stabilized. In this case the $n = \infty$ mode, which is unaffected by the wall because of its highly localized eigenfunction, becomes the more important instability. It is also worth pointing out that with hollow $q(\psi)$ profiles the situation can change. Intermediate n modes, known as "infernal modes" may then set the strictest limits on β (Manickam *et al.*, 1987). Such hollow profiles can occur naturally in the "advanced tokamak" mode of operation which is discussed in the next subsection. For the present discussion, however, attention is focused on standard tokamaks with monotonically increasing $q(\psi)$ profiles.

To begin note that the numerical codes, in addition to treating all relevant parameters as finite, also are used to optimize over profile shape to determine the maximum stable β for a given geometry. Results for such an optimization against $n = \infty$ ballooning modes have already been discussed in Section 12.2.6, the conclusion being the Sykes limit given by

$$\beta_t = \beta_N \frac{I}{aB_0} \qquad \beta_N = 0.044 = 4.4\% \qquad (12.168)$$

where the units are I (MA), B_0 (T), and a (m).

A similar but more comprehensive stability study has been carried out by Troyon and co-workers (Troyon *et al.*, 1984). These studies determine the optimized profiles for stability against the Mercier criterion, ballooning modes, internal modes, and external ballooning-kink modes in a tokamak without a conducting wall over a variety of cross sectional shapes. As stated above, the external ballooning-kink mode is usually the most difficult to stabilize. Like the Sykes limit the result of these studies is remarkably simple. The critical β for stability is known as the "Troyon limit" and can be written as

$$\beta_t = \beta_N \frac{I}{aB_0} \qquad \beta_N = 0.028 = 2.8\% \qquad (12.169)$$

Typically the most stable systems have elongated, outward pointing D-shapes. Also the optimized profiles tend to have peaked current distributions.

The numerical results also show that non-optimized profile effects can be incorporated into the stability criterion by considering β_N to be a function of a current peaking parameter. Specifically, for typical inverse aspect ratios on order of $\varepsilon \sim 1/3$ the numerical studies show that

$$\beta_N \approx 4\, l_i \qquad (12.170)$$

where l_i is the internal inductance per unit length defined by Eq. (6.85). More peaked profiles have larger l_i than broad or hollow profiles.

The most striking feature of the (essentially ballooning-kink) Troyon β limit given by Eq. (12.169) is that it is identical in form to the Sykes limit, except that the numerical value of β_N is slightly reduced from 0.044 to 0.028. There are, however, two additional points to consider. First, the numerical studies are carried out assuming that $q_0 \approx 1$ to avoid internal sawtooth modes. Thus, the expected inverse quadratic scaling of $\beta \propto 1/q^2$ arising from the basic physics probably should be included by the replacement $I/aB_0 \rightarrow I/q_0 aB_0 \propto 1/q_0 q_*$.

More importantly, the pure ballooning mode yields a criterion on the ratio p/I for a fixed geometry which in principle allows high p by operating at high I. Ballooning modes by themselves do not directly limit the size of I. In contrast, the Troyon ballooning-kink stability criterion only applies for $q_a > 2$. This fact is known by most fusion researchers but is often hidden in applications of the Troyon limit. The requirement $q_a > 2$ is frequently introduced as a separate independent criterion but is actually directly coupled to the Troyon limit.

The issue with the current limit has been clarified and sharpened by more recent numerical studies by Menard *et al.* (2004). These authors basically extended the studies of Troyon *et al.* (1984) to include very tight aspect ratios because of their interest in spherical tokamaks. They found that by introducing an appropriately defined β the Troyon limit remains valid even for very tight aspect ratios. In the generalized definition the toroidal β is redefined as follows:

$$\beta_t = \frac{2\mu_0 \langle p \rangle}{B_0^2} \qquad \rightarrow \qquad \frac{2\mu_0 \langle p \rangle}{\langle B^2 \rangle} \qquad (12.171)$$

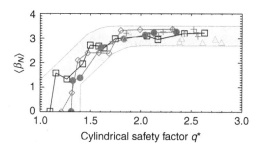

Figure 12.9   Approximately universal curve of $\beta_N = \beta_N(q_*)$ obtained by Menard *et al.* (2002). Reproduced with permission.

Here, $\langle B^2 \rangle$ is the volume average of the total magnetic energy. Interestingly, this definition was originally suggested by Troyon but the definition with B_0^2 has come into wide use because of its simplicity. The two definitions nearly coincide at large aspect ratio but are noticeably different at tight aspect ratio. An equally important contribution of Menard *et al.* is that they calculated the stability limit directly as a function of q_* after carefully defining this quantity in terms of the current. Their definition coincides with the one used in the text, namely

$$q_* = \frac{2\pi a^2 B_0}{\mu_0 R_0 I}\left(\frac{1+\kappa^2}{2}\right) \tag{12.172}$$

Using their generalized definitions one finds that the results of Menard *et al.* can be summarized in the form of an approximately universal β limit, valid for arbitrary aspect ratio. The limit has the form

$$\beta_t = 5\beta_N\left(\frac{1+\kappa^2}{2}\right)\frac{\varepsilon}{q_*} \tag{12.173}$$

where $\beta_N = \beta_N(q_*)$ is illustrated in Fig. 12.9.

Observe that for $q_* > 2$, $\beta_N \approx 0.03 = 3\% = $ constant. In this regime the β_t limit scales as $\beta_t \propto 1/q_* \propto I$ corresponding to the original Troyon result. In contrast, for $q_* \to 1.2$ then $\beta_t \to 0$, indicating the degradation due to too large a current. Now, the numerical data in Fig. 12.9 can be approximated by the following simple formula (in decimal units):

$$\beta_N \approx 0.03\left(\frac{q_*^\nu - 1.2^\nu}{q_*^\nu}\right)^{1/\nu} \tag{12.174}$$

with $\nu = 3$. Using this approximate relation one can see that β_t in Eq. (12.173) has a maximum as a function of q_*. The optimum can easily be calculated leading to a critical q_* and corresponding β_t given by

$$q_* = 1.51$$

$$\frac{\beta_t}{\varepsilon} = 0.079\left(\frac{1+\kappa^2}{2}\right) \tag{12.175}$$

Equation (12.175) with $\kappa = 1$ can be directly compared with the results of the surface current model. The critical q_* values are similar: $q_* = 1.51$ from the Menard results and $q_* = 1.71$ from the surface current model. The maximum β_t values show a wider discrepancy: $\beta_t/\varepsilon = 0.079$ from the Menard relation while $\beta_t/\varepsilon = 0.21$ for the surface current model. The surface current profiles, perhaps unsurprisingly, lead to more optimistic β_t limits than those obtained from realistic diffuse profiles by a factor of about 2.7. The reason is that in addition to external ballooning-kink modes, the diffuse model must also take into account low n internal modes and ballooning modes which are not present in the surface current model.

As a final numerical example consider a tokamak with $\kappa = 1.8$ and $\varepsilon = 0.4$. This yields a no-wall maximum $\beta_t = 0.067 = 6.7\%$. At a toroidal field $B_0 = 6\,\mathrm{T}$ this translates into a total plasma pressure $p = 9.6$ atm, which should be sufficient for a fusion reactor typically requiring $p \sim 5-8$ atm. However, the safety margin is not as large as one might like since there is no guarantee that the optimum profiles can be achieved experimentally. For example, one might need to operate closer to $q_* \approx 2$ to avoid disruptions. Also, the need for a large bootstrap fraction puts further constraints on the profiles. Still, do not lose sight of the forest for the trees.

A tokamak is capable of achieving complete stability against all $n\geq1$ ideal MHD modes without the need of a conducting wall at values of β_t and pressure sufficient for a reactor.

Standard tokamak experiments have achieved β_t values in the range of 5%–10% although at lower toroidal fields than in a reactor (see for instance Lazarus *et al.*, 1991; Strait, 2005). Thus, the actual pressures are lower than ultimately needed. Spherical tokamaks have achieved much higher β_t values, on the order of 30% (see for instance Menard *et al.*, 2002). Here too this performance is achieved at a low toroidal field so the absolute pressures are still below what is needed in a reactor. However, from the point of view of understanding plasma science the MHD model has proven to be a reliable guide to the maximum achievable values of β_t thereby representing a good collaborative theory-experimental success story.

12.9 Stability of tokamaks – advanced tokamak (AT) operation

12.9.1 Introduction

The advanced tokamak (AT) actually refers to a mode of operation of standard tokamaks. It is accessible to some but not all experiments. The AT has the goal of solving one of the basic problems facing the tokamak concept in its quest to become a fusion reactor – the need for steady state operation. To accomplish this a method is needed to non-inductively drive a steady state toroidal current. While RF methods have been known for many years and have demonstrated that they can indeed non-inductively drive current, they are not very efficient. Too many watts (typically 10–20 watts absorbed power) are required to drive one ampere of current. This inefficiency leads to poor reactor power balance. On the order of 300–500 MWe would be required to drive all the current in a 1 GWe tokamak reactor.

A possible resolution to this basic difficulty is to take advantage of the bootstrap current, a naturally driven neoclassical transport current that is automatically generated for "free" in a tokamak. No external drives are required. Reactor studies show that it is necessary for about 75–90% of the total toroidal current to consist of the bootstrap current. Only the remaining 10–25% must be produced by RF current drive, leading to greatly improved power balance.

The purpose of this subsection is to provide some understanding of the requirements and constraints involved in generating 75% bootstrap current. To do this one has to understand the profile shape and magnitude of the bootstrap current. This is accomplished by combining some toroidal and cylindrical tokamak physics.

The overall picture is as follows. Bootstrap current profiles are hollow, not monotonically decreasing as in a standard tokamak. The magnitude of the current is proportional to the pressure. Therefore, high bootstrap current requires high pressure and one must test whether the required pressure exceeds the Troyon limit. Invariably the no-wall Troyon limit is violated, suggesting that wall stabilization is required, which in practice translates into resistive wall mode stabilization. In addition, the profiles are sufficiently different that a new class of ideal MHD modes, known by the tongue-in-cheek name of "infernal modes," can be excited and these too must be stabilized.

Experimentally, the achievement of high bootstrap current requires heating a plasma to sufficiently high temperatures so that the low collisionality bootstrap current physics is applicable. In addition, this must be accomplished at high density so the magnitude of the bootstrap current is large. High bootstrap fractions have been obtained transiently by current programming and RF profile control, but maintaining these high fractions is still far from routine. Additionally, once the profiles can be maintained there will likely be the extra complication of resistive wall stabilization. High-performance AT operation currently remains an important area of tokamak research.

To demonstrate the physics of AT operation, two simple calculations are presented in this section. First, a simplified formula for the bootstrap current obtained from neoclassical transport theory is used to determine the current profiles and amplitude. After some simple analysis the amplitude can be expressed in terms of β_N whose value is determined by requiring the bootstrap fraction f_B to equal 75%. The resulting β_N is shown to exceed the Troyon limit. This motivates the second calculation whose goal is to show how close to place a perfectly conducting wall in order to stabilize the higher value of β_N. This task is carried out using the surface current model.

12.9.2 Bootstrap current profile – the β_N limit

To begin, assume that a high-density plasma has been heated to a sufficient temperature to generate a substantial bootstrap current. Neoclassical transport theory shows that in a large aspect ratio, circular cross section tokamak, the bootstrap current generated has a profile and magnitude given by (see for instance Helander and Sigmar, 2002)

$$J_B(r) = -\frac{1}{B_\theta}\left(\frac{r}{R_0}\right)^{1/2}\left[2.44(T_e + T_i)\frac{dn}{dr} + 0.69\,n\frac{dT_e}{dr} - 0.42\,n\frac{dT_i}{dr}\right] \quad (12.176)$$

This expression can be simplified by assuming plausible profiles for the temperatures and density. A reasonable choice is $T_e = T_i \equiv T$ with a density profile that is flatter than the temperature, which is typical of H-mode tokamaks. The specific profiles chosen are given by

$$\begin{aligned} T(\rho) &= T_0(1 - \rho^2)^{4/3} \\ n(\rho) &= n_0(1 - \rho^2)^{2/3} \\ p(\rho) &= 2n_0T_0(1 - \rho^2)^2 = 3\bar{p}(1 - \rho^2)^2 \end{aligned} \quad (12.177)$$

where $\rho = r/a$ and $\bar{p}$ is the volume averaged pressure. When these profiles are substituted into Eq. (12.176) the expression for J_B reduces to

$$J_B(r) = 10.8\frac{\bar{p}}{a^{1/2}R_0^{1/2}}\frac{\rho^{1/2}}{B_\theta}\left[\rho(1 - \rho^2)\right] \quad (12.178)$$

Now, the total toroidal current flowing in the plasma is $J = J_B + J_{CD}$ with J_{CD} representing the contribution from external current drive. A simple model for J_{CD} assumes that its main purpose is to fill in the current near the axis where $J_B = 0$ and in fact the neoclassical theory breaks down. The model chosen allows the total current to be written as

$$J(r) = 10.8\frac{\bar{p}}{a^{1/2}R_0^{1/2}}\frac{(\rho^{1/2} + \alpha)}{B_\theta}\left[\rho(1 - \rho^2)\right] \quad (12.179)$$

Here, α is a free parameter representing the magnitude of J_{CD}. Its value is obtained by requiring the bootstrap fraction to be 75% of the total current.

To determine f_B one first needs to calculate the $B_\theta(r)$ profile and the corresponding total plasma current $\mu_0 I = 2\pi a B_\theta(a)$. The function $B_\theta(r)$ is determined by a straightforward application of Ampere's law leading to

$$\frac{B_\theta^2(r)}{B_\theta^2(a)} \equiv b_\theta^2(\rho) = 54\beta_N \frac{q_*}{\varepsilon^{1/2}} \left(\frac{2}{9}\rho^{5/2} - \frac{2}{13}\rho^{9/2} + \frac{\alpha}{4}\rho^2 - \frac{\alpha}{6}\rho^4 \right) \qquad (12.180)$$

In this expression $\bar{p}$ has been replaced by β_N by the transformation

$$\frac{2\mu_0 \bar{p}}{B_\theta^2(a)} = \frac{\beta_t q_*^2}{\varepsilon^2} = \left(5\beta_N \frac{\varepsilon}{q_*} \right) \frac{q_*^2}{\varepsilon^2} = 5\frac{q_*}{\varepsilon}\beta_N \qquad (12.181)$$

where β_t is maximized by operating at the MHD stability limit, which is assumed to have a Troyon scaling. However, β_N, at this point, is treated as a free parameter. It's value is directly related to the current drive parameter α and is determined by the requirement $f_B = 0.75$. The resulting β_N must then be compared to the actual Troyon value $\beta_N = 0.028$ to see whether or not the plasma is MHD stable. The explicit relationship between β_N and α is obtained by setting $\rho = 1$ in Eq. (12.180):

$$\beta_N = 0.271 \frac{\varepsilon^{1/2}}{q_*} \left(\frac{1}{1 + 1.22\alpha} \right) \qquad (12.182)$$

With the total $B_\theta(r)$ profile known from Eq. (12.180), one can next integrate Eq. (12.178) over the plasma cross section to determine the total bootstrap current. The bootstrap fraction is then simply the ratio of the bootstrap current to the total current. A short calculation yields

$$f_B = \frac{\mu_0 I_B}{2\pi a B_\theta(a)} = \frac{\mu_0 a}{B_\theta(a)} \int_0^1 J_B(\rho)\rho d\rho = 27\beta_N \frac{q_*}{\varepsilon^{1/2}} \int_0^1 \frac{\rho^{5/2}(1 - \rho^2)}{b_\theta(\rho)} d\rho \qquad (12.183)$$

Equation (12.183) is easily evaluated numerically leading to a relation of the form $f_B = f_B(\alpha)$, as is illustrated in Fig. 12.10. Observe that $f_B = 0.75$ corresponds to $\alpha = 0.24$. Using this value of α one can easily plot the total current and bootstrap current profiles as well as the total $q(r)$ profile. These are shown in Fig. 12.11. The total current profile is indeed hollow while the safety factor profile has an off-axis minimum. The off-axis minimum is often referred to as a "reversed shear" profile since the shear $S(r) \equiv rq'/q$ changes sign. Most AT tokamaks have reversed shear profiles.

Consider next the required value of β_N obtained from Eq. (12.182) with $\alpha = 0.24$. To keep the required β_N as low as possible one sees that small ε and large q_* are desirable. Even so, one cannot make ε too small or q_* too large since the corresponding $\beta_t \propto \beta_N \varepsilon / q_* \propto \varepsilon^{3/2}/q_*^2$ may be too small for reactor

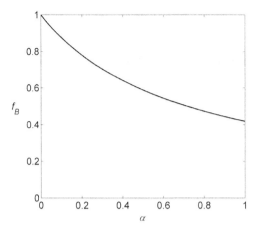

Figure 12.10   Bootstrap fraction f_B as a function of the current drive parameter α.

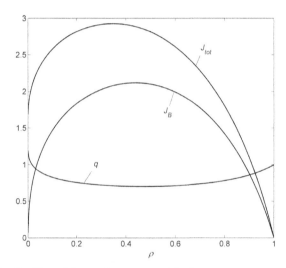

Figure 12.11   Profiles of the total current and the bootstrap current. Also shown is q/q_*. Here, $\rho = r/a$.

applications. For the case $\varepsilon = 1/3$ and $q_* = 3$, which is probably as favorable as possible for a high bootstrap fraction, the required value of β_N has the value

$$\beta_N = 0.04 \tag{12.184}$$

Even this optimistic case leads to a β_N that exceeds the actual Troyon limit $\beta_N = 0.028$. The implication is that a close fitting conducting wall is needed in conjunction with resistive wall stabilization. Also, keep in mind that the required β_N cannot exceed the Troyon value by too large a factor, even assuming wall stabilization is possible. The reason is that internal ballooning modes can be excited which are not affected by wall position.

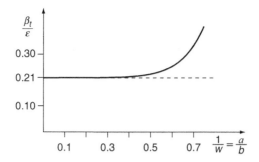

Figure 12.12   The maximum stable β_t/ε vs. normalized wall radius b/a.

12.9.3 Wall stabilization in an advanced tokamak

If a close fitting wall is required for an advanced tokamak then the question that naturally arises is how close must this wall actually be with respect to the plasma? A reasonable estimate can be obtained from the circular cross section sharp boundary model. The idea is to introduce the stabilizing effect of the wall into the evaluation of δW_V and then calculate the critical b/a to produce a gain in critical β equal to the amount necessary for 75% bootstrap operation. Specifically, from the calculation in the previous subsection it follows that the critical β_t for stability must increase by the ratio of $0.04/0.028 = 1.43$ because of the wall.

The calculation proceeds by noting that in the surface current model the vacuum energy in the presence of a wall can be expressed in terms of the no-wall result by a simple transformation of Eq. (12.153) given by

$$\delta \hat{W}_V = 2 \sum_{m,l,p} \frac{G_{ml}G_{mp}}{|m|} \xi_l \xi_p \quad \rightarrow \quad 2 \sum_{m,l,p} \frac{G_{ml}G_{mp}}{|m|} \left[\frac{1 + (a/b)^{2|m|}}{1 - (a/b)^{2|m|}} \right] \xi_l \xi_p \quad (12.185)$$

The maximum stable β_t is found by again assuming that the plasma operates along the equilibrium curve $\beta_t q_*^2/\varepsilon = \pi^2/16$. Next, the full $\delta W = \delta W_P + \delta W_S + \delta W_V$ is evaluated and set to zero to determine the critical $q_* = q_*(a/b)$. For this case the sums involved in the vacuum energy are calculated numerically.

The resulting critical q_* is substituted into the equilibrium limit yielding the desired curve of β_t/ε vs. a/b. This curve is illustrated in Fig. 12.12. One sees that to increase the critical β_t/ε by a factor of 1.43 (i.e., from 0.21 to 0.3) requires the wall to be at a normalized radius $b/a \approx 1.5$. This is a reasonable estimate although it is somewhat optimistic when compared to diffuse profile numerical studies. Still, it is comparable to the wall location needed to stabilize vertical instabilities.

To summarize, AT operation using reversed shear profiles can produce high fractions of bootstrap current. However, to reach $f_B = 75\%$ requires a high value of

β_N that exceeds the Troyon limit. Hence, a conducting wall is required with $b/a < 1.5$. The ideal mode is then transformed into a resistive wall mode requiring feedback stabilization. The goal of simultaneously achieving high f_B and high β_t is a challenging problem and is a major area of current tokamak research.

12.9.4 "Infernal" modes

The last topic of interest related to AT operation concerns infernal modes. The problem arises as follows. Assume that a suitable set of AT profiles exist that are stable against $n = \infty$ ballooning modes and $n = 1$ wall stabilized ballooning-kink modes. For a standard tokamak with a monotonically increasing $q(r)$ profile these are the most dangerous modes. Once their stability is ascertained, the tokamak is stable against all other MHD modes.

The problem with the AT is the existence of a low-shear region around the minimum of $q(r)$. This low-shear region can dramatically alter the stability behavior of the plasma. Qualitatively, when the shear is small, resonant MHD modes localized in this region are only weakly stabilized by line bending. Specifically, in the vicinity of a resonant surface $r = r_s$ line bending stabilization is proportional to $(m - nq)^2 \approx [nq'(r_s)(r - r_s)]^2$. In a low-shear region $q'(r_S)$ is small by definition.

The difficulty that then arises can be understood in the context of the internal mode stability of a straight tokamak. Recall from Eq. (11.176) that the leading-order contribution to the potential energy is of order $\delta W \approx \varepsilon^2 \delta W_2$. Pressure and current profile effects enter as $\varepsilon^4 \delta W_4$ and are normally small for $m \geq 2$. However, when a low-shear region exists then δW_2 itself becomes small since it is proportional to $(m - nq)^2$. When this occurs the distinction between $\varepsilon^2 \delta W_2$ and $\varepsilon^4 \delta W_4$ becomes blurred and a more careful analysis is required.

Such an analysis has been carried out by Manickam *et al.* (1987). They discovered, primarily through numerical calculations, that plasmas that were stable to both $n = \infty$ and $n = 1$ modes could be unstable to intermediate n modes in AT configurations with reversed shear. The instabilities have the form of internal modes that are somewhat but not highly localized about the shear reversal point. The modes are more likely to be excited when there is a substantial pressure gradient in this region. One can almost imagine the authors saying to themselves "We thought we were done after looking at $n = \infty$ and $n = 1$. Now we have to examine these 'infernal' intermediate n modes as well in order to demonstrate AT stability."

The conclusion is that a wide range of n modes must be tested for stability in AT configurations. In addition to the usual requirements related to ballooning and ballooning-kink modes one must make sure that the pressure gradient is not too large in the vicinity of the shear reversal point in order to avoid infernal modes.

There is as yet no simple analytic criterion that defines this critical pressure gradient and so numerical calculations are required.

12.10 Stability of tokamaks – $n = 0$ Axisymmetric modes

12.10.1 Introduction

The final instabilities to consider are the axisymmetric modes. These instabilities correspond to $n = 0$ external perturbations. They have no ϕ dependence and hence at marginal stability can be viewed as neighboring equilibria of the Grad–Shafranov equation. The most unstable mode usually has the form of a nearly rigid vertical shift of the entire plasma. In other words the eigenfunction is predominantly an $n = 0$, $m = 1$ mode.

If excited, the vertical axisymmetric mode is very dangerous, potentially leading to actual physical damage of the first wall, analogous to a major disruption. The mode can be stabilized by feedback and it is an important success story of fusion research that such feedback stabilization is almost routine in modern tokamaks.

This section describes three types of axisymmetric modes. First, vertical stability is investigated for a circular cross section plasma. The resulting stability criterion puts constraints on the curvature of the vertical field holding the plasma in toroidal force balance. With proper curvature, no feedback is required. Second, stability is investigated for a purely horizontal displacement, again for a circular cross section plasma. This analysis leads to further constraints on the vertical field curvature which if satisfied also lead to stability without feedback. Third, the circular cross section assumption is relaxed and the stability of elongated cross sections is investigated. This is an important topic since most modern tokamaks have elongated cross sections. It is shown that even a small amount of elongation leads to a dangerous vertical instability and it is here that feedback is essential.

12.10.2 Vertical instabilities in a circular plasma

A simple model has been studied by several authors that demonstrates the basic nature of the $n = 0$ axisymmetric mode in a circular cross section plasma. In this model the plasma is treated as a thin ($a/R_0 \ll 1$) current carrying loop of wire with a circular cross section embedded in an externally applied vertical field. For simplicity the effects of plasma pressure and the profile of internal magnetic flux are neglected. They are not essential to the discussion.

The goal of the ensuing analysis is to determine the appropriate constraints on the curvature of the vertical field to provide stability against rigid vertical and horizontal displacements. For vertical displacements one can see intuitively that a perfectly uniform vertical field, $\mathbf{B}_V = B_V \mathbf{e}_Z$, $B_V =$ constant, would provide neutral stability against a vertical displacement. If the plasma is moved upward it sees no change in the vertical field force, nor any change in the flux linking the plasma. For positive stability the critical question is then whether the vertical field should curve towards the plasma or away from the plasma.

The analysis proceeds by introducing a potential $\Phi(R, Z)$ such that the equilibrium forces acting on the plasma are given by $\mathbf{F}(R, Z) = -\nabla\Phi$. Equilibrium occurs at the location R_0, Z_0 where $\mathbf{e}_R \cdot \mathbf{F} \equiv F_R(R_0, Z_0) = 0$ and $\mathbf{e}_Z \cdot \mathbf{F} \equiv F_Z(R_0, Z_0) = 0$. If the plasma now undergoes a rigid vertical displacement δZ, then the perturbed vertical force acting on the plasma is given by

$$\delta F_Z = \frac{\partial F_Z(R_0, Z_0)}{\partial Z_0} \delta Z \tag{12.186}$$

The condition that the plasma be stable to a rigid displacement is that

$$\frac{\partial F_Z(R_0, Z_0)}{\partial Z_0} < 0 \qquad \text{for vertical stability} \tag{12.187}$$

In other words, stability occurs when the restoring force is in the opposite direction of the displacement; that is, when δZ is positive then δF_Z is negative. A similar argument holds for a rigid horizontal displacement. In this case the stability condition reduces to

$$\frac{\partial F_R(R_0, Z_0)}{\partial R_0} < 0 \qquad \text{for horizontal stability} \tag{12.188}$$

A key point in the analysis is to recognize that the plasma is a perfect conductor. This implies that the perturbed force must be evaluated subject to the constraint

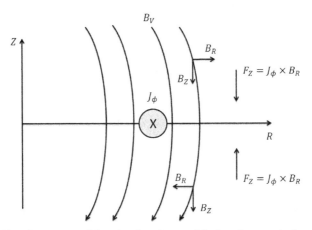

Figure 12.13   Geometry of the simple wire model showing vertical and horizontal stability in a circular cross section wire.

that the poloidal flux Ψ linking the current loop be conserved during the plasma displacement. For the simple model under consideration the appropriate expressions for the potential function and the linked poloidal flux can be written as (see Fig. 12.13 for the geometry)

$$\Phi(R, Z) = \frac{1}{2}LI^2$$

$$\Psi(R, Z) = LI + 2\pi \int_0^R B_Z(R', Z) \, R' dR'$$

(12.189)

Here, $L(R) = \mu_0 R[\ln(8R_0/a) - 2]$ is the external inductance associated with the plasma current I and B_Z (and B_R) are the components of the externally applied vertical field $\mathbf{B}_V = B_R \mathbf{e}_R + B_Z \mathbf{e}_Z$. Note that with the geometry as illustrated in Fig. 12.13 it follows that $B_Z < 0$ for toroidal force balance. The constant flux constraint gives a relationship between $I(R, Z)$ and $L(R)$ as the plasma moves.

With these definitions the first step in the analysis is to calculate the equilibrium forces. Differentiating the potential function leads to

$$F_Z = -LI\frac{\partial I}{\partial Z}$$

$$F_R = -LI\frac{\partial I}{\partial R} - \frac{I^2}{2}\frac{dL}{dR}$$

(12.190)

Constant flux requires setting $d\Psi = (\partial\Psi/\partial R)dR + (\partial\Psi/\partial Z)dZ = 0$. Each coefficient must independently vanish yielding two relations given by

$$\frac{\partial\Psi}{\partial Z} = 0 \rightarrow L\frac{\partial I}{\partial Z} + 2\pi\int_0^R \frac{\partial B_Z}{\partial Z} R' dR' = L\frac{\partial I}{\partial Z} - 2\pi R B_R = 0$$

$$\frac{\partial\Psi}{\partial R} = 0 \rightarrow L\frac{\partial I}{\partial R} + I\frac{\partial L}{\partial R} + 2\pi R B_Z = 0$$

(12.191)

The first of these relations has made use of the $\nabla \cdot \mathbf{B}_V = 0$ relation. The equilibrium forces can thus be written as

$$F_Z(R_0, Z_0) = 0 \rightarrow B_R(R_0, Z_0) = 0$$

$$F_R(R_0, Z_0) = 0 \rightarrow B_Z(R_0, Z_0) = -\frac{I}{4\pi R_0}\frac{dL}{dR_0}$$

(12.192)

Observe that the second of these relations is identical to the term in the Shafranov shift (with a minus sign because of the different coordinate definitions) associated with the external poloidal flux (see Eq. (6.91)). Equation (12.192) describes the required properties for the applied vertical field to produce toroidal force balance.

Consider now stability against a rigid vertical shift. Using the fact that $\nabla \times \mathbf{B}_V = 0$ since $\mathbf{B}_V$ is the externally applied vacuum vertical field, one can easily calculate $\partial F_Z / \partial Z_0$. A short calculation yields

$$\frac{\partial F_Z}{\partial Z_0} = -\frac{I^2}{2R_0} \frac{dL}{dR_0} n \tag{12.193}$$

where, in standard notation,

$$n(R_0, Z_0) = -\left(\frac{R}{B_Z} \frac{\partial B_Z}{\partial R}\right)_{R_0, Z_0} \tag{12.194}$$

is known as the field index, (not to be confused with the mode number). The condition for vertical stability is therefore given by

$$n > 0 \tag{12.195}$$

This condition can be easily understood by examining Fig. 12.13. If the directions of the vertical field and toroidal plasma current are as shown then (1) the vertical field produces an inward force for toroidal force balance and (2) a small upward shift of the plasma gives rise to a downward $\mathbf{J} \times \mathbf{B}$ force of magnitude $2\pi RIB_R$ whose direction is to restore equilibrium. Thus, the curvature of the vertical field as shown in Fig. 12.13 is the correct one for stability and corresponds to the condition $n > 0$. If the vertical field were to bend away from the plasma then $n < 0$ and the plasma would be unstable.

12.10.3 Horizontal instabilities in a circular plasma

The next problem of interest is horizontal stability. Most of the analysis in the previous section applies here as well. The one main difference is that it is now required to calculate $\partial F_R / \partial R_0$. This is a slightly more complicated calculation but is still straightforward. A short calculation leads to

$$\frac{\partial F_R}{\partial R_0} = -\frac{I^2}{2R_0} \frac{dL}{dR_0} \left[-n + 1 + \frac{1}{2}\left(\frac{R_0 L'}{L} - 2\frac{R_0 L''}{L'}\right)\right] \tag{12.196}$$

where $L' = dL / dR_0$. Upon substituting the expressions for L, L' and L'' one finds that the term in the curved parenthesis is approximately equal to unity over the interesting range of aspect ratio $R_0/a > 3$. Within this approximation it then follows that the condition for horizontal stability is

$$n < \frac{3}{2} \tag{12.197}$$

The vertical field must curve inward for vertical stability but not too much or else the plasma becomes horizontally unstable.

There is a simple physical picture that helps provide insight to the horizontal stability criterion. Assume that the plasma is given a small outward shift. At constant flux, the outward hoop force per unit length acting on the plasma decreases because of the smaller current; that is, at $\Psi = LI =$ constant, L increases as R increases implying that I must decrease, thereby reducing the hoop force per unit length. Similarly, for an inward curving vertical field the magnitude of B_V decreases with increasing R because of the larger radius of curvature. Thus the toroidal restoring force also decreases as the plasma moves outward. The stability criterion given by Eq. (12.197) is an expression of the requirement that the curvature of the vertical field be sufficiently weak so that the toroidal restoring force decreases at a slower rate than the hoop force. When this occurs the restoring force dominates and the plasma is stable against horizontal displacements.

To summarize, the conditions on the field index for a circular cross section plasma to be stable against both vertical and horizontal displacements is given by

$$0 < n < \frac{3}{2} \tag{12.198}$$

In general, more accurate numerical calculations show that the axisymmetric stability of circular plasmas is relatively easy to achieve, requiring only a modest shaping of the vertical field.

12.10.4 Vertical instabilities in an elongated plasma

Vertical instabilities in an elongated tokamak are an important issue. Recall that high elongation is a desirable feature in that larger currents can flow with the same q_* as κ increases. This improves the ideal MHD β limit as well as the energy confinement time. Still, increasing elongation makes the plasma more susceptible to $n = 0$ vertical instabilities.

The goal in this subsection is to calculate the critical conducting wall radius required to stabilize more highly elongated plasmas. As in previous cases, since the actual wall is resistive, the instability is transformed into a resistive wall mode. The end result is that when engineering reality is taken into account most tokamak researchers would agree that elongations on the order of $\kappa \sim 1.6$–1.8 can be feedback stabilized. In fact elongations in this range are readily achieved in modern tokamaks, and even higher values are achieved in spherical tokamaks.

The analysis presented here has two components. The first is a simple wire model that shows physically why elongated plasmas are unstable. The second calculates δW as a function of elongation and wall position using the Energy Principle. The calculation is greatly simplified by the use of hindsight, which has shown that the basic instability mechanism does not depend on either β or toroidicity. Thus, the calculation is carried out in a straight cylinder with an elliptic cross section plasma using the tokamak ordering. This calculation leads to an explicit relation between elongation and wall position. The results are then compared to more accurate numerical calculations.

The wire model

A simple physical picture that shows why elongated plasmas tend to be unstable can be obtained from a simple wire model as illustrated in Fig. 12.14. Assume the plasma wire is held in equilibrium by two equally spaced, rigidly mounted wires as shown in the diagram. By symmetry the global forces acting on the plasma wire, when it is centered, cancel and the plasma is in equilibrium. Now, to elongate the plasma vertically the currents in each wire must flow in the same direction as in the plasma wire. The resulting attractive force due to each equilibrium wire "pulls" on the top and bottom of the plasma causing it to become elongated while simultaneously holding the plasma in global force balance.

Assume next that the plasma wire undergoes a small vertical displacement ξ towards the upper wire. Recall that the force of the upper wire on the plasma wire is inversely proportional to the separation distance. Thus, the upward pulling force is increased since the wires are slightly closer together. Similarly, the pulling force of the lower wire is decreased since it is now slightly further away from the plasma wire. The conclusion is that there is a net global force, in addition to the elongation forces, acting on the plasma wire and pointing in the upward direction. The direction of this force causes the plasma to move even further away from its original equilibrium position. This clearly corresponds to an instability.

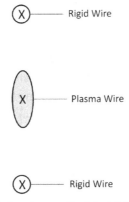

Figure 12.14   Wire model showing vertical instability in an elongated plasma.

The elliptic cylinder model

The properties of the instability, including the important effect of a conducting wall can be quantified by carrying out an Energy Principle analysis of $n = 0$ axisymmetric modes. The calculation is made tractable by neglecting β and toroidicity, neither of which is crucial for the basic excitation of the instability. The analysis is also simplified by using the tokamak expansion for the magnetic fields. The final simplification is to use a trial function, corresponding to a rigid vertical displacement, to evaluate δW. The analysis proceeds as follows.

The configuration of interest is a straight cylindrical plasma with an elliptic cross section of equivalent toroidal length $2\pi R_0$. The geometry is illustrated in Fig. 12.15. The strategy is to evaluate $\delta W = \delta W_F + \delta W_V$ using the $n = 0$ rigid shift trial function given by

$$\boldsymbol{\xi} = \xi_0 \mathbf{e}_y \tag{12.199}$$

where ξ_0 = constant. This is reasonably close to the actual eigenfunction as determined by detailed numerical calculations. As stated above the pressure is set to $p = 0$ and the equilibrium magnetic field is assumed to satisfy the tokamak ordering. Specifically it is assumed that

$$\mathbf{B} = B_0 \mathbf{e}_z + \mathbf{B}_p(x, y) \tag{12.200}$$

with B_0 = constant and $B_p/B_0 \sim \varepsilon$.

Consider first the evaluation of δW_F. If one starts with the intuitive form of δW_F given by Eq. (8.32) then a straightforward calculation leads to

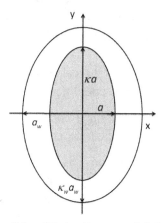

Figure 12.15   Geometry of the elliptic plasma model showing vertical instability in an elongated plasma.

$$\delta W_F = \frac{\pi R_0 \xi_0^2}{\mu_0} \int \left[\left(\frac{\partial B_y}{\partial y} \right)^2 + \frac{\partial B_y}{\partial x} \frac{\partial B_x}{\partial y} \right] dx dy \qquad (12.201)$$

The terms that enter are the line bending and kink contributions. The plasma compressibility and curvature terms vanish because $p = 0$. The magnetic compressibility term is smaller by ε^2 than the other terms in Eq. (12.201) and is thus neglected.

To proceed further an equilibrium magnetic field is required. A convenient choice is the elliptical Solovev model given by Eq. (6.138) with $\beta = 0$. The equilibrium flux and corresponding magnetic fields are given by

$$\psi = \frac{\psi_0}{2} \left(\frac{x^2}{a^2} + \frac{y^2}{\kappa^2 a^2} - 1 \right)$$

$$B_x = -\frac{1}{R_0} \frac{\partial \psi}{\partial y} = -\frac{\psi_0}{R_0 a^2} \frac{y}{\kappa^2} \qquad (12.202)$$

$$B_y = \frac{1}{R_0} \frac{\partial \psi}{\partial x} = -\frac{\psi_0}{R_0 a^2} x$$

Here, ψ_0 is a constant that will scale out of the problem. Equation (12.202) is substituted into Eq.(12.201). A short calculation that makes use of the fact that the elliptical plasma area is equal to $\pi \kappa a^2$ yields the required expression for δW_F

$$\frac{\delta W_F}{W_0} = \delta \hat{W}_F = -\frac{1}{\kappa} \qquad (12.203)$$

where $W_0 = \pi^2 \psi_0^2 \xi_0^2 / \mu_0 a^2 R_0$.

Clearly this term is destabilizing. However, destabilization is not directly due the edge current jump which led to $n \geq 1$ instabilities in a tokamak. It is in fact shown shortly that the circular tokamak ($\kappa = 1$) is stable to the $n = 0$ mode even in the presence of a current jump. In other words, if an instability arises it is due to elongation.

The next step in the analysis is to calculate the vacuum energy. This is accomplished by introducing elliptic coordinates u, v in the vacuum region,

$$x = c \sin u \, \cos v$$
$$y = c \cosh u \, \cos v \qquad (12.204)$$

Here, u is a radial-like variable while v is a poloidal angle-like variable. The parameter c is a constant. Its value is determined by the requirement that the plasma surface correspond to $u = u_p$. The implication is that c and u_p satisfy

$$c \sin u_p = a$$
$$c \cosh u_p = \kappa a \qquad (12.205)$$

which can easily be inverted

$$\coth u_p = \kappa$$
$$c^2 = a^2(\kappa^2 - 1) \tag{12.206}$$

Similarly, the perfectly conducting wall is assumed to be located at $u = u_w$. Note that the wall is not concentric with the plasma. Instead it has the shape of a confocal ellipse which is slightly flattened with respect to the plasma. This wall shape assumption is primarily for mathematical simplicity. The wall radius a_w and elongation κ_w thus satisfy

$$c \sinh u_w = a_w \qquad \kappa_w = \coth u_w$$
$$c \cosh u_w = \kappa_w a_w \qquad a_w^2 = a^2(\kappa^2 - 1)\sinh^2 u_w \tag{12.207}$$

The advantage of introducing elliptic coordinates is that Laplace's equations and the corresponding boundary conditions become quite simple. To see this observe that the perturbed vacuum magnetic field can be written as $\hat{\mathbf{B}}_1 = \nabla V$ with V satisfying $\nabla^2 V = 0$. After some algebra one can show that Laplace's equation in the laboratory coordinate system for $n = 0$ modes transforms to

$$\frac{\partial^2 V}{\partial x^2} + \frac{\partial^2 V}{\partial y^2} = c^2(\cosh^2 u - \sin^2 v)\left(\frac{\partial^2 V}{\partial u^2} + \frac{\partial^2 V}{\partial v^2}\right) = 0 \tag{12.208}$$

The boundary conditions, particularly the one on the plasma surface, also require some algebra. The key point is to recognize that the unit normal vector on any elliptic surface $S(x, y) = x^2 + y^2/\kappa^2 - a^2 = 0$ can be written as

$$\mathbf{n} = \frac{\nabla S}{|\nabla S|} = \frac{1}{(\kappa^4 x^2 + y^2)^{1/2}}(\kappa^2 x \, \mathbf{e}_x + y \, \mathbf{e}_y) \tag{12.209}$$

which can easily be converted into u, v coordinates. Using Eq. (12.209) one can then show that the boundary conditions on $V(u, v)$ are given by

$$\mathbf{n} \cdot \nabla V|_{wall} = 0 \qquad\qquad \rightarrow \qquad \left.\frac{\partial V}{\partial u}\right|_{u_w} = 0$$

$$\mathbf{n} \cdot \nabla V|_{plasma} = \mathbf{n} \cdot \nabla \times (\xi_\perp \times \mathbf{B})|_{plasma} \qquad \rightarrow \qquad \left.\frac{\partial V}{\partial u}\right|_{u_p} = \frac{\psi_0 \xi_0}{a R_0 \kappa} \cos v \tag{12.210}$$

The solution for V satisfying Eqs. (12.208), subject to Eq. (12.210) has the form

$$V(u, v) = \frac{\psi_0 \xi_0}{a R_0 \kappa} \frac{\cosh(u_w - u)}{\sinh(u_w - u_p)} \cos v \tag{12.211}$$

This solution is now substituted into the expression for δW_V,

$$\delta W_V = \frac{1}{2\mu_0} \int_{V_p} |\hat{\mathbf{B}}_1|^2 \, d\mathbf{r} = -\frac{1}{2\mu_0} \int_{S_p} V \, \mathbf{n} \cdot \nabla V \, dS \qquad (12.212)$$

After some standard manipulations one obtains

$$\frac{\delta W_V}{W_0} = \delta \hat{W}_V = \frac{1}{\kappa^2} \coth(u_w - u_p) \qquad (12.213)$$

Equations (12.203) and (12.213) are combined yielding an expression for the total potential energy

$$\delta \hat{W} = \delta \hat{W}_F + \delta \hat{W}_V = -\frac{1}{\kappa} + \frac{1}{\kappa^2} \coth(u_w - u_p) \qquad (12.214)$$

The final step is to eliminate u_p, u_w in terms of more recognizable geometric parameters. The quantity u_p has already been calculated in Eq. (12.206). The quantity u_w can be conveniently expressed in terms of a wall-closeness parameter A which measures the ratio of the area of the wall to the area of the plasma,

$$A \equiv \frac{\pi \kappa_w a_w^2}{\pi \kappa a^2} = \frac{c^2 \sinh u_w \cosh u_w}{\kappa a^2} = \frac{\kappa^2 - 1}{2\kappa} \sinh 2 u_w \qquad (12.215)$$

When the wall is at infinity then $A = \infty$. When it is on the plasma then $A = 1$. A short calculation leads to desired form of the potential energy

$$\delta \hat{W} = -\frac{1}{\kappa} + \frac{1}{\kappa^3} \frac{\kappa^2 + 1 + [(\kappa^2 - 1)^2 + 4\kappa^2 A^2]^{1/2}}{2(A - 1)} \qquad (12.216)$$

Equation (12.216) describes the stability of an elongated plasma against vertical instabilities. There are some useful limits to this expression. First, for a circular plasma corresponding to $\kappa = 1$, $\delta \hat{W}$ simplifies to

$$\delta \hat{W} = \frac{2}{A - 1} > 0 \qquad (12.217)$$

A circular plasma is stable to vertical displacements for any wall radius including placing the wall at infinity. Second, assume a finite ellipticity but move the wall to infinity, $A \to \infty$. The potential energy reduces to

$$\delta \hat{W} = -\frac{1}{\kappa} + \frac{1}{\kappa^2} < 0 \qquad (12.218)$$

In this simple model the plasma is unstable for any elongation with the wall at infinity.

The most interesting relation results from setting $\delta \hat{W} = 0$. Doing this determines how close the wall has to be to stabilize a plasma with a given elongation. A short calculation yields the simple relation

$$A < 2\left(\frac{\kappa^2 + 1}{\kappa^2 - 1}\right) \qquad\qquad \text{for stability} \qquad (12.219)$$

Equation (12.219) implies that an area ratio of 2 or less would stabilize any elongation.

Numerical results

Compared to various numerical studies, the critical wall area given by Eq. (12.219) is somewhat optimistic. The reasons are as follows. First, a trial function is used rather than the actual eigenfunction. Second, the Solovev model has finite edge currents which are closer to the wall, thereby producing a stronger wall effect. Third, the choice of a confocal ellipse for the wall places the wall closer to the top of the plasma. This is exactly where the instability is driving the plasma, the result again being a stronger wall effect than, for instance, with a concentric wall.

On the other hand, numerical solutions indicate that toroidal effects are slightly stabilizing which would tend to make the elliptical cylinder model more pessimistic. When all the effects are included the numerical studies show that the net result is that realistic plasmas need a closer wall than indicated by Eq. (12.219). As stated earlier, even with such a closer wall, the vertical instability still persists but in the form of a resistive wall mode that must be feedback stabilized. When practical engineering constraints are taken into account the conclusion is that elongations in the range of 1.6–1.8 can be feedback stabilized. This range is representative of modern tokamak operation. Spherical tokamaks, because of their very tight aspect ratio, can achieve even higher elongations, on the order of 2–2.5.

The main conclusion to be drawn from this discussion is that axisymmetric instabilities lead to a serious operational limit on tokamak operation. Vertical instabilities in an elongated plasma limit κ to a value less than 2, even though for other MHD and transport reasons it would be desirable to have much larger elongations. Still, it has now become nearly routine to feedback stabilize elongated tokamaks and this in itself is a success story for fusion research.

12.11 Overview of the tokamak

By combining the results from the equilibrium and stability analysis presented in the text one can obtain an overview of the standard tokamak configuration with respect to ideal MHD behavior.

Consider first equilibrium. The tokamak is an axisymmetric toroidal configuration with a large toroidal and a smaller poloidal magnetic field. In the interesting auxiliary heated high β tokamak regime radial pressure balance is provided by poloidal currents which produce a diamagnetic dip in B_ϕ. Toroidal force balance is produced by the interaction of the toroidal current with the applied vertical field, basically the $\mathbf{J}_\phi \times \mathbf{B}_V$ force. The high β tokamak regime is characterized by the following scaling relations: $\beta_t \sim \varepsilon$, $\beta_p \sim 1/\varepsilon$, and $q \sim 1$. A standard tokamak typically has $\varepsilon \approx 1/3$, $q(0) \approx 1$, and $q(a) \sim 2 - 3$. There are equilibrium pressure limits on tokamaks but these limits usually exceed reactor requirements by a substantial margin.

Most tokamaks have elongated cross sections with $\kappa \sim 1.5-2$ and have an outward pointing "D" shape. The maximum elongation is limited by $n = 0$ axisymmetric modes. Interesting variations on the standard tokamak are (1) the spherical tokamak which attempts to achieve high β_t by employing a very tight aspect ratio and (2) the advanced tokamak whose goal is to ease the achievement of steady state operation by means of a high bootstrap current which generates a reverse shear $q(r)$ profile.

The most important MHD limitations on tokamak performance involve both pressure and current limits as set by MHD instabilities. Successful operation of a tokamak requires that neither the pressure nor the current be too large. Current limits exist in both the straight and toroidal tokamak. They arise from the $n = 1$, $m = 1$ internal kink mode, the high n Mercier interchange instability, and the $n = 1$, $m \geq 1$ external kink modes.

The limitations due to these modes are as follows. The $n = 1$, $m = 1$ internal mode is closely connected to the excitation of sawtooth oscillations. This mode basically limits the current density on axis so that $q_0 \approx 1$. The Mercier criterion is most difficult to satisfy near the magnetic axis where the shear is small. To prevent such instabilities and the resulting turbulence, the tokamak must operate with $q_0 \gtrsim 1$. The exact threshold depends on the shape of the cross section. Elongated, outward pointing D-shapes are favorable for Mercier stability. The most dangerous current-driven modes are the $n = 1$ external kinks. The tokamak must operate with $q_a > 1$ to avoid the $m = 1$ Kruskal–Shafranov limit. The actual criterion is more stringent, requiring $q_a \sim 2-3$ to avoid $m \geq 2$ external kinks. If q_a becomes too small then a major disruption occurs. Much higher m instabilities can occur at the plasma surface if the edge current gradient does not vanish sufficiently rapidly. These high m instabilities represent the peeling component of the peeling–ballooning mode believed to be responsible for ELMs.

There are equally important pressure limits in a tokamak that arise from $n = \infty$ ballooning modes and $n = 1$ ballooning-kink modes. The ballooning modes are internal instabilities that set a maximum allowable value of p/I as given by the Sykes limit $\beta_t < 0.044(I/aB_0)$. These modes are the ballooning contribution to the peeling–ballooning modes that may be responsible for ELMs. The ballooning-kink mode also sets a limit on the maximum p/I in this case given by the Troyon limit, $\beta_t < 0.028(I/aB_0)$. However, this maximum is achieved at an optimum current defined by $q_* \approx 1.5$, which when geometric and finite β_t/ε effects are correctly included corresponds to $q_a \approx 2$. Violation of the Troyon limit often leads to major disruptions.

The final word is that from the MHD point of view the standard tokamak is capable of achieving MHD stable profiles without a conducting wall at sufficiently high β_t to meet the requirements of a fusion reactor.

This is clearly a positive conclusion but still, two problems remain. First, a standard tokamak reactor is projected to be a large device implying a large capital cost. By utilizing a very tight aspect ratio the spherical tokamak can achieve much higher β_t in a more compact geometry. Compactness translates into reduced capital costs. One MHD difficulty with the spherical tokamak is that high β_t does not necessarily translate into high p because of the rapid $1/R$ decay of the toroidal field in a tight aspect ratio geometry.

The second problem faced by the standard tokamak is the need for steady state. Because RF current drive is expensive and inefficient, a large portion of the plasma current must be provided by the bootstrap current to achieve a favorable power balance. This has led to the idea of AT operation which is characterized by high bootstrap current and reverse shear profiles. The difficulty here is that the plasma pressure needed to achieve a high bootstrap fraction invariably violates the Troyon limit. The implication is that a close conducting wall is required accompanied by resistive wall feedback.

The standard tokamak, or one of its disciples, may be able to overcome the difficulties discussed above and achieve the fusion dream – a steady state, disruption-free, high-performance tokamak with sufficient pressure and current to meet the requirements of a fusion reactor.

12.12 Stellarator stability

As might be expected, determining the MHD stability of stellarators is one of the most challenging problems in MHD theory. The reasons are self-evident. The geometry is three dimensional. There are a wide variety of different stellarator configurations. Some have a circular magnetic axis and some a helical magnetic axis. There may be small expansion parameters that can serve as the basis for an asymptotic expansion, but often times no such parameters exist.

The result of this complexity is that most stellarator stability results are obtained numerically, a task now made possible by advances in computer modeling and in computers themselves. Carrying out such studies is a critical component in the design of new stellarator experiments as well as interpreting data from existing experiments. Even so, it would be useful if some simple analytic theory was available to shed insight into which geometric properties of stellarators are favorable for MHD stability. In this connection it is worth noting that there does exist some analytic theory, although most of it was developed during the early years of the fusion program. This early work is the basis for much of the material presented below.

As an overview, note that, in general, stellarators are subject to many, but not all, of the same modes found in tokamaks. Specifically, localized interchanges and ballooning mode instabilities are important in stellarators. However, internal and external current-driven kinks are not very important since stellarators typically operate with only a small or zero net toroidal current. As a result, the low n external ballooning-kink mode in a tokamak is essentially transformed into an external ballooning mode which is important in setting β limits. Lastly, as has been stated previously, when an MHD stability limit is violated, stellarators do not disrupt, a consequence of the cage-like confinement resulting from the applied helical field. Instead, performance degrades because of MHD turbulence, which is not desirable but is still better than a major disruption.

The analysis described in this section is primarily focused on high n ballooning and interchange modes. A series of simple calculations is presented that provides some insight into the geometric features needed for favorable stability. It is shown that a straight stellarator always has unfavorable curvature. Shear can provide some stability but only at very low β. Stability at higher values of β requires a combination of toroidicity plus a vertical field. The stability against low n external modes is discussed by reviewing several numerical studies.

12.12.1 High n *modes in a stellarator*

Some initial insight into the high n stability of a stellarator can be obtained by examining the expression for the local shear associated with the s,α diagram of a tokamak. For the simple analytic model the local shear is given by Eq. (12.99), repeated here for convenience:

$$\hat{s}(r,\theta) = s(r) - \alpha(r)\cos\theta \tag{12.220}$$

where $s(r) = rq'(r)/q(r)$ is the average shear and $\alpha(r) = -q^2 R_0 d\beta(r)/dr$ is the normalized pressure gradient.

Now, in a tokamak both s and α are positive. Therefore, in the region of unfavorable toroidal curvature ($\theta = 0$), as the pressure gradient α increases from zero the local shear first decreases and then increases. The value of α for which the local shear vanishes corresponds to a region of vulnerability for ballooning modes. At sufficiently large $\alpha \gg s$ the local shear increases with further increases in pressure in the region of unfavorable curvature, leading to the second region of stability.

The situation is qualitatively different in a stellarator. The reason is that $q(r)$ is usually a decreasing function of r. A stellarator is characterized by a reversed shear profile over the entire plasma: $s(r) < 0$. To the extent that a similar s, α diagram applies to a stellarator, one sees that as the pressure α is increased, the local shear in the unfavorable toroidal field curvature region ($\theta = 0$) also increases monotonically. At the inside of the plasma ($\theta = \pi$) the local shear first decreases and then increases as the pressure rises. However, the region where the local shear vanishes is not that important because it does so in a region of favorable toroidal field curvature.

The conclusion is that the situation with respect to local shear is more favorable in a stellarator than in a tokamak. The local shear in a stellarator increases monotonically in the region where it is needed most, the region of unfavorable curvature. This insight suggests that the balance between interchange and ballooning stability shifts towards interchanges which are usually easier to stabilize.

In order to approximately demonstrate the interchange/ballooning mode properties analytically the stellarator ballooning mode potential energy is evaluated for a simple trial function. This analysis leads naturally to the idea of "average favorable curvature" and the related concept of the "magnetic well." To obtain these results one first needs some subsidiary calculations, specifically a derivation of the parallel current constraint and the relation between average curvature and magnetic well. These and the follow-on stability results are discussed below.

12.12.2 The parallel current constraint

The parallel current constraint is a general relation obtained from the requirement that $\nabla \cdot \mathbf{J} = 0$. The relation is needed in order to simplify the potential energy of the stellarator ballooning mode equation for the trial function used in the analysis.

The derivation of the parallel current constraint proceeds as follows. The current density $\mathbf{J}$ is written as $\mathbf{J} = J_\parallel \mathbf{b} + \mathbf{J}_\perp$ and then substituted into the equation $\nabla \cdot \mathbf{J} = 0$, yielding

$$\mathbf{B} \cdot \nabla \frac{J_\parallel}{B} = -\nabla \cdot \mathbf{J}_\perp \tag{12.221}$$

Next, $\mathbf{J}_\perp$ is substituted from the momentum equation: $\mathbf{J}_\perp = \mathbf{B} \times \nabla p / B^2$. This gives

$$\mathbf{B} \cdot \nabla \frac{J_\parallel}{B} = -\nabla p \cdot \nabla \times \frac{\mathbf{B}}{B^2} \tag{12.222}$$

A simple calculation shows that

$$\left(\nabla \times \frac{\mathbf{B}}{B^2} \right)_\perp = -\frac{\nabla p \times \mathbf{B}}{B^4} - 2 \frac{\boldsymbol{\kappa} \times \mathbf{B}}{B^2} \tag{12.223}$$

Substituting into Eq. (12.222) leads to the parallel current constraint

$$\mathbf{B} \cdot \nabla \frac{J_{\parallel}}{B} = \frac{2}{B}(\mathbf{b} \cdot \nabla p \times \boldsymbol{\kappa}) \tag{12.224}$$

The final desired form is obtained by introducing straight field line flux coordinates with arc length defining the Jacobian. From the analysis in Section 12.4 it then follows that

$$\frac{\partial}{\partial l} \frac{J_{\parallel}}{B} = \frac{2}{B} \frac{dp}{d\psi} \kappa_{\alpha} \tag{12.225}$$

where the curvature has been written as $\boldsymbol{\kappa} = \kappa_{\psi} \nabla \psi + \kappa_{\alpha} \nabla \alpha$.

The parallel current constraint requires that for a given pressure and flux surface shape, a specific amount of parallel current must flow in order to keep the total current density divergence free.

12.12.3 The relation between average curvature and magnetic well

A reasonable guideline for assessing the MHD stability of stellarators can be obtained by substituting a trial function into the ballooning mode potential energy and then making use of the parallel current constraint. The result of these steps is a natural definition of average favorable curvature, which after some additional analysis, leads to the concept of a magnetic well.

The starting point is the ballooning mode potential energy written in terms of arc length as given by Eq. (12.52) and repeated here for convenience

$$\overline{W}(\psi, \alpha, \alpha_0) = \int_{-\infty}^{\infty} \left[k_{\perp}^2 \left| \frac{\partial X}{\partial l} \right|^2 - 2\mu_0 \frac{dp}{d\psi} (k_{\alpha}^2 \kappa_{\psi} - k_{\alpha} k_{\psi} \kappa_{\alpha}) |X|^2 \right] \frac{dl}{B} \tag{12.226}$$

A simple trial function is now substituted into Eq. (12.226). The trial function chosen is given by

$$X(\psi, \alpha, l) = \lim_{\varepsilon \to 0} \left[e^{-\varepsilon^2 l^2} X(\psi, \alpha) \right] \tag{12.227}$$

In the resulting expression for $\overline{W}$ the stabilizing $|\partial X/\partial l|^2$ term is neglected for simplicity. The term tends to be small since X only depends weakly on l. On the other hand, it tends to be large since $k_{\perp}^2$ contains secular terms proportional to l^2. The net effect is a stabilizing contribution of the same order as the pressure gradient term. Still, by neglecting the term one obtains a lower bound on $\overline{W}$ that in the limit $\varepsilon \to 0$ can be written as

$$\overline{W}(\psi, a, a_0) \geq -2\mu_0 \frac{dp}{d\psi} |X|^2 \int_{-\infty}^{\infty} (k_\alpha^2 \kappa_\psi - k_\alpha k_\psi \kappa_\alpha) \frac{dl}{B} \qquad (12.228)$$

Note, however, that since a trial function has been used rather than the true eigenfunction the resulting $\overline{W}$ also must lie above the true minimum value of the potential energy: $\overline{W}(\psi, a, a_0) \geq \overline{W}_{\min}(\psi, a, a_0)$. Since there are two lower bounds on $\overline{W}$ one cannot in general conclude that the expression in Eq. (12.228) is either necessary or sufficient for stability. It does, nonetheless, give a plausible estimate for $\overline{W}$ which can be used to gain some physical insight.

The next step in the analysis is a further simplification of Eq. (12.228) resulting from the application of the parallel current constraint. Focus for the moment on the κ_α term. Since $k_\alpha = \partial S(\psi, \alpha)/\partial \alpha$ and $k_\psi = \partial S(\psi, \alpha)/\partial \psi$ are both independent of l one sees that

$$[\overline{W}]_{\kappa_\alpha} = 2\mu_0 \frac{dp}{d\psi} |X|^2 \int_{-\infty}^{\infty} (k_\alpha k_\psi \kappa_\alpha) \frac{dl}{B} = 2\mu_0 \frac{dp}{d\psi} k_\alpha k_\psi |X|^2 \int_{-\infty}^{\infty} \kappa_\alpha \frac{dl}{B} \qquad (12.229)$$

The quantity κ_α is eliminated by means of the parallel current constraint, leading to

$$[\overline{W}]_{\kappa_\alpha} = 2\mu_0 \frac{dp}{d\psi} k_\alpha k_\psi |X|^2 \int_{-\infty}^{\infty} \kappa_\alpha \frac{dl}{B} = \mu_0 k_\alpha k_\psi |X|^2 \int_{-\infty}^{\infty} \frac{\partial}{\partial l} \frac{J_\parallel}{B} dl = 0 \qquad (12.230)$$

The integrand is a perfect differential implying that the κ_α term makes no contribution to the estimate for $\overline{W}$.

With this simplification, the expression for $\overline{W}$ reduces to

$$\overline{W}(\psi, a, a_0) \geq -2\mu_0 \frac{dp}{d\psi} k_\alpha^2 |X|^2 \int_{-\infty}^{\infty} \kappa_\psi \frac{dl}{B} \qquad (12.231)$$

For a standard negative pressure gradient it thus follows that a good guideline for stability is

$$\langle \kappa_\psi \rangle > 0 \qquad (12.232)$$

where, in normalized form,

$$\langle \kappa_\psi \rangle = \frac{\int_{-\infty}^{\infty} \kappa_\psi \frac{dl}{B}}{\int_{-\infty}^{\infty} \frac{dl}{B}} \qquad (12.233)$$

is the average normal curvature. A positive average curvature is favorable for stability.

A slightly more convenient form of Eq. (12.199) for vacuum fields can be obtained by substituting κ_ψ from Eq. (12.48) after letting $p \to 0$. Specifically one substitutes

$$\kappa_\psi = \frac{1}{B}\frac{\partial B}{\partial \psi} \tag{12.234}$$

into Eq. (12.233), yielding

$$\langle \kappa_\psi \rangle = \frac{\int_{-\infty}^{\infty} \left(\frac{1}{B}\frac{\partial B}{\partial \psi}\right)\frac{dl}{B}}{\int_{-\infty}^{\infty}\frac{dl}{B}} = -\frac{\frac{\partial}{\partial \psi}\int_{-\infty}^{\infty}\frac{dl}{B}}{\int_{-\infty}^{\infty}\frac{dl}{B}} \tag{12.235}$$

Equation (12.235) shows that an equivalent definition of favorable average curvature for a vacuum magnetic field can be written as (Rosenbluth and Longmire, 1957; Lenard, 1964)

$$\frac{\partial}{\partial \psi}\int_{-\infty}^{\infty}\frac{dl}{B} < 0 \tag{12.236}$$

When a magnetic field satisfies this criterion it is said to possess a "vacuum magnetic well." Unfortunately, in terms of intuition, it is not at all obvious what actually lies at the bottom of the "well." One further mathematical transformation is needed to demonstrate in a more intuitive way what quantity the "well" actually describes.

The transformation makes use of the fact that a vacuum magnetic field can be written as a gradient: $\mathbf{B} = \nabla V(\psi, \alpha, l)$. Now, consider the current flowing through the closed loop illustrated in Fig. 12.16. The loop is assumed to lie in a fixed α surface. The upper and lower line segments represent the intersections with the surfaces $\psi = \psi_1$ and $\psi = \psi_2$ respectively. The vertical line segments are chosen to be parallel to the $\nabla \psi$ direction. There are no contributions on these segments since $\mathbf{B} \cdot \nabla \psi = 0$. The current enclosed must be zero since the field is a vacuum field.

This implies that

$$\mu_0 I = \oint \mathbf{B} \cdot d\mathbf{l} = \oint \nabla V \cdot d\mathbf{l} = V_1(\psi_1, \alpha, l)\Big|_{l_1}^{l_2} - V_2(\psi_2, \alpha, l)\Big|_{l_3}^{l_4} = 0 \tag{12.237}$$

The change in the potential function must be the same along each segment. Since this relation must be true for any two arbitrary flux surfaces, it follows that $\delta V = V(\psi, \alpha, l_2) - V(\psi, \alpha, l_1)$ must be independent of ψ: $\partial(\delta V)/\partial \psi = 0$.

The desired transformation is now obtained from the relation

$$\delta V = \int_{l_1}^{l_2} \mathbf{B} \cdot d\mathbf{l} = \int_{l_1}^{l_2} B\,dl \tag{12.238}$$

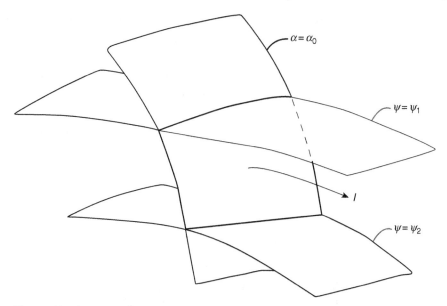

Figure 12.16 Geometry showing the current flowing through a closed loop used in the derivation of the magnetic well criterion.

where l_1 and l_2 are arbitrary points along the field line including $\pm\infty$. Specifically, the transformation makes use of the fact that

$$\frac{\partial}{\partial\psi}\int_{-\infty}^{\infty} B\,dl = 0 \tag{12.239}$$

The more intuitive form of the stability condition is finally obtained by a short calculation that makes use of Eqs. (12.236) and (12.238),

$$\frac{1}{\delta V}\frac{\partial U}{\partial\psi} = \frac{\partial}{\partial\psi}\left(\frac{U}{\delta V}\right) = -\frac{U^2}{(\delta V)^2}\frac{\partial}{\partial\psi}\left(\frac{\delta V}{U}\right) < 0$$

$$U = \int\frac{dl}{B} \tag{12.240}$$

The last term can be rewritten in a form that clearly shows why the stability condition corresponds to a "magnetic well,"

$$\frac{\partial}{\partial\psi}\langle B^2\rangle > 0 \tag{12.241}$$

When $\langle B^2\rangle$ lies at the bottom of a well and increases with flux away from the magnetic axis, the approximate stability criterion is satisfied.

The conclusion from the analysis is that there is a one-to-one relation between average curvature and magnetic well. A short calculation shows that this relationship is given by

$$\langle \kappa_\psi \rangle = \frac{1}{\langle B^2 \rangle} \frac{\partial}{\partial \psi} \langle B^2 \rangle \tag{12.242}$$

In practice one can use whichever form is more convenient to examine stability. Often the curvature expression requires less work to evaluate while the magnetic well expression is more intuitive to understand.

12.12.4 Average curvature of a straight helix

The simplest configuration to investigate is the straight, single helicity stellarator as described in Section 6.7. It is shown below that for a vacuum magnetic field, the configuration always has unfavorable average curvature. The analysis is relatively straightforward and makes extensive use of the results derived in Section 6.7. The calculation is most easily carried out assuming that the amplitude of the helical field δ is small and using the definition of average curvature rather than magnetic well.

The analysis begins by introducing the small δ approximation into the definition of average curvature. The expansion assumes that the magnetic field can be written as $\mathbf{B} = B_0 \mathbf{e}_z + \mathbf{B}_1(r, l\theta + hz)$ with $B_1/B_0 \sim \delta \ll 1$. The first non-vanishing contribution arises in order δ^2. The critical integral quantity required can thus be expanded as

$$\int \frac{dl}{B} = \int \frac{dz}{B_z} = \int \frac{dz}{B_0} \left[1 - \frac{B_{z1}}{B_0} - \frac{1}{B_0} \frac{\partial B_{z1}}{\partial r_0} r_1 - \frac{1}{B_0} \frac{\partial B_{z1}}{\partial \theta_0} \theta_1 + \frac{B_{z1}^2}{B_0^2} + \cdots \right] \tag{12.243}$$

Here, the field line of interest is labeled by r_0, θ_0 and, because of helical symmetry, the integration need only be carried out over one helical period $0 < z < L = 2\pi/h$. All quantities in the integrand are functions of r_0, θ_0, z.

The value of B_{z1} is obtained from Eq. (6.220) and can be written as

$$\frac{B_{z1}}{B_0} = \Delta I_l(u_0) \cos (l\theta_0 + hz)$$

$$\Delta = \frac{ha\delta}{(l^2 + h^2 a^2)^{1/2} I_l(ha)} \tag{12.244}$$

$$u_0 = hr_0$$

where the specific definition of δ is given by

$$\delta \equiv \frac{B_h}{B_0} = \left[\frac{lB_{\theta 1} + hr_0 B_{z1}}{(l^2 + h^2 r_0^2)^{1/2} B_0}\right]_{r_0=a, \theta_0=0, z=0} \tag{12.245}$$

The remaining quantities in the integrand, r_1, θ_1, represent the helical oscillations of the field line trajectory. They are easily found from the field line trajectory equations

$$\frac{dr_1}{dz} = \frac{B_{r1}}{B_0} = \Delta I_l'(u_0) \sin (l\theta_0 + hz)$$

$$\frac{d\theta_1}{dz} = \frac{B_{\theta 1}}{r_0 B_0} = \Delta \frac{l I_l(u_0)}{u_0} \cos (l\theta_0 + hz) \tag{12.246}$$

Integrating these equations yields

$$r_1 = -\frac{\Delta}{h} I_l'(u_0)[\cos (l\theta_0 + hz) - 1]$$

$$\theta_1 = l\Delta \frac{I_l(u_0)}{u_0^2} \sin (l\theta_0 + hz) \tag{12.247}$$

All the quantities in the integrand have now been evaluated. A straightforward calculation now leads to an expression for the renormalized average curvature

$$\overline{\kappa} = u \frac{d\psi_0}{du} \langle \kappa_\psi \rangle = -\frac{u}{U} \frac{\partial U}{\partial u} \tag{12.248}$$

that can be written as

$$\overline{\kappa}(u) = -\frac{\Delta^2}{2} u \frac{d}{du} \left[\left(\frac{dI_l}{du}\right)^2 + \left(\frac{l^2}{u^2} + 1\right) I_l^2\right]$$

$$\approx -2(l-1)\delta^2 \left(\frac{r}{a}\right)^{2(l-1)} \quad l \geq 2$$

$$\approx -h^2 a^2 \delta^2 \left(\frac{r}{a}\right)^2 \quad l = 1 \tag{12.249}$$

$$\approx -2h^2 a^2 \delta^2 \left(\frac{r}{a}\right)^2 \quad l = 0$$

Here for simplicity the "0" subscript has been suppressed. The approximate expressions correspond to the loose helix limit $ha \ll 1$. Observe that for all l the average curvature $\overline{\kappa} < 0$. A straight stellarator always has unfavorable average curvature.

12.12.5 Shear stabilization of a straight helix

A natural question that now arises is whether or not shear can provide stability against the average unfavorable curvature. To answer this question one basically has to include the line bending $|\partial X/\partial l|^2$ term that was neglected in the evaluation of $\overline{W}$. Sinclair *et al.* (1965) have carried out such a calculation, the result being a generalized form of Suydam's criterion. Below, a heuristic derivation is presented of this result. The generalized Suydam's criterion is then applied to a loosely wound $l = 2$ stellarator where it is found that shear can indeed stabilize the unfavorable curvature but only for relatively small values of β_t.

The analysis begins by recalling Suydam's stability criterion (i.e., Eq. (11.109)),

$$rB_z^2 \left(\frac{q'}{q}\right)^2 + 8\mu_0 p' > 0 \tag{12.250}$$

This equation, which is valid for a straight one-dimensional cylinder, can be easily cast in a form more suitable for generalization to a stellarator by replacing q by $2\pi/\iota \approx rB_0/R_0 B_\theta$,

$$r^2 \iota'^2 + 4r\beta'\iota^2 > 0 \tag{12.251}$$

where $\beta = 2\mu_0 p/B_0^2$. Next, the quantity ι^2 is expressed in terms of the actual curvature $\kappa \approx -B_\theta^2/rB_0^2$ which is then expressed in terms of κ_ψ as follows:

$$\frac{\iota^2}{4\pi^2} = \frac{R_0^2 B_\theta^2}{r^2 B_0^2} = -\frac{R_0^2}{r}\kappa = -\frac{R_0^2}{r}\psi_0' \kappa_\psi \tag{12.252}$$

Substituting Eq. (12.252) into Eq. (12.251) yields

$$r^2 \left(\frac{\iota'}{2\pi}\right)^2 - 4R_0^2 \beta' \psi_0' \kappa_\psi > 0 \tag{12.253}$$

The heuristic part of the derivation requires replacing κ_ψ with $\langle \kappa_\psi \rangle$ and then introducing the normalized curvature $\overline{\kappa} = r\psi_0' \langle \kappa_\psi \rangle$. One obtains

$$r^2 \left(\frac{\iota'}{2\pi}\right)^2 - 4R_0^2 \frac{\beta'}{r}\overline{\kappa} > 0 \tag{12.254}$$

Equation (12.254) shows how shear can potentially stabilize unfavorable curvature. To quantify the maximum β that can be shear stabilized assume that Eq. (12.254) is marginally stabilized at every value of radius. The maximum β satisfies

$$\beta' = \frac{1}{4R_0^2}\frac{r^3}{\overline{\kappa}}\left(\frac{\iota'}{2\pi}\right)^2 \tag{12.255}$$

Observe that $\beta' \propto r^3$ for a loosely wound $l = 2$ system implying a flat central pressure. The global β_t that can be stabilized is determined by multiplying Eq. (12.255) by r^2 and integrating over the plasma volume. A short calculation yields

$$\beta_t = -\frac{1}{a^2} \int_0^a \beta' r^2 dr = -\frac{1}{4R_0^2 a^2} \int_0^a r^5 \bar{\kappa} \left(\frac{\iota'}{2\pi}\right)^2 dr \qquad (12.256)$$

The value of β_t can be easily evaluated for a loosely wound $l = 2$ stellarator for which

$$\bar{\kappa} = -2\delta^2 \left(\frac{r}{a}\right)^2$$

$$\frac{\iota}{2\pi} = \frac{R_0 \delta^2}{ha^2} \left(1 + \frac{h^2 r^2}{2}\right) \qquad (12.257)$$

$$\frac{\iota'}{2\pi} = \frac{hR_0\delta^2}{a^2} r$$

where $\bar{\kappa}$ and $\iota/2\pi$ have been obtained from Eqs. (12.249) and (6.243) respectively. The resulting maximum value of β_t is given by

$$\beta_t = \frac{1}{48} h^2 a^2 \delta^2 \qquad (12.258)$$

Equation (12.258) can be rewritten in a more familiar form by noting that $\iota(r)/2\pi$ in a loosely wound $l = 2$ stellarator is approximately constant: $\iota/2\pi \approx \iota_H/2\pi \approx R_0\delta^2/ha^2$. Also, since $2\pi/h$ is the helical wavelength it follows that $ha = N\varepsilon$, where N is the number of helical periods. These substitutions lead to a more stellarator-like form for the β_t limit,

$$\beta_t = \frac{1}{48} (N\varepsilon)^3 \varepsilon \left(\frac{\iota_H}{2\pi}\right) \qquad (12.259)$$

The β_t limit is quite small for LHD and W7-X. For instance, for LHD $N = 10$, $\varepsilon \approx 0.14$, and $\iota_H/2\pi \approx 0.4$ resulting in $\beta_t = 0.3\%$. The W7-X stellarator has an even lower limit. For this case, $N = 5$, $\varepsilon \approx 0.1$, and $\iota_H/2\pi \approx 1$ resulting in $\beta_t = 0.026\%$. The conclusion is that shear is not a very effective way to stabilize a predominantly straight, single helicity, $l = 2$ stellarator. To improve the situation toroidicity combined with a vertical field must be taken into account.

12.12.6 Stabilization of a toroidal stellarator

The average curvature in a stellarator can be made favorable by the addition of an externally applied vertical field with the proper sign and magnitude.

The addition of this field combined with toroidicity can produce a radially outward shift (along R) of the inner flux surfaces with respect to the outer flux surfaces. This relative shift, if large enough, is shown to produce average favorable curvature.

The derivation of the average curvature in a torus is similar to that for the straight stellarator. However, there is one subtlety that complicates the analysis. The issue is as follows. In a straight stellarator the θ trajectory of a field line along ϕ is composed of two contributions, a helical oscillation and a slow precession due to the rotational transform. Because of helical symmetry one need only calculate the transform over one helical period and then multiply by N, the total number of helical periods, to obtain the total transform. In contrast, in a torus as the field line slowly rotates with the transform it samples both the inside ($\theta = \pi$) and the outside ($\theta = 0$) of the torus. Since the toroidal field differs in each region by $(r/R_0)\cos\theta$, this modifies the field line trajectory and in turn, the average curvature. This is the new effect that must be taken into account.

The Greene–Johnson expansion

A convenient way to carry out the analysis is by means of the Greene–Johnson stellarator expansion (Greene and Johnson, 1961), which assumes that (for a vacuum field)

$$B_h^2/B_0^2 \sim \delta^2 \sim 1/hR_0 \sim 1/N \sim \varepsilon \ll 1$$
$$ha \sim \iota/2\pi \sim 1$$
(12.260)

The expansion has the virtue of separating the rapid helical motion from the slow rotational transform motion since the number of helical periods N is assumed to be large. It is this separation of scales that simplifies the analysis. The end goal of the analysis is to determine $\bar{\kappa}$ whose derivation is given below.

The basic quantity of interest U needed to calculate $\bar{\kappa}$ is expanded as follows:

$$U = \int \frac{dl}{B} = \lim_{M\to\infty} \int_{-2\pi M/N}^{2\pi M/N} \frac{R\,d\phi}{B_\phi}$$
$$\approx \lim_{M\to\infty} \frac{R_0}{B_0} \int_{-2\pi M/N}^{2\pi M/N} \left[1 - \frac{B_{\phi 1}}{B_0} - \frac{B_{\phi 2}}{B_0} + \frac{B_{\phi 1}^2}{B_0^2} + \frac{r}{R_0}\cos\theta \right]_1 d\phi$$
(12.261)

Note that the change in ϕ over one helical period is equal to $2\pi/N$ and that the integration is carried out over $2M$ helical periods with $M \to \infty$. This guarantees that the magnetic line will sample the entire flux surface, both inside and outside. Also, the quantity in the square bracket is to be evaluated along the magnetic line.

Now, to evaluate the integrand $I[r(\phi),\theta(\phi),\phi]$ one expands the field line trajectories as

$$\theta(\phi) = \theta_0(\phi) + \theta_1(\theta_0 + N\phi)$$
$$r(\phi) = r_0(\theta_0) + r_1(\theta_0 + N\phi) \tag{12.262}$$

Here, θ_1, r_1 are the small amplitude (of order δ), rapidly oscillating helical contributions to the trajectory. The quantities r_0, θ_0 are slowly varying with ϕ as the field line moves from the outside to the inside of the torus with the rotational transform (of order δ^2 per helical period). During one helical period r_0, θ_0 can be assumed to be constant.

The helical field contribution

Using this representation one can separate U into helical and toroidal contributions given by

$$U = U_h + U_t$$

$$U_h = \lim_{M \to \infty} \frac{R_0}{B_0} \int_{-2\pi M/N}^{2\pi M/N} \left(1 - \frac{B_{\phi 1}}{B_0} - \frac{1}{B_0} \frac{\partial B_{\phi 1}}{\partial r_0} r_1 - \frac{1}{B_0} \frac{\partial B_{\phi 1}}{\partial \theta_0} \theta_1 + \frac{B_{\phi 1}^2}{B_0^2} \right) d\phi$$

$$U_t = \lim_{M \to \infty} \frac{R_0}{B_0} \int_{-2\pi M/N}^{2\pi M/N} \left(-\frac{B_{\phi 2}}{B_0} + \frac{r_0}{R_0} \cos \theta_0 \right) d\phi = \lim_{M \to \infty} \frac{2R_0}{B_0} \int_{-2\pi M/N}^{2\pi M/N} \left(\frac{r_0}{R_0} \cos \theta_0 \right) d\phi$$

$$\tag{12.263}$$

All quantities in the integrand are evaluated at r_0, θ_0. Also, the contribution of $B_{\phi 2}$ to U_t is just the toroidal correction to B_ϕ: $B_{\phi 2}/B_0 = -(r_0/R_0)\cos \theta_0$.

The helical term U_h is identical to the straight stellarator contribution, discussed in the previous subsection as given by Eq. (12.243). Therefore, separating $\bar{\kappa}$ into helical and toroidal components

$$\bar{\kappa} = \bar{\kappa}_h + \bar{\kappa}_t \tag{12.264}$$

leads to

$$\bar{\kappa}_h = -2\delta^2 \left(\frac{r}{a} \right)^2 = -\frac{2}{N} \left(\frac{\iota_H}{2\pi} \right) u^2 \tag{12.265}$$

for a loosely wound $l = 2$ stellarator. Also, from Eq. (12.257), $\iota_H/2\pi \approx R_0 \delta^2/ha^2$ is the nearly constant $l = 2$ rotational transform.

The vertical field contribution

Next consider the toroidal contribution $\bar{\kappa}_t$ which requires the evaluation of U_t. This task can be carried out by switching from ϕ to θ_0 as the integration variable,

$$U_t = \lim_{M\to\infty} \frac{2R_0}{B_0} \int_{-2\pi M/N}^{2\pi M/N} \left(\frac{r_0}{R_0}\cos\theta_0\right) d\phi$$

$$= \lim_{M\to\infty} \frac{2R_0}{B_0} \int \left(\frac{r_0}{R_0}\cos\theta_0\right) \frac{d\phi}{d\theta_0} d\theta_0 \qquad (12.266)$$

$$= \lim_{M\to\infty} \frac{4\pi R_0}{B_0} \int \left[\frac{r_0}{\iota(r_0)R_0}\cos\theta_0\right] d\theta_0$$

Here, use has been made of the fact that as the magnetic line slowly moves around the flux surface due to the rotational transform, its helically averaged trajectory is given by $\theta_0(\phi) = \theta_i + [\iota(r_0)/2\pi]\phi$, where θ_i the starting angle. Therefore, $d\phi/d\theta_0 = 2\pi/\iota(r_0)$.

It is at this point that the vertical field enters the calculation. The vertical field causes the helically averaged flux surfaces to shift outward, along R, by an amount $\sigma(r_0)$. For mathematical self-consistency this shift should be finite corresponding to an ordering $\sigma/r_0 \sim 1$. However, to make the analysis simpler it is assumed that $\sigma/r_0 \ll 1$; the shift is assumed small for evaluation reasons and not because it scales with a small parameter. The small shift assumption implies that the field line trajectory can be approximated by

$$r_0(\theta_0) \approx r_i + \sigma(r_i)\cos\theta_0$$
$$\iota(r_0) \approx \iota(r_i) + \iota'(r_i)\,\sigma(r_i)\cos\theta_0 \qquad (12.267)$$

where r_i is the initial radius and serves as the field line label.

Equation (12.267) is substituted into Eq. (12.266) yielding

$$U_t = \lim_{M\to\infty} \frac{4\pi R_0}{B_0} \int \left[\frac{r_0}{\iota(r_0)R_0}\cos\theta_0\right] d\theta_0$$

$$= \lim_{M\to\infty} \frac{4\pi R_0}{B_0} \left[\frac{r_i}{R_0\iota(r_i)}\right] \int_{-2\pi M\iota/N}^{2\pi M\iota/N} \left[1 + \frac{\sigma(r_i)}{r_i}\cos\theta_0 - \frac{\iota'(r_i)}{\iota(r_i)}\sigma(r_i)\cos\theta_0\right]\cos\theta_0\, d\theta_0$$

$$(12.268)$$

This integral can be easily evaluated. All the oscillating terms average to essentially zero when integrating over many toroidal periods. Only the terms with a non-zero average value (the $\cos^2\theta_0$ terms) survive. The result is

$$U_t = \left(\frac{4\pi R_0 M}{B_0 N}\right)\left(1 - \frac{r}{\iota}\frac{d\iota}{dr}\right)\frac{\sigma}{R_0} \qquad (12.269)$$

Note that the subscript i has been suppressed from r for simplicity.

The value of $\bar{\kappa}_t$ is calculated by noting (from Eq. (12.263)) that the large constant part of U has the value $U_0 \approx 4\pi R_0 M/B_0 N$. This, after setting $u = hr$, leads to

$$\bar{\kappa}_t \approx -\frac{u}{U_0}\frac{\partial U_t}{\partial u} = u\frac{\partial}{\partial u}\left[\frac{u^2\sigma}{\imath R_0}\frac{\partial}{\partial u}\left(\frac{\imath}{u}\right)\right] \tag{12.270}$$

Relation between σ and B_V

One further step is needed to complete the calculation and that is to find the relation between $\sigma(u)$ and the applied vertical field B_V. This relation is found as follows. In the context of the Greene–Johnson expansion, the total normalized flux function can be written as $\psi(r,\theta,\phi) = \overline{\psi}(r,\theta) + \psi_h(r, l\theta + N\phi)$ with $\psi_h/\overline{\psi} \sim \delta$. The toroidally averaged shift $\sigma(r)\cos\theta$ arises from the θ dependence of $\overline{\psi}$ which satisfies the Greene–Johnson equilibrium equation. This equation is given by Eq. (7.57) or (7.62). For a vacuum field both equations overlap and are given by

$$\nabla_\perp^2\left[\overline{\psi} - \left(\sum_1^\infty \frac{i}{n}\mathbf{e}_\phi\cdot\nabla_\perp A_n^* \times \nabla_\perp A_n\right)\right] = 0 \tag{12.271}$$

The solution, including an applied vertical field, which appears as a homogeneous contribution, can be written as

$$\overline{\psi} = \sum_1^\infty \frac{i}{n}\mathbf{e}_\phi\cdot\nabla_\perp A_n^* \times \nabla_\perp A_n + C_1 r\cos\theta \tag{12.272}$$

Here C_1 is known in terms of B_V, the amplitude of the vertical field, once the normalizations are disentangled. Also, the helical field harmonic A_n is known from Eq. (7.60). A short calculation then leads to the following expressions for $\overline{\psi}$ in terms of the normalizations used in this section

$$\overline{\psi}(u,\theta) = \psi_0(u) + \psi_V(u)\cos\theta$$

$$\psi_0(u) = -\left(\frac{l\Delta^2}{2N\varepsilon^2}\right)\frac{I_l I_l'}{u} \tag{12.273}$$

$$\psi_V(u) = \left(\frac{1}{N\varepsilon^2}\frac{B_V}{B_0}\right)u$$

The proper ordering requires $\psi_0 \sim \psi_V \sim 1$ but, as above, ψ_V is assumed to be small for evaluation reasons. This approximation leads to the standard formula for the shift $\sigma(u) = -\psi_V(u)/h\psi'_0(u)$. After another short calculation that makes use of the relation $\imath/2\pi \approx (lN\Delta^2/2u)(I_l I_l'/u)'$ one finds

$$\frac{\sigma(u)}{R_0} \approx \frac{B_V}{B_0} \frac{2\pi}{\iota(u)} \qquad (12.274)$$

Equation (12.274) is substituted into Eq. (12.270) yielding

$$\overline{\kappa}_t(u) = -2\pi \frac{B_V}{B_0} u \frac{d^2}{du^2}\left(\frac{u}{\iota}\right) \qquad (12.275)$$

Observe that for a uniform rotational transform, $\overline{\kappa}_t = 0$. Shear is required in order to generate favorable average curvature.

After an admittedly lengthy calculation one can combine Eqs. (12.265) and (12.275) to obtain the desired expression for the average curvature in a loosely wound $l = 2$ stellarator,

$$\overline{\kappa} = \left[-\frac{2}{N}\left(\frac{\iota_H}{2\pi}\right) + 3\frac{B_V}{B_0}\left(\frac{2\pi}{\iota_H}\right)\right]u^2 \qquad (12.276)$$

Conclusions

Clearly, favorable average curvature is achieved for a sufficiently large vertical field satisfying

$$\frac{B_V}{B_0} > \frac{2}{3N}\left(\frac{\iota_H}{2\pi}\right)^2 \qquad (12.277)$$

There are several further observations that can be made. First, from a practical point of view most stellarators have a set of external vertical field coils whose current can be adjusted to produce average favorable curvature. Second, the rotational transform appears in the numerator of $\overline{\kappa}_h$ and the denominator of $\overline{\kappa}_t$. Therefore, as one moves further out in r, one must maintain the corrections due to the full modified Bessel functions rather than just using the small argument approximations. Since the modified Bessel functions grow exponentially with r, the implication is that the destabilizing helical contribution grows faster than the stabilizing vertical field contribution; it is only possible to create average favorable curvature over a plasma whose outer radius is not too large.

Third, shear can help stabilize unfavorable curvature at the edge of the plasma where the local β is small but this is not a very large effect. Perhaps more importantly, high shear in conjunction with a vertical field increases $\overline{\kappa}_t$, thereby changing the average curvature in the stabilizing direction. The final point is that stellarators whose vacuum magnetic field possesses average favorable curvature should be stable against interchange instabilities. This stability should persist when finite pressure effects are included, up to the point where ballooning effects become important. Ballooning modes will then set the limit on β_t.

12.12.7 Numerical results

The analysis presented above provides some insight into the stability of stellarators. While this is helpful, when one is designing and interpreting actual experiments more accurate predictions are needed. These are obtained from three-dimensional stability codes (see Further reading). There are several features of such codes that are worth discussing.

To test stability it is necessary to begin with an equilibrium. At the time this book is being written most of the equilibria used in stability codes assume closed nested flux surfaces. Equilibria with islands are not considered. This is primarily for numerical convenience since closed surface equilibria are much easier to calculate than equilibria with islands. In essence these codes numerically average over any islands that may exist. Full stability analysis including islands should be forthcoming over the next several years.

Another point of interest concerns configurational optimization. For comparison recall that extensive optimization studies have been carried out for the tokamak that focus on maximizing the MHD stable β as a function of cross section and profile, giving rise to the Sykes and Troyon limits. Optimization in a stellarator is more complicated. Specifically, it is not aimed solely at maximizing MHD β limits but in addition includes neoclassical transport and realistic coil design constraints. In other words, the "best" stellarator is not in general the one with the highest MHD stable β, but the one that makes the best trade-offs between MHD, transport, and engineering constraints.

A further difference between stellarators and tokamaks is as follows. When the MHD β or q limit is violated in a tokamak, the end result is a catastrophic major disruption. Thus, MHD stability boundaries set rather rigid limits on tokamak operation. In contrast, when the maximum β limit in a stellarator is violated, the plasma does not disrupt. Instead, transport starts to gradually increase due to MHD turbulence, behavior often referred to as a "soft landing." This is the reason why MHD does not stand alone in stellarator optimization but can be combined with transport and engineering constraints to arrive at a best design.

Because of this multi-constrained optimization there is no relatively simple scaling relation, such as the Sykes or Troyon limit, that applies to all stellarators. Vastly different stellarator geometries, for example the large aspect ratio isodynamic W7-X, the continuously wound LHD heliotron, and the compact quasi-symmetric NCSX, all have numerically determined critical β values on the order of 5%. Stated differently, to a certain extent stellarator design takes stable $\beta \sim 5\%$ as a constraint and then proceeds to optimize with respect to neoclassical transport and coil design. Overall, $\beta \sim 5\%$ is probably sufficient for energy applications and it is encouraging that a wide variety of stellarators can achieve this goal.

12.13 Summary

In Chapter 12 the Energy Principle has been used to investigate the stability of toroidal configurations, in particular the tokamak and the stellarator. A summary of the main results is given below.

General ballooning mode equation

There is a basic difference between cylindrical and toroidal systems with respect to MHD stability. Cylindrical symmetry implies that the curvature is unidirectional, unfavorable and only a function of r. In toroidal configurations there are usually alternating regions of favorable and unfavorable curvature. The $1/R$ dependence of the toroidal field produces favorable curvature on the inside of the torus ($\theta = \pi$) and unfavorable curvature on the outside ($\theta = 0$).

These alternating regions allow the possibility of ballooning modes. The corresponding eigenfunctions attempt to minimize δW by focusing in the unfavorable region while not causing too much line bending. The most dangerous ballooning modes correspond to $n = \infty$. They have long parallel wavelengths, short perpendicular wavelengths, and are localized about a rational flux surface. By exploiting the high-n limit plus the mode localization, one can reduce the general multidimensional Energy Principle to a one-dimensional differential equation along a field line on each flux surface. Stability can then be tested one field line at a time, representing a large savings in the effort required to test stability. The ballooning mode differential equation describes a competition between the stabilizing effects of line bending versus the destabilizing effects of local unfavorable curvature.

Tokamak stability

Toroidicity makes important modifications to the stability predictions of the straight tokamak. There are quantitative changes in the current-driven kink stability limits and qualitative differences when pressure is included.

In a low beta straight tokamak only current-driven modes are important. For internal modes, $m = 1$, $n = 1$ is usually the only instability and is the initial drive for sawtooth oscillations. The stability condition is $q_0 > 1$. In a torus the situation is more complicated with the results depending sensitively on the shape of the q profile near the axis. For a parabolic q profile the $n = 1$ mode is stable down to $q_0 = 1/2$ as $\beta_p \to 0$. Only when $\beta_p > 0.3$ does stability again require $q_0 > 1$. On the other hand, for a very flat q profile the stability boundary is $q_0 > 1$ even as $\beta_p \to 0$. Experiments have observed sawtooth oscillations with $q_0 \sim 0.75$.

With respect to low beta external kink modes, the stability predictions are similar in both the straight and toroidal models. A value $q_a > 1$ is required to avoid the $m = 1$, $n = 1$ Kruskal–Shafranov limit. However, higher values are required to stabilize higher m number external kinks. Typically $q_a > 2.5$–3 to stabilize external kinks. When the stability criterion is violated, too much current is flowing in the plasma and the usual result is a major disruption.

Another source of major disruptions is due to the Greenwald density limit. Here, at high density a low current edge layer forms, effectively shrinking the size of the plasma. The same current in a smaller plasma is again sufficient to drive a major disruption.

Much higher m external kinks drive the peeling component of the peeling–ballooning mode thought to be responsible for ELMs. These modes are most easily excited when the edge current gradient is large.

Consider now pressure-driven effects. For a straight tokamak pressure-driven modes do not lead to any significant limits on β_t. In contrast, in a torus, interchanges, ballooning modes, and ballooning-kink modes all play an important role in setting stability limits. The toroidal interchange condition is given by the Mercier criterion, a two-dimensional generalization of the Suydam criterion. The $n \to \infty$ interchanges are most important near the magnetic axis where the shear is small. A circular cross section tokamak develops average favorable curvature when $q_0 > 1$. When this condition is satisfied the plasma is stable against localized interchanges across the entire profile. For non-circular tokamaks a combination of elongation and outward pointing triangularity improves stability, allowing operation slightly below $q_0 < 1$, although one must be careful to ensure that the sawtooth condition is also satisfied.

The $n \to \infty$ ballooning modes are more serious than interchanges in a tokamak. These modes set important limits on the maximum allowable pressure gradient, more accurately $\nabla p / I^2$. High shear and elongated outward pointing triangularity are again favorable for stability. Numerical studies by Sykes *et al.* (1983) show that for optimized profiles against ballooning modes the stability limit is given by $\beta_t = 0.11 \, \varepsilon (1 + \kappa^2)/q_*$. For realistic experimental cases one finds $\beta_t \sim 0.1$. Ballooning effects play an important role in the peeling–ballooning modes thought to be responsible for ELMs.

Whereas ballooning modes set one limit on the maximum β_t in a tokamak, external ballooning-kink modes set an even stricter limit. As β_t is increased the current-driven kink mode develops a strong ballooning component, transforming to a ballooning-kink mode. A combination of elongation and outward triangularity also improves stability against these modes. The numerical studies of Troyon *et al.*

(1984) have shown that the maximum beta limit in a profile optimized tokamak is given by $\beta_t = 0.07\,\varepsilon(1 + \kappa^2)/q_*$. For practical cases $\beta_t \sim 0.06$. Higher values can be stabilized with a conducting wall. However, for a finite conductivity wall the mode remains unstable but as a resistive wall mode which requires feedback stabilization. When the ballooning-kink mode does become unstable, it is a very dangerous mode, invariably leading to a major disruption.

A final class of modes that occur in tokamaks results from $n = 0$ axisymmetric perturbations. These are predominantly vertical instabilities which, with a conducting wall, have the form of resistive wall modes. Stability requires feedback control and practical engineering considerations usually limit the allowable elongations to about $\kappa < 1.8$. Excitation of an $n = 0$ axisymmetric mode usually leads to a major disruption.

With care, standard tokamaks can avoid, or limit the number of major disruptions. However, when the need for steady state operation is taken into account, advanced tokamak (AT) operation is required. This invariably results in a violation of the Troyon limit and the excitation of resistive wall modes.

The basic conclusion is that there are a variety of modes that can become unstable in a tokamak. Overall stability can be achieved by a combination of cross sectional shaping, tight aspect ratio, sufficiently low current and pressure, and feedback. The maximum stable values of β_t lie in the range $\beta_t \sim 0.05$–0.1, perhaps slightly low, but still within the regime of reactor interest. In AT operation, the β_t limit will likely be violated, leading to a resistive wall mode that requires feedback stabilization.

Stellarator stability

The stellarator is the most complicated of the configurations investigated. It is a truly three-dimensional system with a wide variety of strategies for optimizing performance. Because of the geometric complexity there are relatively few analytic guidelines to provide intuition. What does exist is largely due to studies carried out in the early days of the program. However, because of advances in computing and computers it has become possible to study the MHD stability of stellarators by means of large-scale codes.

As a general comment note that stability is dominated by pressure-driven modes since stellarators typically have small or zero net toroidal current. Also when the β_t limit is exceeded, stellarators do not suffer major disruptions. This is largely a consequence of the fact that the applied helical field has a fixed magnetic axis as opposed to a tokamak, where the axis can move with the plasma. Therefore, stellarators can sometimes operate above the ideal MHD limit, although once this occurs there is usually an increase in transport due to MHD turbulence. The MHD β_t limit thus sets a reasonable disruption-free limit on high-performance operation.

In terms of intuition, the analytic theory is largely focused on achieving average favorable curvature which is closely related to the existence of a magnetic well. A relatively simple expression is derived that is neither necessary nor sufficient for stability but nonetheless provides some insight into the ways to achieve favorable average curvature.

One key feature of many stellarator configurations is that ballooning modes are often less unstable than in a tokamak. The reason is that the global shear is negative. This in turn implies that as β_t is increased, the local shear increases in the region of unfavorable curvature, exactly the opposite of a tokamak. This is desirable for stability and thereby decreases the relative importance of ballooning instabilities.

The average curvature criterion shows that a straight, single helicity stellarator always has unfavorable average curvature. Shear can provide stability but only at low values of β_t. To improve the situation toroidal effects must be taken into account, specifically the combination of toroidal curvature plus a vertical field. The vertical field affects the toroidally averaged shift of the magnetic axis which has a strong impact on stability.

In the end, there is no simple universal β_t limit for a stellarator because of the wide variety of possible configurations. Three-dimensional numerical studies show that stable β_t values on the order of $\beta_t \sim 0.05$ are possible in MHD-transport optimized stellarators. Higher values are possible but the need to optimize including neoclassical transport effects is a major limiting factor.

12.14 The final word

In the Introduction it is pointed out that a fusion reactor will require a pressure on the order of 5–10 atm for ignited operation. The goal of MHD studies is to discover magnetic geometries that can achieve such pressures with good equilibrium and stability properties. Well, what is the answer?

The tokamak and stellarator appear to be capable of stable operation with a maximum $\beta \sim 5\%$–10%, but without much safety margin. This translates into a pressure on the order of 5–10 atm assuming a reasonable aspect ratio and a maximum field on the (superconducting) coil of about 13 T. Most importantly, $n \geq 1$ stability should be achievable without the need of a conducting wall, a substantial saving in technological complexity. Still the situation is not completely resolved for the following reasons.

In addition to high β, disruption-free operation a tokamak reactor must operate as a steady state device for engineering reasons. This requires non-inductive current drive which because of its low efficiency must rely heavily on the bootstrap current. The resulting AT tokamak profiles that generate the nominal 75% bootstrap fraction required for economics invariably violate the MHD Troyon limit. The implication is that a conducting wall, accompanied by resistive wall stabilization is required. The need for a conducting wall erases some of the original advantages of the tokamak.

Much higher β values are possible in the spherical tokamak, on the order of 30%. However, these do not necessarily translate into high pressures because of the lower maximum field (~ 7–8 T) on the central (copper) coil and the rapid decay of B_ϕ away from the TF coil because of the tight aspect ratio.

Stellarators, on the other hand, are inherently steady state and will likely be disruption free. Theoretical studies and experimental results indicate that stable β values on the order of 5% are possible. This is indeed a favorable result from the MHD point of view. However, because of the 3-D geometry neoclassical transport in a stellarator is in general substantially worse than in a tokamak. Clever magnetic geometries can produce omnigenous behavior which should reduce neoclassical transport to an acceptable level. This is one of the main goals of present stellarator research – achieve transport comparable to H-mode in a tokamak. Still, MHD has done its job. The problem now falls into the domain of the transport community.

Lastly, it is worth noting that the reversed field pinch should be capable of achieving a high stable β, on the order of 20–30%, against all internal ideal MHD modes with a relatively simple coil set, clearly a significant advantage technologically. On the other hand, the RFP is always unstable to a wide range of external MHD modes, which the naysayers believed could never all be simultaneously wall stabilized. Experiments in Europe have shown quite remarkably that a complex feedback system could be designed and built that indeed stabilizes the external modes. This promising result indicates that MHD stability is not a show-stopper for the RFP. Instead, the standard RFP must improve its transport properties, which are substantially poorer than for a tokamak, presumably because of resistive MHD turbulence. Edge profile control and the formation of single helicity states are two possible ideas that address the transport problem. There is also the need for a large amount of current drive since the bootstrap current is small in an RFP.

The final word? MHD has done a plausible job discovering configurations with reactor relevant values of pressure although there is not a lot of safety margin. The transport community still has to discover ways to further reduce cross-field heat conduction. The RF-neutral beam community still needs to develop more efficient ways to drive current in axisymmetric geometries.

References

Bussac, M. N., Pellat, R., Edery, D., and Soule, J. L. (1975). *Phys. Rev. Lett.* **35**, 1638.
Connor, J. W., Hastie, R. J., and Taylor, J. B. (1979). *Proc. R. Soc. London, Ser. A.* **365**, 1.
Coppi, B. (1977). *Phys. Rev. Lett.* **39**, 939.
Dewar, R. L. and Glasser, A. H. (1983). *Phys. Fluids* **26**, 3038.
Freidberg, J. P. and Haas, F. A. (1973). *Phys. Fluids* **16**, 1909.
Goedbloed, J. P., Keppens, R., and Poedts, S. (2010). *Advanced Magnetohydrodynamics*. Cambridge: Cambridge University Press.
Greene, J. M. and Johnson, J. L. (1961). *Phys. Fluids* **4**, 875.
Greene, J. M. and Chance, M. S. (1981). *Nucl. Fusion* **21**, 453.
Hegna, C. C. and Nakajima, N. (1998). *Phys. Plasmas* **5**, 1336.

Helander, P. and Sigmar, D. J. (2002). *Collisional Transport in Magnetized Plasmas.* Cambridge: Cambridge University Press.

Jardin, S. (2010). *Computational Methods in Plasma Physics.* Boca Raton, FL: Chapman and Hall/CRC Press.

Laval, G., Luc, H., Maschke, E. K., *et al.* (1971). In *Plasma Physics and Controlled Nuclear Fusion Research 1970.* Vienna: IAEA, Vol. II, p. 507.

Lazarus, E. A., Chu, M. S., Ferron, J. R. *et al.* (1991). *Phys. Fluids B* **3**, 2220.

Lenard, A. (1964). *Phys. Fluids* **7**, 1875.

Lortz, D. and Nuhrenberg, J. (1973). *Nucl. Fusion* **13**, 821.

Manickam, J., Pomphrey, N., and Todd, A. A. M. (1987). *Nucl. Fusion* **27**, 1461.

Menard, J. E., Bell, M. G., Bell, R. E. *et al.* (2002). *Proc. 19th International Atomic Energy Agency (IAEA) Fusion Energy Conference*, Lyon, France, paper EX/S1 5.

Menard, J. E., Bell, M. G. *et al.* (2004). *Phys. Plasmas* **11**, 639.

Mercier, C. (1960). *Nucl. Fusion* **1**, 47.

Mikhailovskii, A. B. (1974). *Nucl. Fusion* **14**, 483.

Mikhailovskii, A. B. and Shafranov, V. D. (1974). *Sov. Phys. – JETP* **39**, 88.

Rosenbluth, M. N. and Longmire, C. L. (1957). *Annal. Phys. NY* **1**, 120.

Shafranov, V. D. and Yurchenko, E. I. (1968). *Sov. Phys. – JETP* **26**, 682.

Sinclair, R. M., Yoshikawa, S., Harries, W. L., Young, K. M., Weimer, K. E., and Johnson, J. L. (1965). *Phys. Fluids* **8**, 118.

Snyder, P. B., Wilson, H. R., and Xu, X. Q. (2005). *Phys. Plasmas* **12**, 056115.

Solov'ev, L. S., Shafranov, V. D., and Yurchenko, E. I. (1969). In *Plasma Physics and Controlled Nuclear Fusion Research.* Vienna: IAEA, Vol. 1, p. 173.

Strait, E. J. (2005). *Fusion Sci. and Tech.* **48**, 864.

Sykes, A. and Wesson, J. A. (1974). *Nucl. Fusion* **14**, 645.

Sykes, A., Turner, M. F., and Patel, S. (1983). In *Controlled Fusion and Plasma Physics*, 11th European Conference, Aachen, West Germany, p. 363.

Troyon, F., Gruber, R., Saurenmann, H., Semenzato, S., and Succi, S. (1984). *Plasma Phys.* **26**, 209.

Ware, A. A. and Haas, F. A. (1966). *Phys. Fluids* **9**, 956.

Wesson, J. A. and Sykes, A. (1974). In *Plasma Physics and Controlled Nuclear Fusion Research 1974.* Vienna: IAEA, Vol. I, p. 449.

Further reading

Tokamak and stellarator analytic theory

Goedbloed, J. P., Keppens, R., and Poedts, S. (2010). *Advanced Magnetohydrodynamics.* Cambridge: Cambridge University Press.

Kikuchi, M., Lackner, K. and Tran, M. Q., eds. (2012). *Fusion Physics.* Vienna: International Atomic Energy Agency.

Wesson, J. A. (2011). *Tokamaks*, 4th edn. Oxford: Oxford University Press.

Stellarator numerical codes

Anderson, D. V., Cooper, W. A., Gruber, R. *et al.* (1990) *Scient. Comp. Supercomputer* **II**, 159.

Garabedian, P. (2002). *Proc. Nat. Acad. Sci.* **16**, 10257.

Hirshman, S. P. and Whitson, J. C. (1983). *Phy. Fluids* **26**, 3553.

Reiman, A. and Greenside, H. S. (1986). *Comput. Phys. Commun.* **43**, 157.

Problems

12.1 This problem demonstrates that any divergence-free vector field can be written in the form $\mathbf{B} = \nabla\alpha \times \nabla\beta$. Consider two functions $\hat{\alpha}(r)$, $\hat{\beta}(r)$ such that the magnetic lines lie in the contours $\hat{\alpha} = $ constant and $\hat{\beta} = $ constant. The intersection of two contours $\hat{\alpha} = \alpha_0$ and $\hat{\beta} = \beta_0$ defines a given magnetic line.

(a) Show that $\nabla\hat{\alpha} \times \nabla\hat{\beta}$ points in the direction of the magnetic field at every point in space.
(b) Assuming (a) to be true, one can write the magnetic field as $\mathbf{B} = A(\mathbf{r})\nabla\hat{\alpha} \times \nabla\hat{\beta}$. Here, $A(\mathbf{r})$ is a scalar function adjusted appropriately so that $\mathbf{B}$ has the correct magnitude. Show that the condition $\nabla \cdot \mathbf{B} = 0$ implies that $A(\mathbf{r}) = A\left(\hat{\alpha}, \hat{\beta}\right)$.
(c) Define new coordinates α and β as follows:

$$\alpha = \hat{\alpha}$$

$$\beta = \int_0^{\hat{\beta}} A(\hat{\alpha}, \beta')d\beta'$$

Show that

$$\mathbf{B} = \nabla\alpha \times \nabla\beta$$

12.2 Consider an ideal MHD plasma in the presence of a fixed external current density source $\mathbf{J}_e$. Show that δW_F can be written in the following intuitive form:

$$\delta W_F = \frac{1}{2\mu_0} \int d\mathbf{r} \Big[|\mathbf{Q}_\perp|^2 + B^2|\nabla \cdot \boldsymbol{\xi}_\perp + 2\boldsymbol{\xi}_\perp \cdot \boldsymbol{\kappa}_c|^2 + \gamma\mu_0 p_c|\nabla \cdot \boldsymbol{\xi}|^2$$

$$-\mu_0(\mathbf{b} \cdot \mathbf{J}_c)\boldsymbol{\xi}_\perp^* \times \mathbf{b} \cdot \mathbf{Q}_\perp - 2\mu_0(\boldsymbol{\xi}_\perp \cdot \nabla p_c)(\boldsymbol{\xi}_\perp^* \cdot \boldsymbol{\kappa}_c) \Big]$$

where p_c is the plasma pressure, $\mathbf{J}_c = \mathbf{J} - \mathbf{J}_e$ is the plasma current, and

$$\boldsymbol{\kappa}_c = \boldsymbol{\kappa} + \frac{\mu_0}{2}\left(\frac{\mathbf{b} \times \mathbf{J}_e}{B}\right)$$

12.3 Consider the form of δW_F given in Problem 12.2. Introduce an eikonal representation for $\boldsymbol{\xi}_\perp$ and show that in the limit $k_\perp \to \infty$, δW_F can be written as

$$\delta W_F = \frac{1}{2\mu_0} \int d\mathbf{r} \Big[k_\perp^2 |\mathbf{b} \cdot \nabla X|^2 - \frac{2\mu_0}{B^2}(\mathbf{b} \times \mathbf{k}_\perp \cdot \nabla p_c)(\mathbf{b} \times \mathbf{k}_\perp \cdot \boldsymbol{\kappa}_c)|X|^2 \Big]$$

12.4 In this problem the effects of compressibility on the ballooning mode stability of closed line systems is investigated. For simplicity carry out the analysis in a straight 2-D "bumpy" θ-pinch geometry where the coordinates are the flux ψ, the axial arc length l, and the poloidal angle θ (corresponding to the symmetry direction).

(a) Show that the terms giving rise to a zeroth-order contribution in δW proportional to $\mathbf{k}_\perp$ in the eikonal representation are given by

$$\delta W_0 = \frac{1}{2} \int \frac{d\theta d\psi dl}{B} \left[\frac{B^2}{\mu_0} |\nabla \cdot \boldsymbol{\xi}_\perp + 2\boldsymbol{\xi}_\perp \cdot \boldsymbol{\kappa}|^2 + \gamma p |\langle \nabla \cdot \boldsymbol{\xi}_\perp \rangle|^2 \right]$$

where

$$\langle F \rangle = \oint \frac{Fdl}{B} \bigg/ \oint \frac{dl}{B}$$

(b) Write $\boldsymbol{\xi}_\perp = \boldsymbol{\eta}_\perp \exp(iS)$, where $\mathbf{B} \cdot \nabla S = 0$ and $\mathbf{k}_\perp = \nabla S$. Expand $\boldsymbol{\eta}_\perp = \boldsymbol{\eta}_{\perp 0} + \boldsymbol{\eta}_{\perp 1}$. Minimize δW_0 and show that the first non-vanishing contribution is given by

$$\delta W_0 = \frac{1}{2} \int d\theta d\psi W_0$$

Here

$$W_0 = \oint \frac{dl}{B} \left[\frac{B^2}{\mu_0} (U + 2\boldsymbol{\eta}_{\perp 0} \cdot \boldsymbol{\kappa}_0)^2 + \gamma p \langle U \rangle^2 \right]$$

with $U = i\mathbf{k}_\perp \cdot \boldsymbol{\eta}_{\perp 1} + \nabla \cdot \boldsymbol{\eta}_{\perp 0}$ and $\boldsymbol{\eta}_{\perp 0} = (X/B)(\mathbf{b} \times \mathbf{k}_\perp)$.

(c) Note that $\boldsymbol{\eta}_{\perp 1}$ appears only in W_0 and not in the remainder of δW. Minimize W_0 with respect to U. A convenient way to carry out the minimization is to introduce the variable

$$V(l) = \int_0^l U \frac{dl}{B}$$

It immediately follows that

$$U = B \frac{\partial V}{\partial l}$$

and

$$\langle U \rangle = V(L)/K_1$$

where

$$K_n = \oint \frac{dl}{B^n}$$

Show that the minimized W_0 can be written as

$$W_0 = \frac{4\gamma p K_1^2 \langle \boldsymbol{\eta}_{\perp 0} \cdot \boldsymbol{\kappa} \rangle^2}{\mu_0 \gamma p K_3 + K_1}$$

(d) Show that the full contribution to W (for ballooning modes) is given by

$$W = \frac{4\gamma p K_1^2}{\mu_0 \gamma p K_3 + K_1} \left\langle \frac{\kappa_n X}{rB} \right\rangle^2$$

$$+ \oint dl \left[\frac{1}{\mu_0 r^2 B} \left(\frac{\partial X}{\partial l} \right)^2 - 2 \frac{dp}{d\psi} \frac{\kappa_n X^2}{rB^2} \right]$$

where $\kappa_n = \mathbf{n} \cdot \boldsymbol{\kappa}$ is the normal curvature.

(e) Consider the interchange perturbation $X = 1$. Show that the stability condition can be expressed

$$-2\frac{dp}{d\psi} \left\langle \frac{\kappa_n}{rB} \right\rangle + \left[\frac{4\gamma p}{1 + \mu_0 \gamma p \langle 1/B^2 \rangle} \right] \left\langle \frac{\kappa_n}{rB} \right\rangle^2 > 0$$

12.5 This problem demonstrates a practical procedure for calculating the orthogonal flux coordinates used in the tokamak ballooning mode analysis. Assume that the flux function $\psi(R,Z)$ and its R and Z derivatives have been found either numerically or analytically by solving the Grad–Shafranov equation. The goal is to determine the coordinates $R = R(\psi,\chi)$, $Z = Z(\psi,\chi)$ where χ is an angle-like variable with period 2π orthogonal to ψ: $\nabla\psi \cdot \nabla\chi = 0$.

(a) Write the total differentials for $d\psi$ and $d\chi$. Show that for orthogonal coordinates, R and Z satisfy the equations

$$\frac{\partial R}{\partial \psi} = \frac{1}{(\nabla\psi)^2} \frac{\partial \psi}{\partial R}$$

$$\frac{\partial Z}{\partial \psi} = \frac{1}{(\nabla\psi)^2} \frac{\partial \psi}{\partial Z}$$

(b) The equations in part (a) represent a set of coupled, non-linear ordinary differential equations for R and Z with χ appearing as a parameter. R and Z can be obtained by a simple numerical integration once appropriate starting conditions in ψ and a value of χ are specified. The starting conditions are slightly subtle, but can be obtained as follows. Expand all quantities about the magnetic axis:

$$R = R_a + x, \quad Z = y, \quad \psi = \left(\psi_{RR} x^2 + \psi_{ZZ} y^2 \right)/2.$$

Here, ψ_{RR} and ψ_{ZZ} are constants, equal to the derivatives evaluated on the magnetic axis and the arbitrary constant in ψ has been chosen so that $\psi(R_a,0) = 0$. Show that the equations in the vicinity of the axis can be written as

$$\frac{\partial X}{\partial \psi} = 2\psi_{RR}\frac{X}{X+Y}$$

$$\frac{\partial Y}{\partial \psi} = 2\psi_{ZZ}\frac{Y}{X+Y}$$

where $X = \psi_{RR}^2 x^2$, $Y = \psi_{ZZ}^2 y^2$.

(c) Solve the equations in part (b) for X and Y. Show that

$$X = F(\chi)Y^\alpha$$

$$X + \alpha Y = 2\psi_{RR}\psi$$

where $F(\chi)$ is an arbitrary function and $\alpha = \psi_{RR}/\psi_{ZZ}$. For elongated flux surfaces $\alpha > 1$.

(d) The function $F(\chi)$ is not unique; that is, there are an infinite number of ways to label the orthogonal angle coordinate χ. A convenient choice, satisfying the 2π periodicity constraint and possessing the conventional properties that $\chi = 0$ when $y = 0$ and $\chi = \pi/2$ when $x = 0$ is given by

$$F(\chi) = K\cot^2\chi$$

The constant K is arbitrary and its value is of no great significance. Different K values correspond to different labels of the constant χ surfaces as χ varies from 0 to 2π. However, the shape of the constant χ surfaces remains unchanged. Choose $K = \psi_{RR}^2/\psi_{ZZ}^{2\alpha}$. Show that the proper starting conditions for small ψ can be written as a set of coupled transcendental algebraic equations given by

$$\frac{y^2}{\alpha^{1/2}} + y^{2\alpha}\alpha^{1/2}\cot^2\chi = \frac{2\psi}{(\psi_{RR}\psi_{ZZ})^{1/2}}$$

$$x^2 = y^{2\alpha}\cot^2\chi$$

(e) Sketch the $\psi =$ constant and $\chi =$ constant contours in the vicinity of the origin. To calculate the orthogonal coordinates $R(\psi,\chi)$, $Z(\psi,\chi)$ one specifies a value of χ and a small value of ψ (near the magnetic axis). The equations in part (d) are then solved simultaneously for y and x. These values are used as starting conditions for the basic equations in part (a). This process is then repeated for a sequence of χ values.

12.6 Consider the transformation of coordinates $(R,Z) \rightarrow (\psi,\chi)$ used in the discussion of axisymmetric tori. First carry out the transformation in the forward direction

$$R\,dR\,dZ = J\,d\psi\,d\chi$$

Next carry out the transformation in the reverse direction

$$d\psi\,d\chi = J'R\,dR\,dZ$$

By definition $J = 1/J'$. Show that

$$JB_p = \frac{1}{\mathbf{b}_p \cdot \nabla\chi} = \mathbf{b}_p \cdot \frac{\partial \mathbf{r}}{\partial \chi}$$

where $\mathbf{r} = R(\psi,\chi)\mathbf{e}_R + Z(\psi,\chi)\mathbf{e}_Z$ and $\mathbf{b}_p = \mathbf{B}_p/B_p$.

12.7 The maximum value of β_t allowed by ballooning instabilities has been investigated using a simple analytic model. This problem reexamines the analysis using a slightly more realistic model. Assume the entire profile is determined by operation on the marginal stability boundary of the ballooning mode. For simplicity assume the boundary is given by $\alpha = Ks$ where $K \approx 0.6$.

(a) Show that the critical β_t can be written as

$$\beta_t = \frac{\varepsilon K}{2}\left[-\frac{1}{q_a^2} + 3\int_0^1 \frac{x^2}{q^2}\,dx \right]$$

where $x = r/a$.
(b) Assume the q profile has the form

$$q(x) = q_0(1 + \lambda x^3)^\nu$$

Find the relationship between λ and ν such that $J_\phi(a) = 0$.
(c) Calculate $\beta_t = (\varepsilon K/2)F(q_a,q_0)$. Set $q_0 = 1$ and plot $F(q_a)$ versus $1/q_a$. Compare the dependence of F on $1/q_a$ with the predictions of Sykes and Troyon: $F \propto 1/q_a$. Note that there is quite good scaling agreement over the experimentally interesting range $2 < q_a < 6$.

12.8
(a) Write down the general pressure-balance jump condition for the sharp boundary surface current model of an axisymmetric tokamak. Derive an analytic expression for $B_p = B_p(\theta)$ corresponding to an arbitrary plasma cross section $r = r_p(\theta)$. Do not make a large aspect ratio or low β expansion.
(b) Derive the $\beta_t q_*^2/\varepsilon$ equilibrium limit as a function of ε for a circular cross section plasma.
(c) Repeat step (b) for an elliptic cross section plasma with $\kappa = 2$.

12.9 Show that the surface term in δW for an arbitrary three-dimensional sharp boundary equilibrium can be written as

$$\delta W_S = \frac{1}{2} \int dS |\mathbf{n} \cdot \boldsymbol{\xi}_\perp|^2 \left[2p\langle \kappa_n \rangle + \langle B^2/\mu_0 \rangle (\hat{\kappa}_n - \kappa_n) \right]$$

where $\langle \kappa_n \rangle = (\hat{\kappa}_n + \kappa_n)/2$ is the average curvature across the surface and $\langle B^2 \rangle = (\hat{B}^2 + B^2)/2$ is the average magnetic energy. Note that the first term is analogous to the pressure-driven contribution in δW proportional to the product of the pressure gradient and the curvature. The second term is analogous to the current-driven contribution in δW; the quantity $\hat{\kappa}_n - \kappa_n$ is non-zero only when there is a non-zero parallel current.

12.10 In this problem the high β tokamak expansion of δW_F is used to derive a ballooning mode equation. The goal is to see whether it makes any difference in the final equation whether one expands in ε and then takes the ballooning limit or vice versa. Starting with Eq. (12.123) introduce an eikonal representation for $U : U = \hat{U}(r, \theta) \exp[iS_\perp(r, \theta)]$, where $\mathbf{B}_p \cdot \nabla_\perp S_\perp - nB_0/R_0 = 0$ and $n \to \infty$. Derive the ballooning mode equation following the procedure given in Section 12.3. Does the resulting equation agree with the high β limit of the exact ballooning mode equation (i.e., Eq. (12.40)) in the limit of the high β expansion?

12.11 Evaluate the magnetic well for an ohmically heated tokamak as follows. Using the inverse aspect ratio expansion show that on a plasma flux surface $r = r_0 + \Delta(r_0 \cos \theta)$,

$$B^2 \approx B_0^2 \left[1 - 2\frac{r_0}{R_0} \cos \theta - \left(\frac{2\Delta}{R_0} - \frac{3r_0^2}{R_0^2} \right) \cos^2 \theta + \frac{2B_{\phi 2}}{B_0} + \frac{B_{\theta 1}^2}{B_0^2} \right]$$

where $\Delta(r_0) = -\psi_1(r_0)/\psi_0'(r_0)$ is the flux surface shift. From this result show that the magnetic well $W(r_0)$ is given by

$$W = -\frac{B_{\theta 1}^2}{B_0^2} - \frac{r_0}{2R_0} \left[\frac{(r_0^2 \Delta)'}{r_0} \right] + \frac{r^2}{2R_0^2}$$

12.12 Consider a large aspect ratio torsatron in which the magnetic fields are created by a surface current located at $r = a_c$, which is given by

$$\mathbf{K} = K_0 [1 + \cos(l\theta - N\phi)] (\mathbf{e}_\theta \cos \alpha + \mathbf{e}_\phi \sin \alpha)$$

where $\tan \alpha = lR_0/Na_c$. Derive a formula for the rotational transform at $r = a$.

12.13 Show that the parallel current constraint given by Eq. (12.191) is automatically satisfied for an arbitrary one-dimensional screw pinch equilibrium. Repeat the calculation for an arbitrary axisymmetric torus as described by the Grad–Shafranov equation.

Appendix A

Heuristic derivation of the kinetic equation

A.1 Introduction

The basic model describing MHD and transport theory in a plasma is the kinetic-Maxwell equations, which consist of a set of coupled electromagnetic and kinetic equations. The electromagnetic behavior is governed by the full Maxwell's equations (i.e., displacement current and Poisson's equation are included). In the kinetic model each species is described by a distribution function $f_\alpha(\mathbf{r}, \mathbf{v}, t)$, which satisfies a 6-D plus t integro-differential equation including the effect of collisions. The equations are very general and very, very difficult to solve. They accurately describe behavior ranging from the fast $\omega_{c\alpha}$ and $\omega_{p\alpha}$ time scales, down to the slower MHD time scale and the even slower transport time scale.

Since the ideal MHD model is based on the kinetic-Maxwell equations, the first step in the theoretical development of MHD is a derivation of the kinetic equation. A simple heuristic derivation is presented below.

A.2 Heuristic derivation of the kinetic equation

The derivation is based on intuitive common sense applied to the conservation of particles; that is, counting and conserving particles in the 6-D $(\mathbf{r}, \mathbf{v})$ phase space leads to the kinetic equation. The plan of attack is as follows:

1. Derive the conservation of particles relation for a simple fluid in 3-D physical space.
2. Generalize this result to a 6-D phase space assuming only long-range forces are present, which corresponds to neglecting collisions. The result is the Vlasov equation.
3. Add in the effect of short-range collisional forces leading to the kinetic equation. A key assumption is that binary collisions dominate the behavior so that n-body collisions with $n \geq 3$ can be neglected.
4. Discuss the general conservation properties of the kinetic collision operator.

A.3 Conservation of particles in 3-D physical space

For simplicity, consider first the 1-D geometry illustrated in Fig. A.1. Now assume that there are no sources or sinks of particles within the fluid. Then, intuitively, conservation of particles states that the gain in the number of particles in a fixed Eulerian volume element $\Delta x \Delta y \Delta z$ is given by

$$\text{gain in particles} = \text{flow in} - \text{flow out} \tag{A.1}$$

Each of these terms can be evaluated separately as follows:

gain in particles in a time $\Delta t = [n(x, t + \Delta t) - n(x, t)]\Delta x \Delta y \Delta z$

flow in of particles in a time $\Delta t = (\text{flux})(\text{area})\Delta t = \Delta y \Delta z \Delta t (nu_x)\big|_{x - \Delta x/2}$

flow out of particles in a time $\Delta t = (\text{flux})(\text{area})\Delta t = \Delta y \Delta z \Delta t (nu_x)\big|_{x + \Delta x/2}$

$$\tag{A.2}$$

The next step is to Taylor expand assuming Δt, Δx, Δy, Δz are all small quantities

$$\Delta x \Delta y \Delta z \Delta t \frac{\partial n}{\partial t} = \Delta y \Delta z \Delta t \left[nu_x\big|_x - \frac{\Delta x}{2} \frac{\partial (nu_x)}{\partial x}\bigg|_x \right] - \Delta y \Delta z \Delta t \left[nu_x\big|_x + \frac{\Delta x}{2} \frac{\partial (nu_x)}{\partial x}\bigg|_x \right]$$

$$\tag{A.3}$$

which reduces to

$$\frac{\partial n}{\partial t} + \frac{\partial}{\partial x}(nu_x) = 0 \tag{A.4}$$

This is conservation of particles in 1-D. In 3-D it generalizes to

$$\frac{\partial n}{\partial t} + \frac{\partial}{\partial x}(nu_x) + \frac{\partial}{\partial y}(nu_y) + \frac{\partial}{\partial z}(nu_z) = 0$$

$$\frac{\partial n}{\partial t} + \nabla \cdot (n\mathbf{u}) = 0 \tag{A.5}$$

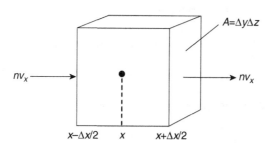

Figure A.1 Geometry in physical space illustrating conservation of particles.

A.4 Kinetic generalization to a 6-D phase space

The same ideas can be used to generalize conservation of particles into a 6-D phase space, with three dimensions corresponding to physical space ($\mathbf{r} \equiv x, y, z$) and three to velocity space ($\mathbf{v} \equiv v_x, v_y, v_z$). The following generalizations are used in the derivation:

$$n(\mathbf{r}, t) \rightarrow f(\mathbf{r}, \mathbf{v}, t) = \text{6-D} \qquad \text{phase space density}$$
$$\Delta x \Delta y \Delta z \rightarrow \Delta x \Delta y \Delta z \Delta v_x \Delta v_y \Delta v_z = \text{6-D} \quad \text{volume element} \qquad \text{(A.6)}$$

Physically, conservation of particles corresponds to particle bookkeeping. It takes into account that particles can flow into or out of the physical part of a volume element with a velocity $\mathbf{v}$. Similarly, particles can accelerate into or out of the velocity part of a volume element with an acceleration $\mathbf{a}$. Note that $\mathbf{v}$ and $\mathbf{r}$ are independent coordinates (variables). At any point $\mathbf{r}$, a particle can have any velocity $\mathbf{v}$. However, in general $\mathbf{a} = \mathbf{a}(\mathbf{r}, \mathbf{v}, t)$. A particle at $\mathbf{r}$ moving with velocity $\mathbf{v}$ will have a known acceleration as determined by Newton's law and the specific force field under consideration.

The conservation of particles relation for the 6-D phase space is similar to the simpler relation for the 3-D physical space discussed above. The one difference is that the relation is generalized to include sources and sinks which, as discussed below, are needed to account for short-range collisions. The generalized relation for each 6-D volume element is given by

$$\text{gain in particles} = \text{flow in} - \text{flow out} + \text{sources} - \text{sinks} \qquad \text{(A.7)}$$

Next, each of these terms is evaluated for the reduced case which allows only 1-D motion in physical space (i.e., the x direction) and 1-D motion in velocity space (i.e., the v_x direction). The extension to 6-D motion is straightforward. The terms in the equation are as follows (see Fig. A.2):

gain in a time $\Delta t = [f(x, v_x, t + \Delta t) - f(x, v_x, t)] \Delta \mathbf{r} \Delta \mathbf{v}$
flow into physical space $= \Delta y \Delta z \Delta t [v_x f \, d\mathbf{v}]_{x - \Delta x/2}$
flow out of physical space $= \Delta y \Delta z \Delta t [v_x f \, d\mathbf{v}]_{x + \Delta x/2}$
acceleration into velocity space $= \Delta v_y \Delta v_z \Delta t [a_x f \, d\mathbf{r}]_{v_x - \Delta v_x/2}$ (A.8)
acceleration out of velocity space $= \Delta v_y \Delta v_z \Delta t [a_x f \, d\mathbf{r}]_{v_x - \Delta v_x/2}$
sources $-$ sink $= s \Delta \mathbf{r} \Delta v \Delta t$

To proceed, Taylor expand as before. This leads to

$$\frac{\partial f}{\partial t} + \frac{\partial}{\partial x}(v_x f) + \frac{\partial}{\partial v_x}(a_x f) = s(x, v_x, t) \qquad \text{(A.9)}$$

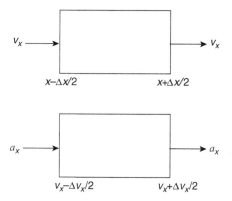

Figure A.2   Geometry in phase space illustrating conservation of particles.

Note that $\partial v_x / \partial x = 0$ (independent coordinates). The generalization to the full 6-D case can thus be written as

$$
\frac{\partial f}{\partial t} + \left(v_x \frac{\partial}{\partial x} + v_y \frac{\partial}{\partial y} + v_z \frac{\partial}{\partial z} \right) f + \left(a_x \frac{\partial}{\partial v_x} + a_y \frac{\partial}{\partial v_y} + a_z \frac{\partial}{\partial v_z} \right) f
$$
$$
= -f \left(\frac{\partial a_x}{\partial v_x} + \frac{\partial a_y}{\partial v_y} + \frac{\partial a_z}{\partial v_z} \right) + s
$$
(A.10)

which in compact form reduces to

$$
\frac{\partial f}{\partial t} + \mathbf{v} \cdot \nabla f + \mathbf{a} \cdot \nabla_v f = -f \, \nabla_v \cdot \mathbf{a} + s
$$
(A.11)

A.5 The Vlasov equation

Now consider the acceleration $\mathbf{a}$ for a plasma. For the cases of interest $\mathbf{a}$ can be divided into two parts:

$$
\mathbf{a} = \mathbf{a}_s (\text{short range}) + \mathbf{a}_l (\text{long range})
$$
(A.12)

The short-range forces act within a volume much smaller than a Debye sphere. These are the Coulomb collisions. Outside this sphere the Coulomb potential is shielded by "Debye shielding." Furthermore, it is a good approximation to assume that each collision occurs at a vanishingly small point in physical space. That is, the particle's position in physical space does not change before and after a collision. In contrast, a collision changes the particle's velocity by a finite amount. For example in a 90° momentum collision a particle's entire x momentum can be lost and converted into y momentum. In terms of the phase space the conclusion is that before and after a collision a particle's position $\mathbf{r}$ in the 6-D volume element remains unchanged while the change in $\mathbf{v}$ of the

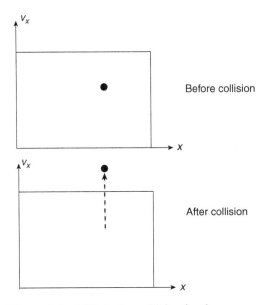

Figure A.3   Effect of a collision in phase space.

particle will in general cause it to leave the volume element by passing through the one of the velocity boundaries. A sketch of this behavior is given in Fig. A.3, showing the projection of the 6-D volume element in (x, v_x) space. The effects of Coulombs collisions are represented by the source/sink term s in Eq. (A.11) and arise from $\mathbf{a}_s$.

The long-range forces act over a distance much greater than a Debye length. By definition these forces behave smoothly over a 6-D volume element. As a result there are no abrupt changes in the position or velocity of a particle within a volume element due to the long-range forces. A sketch of the long-range force behavior is illustrated in Fig. A.4. The long-range forces enter Eq. (A.11) through the terms containing $\mathbf{a}$, which is just an abbreviation for $\mathbf{a}_l$ with the subscript dropped for convenience.

Assume now that collisional effects are negligible. Also assume that the number of particles in a Debye sphere is large so that a statistical description makes sense. Even so, the Debye length is assumed small compared to the gradient lengths of interest. Similarly, the time scales of interest are assumed much slower than ω_{pe}.

In the collisionless limit just described (1) the term s in Eq. (A.11) can be neglected and (2) the long-range acceleration corresponds to the Lorentz force

$$\mathbf{a}(\mathbf{r}, \mathbf{v}, t) = \frac{q}{m}(\mathbf{E} + \mathbf{v} \times \mathbf{B})$$ (A.13)

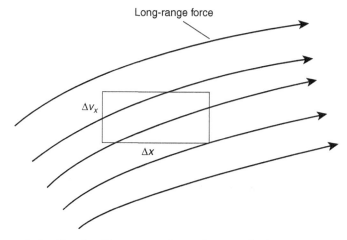

Figure A.4   Sketch of long-range forces in a small phase space volume.

Here $\mathbf{E}(\mathbf{r}, t)$ and $\mathbf{B}(\mathbf{r}, t)$ are the smooth long-range electric and magnetic fields. With this definition the quantity $\nabla_v \cdot \mathbf{a}$ reduces to

$$
\begin{aligned}
\nabla_v \cdot \mathbf{a} &= \frac{q}{m} \nabla_v \cdot (\mathbf{E} + \mathbf{v} \times \mathbf{B}) \\
&= \frac{q}{m} (\mathbf{B} \cdot \nabla_v \times \mathbf{v} - \mathbf{v} \cdot \nabla_v \times \mathbf{B}) \\
&= \frac{q}{m} \mathbf{B} \cdot \nabla_v \times \mathbf{v} \\
&= 0
\end{aligned}
\tag{A.14}
$$

Under these assumptions and conditions the 6-D conservation relation reduces to

$$
\frac{\partial f}{\partial t} + \mathbf{v} \cdot \nabla f + \frac{q}{m} (\mathbf{E} + \mathbf{v} \times \mathbf{B}) \cdot \nabla_v f = 0
\tag{A.15}
$$

This is the Vlasov equation. It has the simple interpretation that

$$
\frac{df}{dt} \equiv \frac{\partial f}{\partial t} + \frac{d\mathbf{r}}{dt} \cdot \nabla f + \frac{d\mathbf{v}}{dt} \cdot \nabla_v f = 0
\tag{A.16}
$$

along the trajectory

$$
\begin{aligned}
\frac{d\mathbf{r}}{dt} &= \mathbf{v} \\
\frac{d\mathbf{v}}{dt} &= \frac{q}{m} (\mathbf{E} + \mathbf{v} \times \mathbf{B})
\end{aligned}
\tag{A.17}
$$

implying that the phase space density is conserved moving with the particle orbits.

The significance of this result is as follows. As f evolves in phase space the volume element $\Delta \mathbf{r} \Delta \mathbf{v}$ varies smoothly as the particles move as shown in Fig. A.5.

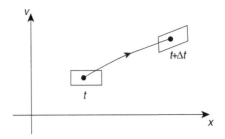

Figure A.5   Evolution of a phase space volume in the absence of collisions.

With only long-range forces acting, no particles suddenly enter or leave the phase space volume element due to collisions. Thus, moving along the trajectory the total number of particles within the volume element is, by definition, conserved: $d(f\Delta\mathbf{r}\Delta\mathbf{v})/dt = 0$. However, the condition $\nabla_v \cdot \mathbf{a} = 0$ leads to the conclusion that $\Delta\mathbf{r}\Delta\mathbf{v}$ is by itself conserved. Thus, not only is the total number of particles $f\Delta\mathbf{r}\Delta\mathbf{v}$ conserved, but the density f as well. This is the significance of the Vlasov equation.

A.6 The kinetic equation

The kinetic equation makes use of the same assumptions as the Vlasov equation concerning the number of particles in a Debye sphere plus the smallness of λ_D and $1/\omega_{pe}$. However, it does not neglect the effect of collisions. The usual approach is to place all terms with a_s on the right-hand side of the equation and simply call them $(\partial f/\partial t)_c$. Since collisions can occur between both like and unlike particles, the collision term is usually written as a sum over all species. Specifically, one writes

$$s \equiv \left(\frac{\partial f_\alpha}{\partial t}\right)_c = \sum_\beta C_{\alpha\beta}(\mathbf{r}, \mathbf{v}, t) \tag{A.18}$$

where $C_{\alpha\beta}$ represents the change in f_α due to collisions with species β.

In this formalism the kinetic equation for species α can be written as

$$\frac{df_\alpha}{dt} = \frac{\partial f_\alpha}{\partial t} + \mathbf{v} \cdot \nabla f_\alpha + \frac{q_\alpha}{m_\alpha}(\mathbf{E} + \mathbf{v} \times \mathbf{B}) \cdot \nabla_v f_\alpha = \left(\frac{\partial f_\alpha}{\partial t}\right)_c \tag{A.19}$$

These kinetic equations are coupled to Maxwell's equation as follows:

$$\nabla \times \mathbf{E} = -\frac{\partial \mathbf{B}}{\partial t}$$

$$\nabla \times \mathbf{B} = \mu_0 \mathbf{J} + \frac{1}{c^2}\frac{\partial \mathbf{E}}{\partial t}$$

$$\nabla \cdot \mathbf{E} = \frac{\sigma}{\varepsilon_0} \tag{A.20}$$

$$\nabla \cdot \mathbf{B} = 0$$

Here, the smoothed charge and current densities determining the long-range electric and magnetic fields are calculated as follows:

$$\sigma = \sum_\alpha q_\alpha n_\alpha = \sum_\alpha q_\alpha \int f_\alpha \, d\mathbf{v}$$

$$\mathbf{J} = \sum_\alpha q_\alpha n_\alpha \mathbf{u}_\alpha = \sum_\alpha q_\alpha \int \mathbf{v} f_\alpha d\mathbf{v} \tag{A.21}$$

The kinetic-Maxwell equations are a set of non-linear, 6-D, time-dependent integro-differential equations – indeed a complex model.

A.7 The collision operater

At this point the collision operators $C_{\alpha\beta}$ have yet to be determined. However, even without giving a specific form for $C_{\alpha\beta}$, it is possible to proceed, at least in a formal manner, and determine a set of "simplified" fluid equations by taking moments of the kinetic equation.

Some of the terms appearing in the moment equations involve velocity integrals of the $C_{\alpha\beta}$. These terms can be partially simplified by invoking certain general conservation relations involving $C_{\alpha\beta}$. These conservation laws arise from the assumption that the collisions characterized by $C_{\alpha\beta}$ are purely elastic – Coulomb collisions to be specific. Inelastic collisions representing ionization, recombination, charge exchange, alpha production, etc. are negligible on the MHD time scale and thus can be ignored.

As previously stated, for purely elastic two-body Coulomb collisions it can be accurately assumed that collisions take place locally, at a single point in space. It then follows that there exists an exact set of conservation laws given by:

- conservation of particles in like and unlike particle collisions

$$\int C_{ee} d\mathbf{v} = \int C_{ii} d\mathbf{v} = \int C_{ei} d\mathbf{v} = \int C_{ie} d\mathbf{v} = 0 \tag{A.22}$$

- conservation of momentum between like particle collisions

$$\int m_e \mathbf{v} C_{ee} d\mathbf{v} = \int m_i \mathbf{v} C_{ii} d\mathbf{v} = 0 \tag{A.23}$$

- conservation of energy between like particle collisions

$$\int \frac{1}{2} m_e v^2 C_{ee} d\mathbf{v} = \int \frac{1}{2} m_i v^2 C_{ii} d\mathbf{v} = 0 \tag{A.24}$$

• conservation of total momentum between unlike particle collisions

$$\int (m_e \mathbf{v} C_{ei} + m_i \mathbf{v} C_{ie}) d\mathbf{v} = 0 \tag{A.25}$$

• conservation of total energy between unlike particle collisions

$$\int \frac{1}{2} \left(m_e v^2 C_{ei} + m_i v^2 C_{ie} \right) d\mathbf{v} = 0 \tag{A.26}$$

These are the desired general conservation properties of the collision operators.

A.8 Specific forms of the collision operators

Even with a knowledge of the general conservation relations one still needs specific forms for the collision operators to estimate some of the transport terms appearing in the moment equations. Since these transport terms turn out to be small compared to the MHD terms, the use of simplified forms for the $C_{\alpha\beta}$ is more than adequate to make the necessary estimations. Below, two forms for the collision operators are given, without proof, in order of increasing simplicity: (1) the Fokker–Planck operator, which is a special limit of the Boltzmann operator; and (2) the Krook operator. The last form makes good intuitive sense. It, therefore, eliminates the need for a detailed derivation while still allowing for a simple estimation of the transport coefficients.

The Fokker–Planck collision operator

$$C_{\alpha\beta} = -\gamma_\alpha \left[\nabla_v \cdot (f_\alpha H_{\alpha\beta}) - \frac{1}{2} \nabla_v \nabla_v : (f_\alpha \nabla_v \nabla_v G_{\alpha\beta}) \right]$$

$$\gamma_\alpha = \frac{e^4 \ln(\Lambda)}{4\pi \varepsilon_0^2 m_\alpha^2} \tag{A.27}$$

$$H_{\alpha\beta} = \frac{m_\alpha + m_\beta}{m_\beta} \int \frac{f_\beta(\mathbf{v}')}{|\mathbf{v} - \mathbf{v}'|} d\mathbf{v}'$$

$$G_{\alpha\beta} = \int |\mathbf{v} - \mathbf{v}'| f_\beta(\mathbf{v}') d\mathbf{v}'$$

This is a rather complicated operator but one that has been successfully used by Braginskii (1965) in his classic paper deriving transport equations and transport coefficients for a collision dominated plasma.

The Krook operator

$$C_{\alpha\beta} = -\nu_{\alpha\beta}\left(f_\alpha - \bar{f}_{\alpha\beta}\right)$$

$$\bar{f}_{\alpha\alpha} = n_\alpha\left(\frac{m_\alpha}{2\pi T_\alpha}\right)^{3/2}\exp\left[-\frac{m_\alpha(\mathbf{v}-\mathbf{u}_\alpha)^2}{2T_\alpha}\right]$$

$$\bar{f}_{\alpha\beta} = n_\alpha\left(\frac{m_\alpha}{2\pi T_\beta}\right)^{3/2}\left[a_0 + a_2\frac{m_\alpha(\mathbf{v}-\mathbf{u}_\beta)^2}{2T_\beta}\right]\exp\left[-\frac{m_\alpha(\mathbf{v}-\mathbf{u}_\beta)^2}{2T_\beta}\right] \qquad (A.28)$$

$$a_0 = 1 - \frac{3}{2}a_2$$

$$a_2 = \frac{(m_\beta - m_\alpha)(T_\alpha - T_\beta)}{m_\beta T_\beta + m_\alpha T_\alpha}$$

In the expression for the Krook operator the "collision frequencies" $\nu_{\alpha\beta}$ are treated as known functions of temperature, density, and magnetic field but are not functions of velocity. It can easily be shown that if one sets $m_i n_i \nu_{ie} = m_e n_e \nu_{ei}$, then the Krook model satisfies the mass, momentum, and energy conservation relations for elastic collisions as described by Eqs. (A.22)–(A.26). The model also yields the correct scaling relations for the transport coefficients necessary in the derivation of the MHD equation.

References

Braginskii, S. I. (1965). In *Reviews of Plasma Physics*, Vol. 1, ed. M. A. Leontovich. New York: Consultants Bureau.

Appendix B

The Braginskii transport coefficients

The two-fluid model given by Eqs. (2.30)–(2.32) is a nearly exact set of moment equations describing conservation of mass, momentum, and energy. Only very high-frequency, short-wavelength information (i.e., $\varepsilon_0 \to 0$) and electron inertia (i.e., $m_e \to 0$) have been neglected, both excellent approximations when considering MHD phenomena. The equations, however, are not closed. To do so requires expressions for the following unknown higher moments: $\mathbf{\Pi}_i$, $\mathbf{\Pi}_e$, $\mathbf{R}_e$, $\mathbf{h}_i$, $\mathbf{h}_e$, Q_i, and Q_e. These quantities have been calculated by Braginskii (1965) for a collision dominated plasma. The calculations are elegant but quite lengthy, beyond the scope of the present book. Analysis and interpretation of the results can be found in several well-known references in addition to Braginskii's original paper (1965). See, for instance, Helander and Sigmar (2002) and Wesson (2011). In Appendix B the moments are simply stated without proof, the goal being to provide reference information that can be used to estimate which terms must be maintained and which can be neglected in the ideal MHD limit.

Qualitatively, Braginskii derives the transport coefficients in the collision dominated limit $\omega\tau_{ee} \ll 1$ and $\omega\tau_{ii} \ll 1$. Even so, he assumes that $\omega \ll \omega_{ca}$ so that $\omega_{ce}\,\tau_{ee} \gg 1$ and $\omega_{ci}\,\tau_{ii} \gg 1$. Particles make many gyrations before undergoing a collision. Braginskii does not assume temperature equilibration. His basic technique is to expand the distribution functions about a local Maxwellian

$$f_\alpha(\mathbf{r}, \mathbf{v}, t) = \frac{n_\alpha}{(2\pi T_\alpha/m_\alpha)^{3/2}} \exp - \left[\frac{m_\alpha}{2T_\alpha} (\mathbf{v} - \mathbf{u}_\alpha)^2 \right] \tag{B.1}$$

with $n_\alpha = n_\alpha\,(\mathbf{r}, t)$, $T_\alpha = T_\alpha\,(\mathbf{r}, t)$, and $\mathbf{u}_\alpha = \mathbf{u}_\alpha\,(\mathbf{r}, t)$. The desired solutions are obtained using the Fokker–Planck form of the collision operator, described in Appendix A. With this as background, the Braginskii transport coefficients are given as follows.

B.1 The pressure tensor

The elements of the anisotropic part of the ion and electron pressure tensors can be written as

$$
\begin{aligned}
\Pi_{zz} &= -\eta_0 W_{zz} \\
\Pi_{xx} &= -\eta_0 (W_{xx} + W_{yy})/2 - \eta_1 (W_{xx} - W_{yy})/2 - \eta_3 W_{xy} \\
\Pi_{yy} &= -\eta_0 (W_{xx} + W_{yy})/2 - \eta_1 (W_{yy} - W_{xx})/2 + \eta_3 W_{xy} \\
\Pi_{xy} &= \Pi_{yx} = -\eta_1 W_{xy} + \eta_3 (W_{xx} - W_{yy})/2 \\
\Pi_{xz} &= \Pi_{zx} = -\eta_2 W_{xz} - \eta_4 W_{yz} \\
\Pi_{yz} &= \Pi_{zy} = -\eta_2 W_{yz} + \eta_4 W_{xz}
\end{aligned}
\tag{B.2}
$$

Here $\mathbf{B}$ is aligned with the z axis and

$$
W_{ij} = \frac{\partial u_i}{\partial x_j} + \frac{\partial u_j}{\partial x_i} - \frac{2}{3} \delta_{ij} \nabla \cdot \mathbf{u}
\tag{B.3}
$$

The viscosity coefficients for ions and electrons are given by

$$
\begin{array}{ll}
\eta_0^i = 0.96 n_i T_i \tau_i & \eta_0^e = 0.73 n_e T_e \tau_e \\[2mm]
\eta_1^i = \eta_2^i/4 = \dfrac{3}{10} \dfrac{n_i T_i}{\omega_{ci}^2 \tau_i} & \eta_1^e = \eta_2^e/4 = 0.51 \dfrac{n_e T_e}{\omega_{ce}^2 \tau_e} \\[4mm]
\eta_3^i = \eta_4^i/2 = \dfrac{1}{2} \dfrac{n_i T_i}{\omega_{ci}} & \eta_3^e = \eta_4^e/2 = -\dfrac{1}{2} \dfrac{n_e T_e}{|\omega_{ce}|}
\end{array}
\tag{B.4}
$$

Here and below,

$$
\begin{aligned}
\tau_e &= 3(2\pi)^{3/2} \frac{\varepsilon_0^2 m_e^{1/2} T_e^{3/2}}{n e^4 \ln \Lambda} = 1.09 \times 10^{-4} \frac{T_k^{3/2}}{n_{20} \ln \Lambda} \ \text{sec} \\[3mm]
\tau_i &= 12\pi^{3/2} \frac{\varepsilon_0^2 m_i^{1/2} T_i^{3/2}}{n e^4 \ln \Lambda} = 6.60 \times 10^{-3} \frac{(m_i/m_p)^{1/2} T_k^{3/2}}{n_{20} \ln \Lambda} \ \text{sec}
\end{aligned}
\tag{B.5}
$$

B.2 Collisional momentum transfer

The collisional momentum transfer is described by $\mathbf{R}_e$, which has two contributions, $\mathbf{R}_e = \mathbf{R}_u + \mathbf{R}_T$, given by

$$
\begin{aligned}
\mathbf{R}_u &= -\frac{m_e n}{\tau_e} (0.51 \mathbf{u}_\| + \mathbf{u}_\perp) = ne \left(\eta_\| \mathbf{J}_\| + \eta_\perp \mathbf{J}_\perp \right) \\[3mm]
\mathbf{R}_T &= -0.71 n \nabla_\| T_e - \frac{3}{2} \frac{n}{|\omega_{ce}| \tau_e} \mathbf{b} \times \nabla T_e
\end{aligned}
\tag{B.6}
$$

B.3 The heat flux vector

The electron heat flux vector $\mathbf{h}_e$ also is composed of two contributions: $\mathbf{h}_e = \mathbf{h}_u^e + \mathbf{h}_T^e$. The ion heat flux has only a single contribution $\mathbf{h}_i = \mathbf{h}_T^i$. These are given by

$$\mathbf{h}_u^e = nT_e \left(0.71\mathbf{u}_\| + \frac{3/2}{|\omega_{ce}|\tau_e} \mathbf{b} \times \mathbf{u} \right)$$

$$\mathbf{h}_T^e = \frac{nT_e\tau_e}{m_e} \left(-3.16\nabla_\| T_e - \frac{4.66}{\omega_{ce}^2\tau_e^2} \nabla_\perp T_e - \frac{5/2}{|\omega_{ce}\tau_e|} \mathbf{b} \times \nabla T_e \right) \qquad \text{(B.7)}$$

$$\mathbf{h}_T^i = \frac{nT_i\tau_i}{m_i} \left(-3.9\nabla_\| T_i - \frac{2}{\omega_{ci}^2\tau_i^2} \nabla_\perp T_i + \frac{5/2}{\omega_{ci}\tau_i} \mathbf{b} \times \nabla T_i \right)$$

B.4 Collisional energy transfer

The collisional energy transfer to and between ions and electrons is characterized by the quantities Q_i and Q_e, which have the form

$$Q_i = 3\frac{m_e}{m_i}\frac{n}{\tau_e} (T_e - T_i)$$

$$Q_e = -\mathbf{R}_e \cdot \mathbf{u} - Q_i \qquad \text{(B.8)}$$

$$= \eta_\| J_\|^2 + \eta_\perp J_\perp^2 + \frac{1}{en}\mathbf{J} \cdot \mathbf{R}_T + 3\frac{m_e}{m_i}\frac{n}{\tau_e} (T_i - T_e)$$

These relationships lead to a closure of the two-fluid equations in the collision dominated regime.

References

Braginskii, S. I. (1965). In *Reviews of Plasma Physics*, Vol. 1, ed. M. A. Leontovich. New York: Consultants Bureau.

Helander, P. and Sigmar, D. J. (2002). *Collisional Transport in Magnetized Plasmas*. Cambridge: Cambridge University Press.

Wesson, J. A. (2011). *Tokamaks*, 4th edn. Oxford: Oxford University Press.

Appendix C

Time derivatives in moving plasmas

C.1 Introduction

A derivation is presented of two basic relations that involve the time derivatives of integrated plasma quantities where the integration domain is moving with the plasma. The desired relations can be obtained from a general differential geometry analysis. However, a simpler derivation is presented here that is less abstract and shows directly how the motion of the plasma enters the relations. The analysis is further simplified by assuming only one-dimensional motion. The generalization to a three-dimensional motion is straightforward but somewhat tedious.

C.2 Time derivatives in a moving volume

Consider a plasma contained within a closed volume V_p. Assume that there is a physical quantity of interest that can be expressed as the volume integral $G(t)$ of a local quantity $g(\mathbf{r},t)$ as follows:

$$G(t) = \int_{V_p} g(\mathbf{r}, t) d\,\mathbf{r} \tag{C.1}$$

The goal is to calculate dG/dt for the general case where the surface S_p bounding V_p is moving with a normal velocity $\mathbf{v}$.

The derivation begins by introducing a set of standard spherical coordinates r, θ, ϕ whose origin is located within V_p as shown in Fig. C.1. For simplicity we assume that while $g(\mathbf{r}, t)$ is a locally 3-D function, its boundary is a moving sphere. Thus, the surface of the moving plasma is written as

$$S_p(r, \theta, \phi, t) \equiv r - r_p(t) = 0 \tag{C.2}$$

The function $r_p(t)$ defines the one-dimensional motion of the surface.

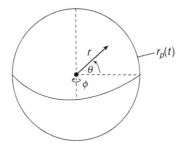

Figure C.1   Geometry used to calculate the time derivative of a quantity in a moving volume.

Using this information one sees that the quantity G can be expressed as

$$G(t) = \int_{V_p} g(\mathbf{r}, t) d\mathbf{r} = \int_0^{2\pi} d\phi \int_0^{\pi} d\theta \int_0^{r_p(t)} g(r, \theta, \phi, t) \, r^2 \sin\theta \, dr \qquad (\text{C.3})$$

To evaluate the desired quantity dG/dt it is necessary to differentiate both the function $g(r, \theta, \phi, t)$ and the integration limit $r_p(t)$. Carrying out this step yields

$$\frac{dG}{dt} = \int_{V_p} \frac{\partial g}{\partial t} d\mathbf{r} + \int_0^{2\pi} \int_0^{\pi} g_p \frac{dr_p}{dt} r_p^2 \sin\theta \, d\theta \, d\phi \qquad (\text{C.4})$$

where $g_p(\theta, \phi, t) = g(r_p, \theta, \phi, t)$ is the value on the surface. The first term is the obvious one and it is the second term that represents the effect of the motion of the surface.

The second term is simplified by noting that the normal velocity of the surface is defined by

$$v_n = \frac{dr_p}{dt} \qquad (\text{C.5})$$

From this relation it follows that Eq. (C.4) can be rewritten as

$$\frac{dG}{dt} = \int_{V_p} \frac{\partial g}{\partial t} d\mathbf{r} + \int_{S_p} g_p v_n \, dS \qquad (\text{C.6})$$

For a three-dimensional motion of the surface this equation generalizes to

$$\frac{dG}{dt} = \int_{V_p} \frac{\partial g}{\partial t} d\mathbf{r} + \int_{S_p} g_p \, \mathbf{n} \cdot \mathbf{v} \, dS \qquad (\text{C.7})$$

which is the desired result.

C.3 Flux change in a moving area

The second relation of interest involves the calculation of the change in magnetic flux passing through a moving area. The derivation is similar to the one for the moving volume although carried out in a cylindrical rather than spherical coordinate system.

The analysis begins with the definition of the magnetic flux passing through an arbitrarily shaped moving surface

$$\psi(t) = \int \mathbf{B} \cdot d\mathbf{S} \tag{C.8}$$

For simplicity the boundary of the surface is assumed to be circular. However, the magnetic field can be three dimensional and its direction is not in general perpendicular to the plane of the circle (see Fig. C.2). Specifically, the boundary at a fixed value of $z = z_0$ is written as

$$l_p(r, \theta, z_0, t) \equiv r - r_p(t) = 0 \tag{C.9}$$

Here $r_p(t)$ represents the shape of the surface which in general need not be planar. This is an open surface whose perimeter

$$r = r_p(t) \tag{C.10}$$

is also allowed to change in time.

Under these the assumptions the expression for the flux reduces to

$$\psi(t) = \int_0^{2\pi} d\theta \int_0^{r_p(t)} B_z(r, \theta, z_0, t) \, r dr \tag{C.11}$$

The time rate of change of the flux is then given by

$$\frac{d\psi}{dt} = \int_{S_p} \frac{\partial \mathbf{B}}{\partial t} \cdot d\mathbf{S} + \int_0^{2\pi} B_z \frac{dr_p}{dt} r_p d\theta \tag{C.12}$$

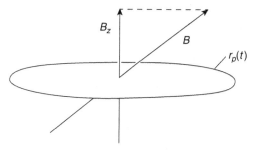

Figure C.2   Geometry used to calculate the time derivative of the magnetic flux through a moving surface.

If one now defines $\mathbf{n} = \mathbf{e}_z$ as the normal to the circular surface, and $\mathbf{e}_\perp = \mathbf{e}_r$ as the direction perpendicular to both the surface normal $\mathbf{n}$ and the boundary arc length $d\mathbf{l} = r_p d\theta\, \mathbf{e}_\theta$, then Eq. (C.12) can be rewritten as

$$\frac{d\psi}{dt} = \int_{S_p} \frac{\partial \mathbf{B}}{\partial t} \cdot d\mathbf{S} + \oint_{l_p} B_n v_\perp dl \qquad\qquad\text{(C.13)}$$

The three-dimensional generalization is straightforward but requires a somewhat tedious calculation. It is given by

$$\begin{aligned}
\frac{d\psi}{dt} &= \int_{S_p} \frac{\partial \mathbf{B}}{\partial t} \cdot d\mathbf{S} + \oint_{l_p} (\mathbf{B} \cdot \mathbf{n})(\mathbf{v} \cdot d\mathbf{l} \times \mathbf{n}) \\
&= \int_{S_p} \frac{\partial \mathbf{B}}{\partial t} \cdot d\mathbf{S} - \oint_{l_p} \mathbf{v} \times \mathbf{B} \cdot d\mathbf{l}
\end{aligned} \qquad\qquad\text{(C.14)}$$

which is the desired result.

Appendix D

The curvature vector

The relation between the curvature vector $\boldsymbol{\kappa} = \mathbf{b} \cdot \nabla \mathbf{b}$ and the radius of curvature vector $\mathbf{R}_c$ is determined as follows. The relevant geometry is illustrated in Fig. D.1. The radius of curvature vector $\mathbf{R}_c$ is defined as the radius vector of an equivalent circle that is tangent to the magnetic field line at the local point of interest. From Fig. D.1 one sees that the magnitude of $\mathbf{R}_c$ is related to the magnitude of the differential arc length by $R_c d\theta = dl$. Now, recall that $\mathbf{b} = \mathbf{B}/B$ is a unit vector tangent to the magnetic field. If the circle is tangent to the field line then from the laws of similar triangles it follows that $|d\mathbf{b}| = |\mathbf{b}|d\theta = d\theta$. Eliminating $d\theta$ yields $R_c = dl/|d\mathbf{b}|$.

Next, observe that the direction of $\mathbf{R}_c$ is anti-parallel to $d\mathbf{b}$ implying that $\mathbf{R}_c = -Cd\mathbf{b}$, where C is a positive quantity. The quantity C is found by noting that $R_c = C|d\mathbf{b}|$. Thus, setting $C = R_c/|d\mathbf{b}| = dl/|d\mathbf{b}|^2$ leads to an expression for $\mathbf{R}_c$ that can be written as

$$\frac{\mathbf{R}_c}{R_c^2} = -\frac{Cd\mathbf{b}}{R_c^2} = -\frac{d\mathbf{b}}{dl} \tag{D.1}$$

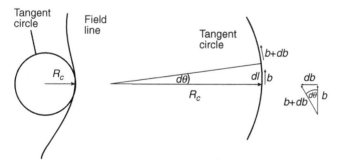

Figure D.1   Geometry used to derive the relationship between the radius of curvature vector $\mathbf{R}_c$ and the curvature vector $\boldsymbol{\kappa}$.

The last step is to recognize that d/dl is the directional derivative parallel to the magnetic field. In other words, $d/dl = \mathbf{b} \cdot \nabla$. Substituting this relation yields an expression for the radius of curvature vector given by

$$-\frac{\mathbf{R}_c}{R_c^2} \equiv \boldsymbol{\kappa} = \mathbf{b} \cdot \nabla \mathbf{b} \tag{D.2}$$

Here, $\boldsymbol{\kappa}$ is known as the curvature vector and is equivalent to and almost always used in place of $\mathbf{R}_c$ in MHD calculations.

Appendix E

Overlap limit of the high β and Greene–Johnson stellarator models

The derivation of the overlap limit of the high β (HBS) and Greene–Johnson (GJ) stellarator models proceeds as follows. The starting equations are given by Eq. (7.56) and are repeated here for convenience:

$$\frac{\partial \beta}{\partial \phi} = -\frac{1}{N^{1/2}} \mathbf{e}_\phi \times \nabla_\perp A \cdot \nabla_\perp \beta$$

$$\frac{\partial \left(\nabla_\perp^2 A\right)}{\partial \phi} = -\frac{1}{N^{1/2}} \mathbf{e}_\phi \times \nabla_\perp A \cdot \nabla_\perp \left(\nabla_\perp^2 A\right) + \frac{1}{N^{3/2}} \mathbf{e}_z \cdot \nabla_\perp \beta \tag{E.1}$$

E.1 The $1/N^{1/2}$ expansion

The analysis begins by first separating each unknown appearing in Eq. (E.1) into two components. The first is independent of ϕ while the second is periodic in ϕ with zero average value,

$$A(\rho, \theta, \phi) = \overline{A}(\rho, \theta) + \tilde{A}(\rho, \theta, \phi)$$
$$\beta(\rho, \theta, \phi) = \overline{\beta}(\rho, \theta) + \tilde{\beta}(\rho, \theta, \phi) \tag{E.2}$$

where

$$\overline{A} = \frac{1}{2\pi} \int_0^{2\pi} A d\phi \qquad \frac{1}{2\pi} \int_0^{2\pi} \tilde{A} d\phi = 0$$

$$\overline{\beta} = \frac{1}{2\pi} \int_0^{2\pi} \beta d\phi \qquad \frac{1}{2\pi} \int_0^{2\pi} \tilde{\beta} d\phi = 0 \tag{E.3}$$

The appropriate $1/N^{1/2}$ expansion for these quantities which leads to the overlap limit is as follows:

$$A = \tilde{A}_0 + \frac{1}{N^{1/2}}\bar{A}_1 + \frac{1}{N}\tilde{A}_2 + \frac{1}{N^{3/2}}\left(\bar{A}_3 + \tilde{A}_3\right) + \cdots$$

$$\beta = \bar{\beta}_0 + \frac{1}{N^{1/2}}\tilde{\beta}_1 + \frac{1}{N}\left(\bar{\beta}_2 + \tilde{\beta}_2\right) + \cdots$$

(E.4)

E.2 The solution

The solution is obtained by substituting Eq. (E.4) in Eq. (E.1) and then setting the coefficient of each power of $1/N^{1/2}$ sequentially to zero. To help keep track of the ordering sequence the first equation in Eq. (E.1) is denoted as the "β equation" and the second as the "A equation." The solution proceeds as follows.

The $(1/N^{1/2})^0$ contribution to the A equation

The leading-order contribution to the vector potential equation is given by

$$\frac{\partial}{\partial\phi}\left(\nabla_\perp^2\tilde{A}_0\right) = 0 \tag{E.5}$$

which requires that

$$\nabla_\perp^2\tilde{A}_0 = 0 \tag{E.6}$$

This, as expected, is just the equation for a loosely wound, vacuum helical field. The solution to Eq. (E.6) can be written as

$$\tilde{A}_0(\rho,\theta,\phi) = \sum_{n\neq0} A_n(\rho,\theta)e^{in\phi}$$

$$A_n(\rho,\theta) = \sum_{l\neq0} \frac{\delta_{nl}}{2N^{1/2}\varepsilon l}\rho^{|l|}e^{il\theta}$$

(E.7)

Here, δ_{nl} is the normalized amplitude of each harmonic of the helical magnetic field at $\rho = 1$: $\left(\tilde{B}_{\rho,nl}^2 + \tilde{B}_{\theta,nl}^2\right)_{\rho=1} = \delta_{nl}^2 B_0^2$. Note that the $l = 0$ contribution is eliminated from the sum because of the long helical wavelength assumption.

The $(1/N^{1/2})^1$ contribution to the β equation

The first non-vanishing contribution to the β equation occurs in order $\left(1/N^{1/2}\right)^1$ and leads to a relation between $\tilde{\beta}_1$ and $\bar{\beta}_0$. The relevant equation is given by

$$\frac{\partial\tilde{\beta}_1}{\partial\phi} = -\mathbf{e}_\phi \cdot \nabla_\perp\tilde{A}_0 \times \nabla_\perp\bar{\beta}_0 \tag{E.8}$$

This equation can be easily integrated with respect to ϕ yielding

$$\tilde{\beta}_1(\rho,\theta,\phi) = \sum_{n\neq 0} \frac{i}{n}\left(\mathbf{e}_\phi \cdot \nabla_\perp A_n \times \nabla_\perp \bar{B}_0\right)e^{in\phi} \tag{E.9}$$

The $(1/N^{1/2})^2$ contribution to the β equation

The next step in the solution procedure involves the $\left(1/N^{1/2}\right)^2$ contribution to the β equation. An analysis of the resulting equation yields an integrability constraint that ultimately relates $\bar{A}_1$ to $\tilde{A}_0$ and $\bar{B}_0$. The equation of interest is given by

$$\frac{\partial \tilde{\beta}_2}{\partial \phi} = -\mathbf{e}_\phi \cdot \left(\nabla_\perp \tilde{A}_0 \times \nabla_\perp \tilde{\beta}_1 + \nabla_\perp \bar{A}_1 \times \nabla_\perp \bar{B}_0\right) \tag{E.10}$$

An explicit expression for $\tilde{\beta}_2$ is not actually needed. Instead, all that is necessary is to ensure that a periodic solution for $\tilde{\beta}_2$ exists. Satisfying this requirement leads to the following integrability constraint on the right-hand side of Eq. (E.10)

$$\frac{1}{2\pi}\int_0^{2\pi} \mathbf{e}_\phi \cdot \left(\nabla_\perp \tilde{A}_0 \times \nabla_\perp \tilde{\beta}_1 + \nabla_\perp \bar{A}_1 \times \nabla_\perp \bar{B}_0\right)d\phi = 0 \tag{E.11}$$

A straightforward calculation in which $\tilde{A}_0$ and $\tilde{\beta}_1$ are eliminated by means of Eqs. (E.7) and (E.9) shows that this condition reduces to

$$\mathbf{e}_\phi \cdot \nabla_\perp \bar{B}_0 \times \nabla_\perp \left(\bar{A}_1 + \sum_1^\infty \frac{i}{n}\mathbf{e}_\phi \cdot \nabla_\perp A_n^* \times \nabla_\perp A_n\right) = 0 \tag{E.12}$$

The solution has the form

$$\bar{A}_1 + \sum_1^\infty \frac{i}{n}\mathbf{e}_\phi \cdot \nabla_\perp A_n^* \times \nabla_\perp A_n = G(\bar{B}_0) \tag{E.13}$$

where $G(\bar{B}_0)$ is an arbitrary free function.

A short calculation presented in "Auxiliary analysis I" at the end of Appendix E shows that $G(\bar{B}_0)$ is related to the average poloidal flux as follows:

$$G(\bar{B}_0) = \psi \tag{E.14}$$

where ψ is defined by

$$\psi = \frac{1}{2\pi a^2 B_0}\psi_p = \frac{1}{2\pi a^2 B_0}\int \mathbf{B}_p \cdot d\mathbf{S} \tag{E.15}$$

Hereafter assume that Eq. (E.14) has been inverted so that $\bar{B}_0 = \bar{B}_0(\psi)$ is a known free function. The final form of the integrability constraint can now be written as

$$\bar{A}_1(r,\theta) = \psi - \sum_1^\infty \frac{i}{n} \mathbf{e}_\phi \cdot \nabla_\perp A_n^* \times \nabla_\perp A_n \qquad (E.16)$$

The $(1/N^{1/2})^2$ contribution to the A equation

The $\left(1/N^{1/2}\right)^1$ contribution to the vector potential equation is trivially satisfied. The next non-vanishing contribution occurs in order $\left(1/N^{1/2}\right)^2$ and has the form

$$\frac{\partial}{\partial\phi}\left(\nabla_\perp^2 \tilde{A}_2\right) = -\mathbf{e}_\phi \cdot \nabla_\perp \tilde{A}_0 \times \nabla_\perp\left(\nabla_\perp^2 \bar{A}_1\right) \qquad (E.17)$$

This equation can be easily integrated with respect to ϕ leading to an expression that relates $\nabla_\perp^2 \tilde{A}_2$ to $\bar{A}_1$ and the A_n:

$$\nabla_\perp^2 \tilde{A}_2 = \sum_{n \neq 0} \frac{i}{n} \left[\mathbf{e}_\phi \cdot \nabla_\perp A_n \times \nabla_\perp \left(\nabla_\perp^2 \bar{A}_1\right)\right] e^{in\phi} \qquad (E.18)$$

In the analysis that follows it will become apparent that only $\nabla_\perp^2 \tilde{A}_2$, and not $\tilde{A}_2$, is required.

The $(1/N^{1/2})^3$ contribution to the A equation

This is the last step in the analysis leading to the helical Grad–Shafranov equation. The equation of interest can be written as

$$\frac{\partial}{\partial\phi}\left(\nabla_\perp^2 \tilde{A}_3\right) = -\mathbf{e}_\phi \cdot \left[\nabla_\perp \tilde{A}_0 \times \nabla_\perp\left(\nabla_\perp^2 \tilde{A}_2\right) + \nabla_\perp \bar{A}_1 \times \nabla_\perp\left(\nabla_\perp^2 \bar{A}_1\right)\right] + \mathbf{e}_Z \cdot \nabla_\perp \bar{B}_0$$

$$(E.19)$$

As above one does not actually need to evaluate $\nabla_\perp^2 \tilde{A}_3$, but only ensure that a periodic solution exists. This leads to the following integrability condition:

$$\frac{1}{2\pi}\int_0^{2\pi}\left\{\mathbf{e}_\phi \cdot \left[\nabla_\perp \tilde{A}_0 \times \nabla_\perp\left(\nabla_\perp^2 \tilde{A}_2\right) + \nabla_\perp \bar{A}_1 \times \nabla_\perp\left(\nabla_\perp^2 \bar{A}_1\right)\right] - \mathbf{e}_Z \cdot \nabla_\perp \bar{B}_0\right\}d\phi = 0$$

$$(E.20)$$

After a straightforward but slightly lengthy calculation the integrability condition reduces to

$$\mathbf{e}_\phi \cdot \nabla_\perp \psi \times \nabla_\perp \left(\nabla_\perp^2 \bar{A}_1 + \frac{d\bar{B}_0}{d\psi}x\right) = 0 \qquad (E.21)$$

with $x = \rho \cos \theta$. The solution to this equation is given by

$$\nabla_\perp^2 \overline{A}_1 = -\frac{d\bar\beta_0}{d\psi} x + H(\psi) \tag{E.22}$$

where $H(\psi)$ is a second free function. A short calculation presented in "Auxiliary analysis II" (Section E.4) shows that $H(\psi)$ is related to the normalized average toroidal current density $J(\psi)$ by

$$H(\psi) = -J(\psi) + \frac{d\bar\beta_0}{d\psi}\langle x \rangle \tag{E.23}$$

where

$$J = -\frac{\mu_0 R_0}{B_0}\langle J_\phi \rangle = -\frac{\mu_0 R_0}{B_0}\frac{\oint J_\phi \frac{dl_p}{|\nabla_\perp \psi|}}{\oint \frac{dl_p}{|\nabla_\perp \psi|}}$$

$$\langle x \rangle = \frac{\oint x \frac{dl_p}{|\nabla_\perp \psi|}}{\oint \frac{dl_p}{|\nabla_\perp \psi|}} \tag{E.24}$$

The notation $\langle Q \rangle$ denotes an average over poloidal angle written in terms of poloidal arc length $l_p = l_p(\theta)$. Also a stellarator with zero net average current density on each flux surface corresponds to $J(\psi) = 0$.

The final equation for loosely wound Greene–Johnson stellarator

The results can now be collected leading to a single non-linear partial differential equation for the flux function. Specifically, the quantity $\overline{A}_1$ is eliminated between Eqs. (E.16) and (E.22). Also, hopefully without increasing the risk of confusion, all subscripts and overbars are eliminated to yield a simple compact form:

$$\nabla_\perp^2 \psi = -J - \frac{d\beta}{d\psi}(x - \langle x \rangle) + \nabla_\perp^2\left(\sum_1^\infty \frac{i}{n}\mathbf{e}_\phi \cdot \nabla_\perp A_n^* \times \nabla_\perp A_n\right) \tag{E.25}$$

This is the desired equation. The basic unknown is ψ and there are two free functions $J(\psi)$ and $\beta(\psi)$. The functions $A_n(r, \theta)$ are the known solutions for the vacuum helical field.

This equation represents a very large step of mathematical simplification in that the original 3-D problem has been reduced to a 2-D problem.

E.3 Auxiliary analysis I: relation between $G(\bar\beta_0)$ and ψ

The relation between $G(\bar\beta_0)$ and ψ is obtained by evaluating the poloidal flux defined by

$$\psi_p = \int \mathbf{B}_p \cdot d\mathbf{S} \tag{E.26}$$

For convenience, the surface area of interest is chosen to be $\theta = 0$ as shown in Fig. E.1 where $d\mathbf{S} = R dr d\hat\phi \mathbf{e}_\theta \approx -(aR_0/N)d\rho d\phi \mathbf{e}_\theta$ The expression for ψ_p reduces to

$$
\begin{aligned}
\psi_p &\approx -\frac{aR_0}{N}\int_0^{-2\pi N} d\phi \int_0^{\rho_0+\sigma_h} d\rho\,(B_\theta)_{\theta=0} \\
&= a^2 B_0 N^{1/2}\int_0^{2\pi} d\phi \int_0^{\rho_0+\sigma_h} d\rho\left(\frac{\partial A}{\partial\rho}\right)_{\theta=0} \\
&= a^2 B_0 N^{1/2}\int_0^{2\pi} d\phi \int_0^{\rho_0+\sigma_h} d\rho\left(\frac{\partial\tilde A_0}{\partial\rho}+\frac{1}{N^{1/2}}\frac{\partial\overline{A}_1}{\partial\rho}\right)_{\theta=0}
\end{aligned}
\tag{E.27}
$$

Here, $\rho_0 = \rho_0(\theta)|_{\theta=0}$ labels the flux surface of interest, and $\sigma_h(\rho_0,\theta,\phi)|_{\theta=0}$ is the helical perturbation to the flux surface at $\theta = 0$. The quantity σ_h, which is evaluated shortly, is required because the leading-order contribution due to the helical fields (i.e., the $\tilde A_0$ term) averages to zero over one period in ϕ. Also, it is assumed for simplicity that $\psi_p = 0$ on the magnetic axis, although this choice is arbitrary.

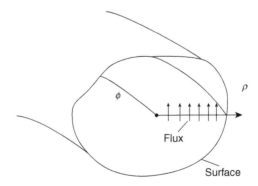

Figure E.1   Surface area used to calculate the poloidal flux.

Consider now the evaluation of the helical term. The first contribution that does not average to zero is found by a simple Taylor expansion of the integral and can be written as

$$
\psi_p^{(1)} = a^2 B_0 N^{1/2} \int_0^{2\pi} d\phi \left[\left(\frac{\partial \tilde{A}_0}{\partial \rho} \right) \sigma_h \right]_{\theta=0}
$$
$$
= a^2 B_0 N^{1/2} \sum_{n \neq 0} \int_0^{2\pi} d\phi \left[\sigma_h \left(\frac{\partial A_n}{\partial \rho} \right) e^{in\phi} \right]_{\theta=0}
$$

(E.28)

The quantity σ_h is easily calculated by noting that the flux surfaces are defined in terms of the pressure contours as follows: $\bar{B}_0(\rho, \theta) + (1/N^{1/2})\tilde{B}_1(\rho, \theta, \phi) = $ constant. Thus, by Taylor expanding $\rho = \rho_0 + \sigma_h$, one finds

$$
\sigma_h(\rho_0, \theta, \phi) = -\frac{1}{N^{1/2}} \frac{\tilde{B}_1}{\partial \bar{B}_0 / \partial \rho_0}
$$
$$
= \frac{1}{N^{1/2}} \frac{1}{\partial \bar{B}_0 / \partial \rho_0} \sum_{n \neq 0} \frac{i}{n\rho_0} \left(\frac{\partial A_n}{\partial \theta} \frac{\partial \bar{B}_0}{\partial \rho_0} + \frac{\partial A_n}{\partial \rho_0} \frac{\partial \bar{B}_0}{\partial \theta} \right) e^{in\phi}
$$

(E.29)

This expression is substituted into Eq. (E.28) and then summed over $\pm n$. The contribution proportional to $(1/n)|\partial A_n / \partial \rho_0|^2$ sums to zero because of odd symmetry in n. The remaining integral is easily evaluated yielding

$$
\psi_p^{(1)} = 2\pi a^2 B_0 \sum_1^\infty \frac{i}{n} \left(\frac{1}{\rho_0} \frac{\partial A_n}{\partial \theta} \frac{\partial A_n^*}{\partial \rho_0} - \frac{1}{\rho_0} \frac{\partial A_n^*}{\partial \theta} \frac{\partial A_n}{\partial \rho_0} \right)_{\theta=0}
$$
$$
= 2\pi a^2 B_0 \sum_1^\infty \frac{i}{n} \mathbf{e}_\phi \cdot (\nabla_\perp A_n^* \times \nabla_\perp A_n)
$$

(E.30)

The second term in Eq. (E.27) is easier to evaluate since the leading-order contribution is already smaller by $1/N^{1/2}$ and does not average to zero. One obtains

$$
\psi_p^{(2)} \approx a^2 B_0 \int_0^{2\pi} d\phi \int_0^{\rho_0} d\rho \left(\frac{\partial \bar{A}_1}{\partial \rho} \right)_{\theta=0}
$$
$$
= 2\pi a^2 B_0 \bar{A}_1
$$

(E.31)

The results from Eqs. (E.30) and (E.31) are combined leading to the following expression for the normalized poloidal flux

$$\psi = \sum_{1}^{\infty} \frac{i}{n} \mathbf{e}_\phi \cdot \left(\nabla_\perp A_n^* \times \nabla_\perp A_n \right) + \overline{A}_1 \qquad (\text{E.32})$$

where $\psi = \psi_p/2\pi a^2 B_0$. A comparison of Eq. (E.32) with Eq. (E.13) shows that

$$G(\overline{B}_0) = \psi \qquad (\text{E.33})$$

As expected, inverting Eq. (E.33) implies that $\overline{B}_0 = \overline{B}_0(\psi)$.

E.4 Auxiliary analysis II: relation between $H(\psi)$ and $J(\psi)$

The relation between $H(\psi)$ and $J(\psi)$ is found by noting that the net toroidal current flowing within any given flux surface is given by

$$I(\psi) = \int J_{\hat{\phi}} \, \mathbf{e}_{\hat{\phi}} \cdot d\hat{\mathbf{A}} = -\int J_\phi \, \mathbf{e}_\phi \cdot d\mathbf{A} = -\int_0^\psi d\psi \oint \frac{dl_p}{|\nabla_\perp \psi|} J_\phi \qquad (\text{E.34})$$

Here, a coordinate transformation (in un-normalized coordinates) has been made from r, θ to ψ, l_p with l_p being poloidal arc length. Next, the average current density within any given flux surface, $\langle J_\phi(\psi) \rangle$, is defined as

$$I(\psi) \equiv -\int \langle J_\phi \rangle \, dA = -\int_0^\psi d\psi \langle J_\phi \rangle \oint \frac{dl_p}{|\nabla_\perp \psi|} \qquad (\text{E.35})$$

Since Eqs. (E.34) and (E.35) are valid on any arbitrary flux surface it follows that the integrands can be equated:

$$\langle J_\phi \rangle = \frac{\oint J_\phi \frac{dl_p}{|\nabla_\perp \psi|}}{\oint \frac{dl_p}{|\nabla_\perp \psi|}} \qquad (\text{E.36})$$

This leads to a relation between $H(\psi)$ and $\langle J_\phi(\psi) \rangle$, obtained by recalling that $\mu_0 J_\phi = -\nabla_\perp^2 A_1$ and that the net (i.e., the cross section averaged) toroidal current arises from the $\overline{A}_1/N^{1/2}$ non-helical contribution to A; that is, since $\tilde{A}_0$ corresponds to a vacuum field the first non-vanishing contribution to the current arises from the $\overline{A}_1/N^{1/2}$ term. Transforming to normalized coordinates one finds

$$\mu_0 \langle J_{\hat{\phi}} \rangle = -\mu_0 \langle J_\phi \rangle = \langle \nabla^2 A_1 \rangle = -\frac{B_0 N^{1/2}}{R_0} \langle \nabla_\perp^2 A \rangle = -\frac{B_0}{R_0} \langle \nabla_\perp^2 \overline{A}_1 \rangle \qquad (\text{E.37})$$

Next, $\langle \nabla_\perp^2 \overline{A}_1 \rangle$ is eliminated by means of Eq. (E.22), finally leading to

$$H(\psi) = -J(\psi) + \frac{d\overline{\beta}_0}{d\psi}\langle x \rangle$$

$$\langle x \rangle = \frac{\oint x \dfrac{dl_p}{|\nabla_\perp \psi|}}{\oint \dfrac{dl_p}{|\nabla_\perp \psi|}} \tag{E.38}$$

where

$$J \equiv \frac{\mu_0 R_0}{B_0}\langle J_{\hat\phi} \rangle = -\frac{\mu_0 R_0}{B_0}\langle J_\phi \rangle \tag{E.39}$$

is the normalized average current density. Equation (E.38) is the required relation.

Appendix F

General form for $q(\psi)$

The goal is to show that $q(\psi) = d\psi_t/d\psi$. This is easily done using the periodicity and straight field line properties of Boozer coordinates. The general definition of safety factor from Chapter 4, Eq. (4.28), is given by

$$q = \frac{\lim_{L \to \infty} \Delta\phi}{\lim_{L \to \infty} \Delta\theta} = \frac{\lim_{L \to \infty} \int_0^L \frac{B_\phi}{RB} dl}{\lim_{L \to \infty} \int_0^L \frac{B_\theta}{rB} dl} \tag{F.1}$$

Now, since χ and ζ change by 2π every time θ and ϕ change by 2π it follows that an equivalent definition of $q(\psi)$ can be written as

$$q = \frac{\lim_{L \to \infty} \Delta\zeta}{\lim_{L \to \infty} \Delta\chi} \tag{F.2}$$

Next, the straight field line property of Boozer coordinates, as was used in Eq. (7.263), can be expressed as

$$\zeta = \frac{d\psi_t}{d\psi}\chi - \alpha_B \tag{F.3}$$

The implication of Eq. (F.3) is that over any specified arc length L, the relation between the change in ζ and the change in χ is given by

$$\Delta\zeta = \frac{d\psi_t}{d\psi}\Delta\chi \tag{F.4}$$

Therefore, substituting into Eq. (F.2) yields the desired relation

$$q(\psi) = \frac{d\psi_t}{d\psi} \tag{F.5}$$

Appendix G

Natural boundary conditions

A classic one-dimensional eigenvalue problem is formulated in terms of a variational principle in order to provide some insight into the use of natural boundary conditions. Two problems are solved, each involving the same differential equation. The first problem has very simple boundary conditions analogous to the plasma surrounded by a conducting wall. The second problem has more complicated boundary conditions similar to the plasma–vacuum system. For the second problem it is shown how the use of a natural boundary condition simplifies the variational formulation.

G.1 The variational principle for simple boundary conditions

Consider the classic eigenvalue problem given by

$$\frac{d}{dx}\left(f\frac{dy}{dx}\right) - (g - \lambda)y = 0$$
$$y(0) = y(1) = 0 \tag{G.1}$$

where $f(x)$, $g(x)$ are known functions, and λ is the eigenvalue. A variational principle is easily derived by multiplying the equation by y and integrating over the region $0 \le x \le 1$. A simple calculation yields

$$\lambda = \frac{\int \left(fy'^2 + gy^2\right)dx}{\int y^2 dx} \tag{G.2}$$

Here and below the limits of integration are $x = 0$ to $x = 1$.

To show that Eq. (G.2) does indeed represent a valid variational principle assume that a trial function $y = y_0(x)$ satisfying the boundary conditions $y_0(0) = y_0(1) = 0$ is substituted into the integral resulting in a corresponding λ_0. Now modify the trial function by a small but arbitrary perturbation, $y = y_0(x) + \delta y(x)$, where $\delta y(x)$ also satisfies the boundary conditions $\delta y(0) = \delta y(1) = 0$. This produces a new $\lambda = \lambda_0 + \delta\lambda$, where $\delta\lambda$ is given by

$$\delta\lambda = \frac{\int \left[f(y'_0 + \delta y')^2 + g\,(y_0 + \delta y)^2 \right] dx}{\int (y_0 + \delta y)^2 dx} - \frac{\int \left(fy_0'^2 + g\,y_0^2 \right) dx}{\int y_0^2 dx} \qquad (G.3)$$

For small δy this expression reduces to

$$\delta\lambda = \frac{2 \int \left[fy'_0 \delta y' + (g - \lambda_0)\,y_0 \delta y \right] dx}{\int y_0^2 dx} \qquad (G.4)$$

The first term can be integrated by parts yielding

$$\delta\lambda = - \frac{2 \int \delta y \left[(fy'_0)' - (g - \lambda_0)\,y_0 \right] dx}{\int y_0^2 dx} \qquad (G.5)$$

Now, for λ_0 to correspond to an extremum then $\delta\lambda = 0$ for arbitrary δy. This can only occur if

$$(f\,y'_0)' - (g - \lambda_0)\,y_0 = 0 \qquad (G.6)$$

which is equivalent to the original eigenvalue problem. Note that in order to show that Eq. (G.2) is a valid variational principle it has been assumed that the trial function y_0 and all arbitrary perturbations δy satisfy the boundary conditions. In practice this is not a heavy mathematical burden to implement since the boundary conditions are very simple.

G.2 The variational principle for more complicated boundary conditions

In the second problem of interest $y(x)$ satisfies the same differential equation but with a more complicated set of boundary conditions

$$y(0) = 0$$
$$y'(1) - Ay(1) = 0 \qquad (G.7)$$

where $A = $ constant.

One can again obtain a variational principle by multiplying Eq. (G.1) by $y(x)$ and integrating. In this case a boundary term appears in the expression for λ

$$\lambda = \frac{\int \left(fy'^2 + gy^2 \right) dx - \left[fy'y \right]_{x=1}}{\int y^2 dx}$$

(G.8)

Setting $y = y_0(x) + \delta y(x)$ and substituting into Eq. (G.8) leads to the following expression for $\delta\lambda$:

$$\delta\lambda = -\frac{2 \int \delta y \left[(fy'_0)' - (g - \lambda_0)y_0 \right] dx + \left[f(y'_0 \delta y - y_0 \delta y') \right]_{x=1}}{\int y_0^2 dx}$$

(G.9)

Assume now that y_0 corresponds to an extremum. For this to be true $\delta\lambda = 0$ for arbitrary δy which can only occur if (1) y_0 satisfies the original differential equation and (2) y_0 and δy satisfy the boundary conditions: $y_0(0) = \delta y(0) = 0$ and $y'_0(1) - Ay_0(1) = 0$, $\delta y'(1) - A\delta y(1) = 0$.

The result is very similar to the simple boundary condition problem. However, there is an important practical drawback to this form of the variational principle. To understand the problem assume that the goal is to extremize λ using a trial function consisting of an expansion in a set of arbitrary amplitude basis functions. It is usually quite difficult practically to choose basis functions in which each one individually satisfies the more complicated boundary condition at $x = 1$. This is an important constraint on the basis functions in order for the total trial function to satisfy the boundary conditions for an arbitrary choice of expansion amplitudes.

Here is where the natural boundary condition enters the analysis. The natural boundary variational principle is obtained by replacing $y'(1)$ in Eq. (G.8) with $Ay(1)$. The modified variational principle becomes

$$\lambda = \frac{\int \left(fy'^2 + gy^2 \right) dx - \left[Afy^2 \right]_{x=1}}{\int y^2 dx}$$

(G.10)

Again setting $y = y_0(x) + \delta y(x)$ yields an expression for $\delta\lambda$ given by

$$\delta\lambda = -\frac{2 \int \delta y \left[(fy'_0)' - (g - \lambda_0)y_0 \right] dx + 2 \left[f\delta y \left(y'_0 - Ay_0 \right) \right]_{x=1}}{\int y_0^2 dx}$$

(G.11)

Consider next the evaluation of $\delta\lambda$ for the class of trial functions that satisfy $y_0(0) = \delta y(0) = 0$ but not the boundary condition at $x = 1$. Both $y_0(1)$ and $\delta y(1)$ are allowed to float freely. In order for Eq. (G.11) to represent a valid variational principle then $\delta\lambda$ must equal zero for arbitrary $\delta y(x)$ including an arbitrary $\delta y(1)$. If a trial function y_0 is found that extremizes λ then this can only occur if (1) $y_0(x)$ satisfies the original differential equation and (2) $y'_0(1) - Ay_0(1) = 0$. In other words, the natural boundary variational principle drives the total trial function in a direction to satisfy the specified boundary condition at $x = 1$ as the number of terms in the expansion increases.

The natural boundary variational principle has the important advantage of allowing the use of trial functions which do not automatically satisfy the boundary conditions.

Appendix H

Upper and lower bounds on δQ_{KIN}

The goal of Appendix H is to derive upper and lower bounds for the kinetic contribution to δW_{KIN} in general toroidal geometry. This contribution is given by

$$\delta Q_{KIN} = -\frac{|\omega|^2}{2} \sum_\alpha \int \frac{\partial \overline{f}_\alpha}{\partial \varepsilon} |s_\alpha|^2 d\mathbf{w}\, d\mathbf{r} \tag{H.1}$$

The bounds are necessary to determine the stability comparison theorems between the various MHD models considered in the main text.

In order to determine the bounds it is necessary to focus on the marginal stability limit $\omega \to 0$ and to make extensive use of Schwartz's inequality. Also, it is helpful distinguish between ergodic and closed line systems. Here closed line systems refer to geometries in which both the equilibrium and perturbations maintain the closed line symmetry. When the perturbations break the closed line symmetry they are included as part of the ergodic systems. The analysis is more transparent for closed line systems and this is the reason for singling them out.

A key point to recognize is that a finite contribution to δQ_{KIN} only arises from terms in $s_\alpha \propto 1/\omega$ as $\omega \to 0$. How do such terms arise? The answer is as follows. Consider first trapped particles which bounce back and forth between the hills and valleys of the equilibrium magnetic field. When integrating along l' these particles only sample part of the flux surface. As is shown, this partial sampling leads to secular-like behavior of the trajectory integral which in turn gives rise to a $1/\omega$ contribution. Passing particles, which sample the entire flux surface, also give rise to a similar $1/\omega$ contribution. The reason is that they sample the surface non-uniformly, spending more time in regions of high field than in low field. Only in a straight cylinder where there are no trapped particles and the surface sampling is uniform do the $1/\omega$ contributions average to zero.

The analysis demonstrating these points is rather complicated, involving considerable algebra and several clever mathematical tricks.

H.1 General theory

The starting point is the Fourier analysis of the orbit integral.

Fourier analysis of s

The general form of the trajectory integral is given in Eq. (10.68) repeated here for convenience,

$$s = e^{-i\omega t} \int_0^\infty e^{i\omega \tau} I[l'(\tau)] d\tau \tag{H.2}$$

Here the species subscript has been suppressed for simplicity. Now, the orbits for trapped particles or passing particles are periodic. For passing particles $\tau_B = 2\pi/\omega_B$ is the period of oscillation while for trapped particles $\tau_B = 2\pi/\omega_B$ is the half period. Note that for a closed line system the period for passing particles corresponds to one toroidal or one poloidal transit of the orbit. For ergodic systems, however, the period can be much, much longer, requiring a large number of toroidal transits, until a particle returns arbitrarily close to its initial position. For all cases one can Fourier expand

$$I[l'(\tau)] \equiv I(\tau) = \sum_n I_n(\varepsilon, \mu, \psi, \chi) e^{in\omega_B \tau}$$

$$I_n = \frac{1}{\tau_B} \int_0^{\tau_B} I(\tau) e^{-in\omega_B \tau} d\tau \tag{H.3}$$

Here, flux coordinates ψ, χ, l have been introduced for the spatial dependence. Also, ε, μ, ψ, χ are all constant along the unperturbed orbit.

Using the Fourier expansion one can easily evaluate s,

$$s = ie^{-i\omega t} \sum_n \frac{I_n}{\omega + n\omega_B} \tag{H.4}$$

It is at this point that the marginal stability assumption becomes very useful. As $\omega \to 0$, the $n = 0$ term dominates the sum. In this limit

$$s = i\frac{I_0}{\omega} = \frac{i}{\omega\tau_B} \int_0^{\tau_B} I(\tau) d\tau \tag{H.5}$$

Slight simplification of I(τ)

The next step in the analysis is to simplify the integrand $I(\tau)$ in the marginal stability limit. The general form of $I(\tau)$ is given by

$$e^{i\omega \tau} I(\tau) = e^{i\omega \tau} \left[q w_\| A_{\|1} + (m w_\perp^2/2) \nabla \cdot \boldsymbol{\xi}_\perp + m \left(w_\perp^2/2 - w_\|^2 \right) (\boldsymbol{\xi}_\perp \cdot \boldsymbol{\kappa}) \right] \tag{H.6}$$

In the limit $\omega \to 0$ the term with $A_{\|1}$ has odd symmetry with respect to $w_{\|}$. Consequently this term averages to zero when integrating over even moments in velocity space. The end result is that s simplifies to

$$s \to s_\perp(\varepsilon, \mu, \psi, \chi) = \frac{i}{\omega \tau_B} \int_0^{\tau_B} I_\perp(\tau) d\tau$$

$$= \frac{i}{\omega \tau_B} \int_0^{\tau_B} \left[(m w_\perp^2/2) \nabla \cdot \boldsymbol{\xi}_\perp + m\left(w_\perp^2/2 - w_\|^2\right)(\boldsymbol{\xi}_\perp \cdot \boldsymbol{\kappa}) \right] d\tau$$

$$\text{(H.7)}$$

The integrand $I_\perp$ now has even symmetry with respect to $w_{\|}$.

Rewriting δQ_{KIN}

The basic quantity of interest δQ_{KIN} can now be rewritten by introducing ψ, χ, l flux coordinates and ε, μ velocity coordinates. The flux coordinate transformation is given by

$$d\mathbf{r} = d\psi d\chi \frac{dl}{B} \qquad \text{(H.8)}$$

The velocity coordinate transformation for integrands with even symmetry in $w_\|$ has the form

$$d\mathbf{w} = (2/m)^{3/2} \pi B \frac{d\varepsilon d\mu}{(\varepsilon - \mu B)^{1/2}} \qquad \text{(H.9)}$$

These transformations are substituted into Eq. (H.1) leading to

$$\delta Q_{KIN} = -\frac{|\omega|^2}{2} \sum_\alpha \int \frac{\partial \overline{f}_\alpha}{\partial \varepsilon} \frac{1}{|\omega|^2 \tau_B^2} \left| \int_0^{\tau_B} I_\perp(\tau) d\tau \right|^2 d\mathbf{w} d\mathbf{r}$$

$$= -\frac{1}{2} \sum_\alpha (2/m_\alpha)^{3/2} \pi \int \frac{d\psi \, d\chi \, dl \, d\varepsilon \, d\mu}{(\varepsilon - \mu B)^{1/2}} \frac{\partial \overline{f}_\alpha}{\partial \varepsilon} |I_0|^2$$

$$\text{(H.10)}$$

In this relation the velocity integrals are to be carried out first followed by the spatial integrals.

Carrying out the l integration

Here, and in the analysis that follows it is often useful to invert the order of integration between l and μ. For instance, by making this inversion one can evaluate the l integration in Eq. (H.10). As a starting point, focus on the l, μ integration including the limits of integration as given by

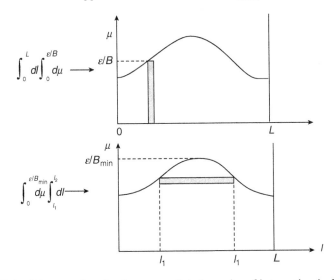

Figure H.1   Diagram showing how to switch the order of integration in l, μ space.

$$\int_0^L dl \int_0^{\varepsilon/B} d\mu \tag{H.11}$$

where $L = L(\psi, \chi)$ and $B = B(\psi, \chi, l)$. To invert the order of integration examine Fig. H.1 (which for simplicity is drawn for an axisymmetric torus with only one region of velocity space corresponding to trapped particles). Observe that the integration domain separates into two regions, one representing the passing particles and the other the trapped particles. From Fig. H.1 one sees that the inversion of the integration order corresponds to

$$\int_0^L dl \int_0^{\varepsilon/B} d\mu = \int_0^{\varepsilon/B_{\min}} d\mu \int_{l_1}^{l_2} dl \tag{H.12}$$

For passing particles $l_{1,2} = 0, L$. For trapped particles $l_{1,2} = l_{1,2}(\varepsilon, \mu, \psi, \chi)$ are the turning points of the orbit defined by $w_{\parallel} = 0$ or equivalently $\varepsilon - \mu B(\psi, \chi, l_{1,2}) = 0$.

Since $I_0 = I_0(\varepsilon, \mu, \psi, \chi)$ is not a function of l the μ, l portion of the integration in Eq. (H.10) can be written as

$$\int_0^{\varepsilon/B_{\min}} d\mu \int_{l_1}^{l_2} \frac{dl}{(\varepsilon - \mu B)^{1/2}} = (2/m)^{1/2} \int_0^{\varepsilon/B_{\min}} d\mu \int_{l_1}^{l_2} \frac{dl}{w_{\parallel}} \tag{H.13}$$

Equation (H.13) can be simplified by recalling that along the unperturbed orbits $dl/dt = w_{\parallel}$, which implies that

$$\tau_B = \int_0^{\tau_B} dt = \int_{l_1}^{l_2} \frac{dl}{w_{\parallel}} \tag{H.14}$$

The end result of these manipulations is that the quantity δQ_{KIN} can be rewritten as

$$\delta Q_{KIN} = -\sum_\alpha \frac{2\pi}{m_\alpha^2} \int d\psi \, d\chi \, d\varepsilon \, d\mu \frac{\partial \bar{f}_\alpha}{\partial \varepsilon} \tau_B |I_0|^2 \tag{H.15}$$

Equation (H.15) is still exact but is in a more convenient form for determining upper and lower bounds.

H.2 A lower bound on δQ_{KIN}

A lower bound on δQ_{KIN} is determined by the use of Schwartz's inequality, which for situations where one function is complex is given by

$$\int g^2 d\mu \int |h|^2 d\mu \geq \left| \int gh d\mu \right|^2 \tag{H.16}$$

If one sets $g = \tau_B^{1/2}$ and $h = \tau_B^{1/2} I_0$ it then follows that

$$\int \tau_B d\mu \int \tau_B |I_0|^2 d\mu \geq \left| \int \tau_B I_0 d\mu \right|^2 \tag{H.17}$$

Therefore, a lower bound on δQ_{KIN} can be written as

$$\delta Q_{KIN} \geq \delta Q_{min} = -\sum_\alpha \frac{2\pi}{m_\alpha^2} \int d\psi \, d\chi \, d\varepsilon \frac{\partial \bar{f}_\alpha}{\partial \varepsilon} \frac{\left| \int \tau_B I_0 d\mu \right|^2}{\int \tau_B d\mu} \tag{H.18}$$

This form is useful because the μ integrals can be evaluated analytically. For example, by again inverting the order of integration, one finds for the integral in the denominator

$$\begin{aligned} \int \tau_B d\mu &= \int_0^{\varepsilon/B_{min}} d\mu \int_{l_1}^{l_2} \frac{dl}{w_\parallel} \\ &= (m/2)^{1/2} \int_0^L dl \int_0^{\varepsilon/B} \frac{d\mu}{(\varepsilon - \mu B)^{1/2}} \\ &= (2m)^{1/2} \varepsilon^{1/2} \int_0^L \frac{dl}{B} \end{aligned} \tag{H.19}$$

Similarly, a short calculation leads to

$$\begin{aligned} \int \tau_B I_0 d\mu &= (m/2)^{1/2} \int_0^L dl \int_0^{\varepsilon/B} \frac{d\mu}{(\varepsilon - \mu B)^{1/2}} [\mu B \, \nabla \cdot \boldsymbol{\xi}_\perp + (3\mu B - 2\varepsilon)(\boldsymbol{\xi}_\perp \cdot \boldsymbol{\kappa})] \\ &= (2^{3/2} m^{1/2}/3) \varepsilon^{3/2} \int_0^L \frac{dl}{B} \nabla \cdot \boldsymbol{\xi}_\perp \end{aligned} \tag{H.20}$$

Equations (H.19) and (H.20) are substituted into Eq. (H.18). The expression for $\delta Q_{\min}$ reduces to

$$
\delta Q_{\min} = -\sum_\alpha \frac{2^{7/2}\pi}{9m_\alpha^{3/2}} \int d\psi \, d\chi \, \frac{dl}{B} |\langle \nabla \cdot \boldsymbol{\xi}_\perp \rangle|^2 \int d\varepsilon \, \varepsilon^{5/2} \frac{\partial \overline{f}_\alpha}{\partial \varepsilon}
$$

$$
\langle \nabla \cdot \boldsymbol{\xi}_\perp \rangle = \frac{\displaystyle\int \frac{dl}{B} \nabla \cdot \boldsymbol{\xi}_\perp}{\displaystyle\int \frac{dl}{B}}
\tag{H.21}
$$

The last step is to recognize that p_α written in ε, μ coordinates is given by

$$
\begin{aligned}
p_\alpha &= \int \frac{m_\alpha}{3}\left(w_\perp^2 + w_\parallel^2\right) f_\alpha d\mathbf{w} \\
&= \left(2^{5/2}\pi/m_\alpha^{3/2}\right) \int_0^\infty \varepsilon f_\alpha \int_0^{\varepsilon/B} \frac{B d\mu}{(\varepsilon - \mu B)^{1/2}} \\
&= \left(2^{7/2}\pi/3m_\alpha^{3/2}\right) \int_0^\infty \varepsilon^{3/2} f_\alpha d\varepsilon \\
&= -\left(2^{9/2}\pi/15m_\alpha^{3/2}\right) \int_0^\infty \varepsilon^{5/2} \frac{\partial f_\alpha}{\partial \varepsilon} d\varepsilon
\end{aligned}
\tag{H.22}
$$

This expression is substituted into Eq. (H.21) leading to the final desired form of $\delta Q_{\min}$

$$
\begin{aligned}
\delta Q_{\min} &= \frac{1}{2}\sum_\alpha \int \frac{5}{3} p_\alpha |\langle \nabla \cdot \boldsymbol{\xi}_\perp \rangle|^2 d\psi \, d\chi \, \frac{dl}{B} \\
&= \frac{1}{2}\sum_\alpha \int \frac{5}{3} p_\alpha |\langle \nabla \cdot \boldsymbol{\xi}_\perp \rangle|^2 d\mathbf{r} \\
&= \frac{1}{2}\int \frac{5}{3} p |\langle \nabla \cdot \boldsymbol{\xi}_\perp \rangle|^2 d\mathbf{r}
\end{aligned}
\tag{H.23}
$$

This expression is identical to δW_C, the compressional energy in ideal MHD.

H.3 An upper bound on δQ_{KIN}

An upper bound on δQ_{KIN} can be obtained by an alternate use of Schwartz's inequality applied to the integral I_0 itself. In this case τ rather than μ is the integration variable. From the definition of I_0 given by Eq. (H.5) and Schwartz's inequality (with $g = 1$ and $h = I_\perp$) it follows that

$$|I_0|^2 \equiv \frac{1}{\tau_B^2} \left| \int_0^{\tau_B} I_\perp \, d\tau \right|^2 \leq \frac{1}{\tau_B^2} \left(\tau_B \int_0^{\tau_B} |I_\perp|^2 \, d\tau \right) \tag{H.24}$$

Therefore, Eq. (H.15) leads to an upper bound for δQ_{KIN} given by

$$\delta Q_{KIN} \leq \delta Q_{max} = -\sum_\alpha \frac{2\pi}{m_\alpha^2} \int d\psi \, d\chi \, d\varepsilon \, d\mu \frac{\partial \bar{f}_\alpha}{\partial \varepsilon} \int_0^{\tau_B} |I_\perp|^2 \, d\tau \tag{H.25}$$

The quantity δQ_{max} can be calculated by switching from τ to l as the integration variable and then inverting the order of integration between l and μ:

$$\delta Q_{max} = -\sum_\alpha \frac{2\pi}{m_\alpha^2} \int d\psi \, d\chi \, d\varepsilon \, \frac{\partial \bar{f}_\alpha}{\partial \varepsilon} \int_0^L dl \int_0^{\varepsilon/B} \frac{|I_\perp|^2}{w_\|} \, d\mu$$

$$I_\perp = \mu B \, \nabla \cdot \boldsymbol{\xi}_\perp + (3\mu B - 2\varepsilon)(\boldsymbol{\xi}_\perp \cdot \boldsymbol{\kappa})$$

$$w_\| = (2/m)^{1/2} (\varepsilon - \mu B)^{1/2} \tag{H.26}$$

The μ integral, in analogy with Eq. (H.19), is straightforward, but slightly tedious to evaluate. The result is

$$\int_0^{\varepsilon/B} \frac{|I_\perp|^2}{w_\|} \, d\mu = \frac{2^{5/2} m^{1/2}}{15} \frac{\varepsilon^{5/2}}{B} G$$

$$G = 2|\nabla \cdot \boldsymbol{\xi}_\perp|^2 + 3|\boldsymbol{\xi}_\perp \cdot \boldsymbol{\kappa}|^2 + (\nabla \cdot \boldsymbol{\xi}_\perp)(\boldsymbol{\xi}_\perp^* \cdot \boldsymbol{\kappa}) + (\nabla \cdot \boldsymbol{\xi}_\perp^*)(\boldsymbol{\xi}_\perp \cdot \boldsymbol{\kappa}) \tag{H.27}$$

Substituting this expression into Eq. (H.26) yields

$$\delta Q_{max} = -\sum_\alpha \frac{2^{7/2}\pi}{m_\alpha^{3/2}} \int G \, d\psi \, d\chi \, \frac{dl}{B} \int \varepsilon^{5/2} \frac{\partial \bar{f}_\alpha}{\partial \varepsilon} \, d\varepsilon \tag{H.28}$$

One now introduces the pressure as defined by Eq. (H.22) and rearranges the terms in G. A short calculation leads to the desired upper bound on δQ_{KIN},

$$\delta Q_{max} = \frac{1}{2} \sum_\alpha \int p_\alpha \left[\frac{5}{3} |\nabla \cdot \boldsymbol{\xi}_\perp|^2 + \frac{1}{3} |\nabla \cdot \boldsymbol{\xi}_\perp + 3(\boldsymbol{\xi}_\perp \cdot \boldsymbol{\kappa})|^2 \right] d\mathbf{r}$$

$$= \frac{1}{2} \int \left[\frac{5}{3} p |\nabla \cdot \boldsymbol{\xi}_\perp|^2 + \frac{p}{3} |\nabla \cdot \boldsymbol{\xi}_\perp + 3(\boldsymbol{\xi}_\perp \cdot \boldsymbol{\kappa})|^2 \right] d\mathbf{r} \tag{H.29}$$

This expression is identical to δQ_{CGL} from the double adiabatic theory. All the information required for the comparison theorems has now been derived.

Index

Printed in the United States
by Baker & Taylor Publisher Services